Brewing Microbiology

Current Research, Omics and Microbial Ecology

https://doi.org/10.21775/9781910190616

Edited by

Nicholas A. Bokulich

Northern Arizona University, Flagstaff, AZ, USA

and

Charles W. Bamforth

University of California, Davis, CA, USA

Caister Academic Press

Caister Academic Press
Norfolk, UK

www.caister.com

British Library Cataloguing-in-Publication Data
A catalogue record for this book is available from the British Library

ISBN: 978-1-910190-61-6 (paperback)
ISBN: 978-1-910190-62-3 (ebook)

Contents

Current Books of Interest

Preface

Research into brewing yeast and other organisms associated with beer and brewing has experienced many important advances in the past decade. These have been nudged in no small way by staggering technological advances in tools fundamental to the investigation of microbes and their metabolism. Cutting-edge approaches, such as highly parallel nucleotide sequencing, genetic modification and mass spectrometry, are leading to new discoveries across the field of brewing microbiology.

The goal of this volume is to survey the most recent discoveries in brewing microbiology, with an emphasis on omics techniques and other modern technologies. The chapters span an array of subjects, including yeast genomics and evolution; the physiology, handling, metabolism and genetic regulation of brewing yeasts; genetic modification; taxonomy of both *Saccharomyces* and non-*Saccharomyces* yeasts; the biology and management of spoilage organisms (both fungal and bacterial); microbial ecology of traditional and 'wild' fermentations; and fungal contamination of barley and malt.

Advances in each of these topics have not only furthered our knowledge of brewing processes, they have yielded applications that touch all aspects of brewing practice, from barley growth and malting to yeast management, strain selection, fermentation control, and quality assurance. Consumer interests and brewing technologies continue to shift, yielding new challenges and research frontiers. For example, trends towards lower-alcohol beers have altered quality assurance demands, and a growing global interest in 'wild' and otherwise sour beers has spurred the need to better understand the ecology of traditional beer fermentations and biology of non-*Saccharomyces* yeasts. Other recent research has revealed that lager yeasts, which are responsible for fermentation of the majority of beers consumed globally, are actually the progeny of hybridization events that occurred only a few centuries ago (likely selected by contemporaneous advances in cave brewing technology), and hence only a few distinct lager strain lineages are available. Now, novel techniques for high-throughput hybridization have yielded dozens more, broadening the range and characteristics of strains available to brewers.

The chapters in this volume aim not only to illuminate recent progress, but also to discuss its impact on brewing practices. We also discuss future research directions, setting out a vision for the next decade of discovery. We are on the cusp of many great innovations, and have only begun to tap the potential of the new tools that pave the way.

Nicholas A. Bokulich
Northern Arizona University, Flagstaff, AZ, USA

Charles W. Bamforth
University of California, Davis, CA, USA

Brewing Yeast Physiology

1

Chris Boulton

Boulton Brewing Consultancy, Burton on Trent, UK.

Correspondence: christopherboulton@gmail.com

https://doi.org/10.21775/9781910190616.01

Abstract

Characterization of the genomes of yeast belonging to the genus *Saccharomyces*, including those used in brewing, has been the subject of intense investigation. Fundamental differences between ale and lager strains have been demonstrated. The linkage of many genes with cellular function has been made. Although essential, this has also highlighted a fundamental lack of knowledge as to how expression of the genome is made manifest in terms of phenotype, in particular how the phenotype is regulated at the level of the metabolome. The intention of this chapter is to review current research that has been directed towards these ends. The chapter addresses how yeast physiology is influenced by the conditions it is exposed to in current modern brewing practice, what are the potential pitfalls, and how can processes be managed to increase the likelihood of a controlled and consistent outcome. Of particular note are the factors that influence passage of cells into and out of quiescence, coincident with fermentation initiation and eventual removal of cropped yeast to storage. Since the majority of published studies have used haploid laboratory strains grown aerobically on semi-defined media, attempts have been made to extrapolate results to encompass brewing strains serially re-pitched in semi-aerobic fermentations using wort as the feedstock.

Introduction

Yeast of the genus *Saccharomyces* is one of the most intensively studied of all eukaryotic organisms and, in consequence, a huge and somewhat daunting body of literature exists that is devoted to the science underlying its activities. It is neither possible nor desirable to attempt to provide an overview of this in its entirety; instead, the object of this chapter is to describe some recent literature that describes the elucidation of the physiological response of brewing yeast to the conditions experienced in modern commercial brewing practice. For a general overview of brewing yeast and fermentation, the reader is referred to Boulton and Quain (2001). Most of the yeast literature describes work that employed haploid laboratory strains of *Saccharomyces cerevisiae*. A much smaller proportion of the literature embraces brewing yeast strains. This presents some difficulties in that lager brewing strains are very different, being hybrid in nature and possessing a polyploid/aneuploid genome (see Chapter 4). It follows that in the case of some of the work described here there is an assumption that it will apply to brewing strains, although, of course, this may not be entirely true.

At the highest level, cells respond to environmental changes by modifications to expression of the genome. Much published work has described how this is accomplished and knowledge of the genome of brewing yeast strains continues to expand and fewer orphan genes remain; however, the hybrid nature of lager strains introduces added complication in that the mosaic of duplicated genes allows for much variation between individual strains (Wendland, 2014). The ways in which genotypic changes are made manifest in terms of the physiological response are not extensively reported and yet it is these responses that underpin much that is of importance in terms of the ability of yeast to transform wort into beer in a manner that is efficient and predictable. Thus, changes in external conditions can prompt large changes in genome expression but of themselves these are insufficient to account for the fine tuning of cellular metabolism required to underpin translational responses. This is accomplished by another hierarchy of responses which include regulation of mRNA stability, allosteric control of enzyme activity and a set of post-translational controls in which the activities of individual enzymes are modified by reactions such as phosphorylation, acetylation and ubiquitination. Predictably, major carbon-metabolizing pathways such as glycolysis, those involved in fermentation and the control of reserve carbohydrates are subject to close regulation via post-translational mechanisms (Tripodi *et al.*, 2014). A measure of the potential importance of these methods of control may be inferred from the finding of Olivera and Sauer (2014) that around half of the enzymes of *S. cerevisiae* involved in metabolic networks are phosphoproteins. Undoubtedly, elucidating the fine details of these cellular regulation methods will pay rich dividends in terms of gaining a better understanding of the behaviour of yeast throughout the brewing process.

The brewing yeast cycle

An obvious characteristic of yeast belonging to the genus *Saccharomyces* is an ability to survive in a wide variety of conditions; perhaps more specifically, an ability to adapt its phenotype rapidly in response to changes in the environment in order that it may grow and proliferate where conditions permit this or simply survive where they do not. Modern brewing practices test these abilities to the full as will be described.

The majority of modern commercial brewers practise a semi-conservative process in which yeast derived from fermentation is recovered from the immature beer, retained in a storage vessel and a proportion used to inoculate a subsequent fermentation. This process of serial fermentation is continued typically for 5–15 successive fermentations (referred to by brewers as 'generations'), after which the yeast is disposed of and replaced by a new culture derived from laboratory stocks and introduced into the brewery via a process of propagation (Fig. 1.1).

The brewing yeast cycle has several consequences. The sequence of events represented by inoculation (pitching), growth, cropping, and storage requires cells to undergo a series of transitions in which they are exposed to rapid and dramatic changes in environmental conditions. In the storage phase, cells are starved of nutrients but usually exposed to low temperature (2–4°C), anaerobic conditions, and relatively high ethanol conditions. In many breweries, before pitching, yeast cells are treated with an acidulant such as phosphoric acid with the aim of killing less acid-tolerant bacteria. In this 'acid washing' step, the pH is reduced from around pH 4.0 to pH 2.1–2.3 and, at least in a properly controlled process, the

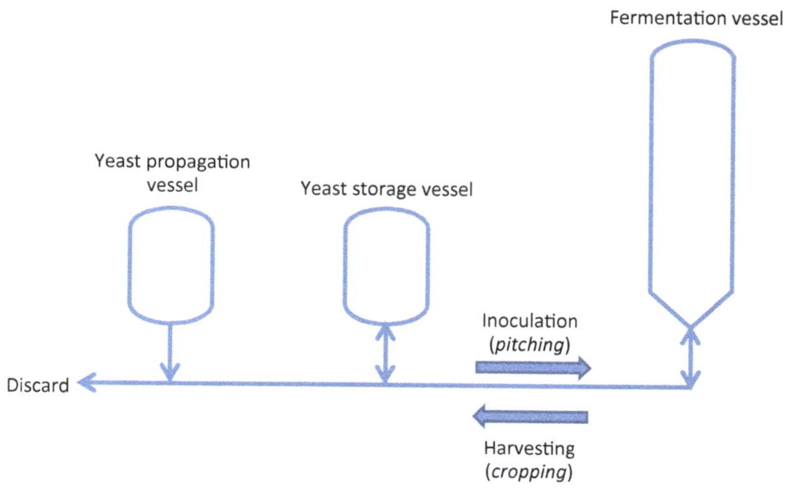

Figure 1.1 The brewing yeast cycle.

cells are held at a temperature of around 3°C for a period of 1–2 hours (Simpson and Hammond, 1989). During pitching, the cells undergo a shift to higher temperature (although one much lower than that optimal for growth) and exposure to oxygenated wort. Thus, cells undergo a transition to aerobic conditions in a complete growth medium, albeit one that is relatively unbalanced, having a very high sugar concentration and comparatively smaller concentrations of other essential nutrients. Transfer to wort is accompanied by a concomitant osmotic shock and in the large capacity and tall vessels most commonly used in modern brewing the cells have considerable hydrostatic pressure exerted on them. As growth proceeds, the yeast exhausts the available oxygen and conditions revert to anaerobic; nutrients are assimilated; and the principal products, ethanol and CO_2, accumulate together with a multitude of other metabolic products, some of which influence beer flavour. Ethanol and CO_2 both exert toxic effects on yeast cells. In modern high-gravity brewing practice, it is common to use highly concentrated worts as a means of increasing fermentation productivity. The concentrated beers produced in this process are then diluted to 'sales gravity' prior to packaging. The osmotic pressures generated as a consequence of high-sugar concentrations in these worts and the consequent high yields of ethanol increase the stress levels experienced by yeast (Udeh and Kgatia, 2013). Commonly, wort strength is increased by the addition of relatively pure sugar syrups since these are usually less expensive compared to malted barley. The effect of this is to dilute the non-sugar components and thereby render the wort an even more unbalanced growth medium.

As fermentation proceeds, the majority of brewing strains settle to the bottom of the vessel to form a sediment in which the cells are subject to the potentially damaging combination of high hydrostatic pressure and high levels of carbonation and ethanol. In those strains capable of doing so, sedimentation is aided by the process of flocculation (Vidgren and Londesborough, 2011). At some stage in the fermentation, yeast, with some entrained beer, is cropped from fermenter, chilled and transferred back into a storage vessel where it may be held for up to 7 days (typically 2 – 5 days). After this time, another cycle of growth is initiated by pitching stored yeast into fresh wort. Cylindroconical fermenters (the type most commonly used in modern brewing practice) are designed to allow removal of yeast crops with a minimum of back-mixing. Thus, there is a spatial relationship between the location of yeast in the crop in the cone of the vessel and the timing of its removal. By inference, the brewer has the opportunity to make a choice over which fraction to retain for re-pitching. This is significant in that Powell *et al.* (2002) noted that there was considerable heterogeneity in terms of location of cells in the cone crop and factors such as cell size, replicative age, flocculation characteristics and levels of storage carbohydrates. Predictably, there was concomitant variability in subsequent fermentation performance depending on which cells were re-pitched.

The modern brewing process is potentially harmful to yeast and prolonged serial fermentation increases the likelihood of contamination and the potential for the emergence of genetic variants. It is for this reason that after a given number of serial fermentations, the precise number chosen by individual brewers on a more or less empirical basis, the yeast culture is disposed of and replaced with a new one of guaranteed identity and purity. The propagation process used by the majority of large- scale commercial brewers involves carrying out a series of batch cultures of ever-increasing size until sufficient biomass is generated to pitch a full-scale production fermentation. In the propagation phase, wort is used as the growth medium and yeast growth, in excess of that observed in fermentation, is encouraged by the provision of continuous oxygenation. Fed-batch approaches, in which biomass yields far in excess of those achievable by conventional brewery propagations on wort are used for the production of active dried yeast, a product used by many of the new generation of craft brewers (Jenkins *et al.*, 2011).

Compared with many other industrial processes and certainly the majority of laboratory studies, brewing fermentation pitching rates are relatively high, typically 15–20 million cells per ml (*ca* 0.8–1.2 g/l dry weight) in a high-gravity wort of 15–20° Plato. During fermentation, growth is modest, usually a 3- to 4-fold increase in cell biomass. A result of this is that the pitched cells not only play an active part in fermentation but may also survive through to cropping and re-pitching. Yeast cells undergo an ageing process that, unless cut short by other causes, culminates in senescence and death (Steinkraus *et al.*, 2008). Depending on the fraction of crop retained for re-pitching, there is the possibility of a gradual increase in the average cell replicative age with each generation.

Yeast physiological response to the brewing process, transitions between growth and non-growth

Most cell types spend the majority of their lifespans in a quiescent state where proliferation is not occurring and metabolism may be in a resting state (Hartwell *et al.*, 1974; Werner-Washburne, 1993). When conditions are suitable, cells may leave this quiescent stage and initiate a new cell cycle. The ability to enter the quiescent phase in a controlled manner, such that metabolism is manipulated in a way that favours survival, is clearly essential as is the requirement for timely re-entry into active growth when conditions allow. In both cases there is a need, as early as possible, to be able to detect changes in external conditions so that the process of phenotypic re-ordering can commence. In the case of an assembly of individual cells, such as is represented by yeast participating in the brewing process, an element of cooperation is also desirable in order to select the fraction of the population deemed most suitable for survival and proliferation.

In the case of yeast participating in the brewing process, passage into and out of quiescence coincides with the ending of one fermentation, the period of intermediate storage, and initiation of the next. Typically, the storage period lasts for 1–5 days and fermentation for 1–2 weeks, and thus brewing yeast cells perhaps spend less time in the quiescent phase and commensurately longer in growth compared with other cell types. The term 'end of fermentation' requires some clarification. For many modern lager fermentations, which encompasses the majority of beer currently produced globally, yeast may be held for a period of days in vessel at a comparatively warm temperature, prior to cropping. In this phase, growth in terms of cell proliferation has ceased; however, outward signs of metabolic activity remain manifest. These include changes that have importance to beer flavour, most notably the assimilation and reduction of vicinal diketones, namely diacetyl and pentanedione (Krogerus and Gibson, 2013).

In terms of the cell cycle, the quiescent state is referred to as G_0 and is characterized as cells that are neither budding nor preparing to bud. Although numerous studies have been published describing the response of the genome to the transitions from growth to quiescence (Wu *et al.*, 2004; Coller *et al.*, 2006; Shimanuki *et al.*, 2007), comparatively few reports discuss this subject at the level of cellular physiology. Clearly the ability of yeast cells to undergo the reversible shifts from quiescence to growth is of critical importance in brewing fermentations. Surprisingly, this subject remains poorly understood for all cell types. Predictably, the majority of studies devoted to yeast have used aerobically grown haploid laboratory strains undergoing diauxic shift where cells arrest after exhausting glucose from the medium and in a second lag phase adjust their metabolism to undergo a second growth phase on ethanol produced in the first. In the absence of studies devoted to brewing fermentation, it is to these investigations that reference must be made. How well they relate to the peculiar conditions of brewing fermentations is a moot point. In particular, the fact that the majority of these laboratory studies have used a carbon-limiting medium, obviously the opposite situation to brewing wort where the ratio of carbon to other nutrients is very high.

Quiescent yeast cells

Compared with actively growing cells, those in the G_0 phase have enhanced thermotolerance, partly a result of a thickened cell envelope and elevated levels of carbohydrate reserves, notably glycogen and trehalose (Gray *et al.*, 2004), as might be expected for cells adopting a strategy for survival. The shift into quiescence occurs in response to nutrient limitation via the concerted action of, amongst other factors, the activities of a number of protein kinases that participate in cellular cascade control pathways (Smets *et al.*, 2010). It has been assumed that the shift to quiescence only takes place in G_1 phase cells (Pardee, 1974); however, other work has shown that cells may enter this phase from any point in the cell cycle (Wei *et al.*, 1993; Laporte *et al.*, 2011). In the latter paper, the authors grew yeast on a carbon-limited medium and noted that within the stationary phase population approximately 90%

of the cells were unbudded and the remaining 10% budded. Examination of the two subpopulations revealed that both groups had the same characteristics in terms of markers of quiescent cells and both sets were capable of resuming proliferation when suitable nutrients were supplied. The authors went on to separate budded and unbudded quiescent cells by micromanipulation and tested their relative abilities to form colonies on solid nutrient medium. Both types were capable of this; however, the budded fraction were much less able (65% of budded cells compared with 95% unbudded). This mirrors observations made in relation to brewing propagation, where it has also been seen that cells in the stationary phase culture after aerobic growth on malt wort had a much higher budding index compared with the same yeast cropped from fermentation (Miller *et al.*, 2012). These authors noted that when the new culture was pitched into the first generation fermentation, there was a lack of synchrony in terms of the move into growth between the budded and unbudded fraction, and it was suggested that this was the basis of the frequent observation that these first fermentations give slower cycle times compared with yeast of an older generation. Conversely, cropped pitching yeast has a much lower budding index, suggesting that in this case the transition to quiescence leads to a more homogeneous population (Miller *et al.*, 2012). Perhaps this is linked to the fact that in a brewery fermentation growth extent may be limited by the quantity of oxygen supplied initially, whereas in propagation it is likely to be another nutrient, in most cases probably amino nitrogen or possibly zinc.

The question has been posed whether the passage into quiescence is the result of following a genetically based programme that, once committed to, must progress to completion, rather like START in the cell cycle; or whether it is simply a number of passive adaptations that occur in response to adverse conditions (Daignan-Fornier and Sagot, 2011). These authors, in reviewing their own work and that of others, make several pertinent points. In yeast, the quiescent stage is entered in response to nutrient limitation; however, profiles of mRNA in quiescent cells are different depending on the limiting nutrient and the transcription of only a few genes are common to different nutrient limitations. This suggests that there is not a signature quiescent gene profile. At the metabolomics level there is little commonality between cells starved for different nutrients, supporting the view that quiescence is an adaptive response to growth at a very low rate. However, it can also be demonstrated that, independent of the nature of the limiting nutrient, cells undergo a common series of cellular adaptations associated with quiescence and these occur in response to sensing systems that are able to detect declining levels of nutrients and predict the onset of starvation. In this sense, there may be a number of different quiescent states depending on the prior history of the cell.

Quiescent yeast cells undergo structural reorganization during the transition to non-growth. The actin cytoskeleton is reorganized to give so-called 'actin bodies', which may act as actin reserves that can be rapidly changed back into a functional form when nutrients become available (Sagot *et al.*, 2006). In addition, the proteasome moves from the nucleus to form a cytoplasmic structure termed the proteasome storage granule (Laporte *et al.*, 2008). This is accompanied by the concentration of many cytosolic metabolic enzymes into discrete structures (Narayanaswamy *et al.*, 2009). Other cellular components are also organized into pools from where they may be rapidly mobilized to facilitate re-entry into proliferation. For example, quiescent cells contain large cytosolic processing bodies where mRNA molecules are held ready to participate in translation when growth-permitting conditions arise. Similarly, it has been demonstrated that, although many genes are repressed in the transition to stationary phase, transcription is held in a state in which it can be rapidly reactivated when the need arises (Radonjic *et al.*, 2006).

Population heterogeneities in stationary phase cells have been reported (Allen *et al.*, 2006). These authors studied differences in cells in the stationary phase as they underwent the diauxic shift. They were able to separate quiescent and non-quiescent yeast cells based on differences in density. A fraction of the population recovered from a glucose-exhausted medium comprised small, dense, unbudded daughter cells. On completion of diauxy, these were able synchronously to re-enter mitotic division. The suggestion was that these daughter cells derived from cell division in the diauxic phase, after which they underwent changes leading to density increase, partly attributable to enhanced trehalose accumulation and a thickened cell wall. A smaller subpopulation of cells apparently did not enter the quiescent phase and these, although viable, rapidly lost the ability to undergo mitosis and became necrotic. A smaller group of cells within this subset exhibited properties of quiescent cells.

Li *et al.* (2013), again working with glucose-grown cells undergoing diauxic shift, used flow cytometry to separate the three distinct cell types described by Allen *et al.* (2006). They confirmed that within these groups was a subpopulation of quiescent cells that were small, had thickened cell walls, and were extremely heat-tolerant. These cells comprised a population of daughter cells formed as a consequence of an asymmetric budding event triggered by the diminishing glucose concentration. Time was required for cells to respond to changes in glucose concentration. Indeed, if the supply was abruptly removed the budding event resulting in the generation of resistant daughter cells was disrupted (Li *et al.*, 2013). The quiescent daughter cells were homogeneous and arrested in the G_1 phase and highly resistant to applied stresses. In the quiescent state they have very long lifespans and are able to begin synchronous proliferation, should conditions change to growth-permitting (Li *et al.*, 2009). The falling glucose concentration triggers a survival response in which there is a lengthening of the G_1 phase, and the quiescent daughter cells have an altered metabolism in which accumulation of both glycogen and trehalose are favoured (Miles *et al.*, 2013). These storage carbohydrates provide a source of carbon and energy to allow survival in the starvation phase and during the subsequent transition from quiescence to growth. In addition, trehalose has a protective role as a stabilizer of membranes and proteins against stresses such as heat and desiccation (Silljé *et al.*, 1999; Elbein *et al.*, 2003). Although these carbohydrates clearly have important roles in survival of potential starvation conditions, they are not determinative of the quiescent state since not all cells with elevated intracellular levels can give rise to quiescent daughters (Li *et al.*, 2013). The second and third subpopulations comprised a much smaller fraction of mother cells that had undergone the shift to quiescence and non-quiescent mother cells.

The transition to quiescence has been shown in the work described above to be related to the ability of cells to sense impending glucose exhaustion (Ozcan *et al.*, 1996; Grose *et al.*, 2007; Conrad *et al.*, 2014). In the case of brewery fermentations, the growth-determining nutrient is usually unknown and probably varies with different types of wort; however, it is never sugar. Where high levels of sugar adjuncts are used, nitrogen-containing nutrients, notably free amino nitrogen, can be low and may be limiting, although this has rarely been subject to proper investigation. In many brewery fermentations, it is often assumed that oxygen is the limiting substrate via its role in anaerobic pitching yeast as substrate for synthesis of sterols and unsaturated fatty acids (see 'Response of yeast to oxygen', below). How does this relate to the transition to quiescence? Using chemostat cultures, Hazelwood *et al.* (2009) investigated the effects of growth under different nutrient limitations on accumulation of trehalose and glycogen. They observed that, in addition to glucose, limitation of ammonia, phosphate, sulfur, and zinc were all influential and by inference storage carbohydrate accumulation was not simply a result of glucose excess. Glucose and ammonia limitation gave 10- to 14-fold more glycogen than growth under conditions of glucose excess. In addition to glycogen, trehalose accumulation was favoured by ammonia limitation. The responses were attributed to post-transcriptional regulation rather than transcriptional regulation alone. Li *et al.* (2013) studied the events occurring in diauxy and concluded that glucose limitation resulted in differentiation into three cell types. One fraction comprised very small quiescent daughter cells that arose from highly asymmetric budding. These cells acquired enhanced thermotolerance and accumulated high levels of reserve carbohydrates. Differentiation between quiescent and non-quiescent cells occurred shortly after diauxy and it was concluded that post-transcriptional regulation of mRNA played a crucial role.

The passage into quiescence, as judged by formation of proteasome storage bodies and actin re-ordering, is seemingly dependent on lack of carbon. Laporte *et al.* (2011) reported that nitrogen-starved yeast cells acquired thermotolerance but did not exhibit the actin skeleton and proteasome modifications. Both starvation of carbon and transfer to distilled water did, presumably indicating the primacy of the glucose signal.

The yeast cell wall is highly plastic in nature and changes in its structure occur in response to environmental changes (Klis *et al.*, 2002, 2006). Remodelling of cell wall structure is regulated by the cell wall integrity (CWI) pathway (Levin, 2011). This system responds to signals received at the cell surface, either as a consequence of normal growth or as environmental challenges, via the appropriate sensors and transmits these to the intracellular targets that mediate the appropriate response. The key element is a G protein, termed Rho1. This switching protein integrates signals from the cell surface and the cell budding cycle, and controls cell wall biogenesis and actin organization.

With regard to entry into quiescence, the increased density of small quiescent daughter cells is in part due to thickening of the cell wall and this change is related to enhanced longevity (Li *et al.*, 2009). Cells become stronger mechanically and less porous. These changes are in part a consequence of the presence of proteins that, based on the observation that they can be removed by treatment with dithiothreitol, are anchored by disulfide bonds (Shimoi *et al.*, 1998; Jansen, 2009). There is a 6- to 7-fold increase in the level of disulfide linkages in stationary phase cells compared with those growing exponentially (de Nobel *et al.*, 1990). The most abundant cell wall protein in quiescent cells is a GPI cell wall protein termed Sed1 (Shimoi *et al.*, 1998). GPI (glycophosphate phosphatidylinositol) cell wall proteins are those in which the lipid moiety is synthesized in the endoplasmic reticulum and from there used to transport proteins to the cell membrane where they may then be further modified and inserted into the cell wall. Here they are covalently bonded to cell wall glucans (Pittet and Conzelmann, 2007). The function of some of these appears to be to deliver mRNA to P bodies, where they are stored and therefore help maintain the quiescent cell wall phenotype. Release of these at the appropriate time facilitates re-entry into growth.

Transition from quiescence to growth

The passage from quiescence to growth occurs in the lag phase of fermentation. It is in the interest of brewers to have as short a lag time as possible, and therefore identification of the factors that regulate its duration is of commercial importance. Just as entry into quiescence is via a programmed sequence of cellular events, so is exit. The process is distinct from growth, as measured in terms of increase in cellular biomass, and proliferation, as manifested by budding. Exit from quiescence must be as tightly regulated as entry and must only occur when conditions are favourable for growth to occur. By inference, premature exit will cause cells to lose their resistant phenotypes and almost certainly result in cell death. Nevertheless, as described in the previous section, the evidence suggests that in the quiescent phase cells have adopted a phenotypic state in which they are poised to pass into the growth phase as soon as conditions allow. Daignan-Fornier and Sagot (2011) summarized the quiescent stage as a convergence between cellular adaption to cope with adversity and preparation for efficient resumption of proliferation.

Laporte *et al.* (2011) used re-ordering of the actin skeleton and disassembly of the proteasome storage granule as markers of passage into growth, and noted that addition of glucose alone was sufficient to trigger these changes. They concluded that the presence of the sugar was sufficient to initiate the early steps of passage from quiescence, and this occurred even when *de novo* protein synthesis was inhibited, and thus independent of growth and proliferation. Activation by glucose required metabolism at least as far as pyruvate but did not require the formation of ATP. This adds weight to the suggestion that there is a two-way interplay between cellular signalling and regulatory systems and the intracellular concentrations of relatively small molecular weight metabolic intermediates. These metabolic activities are not simply required for energy generation (McKnight, 2010).

Cellular signalling systems

Proper regulation of the phenotype requires cells to possess a method of interrogating the external environment and responding in an appropriate manner. This is accomplished by the use of arrays of sensors capable of detecting the nutritional and stress status of the environment, coupled to pathways through which signals are transmitted to targets that adjust metabolism to elicit an appropriate response. The relative simplicity of using *S. cerevisiae* as a model cell has made it a favourite for studying this subject and a great deal is now known, albeit much less in the case of brewing strains. Conrad *et al.* (2014) have published an excellent review summarizing current knowledge.

Several signalling systems are recognized and

they share in common the ability of cells to sense the concentrations of relatively simple nutrients, and to generate multiple responses. The latter is obvious since, for example, starvation of an essential nutrient must not only switch off pathways necessary for growth and proliferation but must also re-engineer metabolism so that the cell acquires a resistant phenotype. Conversely, when conditions become growth-promoting these changes must be reversible. Responses to external cues must be sufficiently rapid to allow timely metabolic shifts and this suggests that many of the changes are likely to involve post-translational modifications. Different levels of control are exquisitely poised to elicit an appropriate response.

Signalling pathways identified in yeast include those for glucose repression and concomitant fermentative metabolism, nitrogen catabolite repression, general amino acid control, phosphate regulation, and regulatory responses triggered by sulfate, metal ions, and some vitamins. Signalling pathways may be implicated in the regulation of uptake of a single or small related class of nutrients; alternatively, general nutrient signalling systems are responsive to the concentrations of multiple classes of nutrients that elicit more global shifts, such as control of biomass formation, progression through the cell cycle, acquisition of stress tolerance, accumulation or mobilization of storage reserves, and apoptosis. Conrad *et al.* (2014) provide the illuminating example that starvation of both zinc and iron will individually trigger the induction of their respective high-affinity uptake systems but with no cross-talk. However, starvation of both ions also elicits a common general response in which growth arrests and there is an increase in stress tolerance. Receptors that are specific to a single or small group of molecules and cause induction of separate specific uptake systems may have themselves, in an earlier evolutionary phase, functioned as transporters, although they have now lost this function (Ozcan *et al.*, 1996).

In order to exert their effects, it is suggested that the triggering nutrients require some degree of metabolic processing. This is consistent with the observation that, for example, the ability of exogenous glucose to cause transition from quiescence requires metabolism at least to pyruvate (McKnight, 2010). The general responses require a target capable of responding to multiple nutrients. Responses may be long or short term and involve different mechanisms. This would be expected to allow for both rapid adjustment of the carbon flux through specific pathways to allow cells to make selective advantage of sudden changes in the environment, and slower global changes such as passage into and out of quiescence. The longer-term responses appear to be via regulation of the genotype, notably by nutrients of ribosomal protein gene expression (Griffioen *et al.*, 1996). The more rapid responses are mediated by cascades of phosphorylation, involving reactions between target enzymes and kinases and phosphatases with regulatory functions. It has been reported that, of 204 enzymes involved in central carbon and amino-acid metabolism, some 35 were observed to change their degree of phosphorylation under five different growth conditions (Oliveira *et al.*, 2012; Oliveira and Sauer, 2012). Signals are transmitted in reactions in which, for example, starvation of several individual nutrients interact reversibly with a common target kinase, and in response the activities of multiple pathways are modulated by the targets of the kinase. In *S. cerevisiae*, some 120 kinases and 40 phosphatases have been identified (Bodenmiller *et al.*, 2010). The effect of phosphorylation on the target protein can result in inactivation or activation by conformational changes, access to the catalytic site, tagging for degradation, intra-organelle translocation, and facilitating association or disassociation with multiprotein complexes (Oliveira *et al.*, 2012).

Assimilation and metabolism of sugars

In order to regulate cell size, a system is required for coordinating growth in terms of biomass formation and the processes of cellular proliferation. Implicit in this is a need to be able to sense cell size and the availability of nutrients (Dungrawala *et al.*, 2012). In yeast the critical point in the cell cycle is START, which once passed commits cells to progress irreversibly from G_1 to S phase. Cells must achieve a critical size before they pass through START; nevertheless, providing conditions are conducive, biomass formation is a continuous process since blocking growth before START restricts proliferation, whereas disruption of *CDC* genes, which regulate cell division, leads to the formation of abnormally large cells (Johnston *et al.*, 1977).

Saccharomyces spp. yeasts exhibit high rates of

glucose uptake and glycolysis coupled to fermentative metabolism even under aerobic conditions, the so-called Crabtree effect (de Deken, 1966). A similar situation occurs in cancer cells although in this case it is termed the Warburg effect (Diaz-Ruiz *et al.*, 2011). These similarities, coupled with the amenability of yeast cells as a model system, have provided the impetus for a huge body of work to elucidate the genetic and biochemical bases of the observations. Fermentative metabolism, which restricts energy generation to substrate-level phosphorylation, is energetically inefficient; however, it probably provides yeast with a selective advantage compared with other cells dependent on oxidative growth since glycolytic rates in Crabtree-positive cells are very high. Under these conditions, yeast cells are able to satisfy their requirements for energy and carbon whilst rapidly depleting carbon nutrients to the detriment of potential competitors.

Wort presents yeast with a complex mixture of assimilable sugars. Maltose is typically the most abundant, with smaller concentrations of others including glucose, fructose, sucrose, and the polymers of maltose, maltotriose, and maltotetrose. Uptake of sugars during fermentation is an ordered process in which the bulk sugar maltose is not used until supplies of sucrose, fructose, and glucose are exhausted. This is explained in that glucose (and the related fructose) is the preferred carbon source for *S. cerevisiae*, and its presence triggers wide-ranging metabolic shifts in which respiratory functions are switched off, as are the uptake and metabolism of other less-preferred carbon sources. This response is independent of and overrides the presence of oxygen (Broach, 2012). The effect has practical ramifications for modern commercial brewing. The use of high proportions of glucose adjuncts will delay maltose uptake, as will prolonged collection times for high-capacity vessels and in continuous fermentation systems.

In terms of signal transduction, the specific effect of glucose acts via a target protein, one of a family of kinases termed snf1/AMPK (sucrose non-fermenting 1/AMP kinase). Predictably, the participation of AMP is indicative that these kinases are involved in sensing energy status (Hardie, 2011). SnfA is a kinase that sits within a complex signalling pathway responsive to changing glucose concentrations, in which metabolism is directed towards eliciting responses associated with repression or derepression as well as inhibiting assimilation of other less preferred carbon sources. Kinase activity allows activation and deactivation responses that are exerted at the level of transcription, such that activation of SnfA by glucose exhaustion causes de-repression of genes required for the assimilation of non-glucose carbon sources as well as activation of pathways for respiration and accumulation of glycogen via gluconeogenesis. It is part of a trimeric protein complex, termed SNF1, which contains the kinase, Snf1, a regulatory subunit Snf4, and a third subunit coded for by three genes, *GAL83*, *SIP1* and *SIP2*. The Snf4 moiety serves to maintain Snf1 activity by preventing autoinhibition. In yeast, glucose exhaustion results in an increase in ADP concentration and the latter binds to Snf4 and the resulting conformational change maintains Snf1 activity. The third subunit regulates the intracellular location of Snf1 activity. Where glucose concentrations are high, all activity is located in the cytosol. However, where glucose is limiting different subcomplexes adopt specific locations; Gal83 and Snf1 Gal83, the β-subunit of SNF1 kinase promotes transfer of the latter to the nucleus. Those associated with Sip1 and Sip2 transfer to the vacuolar membrane and cytosol, respectively. In this way, the various cellular responses associated with the glucose trigger can be mediated (for a fuller explanation see Conrad *et al.*, 2014).

Rapid maltose uptake by brewing yeast is essential for efficient conversion of extract to ethanol. Brewing strains possess a multiplicity of transporters, each with differing specificity. Up to five *MAL* loci are present, each comprising three genes; *MALx1* and *MALx2* code for a transporter and hydrolysing enzymes, respectively, and *MALx3* produces a promoter essential for efficient transcription of the other two (Charron *et al.*, 1989). The Mal system appears relatively specific for maltose and turanose. *MPH2* and *MPH3* (maltose permease homologue) may also be present and these have a wider specificity transporting maltose, maltotriose, α-methylglucoside, and turanose. In some strains, an alternative transporter to *MALx1* is present, coded for by *AGT1* (α-glucosidase transporter), which, in addition to the sugars named already, will also transport trehalose and sucrose (Salema-Oom *et al.*, 2005). Another gene, *MTT1* (*MTY1*), apparently produces a transporter with a higher affinity for maltotriose than maltose.

Efficient utilization of the maltotriose in late fermentation is obviously an important consideration in commercial brewing since in many malt worts it is the third most abundant fermentable sugar present. Ale strains apparently utilize maltotriose less well as compared with lager types. Uptake of both maltose and maltotriose are rate-limiting in terms of fermentation rates, with maltotriose being assimilated more slowly than maltose. Vidgren *et al.* (2005) investigated maltose uptake by both lager and ale strains. They concluded that lager strains predominantly used transporters coded by the *MALx1* genes whereas in ale strains those produced by *AGT1* genes were more important. Once in the cell, maltose and maltotriose are hydrolysed to α-D-glucose which can then enter glycolysis. Vidgren *et al.* (2010), reported that lager strains were more efficient at transporting maltose at low temperatures compared with ale strains, and that this correlated with differential temperature dependence of the Mtt1 transporters. The authors postulate that this may explain the different abilities of the two groups of brewing strains to ferment at different temperatures.

Surprisingly, there are few reports in the literature as to the potential of maltose and other sugars to produce the same repressive effects as glucose. In an early report (Gancedo, 1992), repression is reported for mannose and fructose and to a lesser extent galactose and maltose. The differential effect was reported as either a result of all sugars exerting the same effects on target genes or glucose influencing a greater number of target genes compared with the less effective sugars such as maltose. The report included a description of mutant studies, which showed the presence in *MAL* genes of a binding site for the regulatory protein, Mig1, known to be implicated in glucose repression. Of course, in the case of brewing fermentation using typical all-malt worts, exhaustion of oxygen would normally occur before glucose, which would preclude development of respiratory capacity. However, little information seems to be available describing the regulatory effects of sugars in mid-fermentation under anaerobic conditions. Some unexpected effects of sudden exposure of yeast to high maltose concentrations have been reported. In the phenomenon termed 'maltose-accelerated death' (Postma *et al.*, 1990) it was reported that in chemostat glucose-limited aerobic cultures, sudden exposure to maltose resulted in extensive cell death and lysis. In a similar study, Jansen *et al.* (2004) confirmed the effect of maltose but noted that with time variants that had lost the hypersensitivity and had an increased affinity for maltose could be selected for. Presumably, both of these effects would be of relevance in terms of attempts to isolate strains with improved properties for very-high-gravity brewing. In addition, it may be a point to consider where sugar feeding regimes are used to alleviate osmotic stress in high-gravity brewing.

Global responses to sugars involve two signalling systems in which glucose stimulates fermentation and cell proliferation and represses responses associated with stress resistance and reserve carbohydrate accumulation. Responses are mediated by a cyclic AMP protein kinase pathway (cAMP-PKA). The same response is also elicited by the presence or absence of other essential nutrients (Conrad *et al.*, 2014). Activation of the cAMP-PKA pathway requires glucose to invoke two responses, one extracellular and the other intracellular, both involving G-proteins. The extracellular system senses glucose concentration outside the cell. The intracellular system requires metabolism of glucose via glycolysis in which an intermediate, possibly at the level of phosphofructokinase, is implicated. The consequence of both signalling systems is stimulation of a guanine nucleotide exchange factor, Cdc25, which in turn activates the GTPases, Ras1 and/or Ras2. The Ras enzymes produce cyclic AMP and this activates protein kinase A (PKA). The latter has multiple sites for activation and deactivation, which together allow transmission of the nutrient signal to be coupled to regulation of cellular processes such as acquisition of stress tolerance, growth, and proliferation (Thevelein and de Winde, 1999). The Ras proteins are widely conserved in many cell types and their absence in yeast represents a lethal mutation, which is reversed in other mutants where the cAMP dependence of PKA activation is missing. Interestingly, there is another layer of regulation depending on the intracellular localization of the cAMP-PKA system. Depending on the carbon source and environmental conditions, this may be primarily at the plasma membrane but also internal membranes in the mitochondria, nucleus and endoplasmic reticulum.

Yeast growth and nitrogen assimilation during fermentation

Although glucose can apparently initiate the shift from quiescence, a complete nutrient medium is required for growth proper to begin. In addition to sources of carbon, the nitrogen component of wort has great importance in terms of influencing both yeast growth extent and the development of yeast-derived flavour compounds. The regulation of nitrogen assimilation and utilization is very complex, as would be expected for nutrients that form the building blocks of proteins, notably enzymes, molecules with catalytic activities. Several signalling pathways appear to be involved, which work in parallel and in concert with each other in order to regulate nitrogen assimilation.

Uptake and utilization of wort amino nitrogen during brewery fermentation is an ordered process and by convention four classes (A–D) of amino acids are recognized based on ease and order of assimilation. Amino acids used first include glutamate, glutamine, aspartate, lysine and serine (Class A), followed by isoleucine, valine, leucine and methionine (Class B). Although ammonia is a preferred nitrogen source for *Saccharomyces* yeasts, it is not assimilated until mid to late fermentation (Class C), together with alanine, glycine, tryptophan, tyrosine and phenylalanine. The imino acid proline is the sole member of Class D (Pierce 1987). More recent studies confirm these findings, although with some strain-specific variability (Gibson *et al.*, 2009). The timing of uptake of amino acids has importance for beer flavour since they are precursors in the synthesis of higher alcohols and vicinal diketones (Boulton and Quain, 2001). Proline, in combination with haze-forming polyphenols, is of importance to beer colloidal stability and because its catabolism requires molecular oxygen it is held that it is not assimilated in brewery fermentation to any great degree. This is not entirely the case as demonstrated in Gibson *et al.* (2009), perhaps suggesting that some strains might produce a more colloidally stable beer than others.

The observed patterns of amino acid assimilation are a consequence of the complexity and regulation of families of transporters of both high and low affinity with differing substrate specificity. Among these, at least 12 constitutive and four nitrogen-repressible transporters have been identified and, in addition, a general amino acid permease (Gap1) is present that shows wide specificity (Horák, 1997). Regulation of the transporter systems depends upon the concentrations and spectrum of amino acids present in the medium, and operates by repression of those less preferred by the presence of preferred sources of nitrogen. Gap1 appears to be active only under conditions of nitrogen starvation (Stanbrough and Magasanik, 1995). Regulation occurs at the level of translation (nitrogen catabolite repression) and by post-translational modification (nitrogen catabolite inactivation) as discussed subsequently.

In eukaryotes, including yeast, a nitrogen-responsive signal transduction mechanism, the TOR (so-called 'target of Rapamycin') system has been elucidated. Rapamycin is a bacterial antifungal agent whose ability to prevent growth in many cell types, including those of human tumours in a manner that appeared very similar to passage into G_0, prompted research into its mode of action. Rapamycin growth inhibition was shown to involve a set of kinases that regulate cellular growth in response to nutrient signals. The system sits at the heart of a signalling network and comprises in yeast two large protein complexes termed TOR1 and TOR2 only the first of which is Rapamycin sensitive (Loewith *et al.*, 2002). The TOR1 system can contain both TORC1 and TORC2 complexes, whereas TOR2 contains only TORC2 proteins. TORC1 is considered to regulate cellular growth, possibly in response to external nutrient signals, and the second to control the spatial aspects of cellular polarity associated with growth (Virgilio and Loewith, 2006). The mechanism by which a response to the concentrations of external nutrients operates is not yet fully understood. Conrad *et al.* (2014) suggest that instead the primary signal may be derived from sensing the intracellular nitrogen content of the vacuole. In the same review, the authors ascribe the functions of TORC1, based on the effects of its inactivation, as influencing protein synthesis, ribosome biogenesis, transcription, cell cycle, meiosis, nutrient uptake, and autophagy. The functions of TORC2 include actin cytoskeleton organization, endocytosis, lipid synthesis, and cell survival.

Bearing in mind these global effects, it is perhaps no surprise that there may be linkages between signal transduction via the RAS and TOR systems (Schmelzle *et al.*, 2004). The identification of cellular processes responsive to the TOR system have been probed via identifying the targets of phosphorylation (Huber *et al.*, 2009; Soulard *et al.*, 2010). Such studies have shown that the TOR system exerts its influence at both transcriptional and post-translational levels. The regulation of both sugar and nitrogen metabolism as well as other metabolic pathways has been implicated. With regard to regulation of glycolysis, the pivotal role of phosphofructokinase has long been recognized. Modulation of its activity via allosteric interactions by as many as 20 metabolites has been reported (Sols, 1981). These include ATP, AMP, citrate, and fructose 2,6-bisphosphate. It has now been demonstrated that subunits of the enzyme are phosphorylated in a TOR-dependent manner. Thevelein (2015) has suggested that yeast possess a receptor system used as a sensor of external glucose concentration, which in conjunction with an internal glucose sensor is used to regulate the RAS system. He suggests that fructose 2,6-bisphosphate is a potent activator of the RAS system and control of the concentration of this metabolite via interactions with RAS underpin the observed rapid glycolytic rates characteristic of Crabtree-positive yeast cells and, by inference, cancer cells.

Based on TOR-dependent phosphorylation, several other putative roles for metabolic regulation via this system in yeast have been proposed. The cellular processes involved include control of gluconeogenesis and reserve metabolism via glycogen and trehalose (the latter also presumably being part of a stress response), control of adenylate energy charge, lipid accumulation including sterols, amino acid biosynthesis, folate metabolism, and purine and pyrimidine synthesis (summarized in Table 1.1). Of course, all of these processes are of critical importance to the performance of yeast in brewing fermentation.

In summary, the TOR system is inhibited by conditions that preclude growth. These include starvation and the application of stresses. Conversely, under growth-permitting conditions the TOR system is activated and via its multitude actions mediates passage from quiescence to biomass formation.

The subcellular localization of the TOR system in yeast has importance. The evidence suggests that it is located on the external surface of the tonoplast, the vacuolar membrane (De Viriglio and Loewith, 2006). This is perhaps predictable since this organelle occupies a crossroads, acting as the site for turnover and breakdown of many macromolecules, and storage of their component parts ready for re-use when required (Klionsky *et al.*, 1990). For these reasons, the vacuoles contain high concentrations of many metabolites, including phosphate and amino acids. Yeast vacuoles are highly plastic and the size and number undergo large changes under different physiological conditions. Under starvation or stress conditions, the total vacuosome

Table 1.1 Targets for TOR-mediated phosphorylation (summarized from Loewith, 2011)

Cellular process	Target
Glycolysis	Phosphofructokinase (both Pfk1 and Pfk2 subunits)
Gluconeogenesis	Fructose 1,6-bisphosphate aldolase
Amino acid, purine, and pyrimidine metabolism	5-Phosphoribosyl-1(α)-pyrophosphate synthetase Glutamate dehydrogenase Ser33 (serine and glycine synthesis) Met2, Met12 (methionine synthesis) Hom3 (methionine and threonine synthesis) Lys12 (lysine synthesis)
Energy transduction	Adenine monophosphate deaminase
Nucleic acid synthesis	Uridine kinase
Vitamin synthesis	*FOL1* gene (dihydropteroate synthetase, dihydro-6-hydroxymethylpterin pyrophosphokinase, dihydroneopterin aldolase)
Lipid metabolism	Tgl1 (hydrolysis and mobilization of steryl esters from lipid storage bodies) Tgi5 (regulation of anabolism and catabolism of triacylglycerols)
Reserve carbohydrates	Gph1 (glycogen phosphorylase)
Stress response	Tps3 and Tsl1 (regulatory subunits of trehalose 6-phosphate and trehalose phosphatase complex)

is increased; for example in, very-high-gravity fermentation (Pratt *et al.*, 2012). It is suggested that the linkage between the vacuolar membranes, intracellular protein trafficking membrane systems such as the endoplasmic reticulum and Golgi body, and the plasma membrane allow the TOR system to respond to metabolic cues such as amino acid concentrations in the vacuoles and external medium and manipulate metabolism according to need (Rohde *et al.*, 2008).

TORC2 has received less attention compared with TORC2, largely because its insensitivity to rapamycin has meant more conventional but less tractable mutant studies have had to be used to elucidate its physiological roles. It appears that, as opposed to the regulation of growth by TORC1 in response to nutrient availability, TORC2 is involved in the control of cell division via the cell cycle-dependent polarization of the actin cytoskeleton (Schmidt *et al.*, 1996). Other functions have also been ascribed to TORC2, including regulation of sphingolipid synthesis and the related ceramides (Powers *et al.*, 2010). Sphingolipids are membrane-associated and appear to have several cellular roles, which include extracellular signal receptors, intracellular protein trafficking, and cell cycle regulation. In addition, they appear involved in the formation of eisomes, protein bodies found close to the plasma membrane and implicated in the early stages of endocytosis (Cowart and Obeid, 2008). A key role of TORC2 is to respond to external stresses that could compromise membrane structure and function. A TORC2-dependent protein kinase, Ypk1 has been shown to control membrane lipid content in response to stresses such as heat. Ypk1 activity up-regulates the synthesis of sphingolipid precursors by alleviating negative control of L-serine:palmitoyl-CoA transferase (Muir *et al.*, 2014).

Nitrogen catabolite repression acts, as the name suggests, at the transcription level. In the presence of preferred nitrogen sources such as ammonia, glutamate, and glutamine, the genes required for the utilization of less preferred nitrogen sources are repressed. The mechanism by which the effects are exerted involve transcription activators, whose activities are regulated depending on their intracellular location; thus, repression of genes required for the assimilation of less preferred nitrogen sources is inhibited only if the transcription activators are moved from the cytoplasm to the nucleus. One such key transcription activator, Gln3, is retained in the cytoplasm under nitrogen-rich conditions as a result of it binding to a protein designated Ure2 (Blinder *et al.*, 1996). Under conditions of nitrogen starvation, the binding is reversed and Gln3, together with another transcription activator Gat1, migrate to the nucleus and activate transcription of the genes sensitive to nitrogen catabolite repression. Since the relocation to the nucleus occurs after treatment with rapamycin, the involvement of TORC1 is suggested, acting as a mediator of the phosphorylation state of Gln3, which in turn controls complexing with Ure2. This appears to be so; however, the presence of other non-rapamycin-sensitive routes for the control of Gln3/Ure1 complexing has led to the suggestion that there is probably another non-TORC1 parallel signalling pathway involved in nitrogen catabolite repression (Rai *et al.*, 2013). Yeast cells possess a retrograde signalling pathway so-called because it transmits signals from mitochondria to the nucleus, the reverse of the normal direction (Butow and Avadhani, 2004). The pathway controls adaptions in both carbohydrate and nitrogen metabolism in response to mitochondrial dysfunction in which the TOR system is implicated (see 'Roles of mitochondria', below, for more discussion).

Under conditions of nitrogen starvation (and other stress conditions), the general amino acid control pathway is up-regulated (Straschke *et al.*, 2010). This is yet another example of transcriptional control in response to extracellular cues, in this case, inhibition of amino acid synthetic pathways as means of conserving available resources. The system operates via a transcription initiation factor, elF2 whose active form comprises a ternary complex of elF2, GTP and methionyl initiator tRNA. Under conditions of amino acid starvation, levels of tRNA increase and via a phosphorylation event this precipitates inhibition of the formation of the active GTP-bound form of elF2. Reduction in the concentration of activated elF2 reduces the rate of translation of messenger RNA, decreasing the rate of amino acid synthesis. The pathway is linked to the TORC1-driven intracellular amino acid sensing pathway and glucose sensing via Snf1. In the latter case, the genes coding for pathways involved in ribosome formation are down-regulated (Simpson and Ashe, 2012).

Fermentation growth medium-induced pathway

Yeast growth in brewery fermentation proceeds under conditions of repression as a result of the high sugar concentration. The repressive effects appear to be mediated principally via the activity of targets of PKA. In the exponential phase of growth, as discussed previously, the system is down-regulated and the phenotype is one of low-stress resistance (thinner cell wall, low reserves of trehalose and glycogen, down-regulation of stress-related genes). Under these conditions, exhaustion of an essential nutrient, most likely nitrogen in the case of beer fermentations, causes immediate transition into G_0. The targets of PKA are up-regulated and the aforementioned phenotypic changes are reversed. The fact that the same relationship between PKA and growth on non-repressing carbon sources is absent has led to the suggestion that a specific fermentation growth medium signalling pathway (FMG) exists, which is entered whenever there is a complete growth medium containing repressing concentrations of glucose, or a similar carbon source. Thus, there is a common pathway that can be activated by one of a number of essential nutrients. Based on the fact that mutants lacking the regulatory subunit of PKA show the same response, it is considered independent of this kinase but working in parallel with the Ras-cyclic AMP pathway (Conrad *et al.*, 2014). Activation of the FMG pathway requires activation of a cyclic AMP-dependent protein kinase (cAPK). It has been demonstrated that activation of FMG by provision of nitrogen to a glucose-containing medium requires the presence of another protein kinase, termed Sch9, which in turn regulates the activity of cAPK (Crauwells *et al.*, 1997). Sch9 is itself a target for phosphorylation by TORC1 (Urban *et al.*, 2007); in addition, it appears to act directly as an inhibitor of PKA (Zhang *et al.*, 2011), indicating at least dual functions.

Response of yeast to oxygen

The practice of serial repitching of brewing yeast is responsible for the requirement to add oxygen to wort. Oxygen is considered to be required for synthesis of sterols and unsaturated fatty acids, both essential components of the plasma membrane (Boulton and Quain, 2001). These lipids are synthesized in the aerobic phase of fermentation and become diluted among daughter cells in subsequent growth. The quantity of oxygen supplied is one of the factors that regulate growth extent and this, in conjunction with the pitching rate and attemperation, are the major variables used to control brewery fermentations. Oxygen requirements for lipid synthesis are relatively modest, theoretically much more for sterol synthesis compared to unsaturated fatty acids, and it is likely that more is added to wort than is actually required. This prompts the question as to what is the fate of the excess?

Oxygen can function as both substrate and signalling molecule. Its presence or absence has effects at the genome level, although perhaps less than might be imagined. Based on chemostat studies under conditions of glucose-limitation, Ter Linde *et al.* (1999) observed that, of 6171 open reading frames, 5738 showed detectable transcripts. Of these, based on more than three times higher transcription levels, 219 showed elevated expression under aerobic conditions and 140 under anaerobic conditions. A further subset of genes are up-regulated under hypoxic conditions where oxygen is limiting (Deckert *et al.*, 1995). The presence of these genes would be predicted in that it makes sense that cells would possess systems allowing efficient utilization of oxygen when supplies are restricted. It might be supposed that hypoxia is important for facultative anaerobes such as *S. cerevisiae*, since in the wild this condition would be expected to be more common than absolute anaerobiosis. Hypoxic genes include those coding for oxygen-requiring pathways, such as sterol synthesis and sterol uptake, and isoforms of aerobic enzymes, including those involved in respiration and mitochondrial ATP translocation. Two groups of genes are recognized; those with an aerobic counterpart are repressed at all but very low oxygen concentration ($< 0.5\ \mu M$), and a second subset are active under all oxygen concentrations but activity is stimulated under hypoxic conditions (Becerra *et al.*, 2002; Zitomer *et al.*, 1997; Poyton, 1999). There are further complications since aerobic metabolism has the potential to generate harmful reactive oxygen species and the cell must have mechanisms to deal with these (Morades-Ferreira and Costa, 2000) (for further discussion of oxidative stress in yeast, see Chapter 2).

The response to oxygen requires the existence of a sensing system and the evidence suggests that haem is implicated (Kwast *et al.*, 1999) although sterols have also been proposed (Davies and Rine, 2006), both based on the fact that oxygen is required for synthesis and therefore their intracellular concentrations would be expected to vary in relation to available oxygen levels. In the case of haem, the enzymes for the whole of the synthetic pathway are present under anaerobic conditions (Labbe-Bois and Labbe, 1990). The *ROXI* gene had been shown to produce a haem-induced repressor of yeast hypoxic genes, which exerts its effect by generating a product that binds to the promoter region of receptive genes (Deckert *et al.*, 1995). An additional haem-responsive factor, Hap1, activates the expression of genes involved in respiratory functions. As oxygen levels fall, there is a concomitant reduction in levels of Rox 1 and hypoxic genes are up-regulated. Other mechanisms exist and some of these appear to be related to sterol metabolism. An element, Upc2p, has been identified (Davis and Rine, 2006) that is involved in the regulation of around a third of anaerobically expressed genes. Levels of Upc2p are linked to sterol depletion and, by implication, available oxygen.

Yeast cells respond to exogenous sterol levels in a complex fashion, which is linked to oxygen availability. Under aerobic conditions, sterols are not assimilated, a process termed aerobic sterol exclusion (Rodriguez *et al.*, 1985). This apparently is a result of transcriptional inhibition of the sterol uptake system in which *ROX1* may be implicated (Rosenfeld and Beauvoit, 2003).

The response of anaerobically grown yeast to oxygen is complex, much more so than the simple relationship between oxygen, sterols, and unsaturated fatty acids would suggest (Snoek and Steensma, 2007). The requirement for oxygen for growth can be satisfied by supplementing growth media with sterols and unsaturated fatty acids; however, it has been observed that addition of oxygen at stationary phase to such a medium increases specific fermentation rate, shortens fermentation cycle time, and increases the viability of the crop (Rosenfeld *et al.*, 2003). The stimulation was linked to further sterol synthesis. Kwast *et al.* (2003) provided a description of putative roles of various classes of genes known to be induced under anaerobic conditions. Out of a total of 346 genes, 42 were related to the cell wall, 35 to cellular stress responses, 31 to carbohydrate metabolism, and 28 to the metabolism of lipids, fatty acids, and isoprenoids. In addition, several others coded for enzymes to which there was an aerobically induced isoform. In addition to the haem biosynthetic pathway, anaerobic yeast contain all the enzymes for sterol synthesis. Clearly, neither of these pathways can be active under such conditions, but it does suggest that the cell is primed in such a way that very rapid mobilization will occur should oxygen become available. This is in agreement with the observation that anaerobic yeast contains high levels of squalene, the last step in the non-oxygen requiring sterol biosynthetic pathway, and therefore the pool of this metabolite primes sterol formation should oxygen become available (Jahnke and Klein, 1983). If this is so, it follows that lipid synthesis could proceed in cropped pitching yeast, providing oxygen is supplied. This has been shown to be the case (Boulton *et al.*, 2000; Verbelen *et al.*, 2009). In the work described, oxygenation of concentrated pitching yeast slurries suspended in beer was carried out, care being taken to ensure that exposure to oxygen during handling was minimized. At a temperature of 20°C, both sterols and unsaturated fatty acids were synthesized, maximum concentrations being achieved after 6–8 h. The process also resulted in loss of glycogen. In the latter study, it was shown that trehalose levels increased and the yeast acquired resistance to oxidative stress, based on the up-regulation of genes know to be implicated in these processes. Acquisition of stress resistance occurred after 45–60 min.

On a mass balance basis, the quantity of oxygen used for sterol and unsaturated fatty acid synthesis is small. The conversion of a molecule of squalene to ergosterol requires 12 molecules of oxygen. In the case of unsaturated fatty acid synthesis, desaturation of one molecule of a fatty acid requires just a single molecule of oxygen. The total quantity of sterols in yeast never exceeds around 1% of the dry weight and under brewing conditions is much less (Parks and Casey, 1995). For growth under anaerobic conditions, no more than 5 mg/l is required (Aries and Kirsop, 1978). Unsaturated fatty acids are more abundant than sterols by approximately 5-fold (Rogers *et al.*, 1974). In addition to the requirement for oxygen, yeast cells decrease the ratio of saturated to unsaturated fatty acids as a means of maintaining membrane fluidity at lower temperature; however, not all strains have an equal facility for this (Torija *et al.*, 2003; Beltran *et al.*, 2008; Tronchoni *et al.*, 2012). Although performed largely with oenological yeast strains, these studies possibly suggest that this is implicated in the ability of lager strains to grow at lower temperatures than ale types.

Under aerobic de-repressing conditions, the majority of oxygen is used by yeast as the terminal electron acceptor in oxidative phosphorylation. Under repressing conditions, yeast has an affinity for oxygen in excess of that required for sterol and unsaturated fatty acid synthesis, which prompts the question as to its role. Rosenfeld and Beauvoit (2003) reported a cyanide-resistant oxygen uptake by anaerobic yeast but concluded that this was not due to a functional respiratory chain since it could be observed in both rho^+ (wild-type) and rho^- (petite mutant) cells grown under anaerobic conditions. The authors suggest that synthesis of haem (2.5 molecules of oxygen per molecule), sterol, and unsaturated fatty acids account for some of this. Other oxygen-requiring pathways include the syntheses of nicotinic acid from tryptophan and ubiquinone in a pathway whose early stages are common to that used for sterol formation. The Ferri-reductase system features a plasma membrane oxidase used for high-affinity uptake of ferric ions and a mitochondrial L-proline oxidase, which is repressed under anaerobiosis but induced by proline. Interestingly, in oenological fermentations proline is rapidly metabolized in response to provision of oxygen at the end of the growth phase, presumably when the glucose and nitrogen repression signals are absent (Salmon and Barre, 1998).

Roles of mitochondria

In anaerobically grown yeast mitochondria adopt an undeveloped form termed *promitochondria*. Under non-repressing aerobic conditions these can rapidly adopt a fully functional form (Plattner *et al.*, 1970). Visser *et al.* (1995) noted that under aerobic de-repressed conditions there were numerous small mitochondria, whereas under glucose repression they became much fewer in number but these were larger and branched. The total mitochondriome was similar in each case. Conversely, working with sake yeast, Kitagaki *et al.* (2013) developed procedures to monitor mitochondrial morphology throughout fermentation. They noted that initially the organelles had a filamentous tubular structure and as the fermentation progressed these fragmented and became smaller and non-elongated. Rosenfeld *et al.* (2004) isolated mitochondria from anaerobically grown, repressed yeast cells and reported that they could detect the presence of cyanide-sensitive and non-phosphorylating NADH-dependent oxygen consumption but no antimycin-A-dependent, NADH- or NADPH-dependent oxidase activities and thus they concluded that oxygen consumption by anaerobic mitochondria was negligible. Despite the apparent differences in some reports, the consensus is that the mitochondria are highly dynamic in shape and size and that processes of both fusion and division are of frequent occurrence (Jensen *et al.*, 2000).

The role of mitochondria in brewing yeast under brewing conditions is uncertain; however, it is known that petites produce unsatisfactory fermentation performance, notably slower rates, prolonged vicinal diketone (VDK) stand times, and altered flocculation characteristics (Ernandes *et al.*, 1993). The negative effects on VDK metabolism are perhaps unsurprising, since this organelle is the site of key parts of the ILV pathway in which branched chain amino acids leucine, isoleucine, and valine are synthesized (Kohlhaw, 2003). In modern brewing practice, where large batch sizes and very concentrated worts are commonly used, the increased frequency of occurrence of petites has been observed and ascribed to elevated stress levels associated with these processes (Jenkins *et al.*, 2009; Lawrence *et al.*, 2012, 2013). The quiescent daughter cells produced during the diauxic shift phase discussed earlier (see section 'Quiescent yeast cells') inherit only highly functional mitochondria during the asymmetrical division (Peraza-Hayes *et al.*, 2010; McFaline-Figueroa *et al.*, 2011). Under these conditions, where oxygen is present the quiescent cells adopt a respiratory metabolism and maintain their elevated levels of glycogen and trehalose. The non-quiescent cell fraction retains a repressed metabolism and continues to deplete carbohydrate reserves via glycolysis. This helps to explain the relative differences in longevity between the quiescent and non-quiescent phenotypes under non-growth-permitting conditions.

Interestingly, mitochondria seemingly have a role in sterol uptake and transport in anaerobically grown yeast. Reiner *et al.* (2006) used a range of yeast mutants that were unable to grow under anaerobic conditions. They observed that the largest group of mutants, which were not able to take up exogenous sterols, had disruptions of genes known to have mitochondrial functions. These roles for mitochondria are perhaps unsurprising, since this organelle is the site of several biosynthetic pathways. These include part of those for sterol synthesis, branched-chain amino acids, and the citric acid cycle (Shimizu *et al.*, 1973; Ryan and Kohlhaw, 1974; Jauniaux, 1978). Based on the inhibition of the ADP/ATP transporter by bongkrekic acid, Visser *et al.* (1994) proposed that the energy to fuel these pathways must be derived from substrate-level phosphorylation. In one of the few studies using a brewing yeast strain, Samp (2012) reported that in lager strains there was a link between mitochondrial function and SO_2 production. Respiratory deficient mutants showed reduced conversion rates of sulfate to sulfite compared with the wild type. Conditions that favoured cardiolipin synthesis, an important mitochondrial membrane lipid, also resulted in reduced sulfite.

Mitochondria play a role in apoptosis in yeast. This is the phenomenon of self-programmed suicide. Programmed cell death can occur in response to exposure to external stimuli, such as hydrogen peroxide or acetic acid, and via internal cues, such as cell ageing. In a unicellular organism, such as yeast, it may represent an altruistic cooperative phenomenon whereby unfit cells are eliminated from the population and/or subpopulations are selected that possess enhanced properties, such as improved resistance to external stresses. Several apoptotic pathways have been described and some suicidal processes are mediated by mitochondria. Regulation may involve major signalling pathways, such as Ras and TOR, which serve to link the process to inputs from ageing and nutritional status. A group of proteases termed caspases are involved in the signalling process, and the outward manifestations of mitochondrial involvement are release into the cytosol of cytochrome C and nucleases that migrate to the nucleus and once there disrupt DNA and reactive oxygen species (Peirera *et al.*, 2008). Of course, in the case of yeast subjected to the conditions encountered in brewing, the putative roles of mitochondria in apoptosis are less apparent, since these organelles never become fully developed. For example, Madeo *et al.* (1999) reported that oxygen radicals played an essential role in apoptosis, since conditions that resulted in their depletion, such as anaerobiosis, prevented apoptosis. Conversely, in another study (Aerts *et al.*, 2008) a mutant yeast strain with a much shorter chronological lifespan compared to the wild type was shown, when growing aerobically, to exhibit increased rates of apoptosis and to have dysfunctional mitochondria.

Cell wall plasticity

The yeast cell wall confers shape and has a protective function, providing resistance to mechanical and osmotic stresses. Apart from providing the location for receptors and targets for flocculation, the porosity of the outer protein layer limits entry of larger molecules, such as enzymes that might damage the more fragile plasma membrane (Klis *et al.*, 2002; Free, 2013). The wall must be capable of responding to the requirements of cellular growth and proliferation, and conversely adopt a more rigid protective form under non-growth-permitting conditions. It follows that systems must be available for sensing cell wall structure and making appropriate adjustments in response to signals from the central metabolic pathways. Based on whole-cell transcription studies, some 1200 genes have been implicated in cell wall-related functions, demonstrating how regulation in the structure of this organelle is integral to physiological responses of the cell to the environment (De Groot *et al.*, 2001). Comparisons of stationary phase and exponentially growing walls show many differences. The former are less permeable, show different protein profiles, and are more resistant to cell wall-degrading enzymes. Approximately a 6-fold increase in disulfide bridges in stationary phase cells has been observed (de Nobel *et al.*, 1990). In addition to these changes, anaerobiosis is accompanied by altered levels of transcription of several genes involved in the production of cell wall proteins (Klis *et al.*, 2006).

Remodelling of cell wall structure is accomplished through the action of the cell wall integrity signalling pathway (Levin, 2011). A response to environmental stresses is sensed via a set of cell-wall-located receptors via a G-protein termed Rho1 (Ras-homologous family of GTPases). The same protein is also involved in the regulation of cell wall synthetic reactions associated with progression through the cell cycle. It follows that this process requires sensitivity to the internal cues driving budding. Rho1 coordinates a multitude of functions related to cell wall structure and biogenesis. These include coordination of synthesis of β-glucan at selected sites on the cell wall, the associated endocytosis of the necessary building materials, and organization of the actin cytoskeleton. The cell response to the passage into quiescence, triggered by nutrient starvation and/or application of external stresses, is also mediated by the Rho1-directed signalling pathway (Levin, 2011). More recently, Rho1 has also been implicated in another signalling pathway in yeast involved in regulation of homeostasis of plasma membrane fluidity (Lockshon *et al.*, 2012).

Rho1 is one of a family of six related GTPases that are found in *S. cerevisiae*, located at the surface of the plasma membrane. Two of these are essential, Rho as discussed and Cdc42, the latter being vital for establishment of cell polarity and bud formation. Proper function of the Rho system is dependent on correct location and orientation in the plasma membrane and on the activity of guanosine nucleotide exchange factors and GTPase-activating proteins. The first of these transmit signals from cell-surface receptor proteins of the initiating stimulus (Radico and Heinisch, 2010). Somewhat fascinatingly, some of these sensors have been shown to possess rigid polypeptides that project into the periplasm and possibly the wall, where they may serve as linear nanosprings probing mechanical status (Dupres *et al.*, 2009). Phosphoinositides have been shown to have importance in membrane functionality and it appears that part of this activity involves activation of the Rho sensors (Odorizzi *et al.*, 2000).

Rho1 exerts its effects to the relevant parts of yeast metabolism via a MAP kinase cascade. An important component is the protein kinase, Pkc1, which is itself activated by GTP-bound Rho1 (Kamada *et al.*, 1996). It constitutes one of five MAPK signalling pathways that regulate a multitude of critical cellular functions, which apart from the cell wall integrity sensing pathway include mating response, sporulation, and the morphological changes associated with transitions to pseudohyphal habits associated with those strains capable of this response (Gustin *et al.*, 1998).

Flocculation

Flocculation, the non-sexual aggregation of cells to form clumps, is a key determinant of the suitability of brewing strains for use in brewing. The ability to flocculate in late fermentation assists with separating the yeast crop from green beer. Strains that do not show this property to the appropriate degree lead to high cell counts in processes down-stream of fermentation, with concomitant high loss rates and inefficiencies in primary filtration. Conversely, highly flocculant strains may cause premature separation and thereby increase the risk of stuck fermentation and poor VDK removal.

There is a relationship between sugar availability and the onset of flocculation. This makes perfect sense from the standpoint of yeast. When sugar is plentiful in early fermentation, flocculation is inhibited and this ensures good cell dispersion and equal access to carbon. When sugar becomes exhausted, flocculation can occur and from the standpoint of survival it is advantageous for cells to form flocs, which provide a sheltered environment for those in the interior.

Flocculation is a cell-surface phenomenon and therefore its expression is linked to cell wall structure. It has been established that flocculation involves interactions between lectin-like proteins (flocculins) and cell surface mannans. The former are present on all cells, whereas the flocculins only occur in those strains possessing the appropriate genotype. Calcium ions are required for flocculation to occur and it is thought that these mediate binding by causing an essential conformation change on the lectin structure. Abundant wort sugars such as glucose and maltose preferentially bind to lectins and thereby inhibit the process. In addition to mannan–lectin interactions, floc stabilization may occur via hydrophobic interactions and hydrogen bonding (Soares, 2011). Two main phenotypes are recognized based on the patterns of inhibition by sugars. NewFlo types, which include many brewing strains, do not flocculate in the presence of mannose, glucose, maltose, or sucrose. In Flo1 types, flocculation is inhibited by mannose. A rarer third type, M1, occurs in some strongly top-cropping ale strains and appears to operate via direct protein–protein interactions and does not occur in the absence of ethanol.

Flocculation is conferred by a number of *FLO* genes: *FLO1* and its alleles, *FLO2* and *FLO4*, together with *FLO5*, *FLO9*, and *FLO10*, which show high homology to *FLO1*. Collectively, these are responsible for the Flo1 phenotype. Lager strains alone possess another gene, *lg-FLO1*, which confers the NewFlo phenotype and codes for a flocculin that binds a broader range of sugars than Flo1 types (Verstrepen *et al.*, 2003). Other genes are involved in the regulation of flocculation. The product of *FLO8* is a transcriptional activator of *FLO1* and another gene *FLO11*, which is implicated in a stress response in which in some strains' growth becomes pseudohyphal (Bayly *et al.*, 2005). Regulation of *FLO11* is particularly complex and both transcriptional and post-translational mechanisms are recognized. Several signalling pathways are involved in its expression. These include a mitogen-activated protein kinase pathway, cAMP protein kinase A pathway, and those involved in quorum sensing and nutritional status (Verstrepen and Klis, 2005). In NewFlo strains, flocculation is triggered when exponential growth ceases. Aside from exhaustion of sugars and the consequent availability of Flo-mediated lectins, regulation involves metabolic activities in which nutrient-sensing pathways via intracellular kinases control regulation of *FLO* genes.

Flocculation requires cell-to-cell contact to occur and it follows that mechanical agitation is important. Other factors, apart from the presence of sugars, are the concentrations and range of cations, pH, temperature, oxygen, and ethanol. Some of these effects may be purely physical; for example, lower fermentation temperature will decrease natural agitation rates and therefore lessen the probability of cell-to-cell collision. Several cations (Ba^{2+}, Sr^{2+} and Pb^{2+}) inhibit flocculation, possibly by competing with calcium ions. Others, especially Mn^{2+} and Mg^{2+}, promote flocculation. Optimum pH for flocculation is in the range pH 3.0–5.0 and is inhibited at wider ranging values. This explains the observation that acid washing of pitching yeast is accompanied by slurries adopting a more fluid, less viscous form. The presence of ethanol promotes flocculation by an unknown mechanism but possibly by its influence on cell hydrophobicity (Soares, 2011).

Flocculation impinges in two ways on the practice of serial repitching. It has been observed that older cells are more flocculent compared with those of a younger generational age, presumably a consequence of increased size of the latter and possibly linked to age-related changes in the cell surface, maybe the presence of greater concentrations of chitin (Powell, 2003). Changes in flocculation have also been linked to genetic instability. Some strains appear very stable in this regard, in others an abrupt shift from flocculant to non-flocculant has been observed in the course of serial repitching. Thus, Powell and Diacetis (2007), working with two brewing strains, observed no changes in flocculence behaviour over several generations. Conversely, Sato *et al.* (2001) reported that, for a bottom-fermenting strain, loss of flocculence was the most commonly observed change and this was related to loss of the lg-*FLO1* gene.

Reserve metabolism

Yeast, in common with other cells, must accumulate reserve materials in times of plenty in order to increase the chances of survival through future periods of starvation. In brewing yeast cells, the presence of storage granules of lipids, phosphate (volutin), and glycogen can be observed. In addition, the vacuolar space, the site for amino acid storage, increases significantly in volume during the latter phases of fermentation as a consequence of protein turnover associated with starvation.

Volutin granules comprise linear polyphosphate chains associated with a number of basic proteins (Jacobson *et al.*, 1982). Phosphate regulation in yeast cells is complex, responding to the availability of exogenous phosphate but also linked to the control of many cellular processes, including cell cycle control via the actions of Pho85, a cyclin-dependent kinase (Conrad *et al.*, 2014). The latter report that downstream targets of Pho85 include expression of *PHO* genes, responsible for phosphate sensing and scavenging; Gen4 transcription factor, implicated in regulation of amino acid biosynthetic pathways; Cln3, a cyclin involved in cell cycle control; and Rim15, a protein kinase involved in nutrient sensing.

Lipid granules contain stores of both triacylglycerols and sterol esters, as well as enzyme systems involved in lipid accumulation and its breakdown and mobilization (Rajakumari *et al.*, 2006). Alcohol acetyl transferase (Atf1p), the terminal enzyme of the acetate-forming pathway in yeast, is also localized in lipid particles (Verstrepen *et al.*, 2004). This suggests a linkage between ester formation and lipid metabolism, most likely a mechanism for regulation of the supply of acetyl-CoA depending on the availability of oxygen. Three sterol hydrolases, Tg1p, Yeh1p, and Yeh2p, have been detected in *S. cerevisiae*. These are implicated in liberation of free sterols from the esters, which are subsequently transferred to the endoplasmic reticulum for conversion to ergosterol, and from there incorporated into the plasma membrane (Wagner *et al.*, 2009).

The major carbohydrate reserve materials are glycogen and trehalose, although as discussed later the latter also has important roles as a stress protectant. In brewing yeast cropped from fermenter, glycogen may account for up to 40% of the cell dry weight and its presence can be easily detected by the brownish-red coloration formed when cells are stained with iodine. Mobilization of glycogen in the aerobic phase of fermentation provides energy and possibly some carbon for sterol synthesis (Quain and Tubb, 1982). Glycogen accumulates in yeast when one of several nutrients become limiting. These include carbon, sulfur, nitrogen, and phosphorus (Lillie and Pringle, 1980). In addition, glycogen accumulation is also part of the general stress response. An important part of storage of brewing yeast in the intervals between cropping and repitching is to ensure that glycogen reserves are preserved. Glycogen is synthesized from UDP-glucose, itself derived from glucose-1-phosphate via the action of UDPG pyrophosphorylase. Formation of the glycogen polymer involves the concerted activity of glycogenin, a self-glycosylating initiator protein, glycogen synthase, and a branching enzyme. UDP-glucose is also a substrate for the synthesis of trehalose and cell wall β-glucans. A signalling pathway controls disbursement of UDP-glucose between these three biosyntheses (Grose *et al.*, 2007). A yeast kinase, Psk2p, has been shown to phosphorylate another protein, Ugp1p. This event regulates intracellular localization, either at the plasma membrane (phosphorylated) or the cytoplasm (dephosphorylated), and by this means supply of UDP-glucose for the syntheses of cell wall or glycogen, respectively (Wilson *et al.*, 2010). Transcription of glycogen synthase is up-regulated at the end of the exponential phase of growth under the control of cAMP-dependent protein kinase pathway. Post-translational controls also occur via an inactivating phosphorylation. This can be alleviated by the presence of glucose-6-phosphate. Formation of α-1–4 linkages and α-1–6 branch points are formed by the concerted action of glycogen synthase and the branching enzyme (1,4-α-glucan branching enzyme; EC 2.4.1.18). Mobilization of glycogen is principally via glycogen phosphorylase and a debranching enzyme [4-α-glucanotransferase (EC 2.4.1.25) and amylo-α-glucosidase (EC 3.2.1.33)].

Regulation of glycogen accumulation or mobilization is exceedingly complex. Several overlapping signalling pathways appear to be involved. Deactivation or activation of the PKA pathway is accompanied by concomitant reduced or increased glycogen accumulation, respectively. The transcription factors Msn2p and Msn4p, which are implicated in the general stress response, are involved. Phosphorylation of these controls cellular location between the nucleus, where transcription of sensitive genes may occur, or the cytoplasm. In addition, the SNF1 pathway, associated with glucose repression, the Pho85p kinase linked to phosphate metabolism and cell cycle control, and the TOR system all appear to feed into the control of glycogen accumulation or dissimilation (see Wilson *et al.*, 2010, for an excellent summary).

Part of the response to nutrient starvation is one, termed autophagy, in which cellular structures are re-ordered to improve the chances of survival. It involves the capture of selected cytoplasmic components in vesicles, which eventually fuse with the vacuole where reprocessing occurs (Cebollero and Reggiori, 2009). In this way, cellular components are made available for recycling. The process is used to recycle damaged or excess organelles and under normal growth conditions, the transfer of cytosolic components to the vacuole. The latter process is termed the cytosol-to-vacuole targeting (CVT) pathway. It involves enclosure of the selected cytosolic components in a newly synthesized double-membrane-bound structure called the autophagosome. This fuses with the tonoplast, eventually leading to the formation of an intravacuolar, single-membrane-bound vesicle, the autophagic body. The membrane of this is digested and the contents degraded. The process is completed when selected breakdown products are targeted and returned to the cytosol for recycling.

As might be expected, since nutrient limitation acts as a primary trigger in both, there is a link between autophagy and glycogen metabolism. They also share in common a spatial aspect. It is suggested that there are both cytosolic and vacuolar glycogen pools and that transfer of some of the cytosolic pool to the vacuole is via the autophagy system. This application of cellular compartmentalization is a strategy adopted by the cell to prevent glycogen breakdown at an inappropriate time by ensuring that it does not come into contact with the cytosolic glycogen phosphorylase (Wilson *et al.*, 2010).

Trehalose is a disaccharide comprising two molecules of glucose linked through a α-1–1 bond. The role of trehalose as a stress protectant, as well as reserve carbohydrate, is supported by the fact that elevated levels favour yeast survival in very-high-gravity fermentations and enhanced trehalose and improved ethanol tolerance have been shown to correlate in engineered bioethanol strains (Tao *et al.*, 2012). Similar observations have been made in relation to active dried brewing yeast (Zheng *et al.*, 2013). Trehalose exerts its stress-protecting role via its ability to stabilize membranes (Crowe *et al.*, 1984).

Like glycogen, the trehalose content of cells is dependent on the balance between synthesis and degradation. Synthesis is via a UDP-glucose-dependent trehalose synthase protein complex coded by the genes *TPS1* (trehalose 6-phosphate synthase) and *TPS2* (trehalose 6-phosphate phosphatase). Two further genes, *TPS3* and *TSL1*, code for regulatory proteins. Trehalose degradation is via one of two trehalases, a neutral trehalase (*NTH1*) or an acid trehalase (*ATH1*), their names reflecting their pH optima for activity. A review of trehalose metabolism may be found in François and Parrou (2001). *NTH1* appears to be used for mobilization of intracellular trehalose reserves, whereas *ATH1* may be used for utilization of exogenous trehalose as a carbon source (Parrou *et al.*, 2005).

In addition to these roles, trehalose functions as a regulator of glycolysis via the intermediary of trehalose-6-phosphate and the ability of this intermediate to act as an allosteric inhibitor of hexokinase. Thevelein and Hohmann (1995) suggest that trehalose-6-phosphate synthase occupies a key position at the beginning of multiple glucose signalling pathways.

Yeast responses to stress

A common thread that weaves throughout any discussion of brewing yeast physiology are the numerous stresses to which cells are subjected in the brewing process and the resultant cellular responses. Several of these are alluded to in earlier sections of this chapter. Several external stresses elicit common responses as evidenced by the observation that exposure to a non-lethal stress of one type provides a measure of resistance to others. By implication, common pathways are activated and these are triggered in response to one or more separate receptors. Genomic profiling studies have shown that approximately 900 genes respond to signals generated by the application of environmental stresses (Gasch *et al.*, 2000). Activation of the stress response is mediated in part by the Ras cyclicAMP signalling pathway (Park *et al.*, 2005). Both long-term transcriptional effects and short-term post-translational modifications occur by which the phenotype is adjusted to deal with the applied stress. Of key importance is the presence in the promoter region of a short DNA motif, termed STRE, to which transcription factors bind. The STRE element is found in the promoter region of more than 200 genes known to be implicated in the general stress response of yeast. This subject is not discussed further here since it is dealt with elsewhere in this book (see Chapter 2).

Future perspective

It will be apparent from the material covered in this chapter that yeast cells, in particular *S. cerevisiae*, have been subjected to intense study. Much of the elucidation of the breathtakingly elegant ways in which eukaryotic cells regulate all aspects of cellular function, from birth to self-programmed death, have been performed using this model cell. However, the results of very few of these studies can be used with any degree of certainty to interpret the behaviour of brewing yeast strains growing under brewing conditions. Thus, the combination of growth on an uncharacterized and variable but comparatively unbalanced medium, transient aerobioisis, and yeast recycling coupled to serial fermentation make for a fascinating but perhaps bewildering level of complexity. Nevertheless there are some important lessons to be learnt, which might be used profitably in attempts to make brewing yeast behave in a predictable and more productive manner.

In yeast, the signalling pathways that respond to external nutritional cues do so in a way that multiple nutrients can elicit a common set of responses. The pathways respond to the presence or absence of comparatively simple molecules. Since brewers have yeast ready to pitch in storage vessels, it may be possible to use this as an opportunity to initiate key pathways before pitching has occurred and thereby shorten lag times, improve the performance of newly propagated cultures, remodel the phenotype to a state that is more amenable for very-high-gravity brewing, or simply provide methods for more precise control of the formation of yeast-derived beer flavour metabolites.

An aspect of yeast activity that is not usually considered in normal brewing practice is that of population heterogeneity. For obvious reasons, it is usual to deal with yeast in bulk and by inference there is a tacit assumption that all cells present within that population will behave in a similar fashion. Clearly this is incorrect. The understanding of the relationships between cell age and size and the risks of selection of an inappropriate portion of a bottom crop from fermenter are well-known. It is certain that yeast populations are inherently heterogeneous and there is some evidence that there is some degree of cooperation. The phenomenon of flocculation is considered to be a stress response in which cells within the inner parts of the cell mass are more likely to survive than those on the outside. This may be a purely random process, although it is possible that there is some element of choice in terms of which cells are selected for sacrifice or survival based on their relative positions in the floc. As alluded to already, the phenomenon of apoptosis carries the assumption that members of a population are selected for death. Whether this requires input from all members of the population or it is simply a manifestation of the end of the lifespan of an individual unrelated to potential cooperative behaviour is not known.

It is a sobering thought that the contents of a large fermenter represent an enormous number of individuals. A terminal count of 60 million cells/ml equates to a total of around 1×10^{16} cells at the end of growth in a 2000 hl vessel. This compares with the total human population of the earth for 2015 of approximately 7×10^9. Assuming a pitching rate of 15 million cells/ml and initial and final viabilities of 95%, an additional 4.5×10^{14} dead cells would be present in the crop. Of course, this does not include any cells that would have disappeared via lysis or those that might be approaching senescence. Quorum sensing, the ability of microbial populations to monitor cell densities, is a well-established phenomenon in bacteria. The evidence suggests that a similar system occurs in populations of *S. cerevisiae* and that the process uses aromatic alcohols such as phenylethanol as signalling molecules (Zupan *et al.*, 2013; Wuster and Babu, 2010). In another example, important beer esters such as ethyl acetate and ethyl hexanoate have been shown to function as fruit fly pheromones and by inference may be involved in dispersing natural yeast populations (Siderhurst and Jang, 2006). A greater understanding of the biological basis of, for example, the formation of beer flavour compounds, is likely to lead to the development of more directed control procedures.

References

Aerts, A.M., Zabrocki, P., Govaert, G., Mathys, J., Carmona-Gutierrez, D., Madeo, F., Winderickx, J., Cammue, B.P.A., and Thevissen, K. (2009). Mitochondrial dysfunction leads to reduced chronological lifespan and increased apoptosis in yeast. FEBS Lett., *583*, 113–117. http://dx.doi.org/10.1016/j.febslet.2008.11.028

Allen, C., Büttner, S., Aragon, A.D., Thomas, J.A., Meirelles, O., Jaetao, J.E., Benn, D., Ruby, S.W., Veenhuis, M., Madeo, F., *et al.* (2006). Isolation of quiescent and nonquiescent cells from yeast stationary-phase cultures. J. Cell Biol. *174*, 89–100.

Aries, V., and Kirsop, B.H. (1978). Sterol biosynthesis by strains of *Saccharomyces cerevisiae* in the presence and absence of dissolved oxygen. J. Inst. Brew., *84*, 118–121.

Bayly, J.C., Douglas, L.M., Pretorius, I.S., Bauer, F.F., and Dranginis, A.M. (2005). Characteristics of Flo11-dependent flocculation in *Saccharomyces cerevisiae*. FEMS Yeast Res. 5, 1151–1156.

Becerra, M., Lombardía-Ferreira, L.J., Hauser, N.C., Hoheisel, J.D., Tizon, B., and Cerdán, M.E. (2002). The yeast transcriptome in aerobic and hypoxic conditions: effects of hap1, rox1, rox3 and srb10 deletions. Mol. Microbiol. *43*, 545–555.

Beltran, G., Novo, M., Guillamón, J.M., Mas, A., and Rozès, N. (2008). Effect of fermentation temperature and culture media on the yeast lipid composition and wine volatile compounds. Int. J. Food Microbiol. *121*, 169–177.

Blinder, D., Coschigano, P.W., and Magasanik, B. (1996). Interaction of the GATA factor Gln3p with the nitrogen regulator Ure2p in *Saccharomyces cerevisiae*. J. Bacteriol. *178*, 4734–4736.

Bodenmiller, B., Wanka, S., Kraft, C., Urban, J., Campbell, D., Pedrioli, P.G., Gerrits, B., Picotti, P., Lam, H., Vitek,

O., Brusniak, M.Y., Roschitzki, B., Zhang, C., Shokat, K.M., Schlapbach, R., Colman-Lerner, A., Nolan, G.P., Nesvizhskii, A., Peter, M., Loewith, R., von Mering, C., and Aebersold, R. (2010). Phosphoproteomic analysis reveals interconnected system-wide responses to perturbations of kinases and phosphatases in yeast. Sci Signal., *3*, 153–167. http://dx.doi.org/10.1126/scisignal.2001182

Boulton, C.A., Clutterbuck, V.J., and Durnin, S.C.P. (2000) Yeast oxygenation and storage. In Brewing Yeast Fermentation Performance, K.A. Smart, ed. (Oxford, UK: Blackwell Science), pp. 140–151.

Boulton, C.A., and Quain, D.E. (2001). Brewing Yeast and Fermentation (Oxford, UK: Blackwell Science).

Broach, J.R. (2012). Nutritional control of growth and development in yeast. Genetics *192*, 73–105. http://dx.doi.org/10.1534/genetics.111.135731

Brengues, M., Teixeira, D., and Parker, R. (2005). Movement of eukaryotic mRNAs between polysomes and cytoplasmic processing bodies. Science *310*, 486–489.

Butow, R.A., and Avadhani, N.G. (2004). Mitochondrial signaling: the retrograde response. Mol. Cell *14*, 1–15.

Cebollero, E., and Reggiori, F. (2009). Regulation of autophagy in yeast *Saccharomyces cerevisiae*. Biochim. Biophys. Acta *1793*, 1413–1421. http://dx.doi.org/10.1016/j.bbamcr.2009.01.008

Charron, M.J., Read, E., Haut, S.R., and Michels, C.A. (1989). Molecular evolution of the telomere-associated MAL loci of *Saccharomyces*. Genetics *122*, 307–316.

Coller, H.A., Sang, L., and Roberts, J.M. (2006). A new description of cellular quiescence. PLoS Biol. *4*, e83.

Conrad, M., Schothorst, J., Kankipati, H.N., Van Zeebroeck, G., Rubio-Texeira, M., and Thevelein, J.M. (2014). Nutrient sensing and signaling in the yeast *Saccharomyces cerevisiae*. FEMS Microbiol. Rev. *38*, 254–299. http://dx.doi.org/10.1111/1574-6976.12065

Cowart, L.A., and Obeid, L.M. (2007). Yeast Sphingolipids: Recent developments in understanding biosynthesis, regulation, and function. Biochim. Biophys. Acta, *177*, 421–431. http://dx.doi.org/10.1016/j.bbalip.2006.08.005

Crauwels, M., Donaton, M.C.V., Pernambuco, M.B., Winderikx, J., de Winde, J.H., and Thevelein, J.M. (1997). The Sch9 protein kinase in the yeast *Saccharomyces cerevisiae* controls cAPK activity and is required for nitrogen activation of the fermentable-growth-medium-induced (FGM) pathway. Microbiology *143*, 2627–2637.
http://dx.doi.org/10.1099/00221287-143-8-2627

Crowe, J.H., Crowe, L.M., and Chapman, D. (1984). Preservation of membranes in anhydrobiotic organisms: the role of trehalose. Science *223*, 701–703.

Daignan-Fornier, B., and Sagot, I. (2011). Proliferation/quiescence: the controversial 'aller-retour'. Cell Div. *6*, 10. http://dx.doi.org/10.1186/1747-1028-6-10

Davies, B.S., and Rine, J. (2006). A role for sterol levels in oxygen sensing in *Saccharomyces cerevisiae*. Genetics *174*, 191–201.

De Deken, R.H. (1966). The Crabtree effect: a regulatory system in yeast. J. Gen. Microbiol. *44*, 149–156. http://dx.doi.org/10.1099/00221287-44-2-149

de Groot, P.W., Ruiz, C., Vázquez de Aldana, C.R., Duenas, E., Cid, V.J., Del Rey, F., Rodríquez-Peña, J.M., Pérez, P., Andel, A., Caubín, J., *et al.* (2001). A genomic approach for the identification and classification of genes involved in cell wall formation and its regulation in *Saccharomyces cerevisiae*. Comp. Funct. Genomics. *2*, 124–142. http://dx.doi.org/10.1002/cfg.85

De Nobel, J.G., Klis, F.M., Priem, J., Munnik, T., and Van den Ende, H. (1990). The glucanase-soluble mannoproteins limit cell wall porosity in *Saccharomyces cerevisiae*. Yeast, *6*, 491–499. http://dx.doi.org/10.1002/yea.320060606

De Virgilio, C., and Loewith, R. (2006). The TOR signalling network from yeast to man. Int. J. Biochem. Cell Biol. *38*, 1476–1481.

Deckert, J., Perini, R., Balasubramanian, B., and Zitomer, R.S. (1995). Multiple elements and auto-repression regulate Rox1, a repressor of hypoxic genes in *Saccharomyces cerevisiae*. Genetics *139*, 1149–1158.

Diaz-Ruiz, R., Rigoulet, M., and Devin, A. (2011). The Warburg and Crabtree effects: On the origin of cancer cell energy metabolism and of yeast glucose repression. Biochim. Biophys. Acta *1807*, 568–576. http://dx.doi.org/10.1016/j.bbabio.2010.08.010

Dungrawala, H., Hua, H., Wright, J., Abraham, L., Kasemsri, T., McDowell, A., Stilwell, J., and Schneider, B.L. (2012). Identification of new cell size control genes in *S. cerevisiae*. Cell Div. 7, 24. http://dx.doi.org/10.1186/1747-1028-7-24

Dupres, V., Alsteens, D., Wilk, S., Hansen, B., Heinisch, J.J., and Dufrêne, Y.F. (2009). The yeast Wsc1 cell surface sensor behaves like a nanospring *in vivo*. Nat. Chem. Biol. *5*, 857–862. http://dx.doi.org/10.1038/nchembio.220

Elbein, A.D., Pan, Y.T., Pastuszak, I., and Carroll, D. (2003). New insights on trehalose: a multifunctional molecule. Glycobiology *13*, 17R–27R. http://dx.doi.org/10.1093/glycob/cwg047

Ernandes, J.R., Williams, J.W., Russell, I., and Stewart, G.G. (1993). Respiratory deficiency in brewing yeast strains–effects on fermentation, flocculation, and beer flavor components. J. Am. Soc. Brew. Chem., *51*, 16–20.

François, J., and Parrou, J.L. (2001). Reserve carbohydrates metabolism in the yeast *Saccharomyces cerevisiae*. FEMS Microbiol. Rev. *25*, 125–145.

Free, S.J. (2013). Fungal cell wall organization and biosynthesis. Adv. Genet. *81*, 33–82. http://dx.doi.org/10.1016/B978-0-12-407677-8.00002-6

Gancedo, J.M. (1992). Carbon catabolite repression in yeast. Eur. J. Biochem. *206*, 297–313.

Gasch, A.P., Spellman, P.T., Kao, C.M., Carmel-Harel, O., Eisen, M.B., Storz, G., Botstein, D., and Brown, P.O. (2000). Genomic expression programs in the response of yeast cells to environmental changes. Mol. Biol. Cell, *11*, 4241–4257. http://dx.doi.org/10.3390/biom2010001

Gibson, B.R., Boulton, C.A., Box, W.G., Graham, N.S., Lawrence, S.J., Linforth, R.S., and Smart, K.A. (2009). Amino acid uptake and yeast gene transcription during industrial brewery fermentation. J. Am. Soc. Brew. Chem., *67*, 157–165.

Gray, J.V., Petsko, G.A., Johnston, G.C., Ringe, D., Singer, R.A., and Werner-Washburne, M. (2004). 'Sleeping beauty': quiescence in *Saccharomyces cerevisiae*. Microbiol. Mol. Biol. Rev. *68*, 187–206. http://dx.doi.org/10.1128/MMBR.68.2.187-206.2004

Griffioen, G., Laan, R.J., Mager, W.H., and Planta, R.J. (1996). Ribosomal protein gene transcription in *Saccharomyces cerevisiae* shows a biphasic response to nutritional changes. Microbiology *142*, 2279–2287. http://dx.doi.org/10.1099/13500872-142-8-2279

Grose, J.H., Smith, T.L., Sabic, H., and Rutter, J. (2007). Yeast PAS kinase coordinates glucose partitioning in response to metabolic and cell integrity signaling. EMBO J. *26*, 4824–4830.

Gustin, M.C., Albertyn, J., Alexander, M., and Davenport, K. (1998). MAP kinase pathways in the yeast *Saccharomyces cerevisiae*. Microbiol. Mol. Biol. Rev. *62*, 1264–1300.

Hardie, D.G. (2011). AMP-activated protein kinase: an energy sensor that regulates all aspects of cell function. Genes Dev. *25*, 1895–1908. http://dx.doi.org/10.1101/gad.17420111

Hartwell, L.H., Culotti, J., Pringle, J.R., and Reid, B.J. (1974). Genetic control of the cell division cycle in yeast. Science, *183*, 46–51.

Hazelwood, L.A., Walsh, M.C., Luttik, M.A., Daran-Lapujade, P., Pronk, J.T., and Daran, J.M. (2009). Identity of the growth-limiting nutrient strongly affects storage carbohydrate accumulation in anaerobic chemostat cultures of *Saccharomyces cerevisiae*. Appl. Environ. Microbiol. *75*, 6876–6885. http://dx.doi.org/10.1128/AEM.01464-09

Horák, J. (1997). Yeast nutrient transporters. Biochim. Biophys. Acta *1331*, 41–79.

Huber, A., Bodenmiller, B., Uotila, A., Stahl, M., Wanka, S., Gerrits, B., Aebersold, R., and Loewith, R. (2009). Characterisation of the rapamycin-sensitive phosphoproteome reveals that Sch9 is a central coordinator of protein synthesis. Genes Dev., *23*, 1929–1943. http://dx.doi.org/10.1101/gad.532109

Jacobson, L., Halmann, M., and Yariv, J. (1982). The molecular composition of the volutin granule of yeast. Biochem. J. *201*, 473–479.

Jahnke, L., and Klein, H.P. (1983). Oxygen requirements for formation and activity of the squalene epoxidase in *Saccharomyces cerevisiae*. J. Bacteriol. *155*, 488–492.

Jansen, L.A., Daran-Lapujade, P.D., de Winde, J.H., Piper, M.D.W and Pronk, J.T. (2004). Prolonged maltose-limited cultivation of *Saccharomyces cerevisiae* selects for cells with Improved maltose affinity and hypersensitivity. Appl. Environ. Microbiol., *70*, 1956–1963. http://dx.doi.org/10.1128/AEM.70.4.1956-1963.2004

Jansen, J.M., Wanless, A.G., Seidel, C.W., and Weiss, E.L. (2009). Cbk1 regulation of the RNA-binding protein Ssd1 integrates cell fate with translational control. Curr. Biol. *19*, 2114–2120. http://dx.doi.org/10.1016/j.cub.2009.10.071

Jauniaux, J.C., Urrestarazu, L.A., and Wiame, J.M. (1978). Arginine metabolism in *Saccharomyces cerevisiae*: subcellular localization of the enzymes. J. Bacteriol. *133*, 1096–1107.

Jenkins, C.L., Lawrence, S.J., Kennedy, A.I., Hodgson, J.A and Smart, K.A. (2009). Incidence and formation of petite mutants in lager brewing yeast *Saccharomyces cerevisiae* (*syn. S. pastorianus*) populations. J. Am. Soc. Brew. Chem. *67*, 72–80.

Jenkins, D.M., Powell, C.D., Fischborn, T., and Smart, K.A. (2011). Rehydration of active dry brewing yeast and its effect on cell viability. J. Inst. Brew., *117*, 377–382.

Jensen, R.E., Hobbs, A.E., Cerveny, K.L., and Sesaki, H. (2000). Yeast mitochondrial dynamics: fusion, division, segregation, and shape. Microsc. Res. Tech. *51*, 573–583.

Johnston, G.C., Pringle, J.R., and Hartwell, L.H. (1977). Coordination of growth with cell division in the yeast, *Saccharomyces cerevisiae*. Exp. Cell Res. *105*, 79–98.

Kamada, Y., Qadota, H., Python, C.P., Anraku, Y., Ohya, Y., and Levin, D.E. (1996). Activation of yeast protein kinase C by Rho1 GTPase. J. Biol. Chem. *271*, 9193–9196.

Kitagaki, H., Tan, H., and Jayakody, L.N. (2013). Analysis of the role of mitochondria of Sake yeast during Sake brewing and its applications in fermentation technologies. Agri-BioScience Monographs, *3*, 1–12.

Klionsky, D.J., Herman, P.K., and Emr, S.D. (1990). The fungal vacuole: composition, function, and biogenesis. Microbiol. Rev. *54*, 266–292.

Klis, F.M., Mol, P., Hellingwerf, K., and Brul, S. (2002). Dynamics of cell wall structure in *Saccharomyces cerevisiae*. FEMS Microbiol. Rev. *26*, 239–256.

Klis, F.M., Boorsma, A., and de Groot, P.W.J. (2006). Cell wall construction in *Saccharomyces cerevisiae*. Yeast, *23*, 185–202. http://dx.doi.org/10.1002/yea.1349

Kohlhaw, G.B. (2003). Leucine biosynthesis in fungi: entering metabolism through the back door. Microbiol. Mol. Biol. Rev. *67*, 1–15.

Krogerus, and Gibson, B.J., (2013). 125th Anniversary Review: Diacetyl and its control during brewery fermentation. J. Inst. Brew., *119*, 86 – 97.

Kwast, K.E., Burke, P.V., Staahl, B.T., and Poyton, R.O. (1999). Oxygen sensing in yeast: evidence for the involvement of the respiratory chain in regulating the transcription of a subset of hypoxic genes. Proc. Natl. Acad. Sci. U.S.A. *96*, 5446–5451.

Kwast, K.E., Lai, L.C., Menda, N., *et al.* (2002). Genomic analyses of anaerobically induced genes in *Saccharomyces cerevisiae*: functional roles of Rox1 and other factors in mediating the anoxic response. J. Bacteriol. *184*, 250–265.

Labbe-Bois R., and Labbe, P. (1990). Tetrapyrrole and heme biosynthesis in the yeast *Saccharomyces cerevisiae*. In Biosynthesis of Heme and Chlorophylls, Dailey AH, ed. (New York: McGraw-Hill), pp. 235–285.

Laporte, D., Salin, B., Daignan-Fornier, B., and Sagot, I. (2008). Reversible cytoplasmic localization of the proteasome in quiescent yeast cells. J. Cell. Biol. *181*, 737–745. http://dx.doi.org/10.1083/jcb.200711154

Laporte, D., Lebaudy, A., Sahin, A., Pinson, B., Ceschin, J., Daignan-Fornier, B., and Sagot, I. (2011). Metabolic status rather than cell cycle signals control quiescence entry and exit. J. Cell Biol. *192*, 949–957. http://dx.doi.org/10.1083/jcb.201009028

Lawrence, S.J., Wimalasena, T.T., Nicholls, S.M., Box, W.G., Boulton, C., and Smart, K.A. (2012). Incidence and characterisation of petites isolated from lager brewing yeast *Saccharomyces cerevisiae* populations. J. Am. Soc. Brew. Chem., *70*, 268–274.

Lawrence, S.J., Nicholls, S., Box, W.G., Sbuelz, R., Bealin-Kelly, F., Axcell, B and Smart, K.A. (2013). The influence of yeast cell age and environmental conditions on the distribution of petite mutants in the fermenter cone. J. Am. Soc. Brew. Chem., *71*, 90–96.

Levin, D.E. (2011). Regulation of cell wall biogenesis in *Saccharomyces cerevisiae*: the cell wall integrity signaling pathway. Genetics *189*, 1145–1175. http://dx.doi.org/10.1534/genetics.111.128264

Li, L., Lu, Y., Qin, L.X., Bar-Joseph, Z., Werner-Washburne, M., and Breeden, L.L. (2009). Budding yeast SSD1-V regulates transcript levels of many longevity genes and extends chronological lifespan in purified quiescent cells. Mol. Biol. Cell., *20*, 3851–3864. http://dx.doi.org/10.1091/mbc.E09-04-0347

Li, L., Miles, S., Melville, Z., Prasad, A., Bradley, G., and Breeden, L.L. (2013). Key events during the transition from rapid growth to quiescence in budding yeast require posttranscriptional regulators. Mol. Biol. Cell *24*, 3697–3709. http://dx.doi.org/10.1091/mbc.E13-05-0241

Lillie, S.H., and Pringle, J.R. (1980). Reserve carbohydrate metabolism in *Saccharomyces cerevisiae*: responses to nutrient limitation. J. Bacteriol. *143*, 1384–1394.

Lockshon, D., Olsen, C.P., Brett, C.L., Chertov, A., Merz, A.J., Lorenz, D.A., Van Gilst, M.R., and Kennedy, B.K. (2012). Rho signaling participates in membrane fluidity homeostasis. PLoS One *7*, e45049. http://dx.doi.org/10.1371/journal.pone.0045049

Loewith, R., Jacinto, E., Wullschleger, S., Lorberg, A., Crespo, J.L., Bonenfant, D., Oppliger, W., Jenoe, P., and Hall, M.N. (2002). Two TOR complexes, only one of which is rapamycin sensitive, have distinct roles in cell growth control. Mol. Cell *10*, 457–468.

Loewith, R. (2011). A brief history of TOR. Biochem. Soc. Trans. *39*, 437–442. http://dx.doi.org/10.1042/BST0390437

Madeo, F., Fröhlich, E., Ligr, M., Grey, M., Sigrist, S.J., Wolf, D.H., and Fröhlich, K.U. (1999). Oxygen stress: a regulator of apoptosis in yeast. J. Cell Biol. *145*, 757–767.

McFaline-Figueroa, J.R., Vevea, J., Swayne, T.C., Zhou, C., Liu, C., Leung, G., Boldogh, I.R., and Pon, L.A. (2011). Mitochondrial quality control during inheritance is associated with lifespan and mother–daughter age asymmetry in budding yeast. Aging. Cell *10*, 885–895. http://dx.doi.org/10.1111/j.1474-9726.2011.00731.x

McKnight, S.L. (2010). On getting there from here. Science *330*, 1338–1339. http://dx.doi.org/10.1126/science.1199908

Miles, S., Li, L., Davison, J., and Breeden, L.L. (2013). Xbp1 directs global repression of budding yeast transcription during the transition to quiescence and is important for the longevity and reversibility of the quiescent state. PLoS Genet. *9*, e1003854. http://dx.doi.org/10.1371/journal.pgen.1003854

Miller, K.J., Box, W.G., Boulton, C.A and Smart, K.A. (2012). Cell cycle synchrony of propagated and recycled lager yeast and its impact on lag phase in fermenter. (2012). J. Am. Soc. Brew. Chem. *70*, 1-9.

Moradas-Ferreira, P., and Costa, V. (2000). Adaptive response of the yeast *Saccharomyces cerevisiae* to reactive oxygen species: defences, damage and death. Redox. Rep. *5*, 277–285. http://dx.doi.org/10.1179/135100000101535816

Muir, A., Ramachandran, S., Roelants, F.M., Timmons, G., and Thorner, J. (2014). TORC2-dependent protein kinase Ypk1 phosphorylates ceramide synthase to stimulate synthesis of complex sphingolipids. Elife 3, e03779. http://dx.doi.org/10.7554/eLife.03779

Narayanaswamy, R., Levy, M., Tsechansky, M., Stovall, G.M., O'Connell, J.D., Mirrielees, J., Ellington, A.D., and Marcotte, E.M. (2009). Widespread reorganization of metabolic enzymes into reversible assemblies upon nutrient starvation. Proc. Natl. Acad. Sci. U.S.A. *106*, 10147–10152. http://dx.doi.org/10.1073/pnas.0812771106

de Nobel, J.G., Klis, F.M., Priem, J., Munnik, T., and van den Ende,H. (1990). The glucanase-soluble mannoproteins limit cell wall porosity in *Saccharomyces cerevisiae. Yeast, 6*, 491– 499.

Odorizzi, G., Babst, M., and Emr, S.D. (2000). Phosphoinositide signaling and the regulation of membrane trafficking in yeast. Trends Biochem. Sci *25*, 229–235.

Oliveira, A.P., Ludwig, C., Picotti, P., Kogadeeva, M., Aebersold, R., and Sauer, U. (2012). Regulation of yeast central metabolism by enzyme phosphorylation. Mol. Syst. Biol. *8*, 623. http://dx.doi.org/10.1038/msb.2012.55

Oliveira, A.P., and Sauer, U. (2012). The importance of post-translational modifications in regulating *Saccharomyces cerevisiae* metabolism. FEMS Yeast Res. *12*, 104–117. http://dx.doi.org/10.1111/j.1567-1364.2011.00765.x

Ozcan, S., Dover, J., Rosenwald, A.G., Wölfl, S., and Johnston, M. (1996). Two glucose transporters in *Saccharomyces cerevisiae* are glucose sensors that generate a signal for induction of gene expression. Proc. Natl. Acad. Sci. U.S.A. *93*, 12428–12432.

Quain, D.E., and Tubb, R.S. (1982). The importance of glycogen in brewing yeast. Tech. Quart. Master Brewers Assoc. Am. *19*, 29–33.

Pardee, A.B.(1974). A restriction point for control of normal animal cell proliferation. Proc. Natl. Acad. Sci. USA, *71*, 1286–1290.

Park, J.I., Collinson, E.J., Grant, C.M., and Dawes, I.W. (2005). Rom2p, the Rho1 GTP/GDP exchange factor of *Saccharomyces cerevisiae*, can mediate stress responses via the Ras-cAMP pathway. J. Biol. Chem. *280*, 2529–2535.

Parks, L.W., and Casey, W.M. (1995). Physiological implications of sterol biosynthesis in yeast. Annu. Rev. Microbiol. *49*, 95–116. http://dx.doi.org/10.1146/annurev.mi.49.100195.000523

Parrou, J.L., Jules, M., Beltran, G., and François, J. (2005). Acid trehalase in yeasts and filamentous fungi: localization, regulation and physiological function. FEMS Yeast Res. *5*, 503–511.

Peraza_Reyes, L., Crider, D.G., and Pon, L.A. (2010). Mitochondrial manoeuvres: latest insights and hypotheses on mitochondrial partitioning during mitosis in *Saccharomyces cerevisiae*. BioEssays, *32*, 1040–1049. http://dx.doi.org/10.1002/bies.201000083

Pereira, C., Silva, R.D., Saraiva, L., Johansson, B., Sousa, M.J., and Côrte-Real, M. (2008). Mitochondria-dependent apoptosis in yeast. Biochim. Biophys. Acta *1783*, 1286–1302. http://dx.doi.org/10.1016/j.bbamcr.2008.03.010

Pierce, J.S. (1987). The role of nitrogen in brewing. J. Inst. Brew., *93*, 378–381.

Pittet, M., and Conzelmann, A. (2007). Biosynthesis and function of GPI proteins in the yeast *Saccharomyces cerevisiae*. Biochim. Biophys. Acta *1771*, 405–420.

Plattner, H., Salpeter, M.M., Saltzgaber, J., and Schatz, G. (1970). Promitochondria of anaerobically grown yeast. IV. Conversion into respiring mitochondria. Proc. Natl. Acad. Sci. U.S.A. *66*, 1252–1259.

Postma, E., Verduyn, C., Kuiper, A., Scheffers, W.A., and van Dijken, J.P. (1990). Substrate-accelerated death of *Saccharomyces cerevisiae* CBS 8066 under maltose stress. Yeast, *6*, 149–158. http://dx.doi.org/10.1002/yea.320060209

Powell., C., Quain, D.E., and Smart, K.A. (2002). The impact of yeast sedimentation on the distribution of cells within the cone of a cylindroconical vessel. Proc. Institute and Guild of Brewing – Asia Pacific Section Conv., Adelaide.

Powell, C.D., Quain, D.E., and Smart, K.A. (2003). The impact of brewing yeast cell age on fermentation performance, attenuation and flocculation. FEMS Yeast Res. *3*, 149–157.

Powell, C.D and Diacetis, A.N. (2007). Long term serial repitching and the genetic and phenotypic stability of brewer's yeast. J. Inst. Brew., *113*, 67–74.

Powers, T., Aronova, S., and Niles, B. (2010). TORC2 and sphingolipid biosynthesis and signalling: lessons from budding yeast. Enzymes, *27*, 177–197.

Poyton, R.O. (1999). Models for oxygen sensing in yeast: implications for oxygen-regulated gene expression in higher eucaryotes. Respir. Physiol. *115*, 119–133.

Pratt, P.L., Bryce, J.H., and Stewart, G.G. (2012). The yeast vacuole – a scanning electron microscopy study during high gravity wort Fermentations. J. Inst. Brew., *113*, 55 – 60.

Radico, R., and Heinisch, J.J. (2010). Together we are strong – cell wall integrity sensors in yeasts. Yeast, *27*, 531–540. http://dx.doi.org/10.1002/yea.1785

Radonjic, M., Andrau, J.C., Lijnzaad, P., Kemmeren, P., Kockelkorn, T.T., van Leenen, D., van Berkum, N.L., and Holstege, F.C. (2005). Genome-wide analyses reveal RNA polymerase II located upstream of genes poised for rapid response upon *S. cerevisiae* stationary phase exit. Mol. Cell *18*, 171–183.

Rai, R., Tate, J.J., Nelson, D.R and Cooper, T.G. (2013). gln3 mutations dissociate responses to nitrogen limitation (nitrogen catabolite repression) and rapamycin Inhibition of TorC1. J. Biol. Chem. *288*, 2789 2804. http://dx.doi.org/10.1074/jbc.M113.516740

Rajakumari, S., Grillitsch, K., and Daum, G. (2008). Synthesis and turnover of non-polar lipids in yeast. Prog. Lipid Res. *47*, 157–171. http://dx.doi.org/10.1016/j.plipres.2008.01.001

Reiner, S., Micolod, D., Zellnig, G., and Schneiter, R. (2006). A genomewide screen reveals a role of mitochondria in anaerobic uptake of sterols in yeast. Mol. Biol. Cell *17*, 90–103.

Rodriguez, R.J., Low, C., Bottema, C.D., and Parks, L.W. (1985). Multiple functions for sterols in *Saccharomyces cerevisiae*. Biochim. Biophys. Acta *837*, 336–343.

Rogers, P.J., Yue, S.B., and Stewart, P.R. (1974). Effects of oxygen tension and glucose repression on mitochondrial protein synthesis in continuous cultures of *Saccharomyces cerevisiae*. J. Bacteriol., *118*, 523–533.

Rohde, J.R., Bastidas, R., Puria, R., and Cardenas, M.E. (2008). Nutritional control via Tor signaling in *Saccharomyces cerevisiae*. Curr. Opin. Microbiol. *11*, 153–160. http://dx.doi.org/10.1016/j.mib.2008.02.013

Rosenfeld, E., and Beauvoit, B. (2003). Role of the non-respiratory pathways in the utilization of molecular oxygen by *Saccharomyces cerevisiae*. Yeast, *20*, 1115–1144. http://dx.doi.org/10.1002/yea.1026

Rosenfeld, E., Beauvoit, B., Blondin, B., and Salmon, J.M. (2003). Oxygen consumption by anaerobic *Saccharomyces cerevisiae* under enological conditions: effect on fermentation kinetics. Appl. Environ. Microbiol. *69*, 113–121.

Rosenfeld, E., Schaeffer, J., Beauvoit, B., and Salmon, J.M. (2004). Isolation and properties of promitochondria from anaerobic stationary-phase yeast cells. Antonie Van Leeuwenhoek *85*, 9–21. http://dx.doi.org/10.1023/B:ANTO.0000020268.55350.54

Ryan, E.D., and Kohlhaw, G.B. (1974). Subcellular localization of isoleucine-valine biosynthetic enzymes in yeast. J. Bacteriol. *120*, 631–637.

Sagot, I., Pinson, B., Salin, B., and Daignan-Fornier, B. (2006). Actin bodies in yeast quiescent cells: an immediately available actin reserve? Mol. Biol. Cell *17*, 4645–4655.

Salema-Oom, M., Pinto, V.V., Gonçalves, P., and Spencer-Martins, I. (2005). Maltotriose utilization by industrial *Saccharomyces* strains: characterization of a new member of the α-glucoside transporter family. Appl. Environ. Microbiol., *71*, 5044 – 5049.

Salmon, J.M., and Barre, P. (1998). Improvement of nitrogen assimilation and fermentation kinetics in oenological conditions by derepression of alternative nitrogen-assimilatory pathways in an industrial *Saccharomyces cerevisiae* strain. Appl. Environ. Microbiol., *64*, 3831–3837.

Samp, E.J. (2012). Possible roles of the mitochondria in sulphur dioxide production by lager Yeast. J. Am. Soc. Brew. Chem. *70*, 219–229.

Sato, M., Watari, J., and Shinotsuka, K. (2001). Genetic instability in flocculation of bottom-fermenting yeast. J. Am. Soc. Brew. Chem., *59*, 130–134.

Schmelzle, T., Beck, T., Martin, D.E., and Hall, M.N. (2004). Activation of the RAS/cyclic AMP pathway suppresses a TOR deficiency in yeast. Mol. Cell. Biol. *24*, 338–351.

Schmidt, A., Kunz, J., and Hall, M.N. (1996). TOR2 is required for organization of the actin cytoskeleton in yeast. Proc. Natl. Acad. Sci. U.S.A. *93*, 13780–13785.

Shimanuki, M., Chung, S.Y., Chikashige, Y., Kawasaki, Y., Uehara, L., Tsutsumi, C., Hatanaka, M., Hiraoka, Y., Nagao, K., and Yanagida, M. (2007). Two-step, extensive alterations in the transcriptome from G0 arrest to cell division in *Schizosaccharomyces pombe*. Genes Cells. *12*, 677–692.

Shimizu, I., Nagai, J., Hatanaka, H., and Katsuki, H. (1973). Mevalonate synthesis in the mitochondria of yeast. Biochim. Biophys. Acta *296*, 310–320.

Shimoi, H., Kitagaki, H., Ohmori, H., Iimura, Y., and Ito, K. (1998). Sed1p is a major cell wall protein of *Saccharomyces cerevisiae* in the stationary phase and is involved in lytic enzyme resistance. J. Bacteriol. *180*, 3381–3387.

Siderhurst, M.S., and Jang, E.B. (2006). Female-biased attraction of Oriental fruit fly, bactrocera *dorsalis* (*Hendel*), to a blend of host fruit volatiles from *Terminalia catappa L*. J. Chem. Ecol. 32, 2513–2524. http://dx.doi.org/10.1007/s10886-006-9160-6

Silljé, H.H., Paalman, J.W., ter Schure, E.G., Olsthoorn, S.Q., Verkleij, A.J., Boonstra, J., and Verrips, C.T. (1999). Function of trehalose and glycogen in cell cycle progression and cell viability in *Saccharomyces cerevisiae*. J. Bacteriol. *181*, 396–400.

Simpson, W.J., and Hammond, J.R.M. (1989). The response of brewing yeast to acid washing. J. Inst. Brew., *95*, 347–354.

Simpson, C.E., and Ashe, M.P. (2012). Adaptation to stress in yeast: to translate or not? Biochem. Soc. Trans. *40*, 794–799. http://dx.doi.org/10.1042/BST20120078

Smets, B., Ghillebert, R., De Snijder, P., Binda, M., Swinnen, E., De Virgilio, C., and Winderickx, J. (2010). Life in the midst of scarcity: adaptations to nutrient availability in *Saccharomyces cerevisiae*. Curr. Genet. *56*, 1–32. http://dx.doi.org/10.1007/s00294-009-0287-1

Snoek, I.S and Steensma, Y.H. (2007). Factors involved in anaerobic growth of *Saccharomyces cerevisiae*. Yeast, *24*, 1–10. http://dx.doi.org/10.1002/yea.1430

Soares, E.V. (2011). Flocculation in *Saccharomyces cerevisiae*: a review. J. Appl. Microbiol. *110*, 1–18. http://dx.doi.org/10.1111/j.1365-2672.2010.04897.x

Sols, A. (1981). Multimodulation of enzyme activity. Curr. Top. Cell. Regul. *19*, 77–101.

Soulard, A., Cremonesi, A., Moes, S., Schütz, F., Jenö, P., and Hall, M.N. (2010). The rapamycin-sensitive phosphoproteome reveals that TOR controls protein kinase A toward some but not all substrates. Mol. Biol. Cell *21*, 3475–3486. http://dx.doi.org/10.1091/mbc.E10-03-0182

Stanbrough, M., and Magasanik, B. (1995). Transcriptional and posttranslational regulation of the general amino acid permease of *Saccharomyces cerevisiae*. J. Bacteriol. *177*, 94–102.

Staschke, K.A., Dey, S., Zaborske, J.M., Palam, L.R., McClintick, J.N., Pan, T., Edenberg, H.J., and Wek, R.C. (2010). Integration of general amino acid control and target of rapamycin (TOR) regulatory pathways in nitrogen assimilation in yeast. J. Biol. Chem. *285*, 16893–16911. http://dx.doi.org/10.1074/jbc.M110.121947

Steinkraus, K.A., Kaeberlein, M., and Kennedy, B.K. (2008). Replicative aging in yeast: the means to the end. Annu. Rev. Cell Dev. Biol. *24*, 29–54. http://dx.doi.org/10.1146/annurev.cellbio.23.090506.123509

Tao, X., Zheng, D., Liu, T., Wang, P., Zhao, W., Zhu, M., Jiang, X., Zhao, Y., and Wu, X. (2012). A novel strategy to construct yeast *Saccharomyces cerevisiae* strains for very high gravity fermentation. PLoS ONE 7, e31235. http://dx.doi.org/10.1371/journal.pone.0031235

Ter Linde, J.J.M., Liang, H., Davis, R.W., Steensma, H.Y., van Dijken, J.P., and Pronk, J.T. (1999). Genome-Wide Transcriptional Analysis of Aerobic and Anaerobic Chemostat Cultures of *Saccharomyces cerevisiae*. J. Bacteriol., *181*, 7409 – 7413.

Thevelein, J.M., and Hohmann, S. (1995). Trehalose synthase: guard to the gate of glycolysis in yeast? Trends Biochem. Sci *20*, 3–10.

Thevelein, J.M., and de Winde, J.H. (1999). Novel sensing mechanisms and targets for the cAMP-protein kinase A pathway in the yeast *Saccharomyces cerevisiae*. Mol. Microbiol. *33*, 904–918.

Thevelein, J.M. (2015). A novel conserved mechanism in control of fermentation in brewing yeast and cancer cells. Proc. 35th EBC Congr. Porto, in press.

Torija, M.J., Beltran, G., Novo, M., Poblet, M., Guillamón, J.M., Mas, A., and Rozès, N. (2003). Effects of fermentation temperature and *Saccharomyces* species on the cell fatty acid composition and presence of volatile compounds in wine. Int. J. Food Microbiol. *85*, 127–136.

Tripodi, F., Nicastro, R., Reghellin, V., and Cocetti, P. (2014). Post-translational modifications on yeast carbon metabolism: Regulatory mechanisms beyond transcriptional control. Biochim. Biophys. Acta, *1850*, 620–627. http://dx.doi.org/10.1016/j.bbagen.2014.12.010

Tronchoni, J., Rozès, N., Querol, A., and Guillamón, J.M. (2012). Lipid composition of wine strains of *Saccharomyces kudriavzevii* and *Saccharomyces cerevisiae* grown at low temperature. Int. J. Food Microbiol. *155*, 191–198. http://dx.doi.org/10.1016/j.ijfoodmicro.2012.02.004

Udeh, H.O., and Kgatia, (2013). The role of magnesium ions on yeast performance in very high gravity fermentations. J. Brew. Dist., *4*, 19–45.

Urban, J., Soulard, A., Huber, A., Lippman, S., Mukhopadhyay, D., Deloche, O., Wanke, V., Anrather, D., Ammerer, G., Riezman, H., *et al.* (2007). Sch9 is a major target of TORC1 in *Saccharomyces cerevisiae*. Mol. Cell *26*, 663–674.

Verbelen, P.J., Depraetere, S.A., Winderickx, J., Delvaux, F.R., and Delvaux, F. (2009). The influence of yeast oxygenation prior to brewery fermentation on yeast metabolism and the oxidative stress response. FEMS Yeast Res. *9*, 226–239. http://dx.doi.org/10.1111/j.1567-1364.2008.00476.x

Verstrepen, K.J., Derdelinckx, G., Verachtert, H., and Delvaux, F.R. (2003). Yeast flocculation: what brewers should know. Appl. Microbiol. Biotechnol. *61*, 197–205. http://dx.doi.org/10.1007/s00253-002-1200-8

Verstrepen, K.J., Van Laere, S.D.M., Vercammen, J., Derdelinckx, Dufour, J.P., Pretorius, I.S., Winderickx, J., Thevelein, J.M., and Delvaux, F.R. (2004). The *Saccharomyces cerevisiae* alcohol acetyl transferase Atf1p is localized in lipid particles. Yeast, 21, 367–377. http://dx.doi.org/10.1002/yea.1100

Verstrepen, K.J., and Klis, F.M. (2006). Flocculation, adhesion and biofilm formation in yeasts. Mol. Microbiol. *60*, 5–15.

Vidgren, V., Ruohonen, L., and Londesborough, J. (2005). Characterization and functional analysis of the *MAL* and *MPH* Loci for maltose utilization in some ale and lager yeast strains. Appl. Environ. Microbiol. *71*, 7846–7857.

Vidgren, V., Multanen, J.P., Ruohonen, L., and Londesborough, J. (2010). The temperature dependence of maltose transport in ale and lager strains of brewer's yeast. FEMS Yeast Res. *10*, 402–411. http://dx.doi.org/10.1111/j.1567-1364.2010.00627.x

Vidgren, V., and Londsborough, J. (2011). Yeast flocculation and sedimentation in brewing, J. Inst. Brew. *117*, 475–487.

Visser, V., van der Baan, A.A.,Batenburg-van der Vegte, W., Scheffers, W.A., Kramer, R., and van Dijken, J.P. (1994). Involvement of mitochondria in the assimilatory metabolism of anaerobic *Saccharomyces cerevisiae* cultures. Microbiology, *140*, 3039 – 3046.

Visser, W., van Spronsen, E.A., Nanninga, N., Pronk, J.T., Gijs Kuenen, J., and van Dijken, J.P. (1995). Effects of growth conditions on mitochondrial morphology in *Saccharomyces cerevisiae*. Antonie. Van. Leeuwenhoek. *67*, 243–253.

Wagner, A., Grillitsch, K., Leitner, E., and Daum, G. (2009). Mobilization of steryl esters from lipid particles of the yeast *Saccharomyces cerevisiae*. Biochim. Biophys. Acta *1791*, 118–124. http://dx.doi.org/10.1016/j.bbalip.2008.11.004

Wei, W., Nurse, P., and Broek, D. (1993). Yeast cells can enter a quiescent state through G1, S, G2, or M phase of the cell cycle. Cancer. Res. *53*, 1867–1870.

Werner-Washburne, M., Braun, E., Johnston, G.C., and Singer, R.A. (1993). Stationary phase in the yeast *Saccharomyces cerevisiae*. Microbiol. Rev. *57*, 383–401.

Wilson, W.A., Roach, P.J., Montero, M., Baroja-Fernández, E., Muñoz, F.J., Eydallin, G., Viale, A.M., and Pozueta-Romero, J. (2010). Regulation of glycogen metabolism in yeast and bacteria. FEMS Microbiol. Rev. *34*, 952–985. http://dx.doi.org/10.1111/j.1574-6976.2010.00220.x

Wu, J., Zhang, N., Hayes, A., Panoutsopoulou, K., and Oliver, S.G. (2004). Global analysis of nutrient control of gene expression in *Saccharomyces cerevisiae* during growth and starvation. Proc. Natl. Acad. Sci. U.S.A. *101*, 3148–3153. http://dx.doi.org/10.1073/pnas.0308321100

Wuster, A., and Babu, M.M. (2010). Transcriptional control of the quorum sensing response in yeast. Mol. Biosyst. *6*, 134–141. http://dx.doi.org/10.1039/b913579k

Wendland, J. (2014). Lager yeast comes of age. Eukaryotic. Cell *13*, 1256–1265. http://dx.doi.org/10.1128/EC.00134-14

Zhang, A., Shen, Y., Gao, W., and Dong, J. (2011). Role of Sch9 in regulating Ras-cAMP signal pathway in *Saccharomyces cerevisiae*. FEBS Lett. *585*, 3026–3032. http://dx.doi.org/10.1016/j.febslet.2011.08.023

Zheng, D., Zhang, K., Gao, K., Liu, Z., Zhang, X., Li, O., Sun, J., Zhang, X., Du, F., Sun, P., *et al.* (2013). Construction of novel *Saccharomyces cerevisiae* strains for bioethanol active dry yeast (ADY) production. PLoS ONE *8*, e85022. http://dx.doi.org/10.1371/journal.pone.0085022

Zitomer, R.S., Limbach, M.P., Rodriguez-Torres, A.M., Balasubramanian, B., Deckert, J., and Snow, P.M. (1997). Approaches to the study of Rox1 repression of the hypoxic genes in the yeast *Saccharomyces cerevisiae*. Methods *11*, 279–288.

Zupan, J., Avbelj, M., Butinar, B., Kosel, J., Šergan, M., and Raspor, P. (2013). Monitoring of quorum-sensing molecules during minifermentation studies in wine yeast. J. Agric. Food Chem. *61*, 2496–2505. http://dx.doi.org/10.1021/jf3051363

Yeast Stress and Brewing Fermentations

2

Katherine A. Smart

SABMiller PLC, SABMiller House, Woking, UK.

Correspondence: katherine.smart@sabmiller.com

https://doi.org/10.21775/9781910190616.02

Abstract

A key performance indicator of brewing fermentations is the capacity of the yeast to convert wort into the desired fermentate in an appropriate timescale. Balancing the needs of the brewer and the yeast is more complex than is always appreciated. Brewery fermentations can impose a variety of stresses on the yeast cell, particularly when conducted at scale, and this is exacerbated by the use of serial repitching, in which the yeast is recycled to complete a number of successive fermentations. Brewing yeast strains are routinely exposed to fluctuations in oxygen concentration and the accumulation of carbon dioxide, hyperosmotic stresses which are wort gravity-dependent; pH downshifts during fermentation, which can be extreme if acid washing is applied between successive fermentations; accumulation of acetaldehyde, ethanol, and organic acids; nutrient abundance, imbalance, and scarcity; and temperature shifts from around 21°C to 2°C. This chapter will focus on some of the stresses that may be encountered during lager, ale, and where appropriate wheat beer fermentations conducted at scale.

The origins of brewing yeast

The majority of publications in the field have been focused on lager strains mainly because of the larger global volume associated with these beer styles. Although many of the observations concerning stress responses may also apply to ale and hefe ale yeast strains this is not always the case. One reason for these differences is that lager yeast strains are genetically different from their ale and wheat yeast peers and in fact from each other. Lager yeast strains are interspecific hybrids (meaning hybrids from two different species) but the exact parentage had been a matter of debate (Kodama *et al.*, 2005) (see Chapters 4 and 6). Originally a model involving hybridization of an ale strain of *Saccharomyces cerevisiae* with a strain of the genetically complex species, *Saccharomyces bayanus*, was generally accepted. However, lager yeast also appear to contain DNA not common to either parent (Rainieri *et al.*, 2006; Nakao *et al.*, 2009), suggesting genetic input from an unknown 'parental species'. This remained a matter of much debate until the discovery of a new species, *Saccharomyces eubayanus*, which has been revealed to be an exact genetic match to the non-*S. cerevisiae* parent in the complex (Libkind *et al.*, 2011). The species was discovered associated with Nothofagus (southern Beech) in the forests of Patagonia and unpublished reports suggest it is also found elsewhere in the world but many workers in the field of yeast ancestry and diversity believe this to be but one example habitat from which such species have been derived.

The hybrid condition of *S. pastorianus* (*S. cerevisiae* × *S. eubayanus*) may be the reason for its psychrophilic nature, apparently inherited from the cold-tolerant *S. eubayanus* parent (Libkind *et al.*, 2011). This characteristic has resulted in superior performance in the low-temperature environment of lager fermentations and confers resistance to cold during storage and yeast transfer in breweries.

Lager strains are categorized as Saaz and Frohberg after the locations in Bohemia and Germany in which they were originally used. These two hybrid groups are used by modern brewers but are genetically distinct from one another (Liti *et al.*, 2005; Dunn and Sherlock, 2008).

Ale brewing predates lager brewing, and although the yeast strains have been selected over time, their origins are less well documented or indeed studied. Ale strains belong to the species *S. cerevisiae* but strains can be really quite genetically distinct from one another.

Wheat beer yeast strains also belong to the species *S. cerevisiae* and are characterized by the occurrence of the *PAD1* and *FDC1* genes, encoding phylacrylic acid and ferulic acid decarboxylase, respectively. Both genes must be functional for the characteristic clove-like, phenolic character due to 4-vinylguaiacol (4-VG) to be produced.

Brewing fermentations

Modern brewing practices used by global, regional, and craft brewers are essentially derivatives of those used in traditional, historical brewing. The late nineteenth-century innovations concerning the concept of single-strain brewing and the use of propagation systems eclipsed the practice of continual serial repitching. Although some breweries do practice this even today, they face the unintended consequences of hygiene issues and genetic drift of primary strains. Most global and regional brewers use propagation in specially designed vessels to ensure sufficient inoculating biomass (pitch) of the desired brewing strain. Craft brewers tend to use dried yeast or yeast slurries acquired from Regional or Global neighbouring breweries. Yeast is propagated in wort under aerobic conditions, achieved through the addition of sterile air or oxygen (Boulton *et al.*, 2000). For further discussion of yeast handling, see Chapter 3.

Following propagation, the fresh yeast slurry is transferred (pitched) into aerated or oxygenated wort in a cylindroconical (CC) fermentation vessel. For the most part, these vessels are vertically placed to alleviate footprint constraints but on occasion companies adopt a horizontal position for these vessels presumably to reduce hydrostatic pressure impacts on some strains with the additional consequence of a modified ester (fruity) profile. CC fermenters may, or may not, be mixed using pumped loop systems or internal agitators where appropriate, though this practice is relatively new to the sector. At the beginning of fermentation a brief lag phase occurs. This is partly caused by the requirement for the yeast to exit from G_0 of the yeast cell cycle and commence replication, but is also a consequence of the adjustment of the cells to changes in the nutrient, gaseous, and physical environment upon pitching. Following the lag phase, yeast grows exponentially, rapidly depleting the available oxygen and key nutrients before entering an anaerobic environment. The fermentable wort sugars and assimilable nutrients are rapidly utilized, resulting in carbon and nutrient limitation, typically prompting the cell to enter a quiescent state (see Chapter 1). These carbon- and nutrient-limited conditions coincide with an increase in ethanol concentration (Casey *et al.*, 1984). On completion of fermentation, yeast that has sedimented to the cone at the bottom of the cylindroconical vessel is removed (cropped) from the base of the cone and a portion of this yeast is stored under beer at low temperature (3–4°C) until required for use in subsequent fermentations (Briggs *et al.*, 2004). Before repitching, yeast can be washed with food-grade acids (pH) in order to remove bacterial contaminants (Simpson and Hammond, 1989), though the deployment of this practice is not consistent with best-practice hygiene and yeast management, which should obviate the need for this step. Yeast repitched into a fermentation vessel is subjected to the same procedures and indeed stresses in a cyclic manner. Serial repitching, whereby yeast cropped at the end of the fermentation is re-used in subsequent fermentations, is a process unique to brewery fermentations and should only be conducted a finite number of times to prevent yeast quality deterioration and the consequent fermentation performance compromise.

One of the key drivers for large-scale brewing is the requirement to improve the efficiency of wort fermentation, leading to a reduction in fermentation time and/or an increase in yield. The latter requirement has been addressed by the use of high-gravity worts, produced via the addition of sugar adjuncts that result in higher ethanol concentrations towards the end of fermentations (Casey and Ingledew, 1983; Casey *et al.*, 1984; Stewart *et al.*, 1988; Stewart, 2001). Traditional and more recent innovative developments in practice are usually adopted to improve product quality and/or process efficiency; however, the fermentation of wort by yeast and the handling of yeast between fermentations results in exposure to numerous stresses (Gibson *et al.*, 2007).

The general stress response

Saccharomyces cerevisiae demonstrates two major stress response pathways, the heat shock response (HSR), which is activated in a complex manner by sublethal heat stress (Chatterjee *et al.*, 2000) and mediated by the so-called heat shock transcription factor (HSF) (Bienz and Pelham, 1986; Morimoto *et al.*, 1996). Alternatively, the general (or global) stress response (GSR) is activated by a number of environmental stresses including oxidative, pH, heat, and osmotic stresses, as well as nitrogen starvation (Ruis and Schuller, 1995; Schmitt and McEntee, 1996; Martinez-Pastor *et al.*, 1996). The GSR is believed to be an evolutionary adaptation that allows yeast to respond to adverse environmental conditions in a non-specific manner, in order that cellular fecundity is retained whilst specific responses are activated (Ruis and Schuller, 1995; Martinez-Pastor *et al.*, 1996). The GSR is typified by the up-regulation of approximately 200 genes and their corresponding proteins, which are involved in a diverse array of cellular functions (Gasch *et al.*, 2000; Causton *et al.*, 2001). The expression of these genes has been demonstrated to occur in a process dependent upon the pentameric cis-acting sequence CCCCT within the promoter region of the induced genes. This so-called stress responsive element (STRE) was first identified in reference to the stress-induced expression of the *CTT1* gene encoding cytosolic catalase T (Marchler *et al.*, 1993) and subsequently in control of expression of the *DDR2* gene, which encodes a putative chaperone protein (Kobayashi and McEntee, 1993). It has subsequently been demonstrated that the activation of the STRE element of inducible genes is dependent upon two zinc finger transcriptional activators (Msn2p and Msn4p) (Martinez-Pastor *et al.*, 1996; Schmitt and McEntee, 1996; Treger *et al.*, 1998), which are active during a wide array of stresses (Ruis and Schuller, 1995; Hohmann, 2002), including those associated with yeast handling in the industrial brewery and during the diauxic shift (Boy-Marcotte *et al.*, 1998). This mode of activation explains why exposure to one type of stress often confers resistance to another, unrelated form of stress (Lindquist, 1986). The general stress response is a transient phenomenon and Msn2p is rapidly degraded following the stress response (Bose *et al.*, 2005).

Oxidative stress

Whilst molecular oxygen (O_2) is necessary for the release of energy during respiration, derivative forms of O_2, termed reactive oxygen species (ROS) are produced endogenously by cells under aerobic conditions. These ROS include the superoxide radical ($O_2^{-}\cdot$), hydrogen peroxide (H_2O_2), and the hydroxyl radical (·OH), which can damage cell components, contribute to cellular ageing, and ultimately lead to cell death (Beckman and Ames, 1998). Specific effects include lipid peroxidation (Girotti, 1998), protein inactivation (Cabiscol *et al.*, 2000), and nucleic acid damage (Salmon *et al.*, 2004; Ribeiro *et al.*, 2006), including damage to mitochondrial DNA, which can lead to the generation of respiratory-deficient 'petite' mutants (O'Rourke *et al.*, 2002; Doudican *et al.*, 2005; Gibson *et al.*, 2006). Oxygen's apparently contradictory roles within the cell, that is being essential for aerobic respiration and other metabolic processes, whilst being inherently toxic, have been referred to as the 'oxygen paradox' (Davies, 1995).

Oxygen in the brewery

Interestingly, no definitive work concerning the required levels of oxygen for both propagation and fermentation has been published. In general terms, during fermentation some archaic 'rules of thumb' are applied such that for every °P increase in wort gravity an additional 1 ppm of oxygen is required, which for the purposes of this review will be known as the 1 : 1 theory. The work of Ashraf and Smart, (in preparation), questions this guideline and the premise that the levels of oxygen applied are not attuned to the requirement of yeast generally. In very large-scale fermentations, the point of application of oxygen or air is also not uniform, and in any case oxygen uptake is not a metric for most full-scale fermentations. This is surprising because oxygen plays an essential role in the brewing process despite its potential toxicity. A supply of oxygen is necessary during brewery propagation and early fermentation to generate yeast biomass and ensure that yeast is in optimum physiological condition for effective fermentation (Hammond, 2000; Hulse, 2003). Oxygen is required for lipid synthesis, which is necessary to maintain plasma membrane integrity and function, and consequently for cell replication (Hammond, 2000; Briggs *et al.*, 2004). Sterols are synthesized using carbon devolving from glycolysis via acetyl-CoA. The first part of the synthesis is anaerobic, involving the conversion of acetyl Co-A to squalene; the conversion of squalene to 2,3-epoxysqualene, the precursor to the formation of sterols, requires molecular oxygen. (Boulton and Quain, 2003). A lack of oxygen can lead to an increase in cellular acetyl coenzyme A which can, in turn, lead to increased levels of esters such as ethyl acetate that can affect beer flavour (Briggs *et al.*, 2004). Conversely, overexposure of yeast to oxygen in the fermentation vessel can result in excessive yeast growth at the expense of ethanol production (Briggs *et al.*, 2004). Optimum oxygen levels are therefore necessary for successful beer production.

After fermentation is complete, the yeast cells that have sedimented out of the wort are re-used in subsequent fermentations. Serial repitching, a procedure unique to brewery fermentation, may have important implications for yeast physiological state and fermentation performance (Powell *et al.*, 2003) (see Chapter 1).

Defending against oxygen using antioxidants

The importance of oxidative stress to the cell is illustrated by the number and diversity of antioxidant molecules that are synthesized by *Saccharomyces* spp. (Dawes, 2004). These antioxidant defences include the non-enzymatic molecules L-ascorbic acid, D-erythroascorbic acid, flavohaemoglobin, glutaredoxin, glutathione (GSH), metallothioneins, polyamines, thioredoxin, ubiquinol, trehalose, ergosterol, and enzymatic defences such as catalase, cytochrome c peroxidase, superoxide dismutase, glutathione peroxidase, glutathione reductase, thioredoxin peroxidase, and thioredoxin reductase, (Higgins *et al.*, 2003; Kwon *et al.*, 2003; Dawes, 2004; Saffi *et al.*, 2006). Antioxidants synthesized by brewing strains have been measured in both small-scale and industrial propagations (Martin *et al.*, 2003) and fermentations (Clarkson *et al.*, 1991). However, some of these indicators are regulated by STRE elements and can indicate the occurrence of other stresses (Costa and Moradas-Ferreira, 2001). Interestingly, Swan *et al.* (2003) observed that cellular levels of thioredoxin (TRX), a small, ubiquitous antioxidant protein, remained unchanged throughout an industrial fermentation, although extracellular concentrations increased as fermentation proceeded. It was also observed in laboratory-scale fermentations that the increase in extracellular TRX concentration was greater with higher wort dissolved oxygen concentrations (Swan *et al.*, 2003). Cellular GSH concentrations vary during the brewing process (Gibson *et al.*, 2006), falling in the early stages of both propagation and fermentation before increasing steadily as each of these stages progresses. These results indicate that GSH content of cells may be reduced during exponential growth, followed by an accumulation as cells respond to nutrient depletion and enter stationary phase (Gibson *et al.*, 2006).

It has also been observed that the relative gene expression of several antioxidant-encoding genes is high at the beginning of pilot-scale fermentations (Higgins *et al.*, 2003) and this is followed by a strong up-regulation of oxidative stress response genes in the later stages of a small-scale fermentation (James *et al.*, 2003). One reason for this observation is that antioxidants have been implicated in the protection of cells exposed to reducing environments (Rand and Grant, 2006) and ethanol stress (Costa *et al.*, 1993, 1997; Pereira *et al.*, 2001), both of which may be encountered by cells in the later stages of an industrial brewery fermentation. Furthermore, a number of antioxidant-encoding genes, including *CTT1*, contain the STRE promoter sequence and are involved in the GSR. Another possibility is that the change in antioxidant activity is related to the nutritional status of the wort, with transcription of antioxidant-encoding genes being up-regulated when glucose and other fermentable carbon sources are exhausted (Costa and Moradas-Ferreira, 2001).

The responsiveness of brewing yeast strains to changes in oxygen concentration is likely to be strain-dependent at least in timing (Clarkson *et al.*, 1991) or magnitude (Martin *et al.*, 2003; Gibson *et al.*, 2006). Sensitivity to oxidative stress has also been associated with growth phase and medium composition, both of which influence antioxidant levels and hence the resistance of individual strains (Martin *et al.*, 2003; Gibson *et al.*, 2006).

Antioxidants are not the only defence

Trehalose is an important stress protectant (Thevelein, 1984; Van Laere, 1989; Wiemken, 1990), conferring stability to the plasma membrane and enzymes and additionally acting as a carbon source during starvation (Fales, 1951; Eaton, 1960; Chester, 1963; Panek, 1963). The level of trehalose is thought to be related to stress tolerance and adaptation (Thevelein, 1984; Van Laere, 1989; Hottiger *et al.*, 1987a,b; D'Amore *et al.*, 1991; D'Hautcourt and Smart, 1999; Majara *et al.*, 1996a,b; Conlin and Nelson, 2007). Trehalose is accumulated in response to heat shock (Hottiger *et al.*, 1987a,b; Majara *et al.*, 1996b; Iwashashi *et al.*, 1995), exposure to toxic chemicals (Attfield, 1987), ethanol stress (Eleutherio *et al.*, 1993; Mansure *et al.*, 1994; Majara *et al.*, 1996b), and osmotic stress (MacKenzie *et al.*, 1988; Majara *et al.*, 1996a).

Trehalose has also been implicated in the protection of yeast cells and cellular components against ROS (Benaroudj *et al.*, 2001; Herdeiro *et al.*, 2006). The genes involved in trehalose synthesis (*TPS1*, *TPS2*, *TSL1*, *TPS3*) and degradation (*NTH1*, *NTH2*, *ATH1*) (François and Parrou, 2001) are regulated by STRE elements and are up-regulated in response to various stresses, including oxidative stress (Parrou *et al.*, 1997; Pedreño *et al.*, 2002). Strains lacking the transcriptional activators Msn2p and Msn4p are unable to accumulate trehalose in response to stress (Parrou *et al.*, 1997). Trehalose is known to inhibit the activity of essential enzymes, including glutathione reductase, which is involved in maintaining cellular homeostasis and reducing oxidative damage within the cell (Sebollela *et al.*, 2004). It is thought that this inhibition saves the cell duplicating protective actions against ROS.

How much oxygen should be deployed during fermentation?

Oxygen cannot be eliminated from the brewing process, at least if serial repitching processes are desired. However, the 1 : 1 theory lacks merit and indeed supporting data. Maemura *et al.* (1998) found that the performance of yeast during small-scale wort fermentation was unimpaired when yeast had been propagated with only limited aeration (one hour at the beginning of propagation) compared with yeast exposed to a continuous supply of air. Little difference was seen in terms of cell density, carbohydrate reserves, or unsaturated fatty acid (UFA) level (Maemura *et al.*, 1998).

The presence of O_2 in wort at the beginning of fermentation allows yeast cells to synthesize lipids, thereby revitalizing the sterol-deficient cell population and ensuring that fermentation can proceed efficiently (see Chapter 1). An alternative approach involves oxygenation of the stored yeast prior to pitching, thereby reducing the O_2 concentration necessary in the fermentation wort (Boulton *et al.*, 2000; Depraetere *et al.*, 2003). In this case, UFA synthesis occurs prior to pitching and the pitched yeast, being sterol-replete, has a reduced requirement for wort oxygenation. Trials have found that pre-oxygenated yeast in unoxygenated wort performs as well as normal yeast in oxygenated wort in terms of fermentation profile, ester synthesis, and alcohol production; the only apparent difference in this investigation was a reduced yeast growth in the unoxygenated wort (Boulton *et al.*, 2000). While yeast cells are still exposed to O_2, exposure is more readily controlled in the storage vessels than in larger, industrial-scale fermentation vessels and the use of excess O_2 can be avoided (Boulton *et al.*, 2000). The potential reduction in fermentation rate caused by the reduced cell density may be overcome by adjusting the pitching rate (Boulton *et al.*, 2000). The reduction in cell growth in that investigation may have been due to excessive O_2 consumption, which can result in depletion of trehalose. Optimum aeration of yeast prior to pitching has been shown to increase cell growth in unoxygenated wort (Fujiwara and Tamai, 2003).

It has also been suggested that the cellular requirement for O_2 can be reduced by supplementation of stored yeast or wort with unsaturated fatty acid (UFA) or sterol (David and Kirsop, 1972; Taylor *et al.*, 1979; Moonjai *et al.*, 2003). Moonjai *et al.* (2003) have, for example, proposed the use of linoleic acid supplements as an alternative to wort oxygenation and demonstrated that pre-conditioning yeast in this fashion removed the requirement for wort oxygenation. Viability and fermentation performance of supplemented cells in non-aerated wort were similar to those of unsupplemented cells in aerated wort (Moonjai *et al.*, 2003). Consequently, such supplementations may have potential in industrial fermentations by obviating the requirement for O_2, thereby mitigating the effect of oxidative stress to yeast cells. It should, however, be noted that the β-oxidation of fatty acids within yeast cell's peroxisomes can generate ROS such as H_2O_2. Koercamp *et al.* (2002) detected an oxidative stress response in yeast cells in chemostat cultures when the carbon source within the growth medium was switched from glucose to the fatty acid oleate.

Potential oxidative stress during brewing may be reduced by delaying the introduction of oxygen to the fermentation vessel. It has been found that, at least in small-scale fermentations, the fermentation performance of yeast was improved when oxygenation began 4 hours after pitching compared with oxygenation prior to pitching (Lodolo and Cantrell, 2005). Delayed oxygenation also resulted in a higher UFA and ergosterol synthesis, reduced levels of off-flavour compounds, and reduced free radical activity. It was hypothesized that the improved fermentation performance was due to improved pro-mitochondrial development in yeast cells not exposed to potentially toxic O_2 at an early stage in their lifecycle (Lodolo and Cantrell, 2005). Whether delayed oxygenation is practical during industrial scale (3000 hl) fermentations has yet to be demonstrated.

Osmotic stress

The advent of high- and very-high-gravity brewing as a means of managing brewing capacity constraints has been a focus in the sector since the early 1990s. Two key stresses are associated with its deployment: osmotic and ethanol stress. Osmotic stress can be defined as any situation where there is an imbalance of intracellular and extracellular osmolarities, sufficient to cause a deleterious change in physiology (Csonka and Hanson, 1991). In natural environments, yeasts are continuously subjected to changes in external osmolarity that can be extremely detrimental to cellular functioning (Hounsa *et al.*, 1998; Beney *et al.*, 2000; Tamás and Hohmann, 2003). Indeed, osmotic stress may occur when there is a low external osmotic potential, for example in deionized water, and this is characterized by an influx of water into the cell, resulting in hypo-osmotic stress (Csonka and Hanson, 1991; Dihazi *et al.*, 2001). Conversely, and this is certainly the focus in brewing fermentations, osmotic stress may also result from exposure to environments comprising high solute concentrations, leading to hyperosmotic stress characterized by the loss of cellular water and subsequently turgor (Blomberg and Adler, 1992; Wood, 1999).

Hyperosmotic stress in the brewery

During the brewing process, there are two major sources of osmotic stress. The first involves the practice of acid washing that is employed by some brewers to remove bacterial contaminants (Hammond *et al.*, 2001) and/or fluidise yeast slurries so that dispersal is more efficient following inoculation into the fermentation vessel. Acid washing regimes involve the submersion of yeast slurries in food-grade acids to achieve a pH within the range 2.2–2.4, which imparts osmotic stress mainly due to the abundance of dissociated H^+ ions. The second source of osmotic stress occurs during inoculation into wort, a complex and highly concentrated medium containing high concentrations of sugars (Erasmus *et al.*, 2003). The use of high-gravity (or higher-solute) worts (Panchal and Stewart, 1980) has been suggested to increase external osmotic pressure, resulting in a deterioration of viability, growth, and fermentation performance (D'Amore, 1992). In support of this, Cahill *et al.* (2000) have demonstrated that an increase in gravity of the propagation medium results in decreased viabilities in subsequent high-gravity (17.5°P) fermentations. More recently, Dumont *et al.* (2003) suggested that hyperosmotic stress may reduce the loss of yeast viability incurred in response to ultra-rapid cooling rates during freezing as a consequence of the reduction in intracellular water.

Osmotolerance and osmoadaption

Osmoregulation in yeast is dependent on the capacity to sense external stimuli and the resultant changes in physiology, biochemistry, and other cellular functions to meet the modified needs of the cell in light of that environmental change. There are two forms of 'response' that any cell may apply following exposure to osmotic challenge: osmotolerance and osmoadaption.

Osmotolerance is strain-dependent and refers to the capacity of a strain to withstand osmotic imbalances (Werner-Washburne *et al.*, 1993; Hounsa *et al.*, 1998; Gasch and Werner-Washburne, 2002). In this scenario, tolerance is derived from an innate ability to withstand the deleterious effects of hyperosmotic pressure as a consequence of 'superior' membrane structure, vacuolar functioning, residual trehalose levels, and many other intrinsic factors (Latterich and Watson, 1991; Sharma *et al.*, 1996; Singer and Lindquist, 1998; Nass and Rao, 1999). Indeed, osmotolerance is promoted by the abundance of osmoprotectant macromolecules that stabilize cellular membranes, enzymes, other proteins, and possibly nucleic acids, with little effect on the intracellular water potential (Hernández-Saavedra *et al.*, 1995).

Osmoadaption typically involves the cessation of replication (Poolman and Glaasker 1998) in favour of survival mode. The process involves a highly refined sensing and response system that is activated in either acute or chronic form. Nass and Rao (1999) defined the chronic response (or acquired osmotolerance) as a signal transduction-mediated pathway that alters the levels of specific proteins, whereas the acute response is a rapid response invoked in response to sudden shifts in high external osmolarity. Both involve the accumulation of one or more types of molecule, termed *osmoticum* (pl = *osmotica*), within the cell in order to increase intracellular osmotic potential, and thus prevent cellular water loss (Yancey *et al.*, 1982; Wegmann, 1986; Blomberg and Adler, 1992; Hernández-Saavedra *et al.*, 1995). One subclass of osmotica, the compatible solutes, have very little effect on normal cellular functioning when accumulated at high levels (Poolman and Glaasker 1998) but the accumulation of this solute causes the cell to retain water that would otherwise be effluxed. Their role in stabilizing the cell during hyperosmotic stress is therefore crucial.

Glycerol is the key compatible solute accumulated during osmotic stress (Brown, 1978; Brown *et al.*, 1986, Blomberg and Adler, 1989, 1992, Albertyn *et al.*, 1994; Hohmann, 1997), and deletions in key genes encoding enzymes in the glycerol biosynthetic pathway lead to an inability to survive hyperosmotic conditions (Albertyn *et al.*, 1994; Eriksson *et al.*, 1995; Liden *et al.*, 1996; Ansell *et al.*, 1997; Hounsa *et al.*, 1998). Glycerol has an important secondary role to play in anaerobic stress, as the requirement of NAD^+ in its production enables glycerol to serve as the final product in a 'redox dump' pathway (Ansell *et al.*, 1997) and this is also relevant to brewing yeast fermentations. Cells subjected to high external osmolarity are able to effectively sense this external stimulus using two surface sensor proteins, Sln1p and Sho1p (Maeda *et al.*, 1994, 1995), which results in the activation of the HOG (MAP kinase) pathway. The HOG stimulates the hyperproduction and hyperaccumulation of glycerol as a compatible solute in order to balance the external and internal osmolarities (Albertyn *et al.*, 1994; Remize *et al.*, 2001; Pahlman *et al.*, 2001). It also appears to mediate the activity of the glyceroaquaporin Fps1p (Luyten *et al.*, 1994; Tamás *et al.*, 1999), which reduces the net efflux of glycerol and thus aids intracellular accumulation.

Saccharomyces spp. and other yeasts show a change in cell size concurrent with external osmolarity changes (Hohmann, 1997). An increase in external osmolarity results in a rapid loss of intracellular water and thus cell shrinkage (Morris *et al.*, 1986; Blomberg and Adler, 1992; Hohmann, 1997). Restoration of favourable (isotonic) solute conditions to hyperosmotically stressed cells does not, however, result in cells regaining their former volume (Hohmann, 1997).

pH downshift

Publications concerning fermentation stresses frequently omit pH. This is unfortunate, since wort pH during a lager or ale fermentation is typically reduced from 5.5 to approximately 4.1 (Coote and Kirsop, 1976; Rowe *et al.*, 1994). The practice of using acidified worts changes the extent of the downshift considerably. Any decrease in fermentation pH occurs as a result of production of carbonic acid (CO_2), secretion of organic acids, and the consumption of buffering compounds (basic amino acids and primary phosphates) in the wort (Coote and Kirsop, 1976). Coote and Kirsop (1976) found a similar buffering capacity in wort and beer and calculated that the removal of buffering materials and release of organic acids are insufficient to account for the magnitude of the pH drop typically observed during fermentation. They concluded that the secretion of H^+ by yeast accounts for the discrepancy. The protons that are released potentially cause a decrease in the intracellular pH (pHi) (Ullah *et al.*, 2012), thereby decreasing the activities of metabolic functions. In addition, changes in pHi affect signal transduction (Dechant *et al.*, 2010), protein interaction (Young *et al.*, 2010), and cell division rate (Orij *et al.*, 2012). As a result of their charged state, the anions released also accumulate inside the cell where they may increase the internal turgor pressure and exert growth-inhibitory interactions, depending on the nature of the anion (Ullah *et al.*, 2012). The plasma membrane ATPase is, therefore, important in maintaining the intracellular pH, allowing the cell to adapt to acidic conditions. However, the energy demand for the activity of this enzyme may result in a depletion of the cell's energy reserves in the form of ATP. Very little is known about the impact of lactic acid on brewing yeast during lambic beer fermentations or the production of Berliner Weiss beers; however, it can be hypothesized that during these fermentations the pHi is significantly impacted.

The final pH of the fermentation is dependent upon the buffering capacity of the wort, initial wort pH, and the extent of yeast growth (Heggart *et al.*, 1999; Narziss *et al.*, 1983). Although it has been suggested that brewing yeasts readily tolerate the 1.5 to 2.0 unit down-shift in pH that occurs during fermentation (Boulton and Quain, 2001), recent evidence suggests that brewing lager yeast strains demonstrate some sensitivity to changes in this parameter, manifesting in modified specific growth rates and reduced replicative lifespans (Maskell, 2003).

pH can have a sizeable impact on the production of flavour components by yeast. A lowering of pitching wort pH from pH 5.75 to 5.46 can result in a reduction in dimethyl sulfide production of up to 50% (Anness and Bamforth, 1982). Furthermore, the rate of conversion of α-acetolactate to the detrimental butterscotch flavour diacetyl was increased 4-fold when wort pH was lowered from pH 5.5 to 4.0 (Haukeli and Lie, 1978).

The downshift in pH during a typical lager fermentation also results in a change in gene expression. In laboratory studies, *S. cerevisiae* experiencing a shift from pH 5.5 to 3.5 in minimal media demonstrated an increase in the expression of 36 genes (Kapteyn *et al.*, 2001) involved in the cell wall, carbohydrate metabolism, redox metabolism, and stress.

Acid washing

Exposure to exceptionally low pH can occur if a brewery deploys acid washing. The use of acid washing is, in reality, a compensation for poor hygiene and is not a practice most brewers would opt to use. It may eliminate bacteria, but wild yeast do survive and production strains can be negatively impacted despite retaining viability. Some breweries have noted that the performance of their yeast strains was stimulated during fermentations after the application of acid washing (Brown, 1916; Russell and Stewart, 1995), some suggest that it may be diminished (Bruch *et al.*, 1964; Roessler, 1968), and others (Simpson and Hammond, 1989; Cunningham and Stewart, 1998) suggest that there are no significant differences in the fermentation profiles. Despite this, there is little doubt that acid washing represents a stress to the cell. Trehalose has also been observed to accumulate (Cheng *et al.*, 1999); proteins involved in plasma membrane integrity may be susceptible to denaturation, particularly in the presence of ethanol (Jones, 1988; Simpson and Hammond, 1989); a reduction in viability, vitality, and flocculation may occur (Casey and Ingledew, 1983; Fernandez *et al.*, 1993; Ogden, 1987; Simpson and Hammond, 1989); and newly formed daughter cells tend to exhibit poor survival rates (Maskell and Smart, unpublished data).

The addition of diethylstilboestrol during acid washing dramatically reduces yeast viability (Uchida *et al.*, 1988). Diethylstilboestrol is an inhibitor of plasma membrane H^+-ATPase and it has therefore been suggested that yeast death was most probably due to intracellular acidification as a consequence of the inability of the cell to release protons to maintain its intracellular pH. Evidence to support the hypothesis that plasma membrane H^+-ATPase is essential for survival during acid washing comes from the studies of Eraso and Gancedo (1987), who showed that plasma membrane H^+-ATPase was stimulated by low pH (Eraso and Gancedo, 1987). However, Carmelo *et al.* (1996) observed that *S. cerevisiae* cells grown in media with an initial pH of 2.5–6.0 exhibited H^+-ATPase activities of 30% of the maximum value at pH 2.5, suggesting that overexposure to low pH has a negative effect. Growth impairment was also observed in these cell populations. There have been no reports on the effect of phosphoric acid on the plasma membrane H^+-ATPase, but it is postulated that acid washing is a practice that can be highly damaging to plasma membrane functionality if not applied with due care.

Anaerobic shift and carbon dioxide toxicity

Brewing yeast experiences an anaerobic environment as a consequence of the accumulation of CO_2 during fermentation. In addition, top pressures in storage tanks as a result of carbon dioxide or nitrogen flushing result in a similar effect. Fermentation rate, the rate and extent of yeast growth, and the final concentration of fusel oils are all decreased by increasing CO_2 pressure, whilst the final pH is decreased (Arcay-Ledezma and Slaughter, 1984; Jones and Greenfield, 1982; Kunkee and Ough, 1966). As a consequence, the final concentration of many flavour compounds and esters may be decreased by increasing the CO_2 pressure during fermentation (Drost, 1977). Although its impact has not been extensively elucidated, anaerobiosis influences several cellular functions in yeast, including modifications in cell division (Norton and Krauss, 1972), cell volume (Lumsden *et al.*, 1987), gene expression (Abramova *et al.*, 2001a; James *et al.*, 2003), metabolism (Hammond, 1993; Lewis and Young, 1995; Stewart *et al.*, 1983; Zheng *et al.*, 1994), amino acid uptake (Slaughter *et al.*, 1987), and the cell wall (Abramova *et al.*, 2001a; James *et al.*, 2003; Lawrence *et al.*, 2011).

Increases in the partial pressure of CO_2 (pCO_2) results in decreases in the cell yield and growth rate of *Saccharomyces* yeast in beer fermentation (Knatchbull and Slaughter, 1987; Nakatani *et al.*, 1984), although little change in fermentation activity due to pCO_2 change were observed. Norton and Krauss (1972) showed that cell growth stopped at 280 kPa of CO_2 pressure metabolically produced during ethanol fermentation, but a culture pressurized at 280 kPa using N_2 gas did not reduce the cell growth rate (Norton and Krauss, 1972). Inhibition of budding and cell division by *S. cerevisiae* is therefore a result of increased CO_2 concentration rather than pressure. Although cell division ceases, doubling of the DNA content of the cells still occurs, indicating that the inhibition of cell division is not due to a lack of DNA replication (Norton and Krauss, 1972). Despite the increase in DNA content of cells, the amount of RNA and protein per cell decreases (Lumsden *et al.*, 1987). Lumsden *et al.* (1987) also found that that after 1 h at elevated CO_2 pressures, the mean cell volume of *S. cerevisiae* increased. This indicated that the influence of CO_2 upon cell characteristics may be associated with a change in cell size (Lumsden *et al.*, 1987). Under a CO_2 pressure of 198 kPa, fermentation of malt extract medium by *S. cerevisiae* results in a changed pattern of absorption of amino acids in the first 4 hours, with a general excretion of amino acids thereafter (Slaughter *et al.*, 1987).

Control of gene expression

Lai *et al.* (2005) examined the dynamic remodelling of the transcriptome during acclimatization to short-term anaerobiosis (two generations) under different metabolic states (catabolite-repressed or -derepressed). They determined that shifting cells from aerobic to anaerobic conditions in galactose medium induced an acute, yet transitory induction of Msn2p- and/or Msn4p-regulated genes associated with the retooling of reserve energy and catabolic pathways during the switch from respiro-fermentative to strictly fermentative growth. These changes are involved in balancing energetic supply and demand during this transition. Concomitantly, genes associated with the G_1/S transition phase of the cell cycle were transiently down-regulated, resulting in a temporary arrest in the cell cycle. None of these networks were differentially expressed when cells experienced anaerobic shift in the presence of glucose, suggesting that a metabolically derived signal arising from the abrupt cessation of respiration, rather than O_2 deprivation, elicits this 'stress response'. In both media, under anaerobic conditions, more chronic, haem-dependant effects were observed, including the down-regulation of Hap1p- and possibly Hap2/3/4/5p-regulated networks, derepression of Rox1p-regulated networks, and, following a slight delay, activation of Upc2p-regulated networks. These changes result in functional remodelling of sterol and sphingolipid metabolism, the cell wall, and dissimilatory pathways required for long-term anaerobiosis (Lai *et al.*, 2005).

Ter Linde *et al.* (1999) and James *et al.* (2003) studied the transcription profiles of yeast during anaerobic incubation in chemostat culture and tall tube fermentations, respectively (ter Linde *et al.*, 1999; James *et al.*, 2003). During anaerobiosis, a number of genes, previously shown to respond to hypoxic conditions, were induced. These include *ERG11*, *NCP1*, *AAC3*, *COX5*, *HEM13*, *OLE1*, and the *PAU* gene family (Rachidi *et al.*, 2000), which encodes the seripauperin proteins. Interestingly, 13 ORFs of unknown function, demonstrating homology to the *PAU* genes, are also induced under these fermentation conditions. Members of the anaerobiosis-inducible mannoprotein *DAN/TIR* family are also highly induced under these conditions. Additionally, transcript levels for the hypoxic gene *AAC3*, which is transcribed optimally under anoxic or microaerophilic conditions, were also up-regulated.

Under anaerobic conditions, cells cannot synthesize UFA or sterols (Lorenz and Parks, 1991). An exogenous source of UFA and sterol is essential for long-term anaerobic growth in *S. cerevisiae* (Andreasen and Stier, 1953, 1954). Lai *et al.* (2005) showed that transcriptomic remodelling during anaerobiosis is fairly specific for sterol and sphingolipid pathways, with very few genes specific for phospholipid or fatty acid synthesis (save for *OLE1* and *AAC1*) showing changes in expression. Genes in the early portion of the sterol synthesis pathway exhibit complex expression patterns, with some showing transient (*ERG10*) or chronic up-regulation (*IDI1* and *HMG2*) and others showing transient (*ERG8* and *MVD1*) or chronic down-regulation (*ERG13*, *ERG20* and *HMG1*). Nearly all the genes in the downstream steps of the pathway were chronically up-regulated, as were genes involved in transport (*PDR11* and *AUS1*) and their primary regulator (*UPC2*).

The cell wall responses to anaerobiosis

The yeast cell wall represents a dynamic organelle that modifies in composition and functionality in response to external stimuli (Klis, 1994; Cabib *et al.*, 1997; Caro *et al.*, 1998; Abramova *et al.*, 2001a; Boorsma *et al.*, 2004; Rhymes and Smart, 2001; Smart, 2003; Lesage and Bussey, 2006). Under aerobic growth conditions, Cwp1p and Cwp2p, encoded by the genes *CWP1* and *CWP2*, respectively, are the most abundant cell wall mannoproteins and are involved in cell wall biosynthesis during cell replication (van der Vaart *et al.*, 1995). During environmental stress, the composition of the cell changes, resulting in the increase in cell wall mannoproteins other than Cwp1p and Cwp2p (Abramova *et al.*, 2001a). The most extensive response is the induction of several homologous mannoproteins (Dan1p, Tip1p, Tir1p, and Tir2p) during anaerobic growth. The genes encoding these proteins are referred to as the *DAN/TIR* genes. This group of genes also includes genes for five other mannoproteins, designated Dan2p, Dan3p, Dan4p, Tir3p, and Tir4p. The Tir1p, Tir3p, and Tir4p proteins are required for anaerobic growth, with knockout yeast becoming arrested at G_1 in anaerobic conditions (Abramova *et al.*, 2001a). During N_2-induced anaerobic adaptation, the *CWP1* and *CWP2* genes are down-regulated and the *DAN/TIR* genes are up-regulated, allowing an extensive remodelling of the yeast cell wall (Abramova *et al.*, 2001a). The *DAN* genes *DAN1* and *DAN4* are expressed within an hour of a N_2-induced anaerobic shift, and *DAN2* and *DAN3* are expressed after approximately three hours (Abramova *et al.*, 2001b). Anaerobic expression of the *DAN/TIR* genes depends on the binding of Upc2p to an upstream regulatory element. In addition, Ecm22p plays a role in the expression of *DAN2* and *DAN3* but not in the expression of any other *DAN/TIR* genes. Davies and Rine (2006) identified sterol level as the primary regulator of Upc2p. Furthermore, they showed that *DAN2* and *DAN4* were activated by Upc2p solely in response to sterol depletion rather than haem depletion, whereas *DAN1* responded to both haem and sterols (Davies and Rine, 2006). Repression of the *DAN* genes results from the combined action of at least six repression factors – Mox1p, Mox2p, Rox1p, Rox7p, Tup1p, and Ssn6p (Abramova *et al.*, 2001b). Some of the *DAN* genes are also repressed by Mot3p, a repressor that is induced by haem and repressed in anaerobic cells by Hap1p, in parallel with the Rox1p repressor. Abramova *et al.* (2001a) concluded that Mot3p, which is known to function as an activator or repressor of different genes through the same binding site, is an important activator of *CWP2*, presumably through the three Mot3p sites in the promoter. Dan1p, along with the putative ABC transporter (ATP-binding-cassette transporter) Aus1p, plays an essential role in anaerobic sterol uptake (Alimardani *et al.*, 2004). No functions associated with anaerobiosis have been assigned to the products of *DAN2*, *DAN3* or *DAN4*, despite their responsiveness to this condition.

The remodelling of the cell wall is important in the brewing context, as it has been demonstrated that the flocculation potential of brewing yeast strains is altered by growth in aerobic and anaerobic conditions (Lawrence *et al.*, 2011; Lawrence and Smart, 2011).

Ethanol toxicity

The primary purpose of brewery fermentation is the synthesis of alcohol from fermentable sugars and the production of particular flavour-active compounds by *Saccharomyces* yeast. Under normal fermentation conditions, final ethanol concentrations fall in the range of 3–6%, though under high-gravity fermentation this concentration may be greater than 10% (Briggs *et al.*, 2004). The effects of ethanol toxicity on yeast physiology are diverse, though cellular membranes appear to be the main sites of ethanol damage. Specific effects include growth inhibition, reduced cell size (Canetta *et al.*, 2006), reduced viability particularly for respiratory-deficient mutants of brewing strains (Gibson *et al.*, 2009; Cheung *et al.*, 2012), reduced respiration and glucose uptake (Pascual *et al.*, 1988), reduced fermentation (Fernandes *et al.*, 1997), enzyme inactivation, lipid modification, loss of proton motive force across the plasma membrane (Petrov and Okorokov, 1990; Mizoguchi and Hara, 1997), increased membrane permeability (Marza *et al.*, 2002), lowering of cytoplasmic pH, and the induction of respiratory-deficient mutants (Jiménez *et al.* 1988; Ibeas and Jiménez, 1997; Chi and Arneborg, 1999).

Beers produced through high-gravity fermentation have also been found to have a greater flavour stability and consistency than those produced using normal gravity brewing (McCaig *et al.*, 1992). This improved fermentation efficiency may, however, be at the expense of yeast physiological state and hence performance in subsequent fermentations.

Responses to ethanol

Yeast responding to increased ethanol concentration demonstrates an increased unsaturation index, and fluidity, of their membranes (Beaven *et al.*, 1982; Šajbidor and Grego, 1992; Lloyd *et al.*, 1993; Odumeru *et al.*, 1993; Alexandre *et al.*, 1994a). This relationship is not absolute, and oleic acid (18:1) appears to be a key determinant of ethanol tolerance in *S. cerevisiae*, rather than unsaturation index, *per se* (You *et al.*, 2003). *S. cerevisiae* cells also exhibit an increase in plasma membrane H^+-ATPase activity in response to ethanol exposure (Rosa and Sá-Correia, 1991, 1996; Alexandre *et al.*, 1994b; Monteiro and Sá-Correia, 1997), counteracting the increased influx of protons across the plasma membrane of ethanol-exposed cells (Leão and Van Uden, 1984). Fujita *et al.* (2006), found that homozygous diploid mutant strains of *S. cerevisiae* lacking genes involved in vacuolar H^+-ATPase function were sensitive to ethanol, 1-propanol, and 1-pentanol (Fujita *et al.* 2006).

Mitigating for ethanol

Typically in brewing, strains are selected for high-gravity brewing on the basis of their capacity to ferment according to profile and tolerate higher ethanol concentrations. Dilution of yeast slurries post cropping is a standard protocol used to ensure that cells are not exposed during storage to higher than necessary ethanol levels. However, there have been some studies suggesting other factors that might mitigate for the consequences of ethanol.

Increases in monounsaturated fatty acids, and corresponding decreases in saturated fatty acids have also been observed in ale and lager yeast strains exposed to ethanol, either directly through supplementation or during fermentation (Odumeru *et al.*, 1993). The membrane composition of brewing yeast is influenced by wort composition, and supplementation of high-gravity wort with ergosterol and oleic acid (in the form of Tween 80) has been shown to significantly improve fermentation rate (Casey and Ingledew, 1985) and ethanol productivity (Dragone *et al.*, 2003). However, it is still unclear whether this improved fermentation performance is due to an increase in ethanol tolerance because of changes to lipid membrane composition or simply due to improvement of the nutritional status of the growth medium, as suggested by Casey and Ingledew (1985).

Magnesium ions have a role in maintaining membrane integrity, and reduce the proton, anion, and nucleotide permeability of membranes exposed to ethanol (Salgueiro *et al.*, 1988; Petrov and Okorokov, 1990; Hu *et al.*, 2003). Increasing the bioavailability of Mg prior to or during ethanol shock reduces the synthesis of the heat shock proteins (Birch and Walker, 2000) and increases the viability (Walker, 1998; Birch and Walker, 2000; Hu *et al.*, 2003) and growth (Ciesarová *et al.*, 1996) of cells. In addition, supplementation of fermentation media with magnesium has been shown to increase fermentation rate and ethanol productivity (Dombek and Ingram, 1986; Stewart *et al.*, 1988; D'Amore *et al.*, 1990; D'Amore, 1992; Ciesarová *et al.*, 1996; Walker and Maynard, 1997; Rees and

Stewart, 1999). Impaired fermentation rates in the absence of sufficient concentrations of Mg^{2+} may also be related to its role in regulating the activity of glycolytic enzymes such as pyruvate kinase (Morris *et al.*, 1984).

Exposure of cells to ethanol stress (16%) induces the synthesis of trehalose (Sharma, 1997) and increased accumulation of trehalose has also been observed in ale and lager brewing yeast strains exposed to 10% ethanol (Odumeru *et al.*, 1993). Alexandre *et al.* (2001) demonstrated that genes involved in trehalose synthesis in yeast are up-regulated in response to ethanol stress.

Ethanol can permanently damage the cell

Ethanol is one of the most potent inducers of the petite phenotype in yeast (Jiménez *et al.*, 1988; Ibeas and Jiménez, 1997; Chi and Arneborg, 1999; Castrejón *et al.*, 2002) and it is therefore likely that the generation of petites during brewery handling is strongly influenced by repeated exposure to high ethanol concentrations during fermentation. Despite the association of ethanol toxicity with petite frequency, the actual mechanism of damage is unclear. Ethanol is not a powerful mutagen of yeast mitochondrial DNA, as determined by pulsed-field gel electrophoresis (Ristow *et al.*, 1995). It has been suggested that induction of petites may be related to the effects of ethanol on the mitochondrial membrane rather than on the mitochondrial DNA itself (Ibeas and Jiménez, 1997). It has been shown that lipid composition influences both ethanol sensitivity and susceptibility to petite mutation (Chi and Arneborg, 1999). An ethanol-tolerant yeast strain with a low level of ethanol-induced mitochondrial petite mutation was found to have a relatively high ergosterol/phospholipid ratio, high phosphatidylcholine content, and high long-chain fatty acid component compared with an ethanol-sensitive strain (Chi and Arneborg, 1999), though how these membrane characteristics affect mitochondrial integrity in the presence of ethanol has yet to be elucidated. It may be that long-chain fatty acids have a role in counteracting the fluidizing effect of ethanol on membranes. The important role of the mitochondrial membrane in maintaining mitochondrial DNA integrity has been demonstrated by the fact that strains with reduced ergosterol-synthesizing ability are more susceptible to petite mutation (Jiménez *et al.*, 1988). It has also been suggested that acetaldehyde, the first metabolite of ethanol, may have a direct mutagenic effect on mitochondrial DNA (Ristow *et al.*, 1995).

Cold shock

Low-temperature environments in the brewery

In *S. cerevisiae* (ale and wheat strains), cold shock occurs at temperatures below 20°C (Kondo and Inouye, 1991; Kondo *et al.*, 1992; Kowalski *et al.*, 1995; Abramova *et al.*, 2001b). It is therefore no surprise that ale fermentations are typically conducted at higher temperatures, between 18°C and 25°C. In contrast, lager fermentations are conducted at much lower temperatures and indeed lager strains are better able to grow at 10°C relative to ale strain counterparts. Interestingly, the cold tolerance of lager strains is hybrid group specific. The Saaz strains display greater cold tolerance than the Frohberg strains (Gibson *et al.*, 2013). Of the two yeast hybrids used for lager brewing, the Frohberg group has predominantly retained DNA of the *S. cerevisiae* parent, while the Saaz group has retained proportionally more of the *S. eubayanus* DNA (Dunn and Sherlock, 2008). These genetic differences appear to reflect functional differences, with the Saaz strain apparently inheriting a relatively greater cold tolerance from the characteristically cryotolerant *S. eubayanus* parent.

Temperatures used for yeast storage between successive fermentations typically range from 2°C to 11°C and, where applied, acid washing is also conducted at 4°C. Yeast handling, therefore, consists of a series of low-temperature environments, which may lead to the yeast slurry experiencing cold shock, a phenomenon that is well documented in other organisms (Phadtare *et al.*, 1998; Rodriguez-Vargas *et al.*, 2002), but has not been extensively studied in yeast.

Cold tolerance and brewing

The exact mechanisms controlling cold tolerance in the lager yeast are still unknown. The uptake of the α-glucosides maltose and maltotriose is known to be sensitive to temperature (Vidgren *et al.*, 2010) but different cryotolerances of the yeast groups are not necessarily related to differences in the sensitivity of maltose transporters. Differential maltotriose utilization may explain why Frohberg strains have historically been considered more industrially significant. The lack of active maltotriose transport in Saaz strains suggests that the transporters responsible for uptake are absent or non-functional.

Cold tolerance and membrane fluidity

The principal fatty acyl chains in the plasma membrane of *S. cerevisiae* are oleic acid (18:1) and palmitoleic acid (16:1), with trace amounts of palmitic acid and stearic acid also present. Membrane fluidity is largely determined by the packing of these molecules. Lowering of the temperature leads to a more ordered membrane structure and hence a reduction in fluidity (Shinitsky, 1984). The membrane is essentially modified from a liquid crystalline form to a gel state (Thieringer *et al.*, 1998). This transformation alters various functions of membrane-bound proteins, such as the import and export of metabolites and proteins across the plasma membrane. The mechanisms by which microorganisms tolerate changes in membrane fluidity has been well characterized and is termed 'homeoviscous adaptation' (Sinensky, 1974). Depending on the microorganism, this process involves increasing proportions of unsaturated fatty acids and/or *cis* double bonds into lipids, chain shortening, and methyl branching (Shaw and Ingraham, 1967; Sinensky, 1974; Shinitsky, 1984; McElhaney, 1982; Russell, 1989). Another physiological effect of lowering the temperature is the reduction in the hydrophobic interactions between the carbon skeleton of the polypeptide and the side chains of amino acids, exposing non-polar regions to water and risking protein denaturation (Gounot and Russell, 1999).

Impact of low temperature on genome-wide expression

In response to an abrupt drop in temperature, a change in gene expression has been postulated to allow for adaptation to the low-temperature environment (Sahara *et al.*, 2002; Schade *et al.*, 2004). This response is gene and time specific and involves the differential regulation of certain genes (Zhang *et al.*, 2001, 2003; Sahara *et al.*, 2002; Schade *et al.*, 2004), presumably as a result of the altered physiological state of the cell caused by reduced membrane transport, accumulation of misfolded proteins, and reduced enzyme activity (Sahara *et al.*, 2002; Schade *et al.*, 2004). The early phase of the cold shock response involves adjustments to membrane fluidity and prevents destabilization of RNA secondary structures to allow efficient protein translation. The late phase involves the up-regulation of genes involved in the general stress response, including those encoding heat shock proteins and metabolic enzymes engaged in glycogen and trehalose metabolism (Sahara *et al.*, 2002; Schade *et al.*, 2004). Sahara *et al.* (2002) demonstrated that the genes involved in the general stress response are up-regulated following 4 hours of low-temperature exposure, whilst Schade *et al.* (2004) showed that they are still being up-regulated following 12 and/or 60 hours of low-temperature exposure. The gene expression profile during growth at low temperatures (Homma *et al.*, 2003) has been reported but this is to be differentiated from cold shock gene expression profiles since the former does not involve a sudden downshift in temperature.

Responding to cold

Several genes involved in yeast cold shock have been identified. These include the *NRS1* gene (Lee *et al.*, 1991; Kondo and Inouye, 1992; Kondo *et al.*, 1992), the *LOT* genes (Zhang *et al.*, 2001), and the cell wall-related *TIR* genes (Kowalski *et al.*, 1995; Abramova *et al.*, 2001a). Translation efficiency is greatly reduced at low temperatures due to the formation of secondary structures in RNA molecules and the inactivation of ribosomes (Jones and Inouye, 1996). To combat this, an early response to cold involves an increased expression of the majority of the genes involved in rRNA synthesis and processing (Sahara *et al.*, 2002; Schade *et al.*, 2004).

Mga2p, a cold sensor, (Nakagawa *et al.*, 2002) has been demonstrated to activate *OLE1* transcription in response to low temperature (Nakagawa *et al.*, 2002), which encodes an intrinsic membrane enzyme important for maintaining optimum levels of membrane fluidity and curvature. Interestingly, a large increase in *OLE1* expression has been demonstrated during fermentation (Higgins *et al.*, 2003; James *et al.*, 2003), particularly during the initial stages (Higgins *et al.*, 2003). Brewery fermentation is considered to be a low-temperature environment (Leclaire *et al.*, 2003) where cold-induced Mga2p activation of *OLE1* expression is likely to occur.

Further evidence that the membranes are a focus for cold shock responses comes from the observation that trehalose accumulation occurs at temperatures of 10°C and below, and coincides with induction of the trehalose-synthesizing enzymes Tps1p and Tps2p (Kandror *et al.*, 2004). This is believed to be an adaptive response, which increases tolerance to low temperatures as well as freezing.

A remodelling of the cell wall appears to take place when cells are subjected to cold shock, since a mannoprotein, Cwp1p (cell wall protein), is down-regulated and expression of the *TIP1*, *TIR1*, *TIR2* and *TIR4* genes is induced (Abramova *et al.*, 2001a). Since flocculation of lager yeast is increased at lower temperatures (González *et al.*, 1996) and the temperature of growth has also been reported to affect the final flocculation capacity (Van Iersel *et al.*, 1998), the *TIR* genes may be important for this fermentation characteristic.

Conclusions and future perspectives

The literature concerning yeast stress responses is largely focused on studies utilizing haploid laboratory strains of *S. cerevisiae*. In this chapter, only those findings with relevance to key brewing performance indicators have been discussed. These studies are thorough and provide much evidence concerning the cell biology of stresses incurred during brewing. However, as a cautionary note they may not provide the full picture for lager strains, which are hybrids and do not belong to the same species, and ale and wheat strains, which are much more diverse from an evolutionary perspective than their counterpart lager strains and laboratory strains.

Most laboratory strains originate from ancestors prospected from the environment. These strains in turn were maintained in diploid and haploid forms and crossed in laboratories as scientists attempted to understand replication (sexual and asexual) and genetics. As an example, the strain S288C was isolated as a result of multiple genetic crosses from strains originally derived from lager brewing yeast foam, baking and strains acquired for other laboratories. Several crosses were conducted with the purpose of isolating subsequent biochemical mutants. S288C was isolated because it was non-flocculent, could be grown readily and had a requirement for biotin allowing genetic studies involving selective media to be utilized (Mortimer and Johnstone, 1986). This strain was then shared with the scientific community leading to further crosses, the descendants of which became laboratory strains in their own right.

Lager yeast (*S. pastorianus*) is defined as an allopolyploid hybrid species of *S. cerevisiae* and a non-*cerevisiae* species. Originally this latter parent was assumed to be *S. bayanus*; however, the *S. bayanus* strain was itself shown to be a hybrid of *S. cerevisiae* and a non-*cerevisiae* species now known as *S. eubayanus*. In a recent study, it was demonstrated that *S. pastorianus* Weihenstephan 34/70 and *S. carlsbergensis* CBS-1513 exhibit a markedly distinct genetic distance to *S. cerevisiae* S288c, despite the lager part-ancestry of S288C (Chen *et al.*, 2016). This genetic distance confirms that observations made with laboratory strains may not reflect the responses of a lager strain.

Responses observed with one lager strain may not be indicative for lager strains in general. The reason for this is that lager yeast can be divided into two groups, Group I (Saaz/Carlsberg type) and Group II (Frohberg type). This division is based on the geographic heritage of the strains and has been supported by fermentation characteristics and molecular analyses (Gibson *et al.*, 2013). *S. carlsbergensis* CBS-1513 and *S. pastorianus* Weihenstephan 34/70 are considered typical for strains belonging to Groups I and II, respectively. The main difference between the two groups at the genome level is the chromosome ploidy. The lack of one group of *S. cerevisiae* chromosomes in Group I strains makes the Group I strains allotriploid and distinguishes the Group I strains from the allotetraploid Group II strains such as *S. pastorianus* Weihenstephan 34/70 (Chen *et al.*, 2016). Thus, the portion of chromosomes originating from *S. cerevisiae* in Group I strains is ≈33.3%, while this ratio in Group II strains is around 50.0%. In many brewing publications strain dependent phenomenon have been proposed for findings not conforming to the expected phenotype. It would be interesting to reconsider these findings in light of the current understanding of relatedness and unrelatedness of some key industrial strains.

In addition to these important genetic differences, the translation from brewing fermentations conducted at the laboratory or pilot scale to those occurring at full industrial scale is also problematic.

Primarily this is due to the failure of most laboratory systems to effectively mimic full scale operations. This is a gap that requires closure to fully evaluate the potential of brewing yeast. Despite these issues, the last 15 years has seen much progress in the field of brewing yeast stress research and with every publication understanding improves.

References

Abramova, N., Sertil, O., Mehta, S., and Lowry, C.V. (2001a). Reciprocal regulation of anaerobic and aerobic cell wall mannoprotein gene expression in *Saccharomyces cerevisiae*. J. Bacteriol. *183*, 2881–2887. http://dx.doi.org/10.1128/JB.183.9.2881-2887.2001

Abramova, N.E., Cohen, B.D., Sertil, O., Kapoor, R., Davies, K.J.A., and Lowry, C.V. (2001b). Regulatory mechanisms controlling expression of the *DAN/TIR* mannoprotein genes during anaerobic remodeling of the cell wall in *Saccharomyces cerevisiae*. Genetics *157*, 1169–1177.

Albertyn, J., Hohmann, S., Thevelein, J.M., and Prior, B.A. (1994). *GPD1*, which encodes glycerol-3-phosphate dehydrogenase, is essential for growth under osmotic stress in *Saccharomyces cerevisiae*, and its expression is regulated by the high osmolarity glycerol response pathway. Mol. Cell Biol. *6*, 4135–4144.

Alexandre, H., Rousseaux, I., and Charpentier, C. (1994a). Ethanol adaptation mechanisms in *Saccharomyces cerevisiae*. Biotechnol. Appl. Biochem. *20*, 173–183.

Alexandre, H., Rousseaux, I., and Charpentier, C. (1994b). Relationship between ethanol tolerance, lipid composition and plasma membrane fluidity in *Saccharomyces cerevisiae* and Kloeckera apiculata. FEMS Microbiol. Lett. *124*, 17–21.

Alexandre, H., Ansanay-Galeote, V., Dequin, S., and Blondin, B. (2001). Global gene expression during short-term ethanol stress in *Saccharomyces cerevisiae*. FEBS Lett. *498*, 98–103.

Andreasen, A.A., and Stier, T.J. (1953). Anaerobic nutrition of *Saccharomyces cerevisiae*. I. Ergosterol requirement for growth in a defined medium. J. Cell. Physiol. *41*, 23–36.

Andreasen, A.A., and Stier, T.J. (1954). Anaerobic nutrition of *Saccharomyces cerevisiae*. II. Unsaturated fatty acid requirement for growth in a defined medium. J. Cell. Physiol. *43*, 271–281.

Anness, B.J., and Bamforth, C.W. (1982). Dimethyl sulphide– a review. J. Inst. Brew. *88*, 244–252.

Ansell, R., Granath, K., Hohmann, S., Thevelein, J.M., and Adler, L. (1997). The two isoenzymes for yeast NAD+-dependent glycerol 3-phosphate dehydrogenase encoded by GPD1 and GPD2 have distinct roles in osmoadaptation and redox regulation. EMBO J. *16*, 2179–2187. http://dx.doi.org/10.1093/emboj/16.9.2179

Arcay-Ledezma, G.J., and Slaughter, J.C. (1984). The response of *Saccharomyces cerevisiae* to fermentation under carbon dioxide pressure. J. Inst. Brew. *90*, 81–84.

Attfield, P.V., Kletsas, S., Veal, D.A., van Rooijen, R., and Bell, P.J. (2000). Use of flow cytometry to monitor cell damage and predict fermentation activity of dried yeasts. J. Appl. Microbiol. *89*, 207–214.

Beaven, M.J., Charpentier, C., and Rose, A. (1982). Production and tolerance of ethanol in relation to phospholipid fatty-acyl composition in *Saccharomyces cerevisiae* NCYC 431. J. Gen. Microbiol. *128*, 1447–1455.

Beckman, K.B., and Ames, B.N. (1998). The free radical theory of aging matures. Physiol. Rev. *78*, 547–581.

Benaroudj, N., Lee, D.H., and Goldberg, A.L. (2001). Trehalose accumulation during cellular stress protects cells and cellular proteins from damage by oxygen radicals. J. Biol. Chem. *276*, 24261–24267. http://dx.doi.org/10.1074/jbc.M101487200

Beney, L., Marechal, P.A., and Gervais, P. (2001). Coupling effects of osmotic pressure and temperature on the viability of *Saccharomyces cerevisiae*. Appl. Microbiol. Biotechnol. *56*, 513–516.

Bienz, M., and Pelham, H.R. (1986). Heat shock regulatory elements function as an inducible enhancer in the *Xenopus* hsp70 gene and when linked to a heterologous promoter. Cell *45*, 753–760.

Birch, R.M., and Walker, G.M. (2000). Influence of magnesium ions on heat shock and ethanol stress responses of *Saccharomyces cerevisiae*. Enzyme Microb. Technol. *26*, 678–687.

Blomberg, A., and Adler, L. (1989). Roles of glycerol and glycerol-3-phosphate dehydrogenase (NAD+) in acquired osmotolerance of *Saccharomyces cerevisiae*. J. Bacteriol. *171*, 1087–1092.

Blomberg, A., and Adler, L. (1992). Physiology of osmotolerance in fungi. Adv. Microb. Physiol. *33*, 145–212.

Boorsma, A., de Nobel, H., ter Riet, B., Bargmann, B., Brul, S., Hellingwerf, K.J., and Klis, F.M. (2004). Characterization of the transcriptional response to cell wall stress in *Saccharomyces cerevisiae*. Yeast *21*, 413–427.

Bose, S., Dutko, J.A., and Zitomer, R.S. (2005). Genetic factors that regulate the attenuation of the general stress response of yeast. Genetics *169*, 1215–1226.

Boulton, C. (2000). Trehalose, glycogen and sterol. *Brewing* Yeast *Fermentation Performance*, 1st Edition (Smart, K.A., ed), pp. 10–19. Blackwell Science, Oxford, UK.

Boulton, C., Clutterbuck, V., and Durnin, S. (2000). Yeast oxygenation and storage. In Brewing Yeast Fermentation Performance, 1st edn, K.A. Smart, ed. (Oxford, UK: Blackwell Science). pp. 140–151.

Boulton, C., and Quain, D. (2001). Brewing Yeast and Fermentation. (Oxford, UK: Blackwell Science).

Boy-Marcotte, E., Perrot, M., Bussereau, F., Boucherie, H., and Jacquet, M. (1998). Msn2p and Msn4p control a large number of genes induced at the diauxic transition which are repressed by cyclic AMP in *Saccharomyces cerevisiae*. J. Bacteriol. *180*, 1044–1052.

Briggs, D.E., Boulton, C.A., Brookes, P.A., and Stevens, R. (2004). Brewing: science and practice. (Cambridge, UK: Woodhead)

Brown, H.T. (1916). Fifty years experience of scientific method in brewing practice. J. Inst. Brew. *22*, 267–348.

Brown, A.D. (1978). Compatible solutes and extreme water stress in eukaryotic micro-organisms. Adv. Microb. Physiol. *17*, 181–242.

Bruch, C.W., Hoffman, A., Gosine, R.M., and Brenner, M.W. (1964). Disinfection of brewing yeast with acidified ammonium persulphate. J. Inst. Brew. *70*, 242–246.

Cabib, E., Drgon, T., Drgonová, J., Ford, R.A., and Kollár, R. (1997). The yeast cell wall, a dynamic structure engaged in growth and morphogenesis. Biochem. Soc. Trans. *25*, 200–204.

Cabiscol, E., Piulats, E., Echave, P., Herrero, E., and Ros, J. (2000). Oxidative stress promotes specific protein damage in *Saccharomyces cerevisiae*. J. Biol. Chem. *275*, 27393–27398. http://dx.doi.org/10.1074/jbc.M003140200

Cahill, G., Murray, D.M., Walsh, P.K., and Donnelly, D. (2000). Effect of the concentration of propagation wort on yeast cell volume and fermentation performance. J. Am. Soc. Brew. Chem. *58* (1), 14–20.

Canetta, E., Adya, A.K., and Walker, G.M. (2006). Atomic force microscopic study of the effects of ethanol on yeast cell surface morphology. FEMS Microbiol. Lett. *255*, 308–315.

Carmelo, V., Bogaerts, P., and Sá-Correia, I. (1996). Activity of plasma membrane H+-ATPase and expression of PMA1 and PMA2 genes in *Saccharomyces cerevisiae* cells grown at optimal and low pH. Arch. Microbiol. *166*, 315–320.

Caro, L.H., Smits, G.J., van Egmond, P., Chapman, J.W., and Klis, F.M. (1998). Transcription of multiple cell wall protein-encoding genes in *Saccharomyces cerevisiae* is differentially regulated during the cell cycle. FEMS Microbiol. Lett. *161*, 345–349.

Casey, G., and Ingeldew, W.M. (1983). High gravity brewing: influence of pitching rate and wort gravity on early yeast viability. J. Am. Soc. Brew. Chem. *41*, 148–152.

Casey, G., and Ingeldew, W.M. (1985). Reevaluation of alcohol synthesis and tolerance in brewer's yeast. J. Am. Soc. Brew. Chem. *43*, 75–83.

Casey, G.P., Magnus, C.A., and Ingledew, W.M. (1984). High-gravity brewing: effects of nutrition on yeast composition, fermentative ability and alcohol production. Appl. Environ. Microbiol. *48* (3), 639–646.

Castrejón, F., Codón, A.C., Cubero, B., and Benítez, T. (2002). Acetaldehyde and ethanol are responsible for mitochondrial DNA (mtDNA) restriction fragment length polymorphism (RFLP) in flor yeasts. Syst. Appl. Microbiol. *25*, 462–467.

Causton, H.C., Ren, B., Koh, S.S., Harbison, C.T., Kanin, E., Jennings, E.G., Lee, T.I., True, H.L., Lander, E.S., and Young, R.A. (2001). Remodelling of yeast genome expression in response to environmental changes. Mol. Biol. Cell *12*, 323–337.

Chatterjee, M.T., Khalawan, S.A., and Curran, B.P.G. (2000). Cellular lipid composition influences stress activation of the yeast general stress response element (STRE). Microbiology *146*, 877–844.

Chellappa, R., Kandasamy, P., Oh, C.S., Jiang, Y., Vemula, M., and Martin, C.E. (2001). The membrane proteins, Spt23p and Mga2p, play distinct roles in the activation of *Saccharomyces cerevisiae* OLE1 gene expression. Fatty acid-mediated regulation of Mga2p activity is independent of its proteolytic processing into a soluble transcription activator. J. Biol. Chem. *276*, 43548–43556. http://dx.doi.org/10.1074/jbc.M107845200

Cheung, A.W.Y., Brosnan, J.M., Phister, T., and Smart, K.A. (2012). Impact of dried, creamed and cake supply formats on the genetic variation and ethanol tolerance of three *Saccharomyces cerevisiae* distilling strains. J. Inst. Brew. *111*, 152–162.

Chen, P., Dong, J., Yin, H., Bao, X., Chen, L., He, Y., Chen, R., Wan, X., Zhao, Y., and Hou, X. (2016). Genome comparison and evolutionary analysis of different industrial lager yeasts (*Saccharomyces pastorianus*). J. Inst. Brew. *122*, 42–47.

Cheng, L., Moghraby, J., and Piper, P.W. (1999). Weak organic acid treatment causes a trehalose accumulation in low-pH cultures of *Saccharomyces cerevisiae*, not displayed by the more preservative-resistant *Zygosaccharomyces bailii*. FEMS Microbiol. Lett. *170*, 89–95.

Chester, V.E. (1963). The dissimilation of the carbohydrate reserves of a strain of *Saccharomyces cerevisiae*. Biochem. J. *86*, 153–160.

Chi, Z., and Arneborg, N. (1999). Relationship between lipid composition, frequency of ethanol-induced respiratory deficient mutants, and ethanol tolerance in *Saccharomyces cerevisiae*. J. Appl. Microbiol. *86*, 1047–1053.

Ciesarová, Z., Smogrovicová, D., and Dömény, Z. (1996). Enhancement of yeast ethanol tolerance by calcium and magnesium. Folia. Microbiol. *41*, 485–488.

Clarkson, S.P., Large, P.J., Boulton, C.A., and Bamforth, C.W. (1991). Synthesis of superoxide dismutase, catalase and other enzymes and oxygen and superoxide toxicity during changes in oxygen concentration in cultures of brewing yeast. Yeast *7*, 91–103.

Conlin, L.K., and Nelson, H.C. (2007). The natural osmolyte trehalose is a positive regulator of the heat-induced activity of yeast heat shock transcription factor. Mol. Cell. Biol. *27*, 1505–1515.

Coote, H., and Kirshop, B.H. (1976). Factors responsible for the decrease in pH during beer fermentations. J. Inst. Brew. *82*, 149–153.

Costa, V., Amorim, M.A., Reis, E., Quintanilha, A., and Moradas-Ferreira, P. (1997). Mitochondrial superoxide dismutase is essential for ethanol tolerance of *Saccharomyces cerevisiae* in the post-diauxic phase. Microbiology *143*, 1649–1656.

Costa, V., and Moradas-Ferreira, P. (2001). Oxidative stress and signal transduction in *Saccharomyces cerevisiae*: insights into ageing, apoptosis and diseases. Mol. Aspects Med. *22*, 217–246.

Costa, V., Reis, E., Quintanilha, A., and Moradas-Ferreira, P. (1993). Acquisition of ethanol tolerance in *Saccharomyces cerevisiae*: the key role of the mitochondrial superoxide dismutase. Arch. Biochem. Biophys. *300*, 608–614.

Csonka, L.N., and Hanson, A.D. (1991). Prokaryotic osmoregulation: genetics and physiology. Annu. Rev. Microbiol. *45*, 569–606. http://dx.doi.org/10.1146/annurev.mi.45.100191.003033

Cunningham, S., and Stewart, G.G. (1998). Effects of high-gravity brewing and acid washing on brewers' yeast. J. Am. Soc. Brew. Chem. *56* (1), 12–18.

D'Amore, T. (1992). Improving yeast fermentation performance. J. Inst. Brew. *98*, 375–382.

D'Amore, T., Crumplen, R., and Stewart, G.G. (1991). Activation of yeast trehalase by heat shock. Braz. Biol. Res. *24*, 979–984.

D'Amore, T., Panchal, C.J., Russell, I., and Stewart, G.G. (1990). A study of ethanol tolerance in yeast. Crit. Rev. Biotechnol. *9*, 287–304.

David, M.H., and Kirsop, B.H. (1972). The varied response of brewing yeasts to oxygen and sterol treatment. J. Am. Soc. Brew. Chem. *30*, 14–16.

Davies, K.J.A. (1995). Oxidative stress: the paradox of aerobic life. In Free radicals and oxidative stress: environment, drugs and food additives, C. Rice-Evans, B. Halliwell, and G.G. Lunt, eds. (London, UK: The Biochemical Society), pp 1–32.

Davies, B.S., and Rine, J. (2006). A role for sterol levels in oxygen sensing in *Saccharomyces cerevisiae*. Genetics *174*, 191–201.

Dawes, I.W. (2004). Yeast stress responses. In The Metabolism and Molecular Physiology of *Saccharomyces cerevisiae*, 2nd edn, J.R. Dickinson and M. Schweizer, eds. (Boca Raton, USA: CRC Press), pp. 376–438.

Debourg, A., and Van Nedervelde, L. (1999). The use of dried yeast in the brewing industry. Proc. Congr. Eur. Brew. Conv. *27*, 751–760.

Dechant, R., Binda, M., Lee, S.S., Pelet, S., Winderickx, J., and Peter, M. (2010). Cytosolic pH is a second messenger for glucose and regulates the PKA pathway through V-ATPase. EMBO J. *29*, 2515–2526. http://dx.doi.org/10.1038/emboj.2010.138

Depraetere, S.A., Winderickx, J., and Delvaux, F.R. (2003). Evaluation of the oxygen requirement of lager and ale yeast strains by preoxygenation. Tech. Q. Master Brew. Assoc. Am. *40*, 283–289.

D'Hautcort, O., and Smart, K.A. (1999). Measurement of brewing yeast flocculation. J. Am. Soc. Brew. Chem. *57*, 123–128.

Dihazi, H., Kessler, R., and Eschrich, K. (2001). Phosphorylation and inactivation of yeast 6-phosphofructo-2-kinase contribute to the regulation of glycolysis under hypotonic stress. Biochemistry *40*, 14669–14678.

Dombek, K.M., and Ingram, L.O. (1986). Magnesium limitation and its role in apparent toxicity of ethanol during yeast fermentation. Appl. Environ. Microbiol. *52*, 975–981.

Donzeau, M., Bourdineaud, J.P., and Lauquin, G.J. (1996). Regulation by low temperatures and anaerobiosis of a yeast gene specifying a putative GPI-anchored plasma membrane protein [corrected.] Mol. Microbiol. *20*, 449–459.

Doudican, N.A., Song, B., Shadel, G.S., and Doetsch, P.W. (2005). Oxidative DNA damage causes mitochondrial genomic instability in *Saccharomyces cerevisiae*. Mol. Cell. Biol. *25*, 5196–5204.

Dragone, G., Silva, D.P., de Almeida e Silva, J.B., and de Almeida Lima, U. (2003). Improvement of the ethanol productivity in a high gravity brewing at pilot plant scale. Biotechnol. Lett. *25*, 1171–1174.

Drost, B.W. (1977). Fermentation and storage. Proc. Congr. Eur. Brew. Conv. *16*, 519–532.

Dumont, F., Marechal, P.A., and Gervais, P. (2003). Influence of cooling rate on *Saccharomyces cerevisiae* destruction during freezing: unexpected viability at ultra-rapid cooling rates. *Cryobiology 46*, 33–42.

Dunn, B., and Sherlock, G. (2008). Reconstruction of the genome origins and evolution of the hybrid lager yeast *Saccharomyces pastorianus*. Genome. Res. *18*, 1610–1623. http://dx.doi.org/10.1101/gr.076075.108

Eaton, N.R. (1960). Endogenous respiration of yeast. I. The endogenous substrate. Arch. Biochem. Biophys. *88*, 17–25.

Eleutherio, E.C., Araujo, P.S., and Panek, A.D. (1993). Role of the trehalose carrier in dehydration resistance of *Saccharomyces cerevisiae*. Biochim. Biophys. Acta *1156*, 263–266.

Erasmus, D.J., van der Merwe, G.K., and van Vuuren, H.J. (2003). Genome-wide expression analyses: Metabolic adaptation of *Saccharomyces cerevisiae* to high sugar stress. FEMS Yeast Res. *3*, 375–399.

Eraso, P., and Gancedo, C. (1987). Activation of yeast plasma membrane ATPase by acid pH during growth. FEBS Lett. *224*, 187–192.

Eriksson, P., André, L., Ansell, R., Blomberg, A., and Adler, L. (1995). Cloning and characterization of GPD2, a second gene encoding *sn*-glycerol 3-phosphate dehydrogenase (NAD+) in *Saccharomyces cerevisiae*, and its comparison with GPD1. Mol. Microbiol. *17*, 95–107.

Fales, F.W. (1951). The assimilation and degradation of carbohydrates by yeast cells. J. Biol. Chem. *193*, 113–124.

Fernandes, L., Corte-Real, M., Loureiro, V., Loureiro-Dias, M.C., and Leao, C. (1997). Glucose respiration and fermentation in *Zygosaccharomyces bailii* and *Saccharomyces cerevisiae* express different sensitivity patterns to ethanol and acetic acid. Lett Appl. Microbiol. *25*, 249–253.

Fernandez, S.S., González, M.G., and Sierra, J.A. (1993). Evaluation of the effect of acid washing on the fermentative and respiratory behaviour of yeast by the acidification power test. Tech. Q. Master Brew. Assoc. Am. *30*, 1–8.

François, J., and Parrou, J.L. (2001). Reserve carbohydrates metabolism in the yeast *Saccharomyces cerevisiae*. FEMS Microbiol. Rev. *25*, 125–145.

Fujita, K., Matsuyama, A., Kobayashi, Y., and Iwahashi, H. (2006). The genome-wide screening of yeast deletion mutants to identify the genes required for tolerance to ethanol and other alcohols. FEMS Yeast Res. *6*, 744–750.

Fujiwara, D., and Tamai, Y. (2003). Aeration prior to pitching increases intracellular enzymatic and transcriptional responses under non-nutritional conditions. J. Am. Soc. Brew. Chem. *61*, 99–104.

Gasch, A.P., Spellman, P.T., Kao, C.M., Carmel-Harel, O., Eisen, M.B., Storz, G., Botstein, D., and Brown, P.O. (2000). Genomic expression programs in the response of yeast cells to environmental changes. Mol. Biol. Cell *11*, 4241–4257.

Gasch, A.P., and Werner-Washburne, M. (2002). The genomics of yeast responses to environmental stress and starvation. Funct. Integr. Genomics. *2*, 181–192. http://dx.doi.org/10.1007/s10142-002-0058-2

Gibson, B.R., Smith, J.M., Lawrence, S.J., Shelton, N., Smith, J.M., and Smart, K.A. (2006). Oxygen as toxin: oxidative stress and brewing yeast physiology. Cerevisia *31*, 25–36.

Gibson, B.R., Lawrence, S.J., Leclaire, J.P., Powell, C.D., and Smart, K.A. (2007). Yeast responses to stresses associated with industrial brewery handling. FEMS Microbiol. Rev. *31*, 535–569.

Gibson, B.R., Storgards, E., Krogerus, K., and Vidgren, V. (2013). Comparative physiology and fermentation performance of Saaz and Frohberg lager yeast strains and the parental species Saccharomyces eubayanus. Yeast *30*, 255–266.

Girotti, A.W. (1998). Lipid hydroperoxide generation, turnover, and effector action in biological systems. J. Lipid Res. *39*, 1529–1542.

Gonzáles, M.G., Fernandez, S., and Sierra, J.A. (1996). Effect of temperature in the evaluation of yeast flocculation ability by the Helm's method. J. Am. Soc. Brew. Chem. *54*, 29–31.

Gounot, A.M., and Russell, N.J. (1999). Physiology of cold-adapted micro-organisms. In Cold-adapted organisms, R. Margesin and F. Schinner, eds. (Berlin: Springer), pp. 33–55.

Hammond, J.R.M. (1993). Brewers yeasts. In The Yeasts, Vol. 5, A.H. Rose and J.S. Harrison, eds. (London: Academic Press), pp 7-67.

Hammond, J.R.M. (2000). Yeast growth and nutrition. In Brewing Yeast Fermentation Performance, 1st Edition, K.A. Smart, ed. (Oxford, UK: Blackwell Science), pp. 77–85.

Hammond, J.R.M., Davis, D., Lee, M., and Storey, K. (2001). Does osmotic pressure affect yeast performance in high gravity fermentation? Proc. Congr. Eur. Brew. Conv. *28*, 316–325.

Haukeli, A.D., and Lie, S. (1978). Conversion of α-acetolactate and removal of diacetyl: a kinetic study. J. Inst. Brew. *84*, 85–89.

Heggart, H.M., Margaritis, A., Pilkington, H., Stewart, R.J., Dowhanick, T.M., and Russell, I. (1999). Factors affecting yeast viability and vitality characteristics: a review. Tech. Q. Master Brew. Assoc. Am. *36*, 383–406.

Herdeiro, R.S., Pereira, M.D., Panek, A.D., and Eleutherio, E.C.A. (2006). Trehalose protects *Saccharomyces cerevisiae* from lipid peroxidation during oxidative stress. Biochim Biophys. Acta *1760*, 340–346.

Hernández-Saavedra, N.Y., Ochoa, J.L., and Vazquez-Dulhalt, R. (1995). Osmotic adjustment in marine yeast. J. Plankton Res. *17* (1), 59–69.

Higgins, V.J., Beckhouse, A.G., Oliver, A.D., Rogers, P.J., and Dawes, I.W. (2003). Yeast genome-wide expression analysis identifies a strong ergosterol and oxidative stress response during the initial stages of an industrial lager fermentation. Appl. Environ. Microbiol. *69*, 4777–4787.

Hohmann, S. (1997). Shaping up: the response of yeast to osmotic stress. In Yeast Stress Responses, S. Hohmann and W.H. Mager, eds. (New York: Chapman & Hall), pp. 101–146.

Hohmann, S. (2002). Osmotic stress signaling and osmoadaptation in yeasts. Microbiol. Mol. Biol. Rev. *66*, 300–372.

Homma, T., Iwahashi, H., and Komatsu, Y. (2003). Yeast gene expression during growth at low temperature. Cryobiology *46*, 230–237.

Horsted, M.W., Dey, E.S., Holmberg, S., and Kielland-Brandt, M.C. (1998). A novel esterase from *Saccharomyces carlsbergensis*, a possible function for the yeast *TIP1* gene. Yeast *14*, 793–803.

Hottiger, T., Boller, T., and Wiemken, A. (1987a). Rapid changes of heat and desiccation tolerance correlated with changes of trehalose content in *Saccharomyces cerevisiae* cells subjected to temperature shifts. FEBS Lett. *220*, 113–115.

Hottiger, T., Schmutz, P., and Wiemken, A. (1987b). Heat-induced accumulation and futile cycling of trehalose in *Saccharomyces cerevisiae*. J. Bacteriol. *169*, 5518–5522.

Hounsa, C.G., Brandt, E.V., Thevelein, J., Hohmann, S., and Prior, B.A. (1998). Role of trehalose in survival of *Saccharomyces cerevisiae* under osmotic stress. Microbiology *144*, 671–680. http://dx.doi.org/10.1099/00221287-144-3-671

Hu, C.-K., Bai, F.-W., and An, L.-J. (2003). Enhancing ethanol tolerance of a self-flocculating fusant of *Schizosaccharomyces pombe* and *Saccharomyces cerevisiae* by Mg^{2+} via reduction in plasma membrane permeability. Biotechnol. Lett. *25*, 1191–1194.

Hulse, G.A. (2003). Yeast propagation. Brewing Yeast Fermentation Performance, 2nd Edition, K.A. Smart, ed. (Oxford, UK: Blackwell Science) pp. 249-256.

Ibeas, J.I., and Jimenez, J. (1997). Mitochondrial DNA loss caused by ethanol in *Saccharomyces* flor yeasts. Appl. Environ. Microbiol. *63*, 7–12.

Iwashashi, H., Obuchi, K., Fuiji, S., and Komatsu, Y. (1995). The correlative evidence suggesting that trehalose stabilises membrane structures in the yeast *Saccharomyces cerevisiae*. Appl. Environ. Microbiol. *65*, 5398–5402.

James, T.C., Campbell, S., Donnelly, D., and Bond, U. (2003). Transcription profile of brewery yeast under fermentation conditions. J. Appl. Microbiol. *94*, 432–448.

Jiménez, J., Longo, E., and Benítez, T. (1988). Induction of petite yeast mutants by membrane-active agents. Appl. Environ. Microbiol. *54*, 3126–3132.

Jones, R.P. (1988). Intracellular ethanol – accumulation and exit from yeast and other cells. FEMS Microbiol. Rev. *4*, 239–258.

Jones, P.G., and Inouye, M. (1996). RbfA, a 30S ribosomal binding factor, is a cold-shock protein whose absence triggers the cold-shock response. Mol. Microbiol. *21*, 1207–1218.

Kandror, O., Bretschneider, N., Kreydin, E., Cavalieri, D., and Goldberg, A.L. (2004). Yeast adapt to near-freezing temperatures by STRE/Msn2,4-dependent induction of trehalose synthesis and certain molecular chaperones. Mol. Cell *13*, 771–781.

Kapteyn, J.C., ter Riet, B., Vink, E., Blad, S., De Nobel, H., Van Den Ende, H., and Klis, F.M. (2001). Low external pH induces HOG1-dependent changes in the organization of the *Saccharomyces cerevisiae* cell wall. Mol. Microbiol. *39*, 469–479.

Klis, F.M. (1994). Review: Cell wall assembly in yeast. Yeast *10*, 851–869.

Klis, F.M., Mol, P., Hellingwerf, K., and Brul, S. (2002). Dynamics of cell wall structure in *Saccharomyces cerevisiae*. FEMS Microbiol. Rev. *26*, 239–256.

Knatchbull, F., and Slaughter, J.C. (1987). The effect of low CO_2 pressures on the absorption of amino acids and production of flavour-active volatiles by yeast. J. Inst. Brew. *93*, 420–424.

Kobayashi, N., and McEntee, K. (1993). Identification of cis and trans components of a novel heat shock stress regulatory pathway in *Saccharomyces cerevisiae*. Mol. Cell. Biol. *13*, 248–256.

Kodama, Y., Kielland-Brandt, M.C., Hansen, J. 2005. Lager brewing yeast. In Comparative Genomics: Using Fungi as Models, P. Sunnerhagen and J. Piškur, eds. (Berlin: Springer), pp. 145–160.

Koerkamp, M.G., Rep, M., Bussemaker, H.J., Hardy, G.P., Mul, A., Piekarska, K., Szigyarto, C.A., De Mattos, J.M., and Tabak, H.F. (2002). Dissection of transient oxidative stress response in *Saccharomyces cerevisiae* by using DNA microarrays. Mol. Biol. Cell *13*, 2783–2794. http://dx.doi.org/10.1091/mbc.E02-02-0075

Kondo, K., and Inouye, M. (1991). TIP 1, a cold shock-inducible gene of *Saccharomyces cerevisiae*. J. Biol. Chem. *266*, 17537–17544.

Kondo, K., and Inouye, M. (1992). Yeast NRS1 protein that has structural similarity to mammalian nucleolin is involved in Pre-rRNA processing. J. Biol. Chem. *267*, 16252–16258.

Kondo, K., Kowalski, L.R., and Inouye, M. (1992). Cold shock induction of yeast NSR1 protein and its role in pre-rRNA processing. J. Biol. Chem. *267*, 16259–16265.

Kowalski, L.R., Kondo, K., and Inouye, M. (1995). Cold-shock induction of a family of TIP1-related proteins associated with the membrane in *Saccharomyces cerevisiae*. Mol. Microbiol. *15*, 341–353.

Kwon, M., Chong, S., Han, S., and Kim, K. (2003). Oxidative stresses elevate the expression of cytochrome *c* peroxidase in *Saccharomyces cerevisiae*. Biochim. Biophys. Acta *1623*, 1–5.

Lai, L.C., Kosorukoff, A.L., Burke, P.V., and Kwast, K.E. (2005). Dynamical remodeling of the transcriptome during short-term anaerobiosis in *Saccharomyces cerevisiae*: differential response and role of Msn2 and/or Msn4 and other factors in galactose and glucose media. Mol. Cell. Biol. *25*, 4075–4091.

Latterich, M., and Watson, M.D. (1993). Evidence for a dual osmoregulatory mechanism in the yeast *Saccharomyces cerevisiae*. Biochem. Biophys. Res. Commun. *191*, 1111–1117.

Lawrence, S.J., Gibson, B.R., and Smart, K.A. (2009). Expression of the Cell Wall Mannoprotein Genes *CWP* and *DAN* During Industrial-Scale Lager Fermentations. J. Am. Soc. Brew. Chem., *67*, 58-62.

Lawrence, S.J., and Smart, K.A. (2011). The Impact of the *CWP* and *DAN* gene encoded mannoproteins on cell wall thickness in aerobic and anaerobic conditions. J. Am. Soc. Brew. Chem. *69* (3), 139–143.

Leão, C., and Van Uden, N. (1984). Effects of ethanol and other alkanols on passive proton influx in the yeast *Saccharomyces cerevisiae*. Biochim. Biophys. Acta *774*, 43–48.

Leclaire, J.P.R., Kennedy, A.I., and Smart, K.A. (2003). Occurrence of cold shock genes in ale and lager yeast strains. Proc. Congr. Eur. Brew. Conv. *29*, 575–585.

Lee, S.M., and Park, J.W. (1998). Thermosensitive phenotype of yeast mutant lacking thioredoxin peroxidase. Arch. Biochem. Biophys. *359*, 99–106.

Lee, W.C., Xue, Z.X., and Mélèse, T. (1991). The NSR1 gene encodes a protein that specifically binds nuclear localization sequences and has two RNA recognition motifs. J. Cell Biol. *113*, 1–12.

Lesage, G., and Bussey, H. (2006). Cell wall assembly in *Saccharomyces cerevisiae*. Microbiol. Mol. Biol. Rev. *70*, 317–343.

Lewis, M.G., and Young, T.W. (1995). Brewing. (London: Chapman & Hall).

Libkind, D., Hittinger, C.T., Valério, E., Gonçalves, C., Dover, J., Johnston, M., Gonçalves, P., and Sampaio, J.P. (2011). Microbe domestication and the identification of the wild genetic stock of lager-brewing yeast. Proc. Natl. Acad. Sci. U.S.A. *108*, 14539–14544. http://dx.doi.org/10.1073/pnas.1105430108

Liden, G., Walfridsson, M., Ansell, R., Anderlund, M., Adler, L., and Hahn-Hagerdal, B. (1996). A glycerol-3-phosphate dehydrogenase-deficient mutant of *Saccharomyces cerevisiae* expressing the heterologous *XYL1* gene. Appl. Environ. Microbiol. *62*, 3894–3896.

Lindquist, S. (1986). The heat-shock response. Annu. Rev. Biochem. *55*, 1151–1191. http://dx.doi.org/10.1146/annurev.bi.55.070186.005443

Liti, G., Peruffo, A., James, S.A., Roberts, I.N., and Louis, E.J. (2005). Inferences of evolutionary relationships from a population survey of LTR-retrotransposons and telomeric-associated sequences in the *Saccharomyces sensu stricto* complex. Yeast *22*, 177–192.

Lloyd, D., Morrell, S., Carlsen, H.N., Dean, H., James, P.E., and Rowlands, C.C. (1993). Effects of growth with ethanol on fermentation and membrane fluidity of *Saccharomyces cerevisiae*. Yeast *9*, 825–833.

Lodolo, E.J., and Cantrell, I.C. (2005). Oxygen – friend and foe of yeast metabolism. Proc. Inst. Brew. Dist. Con. (Africa Section). *10*, 42–51.

Lorenz, R.T., and Parks, L.W. (1991). Involvement of heme components in sterol metabolism of *Saccharomyces cerevisiae*. Lipids *26*, 598–603.

Lumsden, W.B., Duffus, J.H., and Slaughter, J.C. (1987). Effects of CO_2 on budding and fission yeasts. J. Gen. Microbiol. *133*, 877–881.

Luyten, K., Albertyn, J., Skibbe, F., Prior, B.A., Ramos, J., Thevelein, J.M., and Hohmann, S. (1994). The FPS1 gene product functions as a glycerol facilitator in the yeast *Saccharomyces cerevisiae*. Folia. Microbiol. *39*, 534–536.

Maeda, T., Takekawa, M., and Saito, H. (1995). Activation of yeast PBS2 MAPKK by MAPKKKs or by binding of an SH3-containing osmosensor. Science *269*, 554–558.

Maeda, T., Wurgler-Murphy, S.M., and Saito, H. (1994). A two-component system that regulates an osmosensing MAP kinase cascade in yeast. Nature *369*, 242–245. http://dx.doi.org/10.1038/369242a0

Maemura, H., Morimura, S., and Kida, K. (1998). Effects of aeration during the cultivation of pitching yeast on its characteristics during the subsequent fermentation of wort. J. Inst. Brew. *104*, 207–211.

Majara, M., O'Connor-Cox, E.S.C., and Axcell, B.C. (1996a). Trehalose – an osmoprotectant and stress indicator compound in high and very high gravity brewing. J. Am. Soc. Brew. Chem. *54*, 149–154.

Majara, M., O'Connor-Cox, E.S.C., and Axcell, B.C. (1996b). Trehalose – a stress protectant and stress indicator compound for yeast exposed to adverse conditions. J. Am. Soc. Brew. Chem. *54*, 539–563.

Mansure, J.J., Panek, A.D., Crowe, L.M., and Crowe, J.H. (1994). Trehalose inhibits ethanol effects on intact yeast cells and liposomes. Biochim. Biophys. Acta *1191*, 309–316.

Mansure, J.J., Souza, R.C., and Panek, A.D. (1997). Trehalose metabolism in *Saccharomyces cerevisiae* during alcoholic fermetentation. Biotechnol. Lett. *19* (12), 1201–1203.

Marchler, G., Schüller, C., Adam, G., and Ruis, H. (1993). A *Saccharomyces cerevisiae* UAS element controlled by protein kinase A activates transcription in response to a variety of stress conditions. EMBO J. *12*, 1997–2003.

Martin, V., Quain, D.E., and Smart, K.A. (2003). Brewing yeast oxidative stress responses: impact of brewery handling. In Brewing Yeast Fermentation Performance, 2nd edn, K.A. Smart, ed. (Oxford, UK: Blackwell Science), pp. 61–73.

Martinez-Pastor, M.T., Marchler, G., Schuller, C., Marchler-Bauer, A., Ruis, H., and Estruch, F. (1996). The *Saccharomyces cerevisiae* zinc finger proteins Msn2p and Msn4p are required for transcriptional induction through the stress response element (STRE). EMBO J. *15* (9), 2227–2235.

Marza, E., Camougrand, N., and Manon, S. (2002). Bax expression protects yeast plasma membrane against ethanol-induced permeabilization. FEBS Lett. *521*, 47–52.

McCaig, R., McKee, J., Pfisterer, E.A., Hysert, D.W., Munoz, E., and Ingledew, W.M. (1992). Very high gravity-laboratory and pilot plant trials. J. Am. Soc. Brew. Chem. *50* (1), 18–26.

McElhaney, R.N. (1982). Effects of membrane lipids on transport and enzymic activities. Curr. Top. Membr. Transp. *17*, 317–380.

Mizoguchi, H., and Hara, S. (1997). Ethanol-induced alterations in lipid composition of *Saccharomyces cerevisiae* in the presence of exogenous fatty acid. J. Ferment. Bioeng. *83*, 12–16.

Moonjai, N., Verstrepen, K.J., Shen, H.-Y., Derdelinckx, G., Verqachert, H., and Delvaux, F.R. (2003). Linoleic acid supplementation of a cropped brewing lager strain: effects on subsequent fermentation performance with serial repitching. J. Inst. Brew. *109*, 262–272.

Morimoto, R.I., Kroeger, P.E., and Cotto, J.J. (1996). The transcriptional regulation of heat shock genes: a plethora of heat shock factors and regulatory conditions. EXS *77*, 139–163.

Morris, C.N., Ainsworth, S., and Kinderlerer, J. (1984). The regulatory properties of yeast pyruvate kinase. Biochem. J. *234*, 691–698.

Morris, G.J., Winters, L., Coulson, G.E., and Clarke, K.J. (1986). Effect of osmotic stress on the ultrastructure and viability of the yeast *Saccharomyces cerevisiae*. J. Gen. Microbiol. *132*, 2023–2034. http://dx.doi.org/10.1099/00221287-132-7-2023

Mortimer, R.K., and Johnston, J.R. (1986). Genealogy of principal strains of the yeast genetic stock center. Genetics *113*, 35–43.

Nakatani, K., Tkahashi, T., Nagami, K., and Kumada, J. (1984). Kinetic study of vicinal diketones in brewing (I). formation of total vicinal diketones. Tech. Q. Master Brew. Assoc. Am. *21*, 73-78.

Nakao, Y., Kanamori, T., Itoh, T., Kodama, Y., Rainieri, S., Nakamura, N., Shimonaga, T., Hattori, M., and Ashikari, T. (2009). Genome sequence of the lager brewing yeast, an interspecies hybrid. DNA. Res. *16*, 115–129. http://dx.doi.org/10.1093/dnares/dsp003

Narziss, L., Miedaner, H., and Gresser, A. (1983). Heferasse und Birgualitat, Exraktabnahme, pH-Aball und faveranderung wahrend der garung. Brauwelt *123*, 478–486.

Nass, R., and Rao, R. (1999). The yeast endosomal Na^+/H^+ exchanger, Nhx1, confers osmotolerance following acute hypertonic shock. Microbiology *145*, 3221–3228. http://dx.doi.org/10.1099/00221287-145-11-3221

Norton, J.S., and Krauss, R.W. (1972). The inhibition of cell division in *Saccharomyces cerevisiae* (Meyen). by carbon dioxide. Plant Cell Physiol. *13*, 139–149.

Odumeru, J.A., D'Amore, T., Russell, I., and Stewart, G.G. (1993). Alterations in fatty acid composition and trehalose concentration of *Saccharomyces* brewing strains in response to heat and ethanol shock. J. Ind. Microbiol. Biotechnol. *11*, 113–119.

Ogden, K. (1987). Cleansing contaminated ptiching yeast with nisin. J. Inst. Brew. *93*, 302–307.

Orij, R., Urbanus, M.L., Vizeacoumar, F.J., Giaever, G., Boone, C., Nislow, C., Brul, S., and Smits, G.J. (2012). Genome-wide analysis of intracellular pH reveals quantitative control of cell division rate by pH(c) in *Saccharomyces cerevisiae*. Genome Biol. *13*, R80. http://dx.doi.org/10.1186/gb-2012-13-9-r80

O'Rourke, T.W., Doudican, N.A., Mackereth, M.D., Doetsch, P.W., and Shadel, G.S. (2002). Mitochondrial dysfunction due to oxidative mitochondrial DNA damage is reduced through cooperative actions of diverse proteins. Mol. Cell Biol. *22*, 4086–4093.

Pahlman, A.K., Granath, K., Ansell, R., Hohmann, S., and Adler, L. (2001). The yeast glycerol-3-phosphatases Gpp1p and Gpp2p are required for glycerol biosynthesis and differentially involved in the cellular responses to osmotic, anaerobic, and oxidative stress. J. Biol. Chem. *6*, 3555–3563.

Panadero, J., Palloti, C., Rodríguez-Vargas, S., Randez-Gil, F., and Prieto, J.A. (2006). A downshift in temperature activates the high osmolarity glycerol (HOG). pathway, which determines freeze tolerance in *Saccharomyces cerevisiae*. J. Biol. Chem. *281*, 4638–4645.

Panchal, C.J., and Stewart, G.G. (1980). The effect of osmotic pressure on the production and excretion of ethanol and glycerol in brewing yeast strains. J. Inst. Brew. *86*, 207–210.

Panek, A. (1963). Function of trehalose in baker's yeast (*Saccharomyces cerevisiae*). Arch. Biochem. Biophys. *100*, 422–425.

Parrou, J.-L., Teste, M.-A., and François, J. (1997). Effects of various types of stress on the metabolism of reserve carbohydrates in *Saccharomyces cerevisiae*: genetic evidence for a stress-induced recycling of glycogen and trehalose. Microbiology *143*, 1891–1900.

Pascual, C., Alonso, A., Garcia, I., Romay, C., and Kotyk, A. (1988). Effect of ethanol on glucose transport, key glycolytic enzymes, and proton extrusion in *Saccharomyces cerevisiae*. Biotechnol. Bioeng. 32, 374–378. http://dx.doi.org/10.1002/bit.260320317

Pedreño, Y., Gimeno-Alcañiz, J.V., Matallana, E., and Argüelles, J.C. (2002). Response to oxidative stress caused by H(2)O(2) in *Saccharomyces cerevisiae* mutants deficient in trehalase genes. Arch. Microbiol. *177*, 494–499. http://dx.doi.org/10.1007/s00203-002-0418-2

Petrov, V.V., and Okorokov, L.A. (1990). Increase of the anion and proton permeability of *Saccharomyces carlsbergensis* plasmalemma by n-alcohols as a possible cause of its de-energization. Yeast 6 (4), 311–318.

Phadtare, S., Alsina, J., and Inouye, M. (1999). Cold-shock response and cold-shock proteins. Curr. Opin. Microbiol. 2, 175–180.

Poolman, B., and Glaasker, E. (1998). Regulation of compatible solute accumulation in bacteria. Mol. Microbiol. *29*, 397–407.

Powell, C.D., Quain, D.E., and Smart, K.A. (2003). The impact of brewing yeast cell age on fermentation performance, attenuation and flocculation. FEMS Yeast Res. *3*, 149–157.

Rachidi, N., Martinez, M.J., Barre, P., and Blondin, B. (2000). *Saccharomyces cerevisiae PAU* genes are induced by anaerobiosis. Mol. Microbiol. *35*, 1421–1430.

Rainieri, S., Kodama, Y., Kaneko, Y., Mikata, K., Nakao, Y., and Ashikari, T. (2006). Pure and mixed genetic lines of *Saccharomyces bayanus* and *Saccharomyces pastorianus* and their contribution to the lager brewing strain genome. Appl. Environ. Microbiol. 72, 3968–3974.

Rand, J.D., and Grant, C.M. (2006). The thioredoxin system protects ribosomes against stress-induced aggregation. Mol. Biol. Cell *17*, 387–401.

Rees, E.M.R., and Stewart, G.G. (1997). The effects of increased magnesium and calcium concentrations on yeast fermentation performance in high gravity worts. J. Inst. Brew. *103*, 287–291.

Remize, F., Barnavon, L., and Dequin, S. (2001). Glycerol export and glycerol-3-phosphate dehydrogenase, but not glycerol phosphatase, are rate limiting for glycerol production in *Saccharomyces cerevisiae*. Metab. Eng. 3, 301–312. http://dx.doi.org/10.1006/mben.2001.0197

Rhymes, M.R., and Smart, K.A. (2001). Effect of storage conditions on the flocculation and cell wall characteristics of an ale brewing yeast strain. J. Am. Soc. Brew. Chem. 59, 32–38.

Ribeiro, G.F., Côrte-Real, M., and Johansson, B. (2006). Characterization of DNA damage in yeast apoptosis induced by hydrogen peroxide, acetic acid, and hyperosmotic shock. Mol. Biol. Cell *17*, 4584–4591.

Ristow, H., Seyfarth, A., and Lochmann, E.R. (1995). Chromosomal damages by ethanol and acetaldehyde in *Saccharomyces cerevisiae* as studied by pulsed field gel electrophoresis. Mutat. Res. *326*, 165–170.

Rodriguez-Vargas, S., Estruch, F., and Randez-Gil, F. (2002). Gene expression analysis of cold and freeze stress in Baker's yeast. Appl. Environ. Microbiol. *68*, 3024–3030.

Roessler, J.G. (1968). Yeast management in the brewery part 2, yeast handling and treatment. Brewers Digest *43*, 94–98.

Rosa, M.F., and Sá-Correia, I. (1991). In vivo activation by ethanol of plasma membrane ATPase of *Saccharomyces cerevisiae*. Appl. Environ. Microbiol. *57*, 830–835.

Rosa, M.F., and Sá-Correia, I. (1996). Intracellular acidification does not account for inhibition of *Saccharomyces cerevisiae* growth in the presence of ethanol. FEMS Microbiol. Lett. *135*, 271–274.

Rowe, S.M., Simpson, W.J., and Hammond, J.R.M. (1994). Intracellular pH of yeast during brewery fermentation. Lett Appl. Microbiol. *18*,135–137.

Ruis, H., and Schüller, C. (1995). Stress signaling in yeast. Bioessays *17*, 959–965. http://dx.doi.org/10.1002/bies.950171109

Russell, I., and Stewart, G.G. (1995). Brewing. In Brewing in Biotechnology, Vol. 9, G. Reed and T.W. Nagathodawithera, eds. (New York: VCH Publishers), pp. 419–462.

Russell, N.J. (1989). Adaptive modifications in membranes of halotolerant and halophilic microorganisms. J. Bioenerg. Biomembr. *21*, 93–113.

Saffi, J., Sonego, L., Varela, Q.D., and Salvador, M. (2006). Antioxidant activity of L-ascorbic acid in wild-type and superoxide dismutase deficient strains of *Saccharomyces cerevisiae*. Redox. Rep. *11*, 179–184. http://dx.doi.org/10.1179/135100006X116691

Sahara, T., Goda, T., and Ohgiya, S. (2002). Comprehensive expression analysis of time-dependent genetic responses in yeast cells to low temperature. J. Biol. Chem. *277*, 50015–50021. http://dx.doi.org/10.1074/jbc.M209258200

Šajbidor, J., and Grego, J. (1992). Fatty acid alterations in *Saccharomyces cerevisiae* exposed to ethanol stress. FEMS Microbiol. Lett. 72, 13–16.

Salgueiro, S.P., Sá-Correia, I., and Novais, J.M. (1988). Ethanol-induced leakage in *Saccharomyces cerevisiae*: kinetics and relationship to yeast ethanol tolerance and alcohol fermentation productivity. Appl. Environ. Microbiol. *54*, 903–909.

Salmon, T.B., Evert, B.A., Song, B., and Doetsch, P.W. (2004). Biological consequences of oxidative stress-induced DNA damage in *Saccharomyces cerevisiae*. Nucleic Acids Res. *32*, 3712–3723. http://dx.doi.org/10.1093/nar/gkh696

Schade, B., Jansen, G., Whiteway, M., Entian, K.D., and Thomas, D.Y. (2004). Cold adaptation in budding yeast. Mol. Biol. Cell *15*, 5492–5502.

Schmitt, A.P., and McEntee, K. (1996). Msn2p, a zinc finger DNA-binding protein, is the transcriptional activator of the multistress response in *Saccharomyces cerevisiae*. Proc. Natl. Acad. Sci. U.S.A. *93*, 5777–5782.

Sebollela, A., Louzada, P.R., Sola-Penna, M., Sarone-Williams, V., Coelho-Sampaio, T., and Ferreira, S.T. (2004). Inhibition of yeast glutathione reductase by trehalose: possible implications in yeast survival and recovery from stress. Int. J. Biochem. Cell Biol. *36*, 900–908. http://dx.doi.org/10.1016/j.biocel.2003.10.006

Sharma, S.C., Raj, D., Forouzandeh, M., and Bansal, M.P. (1996). Salt-induced changes in lipid composition and ethanol tolerance in *Saccharomyces cerevisiae*. Appl. Biochem. Biotechnol. *56*, 189–195.

Sharma, S.C. (1997). A possible role of trehalose in osmotolerance and ethanol tolerance in *Saccharomyces cerevisiae*. FEMS Microbiol. Lett. *152*, 11–15.

Shaw, E., Hill, D.R., Brittain, N., Wright, D.J., Täuber, U., Marand, H., Helm, R.F., and Potts, M. (2003). Unusual water flux in the extracellular polysaccharide of the cyanobacterium *Nostoc commune*. Appl. Environ. Microbiol. *69*, 5679–5684.

Shaw, M.K., and Ingraham, J.L. (1967). Synthesis of macromolecules by *Escherichia coli* near the minimal temperature for growth. J. Bacteriol. *94*, 157–164.

Shinitsky, M. (1984). Membrane fluidity and cellular functions. In Physiology of Membrane Fluidity (Boca Raton, FL: CRC Press), pp 1045–1049.

Simpson, W.J., and Hammond, J.R.M. (1989). The response of brewing yeast to acid washing. J. Inst. Brew. *93*, 347–354.

Sinensky, M. (1974). Homeoviscous adaptation – a homeostatic process that regulates the viscosity of membrane lipids in *Escherichia coli*. Proc. Natl. Acad. Sci. U.S.A. *71*, 522–525.

Singer, M.A., and Lindquist, S. (1998). Multiple effects of trehalose on protein folding in vitro and in vivo. Mol. Cell *1*, 639–648.

Slaughter, J.C., Flint, P.W.N., and Kular, K.S. (1987). The effect of CO_2 on the absorption of amino acids from a malt extract medium by *Saccharomyces cerevisiae*. FEMS Microbiol. Lett. *40*, 239–243.

Smart, K.A. (2003). Differential regulation of yeast cell wall mannoproteins in response to stress during brewing yeast fermentation. Proc. Congr. Eur. Brew. Conv. *29*, 481–489.

Smits, G.J., Kapteyn, J.C., van den Ende, H., and Klis, F.M. (1999). Cell wall dynamics in yeast. Curr. Opin. Microbiol. 2, 348–352.

Stewart, G.G. (2001). Yeast management – the balance between fermentation efficiency and beer quality. Tech. Q. Master Brew. Assoc. Am. *38* (1), 74–53.

Stewart, G.G., D'Amore, T., Panchal, C.J., and Russell, I. (1988). Factors that influence the ethanol tolerance of brewer's yeast strains during high gravity wort fermentations. Tech. Q. Master Brew. Assoc. Am. *25*, 47–53.

Swan, T., Baldwin, B., Clarke, F., Tonissen, K., Anderson, M., and Rogers, P. (2003). The development and application of natural yeast reductants to the stabilisation of beer. Proc. Congr. Eur. Brew. Conv. *29*, 707–722.

Tamás, M.J., and Hohmann, S. (2003). The osmotic stress response of *Saccharomyces cerevisiae*. In Yeast Stress Responses, S. Hohmann and W.H. Mager, eds. (New York: Chapman & Hall), pp. 121–200.

Tamás, M.J., Luyten, K., Sutherland, F.C., Hernandez, A., Albertyn, J., Valadi, H., Li, H., Prior, B.A., Killian, S.G., Ramos, J., Gustafsson, L., Thevelein, J.M., and Hohmann, S. (1999). Fps1p controls the accumulation and release of the compatible solute glycerol in yeast osmoregulation. Mol. Microbiol. *31*, 1087–1104.

Taylor, G.T., Thurston, P.A., and Kirsop, B.H. (1979). The influence of lipids derived from malt spent grains on yeast, metabolism and fermentation. J. Inst. Brew. *85*, 219–227.

ter Linde, J.J., Liang, H., Davis, R.W., Steensma, H.Y., van Dijken, J.P., and Pronk, J.T. (1999). Genome-wide transcriptional analysis of aerobic and anaerobic chemostat cultures of *Saccharomyces cerevisiae*. J. Bacteriol. *181*, 7409–7413.

Thieringer, H.A., Jones, P.G., and Inouye, M. (1998). Cold shock and adaptation. Bioessays *20*, 49–57.

Treger, J.M., Schmitt, A.P., Simon, J.R., and McEntee, K. (1998). Transcriptional factor mutations reveal regulatory complexities of heat shock and newly identified stress genes in *Saccharomyces cerevisiae*. J. Biol. Chem. *273*, 26875–26879.

Uchida, E., Ohsumi, Y., and Anraku, Y. (1988). Purification of yeast vacuolar membrane H^+-ATPase and enzymological discrimination of three ATP-driven proton pumps in *Saccharomyces cerevisiae*. Methods Enzymol. *157*, 544–562.

Ullah, A., Orij, R., Brul, S., and Smits, G.J. (2012). Quantitative analysis of the modes of growth inhibition by weak organic acids in *Saccharomyces cerevisiae*. Appl. Environ. Microbiol. *78*, 8377–8387. http://dx.doi.org/10.1128/AEM.02126-12

van der Vaart, J.M., Caro, L.H., Chapman, J.W., Klis, F.M., and Verrips, C.T. (1995). Identification of three mannoproteins in the cell wall of *Saccharomyces cerevisiae*. J. Bacteriol. *177*, 3104–3110.

Van Iersel, M.F.M., Meersman, E., Arntz, M., Rombouts, F.M., and Abee, T. (1998). Effect of environmental conditions on flocculation and immobilisation of brewer's yeast during production of alcohol-free beer. J. Inst. Brew. *104*, 131–136.

Van Laere, A. (1989). Trehalose, reserve and/or stress metabolite. FEMS Microbiol. Lett. *15*, 133–159.

Vidgren, V., Multanen, J.P., Ruohonen, L., and Londesborough, J. (2010). The temperature dependence of maltose transport in ale and lager strains of brewer's yeast. FEMS Yeast Res. *10*, 402–411. http://dx.doi.org/10.1111/j.1567-1364.2010.00627.x

Walker, G.M. (1998). Magnesium as a stress-protectant for industrial strains of *Saccharomyces cerevisiae*. J. Am. Soc. Brew. Chem. *56*, 109–113.

Walker, G.M., and Maynard, A.I. (1997). Accumulation of magnesium ions during fermentative metabolism in *Saccharomyces cerevisiae*. J. Ind. Microbiol. Biotechnol. *18*, 1–3.

Wegmann, K. (1986). Osmoregulation in eukaryotic algae. FEMS Microbiol. Rev. *39*, 37-43.

Werner-Washburne, M., Becker, J., Kosic-Smithers, J., and Craig, E.A. (1989). Yeast Hsp70 RNA levels vary in response to the physiological status of the cell. J. Bacteriol. *171*, 2680–2688.

Werner-Washburne, M., Braun, E., Johnston, G.C., and Singer, R.A. (1993). Stationary phase in the yeast *Saccharomyces cerevisiae*. Microbiol. Rev. *57*, 383–401.

Wiemken, A. (1990). Trehalose in yeast, stress protectant rather than reserve carbohydrate. Antonie Van Leeuwenhoek *58*, 209–217.

Wood, J.M. (1999). Osmosensing by bacteria: signals and membrane-based sensors. Microbiol. Mol. Biol. Rev. *63*, 230–262.

Yan, C., and Mélèse, T. (1993). Multiple regions of NSR1 are sufficient for accumulation of a fusion protein within the nucleolus. J. Cell Biol. *123*, 1081–1091.

Yancey, P.H., Clark, M.E., Hand, S.C., Bowlus, R.D., and Somero, G.N. (1982). Living with water stress: evolution of osmolyte systems. Science *217*, 1214–1222.

Yoshimoto, H., and Yamashita, I. (1991). The GAM1/SNF2 gene of *Saccharomyces cerevisiae* encodes a highly charged nuclear protein required for transcription of the STA1 gene. Mol. Gen. Genet. *228*, 270–280.

You, K.M., Rosenfield, C.L., and Knipple, D.C. (2003). Ethanol tolerance in the yeast *Saccharomyces cerevisiae* is dependent on cellular oleic acid content. Appl. Environ. Microbiol. *69*, 1499–1503.

Young, B.P., Shin, J.J., Orij, R., Chao, J.T., Li, S.C., Guan, X.L., Khong, A., Jan, E., Wenk, M.R., Prinz, W.A., *et al.* (2010). Phosphatidic acid is a pH biosensor that links membrane biogenesis to metabolism. Science *329*, 1085–1088. http://dx.doi.org/10.1126/science.1191026

Zhang, L., Onda, K., Imai, R., Fukuda, R., Horiuchi, H., and Ohta, A. (2003). Growth temperature downshift induces antioxidant response in *Saccharomyces cerevisiae*. Biochem. Biophys. Res. Commun. *307*, 308–314.

Zhang, L., Ohta, A., Horiuchi, H., Takagi, M., and Imai, R. (2001). Multiple mechanisms regulate expression of low temperature responsive (LOT). genes in *Saccharomyces cerevisiae*. Biochem. Biophys. Res. Commun. *283*, 531–535.

Zhang, S., Skalsky, Y., and Garfinkel, D.J. (1999). MGA2 or SPT23 is required for transcription of the delta9 fatty acid desaturase gene, OLE1, and nuclear membrane integrity in *Saccharomyces cerevisiae*. Genetics *151*, 473–483.

Yeast Supply, Fermentation and Handling – Insights, Best Practice and Consequences of Failure

3

David E. Quain

International Centre for Brewing Science, School of Biosciences, University of Nottingham, Sutton Bonington Campus, Loughborough, UK.

Correspondence: david.quain@nottingham.ac.uk

https://doi.org/10.21775/9781910190616.03

Abstract

The bottom line in brewery fermentation is that consistency is paramount. Whatever the scale of the operation, excellent yeast quality is a fundamental process requirement to ensure good and sustainable beer quality. The recycling of yeast across numerous fermentations adds complexity and biological stress. It is recommended that this be compensated by application of best practice in yeast supply/propagation, pitching, fermentation and storage. Craft brewers operating at a small scale and without recycling should be conscious of best practice for the rehydration of active dried yeast.

Introduction

With the exception of hybrid lager strains (*Saccharomyces pastorianus*), domesticated yeast used in brewing, baking, distilling and most wine making is branded taxonomically as *Saccharomyces cerevisiae.* This loosely translates to 'sugar mould' (saccharomyces) and 'beer' (cerevisiae). Given its on-going contribution, yeast can justifiably claim to be one of humankind's best friends. Indeed, in addition to its role in the fermentation of sugars, budding yeast has in recent decades become the pre-eminent model laboratory microorganism. This combined contribution to humankind has been neatly headlined (Duina *et al.*, 2014) as, 'a kitchen companion for centuries, *S. cerevisiae* has seen exponential growth (pun intended) as a laboratory companion over the past half century'.

A more recent turning point was the groundbreaking sequencing of the genome of *S. cerevisiae* S288C (Goffeau *et al.*, 1996). Not surprisingly, since then *S. cerevisiae* has seen accelerated use as a model organism. So much so that Botstein and Fink (2011) noted 'that yeast has graduated from a position as the premier model for eukaryotic cell biology to become the pioneer organism that has facilitated the establishment of the entirely new fields of study called 'functional genomics' and 'systems biology.' These new fields look beyond the functions of individual genes and proteins, focusing on how these interact and work together to determine the properties of living cells and organisms.'

Stunning contemporary examples of yeast as a 'pioneer organism' include the creation of a synthetic chromosome III (foreshadowing the possibility of entirely synthetic yeast genomes) (Pennisi, 2014), the successful replacement of defective yeast genes with their human counterparts (Kachoo *et al.*, 2015) and diverse 'engineered' yeasts on the road to producing – amongst other things – penicillin, opiates and anti-malarial drugs.

In the world of brewing yeast things have been a little less newsworthy. That said, the genome sequence of the widely used *S. pastorianus* Weihenstephan 34/70 was published in 2009 (Nakao *et al.*, 2009) and that of the first pure culture lager yeast

(Carlsberg Unterhefe No.1) in 2014 (Walther *et al.*, 2014). Recently, there has been a flurry of excitement around the hybrid genome of lager yeast, *S. pastorianus*, comprising *S. cerevisiae* and another (cryotolerant) yeast. The identification of *Saccharomyces eubayanus* from southern beech forests in Patagonia (Libkind *et al.*, 2011) as the 'other' yeast caused something of a stir in the wider news media. In particular, the disconnect between lager brewing originating in Bavaria in the fifteenth century but predating trans-Atlantic trade triggered a hunt for a sources of *S. eubayanus* closer to Germany. Although the route is still by no means clear, it is suggested to 'be the product of multiple long-distance dispersal events' (Gayevskiy and Goddard, 2015), this yeast having been found in Tibet (Bing *et al.*, 2014) and New Zealand (Gayevskiy and Goddard, 2015) (see Chapter 4).

Although a fascinating one-off, there is a wider 'halo effect' of stretching academic research that can be overlaid to aid the understanding of yeast during brewing fermentation and associated handling. This is fitting as globally brewing research is sadly in decline. There are a few hotspots but the halcyon days of the 1980s for applied brewing yeast research are no more. However much can be learned and applied from fundamental studies and related worlds such as biofuel and wine research.

This various process steps reviewed in this article are summarized in Fig. 3.1.

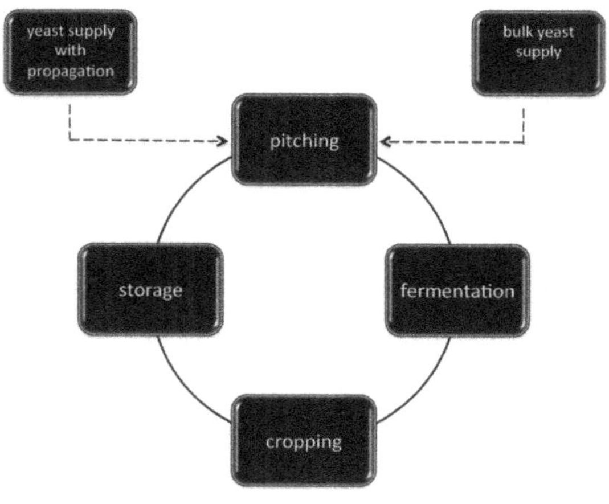

Figure 3.1 Process overview.

Yeast supply

Although very much the norm in brewing, the recycling of yeast from one fermentation to another is unusual. The approach is used in biofuel production (Walker, 2011), but other fermentation-based processes typically adopt a 'single trip' approach with no ambition or opportunity to recover yeast and use it again. The supply of yeast into the fermentation process in breweries can be achieved in a variety of ways that are differentiated by scale and volume. It is rare but not unheard of, to use a yeast strain without replacement for years or decades but is typically capped at so many 'generations' (i.e. fermentation batches). There is no hard and fast rule as to the number of generations, though there is a trend down from 15–20 to 10 to five or fewer generations in recent years. The vast majority of brewers will – as policy – replace their brewing strains periodically with the supply and propagation of pure cultures. The interpretation of this approach can take different shapes where the complexity of supply and level of quality assurance is at its greatest for large brewing companies and groups. This involves the cryopreservation and storage of propriety strains and their quality assured recovery and staged laboratory and plant propagation into fermenter. Whilst this undeniably the best possible practice to achieve strain purity and yeast quality, other approaches are none the less focused on quality and hygiene. Depending on scale and size, some breweries contract with a supplier to store, assure, validate and provide liquid slurries of their strain. Beneath this and typically with the smaller volumes of new and growing craft breweries, a new market for the supply of yeast has grown. Here 'type' yeast strains are provided as liquid cultures or as active dried yeast for pitching directly into fermenter. At around 15 hl or below it is economically viable to adopt a 'pitch and ditch' approach, whereby 'ready to pitch' cultures are used without recycling. Much above these volumes, cropping and re-pitching becomes an increasingly likely process outcome.

Best practice

Irrespective of the number of generations a yeast strain is used, the underpinning process of 'yeast supply' must satisfy the fundamental requirements of strain integrity and purity. The implications of introducing the wrong strain or a contaminated 'right' strain are discussed in 'consequences of failure' below. Accordingly, yeast supply is technically demanding and outside the usual demands of brewery microbiological QA. In larger, typically global operations, the process is managed in-house or increasingly outsourced to a specialist third party. The importance of yeast supply often warrants its inclusion in quality management systems such as ISO 9001. Irrespective of provenance, best practice for the supply of pure yeast cultures is broadly the same, consisting of three key steps; preservation/storage, recovery and supply. A full outline of a successful yeast supply process for the supply of production brewing yeasts has been described (Quain, 1995). This reiterates the importance of the process and the unambiguous requirement to start out with the required yeast strain, a consideration that drives the comparative emphasis of the subject below.

Long-term storage – cryopreservation versus freeze-drying

It goes without saying that the method of long-term storage is fundamental to the success of yeast supply for propagation in the brewery. As noted previously (Quain, 1995), 'the ideal method should be genuinely long-term in that yeast can be stored for many years without compromising viability or genetic stability'. It is generally accepted that the 'gold standard' for preservation of yeasts and other cells (e.g. blood, sperm, ova etc.) is storage in liquid nitrogen at its boiling point of −196°C. Cryopreservation is considered to be the most robust long-term approach, which maintains yeast viability without any genetic change.

There are a number of reports in the brewing literature of the use of liquid nitrogen to store production yeast strains (reviewed in Boulton and Quain, 2001; Quain, 2006b). These publications stem from a different time (1973–2003) and from brewing companies that have now been subsumed into today's global brewers. Irrespective of this, the technology supporting cryopreservation remains essentially unchanged. Suspensions of the yeast 'master culture' are stored in short, coloured polypropylene 'straws' in boxes or tubes immersed in liquid nitrogen. The hardware consists of cryovessels, which must be kept topped up from a storage tank of liquid nitrogen. The health and safety demands of using liquid nitrogen are stringent, a point seemingly overlooked in its use in restaurants and bars!

Although long-term storage in liquid nitrogen requires little more than topping up the cryovessels, the biggest barrier to entry is the need for specialist resources to carefully manage the introduction of yeast into store. For success, there must be strict adherence to detailed methods, which define the physiology of the yeast (oxidative), its concentration (100×10^6 cells/ml) and suspension in fresh media containing a cryoprotectant (glycerol) followed by two-stage phased freezing (room temperature to −30°C and then immersion at −196°C). Recovery is a little more straightforward via removal of the straw from liquid nitrogen and transfer to water at 30°C. Bond (2007b) provides a detailed methodology for the cryopreservation of yeast as used at the National Collection of Yeast Cultures in the UK.

An alternative and older method – freeze-drying or lyophilization – has had a mixed press regarding its impact on yeast viability. Like cryopreservation, freeze-drying also demands expertise and specialist equipment (Bond, 2007a). In a companion article, Bond (2007b) notes that for freeze-drying, 'strain viabilities are generally low, typically between 1 and 30%, as compared with more than 30% for those of yeast preserved in liquid nitrogen'. There have been a number of reports with brewing strains over the decades (Kirsop, 1955, 1974; Hall and Webb, 1975; Russell and Stewart, 1981) that paint a similar picture of a significant loss in viability post drying and then little change during storage over many years. Mindful that methods develop and improve, a similar conclusion was reported by Miyamoto-Shinohara *et al.* (2000). Here the viability of *S. cerevisiae* (95 strains) was 6–10% post drying and 4–8% (43 strains) after 10 years of storage. It is noteworthy that in this study from the Patent Microorganism Depository in Japan, Gram-positive bacteria exhibited a survival rate of around 80% and Gram-negative bacteria around 50%. However, it is clear (Bond, 2007a) that there are opportunities to improve and tune the process through the use of lyoprotectants such as a mix of skimmed milk, trehalose and sodium glutamate.

Yeast culture collections range from 'enormous and well-known' to 'smaller, more specialized boutique collections' (Boundy-Mills, 2012). Some of the industry-specific collections have broadened their focus to embrace other industries and to capture new funding sources. Table 3.1 details some commercial sources of brewing yeasts together with other collections that contain a myriad of diverse yeasts, some of which may have been sourced from breweries.

Although rarely explicit, it is noteworthy that culture collections are ambivalent about the mode of preservation. It would appear that both freeze-drying and cryopreservation are used, with strains stored via both techniques or one or the other. A frank explanation for this is offered by the Agricultural Research Service in the USA, which maintains the biggest collection of yeasts (some 14,500) in the world. They note that 'liquid nitrogen storage seems to cause little or no genetic change in cells. The reasons for not relying on this method exclusively for preservation of cultures are: cost and the fact that lyophilized preparations may be shipped by regular mail whereas strains preserved by liquid nitrogen must first be grown on agar or in liquid medium to avoid the expense of shipping frozen materials'. (http://nrrl.ncaur.usda.gov/TheCollection/AccessionMaintenance.html#Maintenance)

A different and more targeted interpretation of yeast supply is the one that feeds into the vibrant world of craft brewing. Here ale, lager and 'speciality' brewing yeasts are marketed for direct use, either as liquid cultures or as 'active dried' yeasts (see below). Typically, small-scale operations use the yeast culture once without any subsequent handling or recovery. The 'speciality' segment describes a dazzling array of production strains for different beer styles. The less technically onerous approach of supplying liquid cultures is not surprisingly supportive of a greater number of yeast strains.

Recovery and supply

Whilst 'best practice' for yeast preservation and storage from culture collections is ambivalent, it would be argued that cryopreservation is the safer approach. Certainly, if production yeasts are deposited for subsequent supply back to the brewery by a third-party operation, cryopreservation would be strongly recommended as providing a more robust and assured route in maintaining strain integrity and viability. The above concerns regarding the pitfalls of freeze-drying raise real concerns about the viability of the master strain and threat from more robust variants.

Recovery of yeast from storage (in liquid nitrogen) is clearly a 'critical control point' in the supply of yeast for propagation and then production fermentation. Accordingly, the 'recovery' step requires a documented programme of quality assurance to confirm the identity of the yeast strain and its microbiological purity. Application of best practice is mandatory and depending on scale, key points are best performed with two people to eliminate any risk of mix-ups. Whilst the approach detailed in Quain (1995) to confirm strain identity and purity is exhaustive and extensive (Table 3.2), it can be easily justified, as any failure in the process of yeast supply will potentially have a far more dramatic impact on process and product quality. The same principles and group of tests are used to validate the master culture that is stored in liquid nitrogen.

The 'supply' of pure yeast of the required production strain can be either as an agar slope or – increasingly from a third party supplier – a liquid culture. Cutting corners here defeats the object so the entire liquid culture or slope is used to initiate the propagation process.

Table 3.1 Commercial sources of brewing yeasts

Yeasts	Collection	URL	Strains[a]
'Brewing'	Cara Technology	cara-online.com	850
	NCYC	ncyc.co.uk	476
	VTT	culturecollection.vtt.fi	86
	Weihenstephan	hefebank-weihenstephan.de	91
'*Saccharomyces*'	ATCC	lgcstandards-atcc.org	25,299
	ARS	nrrl.ncaur.usda.gov	774
	CBS	cbs.knaw.nl	505

[a]Estimated number of strains as of April 2015.

Table 3.2 Methods to assure strain identity and purity

Method	Detection of
MYGP + ferulic acid (broth)	Phenolic wild yeast
WLN + cycloheximide (agar)	Wild yeast
Raka Ray (agar)	Gram-positive bacteria
MYGP + copper (agar)	Wild yeast
MacConkey (agar)	Gram-negative bacteria
WLN (agar)	Colony colour/ morphology
X-α-Gal (agar)	Lager vs. ale yeast
Genetic fingerprint	Strain identity

Consequences of failure

The consequences of a failure in the yeast supply process can be dramatic. Introduction of contaminated yeast or – far, far worse – the wrong strain can slowly but surely cause accumulated havoc. Scale-up through propagation and subsequent pitching into one fermenter and then two fermenters and beyond can quickly exacerbate a problem. In terms of threat, contamination of the supplied yeast may or may not be an issue depending on the dominance of any contaminants. Bacterial contaminants should hopefully be flagged through routine quality testing. However, from experience, the supply of the wrong yeast results in a much greater challenge to detect. The unfortunate introduction – via a failing yeast supply process – of a phenolic yeast into a major lager brewery caused huge issues that were difficult to manage through blending with 'good beer'. Less dramatic and harder to spot is substitution by similar yeast that behaves atypically in fermenter and triggers a shift in product flavour and aroma.

Propagation

To take a production strain from yeast supply to fermenter requires a progressive scale-up in biomass. The process of pure single culture 'propagation' dates back to Emile Christian Hansen at Gamle Carlsberg brewery in 1883 whose yeast – having been originally sourced from the Spaten brewery in Germany in 1845 – comprised four strains some 38 years later. Hansen developed several techniques to isolate pure yeast strains and – in doing so – paved the way for the philosophy of pure yeast cultures and their propagation.

The first propagator for Hansen's pure brewery yeast (Carlsberg Unterhefe No.1) was built in 1885 in Copenhagen and was made of copper, contained 2.2 hl of wort, a stirrer and pressurized with compressed sterile air (Annemüller *et al.*, 2011). Subsequent developments in what are now stainless steel brewery propagators have rightly focused on the assurance of hygiene, which remains the pre-eminent requirement of the process. As noted succinctly by White and Zainasheff (2010), 'when propagating yeast, the sanitation and oxygen needs are much higher than when brewing'.

Whilst hygiene is a given, attitudes vary in the extent to which oxygen is made available to yeast during propagation. One school of thought is that propagation should be 'aerobic' with the intent of achieving a high cell count. The more traditional approach of 20–30 years ago was to add air intermittently (a few minutes per hour) throughout propagation and to accept a lower yeast cell count (*ca* 50×10^6/ml). Things here have changed such that turnkey, commercial propagators are 'semi-aerobic' with more frequent and efficient aeration and accordingly higher cell counts of *ca* 80–100×10^6/ml (Maca, 2015). Here, the 'fermentation' is often incomplete as actively fermenting yeast ('high krausen') is considered to be better adapted for the first fermentation. Conversely with the aerobic approach, the wort is fully attenuated so as to generate as much yeast as possible. Whichever approach is used, the end game is to ensure that sufficient yeast is available from the propagation process to achieve the required inoculum for the first fermentation.

Propagator designs vary but frequently involve two vessels. These are often differently sized to minimize the extent of scale up, with the final tank being large enough to deliver the required cell count. The complexity of the design is driven by the attitude to oxygen. As yeast has a voracious appetite for oxygen, aerobic propagators require an agitator, sparge ring and baffles. Commercial semi-aerobic brewery propagators are comparatively simpler with an internal aeration lance/air injector, often together with a sinter or stone to aid gas transfer. Formation of foam can be significant and is managed passively by maintaining a large headspace (as much as 100%), top pressure or temporary shutdown of mixing and gas flow (if 'aerobic'), or on/off cycles of aeration (semi-aerobic). Alternatively, foam formation can be reduced by replacing *in situ* aeration with periodic aeration in line via external recirculation of the vessel contents (Maca, 2015).

Although propagation is usually a batch process, the on-going reduction in fermentation 'generations' has increased the pressure on propagation capacity. Realistically, installing a new plant is not always an option, as this requires significant capital expenditure and space. 'Drauflassen' (Kunze, 2014) or 'repeated fed-batch' propagation (Thiele and Back, 2007) can provide a practical solution through cycles of partially pitching-on (80–90% by volume) and topping up the residual 10–20% with fresh wort. As ever, hygiene is paramount, which in such a process must be managed and monitored. Similarly, the threat of selecting for variants of the primary yeast should be considered and the risk assessed. The threat from either route is best managed by limiting the number of top-up cycles.

More recently, though, the philosophy of brewery propagation has moved significantly to being a forcibly aerobic process that achieves better yields of high viability yeast. The drivers and philosophy of aerobic propagation have been outlined at length (Boulton and Quain, 1999, 2001; Quain, 2006b). The basic premise was that 'traditional' propagators had a 'fermentation mindset' and were not geared to growing yeast. Consequently, first-generation fermentations were badly underpitched and excruciatingly slow or required complex protracted arrangements of part-filling and, some days later, topping up. Conversely, 'aerobic propagation' sought to ensure oxygen was not limiting yeast growth and that the biomass yield/cell count was sufficient to pitch at the normal rate and achieve as near as possible standard cycle times. Further, the 'rule' of propagating ale yeast at ale fermentation temperatures and lager yeast at lager fermentation temperatures was parked and replaced with propagation of both lager and ale strains at a fixed temperature of 25°C.

Although labelled 'aerobic', the presence of fermentable sugars (and the associated catabolite repression) results in yeast metabolism remaining fermentative. Accordingly, unlike fed-batch baker's yeast propagation, conversion of sugar to biomass is relatively inefficient, at no more than 7.5%. The key to maintaining detectable but low levels of oxygen throughout propagation is the rate of gas transfer achieved through mixing and increasing gas flow rate. Indeed as described previously (Quain, 2006b), for a 220 hl second-stage propagator with a perforated sparge ring, heavy-duty agitator, and baffles, the maximum agitation speed was 58 rpm with an oxygen flow rate of between 10 and 100 l/min. Dissolved oxygen need not be routinely monitored but is established during commissioning via a blend of agitation and variable gas flow rates. The 'proof of the pudding' is that aerobic propagation delivers a yeast of high viability (>95%) and cell count of *ca* 200×10^6 cells/ml, which is three or four times more than in the traditional brewery yeast propagator. In passing, it is noteworthy that the achievement of such high cell counts requires the addition of zinc at 0.5 mg/l (rather than 0.2 mg/l) (Boulton and Mieleneiewski, unpublished observations; 2006).

In addition to the number of cells, aerobically propagated yeast benefit from being lipid 'replete'. Oxygen is required for the synthesis of unsaturated fatty acids (UFA) and sterols, which play a key role in membrane function (see Chapter 1). The sterols are thought to act as a 'mechanical reinforcer in membranes and also as an antioxidant that protects lipids during dehydration' (Rozenfelde and Rapoport, 2014). In the absence of oxygen and with new cell division, these non-polar lipids are mobilized as sterols and phospholipids to play their role in the membrane bilayers of the cell (Koch *et al.*, 2014). Of course, in the wider context of brewery fermentation, the addition of oxygen to wort is to enable the fresh synthesis of sterols and UFA that are diluted by cell division and found in depleted levels in anaerobic yeast at the end of fermentation (see 'Fermentation').

Best practice

It is important for the propagation process to deliver. After all, the output represents the beginning of a process where yeast will be cycled from fermentation vessel to storage tank and back anywhere between 3 and 15 or more occasions. The success is built on the twin pillars of hygiene and stepwise scale-up in volume (and biomass). It is important that the 'steps' are appropriately sized. Cutting a step and stretching the size of the yeast inoculum can backfire if then a contaminant or variant can 'punch above its weight' and compete with the primary yeast strain. The various steps in a typical large-scale propagation process are detailed in Table 3.3. Whether the yeast is sourced on an agar slope or as a liquid culture, the laboratory and plant steps in yeast propagation are geared to produce sufficient yeast of the correct strain and cleanliness to pitch into fermenter at the normal rate. Depending on scale, provision of liquid culture is beneficial in precluding some (or all) of the laboratory steps.

Irrespective of the size of the first-generation fermenter, propagation should be semi-aerobic – or better still, fully aerobic – so as to maximize yeast count, viability, and levels (stored and free) of sterols and UFA. The propagator should be sized appropriately so that on pitching into the first generation fermenter the yeast pitching rate meets the required specification.

Consequences of failure

Propagation of brewery yeast sets the tone of subsequent fermentations generation by generation. The 'traditional' propagation approach is without focus on cell yield. The addition of oxygen to the propagator via the periodic bubbling of air via a 'lance' achieved little if any gas transfer so that propagation was effectively anaerobic. Yields were poor (*ca* 50×10^6/ml) and little attempt was made to match the required yeast pitching rate. At best the fermentation vessel would be partly filled and then topped up some days later with fresh wort. Typically, the performance of such yeast would be woeful, such that the fermentation charts tracking gravity fall would extend from the usual single page to two or even three pages. Such extended fermentations result in poor-quality yeast of compromised viability. On pitching on, this extends fermentation vessel cycle times, and at best a number of fermentation generations are required before the yeast achieves the required and repeatable cycle times and associated product quality.

The above is the readily visual 'tip of the iceberg'. Cutting corners in terms of hygiene or step-up volumes can add further woe in terms of product

Table 3.3 Yeast propagation from laboratory to the brewery

	Step	Comments
Laboratory	Yeast on agar slope	From 'yeast supply'
	↓	
	2 × 10 ml – static – 24 hours at 25°C	All yeast recovered from slope
	↓	
	2 × 100 ml – shaken – 72 hours at 25°C	Shake flasks in an orbital incubator
	↓	
	3 l – constant aeration – 72 hours at 25°C	Shake flasks in an orbital incubator – entry of 'liquid culture' from a supplier
	↓	
	↓	
	↓	
	20 l – constant aeration – 72 hours at 25°C	'Carlsberg flask' modified to improve gas transfer or entry of 'liquid culture' from a supplier
	↓	
Plant	↓	
	8 hl – oxygenation – 30 hours at 25°C	13 hl seed vessel
	↓	
	140 hl – oxygenation – 30 hours at 25°C	220 hl propagator
	↓	
	1500 hl, 15°P wort	1600 hl FV, pitched at 15×10^6 viable cells/ml

quality and consistency, which is potentially amplified further with successive generations of use.

Yeast supply – active dried yeast

Active dried baker's yeast was developed during the Second World War to avoid the need for refrigeration during its transport and storage. Not surprisingly, the technology has been developed and in recent years extended to include yeasts for brewing, distilling, wine making and biofuels. These days active dried yeast (ADY) is produced via fed-batch fermentations with beet and cane molasses as the feedstock. A review of the process can be found in Ingledew *et al.* (2009).

Unlike aerobic brewery propagation of yeast, the aerobic fed-batch approach is unequivocally focused on the efficient production of yeast biomass such that conversion of sugar to biomass is in the region of 54% (opposed to *ca* 5% from anaerobic fermentation) and with no ethanol formation. Accordingly, the downside is that yeast from this process is physiologically very different (oxidative) from the 'fermentative' yeast produced via aerobic propagation or from brewery fermentation. However, yeast produced in this way is much better equipped for the extreme rigors of drying to approximately 8% moisture via fluidized-bed or airlift dryers. The fed-batch process would be anticipated to result in yeast forming more sterol (4% of the yeast dry weight) than is formed through aerobic propagation (1%) (Quain and Haslam, 1979). Further, as the amount of sterol determines how many rounds of cell division can occur during fermentation, ADY will potentially divide more times. This is helpful should the fermentation be underpitched but, conversely, at normal pitching rates will lead to greater yeast growth than the norm. Either way, with ADY there is no metabolic requirement to aerate or oxygenate the wort as the yeast comes pre-loaded with sterols and unsaturated fatty acids.

In a brewing context, the 'convenience' in terms of long-term, simple storage of ADY is especially beneficial as, in one fell swoop, it removes the need for yeast supply, propagation, cropping, and storage of yeast. Such a single trip or 'pitch and ditch' approach was pushed enthusiastically in the late 1990s as a viable option for 'small scale craft brewers, franchise brewing operations, or in situations where particular beer qualities may be produced infrequently' (Boulton and Quain, 2001). Jump forward to 2015 and ADY has played its part in the dramatic growth of craft brewers around the world with some 1700 breweries in the UK (British Beer and Pub Association, 2015) and *ca* 4200 in the USA (www.brewersassociation.org/statistics/number-of-breweries/).

ADY is used by craft brewers big and small. At the smaller end of the scale, the pitch and ditch approach is very much the norm. Pitching rate is not controlled directly but is based more on marrying the supplied pack size with fermenter volume. However, as output grows this approach becomes increasingly cost-prohibitive and accordingly processes for yeast cropping, storage, and repitching must be introduced.

Concerns over contamination of ADY and its viability have lessened in recent years. As noted by Ingledew *et al.* (2009), 'yeasts grown aerobically in a yeast factory ... will inevitably contain low levels of bacteria and wild yeasts that grow under the same fermentation conditions'. However, hygienic practices have tightened such that the specification for bacteria and wild yeast is a respectable <1 cell per 10^6 viable yeast cells or less. Similarly, there is less focus on viability with suppliers majoring on declarations such as 'viable cells at packaging: $>6\times10^9$/g' or 'living yeast cells at $\geq7\times10^9$ per gram of dry yeast'.

Under the banner of convenience, ADY offers the benefit of protracted storage of up to 2 years, which lends itself to infrequent or seasonal use. However, there are caveats such as storage temperature (4°C for one supplier or <8°C for another) and, upon opening, use within 3 or 7 days. There is some ambiguity on stability with one view that 'activity loss is about 25% per year at 8°C or 50% per year at 22°C' (www.danstaryeast.com/products). A somewhat more coy approach is that ADY can be 'stored at room temperature for periods of time not exceeding 3 months without affecting its performance' (www.fermentis.com/brewing/industrial-brewing/product-range/).

Although yeast 'stress' is a big-ticket item in brewing (Gibson *et al.*, 2007), it is doubtful if any stresses encountered during brewing compare with that of drying and rehydration. There is alarming talk of drying causing protein misfolding and aggregation together with effects on the integrity and functionality of intracellular membranes. On drying and hydration, these 'leaky' membranes allow the loss of low molecular weight compounds (e.g. ions, nucleotides). Going the other way, there is a mandatory need for careful rehydration of ADY. Rehydration of the cell must be performed sympathetically to enable the reconstitution of functional membranes and metabolic homeostasis such that the population is broadly viable for fermentation and cell division to occur.

In preparation for drying, manufacturers use proprietary 'tricks of the trade' to enhance the readiness of yeast for drying. As noted above, the disaccharide trehalose is one of a number of compounds that can be added exogenously to improve the freeze-drying of yeast strains. It – together with the polysaccharide glycogen – are classed as storage or reserve carbohydrates in yeast that accumulate when nutrients (e.g. nitrogen, sulfur, or phosphorus) or energy become limiting (Lillie and Pringle, 1980). More generically, trehalose is present 'in particularly high concentrations in resting or in stressed cells' (Feldmann, 2012). Accumulation of trehalose is reported as 'adding value' to industrial applications such as baker's yeast production and high-temperature bioethanol fermentations in Brazil (Eleutherio *et al.*, 2014). Somewhat multi-talented, trehalose is flagged as conferring general resistance to heat and desiccation. More specifically, trehalose stabilizes and prevents damage to membranes (and proteins) during desiccation by substituting for water through hydrogen bonding. Importantly, to be most effective, trehalose is required on both the inner and outer sides of the membrane (reviewed in Eleutherio *et al.*, 2014). However, after more than 40 years of research the role of trehalose as a generic stress protectant has been questioned. Taking advantage of the same disaccharide transporter to enhance intracellular trehalose, there is direct evidence that trehalose confers desiccation tolerance (Tapia *et al.*, 2015) or that correlation with stress is 'purely coincidental' and is down to a 'moonlighting' protein, a catalytically inactive variant of trehalose-6P synthase (Petitjean *et al.*, 2015).

Rehydration of active dried yeast is very much in the hands of the user. Manufacturers of dried brewing yeasts are prescriptive as to rehydration but differ in the detail. For example, (i) the yeast is sprinkled on 10 times its weight of sterile water at (ii) 27 ± 3°C or 30–35°C after which (iii) stand for 15 or 15–30 minutes and (iv) gently stir for 30 minutes or stir to suspend and stand for 5 minutes at 30–35°C. The yeast is then pitched directly or after cooling to the temperature of the wort. There is a difference of opinion on rehydration of yeast directly into wort, with one manufacturer encouraging it and the other discouraging it.

Rehydration of brewing strains has, by comparison with wine yeasts, attracted little attention. Jenkins *et al.* (2011) noted that 'because of operational constraints this practice (rehydration) is often not adhered to when ADY is employed on an industrial scale'. Further, yeast viability was compromised by incomplete rehydration, rehydration at suboptimal temperatures, or rehydration in wort. The authors suggest that for each commercial dried brewing yeast, there would be merit in establishing the optimum temperature for rehydration and accordingly viability.

Building on this, a variety of 'additives' have been evaluated in rehydration of dried wine yeasts. Given its seasonality of need, wine is a much bigger arena for ADY but reportedly has similar issues, namely 'wineries tend to pay insufficient attention to the rehydration process' (Díaz-Hellín *et al.*, 2013). Well-reasoned additives including lipids (sterols and unsaturated fatty acids), antioxidants, vitamins, and metal ions have been evaluated (Díaz-Hellín *et al.*, 2013; Vaudano *et al.*, 2014) either without effect or where the response varied between strains. Whilst some authors have concluded that there is no 'universal activator' (Díaz-Hellín *et al.*, 2013), others state that 'no relationship was found between viability at the end of rehydration and fermentation performance' (Vaudano *et al.*, 2014).

Best practice

There is a clear, albeit niche opportunity for active dried brewer's yeast. Top of the list of best practices are those actions that are geared to achieving and maintaining viability such as (i) cold storage and use within the 'best before' date, (ii) not storing part-opened packs and (iii) rehydrating to the letter of the manufacturer's instructions. As with wet yeast fermentations, managing viable yeast pitching rate is a given and should be matched with wort gravity. The contribution of dissolved oxygen in wort is of interest, as on the face of it, aerobically fed-batch grown yeast has no immediate requirement for oxygen. However, any dissolved oxygen will be consumed once the yeast population initiates division and thereby may compensate for lower pitching rates than are desirable.

Consequences of failure

The primary consequence of failure in not applying best practice to the use of ADY is inconsistency. This should not be reconfigured as a virtue, as consistent processes deliver consistent products, a quality that the vast majority of beer consumers expect. Fermentations that are too fast or too slow (or incomplete) result in unpalatable and unbalanced beers, which commercially can be 'career threatening'!

Pitching

The inoculation or pitching of yeast into fermenter is one of three primary control factors in brewery fermentations. The other two, oxygen and temperature, are considered below (see section on 'fermentation'). As noted above, brewery fermentations are peculiar in typically transferring yeast between successive fermentations. The efficiency of fermentation (i.e. ethanol production vs. new biomass) is indirectly managed by limiting new yeast growth through a combination of a relatively high pitching rate and availability of the nutrient oxygen. Achieving the desired balance of the two parameters is central to delivering fermentation that is of the required rate and extent and, importantly, is consistent.

Yeast pitching rate is broadly based on the maxim of 1×10^6 viable cells/ml for every 1° Plato of wort (O'Connor-Cox, 1998b). So at a wort gravity of 10° Plato the pitching rate is 10×10^6 viable cells/ml, 15×10^6 viable cells/ml at 15° Plato, and so on. Although by no means carved in stone, this approach – together with management of wort oxygen – provides a good base for optimal fermentation. Expectations include (i) new yeast growth (crop is 3- to 4-fold the amount of yeast pitched), (ii) required fermentation rate and extent (including diacetyl rest) (iii) desired flavour and aroma profile, and (iv) management of fob and cooling demand.

Increasing pitching rate has often been explored (e.g. Edelen *et al.*, 1996; Erten *et al.*, 2007; Verbelen *et al.*, 2009). As ever, the experimental approach, scale, and pitching 'multiple' vary but with consistent oxygen concentration, the process outcomes are broadly consistent. As would be anticipated, increasing the pitching rate results in faster fermentations with correspondingly higher and earlier peak yeast cell counts. The quantity of new yeast growth was unchanged irrespective of pitching rate as this is determined by the (fixed) concentration of oxygen. The combined (pitched + new) biomass obviously increased, resulting in greater losses of bitterness 'bound' to the yeast cell wall. Analysis of beer volatiles from such trials provides no clear, repeatable insights as to the impact of enhanced pitching rate. Given the lack of an experimental baseline, this is no great surprise.

Measurement of pitching rate

There are numerous methods for the measurement of yeast concentration for pitching into a fermenter (Table 3.4). In the past, the complexity, accuracy and cost of these methods was in accordance with the scale of operations. Today, the advent of cost-effective but sophisticated methods has enabled the 'up-selling' of more robust approaches to control yeast pitching rate. For the simplest approaches, the achievement of the desired pitching rate is 'cheap and cheerful'. At its simplest, this may be *x* packs of dried yeast per vessel. Measurement of 'weight' or mass (variously dry, or wet/centrifuged) of 'solids' in yeast slurries enables a more controlled approach to yeast pitching. The dry yeast method is more accurate than wet, reflecting changes in cell size/volume during fermentation (Boulton and Quain, 2001) and during yeast storage (Cahill *et al.*, 1999). However, dry weights are not a practical real-time option unless using a microwave (Rice *et al.*, 1980). However either method is complicated by the distraction of non-yeast solids such as trub and hop material. The quantitative significance of either fraction can be assessed visually (assuming transparent tubes) as distinct bands after centrifugation before weighing or drying. Nevertheless, either approach results in the overestimation of weight and the underestimation of yeast mass. However, whichever approach is used it is strongly recommended that there is correction for yeast viability especially if this is less than 95%. This increases the accuracy of pitching rate and provides a key indication of yeast health and well-being. However, it should be noted that compensation for reduced viability results in the addition of more dead yeast,

Table 3.4 Measurement of pitching yeast concentration

Method	Required equipment	Comments
Wet weight	Centrifuge, analytical balance	'Cheap and cheerful'. Rapid but (even after water washing) inaccurate because of entrained trub and hop material. Requires correction for viability
Dry weight – oven	Centrifuge, oven, analytical balance	As above but slow – 72 hours at 105°C. Takes 2–3 days
Dry weight – microwave	Membrane filter, microwave, analytical balance	As 'wet weight'
Viability	Light microscope, vital stain (e.g. Methylene Blue)	'International method' (ASBC Methods of Analysis, 2011). Rapid but overestimates viability especially <80%. Requires a skilled operator
Viable cell count	Haemocytometer, light microscope, vital stain (e.g. Methylene Blue)	'International method' (ASBC Methods of Analysis, 2012). Rapid but requires a skilled operator
Electronic particle counters	Proprietary equipment	Requires dilution to *ca.* 5×10^4 to 1×10^7 cells/ml and filtration to remove non-yeast particles. Yeast must be deflocculated. Rapid analysis, which includes cell size
Biomass probe using radiofrequency permittivity	Proprietary technology	Real time and representative. Quantifies only viable yeast. Options include in-line, mobile skid and off-line laboratory analyser. Requires calibration for each yeast strain

which – through autolysis – will compromise beer quality.

The issue of non-yeast solids is overcome by measurement of yeast count *as is* or, better still, a viable count in conjunction with a vital stain such as methylene blue, the use of which dates back to 1933 (O'Connor-Cox *et al.*, 1997). Together, they represent the *de facto* reference method by which other methods (below) are compared. The method requires a skilled and trained operator to juggle the use of a haemocytometer (*aka* Thoma chamber) and microscope. This is further compromised by the tedium of counting up to 1,000 cells (ASBC Methods of Analysis, 2011), which results in operator fatigue when analysing numerous yeast samples. Accordingly, an automated slide-based counter using methylene blue has been developed with both enhanced consistency and reduced analysis time compared with the manual method (Thompson *et al.*, 2015).

Whilst both weight and cell numbers have their advocates, especially from the perspective of simplicity, there are some caveats. Whilst trub adds inaccuracy to the measurement of solids, both approaches suffer from errors associated with the difficulties of representative sampling of thick yeast slurries (40% wet solids, *ca* 1.5×10^9 cells/ml) from storage vessels. Considerations include vessel scale, homogeneity (thorough mixing), sampling point and sample size. Such sampling errors can then be further compounded by processing steps required for testing, such as the small slurry volumes involved in serial dilution for cell counts.

As would be anticipated, the driver for the improved control and accuracy of yeast pitching came from the bigger demands of large-scale fermentation management. Ideally, such an approach would be in-line, automated, control the pitching process and better still would 'count' only viable yeast. First up and reported by Riess (1986) was a system using near infra-red turbidometry. Two sensors measured the turbidity of the wort and of the pitched wort, thereby enabling a 'set-point' to be established and constant yeast count to be maintained. The approach was commercialized and met with some success in achieving a tighter range of cell counts compared with the conventional pitching control system. However, the downsides were the lack of correction for viability together with a lack of correction for non-yeast solids entrained in the injected yeast slurry (Boulton and Quain, 2001). Dark beers were also problematic, as haze measurement was more difficult (Noble, 1997).

The issue was neatly summarized by Harris and co-workers (1987) who observed that 'an accurate method for the real-time estimation of microbial biomass during laboratory and industrial fermentations remains an important goal'. The paper then explored the use of radiofrequency (RF) and 'concluded that measurement of RF permittivity provides an extremely powerful and convenient means for the assessment of microbial biomass, *in situ* and in real time'. On application of the radiofrequency electrical field, viable cells (with intact plasma membranes) become charged and the measured capacitance reflects linearly the viable yeast concentration. Two years later, Boulton *et al.* (1989) reported the application of this technology to an automatic yeast pitching system that measured only viable yeast and was independent of any non-yeast solids. However and importantly, because cell size varies, the system required calibration against methylene blue cell counts for each yeast strain in use.

The exploitation of radiofrequency permittivity was undeniably a genuine process innovation and the Aber biomass meter (www.aber-instruments.co.uk) is used in breweries worldwide. Such systems are considered to control pitching rate to better than within ±2% of the target viable count (Boulton, 2006). Accordingly, demonstrable improvements have been reported in target pitching rate (Maca *et al.*, 1994) and fermenter cycle time (Boulton and Quain, 2001; Carvell and Turner, 2003; Boulton, 2006). Unsurprisingly, there have been range extensions to the use of the technology in controlling pitching rate, including an off-line laboratory analyser and a mobile skid-based unit. Other applications have included the control of yeast cropping together with monitoring propagation and fermentation, most notably in the distribution of yeast in cylindroconical vessels (Boulton *et al.*, 2007).

Generation number

The number of times yeast is cycled between fermentations comes up from various perspectives throughout this chapter. Today, a more proactive management (10 to as low as 5 or fewer cycles) has succeeded the laissez-faire approach of 20 or more generations in the past. The drivers for this change are not entirely clear. Whilst hygiene and yeast handling are arguably better than in the past, awareness of genetic change is more pronounced and may contribute to this. A good and quantified explanation for this is reported by Stewart (2015). Over a period of 15 years in a North American brewing group, average wort gravities increased from 12 to 14 to 16 and finally to 18° Plato. In turn this was accompanied by yeast generation numbers reducing from >20 to 16 to 12 and finally to 8 cycles (Stewart, 2015). These changes were introduced over time to avoid 'fermentation difficulties' in rate and/or extent. Another study (Jenkins *et al.*, 2003) reports for a (then) major UK brewer that company policy for maximum generation number of four production yeasts ranged from 10 (two strains) to eight and five.

A word or two on yeast viability

Viability is the key measure of yeast health in brewing and it is used to correct pitching rate, assess fermentation performance and the impact of in-process handling. As noted above, the 'gold standard' is the vital stain methylene blue, which is invariably used to benchmark and calibrate sophisticated methods such as radiofrequency permittivity and other up-and-coming methods. However, despite its position and longevity, methylene blue has its criticisms. Most notably, vital staining infers the cell is a metabolically viable cell but throws no light on whether the cell can divide and replicate. Accordingly, cell division is routinely tested by a functional test involving the plating of samples on nutrient agar. This approach is not 'real time' and takes two or more days incubation to enable a viable cell count. Additionally, plate counts do not necessarily support the growth of all microorganisms in the sample as the media is (intentionally) selective but also the organisms themselves may be physiologically non-culturable for a host of reasons.

Given the debate about measuring yeast viability, it is (with hindsight) somewhat bizarre that an additional concept – vitality – has been added to the mix as a measure of yeast physiological state. Many diverse vitality tests have been proposed to 'assess fitness of individual batches of yeast to pitch, either in a go, no-go approach, or as predictors of subsequent fermentation performance, which preferably allows selection of optimum pitching rates and/or wort oxygenation' (Boulton, 2012). Tellingly, the same author concludes that 'there is little evidence that they provide more information than a simple viability test such as the usual counting of unstained and stained cells treated with the vital dye methylene blue'.

Practically, rightly or wrongly the methylene blue method for yeast viability is firmly embedded in the global brewing industry. Perhaps we should simply pick up on Johnny Mercer's lyrics from 1945 and 'accentuate the positive, eliminate the negative and latch on to the affirmative, do not mess with Mister In-between'.

Acid washing – yes or no?

Rather than beating about the bush, it is hard to offer a balanced, constructive view of acid washing. Its use comes from a different time when yeast hygiene was a concern and 'washing' in (typically) food-grade phosphoric acid (pH 2.2) at 2–4°C with stirring for 2 hours dealt with the problem. For many, rather than being an occasional treatment, acid washing has became hardwired as part of the process irrespective of need or value. In terms of 'need', today's awareness of the importance of hygiene, use of closed vessels and effective, validated cleaning regimes has resulted in a step change in pitching yeast cleanliness. As to 'value', nothing here has really changed. Acid washing is only effective in killing bacteria, or rather brewery bacteria. Extremophile acidophilic bacteria are associated with the mining of metals and coal and have a pH optimum for growth at (or below) pH 3.0 (Johnson, 1988). If wild yeast or the wrong production yeast is the problem, then acid washing will not resolve things as generically 'yeast' is unaffected by the conditions.

Of course, saying that yeast is 'unaffected' by acid washing is debateable. In terms of homeostasis, maintenance of intracellular pH (around pH 7) has been reported in aerobically grown *S. cerevisiae* over an external pH range of 3–8 (Valkonen *et al.*, 2013). More specifically, Simpson and Hammond (1989) noted that brewing yeasts were 'inherently resistant to acid washing treatments but, under some conditions this resistance is diminished'. Interestingly the presence of ethanol (5 and 10% abv) had a marked effect on viability and 'unhealthy' yeast performed poorly. Notably, scanning electron microscopy revealed blistering of the cell wall.

A similar observation – 'the cell surface was deformed' – was made after incubating yeast in rich media at pH 2.5 (de Lucena *et al.*, 2015). This work was triggered from washing yeast with sulfuric acid between fuel alcohol fermentations, as management of (lactic acid) bacteria is important to optimize ethanol production (Walker, 2011). Although a different approach, the conclusions from de Lucena *et al.* (2015) are directional in that acidification with sulfuric acid triggers a host of metabolic stress responses to repair the physical damage to the cell wall and maintain viability. Accordingly, it should be argued that routine or occasional acid washing selectively treats some of the symptoms (not the cause), is unnecessary where hygiene is well managed and is the cause of an additional and unnecessary stress to pitching yeast.

Best practice

Pitching rate along with wort oxygenation are the two pillars of successful and consistent fermentations. Irrespective of process scale, consistently pitching the required number of viable yeast cells is the focus of best practice. How that is delivered is driven by scale and budget. Although the 'radiofrequency permittivity' platform has a range of willing applications that will add value, for some a microscope, haemocytometer and vital stain is sufficient to deliver a consistent yeast pitching rate in fermenter. Other considerations are yeast viability, which directly relates to the management of the other parts of the yeast supply cycle. Generation number – as noted above – should also be proactively managed to 10 or less generations. From a process perspective, pitching of yeast into fermenter must accommodate the reality of lengthy fill times and multiple batches of wort. In terms of consistency, best practice is to pitch – if possible – all the required yeast into the first brewlength. This removes the small risk of wort contaminants taking hold and ensures that the entire yeast population behaves 'as one' in the transition from quiescence to growth.

Consequences of failure

In terms of failure there is a sliding scale of severity for failing to meet best practice recommendations for yeast pitching. Underpitching will result in a sluggish fermentation performance, which – depending on degree – will be readily apparent. If this is due to pitching by mass with no understanding of the contribution of dead yeast, this will result in enhanced cell autolysis and untoward flavours and aromas. A comparable situation can occur running blind with the Aber technology, although increased pitching solids provides an early clue of an increasing dead cell fraction. Although significant overpitching will result in an ageing population of yeast cells, repeated repitching at an elevated rate is unlikely to be a sustainable process. Stretching the maximum generation number is unpredictable and very much specific to the strain in question. The expectation is that this would over time increase the likelihood of sluggish fermentation and selection of variants. Extending yeast pitching from the first brewlength to most or all may be driven by practical considerations or the belief that the approach improves mixing. Whatever the reasoning, extending the pitching profile will result in inconsistency. Here the different batches of yeast will transition from quiescence to growth at different times and (depending how and when oxygen is added) differing availability of oxygen. Competition for oxygen will also not be a 'level playing field' as the uptake rate increases over the time of exposure (Boulton, 2013). Consequently, subsequent batches of yeast will potentially be out-competed for oxygen by the more voracious appetite of the first pitching of yeast.

Fermentation

Fermentation is the heart and – with the exception of 'continental' maturation times (i.e. lagering) – the longest part of the brewing process. At a large scale, the cylindroconical fermenter is far and away the vessel of choice. The roots of the modern cylindroconical vessels (CCV) stem from the remarkable work of Nathan presented at a meeting of the Institute of Brewing in London 86 years ago (Nathan, 1930). Whilst modest in scale (100 hl), these aluminium vessels were designed as unitanks (both fermentation and maturation) with cooling jackets on the vessel. The vessel cone, which was also cooled, was introduced to improve mixing and thereby speed up fermentation as well presumably to catch trub and yeast! Other insights include the aeration of cold water and, after removal of the yeast, gas washing to remove the 'jungbukett' or young beer sulfury aromas. Maule (1986), in an engaging review, describes the evolution of fermenter design pre- and post-Nathan to the many and diverse continuous systems.

Todays CCVs, although rooted in the Nathan's design, have capacities in the range of 2000–10,000 hl and aspect ratios of between 1 : 3 and 1 : 5 (diameter to height). The smaller footprint is of course conducive to the creation of large tank farms. Vessel volumes are nominal, as there is 10–15% freeboard (unfilled headspace) to retain foam within the vessel. Recovery of carbon dioxide from large fermenters is increasingly the norm. These large vessels have a number of distinct cooling jackets on the vessel as well as on the cone. With hygiene being ever more important, fermenters use more sophisticated and effective cleaning-in-place systems than Nathan's use of alcohol vapour. In terms of the big picture, Hoggan (1977) flagged other benefits of CCVs as reduced capital, reduced running costs and losses together with increased flexibility. Good schematics of a typical CCV can be found in Boulton and Quain (2001) and Boulton (2006).

There is no mystery about control; get the simple things right and the big things – i.e. fermentation – look after themselves. The 'simple things' are specific wort gravity, pitching rate and correct wort oxygenation. Yes, yeast viability/physiology is important, as is wort composition and temperature, but the 'simple things' rule routinely. Here, the intent is not to replay excellent reviews on fermentation (Boulton, 2006; Lodolo *et al.*, 2008; Bokulich and Bamforth, 2013; Boulton, 2013), but to focus on three areas; wort oxygen, yeast glycogen and vessel capacity.

In support of the narrative, Fig. 3.2 provides a timeline of some of the key events during fermentation.

Oxygen

As noted above, the generally accepted guideline for pitching rate is 1×10^6 viable cells/ml for every 1° Plato of wort. A similar metric is used for wort oxygen with 1 mg/l (ppm) for every 1° Plato of wort (Barnes, 2006). As with pitching, this is rule of thumb and a useful starting point for optimization. Functionally, as yeast has the requirement for oxygen, pitching rate is the real determinant of oxygen requirement. As ever, it may not be as simple as this! Early reports noted that both ale (Kirsop, 1974) and lager (Jacobsen and Thorne, 1980) strains can vary – from half air saturation to oxygen saturation or more – in the amount of oxygen they require for a successful fermentation.

In terms of process, oxygen is added under backpressure on the cold side of the wort cooler either as air or oxygen. Whilst the proportion of the oxygen in the gas phase is the bigger determinant, pressure together with wort temperature and solids

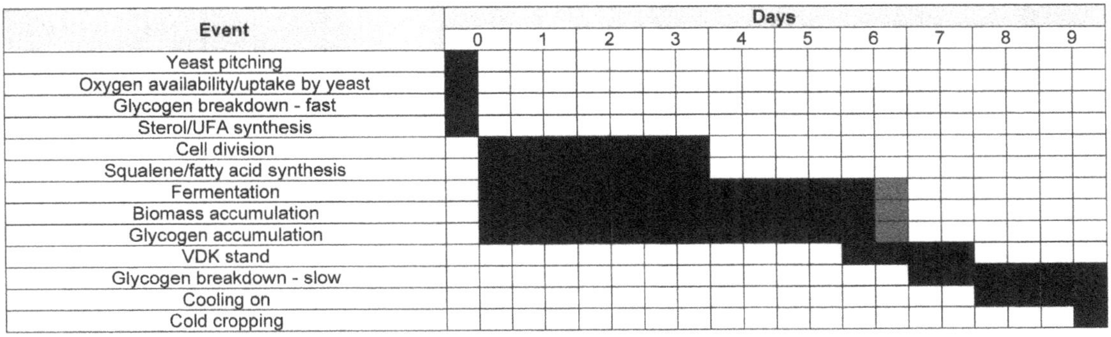

Figure 3.2 Key events during fermentation.

determine solubility. The solubility of oxygen or air in water at 25°C and 1 bar pressure is, respectively, 40 or 8 mg/l. Accordingly, the concentration of oxygen in worts saturated with air at 15°P and 20°P is 6.3/5.5 ppm at 15°C and 5.5/4.9 at 20°C (Kirsop, 1974). Given this, it is not surprising that wort oxygenation – rather than aeration – is the norm with high-gravity wort. Although typically well controlled at the wort cooler, the actual concentration of oxygen in fermenter may be lower due to gas breakout, especially if the vessel is some distance from the point of addition. As with yeast pitching regimes, multiple batches of wort allow different 'tunes' of oxygen addition. Addition to all or some of the wort batches may allow tuning but will inevitably add process complexity and opportunity for error.

Fundamentally, oxygen determines the extent of yeast growth by enabling the synthesis of sterols and mono-unsaturated fatty acids. Too little and there is insufficient lipid synthesis and consequently insufficient yeast growth and incomplete fermentation. Too much oxygen – should it remain available once the cells begin to divide – can result in unrequired additional lipid synthesis and unnecessary yeast growth. As yeast growth impacts negatively on the formation of ethanol, unnecessary yeast growth reduces the fermentation efficiency. These interactions are demonstrated experimentally in Fig. 3.3.

Oxygen and lipid synthesis

Under the anaerobic conditions of fermentation, brewing yeast becomes auxotrophic for sterols and mono-unsaturated fatty acids (UFA). The oxygen added to wort is used by yeast in the synthesis of these key lipids at the beginning of fermentation. During anaerobiosis, these lipids are progressively diluted during cell division so that cells at the end of the process are lipid 'depleted' and once more auxotrophic (Fig. 3.4). The provision of excessive dissolved oxygen does not result in additional synthesis of these lipids as these pathways are highly regulated by various feedback control systems, which conversely are induced when sterol levels are low. Consequentially, it is wise to tune the oxygen concentration to broadly meet the 'needs' of the pitching yeast, as excess and unused oxygen may result in wort oxidative reactions that may damage beer quality. In passing, it is worth noting that in brewery fermentations yeast metabolism is purely fermentative and that respiration and oxidative phosphorylation are not possible. This is a function of the overarching catabolite repression from the fermentable sugars in wort.

The building block for both sterols and fatty acids is acetyl-CoA, the key metabolic intermediary stemming from glycolysis. Of the 20 or so steps from acetyl-CoA to the end product ergosterol, the initial steps to the hydrocarbon squalene do not require oxygen. The steps between squalene and ergosterol require 12 molecules of oxygen for every molecule of ergosterol. In the absence of oxygen, squalene accumulates in anaerobic yeast so as to facilitate sterol synthesis on the provision of oxygen. With UFA, the requirement is one to one. Quantitatively compared to content of sterols, total fatty acids are 5- to 10-fold higher, with UFA content 3-fold or so greater (Rosenfeld and Beauvoit, 2003). For an excellent review on yeast lipid metabolism see Klug and Daum (2014).

Both saturated and unsaturated fatty acids are the basic elements of phospholipids and sphingolipids.

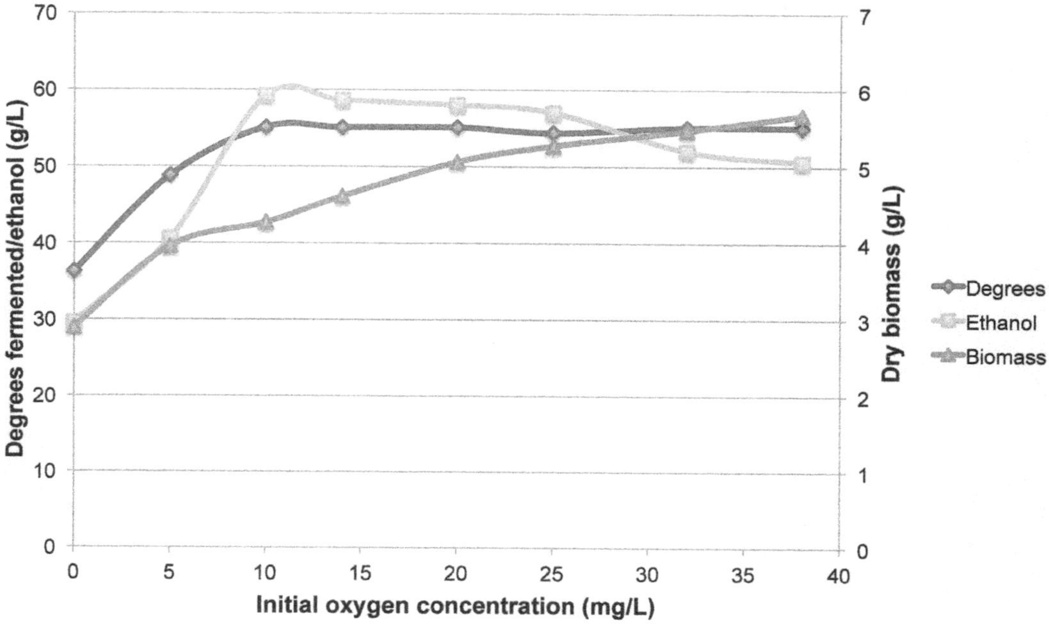

Figure 3.3 Oxygen, yeast growth and fermentation efficiency.

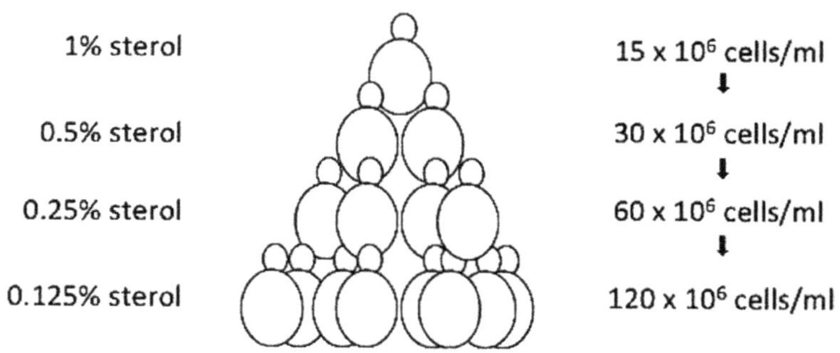

Figure 3.4 Impact of cell division on sterol content.

These are complex amphipathic lipids that are important in biological membrane structure and function. Similarly, sterols are essential for cell viability and are key components of the plasma and organelle membranes where they affect fluidity/rigidity, permeability and protein functionality. In excess, both free sterols and fatty acids can be toxic, so they are sequestered and stored as steryl esters and triacylglycerols in subcellular lipid droplets. With cell division during fermentation (which typically is complete by the mid-point), the sterol esters are converted to free sterols (and UFA) that are transported to the required location in membranes.

It has long been known (Andreasen and Stier, 1953) that anaerobic lipid-depleted yeast will take up exogenous ergosterol dispersed in UFA (oleic acid, C18:1) so as to resume growth and fermentation. Addition of ergosterol in lipid-free detergent to all-malt wort (10°P) supported anaerobic growth (David, 1973), suggesting that anaerobic yeast requires sterol but not UFA for growth. Practically, this is of interest as worts contain lipids, notably stearic acid (C18:0) and linoleic acid (C18:2), which is not made by yeast. However, wort separation/filtration and particularly boiling/clarification remove most of the lipid, with an estimated 0.1–3% of malt lipids surviving through to the pitching wort (Kühbeck *et al.*, 2006). The extent of 'supplementation' by linoleic acid can only be speculated upon and will depend on wort concentration, malt content and clarity (turbid worts usually contain more lipid). Uptake of linoleic acid would be expected to be total, as when added (to suppress ester formation) at 50 mg/l to a 10°P all-malt ale wort it is fully taken up with 18 hours of fermentation (with 70% in 4 hours) (Thurston *et al.*, 1982). Sterol esters account for only 3% of the total lipid reported in pitching worts (Anness and Reed, 1985) so, in accord with David (1973), wort contains insufficient sterol to spare the need for endogenous synthesis.

Quantitatively, total sterol in pitching yeast represents *ca* 0.1% dry biomass and peaks at around 1% dry biomass following oxygenation. With three or four rounds of cell division, the sterol content declines in anaerobic yeast to that found in pitching yeast (Aries and Kirsop, 1978). Qualitatively, the percentage of unsaturated fatty acids in pitching yeast is 20% or so, increasing to *ca* 56% post exposure to oxygen and then diluted out by growth to the initial level (Thurston *et al.*, 1982).

Glycogen

In anaerobic yeast, polysaccharides are quantitatively bigger than the collective protein component. Of these, glucan, mannan and chitin perform a structural role in the cell wall, whereas glycogen is a reserve or storage polysaccharide that can be mobilized when required to provide metabolic energy. Glycogen is a high-molecular-weight branched polymer of α-D-glucose located in the cytosol ('bodies') and vacuole (Wilson *et al.*, 2010) with a 'pool' associated with the cell wall (Gunja-Smith *et al.*, 1977; Deshpande *et al.*, 2011). Regulation of glycogen synthesis and dissimilation is complex and has been reviewed by Wilson *et al.* (2010). Its particular claim to fame has been somewhat dismissively described as 'at the beginning of industrial fermentation procedure stored glycogen is rapidly degraded, while it accumulates once fermentation is complete (Feldman, 2012).

It has long been known that accumulated carbohydrate reserves in anaerobic yeast are rapidly broken down on exposure to oxygen (Chester, 1963). In a brewing context (Quain *et al.*, 1981), glycogen is rapidly broken down after pitching into oxygenated wort, only to reaccumulate during active fermentation before the cycle repeats. The amount of glycogen at the end of fermentation is substantial and is about 30% of the dry biomass (less specific methods of analysis probably over report). On exposure to oxygen, glycogen accounts for < 5% of the dry weight. This rapid mobilization of glycogen has been linked to providing the metabolic energy for the synthesis of new sterols and unsaturated fatty acids. The key argument for this is that the plasma membrane in anaerobic yeast is lipid depleted with limited functionality for transport of wort nutrients (e.g. sugars). Detailed measurement of glucose, sucrose, fructose and specific gravity during the anaerobic-aerobic transition would appear to support this hypothesis (Quain and Tubb, 1982). Subsequent work (Quain *et al.*, 1981) demonstrated a linear relationship between glycogen breakdown and sterol synthesis. The apparent fuelling of sterol synthesis by glycogen in turn triggered an awareness of the need to maintain glycogen levels in pitching yeast. Conditions such as a protracted storage in the fermenter cone or in storage vessel have been shown by many workers to result in glycogen breakdown although at a much slower rate than on exposure to oxygen (e.g. Quain and Tubb, 1982; McCaig and Bendiak, 1985; Powell *et al.*, 2004: Somani *et al.*, 2012).

The 'glycogen story' makes intuitive sense but the experimental techniques and, in particular, analysis was of its time. To the author's knowledge, there has been no more sophisticated analysis to connect more literally glycogen breakdown to sterol synthesis. The expression of the sterol biosynthetic genes (*ERG*) has been reported to be up-regulated shortly after pitching (Higgins *et al.*, 2003; Rautio *et al.*, 2007). However, there are no reports as to the expression of glycogen phosphorylase (*GPH1*) or debranching enzyme (*GDB1*) post-pitching. A more likely explanation is that these enzymes are already 'present and able', as can be inferred from the slow dissimilation of glycogen during storage. Intriguingly, a recent paper (Gsell *et al.*, 2015) has shown that deletion of *GPH1*, the gene for glycogen-mobilizing enzyme, results in decreased levels of sterol esters, triacylglycerols and other lipids. Clearly, this is work in progress and the authors conclude that 'Gph1p may fulfil multiple independent functions which affect carbohydrate metabolism on the one hand and lipid metabolism on the other hand'.

Capacity

Breweries big and small have a 'pinch point' – a combination of capacity and time – that limits end-to-end output. Long-established breweries are often balanced with sufficient capacity across the process to accommodate peaks and troughs. However, commercial pressures such as brand growth, rationalization, brewery takeovers and, pertinently, global consolidation of large companies has driven the need to enhance production capacity without capital investment in new tanks and associated utilities infrastructure. This is not a new concept. Nathan (1930) noted that 'as fewer vessels were needed by his process the installation was about one-third cheaper than the old lager beer system, as a finished beer of 1,048° could be turned out in about 12 days, as compared with 3 or 4 months under the lager system'.

Following the Second World War, the biggest and most sustainable game-changer was the introduction of high-gravity brewing. Rather than brewing at 'sales gravity', worts are collected at a high gravity, the precise definition of which has changed over time. These worts are fermented 'as is' and maturated before dilution to sales gravity post-filtration. Of the various process expectations, one – no alteration in beer flavour or quality – was sacrosanct, although others (no reduction in brewlength, little or no capital expenditure, no additional complexity) would be typically assumed! As can be appreciated, the benefits of growing capacity output without capital expenditure are highly attractive and over time have driven small incremental changes in wort gravity. A modest example over some years would be moving collection gravities (°P) from 9 to 10 to 12, 13, 14.5, 15, 15.5 and 16 and beyond. The holy grail of 20°P fermentations is achievable as a one-off but beyond that massively more challenging. Against a backdrop of consumer acceptance via sales performance or directly through 'hall tests', increasing collection gravity eventually reaches its own 'pinch point'. Typically, brewhouse performance becomes limiting and there is only so far 'extending' worts with sugar syrups can go in terms of yeast performance and beer quality. Indeed, the original wake-up call on high gravity brewing came from matching 10°P and 20°P worts where esters were disproportionality elevated at (very) high gravity. Whilst at the time a concern, the key insight to control ester production at high gravity required that yeast growth to be matched through management of wort oxygenation and yeast pitching rate.

Although the capacity gains through high-gravity brewing have become a fact of brewing life, the law of diminishing returns inevitably begins to play. Whilst this will include numerous factors such yeast strain robustness, malt/sugar ratio, beer complexity and consumer acceptance, there invariably comes a tipping point. This may be manifest in fermentation performance, viability, generation number or beer attributes (appearance, consistency, off-flavours) which taken together suggest that pushing the high-gravity envelope has reached the end of the road. Indeed at this point, reigning back to a lower, high gravity would be advisable.

Gravity is not the only tune to play. Alongside the march to ever-increasing gravities, fermentation temperature has also been questioned. The rules of the game – notably lager fermentations at comparatively low temperatures – have been challenged both in terms of the end game but also the steps in between. Whilst the traditional lagers stay true to their history of low fermentation temperatures and extended maturation times, other lager brands have moved over 10 to 20 years from fermentation at 12°C (or lower) to 15°C, 18°C or higher with concomitantly reduced maturation times. It is noteworthy that increasing the maximum fermentation temperature has not been subject to as much study as increasing collection gravity. Those publications that report the inevitable acceleration of fermentation do so from a laboratory scale rather than production perspective. This most probably reflects concerns and sensitivity around the brands, which is understandable. However, as with extending collection gravities, increasing fermentation temperature brings stress, which combined with high gravity, results in an accumulation of environmental challenges that impact on yeast performance and beer quality. Indeed, it is the long-term consequences of a high-gravity, high-temperature regime that take time to manifest the combined impact on yeast quality. Accordingly, taking time out to reflect on process change and its implications on a periodic basis would be recommended.

Increasing fermentation temperature increases the rate of fermentation. A less controversial approach, which achieves the same objective, is to introduce effective vessel mixing. Whilst making eminent sense, brewery fermentations in CCVs have long been considered as being well mixed during active fermentation. This splendid series of experiments at a production scale as reviewed by Boulton (2013), clearly demonstrate the heterogeneity of fermentations and the advantages of introducing an effective mixing process. Implementation of a commercial mixing system ('Iso-Mix') has an undeniable positive impact on fermentation cycle times, VDK reduction and crash cooling of the vessel.

Best practice

Without fermentation, brewing and other industries would not exist. The cyclical nature of brewery fermentations is unforgiving and requires a more sustainable attention to detail to ensure consistency. Whatever the production scale, a combination of the right viable pitching yeast and the right wort oxygen content will normally assure both a consistent fermentation profile and required beer quality. The onus is on the brewer to tune both variables and, accordingly, to know the normal operating parameters for the plant and yeast. Inevitably an awareness, understanding and implementation of best practice in the up- and downstream processes are part of the bigger fermentation picture.

Consequences of failure

Fermentation is the culmination of a series of events. The back-story of yeast pitching to fermentation and back again exacerbates process and product problems. Failure to manage either yeast pitching rate in concert with wort oxygenation results in a raft of issues. Given that process consistency is a laudable ambition, drift of either parameter will result in atypical fermentation and atypical product quality. Whilst blending provides a short-term solution, the longer-term ambition should be to achieve consistency.

Cropping

At it simplest, the outputs of the cropping process are the recovery of sufficient yeast from the fermenter cone to pitch two or more fermenters. A key requirement is that the cropped yeast is of the required physiological and microbiological quality. Important process considerations include the flocculence of the yeast being recovered, management of cooling, vessel size and shape, cone angle and cropping 'philosophy'. It is a low-key process, a routine recovery of biomass at the end, or near the end, of the transformational conversion of wort to beer. In essence, yeast cropping is a support process and on the face of it not very interesting! First impressions, though, can be deceptive. As with all the steps in the journey of yeast from propagation/pitching through fermentation to storage (and back again), the cropping process can impact positively or – if performed poorly – can cause irreparable harm to yeast quality.

Arguably, visibility is a part of the problem. Irrespective of size and volume, the CCV is nigh on universally the choice of fermenter shape. Whilst undeniably good for hygiene and building footprint, cylindroconical fermenters offer few visual clues (other than overfoaming!) as to what is happening inside the vessel. Perhaps out of sight and out of mind, but despite its importance, yeast recovery from the cone of cylindroconical fermenters has not received a fair press. Sporadic but accumulating evidence paints a persuasive picture that the environment in the vessel cone is highly damaging to yeast physiology and viability. Further, assumptions about mixing (carbon dioxide and convection) are optimistic and as Boulton (2013) has shown, more yeast 'sits' in the cone and arrives there earlier than was thought. Yeast located in the cone is in the wrong place and contributes little or nothing to sugar utilization, diacetyl reduction and the overall progress of fermentation. Conversely, the 'composting' of yeast in the vessel cone together with associated autolysis impacts negatively on beer (and yeast!) quality.

The cropping process

There is not an agreed universal fermenter design; an observation that is frequently reinforced when visiting breweries. Ideally, the CCV can be independently cooled to minimize the development of hostile conditions that damage the cropped yeast. Typically to aid stratification and rundown, the included cone angle is 60 or 70° but flatter or steeper cones are not uncommon. Practically, there are a limited number of processing options yeast cropping from the cone. Firstly and foremost is whether to recover the entire crop or to focus recovery on a particular fraction or 'cut' so as to select the best yeast. Perhaps the most universal approach (Noble, 1997; O'Connor-Cox, 1997) is to recover the 'middle cut' whilst discarding the first runnings, which are rich in trub and dead yeast and (if they are there) enriched with flocculent yeast variants whose selection can cause processing problems (see 'Population heterogeneity and genetic instability'). The first cut or cold break is either removed within a day or so of fermentation commencing or more usually as part of the cropping process itself. The uppermost 'fuzzy' fraction is often specifically not recovered during cropping, as it is lower in solids and enriched in small chronologically young cells together with potentially less flocculent yeast.

The process of yeast recovery can be controlled in a variety of ways ranging from a standard volumetric cut-off, visual assessment, use of turbidity sensors or, better still, application of an in-line viable biomass probe as used to manage pitching yeast (see above). Cropping from the vessel cone is managed via an appropriate pump. Less common is the recovery of powdery, less flocculent yeast by centrifugation. With such yeast this process is unavoidable, but the approach requires capital (for the centrifuge and in-line cooler), adds complexity and is comparatively energy intensive. Cropping from the top of a cylindroconical vessel is unusual and is a demanding process hygienically. Whatever the approach taken to recover cropped yeast, it is preferably stored as slurry in dedicated chilled and mixed yeast storage tanks (see 'Storage', below).

As elsewhere in brewing, time and temperature are key variables of yeast cropping. Typically, where diacetyl reduction is managed in fermenter, yeast is cropped as the final process activity after VDK specification is achieved and the vessel contents have been cooled. Such 'cold' cropping prolongs yeast residence time in fermenter and in doing so inevitably results in more physiologically damaged yeast. An alternative approach with suitably flocculent yeast is to practice 'warm' cropping (Noble, 1997; O'Connor-Cox, 1997; Loveridge *et al.*, 1999). Here, yeast cropping is independent of diacetyl reduction and occurs sooner (rather than later) after fermentation has achieved racking gravity (usually 24 hours) but before cooling is applied. The advantage of this approach is that the yeast is cropped two or more days earlier – with the attendant benefit to yeast quality – but with the additional process complexity of a second crop (that goes to waste) prior to beer transfer. The timelines for cold and warm cropping are compared in Fig. 3.5.

Impact of warm cropping on yeast quality

The benefits of warm cropping were clearly demonstrated in extensive production trials (Loveridge *et al.*, 1999) at 2000 hl and 4000 hl scale. In all, 600 vessels were involved, with 180 warm cropped and the remainder cropped cold. The use of early warm cropping had a marked impact on yeast quality (crop solids, improved viability and consistency) and accordingly reduced fermentation cycle time and faster VDK reduction (Table 3.5). Further work (Quain *et al.*, 2001) focused in greater detail on both warm and cold cropping by sampling '500 kg fractions' during rundown in 2000 hl fermenters. Not unexpectedly, in the yeast cropped 'cold', hotspots were observed with slurry temperature ranging between 1°C and 8°C in the vessel nominally at 4°C. Further, there was a linear

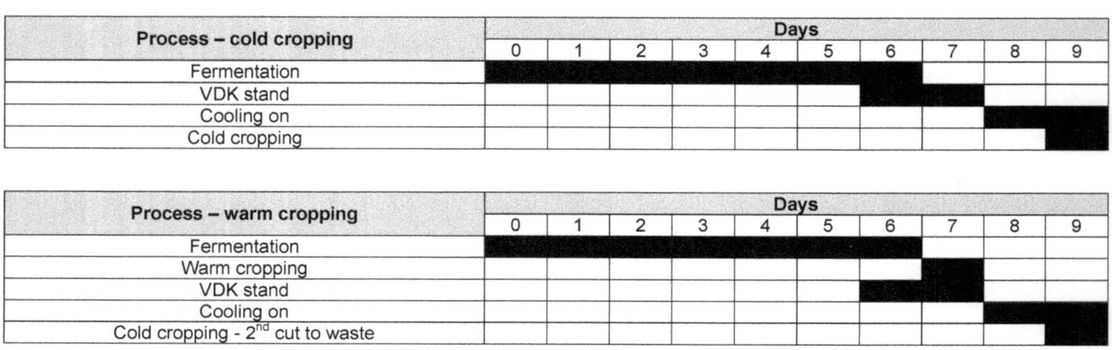

Figure 3.5 Timelines for cold and warm cropping.

Table 3.5 Metrics for cold and warm cropping

	Cold cropping		Warm cropping	
Yeast	Mean	Range	Mean	Range
Viability (%)	90.1	82–94.6	92.3	90.5–93.9
Cell size (µm)	6.54	5.86–7.12	7.23	7.07–7.42
Wet weight (%)	25	9–38	53	43–62

relationship between increasing yeast solids and abv (7% at low solids to *ca* 9% abv at 50% in the early runnings). The most telling comparison was viability, which was markedly higher in the warm cropped yeast (91.1 ± 1.6%) than the control cold crop (83.4 ± 3.3%). Irrespective of which cropping approach was used, there was evidence of stratification and heterogeneity of yeast in the fermenter cone. Parameters such as abv, pH, yeast solids and viability increased towards the base of the cone, whereas beer specific gravity increases towards the top of the cone. Unexpectedly, yeast flocculence (irrespective of cold or warm cropping) peaks at the mid-point of the cone as opposed to the vessel bottom. Prior to this, the perceived wisdom was that that flocculence decreases from bottom to top of the cone.

A sister publication (Powell *et al.*, 2004) provided further insight into yeast quality in the cone, demonstrating that levels on the intracellular storage polysaccharide glycogen increased linearly from the first to last fractions of cropped yeast. Conversely, glycogen breakdown would contribute to the higher abv towards the cone bottom. Analysis of cell size during cropping clearly showed size to reduce across the cropping process. Cell age was also monitored and was found to peak at the mid-point of cropping with yeast of greatest flocculence. As would be expected, small virgin cells predominate at the end of cropping.

Thiele *et al.* (2008) compared warm and cold cropping at a production scale at two different breweries with lower collection gravities (12°P and 11°P) than the 15°P above. This report – which included assessment of yeast vitality or vigour using acidification power – suggested yeast quality to be broadly consistent across the cone. However, crop quality varied within and between breweries and the authors report that 'environmental conditions in the cone cause severe stress to yeast and therefore an early yeast crop can help to maintain healthy yeast'.

Autolysis

The environment in the cone of the fermenter vessel will result in some degree of autolysis by intracellular enzymes in dead yeast cells. A host of environmental factors contribute to autolysis including temperature, pH, osmotic pressure, ethanol, and starvation. Autolysis has been characterized in two stages: (i) the degradation of organelles with the content uniformly distributed within the cell of reduced size; and (ii) hydrolysis of intracellular biopolymers, resulting in the diffusion of hydrolysis products through a porous cell wall into the medium (Babayan and Bezrukov, 1985). The products of autolysis are diverse and include polysaccharides (but not sugars), organic acids, protein, amino acids, fatty acids, nucleic acid products and lipids (Hernawan and Fleet, 1995). The remains of the autolysed cell are primarily cell wall and membrane material called 'ghosts' or 'hulls' in the wine industry (Munoz and Ingledew, 1990).

As noted by Thorn in 1971 'the subject of autolysis is often not a major aspect of publications dealing with fermentation and yeast handling'. Whilst arguably still true in brewing, there has been considerable interest in the world of wine where it is stored on 'lees' consisting mainly of yeast cells (Charpentier, 2010). It is noticeable that in brewing, autolysis 'may lead to off-flavours and should be avoided' (Thorn, 1971) whereas in wine it is positively encouraged in some wine styles and commercial products derived from autolysis are sometimes added. This creates something of a quandary. In terms of negatives, autolysis is linked with short chain fatty acids (Chen *et al.*, 1980), contributing unwelcome 'yeasty' or 'fatty' aromas to beer, as well as hazes via glycogen (Malcorp *et al.*, 2001) or mannan (Siebert *et al.*, 1987). On the other hand, the contribution is encouraged of mannoproteins, peptides, amino acids, nucleotides and nucleosides to the stability and organoleptic properties of wine (Charpentier, 2010). Of particular interest is the release of sterols during autolysis (Hernawan and Fleet, 1995) that, together with the provision of ergosterol and unsaturated fatty acids from yeast hulls (Munoz and Ingledew, 1990) suggests an unexpected upside from cell digestion. As noted above, anaerobic yeast normally forms these lipids in the presence of oxygen but will take them up exogenously should they be available. The insight from the addition of yeast hulls to wine fermentations suggests that these lipids may be recycled from autolysed yeast to the benefit of viable yeast. It would be interesting to test this hypothesis in brewery fermentations.

Best practice

Although the yeast storage process follows cropping, 'storage' effectively begins when yeast sediments (often) early in fermentation into the cone. Accordingly, best practice is to use effective cone cooling (2–4°C), remove the first cut of dead yeast plus trub and practice warm cropping to remove yeast to a more controlled and homogeneous environment in a storage vessel. The importance of early warm cropping increases with high (15°P) and very high (> 15°P) gravity fermentations. A similar argument applies to the use of higher temperature fermentations (to accelerate rate). Although cropping is often managed manually (visually or by time), there is value in the use of 'radiofrequency permittivity' technology to better manage viable yeast solids.

Consequences of failure

Rapid yeast cropping minimizes the inevitable damage to yeast that occurs during this process. Arguably the conditions in the fermenter cone are the most damaging that yeast experience in the fermentation cycle. Prolonged residence in the cone results in metabolic 'hot spots' where the local temperature can increase markedly. Here, the reserve polysaccharide glycogen is dissimilated to glucose, which through glycolysis is converted to ethanol, ATP and heat. As the local temperature rises, the rate of glycogen break down increases, resulting in the formation of more heat. The insulating effect of thick yeast slurries reduces heat dissipation and results in the hot spot getting hotter and bigger. These events result in a spiral of decline encouraging glycogen breakdown, ethanol formation, heat generation and inevitably loss of yeast viability. If this was not bad enough, there are environmental graduations radiating from hotspot(s) where yeast cells retain viability but have reduced levels of glycogen, which may compromise performance in the upcoming fermentation.

Storage

For such an influential process, 'storage' has in recent years had a poor press compared to the other players of propagation, pitching, cropping and the juggernaut of fermentation. The big picture of what best practice looks like for yeast storage is generally agreed. However, practically, it is relatively untouched by new knowledge despite the significant strides in activity and understanding of relevant topics. Indeed as 80% of the earth's environments exhibit temperatures below 5°C, responses of yeast to 'cold' (shock, stress and response) are in scope as is quiescence ('the most common cell state on earth') and near-zero growth rates. Here and where appropriate, these subjects are touched (lightly) on. See Chapter 1 for further discussion of quiescence.

In her seminal series of three articles on 'improving yeast handling in the brewery', Erin O'Connor-Cox (1997, 1998a,b) used the memorable strapline, 'keep it cold, keep it short and keep it simple'. Not surprisingly, as little has changed, from the viewpoint of yeast storage this is as wise a maxim today as it was then. Yeast storage is about maintaining the status quo and minimizing metabolic activity, physiological change or, worse still, death. The bottom line is that yeast storage can – at best – only maintain viability and in reality it will inevitably decline with time!

Yeast storage is a key cog in the wheel of the brewing yeast handling 'cycle'. Hygienic cropping of yeast from fermenter and its hygienic storage as a slurry in a cooled, stirred vessel is a 'given' and is embedded in best practice. Storage of a pressed cake or slurry in 'bins' (hopefully in a cold room) is very much less than ideal and should be avoided or be part of an improvement plan towards the above best practice. The once fashionable option of storage in the fermenter cone cannot be recommended in any shape or form. Indeed, residence time of yeast in the cone should as little as is practical as this environment is most detrimental to yeast quality (see 'Cropping').

Time, temperature and mixing

The object of yeast storage is to minimize the inevitable deterioration of nutrient-limited, stationary phase, lipid-depleted yeast *ex* fermentation. As summarized below (best practice), this is achieved through careful control of temperature (cold but without the risk of freezing), time (the shorter the better), mixing (to assure homogeneity) and minimizing oxygen pick-up (nitrogen gas blanket). Further, with increasingly high collection gravities and consequently high abv, dilution of slurries with sterile water is to be recommended. Whilst beneficial to the maintenance of yeast viability, the barriers to implementation include the unavailability of water of the right microbiological quality and impact on yeast storage capacity of diluted slurries. Despite this, dilution of yeast slurries is increasingly found in the mix of best practice.

It is also worth noting that there are a number of variables specific to the yeast being stored, which add a layer of further complexity to the process. These inevitably include brewing strain variability but also slurry thickness, generation number, cropping technique, cooling and barm-ale quality (including the aforementioned abv). Of the physical inputs, time and temperature are in the front seat of the management of yeast storage. Whilst temperature control and its achievement through mixing are delivered through good process control, 'time management' is more subjective.

All things being equal, there is general acceptance that storage time should be ideally <2 days but no more than 4 days. In routine practice this is more than achievable, although there will be occasions when storage time is stretched beyond four days. The bigger challenge is managing storage time for yeasts used infrequently for niche brands. Such protracted storage, whilst a considered option, will inevitably result in reduced yeast viability and, equally importantly, will also tie up a storage vessel. This scenario has driven workarounds such as the use of dried yeast (which lends itself to 'one-off' fermentations) or, regrettably, culling of niche strains and replacement with a mainstream yeast. Less dramatic approaches have been proposed to successfully prolong the duration of yeast storage. These include the addition of wort (Henson and Reid, 1988; Kunze, 2014) and (memorably in response to the threat of a fuel crisis in 1947), slurrying yeast in potassium dihydrogen phosphate (Bishop *et al.*, 1955).

The recommended temperature range for yeast storage is 2–4°C. There are a number of challenges here as yeast cropped from fermenter is – despite even effective cone cooling – invariably warmer in viscous slurries of 40% or so solids, which in turn are difficult to cool. Overlaid on this is the inaccuracy – also observed in fermenter cones – of temperature probes adjacent to the cooled vessel wall that fail to reflect the temperature of the slurry in the body of the vessel. Ideally, yeast slurries are cooled via an in-line heat exchanger to achieve more homogeneous temperatures on rundown from fermenter. Without such intervention, the transition from fermenter cone temperature to storage at 2–4°C is slower, more demanding and unpredictable.

Whatever the 'prehistory', the norm for yeast storage vessels is for mixing, ideally via an internal top-mounted, off-centred stirrer/agitator or more rarely through a recirculation loop. Whilst the key driver is to minimize slurry and environmental heterogeneity, mixing also degasses entrapped carbon dioxide. This is something of double-edged sword as, with the FV cone, a lack of temperature homogeneity results in 'hot spots' which step-up metabolism through glycogen breakdown resulting in formation of carbon dioxide (and ethanol).

Successful mixing of thick yeast slurries is not straightforward. Vessel design, slurry rheology and location of the mixer require careful consideration. As noted by O'Connor-Cox (1998b) 'many mixers are designed such that they effect slurry movement in a limited area of the tank', such that 'the bulk of the slurry remains unmixed'. In particular, side-entry stirrers in the vessel cone were noted as 'unsuitable', whereas off-centre, top-mounted stirrers were more successful. Other studies are few and far between. Cahill *et al.* (2003) in a landmark publication reported a detailed study of the impact of mixing on temperature, viability and yeast solids in a 10 hl customized storage vessel. Here mixing was evaluated both *in situ* via a hygienically designed off-centred variable-speed agitator or with pumped recirculation from the base of the vessel to the top. Temperature was monitored by a 4×4 array of probes across the central belt of the vessel at different distances from the wall. The conclusions from this work with a top-fermenting ale strain were unequivocal. Mechanical mixing at 200 rpm was key to homogeneity of slurry temperature, which contributed to yeast viability (maintenance and consistency), distribution of solids and vessel volume. By contrast, external recirculation of the vessel (over either 30 minutes or 2 hours) resulted in heterogeneity and could not be recommended with any confidence. Suffice to say, conditions were much worse in the absence of any mixing whatsoever!

Growth, metabolism, stress and quiescence

Cropped yeast at the end of fermentation is metabolically in a non-growing, stationary phase. As a process, yeast storage is akin to the refrigeration of food insomuch that the lower temperature will markedly slow down metabolism and deterioration in 'quality'. Growth is not a realistic consideration at typical storage temperatures of 2–4°C, *Saccharomyces* (as a mesophile) is at or below the minimum temperature for growth. Indeed, even under optimum growth conditions – which yeast storage in beer clearly fails to achieve – doubling times are 50 hours or more at 4°C (Homma *et al.*, 2003; Murata *et al.*, 2006) and 42–63 hours at the higher temperature of 6°C (Walsh and Martin, 1977). This would, in turn, be further extended by the presence of significant levels of ethanol, which adversely impacts on the growth rate of yeast (Sá-Correia and Van Uden, 1983).

Yeast metabolism is muted during storage, as extracellular, assimilable nutrients are few and far between in slurry held in beer or 'barm ale'. Indeed not surprisingly, the viability of yeast declines during storage. Temperature is key with viability declining progressively more rapidly as storage temperatures increase from between 1°C to 5°C, from 5°C to 10°C, 10°C to 15°C and so on (McCaig and Bendiak, 1985). Overlaid on this, viability is linked to the concentration of extracellular ethanol, and is increasingly compromised above 7% abv (Loveridge *et al.*, 1999). Breakdown of glycogen – the quantitatively important reserve polysaccharide – is also an inevitable consequence of storage (Quain and Tubb, 1982; McCaig and Bendiak, 1985), the rate of which is accelerated by increasing temperature. As with storage in the fermenter cone, glycogen dissimilation via fermentation results in the formation of ethanol, which adds to that present in the barm ale.

Analysis (Gibson *et al.*, 2007) of the various stresses that brewing yeast are exposed to during fermentation and handling identified 'cold shock' (and ethanol toxicity) as a major concern during yeast storage at 2–4°C. Of course, the size of the shift in temperature and rate of change will vary depending on fermentation temperature, cropping regime, efficacy of cone cooling and in-line cooling on transfer to storage vessel. As noted by Somani *et al.* (2012), 'the impact of thermal downshift on brewing yeast has not been the subject of extensive investigation. This is surprising in light of the routine application of low-temperature storage to brewing yeast during industrial handling'. What little work has been published is directional in terms of the yeast storage process, as the insight comes from laboratory experiments with haploid yeast (Homma *et al.*, 2003). Importantly in terms of relevance to storage, this work involved the growth – albeit slowly – of yeast aerobically at 4°C. As is the way with DNA microarray, a host of genes were either up- or down-regulated. Although reflecting growth and not a shift from (say) 30°C to 10°C, cold shock genes were typically (although not exclusively) up-regulated. Tellingly, the authors note that their work 'suggests that long-term storage without growth would be preferable to growth at low temperature'.

In a follow-up study (Murata *et al.*, 2006), yeast was cultured at 25°C then transferred to 4°C after which gene expression was monitored over 6–48 hours. Again, the work was with a haploid laboratory strain and was aerobic. Responses to the downshift in temperature to 4°C include an upshift in the synthesis of phospholipid, trehalose, glycogen and cold shock mannoproteins located in the yeast cell wall. In response to the switch in temperature, the authors conclude that 'trehalose and glycogen were synthesized for cold tolerance and energy preservation, and the synthesis of phospholipid and mannoproteins was for maintenance of cell membranes and permeability of cell wall'.

Stored yeast is in stationary/resting phase or in cell cycle terms G_0 or G zero. Such cells are also 'quiescent', which is described as the 'most common and, arguably, most poorly understood cell cycle state' (Allen *et al.*, 2006). Quiescent cells are characterized as maintaining viability under growth-arrested conditions, having a decreased metabolic rate, exhibiting greater stress resistance and resuming growth when growth-promoting conditions return (Klosinska *et al.*, 2011). Candidates for growth arresting conditions include starvation, stress and the accumulation of growth-inhibiting metabolites. Of course, such conditions sum-up those found in the stationary phase environment at the end of brewery fermentations.

However, once again, insight is at best directional as studies on quiescence in yeast are invariably in aerobic liquid cultures undergoing nutrient starvation (nitrogen, carbon and less detrimentally sulfur and phosphate) or diauxic shift from the metabolism of glucose to ethanol. Despite this there are numerous fascinating and stimulating insights into the world of quiescence. Although from the related world of an aerobic culture, it has been shown that not all yeast cells in the stationary phase of (non) growth are quiescent (Allen *et al.*, 2006). Using density gradient centrifugation, the population was separated into (i) quiescent dense, unbudded virgin daughter cells, which are able to synchronously reproduce, and (ii) less dense, heterogeneous (budded/unbudded, mother/daughter) non-quiescent cells which lose the ability to reproduce. Further, the quiescent population contained significantly more glycogen and markedly higher viability during 28 days storage.

Population heterogeneity and genetic instability

Whilst the direct relevance of the quiescence story is not clear, it is a further reminder of the heterogeneity of microbial populations. This has not always been recognized. On reflection, the vast majority of studies of yeast physiology and metabolism in the brewing world have been performed at a 'bulk' level yielding an 'average' measurement. Indeed, rather than being homogeneous, the dawning realization is that populations of single-strain, pure-culture yeast in fermenter, handling and storage are heterogeneous. This is a generic microbiological principle, such that microorganisms exhibit cellular individuality that 'even when genetic and environmental differences between cells are reduced as much as possible, single cells differ from each other with respect to gene expression and other phenotypic traits' (Ackerman, 2015). The bottom line is that heterogeneity of genetically uniform populations is considered to confer a selective advantage during stress and in response to fluctuating environments.

As a subject, 'heterogeneity' has been flagged most notably at the macro scale, where yeast distribution in fermenter is heterogeneous, the extent of which varies with time and yeast strain/flocculence (Boulton *et al.*, 2007). Heterogeneity is increasingly recognized as the size of individual yeast cells varies with chronological age/bud scars (Barker and Smart, 1996), growth rate (Johnston *et al.*, 1979) and ploidy (Galitski *et al.*, 1999). Better recognition of phenotypic heterogeneity has come through cell sorting and flow cytometry (for a review see Davey and Kell, 1996), which has demonstrated heterogeneity in brewing yeast for internal pH (Imai and Ohno, 1995). Heterogeneity of cellular glycogen content during fermentation has been shown with 'individual cell spectroscopy' (Cahill *et al.*, 2000).

Another source of phenotypic heterogeneity is genetic instability. This reflects the pressures of adaptive evolution 'where populations must secure a margin of genome variability that allows for the adjustment to new environmental conditions' (Skoneczna *et al.*, 2015). Instability is 'a given' occurring via a variety of routes and driven by a number of factors. Whatever their shape, mutations are spontaneous during DNA replication and 'if a mutation is beneficial in the environment, allowing the cell and its descendants to proliferate more rapidly, that lineage will begin to increase in relative abundance in the population' (Gresham, 2015). As populations evolve, lineages with beneficial mutations increase at the expense of cells without such mutations. Less predictably, lineages with greater 'fitness' then predominate, winning out and over time dominating the population. The hitherto difficulty in estimating the extent and complexity of spontaneous mutation has underplayed its importance, scale and significance (Zhu *et al.*, 2014).

In brewing, genetic instability has had a somewhat chequered history, which has undermined widespread acceptance. However, without (typically) being mentioned explicitly, it is one of the drivers that has led to the management of the number of yeast cycles or generations that are replenished through the yeast supply process. Early sightings of instability stem from a series of publications from Guinness between 1963 and 1996 (for details see Boulton and Quain, 2001) that detailed 'spontaneous' changes in maltotriose utilization and marked switches in flocculence. Whether this reflects a benefit in terms of fitness or simply process visibility and detection is not clear. A landmark paper (Casey, 1996), using the then new technique of karyotyping reported changes in the fingerprints of production lager yeast sampled and stored between 1958 and 1985. More specifically, two publications from different Japanese brewers reported the detection of flocculence changes during serial repitching in production (Jibiki *et al.*, 2001; Sato *et al.*, 2001). Another report (Quain, 2006a) detailed the tortuous journey from the observation of cropped yeast with enhanced flocculence, demonstration of genetic change through DNA fingerprinting, the isolation of seemingly the same genetic variant some 7 years later and in a brewery more than 400 km away, and culminating in the demonstration that both variants (of the same strain) carried an additional copy of chromosome VII (Table 3.6). Conversely, Powell and Diacetis (2007) – in an extended production-scale study with an ale and lager yeast for, respectively, 98 and 135 generations – showed no genetic changes or changes in fermentation performance.

Chromosomal loss (or addition) is one of a number or routes that are associated with genetic instability in aneuploid and polyploid yeasts. There are different perspectives whether this results in reduced fitness (Storchova, 2014) or 'large fitness gains' which 'accelerate evolutionary adaptation' (Selmecki *et al.*, 2015). Although beyond the scope of this chapter, it is tempting to build on the anecdotal preponderance of lager strains exhibiting genetic instability as being a consequence of the hybrid genome as well as aneuploidy. Of course this may reflect the relatively young evolutionary 'age' of lager yeasts or the enhanced visibility through the dominance of lager-type beer production volumes over ale.

Less heralded but long recognized as part of the 'genetic variants' story is the petite mutation. First identified in France by Ephrussi *et al.* (1949), and named to reflect the atypically phenotypic small size of colonies on agar plates. Petites are respiratory deficient as a consequence of alterations or part deletion of mitochondrial DNA. Such *Rho*$^-$ or p^- mutants grow more slowly on fermentable sugars and are unable to grow aerobically on non-fermentable carbon sources (e.g. glycerol, ethanol etc.) Mutants without mitochondrial DNA (p^0), although rare, have been found in cropped yeast (Lawrence *et al.*, 2013) but are often lethal in cells growing aerobically. Deletion or rearrangement of mitochondrial DNA (p^-) impacts on the genes for transfer and ribosomal RNAs, thus blocking the synthesis of proteins in mitochondria. Further, the mitochondria contribute to the biosynthesis of haem, amino acids, nucleotides and fatty acid metabolism.

Generically, the petite mutation is considered to be 'spontaneous' and 'natural' and is present in brewery fermentations at a frequency of *ca.* 0.1–1 to 6% or more (Šilhánková *et al.*, 1970; Smart, 2007; Lawrence *et al.*, 2013) of the yeast cell population. Strain susceptibility is (as always) a variable that is associated with many and various causes, including 'stress' (e.g. oxidative), cone storage and serial repitching (Lawrence *et al.*, 2013). The frequency of petites is estimated 'to be 10^6–10^8 times more frequent than other mutational types' (Šilhánková *et al.*, 1970).

In passing, another spontaneous (and stable) mutation of brewing yeasts results in glycogen deficiency at a frequency of ≤ 0.8%. Growth of these

Table 3.6 Unravelling genetic change in a production yeast

Observation	Reference
During process trials on yeast oxygenation, a heavily flocculent line was observed on yeast cropping. Pumped solids, typically at 40%, were at 60–75%. Laboratory studies showed the variant (BB11 '56') to be more sensitive to calcium which promoted greater flocculence than the 'parent' yeast (BB11)	Boulton and Quain (2001)
DNA (RFLP) fingerprinting demonstrated small but definite genetic differences between BB1 and BB11 '56'	Wightman *et al.* (1996)
Seven years after the observation of BB 11 '56', a similarly heavily flocculent yeast (BB11 'W') of the same strain was recovered from a different brewery in the Group located 444 km away	–
Genomic analysis shows both BB11 '56' and 'W' to have an additional copy of chromosome VII compared to BB11	Lockhart (2003)

mutants was linked with elevated levels of petites (2–9%) (Chester, 1967) and conversely analysis of 'normal' petites were found to have 'appreciably less glycogen than those of wild-type yeast (Chester, 1968).

Whatever the spontaneous mutation, it is salutary to reflect that a frequency of 1%, pitching yeast at 15×10^6/ml contains a substantial petite population of 1.5×10^5/ml. Whilst rare, there are reports that the frequency of petites can be much higher. For example, reported issues in multisite brand matching were associated with a brewery where *ca* 4% of yeast harvested from fermenters and up to 50% of stored yeast was respiratory deficient (Morrison and Suggett, 1983). Beers from the problem brewery were characterized by elevated diacetyl and reduced levels of esters together with reduced yeast viability. The issue was seemingly associated with protracted yeast storage (4–7 days at 4°C) that, when replaced with 24 hours at 4°C, reduced the occurrence of petites to 1% or less. Whilst an extreme and somewhat surprising demonstration from some 40 years ago – which sadly received no subsequent elaboration – petites remain a concern during brewery yeast processing.

Today the generic threat of petites are arguably a greater concern, as they are much less likely to be detected through testing or established through problem-solving or root-cause analysis. This sad state of affairs reflects reduced technical resources, a diluted knowledge base and a lack of appreciation of what is a relatively niche element of brewery yeast performance. Indeed, our understanding of petites in brewery fermentations and handling is something of a 'grey' area as – under anaerobic, catabolite-repressed conditions – respiratory capability is not possible and the organelle is cytologically underdeveloped (and often termed 'promitochondria'). However, despite this, petites with compromised mitochondrial function and performance are not fit for purpose. This is intriguing and is counter to the aforementioned selection of genetic variants in terms of enhanced fitness.

As with chromosomal instability, new insights from applied yeast genomics pose new questions as to the formation of petites in the two distinct lineages of *S. pastorianus*. These investigations build on the observation that the Group II 'strain generated a lower amount of respiratory deficient 'petite' cells' (Walther *et al.*, 2014). Further, with the mitochondria of lager yeast being derived from the *S. eubayanus* parent (Dunn and Sherlock, 2008) – and not *S. cerevisiae* – there is a case for a fundamental difference in frequency of petite mutations in ale and lager yeasts.

Best practice

It is thought-provoking that 'best practice' for brewery yeast storage is based on a combination of common sense and practical good design. Of course, as with all aspects of yeast handling and management, hygienic design of vessels and mains together with effective clean-in-place procedures is a given. More specifically in terms of yeast storage and minimizing damage, key parameters include storage temperature (2–4°C), homogeneity through 'gentle' mixing of yeast, time (ideally < 2 days but no more than 4 days) and yeast solids at *ca* 40%. Ideally, there are two further considerations that 'add value'. As noted above, barm ale abv has an overarching impact on yeast viability during storage and ideally should be reduced with sterile water to < 5%. Further, there is benefit in avoiding any exposure to air/oxygen by the application of an inert headspace gas to the storage vessel.

Consequences of failure

As a process, yeast storage is assured through control of temperature and management of time. Should either of these parameters drift or hygienic practices fail, then yeast in storage vessel will be compromised in viability or microbial contamination. In such an event, the yeast would probably go forward without notice and any subsequent issues of fermentation performance or beer quality would possibly be flagged after the event. It is noteworthy that where viability has been compromised, traditional yeast viability measurement prior to pitching adds value in flagging deterioration. Conversely, automated pitching systems that pitch on viable cell number would correct for reduced viability and pitch more yeast biomass to achieve the desired pitching rate. However, in doing so, more dead yeast would be pitched, to the detriment of beer quality.

Acknowledgments

Thanks to the Quain family for putting up with the yeast obsession.

Sadly, as has become the way, I acknowledge Ellie the Labrador (December 2001–March 2015), who, through years of walking and companionship, gave me time to reflect, think and conjure up the next steps for projects like this.

References

Ackermann, M. (2015). A functional perspective on phenotypic heterogeneity in microorganisms. Nat. Rev. Microbiol. *13*, 497–508. http://dx.doi.org/10.1038/nrmicro3491

Allen, C., Buttner, S., Aragon, A.D., Thomas, J.A., Meirelles, O., Jaetao, J.E., Benn, D., Ruby, S.W., Veenhuis, M., Madeo, F., and Werner-Washburne. M. (2006). Isolation of quiescent and nonquiescent cells from yeast stationary-phase cultures. J. Cell. Biol. *174*, 89–100.

Andreasen, A.A., and Stier, T.J. (1953). Anaerobic nutrition of *Saccharomyces cerevisiae*. I. Ergosterol requirement for growth in a defined medium. J. Cell. Comp. Physiol. *41*, 23–36.

Annemüller, G., Manger, H.-J., and Lietz, P. (2011). The Yeast in the Brewery (Berlin, Germany: VLB).

Anness, B.J., and Reed, R.J.R. (1985). Lipids in the brewery – a material balance. J. Inst. Brew. *91*, 82–87.

Aries, V., and Kirsop, B.H. (1978). Sterol biosynthesis by strains of *Saccharomyces cerevisiae* in the presence and absence of dissolved oxygen. J. Inst. *Brew. 84*, 118–122.

ASBC Methods of Analysis (2010). Yeast Method 3. Yeast Stains. Approved (1958), rev. (1978, 1990, 2010). American Society of Brewing Chemists, St. Paul, MN, U.S.A. doi: 10.1094/ASBCMOA-Yeast-3

ASBC Methods of Analysis. (2011). Yeast Method 4. Microscopic yeast cell counting. Approved (1958), rev. (1978, 1998, 2011). American Society of Brewing Chemists, St. Paul, MN, U.S.A. doi: 10.1094/ASBCMOA-Yeast-4

Babayan, T.L., and Bezrukov, M.G. (1985). Autolysis in yeasts. Acta. Biotechnol. *2*, 129–136.

Barker, M.G., and Smart, K.A. (1996). Morphological changes associated with the cellular aging of a brewing yeast strain. J. Am. Soc. Brew. Chem. *54*, 121–126.

Barnes, Z. (2006). Brewing process control. In Handbook of Brewing, F.G. Priest and G.G Stewart, eds, 2nd edition (Boca Raton, USA; CRC Press), pp. 447–486.

Bing, J., Han, P.-J., Liu, W.-Q., Wang, Q.-M., and Bai, F.-Y. (2014). Evidence for a Far East Asian origin of lager beer yeast. Curr. Biol. 24, R380–R381.

Bishop, L.R., Alexander, W.H.A., and Cory, F.H.M. (1955). The storage of yeast. J. Inst. Brew. *61*, 150–160.

Bokulich, N.A., and Bamforth, C.W. (2013). The microbiology of malting and brewing. Microbiol. Mol. Biol. Rev. *77*, 157–172. http://dx.doi.org/10.1128/MMBR.00060-12

Bond, C. (2007a). Freeze-drying of yeast cultures. In Cryopreservation and freeze-drying protocols. Methods in Molecular Biology 368, 2nd edition, J.G. Day and G.N. Stacey eds (New Jersey, USA; Humana Press), pp 99–107.

Bond, C. (2007b). Cryopreservation of yeast cultures. In Cryopreservation and freeze-drying protocols. Methods in Molecular Biology 368, 2nd edition, J.G. Day and G.N. Stacey eds (New Jersey, USA; Humana Press), pp. 109–117.

Botstein, D., and Fink, G.R. (2011). Yeast: an experimental organism for 21st Century biology. Genetics *189*, 695–704. http://dx.doi.org/10.1534/genetics.111.130765

Boulton, C.A., Maryan, P.S., Loveridge, D., and Kell, D.B. (1989). The application of a novel biomass sensor to the control of yeast pitching rate. Proc. 22nd Cong. Eur. Brew. Conv. *Zürich*, 653–661.

Boulton, C.A., and Quain, D.E. (1987). Yeast, oxygen and the control of brewery fermentation. Proc. 21st Cong. Eur. Brew. Conv. *Madrid*, 401–408.

Boulton, C.A., and Quain, D.E. (1999). A novel system for propagation of brewing yeast. Proc. 27th Cong. Eur. Brew. Conv. *Cannes*, 647–654.

Boulton, C., and Quain, D. (2001). Brewing Yeast and Fermentation (Oxford, UK: Wiley-Blackwell).

Boulton, C. (2006). Fermentation of beer. In Brewing: new technologies, C.W Bamforth, ed (Cambridge, UK; Woodhead Publishing), pp. 228–253.

Boulton, C., Price, S., and Peters, A. (2007). Yeast distribution in cylindroconical vessels, new insights into fermentation performance and management. Proc. 31st Cong. Eur. Brew. Conv. *Venice*. 267–279.

Boulton, C. (2012). 125th Anniversary Review: Advances in analytical methodology in brewing. J. Inst. Brew. *118*, 255–263.

Boulton, C. (2013). Practical aspects of fermentation management – a review. MBAA Tech. Quart. *50*, 53–61.

Boundy-Mills, K. (2012). Yeast culture collections of the world: meeting the needs of industrial researchers. J. Ind. Microbiol. Biotechnol. 39, 673–680. http://dx.doi.org/10.1007/s10295-011-1078-5

British Beer and Pub Association. (2015). Statistical Handbook (Tettenborn, M. Ed), Brewing Publications Ltd, London.

Cahill, G., Walsh, P.K., and Donnelly, D. (1999). Improved control of brewery yeast pitching using image analysis. J. Am. Soc. Brew. Chem. 57, 72–78.

Cahill, G., Walsh, P.K., and Donnelly, D. (2000). Determination of yeast glycogen content by individual cell spectroscopy using image analysis. Biotechnol. Bioeng. 69, 312–322.

Cahill, G., Walsh, P.K., and Donnelly, D. (2003). A study of the variation in temperature, solids concentration and yeast viability in agitated stored yeast. Proc. 29th Cong. Eur. Brew. Conv. Dublin., pp.454–469.

Carvell, J.P., and Turner, K. (2003). New applications and methods utilizing radio-frequency impedance measurements for improving yeast management. MBAA Tech. Quart. *40*, 30–38.

Casey, G. (1996). Practical applications of pulsed field electrophoresis and yeast chromosome fingerprinting in brewing QA and R&D. MBAA Tech. Quart. 33, 1–10.

Charpentier, C. (2010). Ageing on lees *(sur lies)* and the use of specialty inactive yeasts during wine fermentation. In Managing wine quality, vol 2. Oenology and wine quality. A.G. Reynolds, ed (Cambridge, UK; Woodhead Publishing), pp. 164–187.

Chen, E.C.-H., Jamieson, A.M., and Van Gheluwe, G. (1980). The release of fatty acids as a consequence of yeast autolysis. J. Am. Soc. Brew. Chem. *38*, 13–18.

Chester, V.E. (1963). The dissimilation of the carbohydrate reserves of a strain of *Saccharomyces cerevisiae*. Biochem. J. *86*, 153–160.

Chester, V.E. (1967). Spontaneous mutants of yeast deficient in glycogen. Nature *214*, 1237–1238.

Chester, V.E. (1968). Heritable glycogen-storage deficiency in yeast and its induction by ultra-violet light. J. Gen. Microbiol. *51*, 49–56. http://dx.doi.org/10.1099/00221287-51-1-49

Davey, H.M., and Kell, D.B. (1996). Flow cytometry and cell sorting of heterogeneous microbial populations: the importance of single-cell analyses. Microbiol. Rev. *60*, 641–696.

David, M.H. (1974). Absence of nutritional requirement for unsaturated fatty acids by brewer's yeast in malt wort. J. Inst. Brew. *80*, 80–81.

de Lucena, R.M., Elsztein, C., de Barros Pita, W., de Souza, R.B., de Sá Leitão Paiva Júnior, S., and de Morais Junior, M.A. (2015). Transcriptomic response of *Saccharomyces cerevisiae* for its adaptation to sulphuric acid-induced stress. Antonie Van Leeuwenhoek *108*, 1147–1160. http://dx.doi.org/10.1007/s10482-015-0568-2

Deshpande, P.S., Sankh, S.N., and Arvindekar, A.U. (2011). Study of two pools of glycogen in *Saccharomyces cerevisiae* and their role in fermentation performance. J. Inst. Brew. *117*, 113–119.

Díaz-Hellín, P., Úbeda. J., and Briones, A. (2013). Improving alcoholic fermentation by activation of *Saccharomyces* species during the rehydration stage. LWT. Food. Sci. Technol. *50*, 126–131.

Duina, A.A., Miller, M.E., and Keeney, J.B. (2014). Budding yeast for budding geneticists: a primer on the *Saccharomyces cerevisiae* model system. Genetics *197*, 33–48. http://dx.doi.org/10.1534/genetics.114.163188

Dunn, B., and Sherlock, G. (2008). Reconstruction of the genome origins and evolution of the hybrid lager yeast *Saccharomyces pastorianus*. Genome. Res. *18*, 1610–1623. http://dx.doi.org/10.1101/gr.076075.108

Edelen, C.L., Miller, J.L., and Patino., H. (1996). Effects of yeast pitch rates on fermentation performance and beer quality. MBAA Tech. Quart. *33*, 30–32.

Eleutherio, E., Panek, A., De Mesquita, J.F., Trevisol, M E., and Magalhães, R. (2014). Revisiting yeast trehalose metabolism. Curr. Genet. doi: 10.1007/s00294-014-0450-1.

Ephrussi, B., Hottinguer, H., and Tavlitzki, J. (1949). Action de l'acriflavine sur les levures. II. Etude genetique du mutant 'petite colonies'. Ann. Inst. Pasteur. 79, 419–50.

Erten, H., Tanguler, H., and Cakıroz, H. (2007). The effect of pitching rate on fermentation and flavour compounds in high gravity brewing. J. Inst. Brew. *113*, 75–79.

Feldmann, H. (2012). Yeast metabolism. In Yeast. Molecular and Cell Biology, 2nd edition, H. Feldmann ed (Weinheim, Germany; Wiley-VCH Verlag), pp 25–58.

Galitski, T., Saldanha, A.J., Styles, C.A., Lander, E.S., and Fink, G.R. (1999). Ploidy regulation of gene expression. Science *285*, 251–254.

Gibson, B.R., Lawrence, S.J., Leclaire, J.P.R., Powell, C.D., and Smart, K.A. (2007). Yeast responses to stresses associated with industrial brewery handling. FEMS Microbiol. Rev. *31*, 535–569

Goffeau, A., Barrell, B.G., Bussey, H., Davis, R.W., Dujon, B., Feldmann, H., Galibert, F., Hoheisel, J.D., Jacq, C., Johnston, M., *et al.* (1996). Life with 6000 genes. Science *274*, 546–567.

Gayevskiy, V., and Goddard, M.R. (2015). *Saccharomyces eubayanus* and *Saccharomyces arboricola* reside in North Island native New Zealand Forests. Environ. Microbiol. DOI: 10.1111/1462-2920.13107

Gresham, D. (2015). Evolution: Fitness tracking for adapting populations. Nature *519*, 164–165. http://dx.doi.org/10.1038/nature14207

Gsell, M., Fankl, A., Klug, L., Mascher, G., Schmidt, C., Hrastnik, C., Zellnig, G., and Daum, G. (2015). A yeast mutant deleted of GPH1 bears defects in lipid metabolism. PLoS ONE *10*, e0136957. http://dx.doi.org/10.1371/journal.pone.0136957

Gunja-Smith, Z., Patil, N.B., and Smith, E.E. (1977). Two pools of glycogen in *Saccharomyces*. J. Bacteriol. *130*, 818–825.

Hall, J.F., and Webb, T.J.B. (1975). Factors affecting the survival of lyophilized brewery yeast strains. J. Inst. Brew. *81*, 471–475.

Harris, C.M., Todd, R.W., Bungard, S.J., Lovitt, R.W., Morris, J.G., and Kell, D.B. (1987). Dielectric permittivity of microbial suspensions at radio frequencies: a novel method for the real-time estimation of microbial biomass. Enz. Microb. Tech. *9*, 181–186.

Henson, M.G., and Reid, D.M. (1988). Practical management of lager yeast. Brewer. *74*, 3–8.

Hernawan, T., and Fleet, G. (1995). Chemical and cytological changes during the autolysis of yeasts. J. Ind. Microbiol. *14*, 440–450.

Higgins, V.J., Beckhouse, A.G., Oliver, A.D., Rogers, P.J., and Dawes., I.W. (2003). Yeast genome-wide expression analysis identifies a strong ergosterol and oxidative stress response during the initial stages of an industrial lager fermentation. App. Environ. Microbiol. *69*, 4777–4787.

Hoggan, J. (1977). Aspects of fermentation in conical vessels. J. Inst. Brew. *83*, 133–138.

Homma, T., Iwahashi, H., and Komatsu, Y. (2003). Yeast gene expression during growth at low temperature. Cryobiology *46*, 230–237.

Ingledew, W.M., Austin, G.D., and Kraus, J.K. (2009). Commercial yeast production for the fuel ethanol and distilled beverage industries. In The Alcohol Textbook, 5th edition, W.M. Ingledew, D.R. Kelsall, G.D. Austin and C. Kluhspies eds (Nottingham, UK; Nottingham University Press), pp 127–144.

Imai, T., and Ohno, T. (1995). The relationship between viability and intracellular pH in the yeast *Saccharomyces cerevisiae*. App. Environ. Microbiol. *61*, 3604–3608.

Jakobsen, M., and Thorne, R.S.W. (1980). Oxygen requirement of brewing strains of *Saccharomyces uvarum (carlsbergensis)* – bottom fermenting yeast. J. Inst. Brew. *86*, 284–287.

Jenkins, C.L., Kennedy, A.I., Hodgson, J.A., Thurston, P.A., and Smart, K.A. (2003). Impact of serial repitching on lager brewing yeast quality. J. Am. Soc. Brew. *61*, 1–9.

Jenkins, D.M., Powell, C.D., Fischborn, T., and Smart, K.A. (2011). Rehydration of active dry brewing yeast and its effect on cell viability. J. Inst. Brew. *117*, 377–382.

Jibiki, M., Ishibiki, T., Yuuki, T., and Kagami, N. (2001). Application of polymerase chain reaction to determine the flocculation properties of brewer's lager yeast. J. Am. Soc. Brew. Chem. *59*, 107–110.

Johnson, D.B. (1998). Biodiversity and ecology of acidophilic microorganisms. FEMS. Microbiol. Ecol. *27*, 307–317.

Johnston, G.C., Ehrhardt, C.W., Lorincz, A., and Carter, B.L. (1979). Regulation of cell size in the yeast *Saccharomyces cerevisiae*. J. Bacteriol. *137*, 1–5.

Kachroo, A.H., Laurent, J.M., Yellman, C.M., Meyer, A.G., Wilke, C.O., and Marcotte, E.M. (2015). Evolution. Systematic humanization of yeast genes reveals conserved functions and genetic modularity. Science *348*, 921–925. http://dx.doi.org/10.1126/science.aaa0769

Kirsop, B. (1955). Maintenance of yeasts by freeze-drying. J. Inst. Brew. *61*, 466–471.

Kirsop, B. (1974). The stability of biochemical, morphological and brewing properties of yeast cultures maintained by subculturing and freeze-drying. J. Inst. Brew. *80*, 565–570.

Kirsop, B.H. (1974). Oxygen in brewery fermentation. J. Inst. Brew. *80*, 252–259.

Klosinska, M.M., Crutchfield, C.A., Bradley, P.H., Rabinowitz, J.D., and Broach, J.R. (2011). Yeast cells can access distinct quiescent states. Genes. Dev. *25*, 336–349. http://dx.doi.org/10.1101/gad.2011311

Klug, L., and Daum, G. (2014). Yeast lipid metabolism at a glance. FEMS Yeast Res. *14*, 369–88. http://dx.doi.org/10.1111/1567-1364.12141

Koch, B., Schmidt, C., and Daum, G. (2014). Storage lipids of yeasts: a survey of nonpolar lipid metabolism in *Saccharomyces cerevisiae*, *Pichia pastoris*, and *Yarrowia lipolytica*. FEMS Microbiol. Rev. *38*, 892–915. http://dx.doi.org/10.1111/1574-6976.12069

Kühbeck, F., Back, W., and Krottenthaler, M. (2006). Influence of lauter turbidity on wort composition, fermentation performance and beer quality – a review. J. Inst. Brew. *112*, 215–221.

Kunze, W. (2014). Technology Brewing and Malting. 5th edition. (Berlin, Germany: VLB).

Lawrence, S.J., Nicholls, S., Box, W.G., Sbuelz, R., Bealin-Kelly, F., Axcell, B., and Smart, K.A. (2013). The relationship between yeast cell age, fermenter cone environment, and petite mutant formation in lager fermentations. J. Am. Soc. Brew. Chem. *71*, 90–96.

Libkind, D., Hittinger, C.T., Valério, E., Gonçalves, C., Dover, J., Johnston, M., Gonçalves, P., and Sampaio, J.P. (2011). Microbe domestication and the identification of the wild genetic stock of lager-brewing yeast. Proc. Natl. Acad. Sci. U.S.A. *108*, 14539–14544. http://dx.doi.org/10.1073/pnas.1105430108

Lillie, S.H., and Pringle, J.R. (1980). Reserve carbohydrate metabolism in *Saccharomyces cerevisiae*: responses to nutrient limitation. J. Bacteriol. *143*, 1384–1394.

Lockhart, L. (2003). Genomic analysis of variability in brewers' yeast, Ph.D thesis, University of Manchester, Faculty of Science and Engineering.

Lodolo, E.J., Kock, J.L., Axcell, B.C., and Brooks, M. (2008). The yeast *Saccharomyces cerevisiae*- the main character in beer brewing. FEMS Yeast Res. *8*, 1018–1036. http://dx.doi.org/10.1111/j.1567-1364.2008.00433.x

Loveridge, D., Ruddlesden, J.D., Noble, C.S., and Quain, D.E. (1999). Improvements in brewery fermentation performance by 'early' yeast cropping and reduced storage time. Proc. 7th Convention. Inst. Brew. Africa Sect. 95–99.

Maca, H.W., Barney, M., Goetzke, G., Daniels, D., and Ryder, D. (1994). The use of radio frequency capacitance for the measurement of yeast viable biomass and its use in the automatic pitching of fermentations. MBAA Tech. Quart. *31*, 146–148.

Maca, W., (2015). A unique system for propagating high fermentative capacity yeast, enhanced with yeast propagation management capabilities. MBAA Tech. Quart. *52*, 13–21.

Malcorp, P., Haselaars, P., Dupire, S., and Van den Eynde, E. (2001). Glycogen released by yeast as a cause of unfilterable haze in the beer. MBAA Tech. Quart. *38*, 95–98.

Maule, D.R. (1986). A century of fermenter design. J. Inst. Brew. *92*, 137–145.

McCaig, R., and Bendiak, D.S. (1985). Yeast handling studies. II. Temperature of storage of pitching yeast. MBAA Tech. Quart. *22*, 177–180.

Miyamoto-Shinohara, Y., Imaizumi, T., Sukenobe, J., Murakami, Y., Kawamura, S., and Komatsu, Y. (2000). Survival rate of microbes after freeze-drying and long-term storage. Cryobiology *41*, 251–255. http://dx.doi.org/10.1006/cryo.2000.2282

Morrison, K.B., and Suggett, A. (1983). Yeast handling, petite mutants and lager flavour. Proc. 19th Cong. Eur. Brew. Conv. *London*. 489–496.

Munoz, E., and Ingledew, W.M. (1990). Yeast hulls in wine fermentations – a review. J. Wine. Res. *1*, 197–209.

Murata, Y., Homma, T., Kitagawa, E., Momose, Y., Sato, M.S., Odani, M., Shimizu, H., Hasegawa-Mizusawa, M., Matsumoto, R., Mizukami, S., *et al.* (2006). Genome-wide expression analysis of yeast response during exposure to 4°C. Extremophiles *10*, 117–128. http://dx.doi.org/10.1007/s00792-005-0480-1

Nakao, Y., Kanamori, T., Itoh, T., Kodama, Y., Rainieri, S., Nakamura, N., Shimonaga, T., Hattori, M., and Ashikari, T. (2009). Genome sequence of the lager brewing yeast, an interspecies hybrid. DNA. Res. *16*, 115–29. http://dx.doi.org/10.1093/dnares/dsp003

Nathan, L. (1930). Improvements in the fermentation and maturation of beers. J. Inst. Brew. *36*, 544–550.

Noble, S. (1997). Practical aspects of fermentation management. Brew. Guard. *June*, 253 – 261.

O'Connor-Cox. E. (1997). Improving yeast handling in the brewery. Part 1: yeast cropping. Brew. Guard. December, 26–32.

O'Connor-Cox, E., Mochaba, F.M., Lodolol, E.J., Majara, M., and Axcell, B. (1997). Methylene blue staining: Use at your own risk. MBAA Tech. Quart. *34*, 308–312.

O'Connor-Cox. E. (1998a). Improving yeast handling in the brewery. Part 2: yeast collection. Brew. Guard. February, 22–34.

O'Connor-Cox. E. (1998b). Improving yeast handling in the brewery. Part 3: yeast pitching and measurement of yeast quality. Brew. Guard. *March*, 20–26.

Pennisi, E. (2014). Building the ultimate yeast genome. Science *343*, 1426–1429. http://dx.doi.org/10.1126/science.343.6178.1426

Petitjean, M., Teste, M.A., François, J.M., and Parrou, J.L. (2015). Yeast tolerance to various stresses relies on the trehalose-6P synthase (Tps1) protein, not on trehalose.

J. Biol. Chem. *290*, 16177–16190. http://dx.doi.org/10.1074/jbc.M115.653899

Powell, C.D., Quain, D.E., and Smart, K.A.. (2004). The impact of sedimentation on cone yeast heterogeneity. J. Am. Soc. Brew. Chem. *62*, 8–17.

Powell, C.D., and Diacetis, A.N. (2007). Long term serial repitching and the genetic and phenotypic stability of brewer's yeast. J. Inst. Brew, *113*, 67–74.

Quain, D.E., and Haslam, J.M., (1979). The effects of catabolite derepression on the accumulation of steryl esters and the activity of β-hydroxymethylglutaryl-CoA reductase in *Saccharomyces cerevisiae*. J. Gen. Microbiol. *111*, 343–351.

Quain, D.E., Thurston, P.A., and Tubb, R.S. (1981). The structural and storage carbohydrates of *Saccharomyces cerevisiae*: Changes during fermentation of wort and a role for glycogen catabolism in lipid biosynthesis. J. Inst. Brew. *87*, 108–111.

Quain, D.E., and Tubb, R.S., (1982). The importance of glycogen in brewing yeast. MBAA Tech. Quart. *19*, 29–33.

Quain, D.E. (1995). Yeast supply – the challenge of zero defects. Proc. 25th Cong. Eur. Brew. Conv. Brussels, 309–318.

Quain, D., Powell, C., Hamilton, A., Ruddlesden, D., and Box, W., (2001). Why warm cropping is best. Proc. 28th Cong. Eur. Brew. Conv. *Budapest*. 388–396.

Quain, D.E. (2006a). Yeast genetics in brewing: new insights and opportunities. In Brewing: new technologies, C.W Bamforth, ed (Cambridge, UK; Woodhead Publishing), pp. 149–166.

Quain, D.E. (2006b). Yeast supply and propagation. In Brewing: new technologies, C.W Bamforth, ed (Cambridge, UK; Woodhead Publishing), pp. 167–182.

Rautio, J.J., Huuskonen, A., Vuokko, H., Vidgren, V., and Londesborough, J. (2007). Monitoring yeast physiology during very high gravity wort fermentations by frequent analysis of gene expression. Yeast. *24*, 741–760.

Rice, J.F., Sullivan, T.R., and Helbert, J.R. (1980). A rapid method for the determination of yeast dry weight concentration. J. Am. Soc. Brew. Chem. *38*, 142–145.

Riess, S. (1986), Automatic control of the addition of pitching yeast. MBAA Tech. Quart. *23*, 32–35.

Rosenfeld, E., and Beauvoit, B. (2003), Role of the non-respiratory pathways in the utilization of molecular oxygen by *Saccharomyces cerevisiae*. Yeast, *20*, 1115–1144.

Rozenfelde, L., and Rapoport, A. (2014). Anhydrobiosis in yeast: is it possible to reach anhydrobiosis for yeast grown in conditions with severe oxygen limitation? Antonie Van Leeuwenhoek *106*, 211–217. http://dx.doi.org/10.1007/s10482-014-0182-8

Russell, I., and Stewart, G.G. (1981). Liquid nitrogen storage of yeast cultures compared to more traditional storage methods. J. Am. Soc. Brew. Chem. *39*, 19–23.

Sá-Correia, I., and Van Uden, N. (1983). Temperature profiles of ethanol tolerance: Effects of ethanol on the minimum and the maximum temperatures for growth of the yeasts *Saccharomyces cerevisiae* and *Kluyveromyces fragilis*. Biotechnol. Bioeng. 25, 1665–1667. http://dx.doi.org/10.1002/bit.260250620

Sato, M., Watari, J., and Shinotsuka, K. (2001). Genetic instability in flocculation of bottom-fermenting yeast. J. Am. Soc. Brew. Chem. *59*, 130–134.

Selmecki, A.M., Maruvka, Y.E., Richmond, P.A., Guillet, M., Shoresh, N., Sorenson, A.L., De, S., Kishony, R., Michor, F., Dowell, R., *et al.* (2015). Polyploidy can drive rapid adaptation in yeast. Nature *519*, 349–352. http://dx.doi.org/10.1038/nature14187

Siebert, K.J., Stenroos, L.E., Reid, D.S., and Grabowski, D. (1987). Filtration difficulties from damage to yeast during centrifugation. MBAA Tech. Quart. *24*, 1–8.

Šilhánková, L., Šavel, J., and Moštek, J. (1970). Respiratory deficient mutants of bottom brewer's yeast. 1. Frequencies and types of mutant in various strains. J. Inst. Brew. *76*, 280–288.

Simpson, W.J., and Hammond, J.R.M. (1989). The response of brewing yeasts to acid washing. J. Inst. Brew. *95*, 347–354.

Skoneczna, A., Kaniak, A., and Skoneczny, M., (2105). Genetic instability in budding and fission yeast—sources and mechanisms. FEMS. Microbiol. Rev. *39*, 917–967. http://dx.doi.org/10.1093/femsre/fuv028

Smart, K.A. (2007). Brewing yeast genomes and genome-wide expression and proteome profiling during fermentation. Yeast, *24*, 993–1013.

Somani, A., Bealin-Kelly. F., Axcell, B., and Smart, K.A. (2012). Impact of storage temperature on lager brewing yeast viability, glycogen, trehalose, and fatty acid content. J. Am. Soc. Brew. Chem. *70*, 123–130.

Stewart, G.G. (2015). Yeast quality assessment, management and culture maintenance. In Brewing Microbiology: Managing microbes, ensuring quality and valorising waste. A.E. Hill ed (Cambridge, UK; Woodhead Publishing), pp. 11–29.

Storchova, Z. (2014). Ploidy changes and genome stability in yeast. Yeast. *31*, 421–430.

Tapia, H., Young, L., Fox, D., Bertozzi, C.R., and Koshland, D. (2015). Increasing intracellular trehalose is sufficient to confer desiccation tolerance to *Saccharomyces cerevisiae*. Proc. Natl. Acad. Sci. U S A. *112*, 6122–6127.

Thiele, F., and Back, W. (2007). Assimilation technology as a basis for optimum yeast management. Brauwelt. Int. *2*, 82–85.

Thiele, F., Hartwig. A., and Back. W. (2008). Yeast quality distribution in the cone of cylindroconical tank. Brewing Science, July/August, 148–161.

Thomson, K., Bhat, A., and Carvell.J. (2015). Comparison of a new digital imaging technique for yeast cell counting and viability assessments with traditional methods. J. Inst. Brew. *121*, 231–237.

Thorn, J.A. (1971). Yeast autolysis and its effect on beer. Brewers Digest. October, 110–113.

Thurston, P.A., Quain, D.E., and Tubb, R.S. (1982). Lipid metabolism and regulation of volatile ester synthesis in *Saccharomyces cerevisiae*. J. Inst. Brew. *88*, 90–94.

Valkonen, M., Mojzita, D., Penttilä, M., and Bencina, M. (2013). Noninvasive high-throughput single-cell analysis of the intracellular pH of *Saccharomyces cerevisiae* by ratiometric flow cytometry. Appl. Environ. Microbiol. *79*, 7179–7187. http://dx.doi.org/10.1128/AEM.02515-13

Vaudano, E., Noti, O., Costantini. A., and Garcia-Moruno, E. (2014). Effect of additives on the rehydration of *Saccharomyces cerevisiae* wine strains in active dry form: influence on viability and performance in the early fermentation phase. J. Inst. Brew. *120*, 71–76.

Verbelen, P.J., Dekoninck, T.M., Saerens, S.M., Van Mulders, S.E., Thevelein, J.M., and Delvaux, F.R. (2009). Impact of pitching rate on yeast fermentation performance and beer flavour. Appl. Microbiol. Biotechnol. *82*, 155–167. http://dx.doi.org/10.1007/s00253-008-1779-5

Walsh, R.M., and Martin, P.A. (1977). Growth of *Saccharomyces cerevisiae* and *Saccharomyces uvarum* in a temperature gradient incubator. J. Inst. Brew. *83*, 169–172.

Walker, G.M. (2011). 125th Anniversary Review: Fuel alcohol: Current production and future challenges. J. Inst. Brew. *117*, 3–22.

Walther, A., Hesselbart, A., and Wendland, J. (2014). Genome sequence of *Saccharomyces carlsbergensis*, the world's first pure culture lager yeast. G3 *4*, 783–793. http://dx.doi.org/10.1534/g3.113.010090

White, C., and Zainasheff, J. (2010). Yeast. The Practical Guide to Beer Fermentation. (Boulder, USA; Brewers Publications).

Wightman, P., Quain, D.E., and Meaden, P.G. (1996). Analysis of production brewing strains of yeast by DNA fingerprinting. Lett. Appl. Microbiol. *22*, 90–94.

Wilson, W.A., Roach, P.J., Montero, M., Baroja-Fernández, E., Muñoz, F.J., Eydallin, G., Viale, A.M., and Pozueta-Romero, J. (2010). Regulation of glycogen metabolism in yeast and bacteria. FEMS Microbiol. Rev. *34*, 952–985. http://dx.doi.org/10.1111/j.1574-6976.2010.00220.x

Zhu, Y.O., Siegal, M.L., Hall, D.W., and Petrov, D.A. (2014). Precise estimates of mutation rate and spectrum in yeast. Proc. Natl. Acad. Sci. U.S.A. *111*, E2310–8. http://dx.doi.org/10.1073/pnas.1323011111

Taxonomy, Diversity, and Typing of Brewing Yeasts

4

José Paulo Sampaio[1]*, Ana Pontes[1], Diego Libkind[2] and Mathias Hutzler[3]

[1]UCIBIO-REQUIMTE, Departamento de Ciências da Vida, Faculdade de Ciências e Tecnologia, Universidade Nova de Lisboa, Caparica, Portugal.
[2]Laboratorio de Microbiología Aplicada, Biotecnología y Bioinformática de levaduras, Instituto Andino-Patagónico de Tecnologías Biológicas y Geoambientales (IPATEC), Consejo Nacional de Investigaciones en Ciencia y Técnica (CONICET), Universidad Nacional del Comahue, Bariloche, Argentina.
[3]Research Center Weihenstephan for Brewing and Food Quality, TU München, Freising, Germany.

*Correspondence: jss@fct.unl.pt

https://doi.org/10.21775/9781910190616.04

Abstract

The taxonomy and systematics of brewing yeasts have been a matter of debate and controversy since the early days of microbiology in the 1800s, when *Saccharomyces cerevisiae* and *Saccharomyces carlsbergensis* were first cultivated. The turbulent history of beer yeast systematics epitomizes the endeavours of yeast taxonomy since its origins when researchers used morphological characters and physiological traits to distinguish and classify species. The molecular revolution initiated in the 1980s exposed limitations of phenotypic methods, revealing numerous species synonyms and misclassifications. Today, DNA sequence data provide the means for accurate species identifications, strain typing, and phylogenetic classifications. Progress in the scientific knowledge of beer yeasts was also delayed by another level of complexity, which included inter-species hybridizations occurring in the brewing environment. Inter-species hybridizations created a plethora of chimeric genomes that could only be completely resolved when genomics entered the scene in the last two decades. Indeed, many key beer genotypes like *S. pastorianus*, the lager yeast, and *S. bayanus*, a beer contaminant, are complex multi-*Saccharomyces* species hybrids whose life history and ancestry are only now being revealed. Recently, a combination of novel genome sequencing approaches and microbial ecology studies solved decades-long disputes and revealed the wild genetic stocks of domesticated beer lineages. Here, we give an historical perspective of brewer's yeast taxonomy including also non-*Saccharomyces* yeast species and review available phenotypic and genetic-based typing methods for species and strain discrimination.

Introduction

Yeasts together with malt, water, and hops are an essential ingredient for beer. Their primary role is to convert sugars extracted from cereals into ethanol and carbon dioxide. Myriad other products of yeast metabolism are also essential for the unique sensorial characteristics that contribute to the final product. Studies on the history of beer and brewing have traced the ancestor of modern beers to cereal-based beverages produced in Ancient Egypt and in Mesopotamia circa 6,000 BC (Hornsey, 2003; McGovern, 2009). It is likely that the microorganisms that fermented those primeval beers were not radically different from those that we use today. For example, yeast cells were detected in Egyptian leavened bread dough dated 3.000 BC (Samuel, 1996;

2000) and molecular evidence for the presence of *S. cerevisiae* in wine fermentation has been obtained from pottery jars of the same period (Cavalieri *et al.*, 2003).

The most relevant microorganisms for brewing are by far yeasts of the genus *Saccharomyces*. This genus includes also the yeast species that are essential for all the other processes that involve alcoholic fermentation. The best examples of such products are bread, wine and sake, but many other foods and beverages with a more local range of production and consumption, like palm wine or sorghum beer (Tapsoba *et al.*, 2015), also rely on yeasts of the genus *Saccharomyces*.

Yeasts form a heterogeneous and artificial group of fungi that live predominantly in the unicellular stage and, similarly to other groups of microorganisms, lack distinct morphological traits (Kreger van Rij, 1987). Therefore, contrary to most plants or animals, yeasts are not only normally invisible to the naked eye but, when observed with a microscope, lack the distinctive morphological features that could allow the recognition of different species. Invisibility and lack of distinguishable micro-morphologies or other stable and phylogenetically coherent phenotypic markers, such as species-specific physiological properties, condemned yeasts, and microbes in general, to an unstable taxonomy. This contributed to consolidate, especially among the non-specialists, the belief that the recognition of the relevant yeasts for brewing was a daunting task, shadowed by disputed and conflicting taxonomic criteria and obscure nomenclature rules. In the period of more than 150 years from the first scientific records of yeasts by Cagniard-Latour, Kützing and Schwann in the 1830s (Barnett, 2004) to the generalization of DNA sequence data for taxonomy in the 1990s, a combination of methodologies that did not have the power to clearly differentiate species was employed in successive attempts at understanding yeast diversity and the species and lineages underlying it. Different approaches and different philosophies lead to frequent name changes that, for the most part, confounded those in industry or other applied fields that required at best a stable and logical nomenclature system. The age of enlightenment for microbial taxonomy was initiated when molecular sequence data were used to recognize species and to infer phylogenetic relationships, first for bacteria and archaea (Woese and Fox, 1977) and then for microscopic eukaryotes, with fungi being among the first to be investigated (e.g. Walker and Doolittle, 1982).

Yeasts – significance and relevance

Yeasts represent an artificial group because the predominant or exclusive unicellular stage arose by convergent evolution in distinct and not directly related evolutionary lineages of fungi. As a consequence, the filamentous growth mode is reduced or absent and cell division takes place normally by budding, although other processes can also occur, e.g. fission or the formation of forcibly discharged propagules at the end of sterigmata. Based on the available molecular phylogenetic evidence, it can be inferred that the yeast stage evolved multiple times in the two main lineages of fungi, the phyla Ascomycota and Basidiomycota (Taylor *et al.*, 2004).

The scientific understanding of the relevance of yeasts started with Pasteur (1866, 1876), who demonstrated that living cells (yeasts) were responsible for the alcoholic fermentation in wine and beer. Another aspect of major importance in those early days was the development of the procedures for the cultivation of yeasts as pure cultures by Hansen. Since yeast cultures could be easily obtained from wine or beer and were amenable to propagation and manipulation in the laboratory, *Saccharomyces*, notably the beer and wine yeasts, played a central role in the early studies of biochemistry, microbiology, cell biology and genetics. It is therefore natural that *S. cerevisiae* became the first model eukaryotic organism, being also the first eukaryote to have the full extent of its genome sequence determined (Goffeau *et al.*, 1996).

Yeast taxonomy

Yeast taxonomy arose with the pioneering works of Meyen, who created the name *Saccharomyces* in 1838, and Kützing, who in 1833 had given rise to *Cryptococcus*, although some of the species he described were in fact algae. Subsequent milestones were the recognition of the sexual features of the ascomycetes in some yeasts, by Schwann and Reess; the discovery of the fission yeast

Schizosaccharomyces by Lindner in 1893; and of conjugating yeasts described as *Zygosaccharomyces* by Baker in 1901 (reviewed in Barnett, 2004). Guilliermond published the first treatise on yeasts in 1912, followed in 1928 by a dichotomous key for yeast identification that included 22 genera. The genus *Candida*, which included clinically relevant asexual yeasts, was created in 1923 by Berkhout, and Harrison created in 1927 the genus *Rhodotorula* for yeasts that formed pink to red colonies. Starting in 1931, a series of monographs was published by Dutch researchers: Stelling-Dekker (1931), Lodder (1934), Diddens and Lodder (1942), Lodder and Kreger van Rij (1952), Lodder (1970) and Kreger-van Rij (1984). This series contributed to unifying the taxonomic study of yeasts and the publications of 1952, 1970, and 1984 constituted the first three editions of the most comprehensive monograph on yeasts, *The Yeasts: a Taxonomic Study*. The last two editions of this monograph were published in 1998 and 2011. Whereas the first edition included 180 species, subsequent editions incorporated an increasing number of species and genera so that the fifth edition listed nearly 1500 species.

The taxonomic criteria used to study yeasts changed over the years in tandem with our understanding of their biology and with the availability of new methodologies for scientific inquiry. In the early days, a strong emphasis was given to morphology. The shape of vegetative cells and mode of cell division, formation of pseudomycelium or of mycelium in certain media, and details of the sexual stage were considered key criteria for species and genus recognition. The study of the sexual stage evaluated the presence or absence of cell-to-cell conjugation prior to ascospore (meiospore) formation, the number and shape of ascospores, and the persistence or dehiscence of the ascus, the cell that contained the ascospores. Another line of taxonomic criteria that were gradually implemented in yeast taxonomy consisted of comparing physiological and biochemical properties of cultures. This approach, aimed at distinguishing species based on their physiological profiles, was obviously related to the applied concerns of researchers working close to the brewing and wine industries. In fact, yeast taxonomy has an applied origin, with strong connections to the fermentation industries of the late nineteenth and early twentieth centuries. Logically, the first tests consisted of determining which sugars could be fermented. As the number of known yeast species expanded, so the number of physiological tests had to be increased. In 1928, Guilliermond's scheme included testing for the fermentation of glucose, mannose, fructose, galactose, sucrose, lactose, and inulin, with Stelling-Dekker adding maltose and raffinose in 1931. In 1934, Lodder tested also the ability of each yeast to grow aerobically on sugars as sole carbon sources. In 1951, Wickerham used chemically defined media for obtaining the profiles of utilization of different sources of carbon and nitrogen (Wickerham, 1951), creating the standards for yeast identification: the 'nitrogen base' and 'carbon base' media. These formulations were so successful that they have been produced commercially ever since. Unfortunately, all these types of taxonomic criteria had a major weakness. The simple 'positive' or 'negative' results given by of many of these phenotypic characters and their rapid evolution could give rise to homoplasies, i.e. to similarity not caused by a common evolutionary history, thus introducing erroneous assessments in taxonomic schemes.

Starting in the 1960s, new methods were applied in yeast taxonomy in order to complement the macro- and micro-morphological criteria and physiological data. Electron microscopy was used to study yeast cytology and addressed details of the septal pore or ascospore ornamentation among other aspects of ultrastructure (Barnett and Robinow, 2002). The 1980s witnessed the gradual implementation of molecular methods aiming at analysing nucleotide sequences. DNA–DNA reassociation studies were used to test conspecific relationships among phenotypically similar strains (Kurtzman, 1987). Chemotaxonomy played a role in genus-level analyses, for example with the determination and comparison of the number of Coenzyme Q isoprene units (Yamada *et al.*, 1987) and for assessment of more distant relationships with the analyses of major cell-wall carbohydrates (Prillinger *et al.*, 1993).

However, none of these advances compares with the revolution caused by the use of DNA sequence data for yeast identification and for inferring phylogenetic relationships among species or genera. A first necessary step was the determination and archiving of sequence data from the type strains of known yeast species. This was achieved in the late 1990s for ascomycetous yeasts (Kurtzman and Robnett, 1998) and 2 years later for basidiomycetous yeasts (Fell *et al.*, 2000). The sequence most amenable was the variable D1/D2 domain of large subunit (26S) ribosomal RNA (rRNA) gene. Availability of sequence data for the entire spectrum of yeast species fuelled a major advance in yeast taxonomy. In spite of notable exceptions, a pattern of sequence uniformity among strains of a given species as well as sequence divergence among different species paved the way for the molecular identification of yeasts. As a consequence, the phenotypic criteria that were at the base of identification schemes, employing traditional dichotomous keys or computer-assisted approaches (Barnett *et al.*, 2000), were relegated to a secondary and mostly complementary role in species identifications and also in descriptions of new species. Sequence data provided species-specific DNA barcodes and was easily archived in public databases from which it could be searched and retrieved through the increasingly generalized use of the worldwide web. Finally, and contrary to most of the previously developed chemotaxonomic or molecular methods, the protocols for obtaining DNA sequence data were relatively simple to implement and were not dependent on particular local laboratory optimizations. At last, one taxonomic criterion provided data that could be universally interchanged among laboratories and could feed global databases like DDBJ (DNA Data Bank of Japan), GenBank (National Center for Biotechnology Information, USA) and EMBL-ENA (European Molecular Biology Laboratory European Nucleotide Archive). As the availability of sequencing facilities increased, the methods for analysing sequence data became easier to implement. Since sequence analyses encompassed more objective criteria for the recognition of species boundaries than the previous phenotype-based methods, the number of yeast species descriptions increased dramatically from 500 species in 1984, the beginning of the sequencing era, to 1500 species, that are included in the latest edition of '*The Yeasts*' monograph published in 2011 (Kurtzman *et al.*, 2011). Although in some cases sequence data provided arguments for merging species that were originally defined based on phenotypic criteria, the reverse situation, i.e. the detection of cryptic species only (or mostly) recognizable based on sequence data, was much more frequent. Another major advantage of the generalization of sequence data to support yeast classification schemes was the gradual abolishment of the dichotomy between sexual (teleomorphic) and asexual (anamorphic) yeast species. The dual nomenclature system, one for sexual taxa and the other for asexual ones, that had been in place since the beginning of yeast taxonomy, could be merged into a unified one, a change that had to await the necessary alterations on the Code for Nomenclature of Plants Algae and Fungi (Hawksworth, 2011; McNeill and Turland, 2011; McNeill *et al.*, 2011) and that is currently being implemented for yeasts (e.g. Kurtzman and Robnett, 2013; Liu *et al.*, 2015; Wang *et al.*, 2015). This opened the possibility of reducing the complexity of fungal nomenclature by classifying sexual and asexual yeast stages in the same genus. Another important aspect of the new Code was that no longer did names of sexual genera have priority over names of asexual genera. Therefore, the name to be retained could be the oldest name, according to the traditional rules of nomenclature, or a more recent one if other aspects such as, for example, the frequency of name usage were considered to be more relevant (Kurtzman, 2015).

The limitations of single-gene analyses to infer more distant phylogenetic relationships were made evident with the accumulation of studies of different fungal groups. As a consequence, multigene phylogenies have become prevalent. In recent years, yeast genomics, i.e. the use of whole genome data for the investigation of evolutionary, developmental, and functional questions, has gained momentum. For many yeast species, complete genome data has been obtained with next generation sequencing (NGS) approaches, which allows a deeper level of scrutiny on the evolution of individual species and of lineages encompassing species with different levels of phylogenetic relatedness (Liu *et al.*, 2009). For example, cases of horizontal gene transfer or of interspecies hybridization could only be detected with the implementation of comparative genomics, functional genomics and phylogenomics (Coelho *et al.*, 2013). Moreover, the prospect of having whole-genome sequences for all described yeast species appears feasible in the next few years (e.g. Hittinger *et al.*, 2015).

Taxonomy of the genus *Saccharomyces*

As mentioned in the previous section, the creation of the genus *Saccharomyces* dates back to the beginning of yeast taxonomy. The changes that it endured epitomize the turbulent history of yeast taxonomy. Reess (1870) provided the first comprehensive diagnosis of the genus that emphasized multipolar budding, asci with one to four ascospores, and absence of mycelium. While Meyen described three species in 1838, *S. cerevisiae* for beer strains, *S. pomorum* for cider strains, and *S. vini* for wine strains, Reess (1870) recognized two species, the beer yeast *S. cerevisiae* and *S. ellipsoideus*, which fermented fruit juices. Later, Hansen (1908) retained *S. cerevisiae* for top-fermenting (ale) beer yeasts and described *S. carlsbergensis* for bottom-fermenting (lager) beer yeasts. Stelling-Dekker (1931) accepted 23 species in the genus, collapsing *Zygosaccharomyces* into *Saccharomyces*. Moreover, *S. ellipsoideus* was considered to represent a variety of *S. cerevisiae* due to the absence of relevant differences between them. The tendency to enlarge the genus *Saccharomyes* continued in subsequent years and in 1952, in the first edition of *The Yeasts: a Taxonomic Study*, Lodder and Kreger-van Rij (1952) incorporated species of *Torulaspora* and *Debaryomyces* while merging several of the existing *Saccharomyces* species. After these changes, the genus contained 30 species. In the second edition of *The Yeasts* monograph, van der Walt (1970) accepted 41 species in *Saccharomyces*. He also merged some species and *S. carlsbergensis* was included in *S. uvarum*. van der Walt referred to this broad circumscription of the genus *Saccharomyces*, which included species previously and/or afterwards classified in *Zygosaccharomyces*, *Torulaspora*, and *Debaryomyces*, as *Saccharomyces sensu lato* (van der Walt 1970). Within the genus, the species closely related to the type species (*S. cerevisiae*) constituted the so-called *Saccharomyces sensu stricto*. At that time notable members of the *sensu stricto* group were *S. cerevisiae*, *S. bayanus*, and *S. uvarum*.

The next chapter of the history of *Saccharomyces* witnesses the arrival of the first generation of DNA-based methods: the determinations of base composition of nuclear DNA and DNA–DNA reassociation assays. Yarrow and Nakase (1975) compared the GC content of the DNA of species classified in *Saccharomyces* and confirmed the four groups previously established by van der Walt (1970). Moreover, because of the narrow range of the GC content of the species of the *sensu stricto* group, they defended that they represented 'populations or physiological variants of a single species, *S. cerevisiae*'. Along the same lines, results that were obtained with DNA–DNA reassociation assays suggested several *Saccharomyces* species were conspecific (Rosini *et al.*, 1982). For example, *S. chevalieri* and *S. italicus* were considered to be equivalent to *S. cerevisiae*. Therefore in the third edition of *The Yeasts*, Yarrow (1984) considered that *S. cerevisiae* was undistinguishable from 17 other species, including *S. bayanus*, *S. carlsbergensis*, and *S. uvarum*. As a result, the entire *sensu stricto* group was merged into *S. cerevisiae* and the genus was reduced to seven species.

However, DNA–DNA reassociation studies published shortly after the radical reduction of *Saccharomyces sensu stricto* to a single species told a different story. DNA–DNA reassociation values between *S. cerevisiae* and *S. bayanus* were low, thus suggesting two separate species (Martini and Kurtzman, 1985). Moreover, *S. carlsbergensis* showed intermediate relatedness with both *S. cerevisiae* and *S. bayanus*. This and the higher genome size of *S. carlsbergensis* led Martini and Kurtzman (1985) to hypothesize that it was an amphidiploid hybrid between *S. cerevisiae* and *S. bayanus*. Martini and Martini (1987) confirmed these results using a larger set of strains. They also replaced the name *S. carlsbergensis* with *S. pastorianus* for reasons of nomenclatural priority since Hansen had described *S. pastorianus* in 1904 and *S. carlsbergensis* in 1908. By the end of the 1980s, *S. paradoxus* was resurrected from synonymy with *S. cerevisiae* after passing the test of DNA–DNA reassociation (Martini, 1989). Therefore, four species (*S. cerevisiae, S. bayanus, S. pastorianus and S. paradoxus*) had been confirmed by the latest techniques available at the time to comprise *Saccharomyces*, i.e. to the *sensu stricto* group. In the fourth edition of '*The Yeasts*', Vaughan-Martini and Martini (1998) considered a total of 14 species in *Saccharomyces*, an expansion from the previous edition but that did not affect the *sensu stricto* group. In subsequent years, the use of molecular sequence data confirmed that these four species were well separated from the other members of the genus (James *et al.*, 1997; Kurtzman and Robnett, 1998), thus supporting the view that the genus *Saccharomyces* should be restricted to the *sensu stricto* species. This monophyletic group was enlarged by Naumov *et al.* (2000) who described, based on molecular sequence data from strains originally deposited in culture collections in Brazil and Japan, *S. cariocanus*, *S. kudriavzevii*, and *S. mikatae*. One of these species (*S. cariocanus*) is currently viewed as representing one of the main populations of *S. paradoxus* and is therefore is not recognized as a separate species. A comprehensive multigene phylogenetic analysis of several genera of ascomycetous yeasts more related to *Saccharomyces* was performed by Kurtzman and Robnett (2003) showing once more the monophylly of the *sensu stricto* group and justifying the assignment of the species classified in the *sensu lato* group to the new genera *Kazachstania*, *Naumovia* and *Lachancea* (Kurtzman, 2003). Therefore, 165 years after its birth, the genus *Saccharomyces* appeared to have found stable waters after being rescued from its turbulent past by the awesome power of molecular phylogenetics. However, as a probable reminiscence of the trauma of those times of unrest, many researchers seem to ignore that *Saccharomyces* has a well-defined phylogenetic circumscription and still prefer to use the outdated *sensu stricto* label (Replansky *et al.*, 2008) or to overlook the genera that now accommodate the species previously classified in the *sensu lato* group (e.g. Roop *et al.*, 2016). The genus received a new species in 2008 when researchers in China isolated strains associated with oak trees that they named *S. arboricola* (Wang and Bai, 2008). Therefore, in the fifth and latest edition of *The Yeasts*, Vaughan-Martini and Martini (2011) adopted the phylogenetic circumscription proposed in 2003 and recognized eight species in the genus (*S. arboricola*, *S. bayanus*, *S. cariocanus*, *S. cerevisiae*, *S. kudriavzevii*, *S. mikatae*, *S. paradoxus* and *S. pastorianus*). A point of much debate was the status of *S. uvarum*. High DNA–DNA reassociation values between the type strains of the two species (Rosini *et al.*, 1982) led Naumov (1996) to recommend that *S. uvarum* should be considered a synonym of *S. bayanus*. Later, Naumov (2000) and Naumova *et al.* (2005) observed that hybrids between the two type strains had intermediate fertility (27–37%) and recognized two varieties, *S. bayanus* var. *bayanus* and *S. bayanus* var. *uvarum* (Naumov, 2000). Emphasizing earlier DNA–DNA reassociation results and disregarding more recent arguments indicating that *S. uvarum* represented a species distinct from *S. bayanus* (e.g. Pulvirenti *et al.*, 2000; Nguyen and Gaillardin, 2005), Vaughan-Martini and Martini (2011) maintained *S. uvarum* as a synonym of *S. bayanus*. Libkind *et al.* (2011) detected sympatric *Saccharomyces* populations in *Nothofagus* (Southern beech) forests in Patagonia and characterized them using whole-genome data. One of the populations was identified as *S. uvarum*, which was definitively shown to be distinct from *S. bayanus*. The second population represented an undescribed species that was designated *S. eubayanus*. This name means the 'true *S. bayanus*' and refers to the complex genome of *S. bayanus*, shown by Libkind *et al.* (2011) to be a triple hybrid (see also section below for more discussion of *S. eubayanus* and *S. bayanus*).

In conclusion, and as summarized in Fig. 4.1, presently the genus includes seven natural species (*S. cerevisiae*, *S. paradoxus*, *S. mikatae*, *S. kudriavzevii*, *S. arboricola*, *S. eubayanus* and *S. uvarum*), together with two artificial and hybrid species exclusively associated with human-made fermentative environments (*S. pastorianus* and *S. bayanus*).

Brewing yeasts

In the previous section, we briefly illustrated the taxonomic changes associated with the genus *Saccharomyces* during its approximate one and a half centuries of existence. A direct consequence of those changes was the modification, multiple times, of the names and concepts of species. This has made the subject of understanding the circumscription of the different species, their most relevant differences, and their participation in human-driven fermentations particularly complex, even for those more familiar with the topic. *Saccharomyces* species names and corresponding species concepts are very dependent on the scientific context of a particular period and have to be viewed and discussed in light of their historical background. In many cases, new data promoted a change of the classification criteria and an almost inevitable consequence was the change of the nomenclature. In the next sections we discuss the relevance for brewing of various *Saccharomyces* species, using the most up-to-date circumscription of each species. In Table 4.1 we list the most relevant *Saccharomyces* species for brewing, including old species names not currently

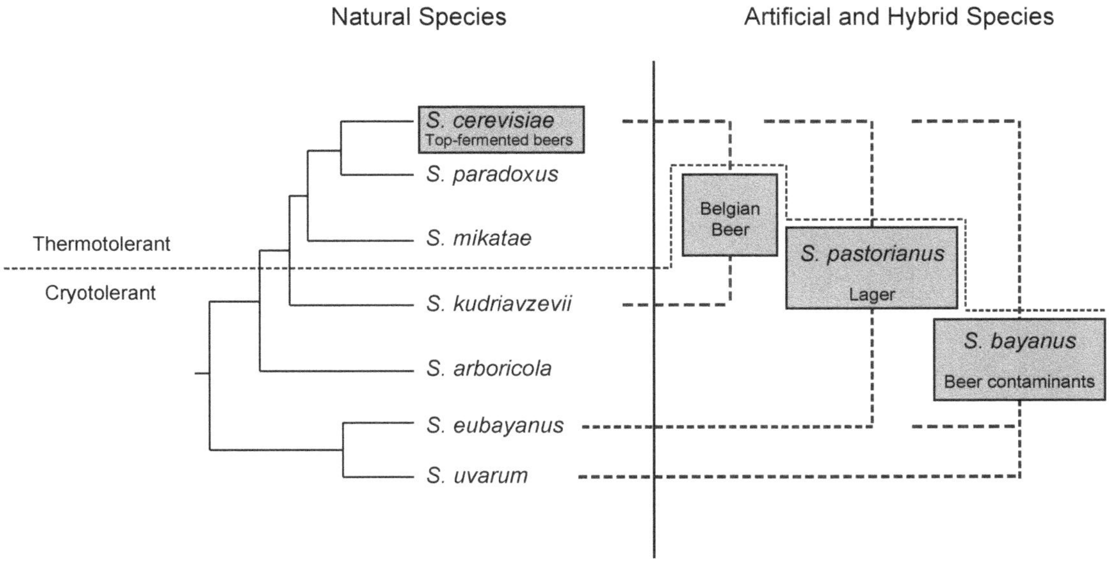

Figure 4.1 Schematic phylogeny of *Saccharomyces* natural species (left) and derived relevant brewing hybris (right) with reference to temperature adaptations. Shadowed boxes highlight species relevant to beer brewing.

in use. We also discuss the use of yeasts other than *Saccharomyces* for brewing.

Saccharomyces cerevisiae

S. cerevisiae, the flagship species of the genus, is the most industrially exploited microorganism. Strains of this species are used for the production of industrial ethanol and in the fermentation of alcoholic beverages, including top-fermented beer, wine, sake, and distilled spirits. This species is also used for bread dough leavening and probably is the species used in most cases of yeast fermentation of artisanal foods and beverages in Africa (Jespersen, 2003; Ludlow *et al.,* 2016), Asia (Pathania *et al.,* 2010; Shrestha *et al.,* 2002) and America (Vallejo *et al.,* 2013; Ludlow *et al.,* 2016).

Because *S. cerevisiae* dominates many spontaneous fermentations but is rarely found in more natural systems such as vineyards, the view that it represents a purely domesticated species that is exclusively associated with man-driven fermentations was defended by Martini (1993) and Vaughan-Martini and Martini (1995) and received generalized acceptance (e.g. Boulton and Quain, 2001). In the view of many researchers, *S. cerevisiae* did not occur in natural habitats and was considered to be associated with wineries and their equipment. This perspective was challenged by the finding of *S. cerevisiae* in wild habitats. The most consistent studies relied on two key aspects: the use of an enrichment medium with ethanol, thus favouring the growth of ethanol-resistant yeasts such as those of the genus *Saccharomyces;* and the sampling of oak trees (*Quercus* spp., family *Fagaceae*), namely bark and soil underneath the trees. In the last two decades, laborious field-work studies have revealed that *S. cerevisiae* has a natural distribution in temperate regions of the northern hemisphere (e.g. Sniegowski *et al.,* 2002; Sampaio and Gonçalves, 2008; Wang *et al.,* 2012), being therefore not limited to artificial environments created or modified by humans like wineries, vineyards, or grapes. More recently, the same approach was used to investigate the natural distribution of *S. cerevisiae* in a tropical region, not colonized by oaks, and new populations and a new candidate tree-habitat were uncovered (Barbosa *et al.,* 2016). The finding of these wild populations has shown that the natural history of *S. cerevisiae* goes far beyond its association with humans.

The degradation of glucose to ethanol without the complete oxidation to CO_2, even when oxygen is present, results in the build-up of toxic ethanol levels, an ecological strategy that enables *S. cerevisiae* to outcompete other species (Piskur *et al.,* 2006). Although other yeasts are also capable of fermenting sugars to ethanol and CO_2, the vast

Table 4.1 *Saccharomyces* spp. nomenclature with relevance for brewing

Species	Older designations and synonymy	Type or reference strain	Main features	Brewing value
S. bayanus	*S. bayanus* var. *bayanus*	CBS 380^T	Complex triple hybrid of *S. cerevisiae* × *S. eubayanus* × *S. uvarum*	Originally found in turbid beer. Probably only relevant as beer contaminant
S. carlsbergensis	Synonym of *S. pastorianus*	CBS 1513^T	See *S. pastorianus* CBS 1538^T	Bottom-fermenting yeast of the Saaz group (Group I)
S. cerevisiae		CBS 1171^T	Wine, beer, and bread yeast	Top-fermenting yeast. Other strains are currently used for brewing
S. diastaticus/*S. cerevisiae* var. *diastaticus*	Synonym of *S. cerevisiae*	CBS 1782^T	Degrades dextrins, resulting in over-attenuated beers	Beer contaminant
S. ellipsoideus/*S. cerevisiae* var. *ellipsoideus*	Synonym of *S. cerevisiae*	CBS 1395^T	Ellipsoid cell morphology	Beer contaminant
S. eubayanus		CBS 12357^T	Cryotolerant wild yeast representing one of the ancestors of lager yeasts	Experimental stage. Parental stock for controlled hybridizations.
S. kudriavzevii		IFO 1802^T	Cryotolerant wild yeast forming hybrids in some brewing environments	*S. cerevisiae* × *S. kudriavzevii* hybrids in some Trappist beers
S. monacensis	Synonym of *S. pastorianus*	CBS 1503^T	See *S. pastorianus* CBS 1538^T	Bottom-fermenting yeast of the Saaz group (Group I)
S. pastorianus		CBS 1538^T	Allopolyploid hybrid of *S. cerevisiae* × *S. eubayanus*	Bottom-fermenting yeast of the Saaz group (Group I)
S. pastorianus		Weihenstephan TUM 34/70	tetraploid with roughly two copies of chromosomes from *S. cerevisiae* and two from *S. eubayanus*	Bottom-fermenting yeast of the Frohberg group (Group II) and most used lager strain in the world
S. uvarum	*S. bayanus* var. *uvarum*	CBS 395^T	Cryotolerant wild yeast forming hybrids in some brewing environments	Associated with the brewing environment after hybridization (see *S. bayanus*)
S. uvarum var. *carlbergensis*	Synonym of *S. pastorianus*	CBS 1513^T	See *S. pastorianus* CBS 1538^T	

Names of currently recognized species are shown in bold. ^{T} indicates the type strain of the species (note that some reference strains are not type strains).

wealth of knowledge that has been gathered on *S. cerevisiae* sugar metabolism has revealed a combination of several properties that together give rise to a unique phenotype. These include a fast growth rate under a fermentation regime; efficient repression by glucose of respiration, thus promoting fermentation and ethanol production; good ability to produce ethanol; tolerance to the resulting environmental stresses, such as high ethanol concentration and low oxygen levels; and, most importantly, a fine-tuned regulatory network that optimizes the strategy of sugar utilization and ethanol formation through the different fermentation stages (Ihmels *et al.*, 2005). Therefore, the confirmed association of wild populations of *S. cerevisiae* with oaks hints at a natural ecology not entirely dominated by the colonization of sugar-rich substrates. In fact tree bark and soils are typically nutrient-poor environments and support small populations. Together, these new findings challenge the traditional concept of *S. cerevisiae* as a yeast exclusively associated with sugar-rich substrates and suggest that we are far from understanding its life cycle in nature. Reconciling the new evidence with the fundamental physiological characteristics of *S. cerevisiae* is the task that lies ahead.

If the natural ecology of *S. cerevisiae* remains, for the most part, a mystery, the population structure of this species is also far from being understood. Using multilocus sequencing, Fay and Benavides (2005) showed that *S. cerevisiae* included both wild and domesticated strains and suggested two independent domestication events, one for wine and the other for sake. They also demonstrated that domesticated strains derive from wild relatives and not the opposite. Population genomics is allowing researchers to disentangle the relationships among wild populations of *S. cerevisiae* and putatively domesticated groups. Liti *et al.* (2009) revealed a complex population structure composed of 5 genetically clean lineages. Two of them represented domesticated groups and corresponded to wine and sake yeasts. Another two populations were natural (wild) and consisted of an oak-associated North American population and a mostly clonal Malaysian population associated with inflorescences of Bertram palms. A fifth population was found in West African artisanal fermented beverages and it is not yet clear if it represents a wild or domesticated population. Another important finding was that many strains were genetically admixed and had contributions from two or more groups, probably because of outcrossing involving ancestors from different populations. A recent study expanded this analysis and included *S. cerevisiae* strains isolated as opportunistic pathogens in humans, most of which had mosaic recombinant genomes (Strope *et al.*, 2015). Very likely the entire sequence space of *S. cerevisiae* remains unknown and new wild and domesticated populations will be revealed in the future. For example, divergent populations from primary forests in China have been reported (Wang *et al.*, 2012) and, although the full extent of their genome sequence has not yet been analysed, it is likely that they represent additional wild populations. In addition, Almeida *et al.* (2015) showed that an oak-associated population in southern Europe was markedly distinct from similar populations in North America and Japan, thus representing a distinct wild lineage. Moreover, this Mediterranean population was identified as the closest wild relative of domesticated wine yeasts (Almeida *et al.*, 2015).

Whereas the genetic diversity and domestication signatures of wine yeasts has been explored, even if incompletely (Marsit and Dequin, 2015), comparatively much less is known for *S. cerevisiae* beer yeasts. Traditionally, two kinds of brewing yeasts have been recognized – ale and lager yeasts, the top-fermenting and bottom-fermenting yeasts, respectively. Presently, it is evident that only ale yeasts belong to *S. cerevisiae* since lager yeasts belong to *S. pastorianus* (see below). Classically, these two yeast groups were distinguished based on the ability to ferment the disaccharide melibiose. Ale yeasts, contrary to lager yeasts, did not ferment melibiose. Whereas the genetics and the genomes of lager yeasts have been intensively studied, top-fermenting ale-type yeasts have only recently been investigated with population genomics tools. Casey (1990, 1996) analysed ale and lager strains with Pulsed Field Gel Electrophoresis in order to generate chromosome fingerprints and reported a higher diversity for ale yeasts, which was hypothesized to be a consequence of the multiple locations of origin of top-fermenting yeasts, with breweries in different regions selecting unique yeast strains for each location. A genotyping study employing microsatellite markers and a large collection of industrial *S. cerevisiae* strains revealed distinct genotypes organized according to their usage, i.e. bread, beer, wine, and sake. Although wine strains were much more numerous than any other group, this study provided the first indication that wine and ale beer strains were genetically distinct, at least at neutral markers (Legras *et al.*, 2007). Borneman *et al.* (2011) analysed the genomes of two ale yeast strains, FostersB and FostersO. They observed a much higher level of heterozygosity in the two ale strains than in four wine strains, which was attributed to a ploidy level higher than 2*n* for beer strains, whereas wine strains were either haploid or diploid. Moreover, a considerable genetic distance was detected between brewing strains and other industrially relevant strains, such as wine strains and those used for bioethanol production. This remained the first clear indication that top-fermenting brewing strains appeared distinct, at the genome level, from other strains of *S. cerevisiae* until Gallone *et al.* (2016) and Gonçalves *et al.* (2016) investigated a comprehensive collection of top-fermenting beer strains. These studies revealed that ale-type beer yeasts are fundamentally distinct from other industrially relevant strains and much more genetically diverse than wine yeasts. Moreover, beer yeasts have a high incidence of polyploidy and aneuploidy and, probably as a consequence of this, a low or null sporulation ability. The phylogeny presented in Fig. 4.2 combines genome sequences of Gallone *et al.* (2016) and Gonçalves *et al.* (2016). Besides a main cluster of beer yeasts related with various types of German, British, Belgian and American beers, beer strains were also found to cluster in the wine, bread and sake clades and in an independent clade sister to the wine clade. Large-scale phenotyping showed industry-specific traits and genome analyses revealed domestication signatures of brewing yeasts. For example, a three-gene cluster that included *RTM1*, a subtelomeric gene associated with sucrose utilization that provides resistance to the inhibitory compounds present in molasses (Borneman *et al.*, 2011), was prevalent in the strains of the main beer clade but not in wine strains (Gonçalves *et al.*, 2016). Moreover, beer yeasts show a significant higher capacity to metabolize maltotriose, and abundant substrate in beer wort (Gallone *et al.*, 2016). Another characteristic that appears to have been selected during brewing yeast domestication is the low or null production of undesirable volatile compounds designated as 'phenolic off flavours' (POF) (Thurston *et al.*, 1981). These compounds, exemplified by 4-vinyl guaiacol, are formed when ferulic or other phenylacrylic acids are decarboxylated through enzymes encoded by two paralogous genes, *FDC1* (ferulic acid decarboxylase) and *PAD1* (phenylacrylic acid decarboxylase) (Mukai *et al.*, 2010). Mukai *et al.* (2014) showed that in most ale strains these genes contained inactivating mutations and consequently functional proteins could not be produced, which prevented POF formation. Genomic analyses showed that different lineages have acquired distinct inactivating mutations hinting at convergent evolution during domestication (Gallone *et al.*, 2016; Gonçalves *et al.*, 2016). However, for some beer styles, such as the Bavarian wheat beers and some Belgian beers, POFs are desired flavour components since they impart a characteristic clove-like flavour due to the desired decarboxylation of ferulic or cinnamic acids (Dufour *et al.*, 2003; Schneiderbanger *et al.*, 2013; Hutzler *et al.*, 2015). Richard *et al.* (2015) demonstrated that the overexpression of both genes resulted in a significant increase of the synthesis of phenolic compounds. *S. cerevisiae* strains that ferment these beer types have *PAD1* and *FDC1* functional, similarly to what was observed in other industrial variants like wine yeasts, as well as wild strains (Gonçalves *et al.*, 2016).

Saccharomyces pastorianus and *Saccharomyces carlsbergensis*

As indicated in the previous section, the two main types of beers are fermented by two distinct and presently clearly recognizable *Saccharomyces* groups. Top-fermenting ale-style strains belong to *S. cerevisiae* and represent the prototypical brewing yeasts, whereas bottom-fermenting lager yeasts, which are hybrids of two *Saccharomyces* species, are presently classified as *S. pastorianus*. In the past, the ability to ferment melibiose was an essential element for the taxonomy of beer yeasts and melibiase-negative yeasts were classified as *S. cerevisiae*, whereas melibiase-positive strains were classified as *S. carlsbergensis*, a name that was for some time considered a synonym of another key melibiose-fermenting species, *S. uvarum*.

Older literature (pre-1987) refers to lager yeasts as *S. carlsbergensis* instead of *S. pastorianus*. Although the two names refer to the same group of yeasts, two distinct type strains have to be considered. CBS 1538 is the type strain of *S. pastorianus* and CBS 1513 is the type strain of *S. carlsbergensis*. As explained above, the nomenclatural change concerned the belated recognition that *S. pastorianus* was an older name than *S. carlsbergensis*. This hybrid species is by far the most relevant brewing yeast, since lager beer accounts for 94% of the world market (Riese and Esslinger, 2009). However, *S. pastorianus* has never been isolated outside the brewing environment; therefore, it appears to be an exclusively domesticated species. Although the precise origin of lager yeasts is unknown, it is logical to assume that their emergence is intimately related with the origin and development of lager-brewing in the fifteenth century in Bavaria (Meussdoerffer, 2009). Although *S. pastorianus* was described by Hansen in 1904, the hybrid nature of lager yeasts was first hypothesized much later in the 1980s (Nilsson-Tillgren *et al.*, 1981; Martini and Kurtzman, 1985; Martini and Martini, 1987). Additional proof was obtained when two different sets of complete chromosomes were detected in lager yeast (Tamai *et al.*, 1998), thus confirming its alloploid nature. Although one of the progenitors of *S. pastorianus* was readily identified as *S.*

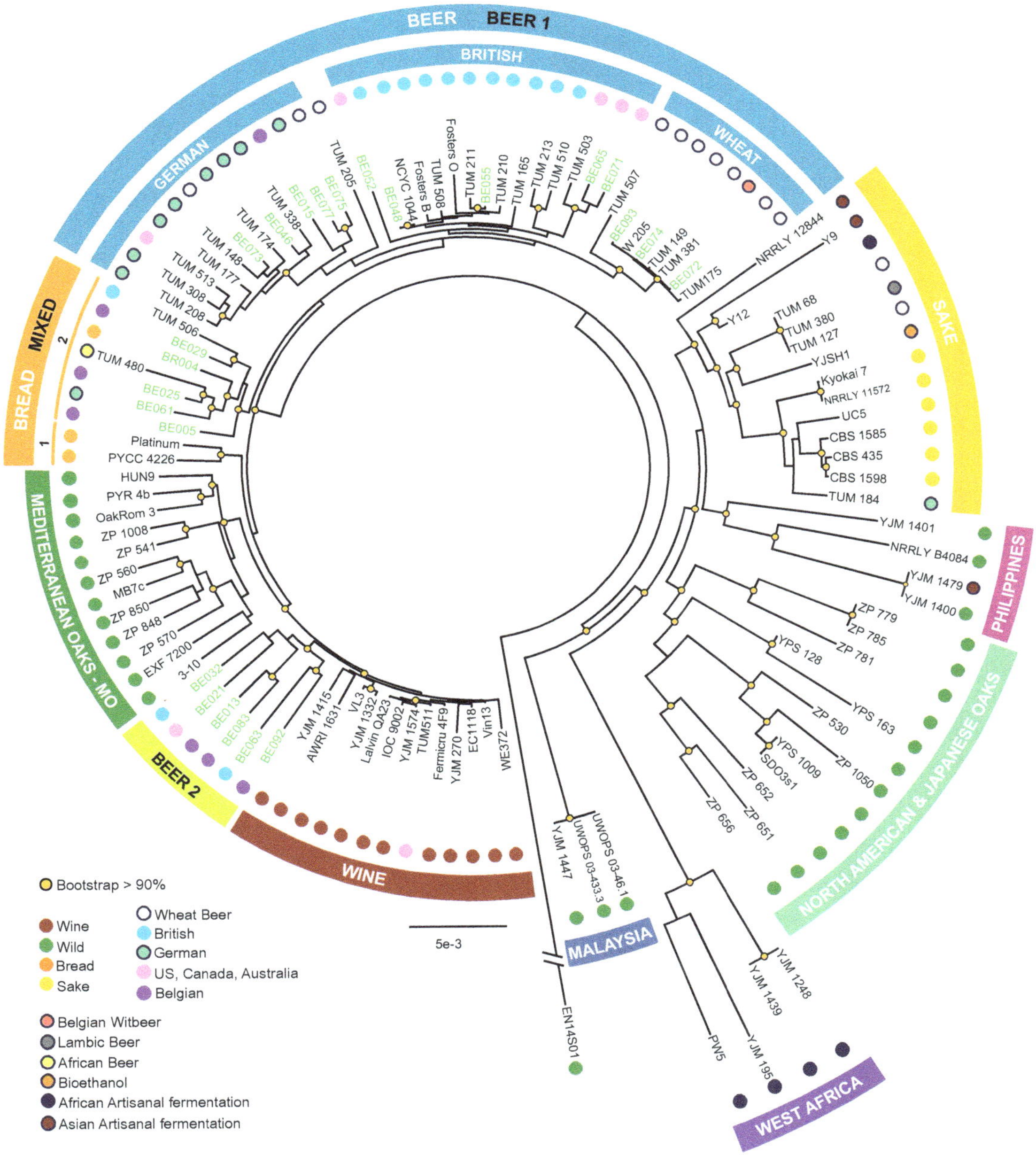

Figure 4.2 Whole-genome phylogeny of *S. cerevisiae* showing the diversity of top-fermenting beer strains. Phylogeny of 114 strains, inferred from 174991 SNPs, using the maximum likelihood method as implemented in RAxML with the GTRGAMMA model of sequence evolution and rooted with *S. paradoxus*. Branch lengths correspond to the expected number of substitutions per site. Support values from bootstrap replicates above 90% are shown. The analysis includes representatives of different types of beer starter cultures and of previously described populations (Wine, Mediterranean Oaks, Sake, Philippines, North American and Japanese oaks, West Africa and Malaysia). The ecology or industrial application of each strain is indicated in colour-coded circles after strain designation. Strains from Gallone *et al*. (2016) are indicated in green and strains from Gonçalves *et al*. (2006) are indicated in black. The black designations of beer groups follow Gallone *et al*. (2016) and those in white follow Gonçalves *et al*. (2006). Where only finished genome sequences were available, the corresponding error-free Illumina reads were simulated using dwgsim (http://sourceforge.net/apps/mediawiki/dnaa/).

cerevisiae, the non-cerevisiae progenitor remained contentious until the discovery in 2011 of a new yeast species, *S. eubayanus*, whose genome is an almost exact genetic match of the non-cerevisiae subgenome of *S. pastorianus* (Libkind *et al.*, 2011). Until that time, the non-cerevisiae contributor to the genome of *S. pastorianus* was thought to be *S. bayanus* (Martini and Kurtzman, 1985; Turakainen *et al.*, 1994; Montrocher *et al.*, 1998; Nakao *et al.*, 2009; Bond, 2009), now known to represent also an artificial and hybrid species only associated with anthropic environments, or *S. monacensis* (Pedersen, 1994; Hansen and Kielland-Brandt, 1994; Kielland-Brandt *et al.*, 1995), presently considered a synonym of *S. pastorianus*. Given the historical and technological context of the development of lager beer, one would suspect that the *S. cerevisiae* ancestor would be an ale yeast, a hypothesis that was validated with array-comparative genomic hybridization data (Dunn and Sherlock, 2008).

Early studies that analysed the diversity of lager yeasts were based on chromosome fingerprints and suggested two groups, here designated I and II, the Carlsberg and Tuborg types, respectively (Casey, 1996). Molecular analyses of transposon distribution also confirmed this division (Liti *et al.*, 2005) and more recent studies have contributed to the elucidation of the composition of these two groups (Dunn and Sherlock, 2008; Gibson *et al.*, 2013; Walther *et al.*, 2014). Group I includes well-known strains from the Carlsberg brewery in Denmark, like CBS 1538, the type strain of *S. pastorianus*; CBS 1513, the type strain of *S. carlsbergensis*; and CBS 1503, the type strain of *S. monacensis*. It includes also strains used to produce the Saaz-type beer, typically brewed in regions now in the Czech Republic. Group II contains strains used to produce the Frohberg-type beer, strains from the Netherlands (Heineken and other breweries), and the most used lager strain in the world, Weihenstephan TUM 34/70 from Bavaria. These two lineages of lager yeasts suggest distinct geographic distributions, corresponding to different breweries. Group I (Saaz strains) is associated with the Bohemian region in Czech Republic and with the Carlsberg brewery in Denmark, whereas group II (Frohberg strains) is associated with the Netherlands and with the Weihenstephan brewery in Germany. The analysis of the genome sequence of *S. carlsbergensis* CBS 1513 showed that it is substantially larger than that of *S. cerevisiae* (19.5Mb instead of 12 Mb) and is basically triploid, with a diploid *S. eubayanus* and haploid *S. cerevisiae* genome content (Walther *et al.*, 2014). The genome content of the best-known representative of group II, the Weihenstephan strain, is tetraploid, with two diploid *S. eubayanus* and *S. cerevisiae* genomes (Nakao *et al.*, 2009; Walther *et al.*, 2014). This strain has 36 chromosomes, with eight chromosome translocations occurring between the two subgenomes (Nakao *et al.*, 2009). For *S. carlsbergensis* CBS 1513, 29 chromosomes were detected, together with seven unique translocations between chromosomes of both parental genomes (Walther *et al.*, 2014). Besides distinct genetic compositions, the two lineages of lager yeasts have different flavour and brewing profiles. Saaz strains grow better at low temperatures (10°C), have a relatively poor fermentation performance compared with Frohberg strains at 22°C, do not use maltotriose, and produce lower amounts of esters (Gibson *et al.*, 2013; Walther *et al.*, 2014). Therefore, while both groups evidence their alloploid hybrid nature by combining the cryotolerant phenotype of *S. eubayanus* with the good fermentation performance of *S. cerevisiae*, detectable differences both in cryotolerance and in fermentation have been linked to the proportional amount of *S. eubayanus* DNA retained in their respective genomes. Moreover, the different phenotypes offer an explanation for the dominance of Frohberg strains in modern industrial brewing.

Multiple lines of evidence point to two independent origins of lager yeasts. It appears that the Saaz and Frohberg lineages were generated by at least two distinct hybridization events between very similar strains of *S. eubayanus* with more divergent ale strains of *S. cerevisiae* (Dunn and Sherlock, 2008; Baker *et al.*, 2015; Monerawela *et al.*, 2015). From the strict biological point of view, assigning the rank of species to a hybrid is questionable. In fact, lager yeasts form a heterogeneous assemblage of alloploid hybrids lacking a natural ecological niche, natural populations, and sexual reproduction, contrary to natural *Saccharomyces* species. However, although modern *Saccharomyces* taxonomy is strongly rooted on phylogenetics, it should in parallel provide a useful and understandable system for non-specialists, especially those in the industry. Therefore, even if in the strict sense *S. pastorianus* does not deserve the rank of species it is, from the practical view, useful to have a simple descriptor for lager yeasts, given their importance and easily recognizable attributes both in the brewery and in the laboratory. Moreover, the recognition of two groups of lager yeasts has led some authors to propose to distinguish them taxonomically. Wendland (2014) suggested to use *S. pastorianus* for the strains of group I (Saaz) and *S. carlsbergensis* for group II (Frohberg). Following the considerations above, the recognition in the nomenclature of the two groups might be adopted, especially if the users of taxonomy, e.g. brewers, feel that this recognition is desirable. However, whereas naming group I as *S. pastorianus* is logical, since the type strain of *S. pastorianus* belongs to this group, naming group II as *S. carlsbergensis* is not because the type strain of *S. carlsbergensis* also belongs to group I. This resolution would go against taxonomic rules and, more importantly, would be a confusing designation. A better description of group II would be, perhaps, '*S. weihenstephanii*'. An alternative solution would be to consider these groups as separate races of *S. pastorianus*. Although typically they are not used in yeast nomenclature, races constitute an informal rank in the taxonomic hierarchy equivalent to breeds and cultivars, and one example could be *S. pastorianus* 'race Weihenstephan'.

Saccharomyces eubayanus and *Saccharomyces bayanus*

A yeast isolated by Will in 1891 from turbid beer, in which it caused a disagreeable flavour, was studied by Bay in 1893 and formally described as *S. bayanus* by Saccardo in 1895. The original strain was lost and one strain originated from the Institute of Brewing in Tokyo and sent to CBS from the Central Laboratory of the South Manchuria Railway Company (Dairen) in 1927, was adopted as the type strain (Lodder and Kreger-van Rig, 1952). The relationship between *S. bayanus* and *S. uvarum*, a contemporary species described in 1898 by Beijerinck, has been contentious and was settled only recently (Libkind *et al.*, 2011). Although as discussed earlier (section 'Taxonomy of the genus *Saccharomyces*'), *S. uvarum* was considered a synonym of *S. bayanus*, evidence from molecular studies (e.g. Nguyen and Gaillardin, 1997; Rainieri *et al.*, 1999) argued for the reinstatement of two independent species. Whereas *S. bayanus* encompassed a small and heterogeneous assemblage of beer contaminants with low frequency of sporulation, *S. uvarum* included a larger, less variable, and more fertile group of wine and cider yeasts, with the wine isolates originating normally from low-temperature fermentations. A key point to the confusion was the nomenclature of CBS 7001, a *S. uvarum* strain selected for genome sequencing (Cliften *et al.*, 2001; Kellis *et al.*, 2003) but still nowadays referred to as *S. bayanus* in the literature and SGD and NCBI databases. The hybrid nature of *S. bayanus* was anticipated in several publications (e.g. Nguyen and Gaillardin, 2005) and has now been fully demonstrated (Nguyen *et al.*, 2011; Libkind *et al.*, 2011). This species shares with *S. pastorianus* a preferential occurrence in the brewing environment (often as a beer contaminant), the apparent absence of a natural habitat, and a complex genome composition. The hybrid nature of the genome of *S. bayanus* is even more complex than that of *S. pastorianus* since, in addition to having contributions from *S. eubayanus* and *S. cerevisiae*, it also contains genomic regions from *S uvarum*. Similarly to *S. pastorianus*, each strain of *S. bayanus* has a unique genomic composition. For example, the subgenomes of the type strain, CBS 380, are 67% *S. uvarum* and 33% *S. eubayanus*, whereas strain NBRC 1948, considered in the pre-genomic era to represent a pure *S. bayanus* line (Rainieri *et al.*, 2006), is 37% *S. uvarum* and 63% *S. eubayanus* (Libkind *et al.*, 2011). Although, as referred in the previous section, *S. bayanus* has been hypothesized as one of the parents of the interspecies hybridization that gave rise to *S. pastorianus*, it is now postulated that *S. pastorianus* may have served as progenitor of the various *S. bayanus* hybrid strains through additional hybridization events with *S. uvarum* (Libkind *et al.*, 2011; Nguyen *et al.*, 2011). In conclusion, *S. bayanus* is now viewed as an artificial hybrid species whose origin is intimately related with the particular environmental conditions of the brewing environment.

S. eubayanus, the latest addition to the genus, was originally found in South America (Libkind *et al.*, 2011). Five populations have been detected so far in *S. eubayanus*, two in South America (Peris *et al.*, 2014) and three in Asia, in Tibet, Sichuan, and western China (Bing *et al.*, 2014). This species was also found, albeit infrequently, in Wisconsin (Peris *et al.*, 2014) and New Zealand (Gayevskiy and Goddard, 2016). Interestingly, one population from the Tibetan plateau is apparently the closest relative of the *S. eubayanus* portion of *S. pastorianus*, with 99.8% sequence similarity based on multilocus sequencing. For Patagonian *S. eubayanus*, an equivalent comparison using a whole-genome estimate is slightly lower at 99.5% (whole genome sequences are not yet available for Tibetan *S. eubayanus*). Since natural populations of *S. eubayanus* have not been detected yet in Europe, it has been suggested that the non-cerevisiae subgenome of lager yeast is of Asian origin (Bing *et al.*, 2014). However, recent genomic studies using novel *S. eubayanus* isolates have shed new light into this issue, suggesting that a primary dispersal from South America into the Holarctic may be more likely based on the relative diversities of the Holarctic and one of the two Patagonian subpopulations and the confinement of a signature of recent demographic expansion to the Tibetan subpopulation (Peris *et al.*, 2016).

The brewing properties of *S. eubayanus* (type strain) have been analysed and found to be more similar to those of Saaz strains than to those of Frohberg strains (Gibson *et al.*, 2013). *S. eubayanus*-Saaz were least sensitive to cold (10°C) and unable to use maltotriose. At 22°C *S. eubayanus* performed poorly, producing less ethanol, even when compared with Saaz strains. Besides being investigated as a brewing yeast, *S. eubayanus* was also used to generate de novo lager hybrids in crosses with an ale strain (Krogerus *et al.*, 2015; Hebly *et al.*, 2015; Mertens *et al.*, 2015). The hybrids inherited relevant brewing properties from both parents and showed apparent hybrid vigour, fermenting faster and producing more ethanol than the parents, thus validating this strategy for the production of new lager strains. Therefore, it is possible that in the future, *S. eubayanus* or new hybrids containing portions of its genome, will be used industrially as brewing yeasts.

Saccharomyces kudriavzevii hybrids

S. kudriavzevii was originally known from a few strains obtained from decayed leaves and soil in Japan (Naumov *et al.*, 1995, 2000). Subsequently, a larger set of isolates was collected from oak trees in Europe (Sampaio and Gonçalves, 2008; Lopes *et al.*, 2010; Erny *et al.*, 2012) and from soil, leaves, and mushrooms in Taiwan (Naumov *et al.*, 2013). Among the natural species of *Saccharomyces*, *S. kudriavzevii* can be considered a cryotolerant yeast since it can grow at low temperatures (≈5°C) and grows poorly at 30°C (Belloch *et al.*, 2008; Sampaio and Gonçalves, 2008; Arroyo-López *et al.*, 2009). This is consistent with the observation that enrichment media incubated at low temperatures (10°C) readily yield *S. kudriavzevii* from oaks across Eurasia (Hittinger, 2013). By studying the glycolytic flux and the level of activity of individual enzymatic steps of the glycolysis, Gonçalves *et al.* (2011) concluded that *S. kudriavzevii* metabolism evolved towards a better performance at low temperatures, particularly in ethanol production, at the cost of heat resistance. Later, the metabolome at low temperatures was compared with the thermotolerant species *S. cerevisiae* and the main differences found were in carbohydrate metabolism, mainly in fructose metabolism (López-Malo *et al.*, 2013). Other metabolic data in favour of the cryotolerant nature of this species were a higher glycerol level and a lower ethanol content of *S. kudriavzevii* compared with *S. cerevisiae* at low temperatures (14°C), which was due to a higher glycerol-3-phosphate dehydrogenase activity (*Arroyo-López et al., 2010*). Strains of *S. kudriavzevii* showed also higher percentages of medium-chain fatty acids and squalene regardless of the growth temperature in comparison to *S. cerevisiae* (Tronchoni *et al.*, 2012). This differential lipid composition may partially explain the better adaptation of S. kudriavzevii at low temperatures. A similar comparison, made at the transcriptome level, led Tronchoni *et al.* (2014) to conclude that the cryotolerance of *S. kudriavzevii* is due to an enhanced ability to initiate a quick and efficient translation of crucial genes for cold adaptation (i.e. the cold stress marker gene *NSR1* and lipid metabolism related genes).

The ethanol tolerance of *S. kudriavzevii* is relatively low in comparison with other *Saccharomyces* species, given that *S. kudriavzevii* shows weak or null growth above 5% ethanol (Belloch *et al.*, 2008). Competitive exclusion of *S. kudriavzevii* by other mesophilic and/or more ethanol-tolerant *Saccharomyces* species has been experimentally demonstrated in laboratory mixed cultures (Sampaio and Gonçalves, 2008; Arroyo-López *et al.*, 2011). The low stress tolerance and thus low competitiveness exhibited by *S. kudriavzevii* probably explains the lack of reports on the occurrence of this yeast in human-related fermentations or environments, with one exception – a few isolates from a brewery in New Zealand (González *et al.*, 2006). Hence, the absence of *S. kudriavzevii* in most brewing monographs is understandable. However, as already mentioned for *S. eubayanus*, *S. kudriavzevii* might be relevant to brewing as a contributor to hybrid strains rather than as a pure lineage.

Indeed, hybrid strains combining the genomes of *S. cerevisiae* and *S. kudriavzevii* have been isolated and characterized from fermenting environments, mostly from wine and cider fermented at low temperatures (Groth *et al.*, 1999; Lopandic *et al.*, 2007). More recently, similar hybrids have been associated to beer and seem to be common in Belgian-style beers (González *et al.*, 2008). With the implementation of genome sequencing studies, strains originally assumed to be *S. cerevisiae* are being recognized as *S. cerevisiae* × *S. kudriavzevii* hybrids. For example, González *et al.* (2008) analysed the genomic composition of 24 brewing strains labelled as *S. cerevisiae* and found that 25% of them were in fact *S. cerevisiae* × *S. kudriavzevii* hybrids. Even triple hybrids of *S. cerevisiae*, *S. uvarum*, and *S. kudriavzevii* have been identified in cider and wine (Naumova *et al.*, 2005; González *et al.*, 2006). On the other hand, Lopes *et al.* (2010) performed several genome-wide molecular analyses with European wild *S. kudriavzevii* strains and found intraspecific differences with respect to the Japanese population, in line with Hittinger *et al.* (2010), who showed that the two populations differed considerably in their GAL gene network. A European/Mediterranean origin for the brewing hybrids has been hypothesized (González *et al.*, 2008), and molecular analyses of independently isolated hybrids lead to the conclusion that they are the result of multiple independent hybridization events (Erny *et al.*, 2012; Peris *et al.*, 2012). Together, these results suggest that an important fraction of brewing strains may correspond to *S. cerevisiae* × *S. kudriavzevii* hybrids.

In contrast with *S. pastorianus*, whose origin is directly connected with the brewing environment (see '*Saccharomyces pastorianus* and *Saccharomyces carlsbergensis*', above), some authors believe that *S. kudriavzevii* hybrids might have originated in wild environments due to the low resistance of this species to the stress conditions of human-driven fermentations (Sipiczki, 2008; Arroyo-López *et al.*, 2011). However, despite the repeated isolation of *S. kudriavzevii* from natural sources, hybrid strains are only known from human-driven fermentations. Nevertheless, the role of *S. eubayanus* and *S. kudriavzevii* subgenomes in the hybrids seems to be similar, guaranteeing a good fermentative performance at low temperatures. The hybrids retained by the industry exhibit the best properties of both parental species, such as the low-temperature fermentation abilities of *S. kudriavzevii* and the high ethanol resistance of *S. cerevisiae* (Giudici *et al.*, 1998; Belloch *et al.*, 2009; Arroyo-López *et al.*, 2009). The genomic contribution of *S. kudriavzevii* adds to the complexity of attributes provided by the genomes of *S. cerevisiae* and *S. eubayanus* in brewing environments and most probably to the properties of the final product. The limited data available on *S. kudriavzevii* × *S. cerevisiae* hybrids in brewing, starting from the low number of isolates so far analysed, precludes drawing conclusions about the prevalence of these newly discovered hybrids in beer, especially their actual contribution to the properties of the final product. Interestingly, half of the *S. kudriavzevii* × *S. cerevisiae* strains detected by González *et al.* (2008) were recovered from Belgian specialty beers from Trappist monasteries (Trappist beers). Bottle re-fermentation or conditioning is a common practice in the production of these types of beers (van Landschoot *et al.*, 2005), which allows adjusting and/or modifying the final flavour of beer, also known as bioflavouring (Vanderhaegen *et al.*, 2003a). Further studies are needed to elucidate whether *S. kudriavzevii* hybrids have a role in the primary and/or secondary fermentation stages and to determine what is their actual contribution to the typical flavour complexity of Belgian specialty beers.

Brettanomyces and *Dekkera*

Yeasts presently classified in the genus *Brettanomyces* were first discovered around 1903 in Great Britain by Clausen (Carlsberg Laboratories) in English old stock beers. Their name came from misspelling of the name 'Brittanomyces', and intended to mean 'a fungus from British beer' (Verarchtert and Derdelinckx, 2014). These yeasts were initially classified in the genus *Torula* and their assignment to *Brettanomyces bruxellensis* (synonym of *B. lambicus*) occurred when another yeast with similar characteristics was isolated from lambic beer by Kufferath and van Laer (1921). At the same time, these authors suggested that this new yeast was responsible for the characteristic taste of lambic beer. The relationship of *Brettanomyces* yeasts with beer was consolidated when Custers (1940) characterized multiple strains recovered from English and Belgian beers.

The taxonomy of *Brettanomyces* has endured many changes over the years. The most relevant was triggered by the discovery of ascospores, i.e. the sexual stage, in some strains. Therefore, sexual (teleomorphic) species were classified in a new genus, *Dekkera*, whereas asexual (anamorphic) stages were kept in *Brettanomyces* (van der Walt, 1984). As discussed above (section 'Yeast Taxonomy'), *Dekkera* and *Brettanomyces* represent the same genus since there is no longer a taxonomic distinction between sexual and asexual states (Oelofse *et al.*, 2008). Since the name *Brettanomyces* is older, better known, and used more commonly in the food and beverage industries, it is expected to be prioritized over *Dekkera* and will be used in this sense hereafter.

Currently, five species of *Brettanomyces* are recognized: *B. anomalus* (*D. anomala*), *B. bruxellensis* (*D. bruxellensis*), *B. custersianus*, *B. naardenensis*, and *B. nanus*. Except for *B. naardenensis*, all other species have been isolated from beer mainly as contaminants. Relevant data on each *Brettanomyces* species are summarized in Table 4.2. All five species can be easily distinguished using 26S rRNA gene sequences (Crauwels *et al.*, 2014). In general, low sequence divergence was observed between strains belonging to the same species (0% to 1.4%), indicating that this molecular marker is relatively well conserved and a reliable tool for identification at the species level. By far, *B. bruxellensis* is the most relevant species for brewing either as a contaminant or as non-conventional fermentation agent.

In spite of their bad reputation as wine- and beer spoilage agents, yeasts of the genus *Brettanomyces* are a desired component of certain beers due to their potent bioflavouring activity. The positive influence of *Brettanomyces* in brewing is mostly related to spontaneously fermented beers, such as Trappist beers or similar, and in sour beers intentionally inoculated with these type of yeasts and sometimes other microorganisms (lactic acid bacteria and/or other yeasts) (Bokulich and Bamforth, 2013). The impact of *Brettanomyces* on the beer organoleptic characteristics is multiple, thus explaining why they are the most well studied non-*Saccharomyces* yeasts for flavour modulation. A revision on this issue has been recently published (Steensels *et al.*, 2015).

One of the most distinctive contributions of *Brettanomyces* to beer flavouring is the production of highly volatile ethylphenols. Two compounds, 4-ethylphenol (barnyard, medicinal, and mousy aromas) and 4-ethylguaiacol (4EG; bacon, spice, or smoky aromas), derive from two vinylphenols previously referred to as POF, 4-vinylguaiacol (4VG) and 4-vinylphenol (4VP) (see '*Saccharomyces cerevisiae*', above), as a result of the activity of a vinylphenol reductase (Chatonnet *et al.*, 1992; Edlin *et al.*, 1998). This activity is absent in *S. cerevisiae* but seems to be present in all *Brettanomyces* species. A screening of 13 strains of *B. bruxellensis*, *B. custersii*, and *B. anamolus* isolated from lambic beer showed that all were efficiently able to convert ferulic acid to 4VG and then to 4EG, the latter being the most dominant volatile phenol present (Vanbeneden *et al.*, 2008). However, the decarboxylation of hydroxycinnamic acid, which is mediated

Table 4.2 *Brettanomyces*: nomenclature and ecology

Species	Older designations and synonymy	Type strain	Ecology
B. anomalus	*B. abstinens*, *B. claussenii*, *Dekkera anomala*	CBS 8139^T	Beer, cider, spoiled soft drinks
B. bruxellensis	*B. intermedia*, *B. intermedium*, *B. lambicus*, *B. custersii*, *D. bruxellensis*	CBS 74^T	Belgian stout, lambic beer, beer, equipment beer brewery, grape must, wine, sour wine, ginger ale
B. custersianus		CBS 4805^T	Equipment at a Bantu beer brewery (South Africa)
B. naardenensis		CBS 6042^T	Carbonated water and fruit drinks with low pH
B. nanus	*Eeniella nana*	CBS 1945^T	Bottled beer in Sweden

^{T} indicates the type strain of the species.

by POF genes, seems to be the rate-limiting step in the formation of ethylphenols and *Brettanomyces* shows higher activity than *S. cerevisiae* (Vanbeneden *et al.*, 2008). However, for *B. bruxellensis*, the most studied species, strain heterogeneity at the 4EG production level was recorded, together with variation in terms of ploidy (2*n* or 3*n*), PCR fingerprinting, genome composition, microsatellite markers, and SO_2 tolerance (Vigentini *et al.*, 2013; Albertin *et al.*, 2014; Crauwels *et al.*, 2014, 2015). A correlation between genotype groups of *B. bruxellensis* and their source of isolation has been shown (Vigentini *et al.*, 2012; Crauwels *et al.*, 2014). Employing genomic analyses, the lack of a β-glucosidase due to a specific deletion was found to be trait distinguishing beer strains from strains from other sources. This aspect may have direct impact on the ability of these strains to compete with other microbes present in the brewing environment and/or on flavour production (Crauwels *et al.*, 2015). Since wine isolates of *B. bruxellensis* outnumber by far those isolated from beer, current knowledge on the brewing attributes of these yeasts is still fragmentary. Additional isolations of *B. bruxellensis* from different beer sources are crucial to further illuminate our understanding of the prevalence of this yeast in specific beers.

It has been suggested that *B. bruxellensis* may have the ability to rapidly adhere to surfaces in low-nutrient environments, thus allowing the formation of strong biofilms (Joseph *et al.*, 2007) that would generate protection against several stress factors such as the disinfectants used in breweries. This is consistent with the hypothesis that this yeast may persist in brewery environments for extended periods of time (Suzuki *et al.*, 2008). However, little is known about the dynamics of this yeast within the brewery environment, and even less is known about the ecology and natural habitats of *Brettanomyces* outside the brewery. It was recently hypothesized that some of the potent flavouring compounds synthesized by *Brettanomyces* (ethyl phenols) play a crucial role in the dispersal through insect vectors in nature (Dweck *et al.*, 2015). We anticipate that finding the wild genetic stocks of domesticated microbes such as members of the genus *Brettanomyces* will contribute not only to understanding their natural history but also to identifying and manipulating the genetic changes that are relevant for brewing.

The microbiota of Belgian specialty beers and other non-conventional brewing yeasts

Fermentation of most beers worldwide depends on a monoculture process using either *S. cerevisiae* or *S. pastorianus*. In contrast, the fermentation and/or maturation of some Belgian beers is determined by non-conventional yeasts belonging to the autochthonous non-starter community or to uncharacterized mixed starter cultures (Verachtert and Iserentant, 1995; Verachtert and Derdelinckx, 2005; see Chapter 7). These beers represent culturally important alcoholic beverages for which complex microbial communities play critical roles, both for production and for quality aspects. Indeed, the end products of these fermentations have unique flavours that are gaining popularity worldwide (Snauwaert *et al.*, 2016). It is clear that, either as hybrids or as pure lineages, non-conventional yeasts have a major role in flavour modulation of specialty beers, members of the genus *Brettanomyces* being those with greater impact in the beer industry (Vanderhaegen *et al.*, 2003a).

Belgian beers are peculiar in several respects. Unique herbs, spices, and fruits are often used, as well as pale or dark candy sugars, caramelized or aromatic malts, and honey. Also, the use of wooden vessels and the practice of bottle conditioning are typical of Belgian-style brewing. Most top-fermented Belgian specialty beers experience a secondary fermentation in the bottle (van Landschoot *et al.*, 2005). To achieve bottle conditioning, mature beer is inoculated with yeast (not necessary the same used for primary fermentation) and a fermentable sugar is added. As a consequence, a re-fermentation takes place, normally lasting less than 2 weeks. The secondary fermentation in the bottle results in a fully saturated beer with an enriched flavour perception and prolonged flavour stability. Phenolic compounds such as 4VG and 4VP, and in some cases 4EG and 4EP, give Belgian-style beers their unique flavour complexity, dominated by spicy clove-like and medicinal aromas (Vanderhaegen *et al.*, 2003b). These have been historically catalogued as POF. As indicated previously (see '*Saccharomyces cerevisiae*'), POF are atypical for most beers, though they are considered a trait marker of Belgian beers such as Trappist and abbey beers, and witbier (brewed with unmalted wheat), as well as German weizen (brewed with malted wheat) and rauchbier (brewed with smoked malts) (Coghe *et al.*, 2004). Several *S. kudriavzevii* × *S. cerevisiae* strains were found in Trappist beer with POF (Gonzalez *et al.*, 2008), thus questioning the belief that all POF-positive yeasts used in brewing were *S. cerevisiae*. Whether the POF character of these hybrids comes from both parents or only from one of them remains unclear, given the lack of complete genome sequences or gene expression studies of such hybrids.

Trappist breweries are run by a monastic order of Cistercian monks, the Trappists. Trappist monks are specialized in the production Belgian-style beers, particularly the more complex ones. Currently, there are twelve recognized and certified Trappist breweries in the world, most of them in Belgium. The designation 'abbey beers' (*Bières d'Abbaye* or *Abdijbier*) was originally applied to any monastic or monastic-style beer, although it is now used for products similar in style or presentation to Trappist beers. Besides the use of *S. cerevisiae* × *S. kudriavzevii* hybrids, it is known that in the production of Orval, a Trappist beer, a strain of *B. bruxellensis* is intentionally used together with *Saccharomyces* spp. for bioflavouring (Vanderhaegen *et al.*, 2003a). Moreover, many breweries mimicking Trappist beers (mostly US craft breweries) use one or more *Brettanomyces* strains in their fermentation process.

Another special type of Belgian beer that employs non-conventional yeasts is lambic beer, generally known as acidic or sour beer. This beer is characterized by a spontaneous fermentation process conduced by an autochthonous microbiota and involving a succession of several bacteria and yeasts (Spitaels *et al.*, 2014, 2015a; see Chapter 7). Gueuze is another type of acidic beer prepared by mixing young and old lambic beers that, for ageing, are re-fermented spontaneously in bottle (Spitaels *et al.*, 2015b). In the American craft-brewing sector, American coolship ales mimic some aspects of the lambic beer production method, although beer is fermented from a single strain and growth of lactic acid bacteria is not encouraged. Bokulich *et al.* (2012) showed there is a predictable ecological microbial succession for these types of spontaneous fermentations although subtle differences at the species-level composition of the various communities were detected. In lambic beer, microbial growth starts during the overnight cooling of the cooked wort in a shallow open vessel, called the cooling tun or coolship and the spontaneous fermentation lasts for 13 years due to a prolonged maturation process (de Keersmaecker, 1996). Lambic brewing is traditionally initiated during the colder months of the year (October to March), since enough coldness is needed to lower the wort temperature to about 20°C in a single night. After this, the wort, inoculated through contact with the air of the brewery, is transferred into wooden casks that are stored at 15–20°C. Lambic beer is fermented and matured in the same casks. Normally, a dominance of *Enterobacteriaceae* is observed in the first month. These bacteria are replaced in the second month by *Pediococcus damnosus* and *Saccharomyces* spp., the latter being replaced after 6 months of fermentation by *B. bruxellensis* (Spitaels *et al.*, 2014). The end product is a non-carbonated, sour beer that mainly serves as a base for gueuze or fruit lambic beers. Lambic sour beers are among the oldest types of beers still brewed and are traditionally produced in or near the Senne river valley, an area near Brussels, Belgium. Whereas a diverse range of yeasts are present in lambic beers, *B. bruxellensis* is the most dominant species and exerts a significant impact on the aroma (Vanoevelen *et al.*, 1977).

Besides the most relevant species for brewing, *S. cerevisiae* and *S. pastorianus*, and those already mentioned in previous sections such as *Brettanomyces* and *S. kudriavzevii*, additional non-conventional yeast species have received increased interest. For example, *Saccharomycodes ludwigii* has been evaluated for the production of low- or non-alcoholic beer types (de Francesco *et al.*, 2015), *Torulaspora delbrueckii* for the production of beer with a special fruity flavour (Canonico *et al.*, 2016; Michel *et al.*, 2016), and *Schizosaccharomyces pombe* for the fermentation of some indigenous African beer styles (Jespersen, 2003). A modern trend in the brewing industry concerns the development of new varieties of beer with novel flavours. Therefore, the use of less-conventional strains and species and the bioprospection of wild yeasts from nature will certainly be intensified in order to add interesting candidates to the pool of brewing yeasts.

Typing and differentiation

As discussed above, brewing strains can be regarded as the main driving force in the production of beer. In addition to the other main raw materials (e.g. barley malt, hops, and water), yeast determines the flavour and character of beer. For example, wort pitched and fermented with three different brewing yeast strains will result in three different beers. Therefore, depending on the degree of strain-related genetic and phenotypic attributes, beer profiles can vary considerably. A practical way of comparing and differentiating strains is to perform standardized small-scale pilot fermentations. These can be followed up by sensorial and chemical analyses, which can be tailored according to the specific aim of yeast performance, optimization, aroma, and flavour characteristics of the final beer. General screening programs of brewing yeast strains usually follow a much broader approach, with a logical first step aiming at species identification, followed by the implementation of a DNA-based typing method that generates a strain fingerprint appropriate for strain discrimination. Afterwards or in parallel, a broad variety of phenotypic methods can be applied to characterize strain-specific fermentation behaviour and to highlight strain-specific key metabolic components. Nowadays, there is a tendency to use high throughput systems (Steensels *et al.*, 2014) and miniature platforms (e. g. multiwell plate systems) to increase screening efficiency (Greetham, 2014). Those systems can be used to quickly generate an overall picture of yeast strain performance and to identify important strain-specific metabolic pathways and key products. Selected phenotypic and DNA-based methods for yeast species and strain discrimination are listed in Table 4.3 and are evaluated according to their practical importance for brewing.

Phenotypic methods

As mentioned above, a historic, traditional, and valuable approach to characterize brewing strains consists of making a small-scale pilot fermentation with subsequent beer tasting. This phenotypic approach gives indications about yeast contribution to beer flavour and the fermentation performance under the specified conditions. It also can be combined with chemical measurements over defined time intervals to record kinetic profiles. In recent years, many new microbiological, biochemical, chemical, and optical methods have been developed to characterize yeast phenotypes during the different stages of the brewing process. Basic microbiological tests such as the 37°C incubation method or the Melibiose-Agar Test were implemented to easily differentiate *S. pastorianus* lager strains from *S. cerevisiae* strains (Röcken and Schulte, 1986; Back, 1987; 1994; 2006; Jespersen and Jakobsen, 1996; Hutzler, 2010). Other agar plate or cultivation broth-based tests, such as the POF test (Table 4.3), were designed to interrogate specific metabolic key activities. As already mentioned, for some German and Belgian beer types, POFs are desired flavour components and POF+ or POF– phenotypes can be determined and used to differentiate brewing strains. The determination of other phenotypic characteristics can be used to generate quantitative differences between strains, such as flocculation tests or measurement of fermentation speed. Other phenotypic key characteristics are the ability to ferment maltotriose and to completely ferment maltose (Gibson and Liti, 2015; Vidgren *et al.*, 2011; Hutzler *et al.*, 2015; Schneiderbanger *et al.*, 2013). Hence, the specific sugar utilization pattern of each strain determines the final degree of attenuation of the beer and therefore the final alcohol concentration, which in turn contributes to mouthfeel and sweetness besides influencing the complete flavour matrix.

Flocculation is also a crucial phenotypic characteristic that determines the yeast handling strategy for serial re-pitching (Verstrepen *et al.*, 2003; Holle *et al.*, 2012; Soares, 2011). Strong flocculating yeast strains can be cropped quite quickly from the bottom of a cylindroconical fermentation tank in order to ensure high yeast vitality. Low flocculating strains, so-called 'dusty' strains, stay longer in suspension and often require other cropping strategies (e.g. later cropping, enforced cooling). Strong top-fermenting strains can even be cropped from special open tanks from above (manually or via a slide).

The strain-dependent requirements of oxygen, nitrogen sources, and zinc-ions are also crucial factors that determine yeast vitality and the 'brewing phenotype'. Most data for these physiological requirements in the literature refer to specific *S. pastorianus* lager brewing strains. From practical experience, we know that there are pronounced strain- dependent concentration differences in the

Table 4.3 Selected phenotypic and DNA-based methods for species and strain discrimination (+, high importance for brewing strains; +/–, medium importance for brewing strains or single studies; –, low importance for brewing strains)

Method	Relevance
Phenotypic methods	+
Growth and colony appearance on specific culture media (e. g. Melibiose-Agar, WLN-Agar)	+
Strong maltose fermentation at 6°C	+/–
Sugar fermentation profiling (e.g. API 50 C AUX)	+
Maltose fermentation at 28°C	+
Maltotriose fermentation 28°C	+
Fermentation performance utilizing high gravity worts (osmotic stress)	+
Growth at 37°C	+
Decarboxylation of coumaric acid and production of 4-vinylphenol (POF)	+/–
Yeast viability and vitality methods	
Flocculation behaviour	+
Decarboxylation of ferulic acid and production of 4-vinylguaiacol (POF)	+
Decarboxylation of cinnamic acid and production of styrene (POF)	+
Amino acid uptake pattern	+/–
Ethanol production	+
Ethanol resistance	+
Screening of fermentation by-products after fermentation in standardized wort	+
Head space SPME for ester production	+/–
Screening of fatty acids after fermentation in standardized wort	+
Screening for flocculation	+
Fermentation kinetics in standardized wort analysing cell concentration, pH and ethanol	+
Sensorial beer analysis after fermentation in standardized wort	+
Propagation characteristics under standardized aeration in standardized wort (generation times during propagation)	+
Total fatty acids analysis (FAME = determination of fatty acid methyl ester compounds)	–
Protein fingerprinting (e.g. 2D protein map)	–
MALDI-TOF and other mass-spectrometry-based methods (protein fingerprint)	+/–
FTIR-spectroscopy (chemotaxonomic fingerprint)	+/–
Phenotype Microarray platforms (e. g. BIOLOG system)	+/–
Flow cytometry and cell sorting	+/–

Table 4.3 Continued

Method	Relevance
DNA-based methods	
Karyotyping	+
Standard PCR (specific primers and targets)	+
rRNA gene sequencing	+
Sequencing of specific genes/house-keeping genes	+
Sequencing of mtDNA	+/–
PCR-RFLP of the 5.8s internal transcribed spacer region and other genes	+/–
RFLP mtDNA	+/–
Fluorescence/chemoluminescence *in situ* hybridization (FisH/CisH)	–
Real-time PCR (specific primers and targets)	+
PCR-DGGE, PCR TGGE	+/–
PCR-DHPLC	+/–
RAPD-PCR	+/–
SAPD-PCR	+/–
Microsatellite PCR	+/–
AFLP-PCR	–
δ-Sequence PCR	+/–
Next generation sequencing (NGS)-based methods (whole genome sequencing)	+
RNA-based transcriptome analysis (miccroarray, *de novo* approaches)	+/–

requirements for essential substrates, especially within *S. cerevisiae* ale, kölsch, alt, Bavarian wheat beer, and Belgian strains.

Aside from the above-mentioned crucial key phenotypic traits and correspondent methods to analyse them, there is an additional broad variety of phenotypic methods aimed at further understanding strain performance at a certain time point during the production process. These include ester profiling, aroma component profiling, profiling of main fermentation products and of by-products, protein fingerprint typing/profiling, viability and vitality methods, fatty acid profiling, and others (Table 4.3). Based on a broad phenotypic understanding of brewing strains over fermentation, maturation, lagering, and also the understanding of the effects of yeast storage, yeast revitalization, and yeast re-pitching, the most important aspects of yeast performance and handling can be optimized according to specific production aims, e.g. faster process, enhancement of specific aroma compounds, enhancement of compounds that increase flavour stability, increased mouthfeel, and other goals. Presently, knowledge about the optimization of strain-dependent processes is only available for a few 'high-performance' *S. pastorianus* lager strains like TUM 34/70 and for some *S. cerevisiae* strains that ferment special beer types, like TUM 68 used for Bavarian wheat beer (Hutzler *et al.*, 2015; Annemueller *et al.*, 2011). Nowadays, a quick optimization of poorly characterized strains can be achieved using a deliberately designed scheme employing phenotypic methods (Table 4.3). Additionally, phenotypic methods can be implemented to control quality and performance of a specific yeast strain.

DNA-based methods

Control of species and strain identity of industrial yeasts is of crucial importance in order to maintain beer quality. As discussed above, the current gold standard for species identification is sequence analysis of partial sequence fragments from rRNA gene domains, such as the D1/D2 domains of 26S rRNA genes and internal transcribed spacer regions (ITS) (Kurtzman, 2015; Kurtzman *et al.*, 2011; Kurtzman and Robnett, 1998, 2003). The resulting sequences can be compared to sequences of public or individually tailored databases (e.g. NCBI, EBI, Yeast IP) by using bioinformatics search and identification tools (e.g. BLAST and Yeast IP identification tool) (Weiss *et al.*, 2013; Fernandez-Espinar *et al.*, 2006). Also, other genes can be used for identification purposes (e.g. *ACT1*, *HIS4*, *RBP1*, *RPB2*, mt*COXII*) (Weiss *et al.*, 2013; Kopecká *et al.*, 2016). The hybrid nature of lager yeasts must also been taken into consideration. As already mentioned, these yeasts can have two distinct rRNA gene sequences, each corresponding to one of the parental species. This is further complicated because there are two main lager strain types (see section '*Saccharomyces eubayanus* and *Saccharomyces bayanus*'), each with distinct relative proportions of the two subgenomes (Hutzler *et al.*, 2015; Brandl *et al.*, 2005), which could lead to the preferential amplification of one of the subgenomes in standard Sanger sequencing. For such cases, the use of specific primer pair combinations can be considered. Moreover, these or other kinds of hybrids can be adequately dealt with using NGS approaches.

Real-time PCR systems for qualitative identification of the most used brewing yeast species are shown in Table 4.4. Such a system was developed for *S. eubayanus* because of its recent use in brewing trials (see '*Saccharomyces eubayanus* and *Saccharomyces bayanus*'). The measurement of the real-time PCR matrix in Table 4.4 enables a quick and easy identification of brewing strains at the species level and can also reveal hybrids. As mentioned above, after species identification it is important to confirm strain identity.

In addition, monitoring the genetic purity of a strain is important for beer flavour consistency. Karyotyping, i.e. generating a chromosomal pattern, is regarded as the reference method for strain typing. However, PCR-based typing techniques are less laborious and easier to set up, particularly for brewing microbiology laboratories. Table 4.3 shows some of the relevant strain typing methods. It has been mentioned already that *S. cerevisiae* strains are genetically more heterogeneous than *S. pastorianus* lager strains. For the differentiation of strains of *S. cerevisiae*, one PCR-based typing method is sufficient proof of strain identity, in most cases (Hutzler *et al.*, 2015). For closely related *S. cerevisiae* strains, for example from the same beer type, one typing method often is not enough, as in the case of discrimination between Kölsch beer and Alt beer strains. It should be noted that strain evaluation methods should desirably be able to discriminate different performance types, because this is a common problem encountered by brewers. Fig. 4.3 shows a fingerprint pattern of PCR-products of a partial sequence of the IGS2 intergenic spacer region (IGS2-r314 system), with subsequent capillary electrophoresis and phylogenetic analysis. Note that the *S. cerevisiae* Kölsch strain TUM 177 and *S. cerevisiae* Alt strain TUM 184 have very similar banding patterns. Additionally, the two *S. pastorianus* lager yeast strains TUM 34/70 and TUM 194 also show similar patterns. For the discrimination of strains within the Kölsch/Alt and the lager beer groups, at least one further typing method is recommended. The wheat beer yeast and the ale yeast strains have unique patterns in Fig. 4.3. In accordance with these results, van Zandycke *et al.* (2007) suggested that for most lager strains at least two typing techniques are needed. There are, however, rare cases where one strain can be typed by a single method. Kopecká *et al.* used a combination of RAPD (primer 21), RFLP mtDNA, karyotyping, and other methods on a brewing yeast strain set (Kopecká *et al.*, 2016). A combination of the methods enabled a discrimination of most strains, in which RAPD (primer 21) showed a high discrimination level (Kopecká *et al.*, 2016). Hutzler *et al.* (2010) used a PCR-DHPLC (denaturing HPLC) approach that is similar to DGGE or TGGE. DHPLC employed a partial sequence of the IGS2 intergenic spacer to differentiate lager yeast strains. This method could be adjusted to separate PCR products under temperature-dependent denaturing conditions, with a higher resolution than electrophoresis approaches (Hutzler *et al.*, 2010). The selection of the appropriate type of strain differentiation tools depends mostly on the strains that are used in a specific brewery and should be tested with the whole strain set. The combination of methods should be able to discriminate all yeast strains within the strain set. Nowadays, besides classical sequencing approaches aiming at genes or DNA-spacers with strain specific polymorphisms, NGS approaches based on whole genome sequencing may also be useful for strain differentiation and strain quality control. Such techniques are also the basis for the design of strain-specific PCR or real-time PCR systems that target strain-specific

Table 4.4 Qualitative real-time PCR systems for the differentiation of brewing yeast species

		Qualitative real-time PCR detection specificity									
Real-time PCR systems, primer and probe sequences (5′–3′)	Reference	*B. ano.*	*B. brux.*	*S. cer.*	*S. c. dia.*	*S. eub.*	*S. kud.*	*S. past.*	*Smy. lud.*	*Sch. pom*	*Tor. del.*
Da-f ATTATAGGGAGAAATCCATATAAAACACG Da-r CACATTAAGTATCGCAATTCGCTG Y58 FAM- AACGGATCTCTTGGTTCTCGCATCGAT-BHQ1	Brandl (2006), Brandl *et al*. (2005)	+	–	–	–	–	–	–	–	–	–
Db-f TGCAGACACGTGGATAAGCAAG Db-r CACATTAAGTATCGCAATTCGCTG Y58 FAM- AACGGATCTCTTGGTTCTCGCATCGAT-BHQ1	Brandl (2006), Brandl *et al*. (2005)	–	+	–	–	–	–	–	–	–	–
Sbp-f CTTGCTATTCCAAACAGTGAGACT Sbp-r1 TTGTTACCTCTGGGCGTCGA Sbp-r2 GTTTGTTACCTCTGGGCTCG Sbp ACTTTTGCAACTTTTTCTTTGGGTTTCGAGCA	Josepa *et al*. (2000), Brandl (2006), Brandl *et al*. (2005)	–	–	–	–	+	–	+	–	–	–
Sc-f CAAACGGTGAGAGATTTCTGTGC Sc-r GATAAAATTGTTTGTGTTTGTTACCTCTG Scer FAM-ACACTGTGGAATTTTCATATCTTTGCAACTT-BHQ1	Josepa *et al*. (2000), Brandl (2006), Brandl *et al*. (2005)	–	–	+	+	–	–	+	–	–	–
Sc-GRC-f CACATCACTACGAGATGCATATGCA Sc-GRC-r GCCAGTATTTTGAATGTTCTCAGTTG Sc-GRC FAM-TCCAGCCCATAGTCTGAACCACACCTTATCT-BHQ1	Hutzler (2010)	–	–	+	+	–	–	+	–	–	–
TF-f TTCGTTGTAACAGCTGCTGATGT TF-r ACCAGGAGTAGCATCAACTTTAATACC TF-MGB FAM-ATGATTTTGCTATCCCAAGTT-BHQ1 (MGB probe)	Hutzler (2010)	–	–	+	+	–	–	–	–	–	–
SCF1 GGACTCTGGACATGCAAGAT SCR1 ATACCCTTCTTAACACCTGGC SC FAM-CCCTTCAGAGCGTTTTCTCTAAATTGATAC-BHQ1	Salinas *et al*. (2009)	–	–	+	+	–	–	–	–	–	–
Sd-f TTCCAACTGCACTAGTTCCTAGAGG Sd-r GAGCTGAATGGAGTTGAAGATGG Sdia FAM-CCTCCTCTAGCAACATCACTTCCTCCG-BHQ1	Scherer, (2002),, Brandl (2006), Brandl *et al*. (2005)	–	–	–	+	–	–	–	–	–	–
Seub F3 GTCCCTGTACCAATTTAATATTGCGC Seub R2 TTTCACATCTCTTAGTCTTTTCCAGACG Seub-probe FAM-CGAATTGTTTGGTTATATTGTTGGTGTGATTTTC TTTG-BHQ1	Primers: Pengelly and Wheals (2013) Probe: Hutzler *et al*. (2016)	–	–	–	–	+	–	+/–	–	–	–
Sk-f TCCTTACCTTATTCATCATATTCTCCAC Sk-r CGATATTTGGTAAGGGGAGGTAGA Skud FAM-TGCTATTACTTTTGCTTTTTCACTCACCACACCCT-BHQ1	Hutzler (2010)	–	–	–	–	–	+	–	–	–	–
BF300E CTCCTTGGCTTGTCGAA BF300M GGTTGTTGCTGAAGTTGAGA BF300 FAM-TGCTCCACATTTGATCAGCGCCA-BHQ1	Brandl (2006)	–	–	–	–	+	–	+	–	–	–

Table 4.4 Continued

Real-time PCR systems, primer and probe sequences (5′–3′)	Reference	Qualitative real-time PCR detection specificity									
		B. ano.	*B. brux.*	*S. cer.*	*S. c. dia.*	*S. eub.*	*S. kud.*	*S. past.*	*Smy. lud.*	*Sch. pom*	*Tor. del.*
BF-LRE-f ACTCGACATTCAACTACAAGAGTAAAATTT BF-LRE-r TCTCCGGCATATCCTTCATCA BF-LRE FAM-ATCTCTACCGTTTTCGGTCACCGGC-BHQ1	Hutzler (2010)	–	–	–	–	+	–	+	–	–	–
Sl-f GACGAGCAATTGTTCAAGGGTC Sl-r ACTTATCGCAATTCGCTACGTTC Y58 FAM-AACGGATCTCTTGGTTCTCGCATCGAT-BHQ1	Brandl (2006), Brandl *et al.* (2005)	–	–	–	–	–	–	–	+	–	–
Sipo-f AAAAGTGTTAGAAAGAGAAAGACGAAAAAA Sipo-r AAAAGTAATGATATGCTTGGCATGC Sipo FAM-ACGAGTGGATGATTTTTGTTTGGTGTGTTTC-BHQ1	Hutzler (2010), Brandl (2006), Brandl *et al.* (2005)	–	–	–	–	–	–	–	–	+	–
Td-f AGATACGTCTTGTGCGTGCTTC Td-r GCATTTCGCTGCGTTCTT Y58 FAM-AACGGATCTCTTGGTTCTCGCATCGAT-BHQ1	Hutzler (2010)	–	–	–	–	–	–	–	–	–	+

B. ano. = *Brettanomyces/Dekkera anomalus/anomala*; *B. brux.* = *Brettanomyces/Dekkera bruxellensis*; *S. cer.* = *Saccharomyces cerevisiae*; *S. c. dia* = *Saccharomyces cerevisiae* var. *diasticus*; *S. eub.* = *Saccharomyces eubayanus*; *S. kud.* = *Saccharomyces kudriavzevii*: *S. past.* = *Saccharomyces pastorianus* (lager strains); *Smy. lud.* = *Saccharomyceodes ludwigii*; *Sch. pom.* = *Schizosaccharomyces pombe*; *Tor. del.* = *Torulaspora delbrueckii*.

FAM, 6-carboxyfluorescein; BHQ1, black hole quencher 1; MGB, minor groove binder probe.

+, positive; –, negative; +/–, positive and negative strains.

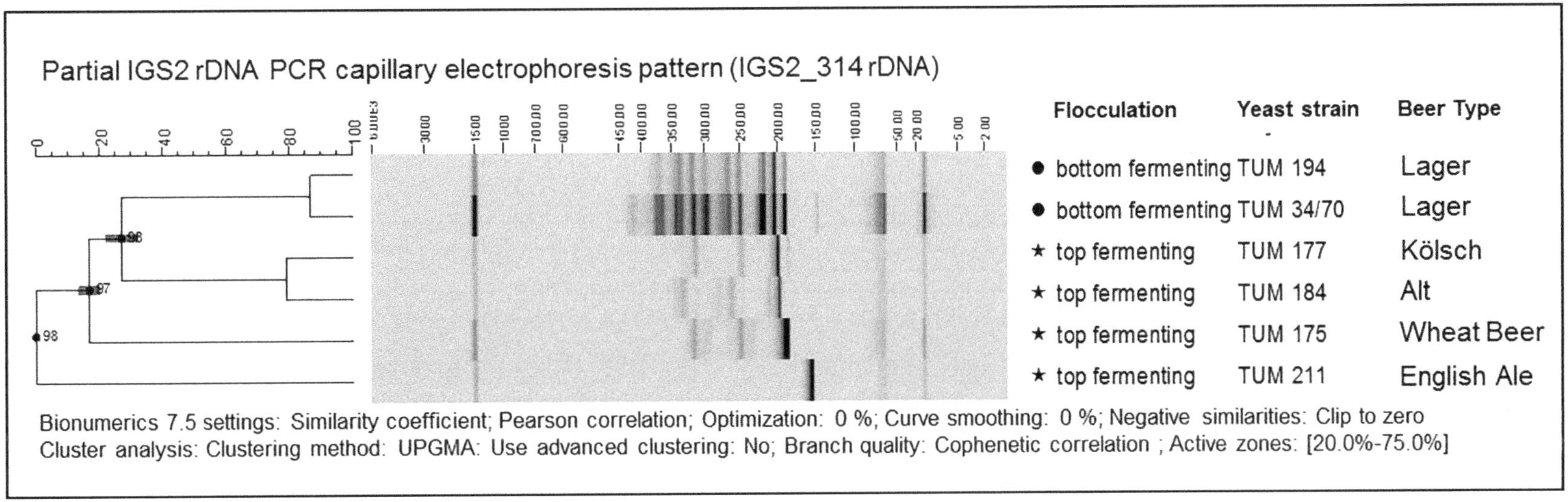

Figure 4.3 Capillary electrophoresis patterns of IGS2 intergenic spacer sequences for brewing strains.

polymorphisms. A future development might consist in a specific real-time PCR system for each production strain within a brewery so that the identity of production strains and of potential cross-contaminations can be controlled easily over the whole brewing process.

Acknowledgments

This work was supported by FCT Portugal grant UID/Multi/04378/2013 (JPS), ANPCyT PICT3677 and CONICET PIP0392 Argentina (DL). We are grateful to Steve Wagner (Central Washington State University) for critical reading of the manuscript.

References

Albertin, W., Panfili, A., Miot-Sertier, C., Goulielmakis, A., Delcamp, A., Salin, F., Lonvaud-Funel, A., Curtin, C., and Masneuf-Pomarede, I. (2014). Development of microsatellite markers for the rapid and reliable genotyping of *Brettanomyces bruxellensis* at strain level. Food Microbiol. *42*, 188–195. http://dx.doi.org/10.1016/j.fm.2014.03.012

Almeida, P., Barbosa, R., Zalar, P., Imanishi, Y., Shimizu, K., Turchetti, B., Legras, J.L., Serra, M., Dequin, S., Couloux, A., *et al.* (2015). A population genomics insight into the Mediterranean origins of wine yeast domestication. Mol. Ecol. *24*, 5412–5427. http://dx.doi.org/10.1111/mec.13341

Annemueller, G., Manger, H.-J., and Lietz, P. (2011). The Yeast in the Brewery (Berlin, VLB: Berlin).

Arroyo-López, F.N., Orlić, S., Querol, A., and Barrio, E. (2009). Effects of temperature, pH and sugar concentration on the growth parameters of *Saccharomyces cerevisiae, S. kudriavzevii* and their interspecific hybrid. Int. J. Food Microbiol. *131*, 120–127. http://dx.doi.org/10.1016/j.ijfoodmicro.2009.01.035

Arroyo-López, F.N., Pérez-Torrado, R., Querol, A., and Barrio, E. (2010). Modulation of the glycerol and ethanol syntheses in the yeast *Saccharomyces kudriavzevii* differs from that exhibited by *Saccharomyces cerevisiae* and their hybrid. Food Microbiol. 628–637.

Arroyo-López, F.N., Pérez-Través, L., Querol, A., and Barrio, E. (2011). Exclusion of *Saccharomyces kudriavzevii* from a wine model system mediated by *Saccharomyces cerevisiae*. Yeast *28*, 423–435.

Back, W. (1987). Nachweis und Identifizierung von Fremdhefen in der Brauerei. Brauwelt *127*, 735-737.

Back, W. (1994). Farbatlas und Handbuch der Getränkemikrobiologie (Nürnberg, Hans Carl Fachverlag).

Back, W. (2006). Colour Atlas and Handbook of Beverage Biology (Nürnberg, Hans Carl Fachverlag).

Baker, E., Wang, B., Bellora, N., Peris, D., Hulfachor, A.B., Koshalek, J.A., Adams, M., Libkind, D., and Hittinger, C.T. (2015). The genome sequence of *Saccharomyces eubayanus* and the domestication of lager-brewing yeasts. Mol. Biol. Evol. *32*, 2818–2831. http://dx.doi.org/10.1093/molbev/msv168

Barbosa, R., Almeida, P., Safar, S.V., Santos, R.O., Morais, P.B., Nielly-Thibault, L., Leducq, J.B., Landry, C.R., Gonçalves, P., Rosa, C.A., *et al.* (2016). Evidence of Natural Hybridization in Brazilian Wild Lineages of *Saccharomyces cerevisiae*. Genome. Biol. Evol. *8*, 317–329. http://dx.doi.org/10.1093/gbe/evv263

Barnett, J.A. (2004). A history of research on yeasts 8: taxonomy. Yeast *21*, 1141-1493.

Barnett, J.A., and Robinow, C.F. (2002). A history of research on yeasts 4: citology part II, 1950-1990. Yeast *19*, 745-772.

Barnett, J.A., Payne, R.W., and Yarrow, D. (2000). Yeasts: Characteristics and Identification, 3rd edn (Cambridge University Press: Cambridge).

Belloch, C., Orlic, S., Barrio, E., and Querol, A. (2008). Fermentative stress adaptation of hybrids within the *Saccharomyces sensu stricto* complex. Int. J. Food Microbiol. *122*, 188–195. http://dx.doi.org/10.1016/j.ijfoodmicro.2007.11.083

Belloch, C., Pérez-Torrado, R., González, S.S., Pérez-Ortín, J.E., García-Martínez, J., Querol, A., and Barrio, E. (2009). Chimeric genomes of natural hybrids of *Saccharomyces cerevisiae* and *Saccharomyces kudriavzevii*. Appl. Environ. Microbiol. *75*, 2534–2544. http://dx.doi.org/10.1128/AEM.02282-08

Bing, J., Han, P.J., Liu, W.Q., Wang, Q.M., and Bai, F.Y. (2014). Evidence for a Far East Asian origin of lager beer yeast. Curr. Biol. *24*, R380–381. http://dx.doi.org/10.1016/j.cub.2014.04.031

Bokulich, N.A., and Mills, D.A. (2012). Next-generation approaches to the microbial ecology of food fermentations. BMB. Rep. *45*, 377–389.

Bokulich, N.A., and Bamforth, C.W. (2013). The microbiology of malting and brewing. Microbiol. Mol. Biol. Rev. *77*, 157–172. http://dx.doi.org/10.1128/MMBR.00060-12

Bokulich, N.A., Bamforth, C.W., and Mills, D.A. (2012). Brewhouse-resident microbiota are responsible for multi-stage fermentation of American coolship ale. PLOS ONE *7*, e35507. http://dx.doi.org/10.1371/journal.pone.0035507

Bond, U. (2009). Chapter 6: The genomes of lager yeasts. Adv. Appl. Microbiol. *69*, 159–182. http://dx.doi.org/10.1016/S0065-2164(09)69006-7

Borneman, A.R., Desany, B.A., Riches, D., Affourtit, J.P., Forgan, A.H., Pretorius, I.S., Egholm, M., and Chambers, P.J. (2011). Whole-genome comparison reveals novel genetic elements that characterize the genome of industrial strains of *Saccharomyces cerevisiae*. PLOS Genet. *7*, e1001287. http://dx.doi.org/10.1371/journal.pgen.1001287

Borneman, A.R., Desany, B.A., Riches, D., Affourtit, J.P., Forgan, A.H., Pretorius, I.S., Egholm, M., and Chambers, P.J. (2012). The genome sequence of the wine yeast VIN7 reveals an allotriploid hybrid genome with *Saccharomyces cerevisiae* and *Saccharomyces kudriavzevii* origins. FEMS Yeast Res. *12*, 88–96. http://dx.doi.org/10.1111/j.1567-1364.2011.00773.x

Boulton, C., and Quain, D. 2001. Brewing Yeast and Fermentation (Blackwell Science, U.K).

Brandl, A. (2006). Entwicklung und Optimierung von PCR-Methoden zur Detektion und Identifizierung von brauereirelevanten Mikroorganismen zur

Routine-Anwendung in Brauereien. Dissertation, TU München.

Brandl, A., Hutzler, M., and Geiger, E. (2005). Optimisation of brewing yeast differentiation and wild yeast identification by Real-Time PCR. Proc. 30th EBC Congr. Prague. (Fachverlag Hans Carl [CD-ROM], Nürnberg).

Canonico, L., Agarbati, A., Comitini, F., and Ciani, M. (2016). *Torulaspora delbrueckii* in the brewing process: A new approach to enhance bioflavour and to reduce ethanol content. Food Microbiol. *56*, 45–51. http://dx.doi.org/10.1016/j.fm.2015.12.005

Casey, G.P. (1990). Yeast selection in brewing. In Yeast Strain Selection, C.J. Panchal, ed., (Marcel Dekker, New York, NY), pp. 65-112.

Casey, G.P. (1996). Practical applications of pulsed field electrophoresis and yeast chromosome fingerprinting in brewing QA and R&D. Tech. Q. Master Brew. Assoc. Am. *33*, 1-10.

Cavalieri, D., McGovern, P.E., Hartl, D.L., Mortimer, R., and Polsinelli, M. (2003). Evidence for *S. cerevisiae* fermentation in ancient wine. J. Mol. Evol. *57* (Suppl. 1), S226–32. http://dx.doi.org/10.1007/s00239-003-0031-2

Chatonnet, P., Dubourdieu, D., Boidron, J.N., and Pons, M. (1992). The origin of ethylphenols in wines. J. Sci. Food Agric. *60*, 165–178.

Cliften, P.F., Hillier, L.W., Fulton, L., Graves, T., Miner, T., Gish, W.R., Waterston, R.H., and Johnston, M. (2001). Surveying Saccharomyces genomes to identify functional elements by comparative DNA sequence analysis. Genome Res. *11*, 1175-1186.

Coelho, M.A., Gonçalves, C., Sampaio, J.P., and Gonçalves, P. (2013). Extensive intra-kingdom horizontal gene transfer converging on a fungal fructose transporter gene. PLOS Genet. *9*, e1003587. http://dx.doi.org/10.1371/journal.pgen.1003587

Coghe, S., Benoot, K., Delvaux, F., Vanderhaegen, B., and Delvaux, F.R. (2004). Ferulic acid release and 4-vinylguaiacol formation during brewing and fermentation: Indications for feruloyl esterase activity in *Saccharomyces cerevisiae*. J. Agric. Food Chem. *52*, 602-608.

Crauwels, S., Zhu, B., Steensels, J., Busschaert, P., De Samblanx, G., Marchal, K., Willems, K.A., Verstrepen, K.J., and Lievens, B. (2014). Assessing genetic diversity among *Brettanomyces* yeasts by DNA fingerprinting and whole-genome sequencing. Appl. Environ. Microbiol. *80*, 4398–4413. http://dx.doi.org/10.1128/AEM.00601-14

Crauwels, S., van Assche, A., de Jonge, R., Borneman, A.R., Verreth, C., Troels, P., de Samblanx, G., Marchal, K., van de Peer, Y., Willems, K.A., Verstrepen, K.J., Curtin, C.D., and Lievens, B. (2015). Comparative phenomics and targeted use of genomics reveals variation in carbon and nitrogen assimilation among different Brettanomyces bruxellensis strains. Appl. Microbiol. Biotechnol. *99*, 9123-9134.

Custers, M.T.J. (1940). Onderzoekingen over het gistgeslacht *Brettanomyces*. Thesis, Delft.

de Francesco, G., Turchetti, B., Sileoni, V., Marconi, O., and Perretti, G. (2015). Screening of new strains of *Saccharomycodes ludwigii* and *Zygosaccharomyces rouxii* to produce low-alcohol beer. J. Inst. Brewing *21*, 113-121.

de Keersmaecker, J. (1996). The mystery of lambic beer. Sci. Am. *275*: 74–81.

Diddens, H.A., and Lodder, J. (1942). Die anaskosporogenen Hefen, II. (Halfte. North-Holland, Amsterdam).

Dufour, J.-P., Verstrepen, K., and Derdelinckx, G. (2003). Brewing yeasts. In Yeasts in Food, T. Boekhout and V. Robert, eds. (Hamburg: Behr's Verlag) pp 347-388.

Dunn, B., and Sherlock, G. (2008). Reconstruction of the genome origins and evolution of the hybrid lager yeast *Saccharomyces pastorianus*. Genome. Res. *18*, 1610–1623. http://dx.doi.org/10.1101/gr.076075.108

Dweck, H.K., Ebrahim, S.A., Farhan, A., Hansson, B.S., and Stensmyr, M.C. (2015). Olfactory proxy detection of dietary antioxidants in *Drosophila*. Curr. Biol. *25*, 455–466. http://dx.doi.org/10.1016/j.cub.2014.11.062

Edlin, D.A.N., Narbad, A., Gasson, M.J., Dickinson, J.R., and Lloyd, D. (1998). Purification and characterization of hydrocinnamate decarboxylase from *Brettanomyces anomalus*. Enzyme Microb. Technol. *22*, 232–239.

Erny, C., Raoult, P., Alais, A., Butterlin, G., Delobel, P., Matei-Radoi, F., Casaregola, S., and Legras, J.L. (2012). Ecological success of a group of *Saccharomyces cerevisiae /Saccharomyces kudriavzevii* hybrids in the northern european wine-making environment. Appl. Environ. Microbiol. *78*, 3256–3265.

Fay, J.C., and Benavides, J.A. (2005). Evidence for domesticated and wild populations of *Saccharomyces cerevisiae*. PLOS Genet. *1*, 66–71. http://dx.doi.org/10.1371/journal.pgen.0010005

Fell, J.W., Boekhout, T., Fonseca, A., Scorzetti, G., and Statzell-Tallman, A. (2000). Biodiversity and systematics of basidiomycetous yeasts as determined by large-subunit rDNA D1/D2 domain sequence analysis. Int. J. Syst. Evol. Microbiol. *50*, 1351–1371. http://dx.doi.org/10.1099/00207713-50-3-1351

Fernandez-Espinar, M.T., Martorell, P., de Llanos, R., and Querol, A. (2006). Molecular methods to identify and charcterize yeasts in foods and beverages. In Yeasts in food and beverages, A. Querol and G.H Fleet eds. (New York: Springer Inc.).

Gallone, B., Steensels, J., Prahl, T., Soriaga, L., Saels, V., Herrera-Malaver, B., Merlevede, A., Roncoroni, M., Voordeckers, K., Miraglia, L., *et al.* (2016). Domestication and divergence of *Saccharomyces cerevisiae* beer yeasts. Cell *166*, 1397–1410.e16. http://dx.doi.org/10.1016/j.cell.2016.08.020

Gayevskiy, V., and Goddard, M.R. (2016). *Saccharomyces eubayanus* and *Saccharomyces arboricola* reside in North Island native New Zealand forests. Environ. Microbiol. *18*, 1137–1147. http://dx.doi.org/10.1111/1462-2920.13107

Gibson, B., and Liti, G. (2015). *Saccharomyces pastorianus*: genomic insights inspiring innovation for industry. Yeast *32*, 17-27.

Gibson, B.R., Storgårds, E., Krogerus, K., and Vidgren, V. (2013). Comparative physiology and fermentation performance of Saaz and Frohberg lager yeast strains and the parental species *Saccharomyces eubayanus*. Yeast *30*, 255–266.

Giudici, P., Caggia, C., Pulvirenti, A., and Rainieri, S. (1998). Karyotyping of Saccharomyces strains with different temperature profiles. J. Appl. Microbiol. *84*, 811–819.

Goffeau, A., Barrell, B.G., Bussey, H., Davis, R.W., Dujon, B., Feldmann, H., Galibert, F., Hoheisel, J.D., Jacq, C., Johnston, M., *et al.* (1996). Life with 6000 genes. Science *274*, 546, 563–567.

Gonçalves, P., Valério, E., Correia, C., de Almeida, J.M., and Sampaio, J.P. (2011). Evidence for divergent evolution of growth temperature preference in sympatric *Saccharomyces* species. PLOS ONE *6*, e20739. http://dx.doi.org/10.1371/journal.pone.0020739

Gonçalves, M., Pontes, A., Almeida, P., Barbosa, R., Serra, M., Libkind, D., Hutzler, M., Gonçalves, P., and Sampaio, J.P. (2016). Distinct domestication trajectories in top-fermenting beer yeasts and wine yeasts. Curr. Biol. *26*, 2750–2761.

González, S.S., Barrio, E., Gafner, J., and Querol, A. (2006). Natural hybrids from *Saccharomyces cerevisiae, Saccharomyces bayanus* and *Saccharomyces kudriavzevii* in wine fermentations. FEMS Yeast Res. *6*, 1221–1234.

González, S.S., Barrio, E., and Querol, A. (2008). Molecular characterization of new natural hybrids of *Saccharomyces cerevisiae* and *S. kudriavzevii* in brewing. Appl. Environ. Microbiol. *74*, 2314–2320. http://dx.doi.org/10.1128/AEM.01867-07

Greetham, D. (2014). Phenotype microarray technology and its application in industrial biotechnology. Biotechnol. Lett *36*, 1153–1160. http://dx.doi.org/10.1007/s10529-014-1481-x

Groth, C., Hansen, J., and Piskur, J. (1999). A natural chimeric yeast containing genetic material from three species. Int. J. Syst. Bacteriol. *49*, 1933–1938. http://dx.doi.org/10.1099/00207713-49-4-1933

Hansen, E.C. (1908). Recherches sur la physiologie et la morphologie des ferments alcooliques. XIII. Nouvelles etudes sur les levures de brasserie a fermentation basse. C.R. Trav. Lab. Carlsberg *7*, 179–217.

Hansen, J., and Kielland-Brandt, M.C. (1994). *Saccharomyces carlsbergensis* contains two functional MET2 alleles similar to homologues from *S. cerevisiae* and *S. monacensis*. Gene *140*, 33–40.

Hawksworth, D.L. (2011). A new dawn for the naming of fungi: impacts of decisions made in Melbourne in July 2011 on the future publication and regulation of fungal names. IMA. Fungus. *2*, 155–162. http://dx.doi.org/10.5598/imafungus.2011.02.02.06

Hebly, M., Brickwedde, A., Bolat, I., Driessen, M.R., de Hulster, E.A., van den Broek, M., Pronk, J.T., Geertman, J.M., Daran, J.M., and Daran-Lapujade, P. (2015). *S. cerevisiae* × *S. eubayanus* interspecific hybrid, the best of both worlds and beyond. FEMS Yeast Res. *15*, fov005. http://dx.doi.org/10.1093/femsyr/fov005

Hittinger, C.T., Gonçalves, P., Sampaio, J.P., Dover, J., Johnston, M., and Rokas, A. (2010). Remarkably ancient balanced polymorphisms in a multi-locus gene network. Nature *464*, 54–58. http://dx.doi.org/10.1038/nature08791

Hittinger, C.T. (2013). *Saccharomyces* diversity and evolution: a budding model genus. Trends Genet. *29*, 309–317. http://dx.doi.org/10.1016/j.tig.2013.01.002

Hittinger, C.T., Rokas, A., Bai, F.Y., Boekhout, T., Gonçalves, P., Jeffries, T.W., Kominek, J., Lachance, M.A., Libkind, D., Rosa, C.A., *et al.* (2015). Genomics and the making of yeast biodiversity. Curr. Opin. Genet. Dev. *35*, 100–109. http://dx.doi.org/10.1016/j.gde.2015.10.008

Holle, A.V., Machado, M.D., and Soares, E.V. (2012). Flocculation in ale brewing strains of *Saccharomyces cerevisiae*: re-evaluation of the role of cell surface charge and hydrophobicity. Appl. Microbiol. Biotechnol. *93*, 1221–1229. http://dx.doi.org/10.1007/s00253-011-3502-1

Hornsey, I.S. (2003). A History of Beer and Brewing (Royal Society of Chemistry paperbacks, Cambridge, UK).

Hutzler, M. (2010). Getränkerelevante Hefen - Identifizierung und Differenzierung (Saarbrücken, SVH-Verlag).

Hutzler, M., Geiger, E., and Jacob, F. (2010). Use of PCR-DHPLC (Polymerase Chain Reaction–Denaturing High Performance Liquid Chromatography) for the rapid differentiation of industrial *Saccharomyces pastorianus* and *Saccharomyces cerevisiae* strains. J. Inst. Brew. *116*, 464–474.

Hutzler, M., Koob, J., Riedl, R., Schneiderbanger, H., Mueller-Auffermann, K.,and Jacob, F. (2015). Yeast identification and characterization. In Brewing Microbiology - Managing Microbes, Ensuring Quality and Valorising Waste. 1st ed., A. Hill, ed. (London: Woodhead Publishing).

Hutzler, M., Methner, F.-J., and Jacob, F. (2016). Multi-Screening Method to Identify Potential Novel Brewing Yeast in Environmental Samples. 12th Trends in Brewing. (Ghent: KU Leuven).

Ihmels, J., Bergmann, S., Gerami-Nejad, M., Yanai, I., McClellan, M., Berman, J., and Barkai, N. (2005). Rewiring of the yeast transcriptional network through the evolution of motif usage. Science *309*, 938–940.

James, S.A., Cai, J., Roberts, I.N., and Collins, M.D. (1997). A phylogenetic analysis of the genus *Saccharomyces* based on 18S rRNA gene sequences: description of *Saccharomyces kunashirensis* sp. nov. and *Saccharomyces martiniae* sp. nov. Int. J. Syst. Bacteriol. *47*, 453–460.

Jespersen, L. (2003). Occurrence and taxonomic characteristics of strains of *Saccharomyces cerevisiae* predominant in African indigenous fermented foods and beverages. FEMS Yeast Res. 3,191-200.

Jespersen, L., and Jakobsen, M. (1996). Specific spoilage organisms in breweries and laboratory media for their detection. Int. J. Food Microbiol. *33*, 139–155.

Josepa, S., Guillamon, J.M., and Cano, J. (2000). PCR differentiation of *Saccharomyces cerevisiae* from *Saccharomyces bayanus/Saccharomyces pastorianus* using specific primers. FEMS Microbiol. Lett. *193*, 255–259.

Joseph, C.M.L., Kumar, G., Su, E., and Bisson. L.F. (2007). Adhesion and biofilm production by wine isolates of *Brettanomyces bruxellensis*. Am. J. Enol. Vitic. *58*, 373-378.

Kellis, M., Patterson, N., Endrizzi, M., Birren, B., and Lander, E.S. (2003). Sequencing and comparison of yeast species to identify genes and regulatory elements. Nature *423*, 241–254. http://dx.doi.org/10.1038/nature01644

Kielland-Brandt, M.C., Nilsson-Tillgren, T., Gjermansen, C., Holmberg, S., and Pedersen, M.B. (1995). Genetics of brewing yeasts, In The Yeasts, 2nd ed., vol. 6, A.H. Rose, A.E. Wheals and J.S. Harrison eds. (Academic Press: London) pp. 223–254.

Kopecká, J., Němec, M., and Matoulková, D. (2016). Comparison of DNA-based techniques for differentiation of production strains of ale and lager brewing yeast. J. Appl. Microbiol. *120*, 1561–1573. http://dx.doi.org/10.1111/jam.13116

Kreger-van Rij, N.J.W. (1984). The Yeasts, A Taxonomic Study, 3rd edn. (Elsevier, Amsterdam).

Kreger-van Rij, N.J.W. (1987). Classification of yeasts. In The Yeasts, volume 1, second edition, A.H. Rose and J.S. Harrison, eds (London: Academic Press).

Krogerus, K., Magalhães, F., Vidgren, V., and Gibson, B. (2015). New lager yeast strains generated by interspecific hybridization. J. Ind. Microbiol. Biotechnol. *42*, 769–778. http://dx.doi.org/10.1007/s10295-015-1597-6

Kufferath, H., and Van Laer, M. (1921). Études sur les levures du Lambic. Bull. Soc. Chim. Belgique *30*, 270–276.

Kurtzman, C.P. (1987). Prediction of biological relatedness among yeasts from comparisons of nuclear DNA complementarity. Stud. Mycol. *30*, 459-468.

Kurtzman, C.P. (2003). Phylogenetic circumscription of *Saccharomyces, Kluyveromyces* and other members of the Saccharomycetaceae, and the proposal of the new genera *Lachancea, Nakaseomyces, Naumovia, Vanderwaltozyma* and *Zygotorulaspora*. FEMS Yeast Res. *4*, 233-245.

Kurtzman, C.P. (2015). Identification of food and beverage spoilage yeasts from DNA sequence analyses. Int. J. Food Microbiol. *213*, 71–78. http://dx.doi.org/10.1016/j.ijfoodmicro.2015.05.023

Kurtzman, C.P., and Robnett, C.J. (1998). Identification and phylogeny of ascomycetous yeasts from analysis of nuclear large subunit (26S) ribosomal DNA partial sequences. Antonie Van Leeuwenhoek *73*, 331–371.

Kurtzman, C.P., and Robnett, C.J. (2003). Phylogenetic relationships among yeasts of the 'Saccharomyces complex' determined from multigene sequence analyses. FEMS Yeast Res. *3*, 417–432.

Kurtzman, C.P., and Robnett, C.J. (2013). Description of *Ambrosiozyma oregonensis* sp. nov., and reassignment of *Candida* species of the *Ambrosiozyma* clade to *Ambrosiozyma kashinagacola* f.a., comb. nov., *Ambrosiozyma llanquihuensis* f.a., comb. nov., *Ambrosiozyma maleeae* f.a., comb. nov., *Ambrosiozyma pseudovanderkliftii* f.a., comb. nov., and *Ambrosiozyma vanderkliftii* f.a., comb. nov. Int. J. Syst. Evol. Microbiol. *63*, 3877-3883.

Kurtzman, C.P., Fell, J.W., and Boekhout, T. (2011). The Yeasts, a Taxonomic Study, (Amsterdam, Elsevier).

Legras, J.L., Merdinoglu, D., Cornuet, J.M., and Karst, F. (2007). Bread, beer and wine: *Saccharomyces cerevisiae* diversity reflects human history. Mol. Ecol. *16*, 2091–2102.

Libkind, D., Hittinger, C.T., Valério, E., Gonçalves, C., Dover, J., Johnston, M., Gonçalves, P., and Sampaio, J.P. (2011). Microbe domestication and the identification of the wild genetic stock of lager-brewing yeast. Proc. Natl. Acad. Sci. U.S.A. *108*, 14539–14544. http://dx.doi.org/10.1073/pnas.1105430108

Liti, G., Peruffo, A., James, S.A., Roberts, I.N., and Louis, E.J. (2005). Inferences of evolutionary relationships from a population survey of LTRretrotransposons and telomeric-associated sequences in the *Saccharomyces sensu stricto* complex. Yeast 22, 177–192.

Liti, G., Carter, D.M., Moses, A.M., Warringer, J., Parts, L., James, S.A., Davey, R.P., Roberts, I.N., Burt, A., Koufopanou, V., Tsai, I.J., Bergman, C.M., Bensasson, D., O'Kelly, M.J., van Oudenaarden, A., Barton, D.B., Bailes, E., Nguyen, A.N., Jones, M., Quail, M.A., Goodhead, I., Sims, S., Smith, F., Blomberg, A., Durbin, R., and Louis, E.J. (2009). Population genomics of domestic and wild yeasts. Nature *458*, 337–341.

Liu, X.Z., Wang, Q.M., Göker, M., Groenewald, M., Kachalkin, A.V., Lumbsch, H.T., Millanes, A.M., Wedin, M., Yurkov, A.M., Boekhout, T., *et al.* (2015). Towards an integrated phylogenetic classification of the Tremellomycetes. Stud. Mycol. *81*, 85–147. http://dx.doi.org/10.1016/j.simyco.2015.12.001

Liu, Y., Leigh, J.W., Brinkmann, H., Cushion, M.T., Rodriguez-Ezpeleta, N., Philippe, H., and Lang, B.F. (2009). Phylogenomic analyses support the monophyly of Taphrinomycotina, including *Schizosaccharomyces* fission yeasts. Mol. Biol. Evol. 26, 27-34.

Lodder, J. (1934). Die Anaskosporogenen Hefen, I. Halfte. Verh. K. Ned. Akad. Wet., Afd. Natuurk., Sect. II. *32*, 1-256.

Lodder, J. (1970). The Yeasts, A Taxonomic Study, 2nd edn. (North-Holland, Amsterdam).

Lodder, J., and Kreger-van Rij, N.J.W. (1952). The Yeasts, A Taxonomic Study (North-Holland, Amsterdam).

Lopandic, K., Gangl, H., Wallner, E., Tscheik, G., Leitner, G., Querol, A., Borth, N., Breitenbach, M., Prillinger, H., and Tiefenbrunner, W. (2007). Genetically different wine yeasts isolated from Austrian vine-growing regions influence wine aroma differently and contain putative hybrids between *Saccharomyces cerevisiae* and *Saccharomyces kudriavzevii*. FEMS Yeast Res. *7*, 953-965.

Lopes, C.A., Barrio, E., and Querol, A. (2010). Natural hybrids of *S. cerevisiae* x *S. kudriavzevii* share alleles with European wild populations of *Saccharomyces kudriavzevii*. FEMS Yeast Res. *10*, 412–421. http://dx.doi.org/10.1111/j.1567-1364.2010.00614.x

López-Malo, M., Querol, A., and Guillamon, J.M. (2013). Metabolomic comparison of *Saccharomyces cerevisiae* and the cryotolerant species *S. bayanus* var. *uvarum* and *S. kudriavzevii* during wine fermentation at low temperature. PLOS ONE *8*, e60135. http://dx.doi.org/10.1371/journal.pone.0060135

Ludlow, C.L., Cromie, G.A., Garmendia-Torres, C., Sirr, A., Hays, M., Field, C., Jeffery, E.W., Fay, J.C., and Dudley, A.M. (2016). Independent Origins of Yeast Associated with Coffee and Cacao Fermentation. Curr. Biol. *26*, 965–971. http://dx.doi.org/10.1016/j.cub.2016.02.012

Marsit, S., and Dequin, S. (2015). Diversity and adaptive evolution of *Saccharomyces* wine yeast: a review. FEMS Yeast Res. *15*, fov067. http://dx.doi.org/10.1093/femsyr/fov067

Martini, A.V. (1989). *Saccharomyces paradoxus* comb. nov., a newly separated species of the *Saccharomyces sensu stricto* complex based upon nDNA/nDNA homologies. System Appl. Microbiol. *12*, 179-182.

Martini, A.V. (1993). Origin and domestication of the wine yeast *Saccharomyces cerevisiae*. Journal of Wine Research. *4*, 165-176.

Martini, A.V., and Kurtzman, C.P. (1985). Deoxyribonucleic acid relatedness among species of the genus

Saccharomyces sensu stricto. International Journal of Systematic Bacteriology *35*, 508-511.

Martini, A.V., and Martini, A. (1987). Three newly delimited species of *Saccharomyces sensu stricto*. Antonie Van Leeuwenhoek *53*, 77–84.

McGovern, P.E. (2009). Uncorking the Past: the Quest for Wine, Beer, and other Alcoholic Beverages (University of California Press, Berkeley, CA).

McNeill, J., and Turland, N.J. (2011). Major changes to the Code of Nomenclature-Melbourne. Taxon *60*, 1495–1497.

McNeill, J., Barrie, F.R., Buck, W.R., Demoulin, V., Greuter, W., Hawksworth, D.L., Herendeen, P.S., Knapp, S., Marhold, K., Prado, J., Prud'Homme Van Reine, W.F., Smith, G.F., and Wiersema, J.H. (2011). International Code of Nomenclature for Algae, Fungi, and Plants, Melbourne Code (Koeltz Scientific, Konigstein).

Mertens, S., Steensels, J., Saels, V., De Rouck, G., Aerts, G., and Verstrepen, K.J. (2015). A large set of newly created interspecific Saccharomyces hybrids increases aromatic diversity in lager beers. Appl. Environ. Microbiol. *81*, 8202–8214. http://dx.doi.org/10.1128/AEM.02464-15

Meussdoerffer, F.G. (2009). A comprehensive history of beer brewing. In Handbook of Brewing Processes, Technology, Markets, H.M. Esslinger, ed. (Weinheim, Germany: Wiley-VCH), pp. 1–42.

Michel, M., Kopecka, J., Meier-Doernberg, T., Zarnkow, M., Jacob, F., and Hutzler, M. (2016). Screening for new brewing yeasts in the non-*Saccharomyces* sector with *Torulaspora delbrueckii* as model. Yeast *33*, 129-144.

Monerawela, C., James, T.C., Wolfe, K.H., and Bond, U. (2015). Loss of lager specific genes and subtelomeric regions define two different *Saccharomyces cerevisiae* lineages for *Saccharomyces pastorianus* Group I and II strains. FEMS Yeast Res. *15*, fou008.

Montrocher, R., Verner, M.C., Briolay, J., Gautier, C., and Marmeisse, R. (1998). Phylogenetic analysis of the *Saccharomyces cerevisiae* group based on polymorphisms of rDNA spacer sequences. Int. J. Syst. Bacteriol. *48*, 295–303. http://dx.doi.org/10.1099/00207713-48-1-295

Mukai, N., Masaki, K., Fujii, T., Kawamukai, M., and Iefuji, H. (2010). *PAD1* and *FDC1* are essential for the decarboxylation of phenylacrylic acids in *Saccharomyces cerevisiae*. J. Biosci. Bioeng. 564–569.

Mukai, N., Masaki, K., Fujii, T., and Iefuji, H. (2014). Single nucleotide polymorphisms of *PAD1* and *FDC1* show a positive relationship with ferulic acid decarboxylation ability among industrial yeasts used in alcoholic beverage production. J. Biosci. Bioeng *118*, 50-55.

Nakao, Y., Kanamori, T., Itoh, T., Kodama, Y., Rainieri, S., Nakamura, N., Shimonaga, T., Hattori, M., and Ashikari, T. (2009). Genome sequence of the lager brewing yeast, an interspecies hybrid. DNA. Res. *16*, 115–129. http://dx.doi.org/10.1093/dnares/dsp003

Naumov, G.I. (1996). Genetic identification of biological species in the *Saccharomyces sensu stricto* complex. J. Indust. Microbiol. *17*, 295-302.

Naumov, G.I. (2000). [*Saccharomyces bayanus* var. Uvarum comb.nov., a new variety established by genetic analysis.] Mikrobiologiia *69*, 410–414.

Naumov, G.I., Naumova, E.S., and Louis, E.J. (1995). Two new genetically isolated populations of the *Saccharomyces sensu stricto* complex from Japan. J. Gen. Appl. Microbiol. *41*, 499–505.

Naumov, G.I., James, S.A., Naumova, E.S., Louis, E.J., and Roberts, I.N. (2000). Three new species in the *Saccharomyces sensu stricto* complex: *Saccharomyces cariocanus*, *Saccharomyces kudriavzevii* and *Saccharomyces mikatae*. Int. J. Syst. Evol. Microbiol. *50*, 1931–1942.

Naumov, G.I., Lee, C.-F., and Naumova, E.S. (2013). Molecular genetic diversity of the *Saccharomyces* yeasts in Taiwan: *Saccharomyces arboricola*, *Saccharomyces cerevisiae* and *Saccharomyces kudriavzevii*. Antonie van Leeuwenhoek *103*, 217–228.

Naumova, E.S., Naumov, G., Masneuf-Pomarède, I., Aigle, M., and Dubourdieu, D. (2005). Molecular genetic study of introgression between *Saccharomyces bayanus* and *S. cerevisiae*. Yeast *22*, 1099-1115.

Nguyen, H.V., and Gaillardin, C. (1997). Two subgroups within the *Saccharomyces bayanus* species evidenced by PCR amplificatiom and restriction polymorphism of the non-transcribed spacer 2 in the ribosomal DNA unit. Syst. Appl. Microbiol. *20*, 286-294.

Nguyen, H.V., and Gaillardin, C. (2005). Evolutionary relationships between the former species *Saccharomyces uvarum* and the hybrids *Sacharomyces bayanus* and *Saccharomyces pastorianus*; reinstatement of *Sacccharomyces uvarum* (Beijerinck) as a distinct species. FEMS Yeast Res. *5*, 471-483.

Nguyen, H.V., Legras, J.L., Neuvéglise, C., and Gaillardin, C. (2011). Deciphering the hybridisation history leading to the Lager lineage based on the mosaic genomes of *Saccharomyces bayanus* strains NBRC1948 and CBS380. PLOS ONE *6*, e25821. http://dx.doi.org/10.1371/journal.pone.0025821

Nilsson-Tillgren, T., Gjermansen, C., Kielland-Brandt, M.C., Petersen, J.G.L., and Holmberg, S. (1981). Genetic differences between *Saccharomyces carlsbergensis* and *S. cerevisiae*. Analysis of chromosome III by single chromosome transfer. Carlsberg Res. Commun. *46*, 65–76.

Oelofse, A., Pretorius, I.S., and du Toit, M. (2008). Significance of *Brettanomyces* and *Dekkera* during winemaking: a synoptic review. S. Afr. J. Enol. Vitic. *29*, 128–144.

Pasteur, L. (1866). Études sur le Vin (Imprimeurs Impérials, Paris).

Pasteur, L. (1876). Études sur la Bière (Gauthier-Villars, Paris).

Pathania, N., Kanwar, S.S., Jhang, T., Koundal, K.R., and Sharma, T.R. (2010). Application of different molecular techniques for deciphering genetic diversity among yeast isolates of traditional fermented food products of Western Himalayas. World J. Microbiol. Biotechnol. *26*, 1539-1547.

Pedersen, M.B. (1994). Molecular analysis of yeast DNA - tools for pure yeast maintenance in the brewery. J. Am. Soc. Brew. Chem. *52*, 23-27.

Pengelly, R.J., and Wheals, A.E. (2013). Rapid identification of *Saccharomyces eubayanus* and its hybrids. FEMS Yeast Res. *13*, 156–161. http://dx.doi.org/10.1111/1567-1364.12018

Peris, D., Lopes, C.A., Arias, A., and Barrio, E. (2012). Reconstruction of the evolutionary history of *Saccharomyces cerevisiae* X *S. kudriavzevii* hybrids based on multilocus sequence analysis. PLOS ONE 7, e45527.

Peris, D., Sylvester, K., Libkind, D., Gonçalves, P., Sampaio, J.P., Alexander, W.G., and Hittinger, C.T. (2014). Population structure and reticulate evolution of *Saccharomyces eubayanus* and its lager-brewing hybrids. Mol. Ecol. *23*, 2031–2045. http://dx.doi.org/10.1111/mec.12702

Peris, D., Langdon, Q., Moriarty, R.V., Sylvester, K., Bontrager, M., Charron, G., Leducq, J.-B., Landry, C.R., Libkind, D., and Hittinger, C.T. (2016). Complex origins of lager-brewing hybrids were shaped by standing variation in the wild yeast *Saccharomyces eubayanus*. PLOS Genetics (submitted).

Piskur, J., Rozpedowska, E., Polakova, S., Merico, A., and Compagno, C. (2006). How did *Saccharomyces* evolve to become a good brewer? Trends Genet. *22*, 183–186.

Prillinger, H., Oberwinkler, F., Umile, C., Tlachac, K., Bauer, R., Dorfler, C., and Taufratzhofer, E. (1993). Analysis of cell wall carbohydrates (neutral sugars) from ascomycetous and basidiomycetous yeasts with and without derivatization. J. Gen. Appl. Microbiol. *39*, 1-34.

Pulvirenti, A., Nguyen, H., Caggia, C., Giudici, P., Rainieri, S., and Zambonelli, C. (2000). *Saccharomyces uvarum*, a proper species within *Saccharomyces sensu stricto*. FEMS Microbiol. Lett. *192*, 191–196.

Rainieri, S., Zambonelli, C., Hallsworth, J.E., Pulvirenti, A., and Giudici, P. (1999). *Saccharomyces uvarum*, a distinct group within *Saccharomyces sensu stricto*. FEMS Microbiol. Lett. *177*, 177–185.

Rainieri, S., Kodama, Y., Kaneko, Y., Mikata, K., Nakao, Y., and Ashikari, T. (2006). Pure and mixed genetic lines of *Saccharomyces bayanus* and *Saccharomyces pastorianus* and their contribution to the lager brewing strain genome. Appl. Environ. Microbiol. *6*, 3968-3674.

Reess, M. (1870). Botanische Untersuchungen über die Alkoholgärhungspilze (Felix, Leipzig).

Replansky, T., Koufopanou, V., Greig, D., and Bell, G. (2008). *Saccharomyces sensu stricto* as a model system for evolution and ecology. Trends Ecol. Evol. *23*, 494–501. http://dx.doi.org/10.1016/j.tree.2008.05.005

Richard, P., Viljanen, K., and Penttilä, M. (2015). Overexpression of PAD1 and FDC1 results in significant cinnamic acid decarboxylase activity in *Saccharomyces cerevisiae*. AMB. Express. *5*, 12. http://dx.doi.org/10.1186/s13568-015-0103-x

Riese, J.C., and Eßlinger, H.M. (2009). World beer market. In Handbook of Brewing Processes, Technology, Markets, H.M. Esslinger, ed. (Weinheim, Germany: Wiley-VCH), pp. 497–513.

Röcken, W., and Schulte, S. (1986). Nachweis von Fremdhefen. Brauwelt *41*, 1921-1927.

Roop, J.I., Chang, K.C., and Brem, R.B. (2016). Polygenic evolution of a sugar specialization trade-off in yeast. Nature *530*, 336–339. http://dx.doi.org/10.1038/nature16938

Rosini, G., Federici, F., Vaughan, A.E., and Martini, A. (1982). Systematics of species of the yeast genus *Saccharomyces* associated with the fermentation industry. Eur. J. Appl. Microbiol. Biotechnol. *15*, 188–193.

Salinas, F., Garrido, D., Ganga, A., Veliz, G., and Martínez, C. (2009). Taqman real-time PCR for the detection and enumeration of *Saccharomyces cerevisiae* in wine. Food Microbiol. *26*, 328–332. http://dx.doi.org/10.1016/j.fm.2008.12.001

Sampaio, J.P., and Gonçalves, P. (2008). Natural populations of *Saccharomyces kudriavzevii* in Portugal are associated with oak bark and are sympatric with *S. cerevisiae* and *S. paradoxus*. Appl. Environ. Microbiol. *74*, 2144–2152. http://dx.doi.org/10.1128/AEM.02396-07

Samuel, D. (1996). Investigation of Ancient Egyptian Baking and Brewing Methods by Correlative Microscopy Science 273, 488–490.

Samuel, D. (2000). Brewing and baking. In: Ancient Egyptian Materials and Technology, (Cambridge University Press, Cambridge; New York), pp. 537–576.

Scherer, A. (2002). Entwicklung von PCR-Methoden zur Klassifizierung industriell genutzter Hefen (TU München).

Schneiderbanger, H., Strauß, M., Hutzler, M., and Jacob, F. (2013). Aroma profiles of selected wheat beer yeast strains. Brauwelt International *31*, 219-222.

Shrestha, H., Nand, K., and Rati, E.R. (2002). Microbiological profile of murcha starters and physico-chemical characteristics of poko, a rice based traditional fermented food product of Nepal. Food Biotechnol. *16*, 1-15.

Sipiczki, M. (2008). Interspecies hybridization and recombination in Saccharomyces wine yeasts. FEMS Yeast Res. *8*, 996–1007. http://dx.doi.org/10.1111/j.1567-1364.2008.00369.x

Snauwaert, I., Roels, S.P., Van Nieuwerburg, F., Van Landschoot, A., De Vuyst, L., and Vandamme, P. (2016). Microbial diversity and metabolite composition of Belgian red-brown acidic ales. Int. J. Food Microbiol. *221*, 1–11. http://dx.doi.org/10.1016/j.ijfoodmicro.2015.12.009

Sniegowski, P.D., Dombrowski, P.G., and Fingerman, E. (2002). *Saccharomyces cerevisiae* and *Saccharomyces paradoxus* coexist in a natural woodland site in North America and display different levels of reproductive isolation from European conspecifics. FEMS Yeast Res. *1*, 299–306.

Soares, E.V. (2011). Flocculation in *Saccharomyces cerevisiae*: a review. J. Appl. Microbiol. *110*, 1–18. http://dx.doi.org/10.1111/j.1365-2672.2010.04897.x

Spitaels, F., Wieme, A.D., Janssens, M., Aerts, M., Daniel, H.M., Van Landschoot, A., De Vuyst, L., and Vandamme, P. (2014). The microbial diversity of traditional spontaneously fermented lambic beer. PLOS ONE *9*, e95384. http://dx.doi.org/10.1371/journal.pone.0095384

Spitaels, F., Wieme, A.D., Janssens, M., Aerts, M., Van Landschoot, A., De Vuyst, L., and Vandamme, P. (2015a). The microbial diversity of an industrially produced lambic beer shares members of a traditionally produced one and reveals a core microbiota for lambic beer fermentation. Food Microbiol. *49*, 23–32. http://dx.doi.org/10.1016/j.fm.2015.01.008

Spitaels, F., Van Kerrebroeck, S., Wieme, A.D., Snauwaert, I., Aerts, M., Van Landschoot, A., De Vuyst, L., and Vandamme, P. (2015b). Microbiota and metabolites of aged bottled gueuze beers converge to the same composition. Food Microbiol. *47*, 1–11. http://dx.doi.org/10.1016/j.fm.2014.10.004

Steensels, J., Daenen, L., Malcorps, P., Derdelinckx, G., Verachtert, H., and Verstrepen, K.J. (2015). *Brettanomyces* yeasts – From spoilage organisms to valuable contributors to industrial fermentations. Int. J. Food Microbiol. *206*, 24–38. http://dx.doi.org/10.1016/j.ijfoodmicro.2015.04.005

Steensels, J., Meersman, E., Snoek, T., Saels, V., and Verstrepen, K.J. (2014). Large-scale selection and breeding to generate industrial yeasts with superior aroma production. Appl. Environ. Microbiol. *80*, 6965–6975. http://dx.doi.org/10.1128/AEM.02235-14

Stelling-Dekker, N.M. (1931). Die sporogenen Hefen. Verh. K. Ned. Akad. Wetensch., Afd. Natuurk., Sect. II, *28*, 1-547.

Strope, P.K., Skelly, D.A., Kozmin, S.G., Mahadevan, G., Stone, E.A., Magwene, P.M., Dietrich, F.S., and McCusker, J.H. (2015). The 100-genomes strains, an *S. cerevisiae* resource that illuminates its natural phenotypic and genotypic variation and emergence as an opportunistic pathogen. Genome Res. *25*, 762-774.

Suzuki, K., Asano, S., Iijima, K., Ogata, T., and Kitagawa, Y. (2008). Effects of beer adaptation on culturability of beer-spoilage *Dekkera/Brettanomyces* yeasts. J. Am. Soc. Brew. Chem. *66*, 239-244.

Tamai, Y., Momma, T., Yoshimoto, H., and Kaneko, Y. (1998). Co-existence of two types of chromosome in the bottom fermenting yeast, *Saccharomyces pastorianus*. Yeast *14*, 923–933.

Tapsoba, F., Legras, J.L., Savadogo, A., Dequin, S., and Traore, A.S. (2015). Diversity of Saccharomyces cerevisiae strains isolated from Borassus akeassii palm wines from Burkina Faso in comparison to other African beverages. Int. J. Food Microbiol. *15*, 128-133.

Taylor, J.W., Spatafora, J., O'Donnell, K., Lutzoni, F., James, T., Hibbett, D.S., Geiser, D., Bruns, T.D., and Blackwell, M. (2004). The Fungi. In Assembling the Tree of Life, J. Cracraft and M.J. Donoghue, eds. (Oxford University Press) pp. 171-194.

Thurston, P.A., and Tubb, R.S. (1981). Screening yeast strains for their ability to produce phenolic off-flavours: A simple method for determining phenols in wort and beer. J. Inst. Brew. *87*,177-179.

Tronchoni, J., Rozès, N., Querol, A., and Guillamón, J.M. (2012). Lipid composition of wine strains of *Saccharomyces kudriavzevii* and *Saccharomyces cerevisiae* grown at low temperature. Int. J. Food Microbiol. *155*, 191–198. http://dx.doi.org/10.1016/j.ijfoodmicro.2012.02.004

Tronchoni, J., Medina, V., Guillamón, J.M., Querol, A., and Pérez-Torrado, R. (2014). Transcriptomics of cryophilic *Saccharomyces kudriavzevii* reveals the key role of gene translation efficiency in cold stress adaptations. BMC Genomics *15*: 432.

Turakainen, H., Kristo, P., and Korhola, M. (1994). Consideration of the evolution of the *Saccharomyces cerevisiae* MEL gene family on the basis of the nucleotide sequences of the genes and their flanking regions. Yeast *10*, 1559-1568.

Vallejo, J.A., Miranda, P., Flores-Félix, J.D., Sánchez-Juanes, F., Ageitos, J.M., González-Buitrago, J.M., Velázquez, E., and Villa, T.G. (2013). Atypical yeasts identified as *Saccharomyces cerevisiae* by MALDI-TOF MS and gene sequencing are the main responsible of fermentation of chicha, a traditional beverage from Peru. Syst. Appl. Microbiol. *36*, 560-564.

van der Walt, J.P. (1970). *Saccharomyces* (Meyen) emend. Reess. In The Yeasts, A Taxonomic Study, 2nd edn., J. Lodder, ed. (North-Holland: Amsterdam) pp. 555–718.

van der Walt, J.P. (1984). Dekkera van der Walt. In: The Yeasts, A Taxonomic Study, 3rd edn. N.J.W. Kreger-van Rij, ed. (Elsevier: Amsterdam) pp. 146–150.

van Landschoot, A., Vanbeneden, N., Machtelinckx, M., Stals, I., and Claeyssens, M. (2005). Peculiarities of seven referemented Belgian strong ales and their corresponding industrial yeasts. Cerevisiae *30*, 181-188.

van Zandycke, S.M., Bertrand, D.,and Powell, C. (2007). The use of genetic tools to differentiate related lager yeast strains. EBC Congress, 6-10 May 2007 Venice, Italy.

Vanbeneden, N., Gils, F., Delvaux, F., and Delvaux, F.R. (2008). Formation of 4-vinyl and 4-ethyl derivatives from hydroxycinnamic acids: Occurrence of volatile phenolic flavour compounds in beer and distribution of Pad1 activity among brewing yeasts. Food Chem. *107*, 221-230.

Vanderhaegen, B., Neven, H., Coghe, S., Verstrepen, K.J., Derdelinckx, G., and Verachtert, H. (2003a). Bioflavoring and beer refermentation. Appl. Microbiol. Biotechnol. *62*, 140–150. http://dx.doi.org/10.1007/s00253-003-1340-5

Vanderhaegen, B., Neven, H., Coghe, S., Verstrepen, K.J., Verachtert, H., and Derdelinckx, G. (2003b). Evolution of chemical and sensory properties during aging of top-fermented beer. J. Agric. Food. Chem. *51*, 6782–6790. http://dx.doi.org/10.1021/jf034631z

Vanoevelen, D., Spaepen, M., Timmermans, P., and Verachtert, H. (1977). Microbial aspects of spontaneous fermentation in production of lambic and gueuze. J. Inst. Brew. *83*: 356–360.

Vaughan-Martini, A., and Martini, A. (1995). Facts, myths and legends on the prime industrial microorganism. J. Ind. Microbiol. *14*, 514–522.

Vaughan-Martini, A., and Martini, A. (1998). *Saccharomyces* Meyen ex Reess In The Yeasts, A Taxonomic Study, 4th edn., C.P. Kurtzman and J.W. Fell eds. (Elsevier, Amsterdam), pp. 358-371.

Vaughan-Martini, A., and Martini, A. (2011). *Saccharomyces* Meyen ex Reess In The Yeasts, A Taxonomic Study, 5th edn., C.P. Kurtzman, J.W. Fell and T. Boekhout, eds. (Elsevier, Amsterdam), pp. 733-746.

Verachtert, H., and Iserentant, D. (1995). Properties of Belgian acid beers and their microflora. Part I. The production of gueuze and related refreshing acid beers. Cerevisia *20*, 37–41.

Verachtert, H., and Derdelinckx, G. (2005). Acidic beers: enjoyable reminiscences of the past. Cerevisia *30*, 38–47.

Verachtert, H., and Derdelinckx, G. (2014). Belgian acidic beers. Daily reminiscences of the past. Cerevisia *38*, 121–128.

Verstrepen, K.J., Derdelinckx, G., Verachtert, H., and Delvaux, F.R. (2003). Yeast flocculation: what brewers should know. Appl. Microbiol. Biotechnol. *61*, 197–205. http://dx.doi.org/10.1007/s00253-002-1200-8

Vidgren, V., Kankainen, M., Londesborough, J., and Ruohonen, L. (2011). Identification of regulatory elements in the AGT1 promoter of ale and lager strains of brewer's yeast. Yeast *28*, 579-594.

Vigentini, I., de Lorenzis, G., Picozzi, C., Imazio, S., Merico, A., Galafassi, S., Piškur, J., and Foschino, R. (2012). Intraspecific variations of *Dekkera/Brettanomyces bruxellensis* genome studied by capillary electrophoresis separation of the intron splice site profiles. Int J Food Microbiol. *157*, 6-15.

Vigentini, I., Joseph, C.M., Picozzi, C., Foschino, R., and Bisson, L.F. (2013). Assessment of the *Brettanomyces bruxellensis* metabolome during sulphur dioxide exposure. FEMS Yeast Res. *13*, 597–608. http://dx.doi.org/10.1111/1567-1364.12060

Walker, W.F., and Doolittle, W.F. (1982). Redividing the basidiomycetes on the basis of 5S rRNA sequences. Nature *299*, 723–724.

Walther, A., Hesselbart, A., and Wendland, J. (2014). Genome sequence of *Saccharomyces carlsbergensis*, the world's first pure culture lager yeast. G3 *4*, 783–793. http://dx.doi.org/10.1534/g3.113.010090

Wang, Q.M., Liu, W.Q., Liti, G., Wang, S.A., and Bai, F.Y. (2012). Surprisingly diverged populations of *Saccharomyces cerevisiae* in natural environments remote from human activity. Mol. Ecol. *21*, 5404–5417. http://dx.doi.org/10.1111/j.1365-294X.2012.05732.x

Wang, Q.M., Yurkov, A.M., Göker, M., Lumbsch, H.T., Leavitt, S.D., Groenewald, M., Theelen, B., Liu, X.Z., Boekhout, T., and Bai, F.Y. (2015). Phylogenetic classification of yeasts and related taxa within Pucciniomycotina. Stud. Mycol. *81*, 149–189. http://dx.doi.org/10.1016/j.simyco.2015.12.002

Wang, S.A., and Bai, F.Y. (2008). *Saccharomyces arboricolus* sp. nov., a yeast species from tree bark. Int. J. Syst. Evol. Microbiol. *58*, 510–514. http://dx.doi.org/10.1099/ijs.0.65331-0

Weiss, S., Samson, F., Navarro, D., and Casaregola, S. (2013). YeastIP: a database for identification and phylogeny of Saccharomycotina yeasts. FEMS Yeast Res. *13*, 117–125. http://dx.doi.org/10.1111/1567-1364.12017

Wendland, J. (2014). Lager yeast comes of age. Eukaryotic. Cell *13*, 1256–1265. http://dx.doi.org/10.1128/EC.00134-14

Wickerham, L.J. (1951). Taxonomy of yeasts. Techn. Bull. 1029, U.S. Dept. Agr., Washington, D.C.

Woese, C.R., and Fox, G.E. (1977). Phylogenetic structure of the prokaryotic domain: the primary kingdoms. Proc. Natl. Acad. Sci. U.S.A. *74*, 5088–5090.

Yamada, Y., Banno, I., von Arx, J.A., and van der Walt, J.P. (1987). Taxonomic significance of the coenzyme Q system in yeasts and yeast-like fungi. Stud. Mycol. *30*, 299-308.

Yarrow, D. (1984). *Saccharomyces* Meyen ex Reess. In The Yeasts, A Taxonomic Study, 3rd edn., N.J.W. Kreger-van Rij, ed. (Elsevier, Amsterdam), pp. 379–395.

Yarrow, D., and Nakase, T. (1975). DNA base composition of species of the genus *Saccharomyces*. Antonie Van Leeuwenhoek *41*, 81–88.

Genetic Manipulation of Brewing Yeasts: Challenges and Opportunities

5

Barbara Dunn[1]*, Daniel J. Kvitek[2] and Gavin Sherlock[1]

[1]Department of Genetics, Stanford University, Stanford, CA, USA.
[2]Invitae Corporation, San Francisco, CA, USA.

*Correspondence: bdunn@stanford.edu

https://doi.org/10.21775/9781910190616.05

Abstract

Brewing yeasts are notoriously difficult to work with genetically due to their complex genomes, which are often polyploid, aneuploid, and/or derived from interspecific hybridization events. We discuss the possibilities for both traditional and non-traditional genetic manipulations of brewing yeasts as a way to combine or enhance beneficial traits already present in such yeasts, or to possibly identify and introduce novel traits from non-brewing yeasts. For genetically tractable yeast strains, classic genetic breeding via meiosis and direct mating of spores can increase genetic variability and combine desired traits. For intractable strains, non-traditional breeding methods such as rare-mating, mating by transient HO induction, cytoduction, and protoplast fusion can be utilized. These various techniques can also be performed using mixed populations in a 'mass-mating' manner and/or to create interspecific hybrids. Once cells or populations with increased genetic variability are obtained, genome shuffling can create novel combinations of traits; adaptive evolution of shuffled populations allows the eventual selection of strains that exhibit desired fermentation or other selectable characteristics, while high-throughput quantitative trait loci (QTL) analysis can lead to insights about the actual genes that contribute to the traits of interest. The above techniques yield non-genetically modified (non-GM) yeasts. However, recent advances in 'minimally invasive' GM techniques that result in precise genome modifications with no remaining foreign DNA may eventually be deemed as acceptable by consumers and the brewing industry as a way to obtain brewing yeasts with desired traits. Overall, there is a wide variety of tools available for the genetic manipulation of brewing yeasts to alter or enhance any of a number of characteristics, from fermentation behaviour to sensory profile.

Life cycle of *Saccharomyces* yeasts; genomes, ploidy, aneuploidy, and interspecific hybrids

Budding yeasts, which are members of the fungal genus *Saccharomyces* (Latin: 'sugar fungus'), are minuscule single-celled organisms, yet they are ultimately responsible for producing almost all of the alcoholic beverages consumed in the world, including beer, wine, sake, and distilled spirits. In the case of beer brewing, *Saccharomyces* yeasts are the primary biological agents that transform wort into beer, by metabolizing sugars present in the wort (mostly maltose, glucose, and maltotriose) and converting them into ethanol and carbon dioxide; these organisms are thus literally 'sugar (-eating) fungi'. Yeasts, as are all fungi, are eukaryotic organisms, with their genomes organized into linear chromosomes contained within a nucleus. Yeasts also contain cytoplasmically located mitochondria, which have their own separate genome, but this chapter will not discuss mitochondria and their genomes in further detail. For *Saccharomyces* yeasts, the typical strains found in nature have a diploid genome containing two copies of each of 16 different chromosomes (16 pairs of chromosomes, hence 32 chromosomes total in the diploid genome). Haploid cells (containing just one copy of each of the chromosomes) can exist briefly within the sexual mating cycle, derived by sporulation of the diploid cell, or can exist indefinitely as free-living cells if they are unable to mate successfully, for example due to mutations in the mating system.

As illustrated in Fig. 5.1, the life cycle of typical *Saccharomyces* yeasts includes both asexual and sexual phases, with both diploid and, under some conditions, haploid cells able to undergo mitosis (i.e. to divide asexually, also called 'clonally') in an unlimited manner when there are sufficient nutrients. However, when nutrients, especially nitrogen, become limiting, a diploid cell – but not a haploid – is able to progress through meiosis and produce 4 haploid spores (two spores each of two different, or 'opposite', mating types, called 'a' and 'α'), which are resistant to desiccation and other environmental stresses (see Herskowitz, 1988, for a review of the yeast life cycle). Upon resumption of nutrient availability and beneficial environmental conditions, the haploid spores can germinate (become metabolically active) and then divide and continue to grow asexually, or – when and if they come into contact with a haploid cell of the opposite mating type – they can undergo cell fusion and mating

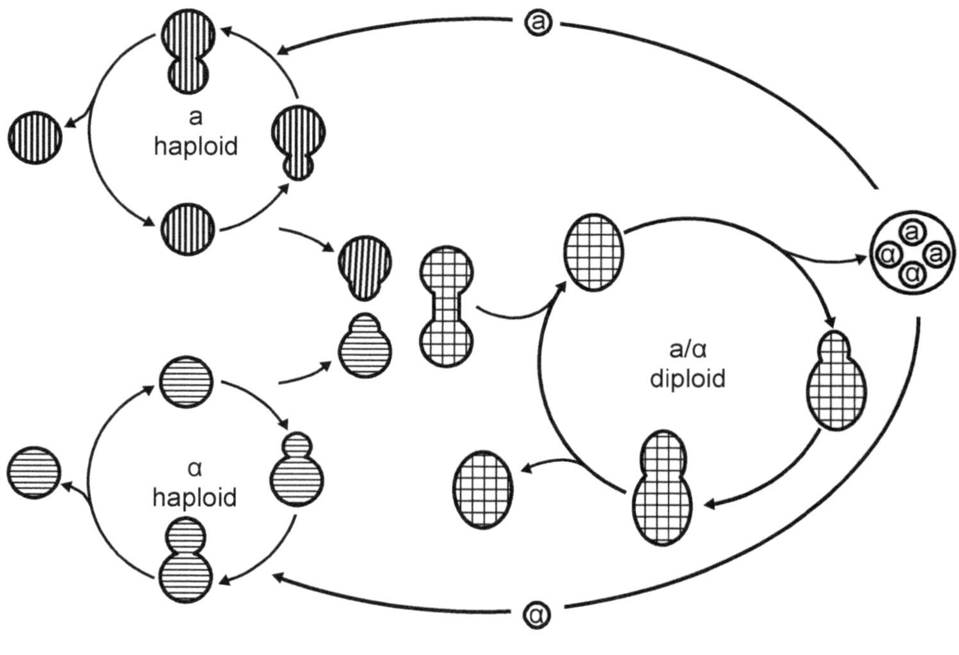

Figure 5.1 Life cycle of *S. cerevisiae*. Yeast cells can exist in both a haploid and diploid state. Haploid cells, shown on the left side of the figure, are either mating type 'a' or mating type 'α'; haploids are capable of fusing to and mating with a cell of the opposite mating type (centre-left of figure; see also Fig. 5.3A), resulting in a diploid cell that now contains the genetic material (chromosomes) from both parental haploids. The resulting diploid cell is heterozygous for the mating-type locus, and thus is called 'a/α', a situation that makes them incapable of mating. In nutrient-rich conditions, HO-deficient haploid cells (see main text), and all diploid cells, can proliferate by asexually (i.e. clonally) 'budding off' new daughter cells by mitotic division, as indicated by the circular arrowed cycles. When exposed to certain nutrient-poor conditions, diploids can undergo sporulation (meiosis followed by spore formation; right side), resulting in the conversion of a diploid cell into four haploid spores: two spores possessing mating type 'a' and two possessing mating type 'α'. The spores can germinate into haploid cells when conditions improve (top and bottom arrows leading back to haploid cycles). The 'sexual phase' of yeast reproduction encompasses both the mating and fusion of the opposite-mating-type haploid cells that results in a diploid cell containing the genetic material of the 2 parents in a single nucleus, as well as the subsequent meiotic division of the diploid cell, which results in recombination, or genetic exchange, between the two parental sets of chromosomes, and yields 4 haploid cells whose chromosomes are recombinant, i.e. containing portions of each parental DNA (note; meiotic recombination is not indicated in this figure).

(conjugation) with that other cell to form a new diploid cell (see Merlini *et al.*, 2013, for review). Normal ('wild-type') haploid cells contain an enzyme, HO endonuclease, that causes a 'mother' cell (i.e. a cell that has given rise to a newly budded 'daughter' cell formed after mitotic, i.e. asexual, cell division) to switch its mating type to the opposite of that of the daughter cell (Cosma, 2004; Haber, 2012). This allows the mother and daughter haploid cells, which are in close proximity and of opposite mating types, to mate and regenerate the diploid genome very quickly after sporulation. This behaviour has been postulated as not merely acting to restore diploidy, *per se*, but instead to quickly restore to cells that are germinating in uncertain or variable environments the ability to make stress-resistant spores if nutrient levels fall again (Knop, 2006; Hanson *et al.*, 2014). After mating, the resulting diploid cells are a/α – in this state they do not express either mating type and therefore do not mate with other cells, and the *HO* gene (which encodes HO endonuclease) is turned off. However, haploid cells with a disabling mutation in the *HO* gene, such as laboratory strains with a natural *HO* gene mutation, or any strain that has been genetically modified to introduce an *HO* gene mutation, will produce daughter cells that do not switch their mating type; these types of yeast strains can exist indefinitely as free-living, mitotically dividing, mating-competent haploids of a particular mating type. Note that if these cells do encounter a cell of the opposite mating type, they are able to mate with it and form a diploid cell. However, since virtually all brewing-related yeast strains do not have *HO* gene mutations and cannot exist as free-living, mating-competent haploids, this chapter will not discuss any types of yeast breeding and/or genetic manipulations that require such cells.

Members of the *Saccharomyces* genus; interspecific hybridization

Within the *Saccharomyces* genus, there are at least seven closely related, naturally occurring species, as defined in recent reviews (Hittinger, 2013; Borneman and Pretorius, 2015); the two species most relevant to brewing yeasts are *S. cerevisiae* and *S. eubayanus*. All seven of the *Saccharomyces* species have very similar genomes: a haploid complement of 16 chromosomes, with most genes (and gene order) shared among all strains; additionally, the genomes are very similar at the DNA level. All *Saccharomyces* species also share the same basic life cycle and mating systems (Fig. 5.1). Interestingly, these species are able to mate with each other, i.e. haploid spores of one *Saccharomyces* species are able to mate with haploid spores of the opposite mating type of any of the other *Saccharomyces* species to form a 'pseudo-diploid' interspecific hybrid; this occurs both in the wild and in human-related environments (reviewed by Morales and Dujon, 2012). These interspecific hybrids are unable to efficiently proceed through meiosis, but are able to indefinitely reproduce in an asexual (clonal) manner. Other mechanisms, such as multiple rounds of spontaneous genome duplication, or aberrant mating between diploids, can lead to polyploidy (more than 2 copies of each of the basic haploid set of 16 chromosomes) within a species, and similar aberrant mating of higher ploidy cells between different species can give rise to interspecific hybrids of varying ploidy levels (Hittinger, 2013; Borneman and Pretorius, 2015). Finally, if a yeast cell is of diploid or higher ploidy, it can often tolerate the loss or gain of a single chromosome, or even several chromosomes, leading to a state called 'aneuploidy', where within a cell different chromosomes may be present at different copy numbers (Barrio *et al.*, 2006; Storchová, 2014).

For brewing yeasts, these concepts of ploidy and aneuploidy (i.e. chromosome numbers within a cell), as well as interspecific hybridization, are very important. Ale yeasts are typically strains of *S. cerevisiae*, and are diploid, or more often, are of even higher ploidy and/or are aneuploid, depending on the particular strain (Legras *et al.*, 2007; Borneman *et al.*, 2011; Gallone *et al.*, 2016; Gonçalves *et al.*, 2016). In contrast, lager yeasts are interspecific hybrids that were formed by mating between *S. cerevisiae* and *S. eubayanus* (Libkind *et al.*, 2011; Wendland, 2014; Gibson and Liti, 2014; Baker *et al.*, 2015; Peris *et al.*, 2016) (see Chapter 4). Note that, because these lager yeast strains are interspecific hybrids, they are not 'true' species. However, this was not clear when they were first studied; instead they were thought to represent unique species and were given names such as *Saccharomyces pastorianus* and/or *Saccharomyces carlsbergensis*, and for convenience they are still often referred to by these names. Similarly, *Saccharomyces bayanus* is the name originally given to what is now known to be a collection of similar cold-tolerant interspecific hybrid strains, most involving *Saccharomyces uvarum* and *S. eubayanus* as the component genomes. *S. bayanus* was also once thought to be a 'pure' *Saccharomyces* species, while *S. uvarum*, now known to be a free-living, true species, was thought to be a subspecies of *S. bayanus* [see Nguyen *et al.* (2011) and Pérez-Través *et al.* (2014) for detailed descriptions of the complex genomes and history of *S. bayanus* yeasts]. It had been known for years that the non-*S. cerevisiae* genome component of the lager yeast was closely related to *S. bayanus* and/or *S. uvarum* (Rainieri *et al.*, 2006); thus, prior to the discovery in 2011 of *S. eubayanus* as the 'pure', free-living non-*S. cerevisiae* component of the lager yeast genome (Libkind *et al.*, 2011), much of the lager yeast literature uses the name *S. bayanus* for the non-*S. cerevisiae* component.

Among the lager hybrids, ploidy levels and aneuploidy also vary from strain to strain (de Barros Lopes *et al.*, 2002; Rainieri *et al.*, 2006; Legras *et al.*, 2007; Dunn and Sherlock, 2008; Hewitt *et al.*, 2014; Wendland, 2014; Gibson and Liti, 2014; Monerawela *et al.*, 2015; Okuno *et al.*, 2016; van den Broek *et al.*, 2015). It is possible that the widespread occurrence of aneuploidy and higher ploidies seen among brewing yeasts reflects selective events that give advantage to these cells in the brewing environment; indeed, polyploidy and aneuploidy have been observed to be adaptive in laboratory yeasts under some growth environments (Dunham *et al.*, 2002; Storchová, 2014; Selmecki *et al.*, 2015; Sunshine *et al.*, 2015). These altered ploidy levels can presumably be tolerated, and even selected for, because brewing yeasts are not required to pass through a sexual phase as part of their lifestyle, instead being propagated asexually in perpetuity during the brewing process. Other chapters in this book (see Chapters 4 and 6) describe the detailed characterization of the genomes of ale yeasts and hybrid lager yeasts, as well as the discovery of the *S. eubayanus* yeast in the wild, and how the formation of the interspecific hybrid lager yeasts may have occurred; the present chapter instead focuses on the challenges and possible benefits or drawbacks of performing directed breeding and genetic analyses – and possibly even genomic manipulations – on brewing yeasts. Note that in this chapter we will focus only on *Saccharomyces* yeasts, and not on various non-*Saccharomyces* yeasts or bacteria that are used in fermentations to create lambic or other 'wild' or sour-type beers. Some aspects of the biology, genomics, and brewing relevance of these yeasts are discussed in other chapters of this book (see Chapters 6 and 7).

Genetic traits of interest to brewers

As we describe below the different genetic manipulations and genetic/genomic modification techniques that might be used to construct new brewing yeasts, keep in mind that these are techniques that will allow brewers to isolate, change, or introduce various traits into pre-existing production brewing yeast strains; however, it is also possible that yeast strains from other industries or from natural environments, even non-*Saccharomyces* yeasts, may be commandeered for use in creating new brewing strains (Steensels and Verstrepen, 2014; Steensels *et al.*, 2014b). Genetic traits ('phenotypes') are growth behaviours and other qualities (such as production of flavour or aroma molecules) that are exhibited by particular yeast strains and that are encoded for by the genes found in the yeast genome. The different traits seen among different *Saccharomyces* strains are thus the result of genetic variation found between different yeast strains. *Saccharomyces* yeasts play a role in, and have been isolated from, many human-associated food, beverage, and industrial fermentations and environments: beer, bread, wine, distilled spirits, biofuel, sake, chocolate, coffee and many others (see, for example, Ludlow *et al.*, 2016); *Saccharomyces* yeasts also exist away from humans and have been found in many 'wild' or natural (e.g. tree- and soil-associated) habitats across the world (Robinson *et al.*, 2016; Fig. 5.2A). The list of traits that brewers are interested in as targets for modification, i.e. the beer-related traits that almost certainly have a genetic basis and are able to be monitored or investigated during various genetic screens and manipulations, can also be kept in mind throughout the rest of this chapter. Many of the qualities and attributes of brewing yeast strains that are of particular interest to brewers include:

- improved resilience and stress tolerance
- improved efficiency of sugar utilization
- altered sugar substrate utilization
- improved efficiency of nitrogen utilization
- improved ethanol tolerance

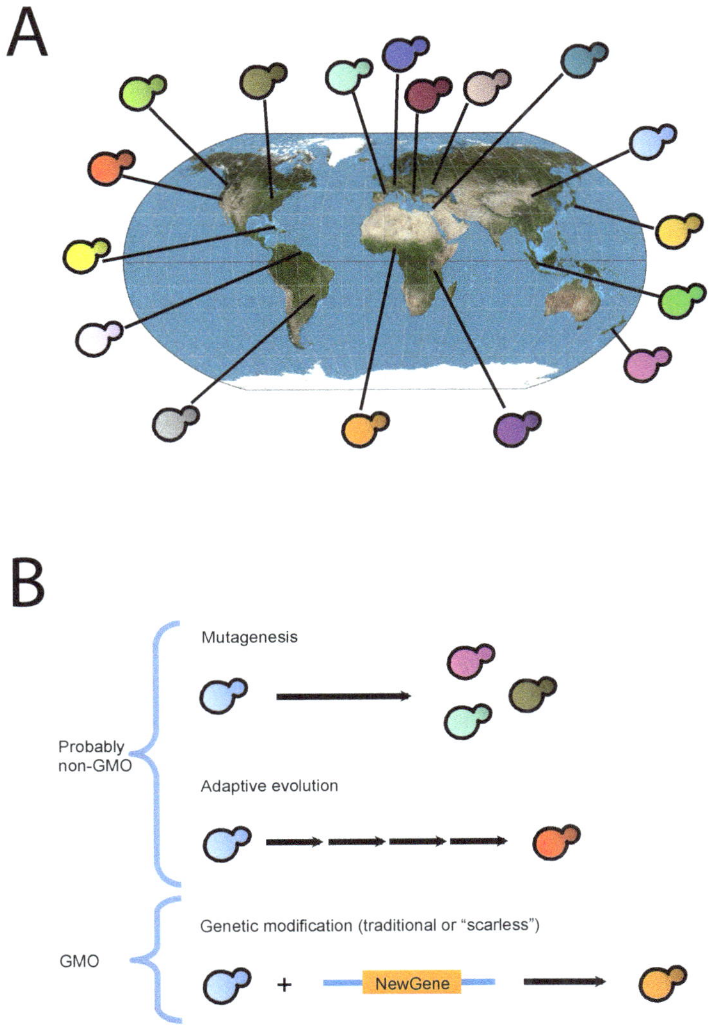

Figure 5.2 Genetic diversity in *Saccharomyces* yeasts: natural and human-made. (A) Idealized view of the world distribution of natural ('standing') variation within the *S. cerevisiae* species. Natural ('standing') genetic variation among yeast strains exists due to long term separation by geography and/or industrial setting; different coloured yeast cells indicate strains that differ genetically from each other (see Fig. 4 of Robinson *et al*., 2016 for a detailed world map of locations where *S. cerevisiae* strains have been isolated). Hundreds of *S. cerevisiae* strains, collected from around the world from both natural and human-associated (food/beverage/industrial) settings have now had their whole genomes sequenced, giving a good sense of the amount of the natural genetic diversity that exists among this entire species (see main text). (B) Human-directed methods to generate genetic diversity and/or to introduce specific genomic changes into a single starting yeast strain. Starting with one yeast strain of interest – such as a brewing strain displaying some beneficial traits, but to which new traits are desired to be added – there are many different ways to introduce genetic changes into the strain. As detailed in the main text, some methods, such as mutagenesis or adaptive evolution, will result in new strains that are probably not considered by consumers or government agencies to be 'genetically modified organisms' (GMOs), while other methods, such as 'traditional' genetic modification approaches that result in strains carrying foreign DNA (shown as 'NewGene' in the figure) in their chromosomes or on a plasmid, will surely result in GMOs. Finally, there are new methods that result in precise 'scarless' genetic modifications to the strain (i.e. with no foreign DNA remaining in the strain); in this case 'NewGene' in the figure could refer to something as simple yet precise as a single nucleotide change in a gene within the strain that would confer a new function or trait to the strain. Such strains may reside in a 'grey zone' with regard to GMO status, being considered GMO by some consumers or governments, and not GMO by others; it is likely there will be much debate about these techniques and the status of the resulting organisms in the coming years.

- decreased ethanol and/or caloric production
- increased tolerance to antimicrobial compounds
- altered fermentation temperature optimum
- controlling cell sedimentation and flocculation
- enhanced or modified flavour attributes
- reduced vicinal diketone production
- reduced production of hydrogen sulfide
- avoidance of ethyl carbamate aka urethane, a carcinogen (although more of a problem in wine)
- modified production of phenolic compounds.

It is essential that for any trait that a yeast breeder would like to investigate, there must be a way to first generate genetic variation either across the whole genome (or if possible, just for the specific trait) within the yeast cells that will be genetically manipulated. Afterward, one must either (1) 'select' for yeast cells bearing the desired trait, i.e. to grow or subject a population of yeast cells in a certain environment such that only the yeast cells bearing the trait of interest will survive, and all other cells will be killed, or (2) to 'screen' for yeast cells bearing that trait, i.e. to grow or subject a population of yeast cells in a certain environment in which all cells grow, but the cells displaying the trait of interest can somehow be identified and isolated away from the non-trait-bearing cells, and can then be studied further. Note that the 'screening' approach is much more laborious than a 'selection' approach, and if the proportion of trait-bearing cells within a population is very low, a screening approach may be impossible. In most cases, selection or screening is carried out on plates, which allows the isolation of single (i.e. clonal) colonies. Typically, several colonies are selected and tested not for only the presence of the desired trait(s), but also tested to assure that the clones retain the necessary overall growth properties of the parent yeast; in the case of brewing yeast, this would mean making sure that the selected clones still ferment wort and produce beer to the same (or superior) specifications as the parent yeast strain.

Using classical genetics alone and/or in combination with molecular biology or genomics approaches, the genes or pathways encoding some of these beer-related traits have been partially or wholly elucidated, but many remain completely uncharacterized genetically. Determining the genetic underpinnings of these traits could eventually allow brewers to custom-design yeast strains to their own specifications and needs, using any of the variety of techniques described below.

Breeding and hybridization strategies currently used with brewing yeasts

As discussed above, most ale and lager yeasts are not 'pure' diploid organisms, but instead exhibit various levels of aneuploidy, polyploidy, and/or interspecific hybridization. This means that these strains are generally unable to proceed through meiosis to give viable haploid spores, probably due to problems with chromosome pairing during meiosis and/or other incompatibility issues that specifically arise as a result of interspecific hybridization (Greig *et al.*, 2002; Greig, 2007; Lee *et al.*, 2008; Kao *et al.*, 2010; Hou *et al.*, 2015). This makes it difficult or impossible to perform typical genetic analyses, and thus typical breeding strategies, with most brewing yeasts. Recently, however, various directed breeding techniques using brewing and other industrial yeasts have been carried out with success, potentially leading to useful ways to generate novel brewing yeasts.

Directed breeding, also called selective or controlled breeding or artificial selection (Darwin, 1859, 1868), is the process by which desired phenotypic characteristics from different lineages are combined by humans, through the controlled mating of selected individuals from these lineages. For millennia, despite lacking an understanding of the underlying genetic mechanisms, humans both intentionally and unintentionally domesticated a variety of different organisms, including plants, animals, and microorganisms, by continually selecting individuals that displayed the phenotypic characteristics of interest and allowing only those chosen individuals to reproduce. There is evidence from many regions across the world that humans have been interacting with yeasts for more than 9000 years, almost certainly at first as a way to transform fruits and grains into intoxicating beverages (McGovern, 2003), possibly aiding in societal development (McGovern, 2009; Kahn, 2015). These early humans were not only unintentionally selecting for yeast strains that would merely survive and thrive in the somewhat harsh conditions of fruit juices or soaked grains, but were also – most likely by transferring foam from only the 'best' fermentations to subsequent batches – repeatedly enriching and selecting for those yeasts that made a pleasing flavour while still providing an intoxicating experience (McGovern, 2009), eventually selecting for different strains of yeasts specialized for various different types of fermentation (Legras *et al.*, 2007). In fact virtually all of the yeast strains used in brewing today, possibly even including lambic and 'wild' beers that are not deliberately inoculated and are fermented by organisms existing in the brewery environment (see Chapter 7), are found only in association with human activity, having become totally 'domesticated'. On the other hand, some would say (albeit in semi-jest – or possibly not), that humans have been 'domesticated' by yeasts to become their unwitting caretakers (Katz, 2010; Dawson, 2013).

However, after the concepts of natural selection (Darwin and Wallace, 1858) and genetic inheritance (Mendel, 1865) were introduced and understood, it became even easier to carry out directed breeding of organisms. In many cases, the inheritance patterns of different traits could be easily elucidated, leading to precise predictions of the phenotypes of offspring relative to the parents. These calculations are very simple if a trait is controlled by one or two genes, but the vast majority of traits are 'quantitative', meaning that the particular trait is influenced by many genes, also called a 'polygenic' trait. The genes controlling such traits are called quantitative trait loci (QTL). However, statistical calculations can be used in these cases to accurately predict offspring phenotypes. Today, directed breeding is carried out in all areas of agriculture and animal husbandry.

Yeasts capable of sexually reproducing can also be bred through this process (see Fig. 5.3A and 'Direct mating', below). However, as mentioned above, many brewing yeasts are essentially sterile, i.e. they do not produce viable spores. Although cases of successful ale and lager yeast breeding using rare viable spores have been reported and will be described in more detail below, in general both lager and ale yeasts are very difficult to selectively breed using traditional spore mating methodology. Furthermore, the rare surviving spores that arise after such defective meiotic events are, by definition, selected to carry traits correlated with spore survival, but in fact these traits may not be correlated (and may even be anti-correlated) with desired brewing traits.

Several methods have been adapted to overcome the sexual limitation of brewing yeasts, with most relying on the fusion of two parental strains (usually diploid, aneuploid, or polyploid) to form a higher ploidy hybrid line via 'rare mating' events or 'protoplast fusion' (described in more detail below). These methods avoid sporulation – and thus the chromosomal reduction step that occurs with meiosis – therefore sidestepping the problems of spore inviability and selection of rare viable spores that can occur with aneuploid or interspecific hybrid yeasts. The resulting higher ploidy organism (tetraploid in the case of two diploid parental strains) ideally combines the phenotypic characteristics of both parental strains, similar to the way that standard mating of haploid spores combines those characteristics in the diploid zygote. This hybrid line may even be more genetically tractable than the aneuploid parental strains, since its increased gene complement may prevent inviability upon meiosis (Morales and Dujon, 2012). An additional benefit of producing hybrids between already domesticated lines is increased fitness and vigour during fermentation (Plech *et al.*, 2014; Steensels *et al.*, 2014a). Conversely, strains resulting from forced interspecific hybridization have been shown to experience an increased rate of chromosome loss (Marinoni *et al.*, 1999), although the magnitude of this effect is debated (Kumaran *et al.*, 2013).

The following sections describe several different directed breeding techniques for yeast that are genetically tractable (i.e. undergo meiosis and produce viable spores), as well as techniques that circumvent the need for 'genetically well-behaved strains' (Figs. 5.3 and 5.4). For comprehensive treatments of directed and other types of breeding methods in yeasts, see Chambers *et al.* (2009) and Steensels *et al.* (2014b).

Direct mating

Classical direct mating (Fig. 5.3B), accomplished via controlled spore-to-spore mating, requires two parental strains that produce viable spores with high efficiency, a characteristic that is generally not found in brewing yeasts, especially not in lager yeasts. Ale yeasts vary in their spore viability, with some strains more capable of classical

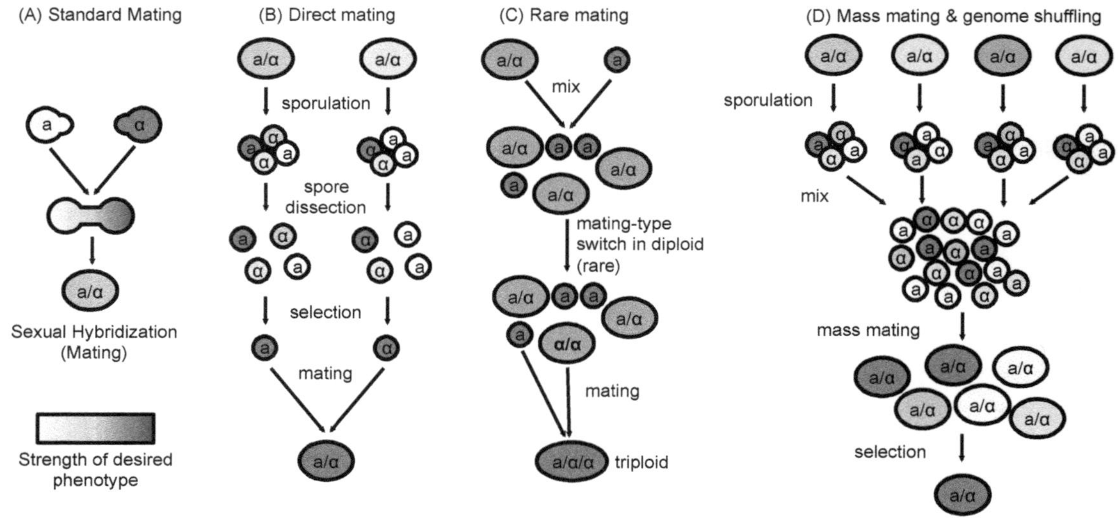

Figure 5.3 'Natural' mating-based strain improvement techniques. Sexual hybridization techniques are often used as a way to generate artificial diversity in yeasts that are capable of (1) mating with each other and subsequently going through successful meiosis and sporulation (e.g. genetically well-behaved yeasts within the same species, e.g. diploid strains of *S. cerevisiae*), or (2) mating with each other through the natural ability of cells or spores of opposite mating types to fuse with each other (e.g. between haploid or spore-producing strains of the various *Saccharomyces* species). Due to the sometimes complex genetics of yeast, different techniques have been developed. Most techniques start from two parental strains (i.e. the cells shown in the top row of each panel) that have been selected for target phenotype(s), and result in diploid or higher ploidy strains that can be selected or screened for the desired phenotype. The greyscale bar at lower left indicates the strength of the phenotype, for example, dark=strong ethanol tolerance, light=weak ethanol tolerance. (A) Standard mating. For 'standard mating', two haploid yeast strains of opposite mating type ('a' and 'α', respectively) and which are capable of continually asexually (mitotically) dividing as haploids (due to inactivation of the HO-endonuclease; see main text) can be pre-screened for desired traits. If selection for diploid cells is possible for this combination of strains, it is very simple to mix a large mass of cells from each parent strain together, whereupon cells of opposite mating type, when located close to each other, can fuse and form 'a/α' diploid cells (Fig. 5.1); these can then be evaluated for presence and strength of the desired trait. If there is no selection available for diploid cells, then individual cells from each parent can be microscopically manipulated to be adjacent to each other on a petri plate (at a marked location), and the colony appearing there will be composed of the diploid cells arising from the original mating of the two cells. This technique can be performed with haploid parents of the same species, or haploid parents of different species within the *Saccharomyces* genus. (B) Direct mating. When one or both of the parental strains contain an active *HO* gene, the strains are essentially always diploid and instead of standard haploid-haploid mating, direct spore-to-spore mating must be used. Two parental strains (shown as intermediate in phenotype strength, but in the case shown, are heterozygous) are each sporulated, then the haploid spores (before they can germinate and self-mate to form diploids) are microscopically manipulated to be placed adjacent to each other on a petri plate as described above; the diploid colony of cells arising from the mating of the two spores can then be seen on the plate and chosen for further evaluation. Because the parents are heterozygous, their spores are genetically diverse as shown by the differences in colouring, and randomly choosing spores for mating will result in a wide range of the phenotype of interest. However, if a screening or selection step is applied to the spores before mating, then only those spores with the desired phenotype can be chosen for mating, leading to enhancement of the phenotype; if a DNA-based screening of a single spore is used to identify spores with known desired gene variants, this method becomes a type of 'marker-assisted breeding'. (C) Rare mating. For 'rare mating', strains are crossed without a sporulation step. This is possible because normal 'α/α' diploid yeasts will very rarely undergo a mating-type switch on one chromosome, yielding an 'a/a' or 'a/α' diploid cell. These cells can subsequently mate with a haploid cell of the opposite mating type as shown. It is important to note that rare mating is not limited to the development of triploid yeasts. For example, tetraploid hybrids can be obtained if the parent on the right was an 'a/a' diploid cell instead of an 'a' haploid. (D) Mass mating and genome shuffling. For 'mass mating', multiple parental strains, or a heterogeneous population (e.g. after mutagenesis) of the same parental strain, can be used. After mass sporulation and mixing of the resulting spores, mass mating will occur. These rounds of mass sporulation and mass mating can be repeated multiple times, a process which is one way to perform so-called genome shuffling. In genome shuffling, the mass-sporulation and mass-mating steps can also be replaced by protoplast fusion (Fig. 5.4B).

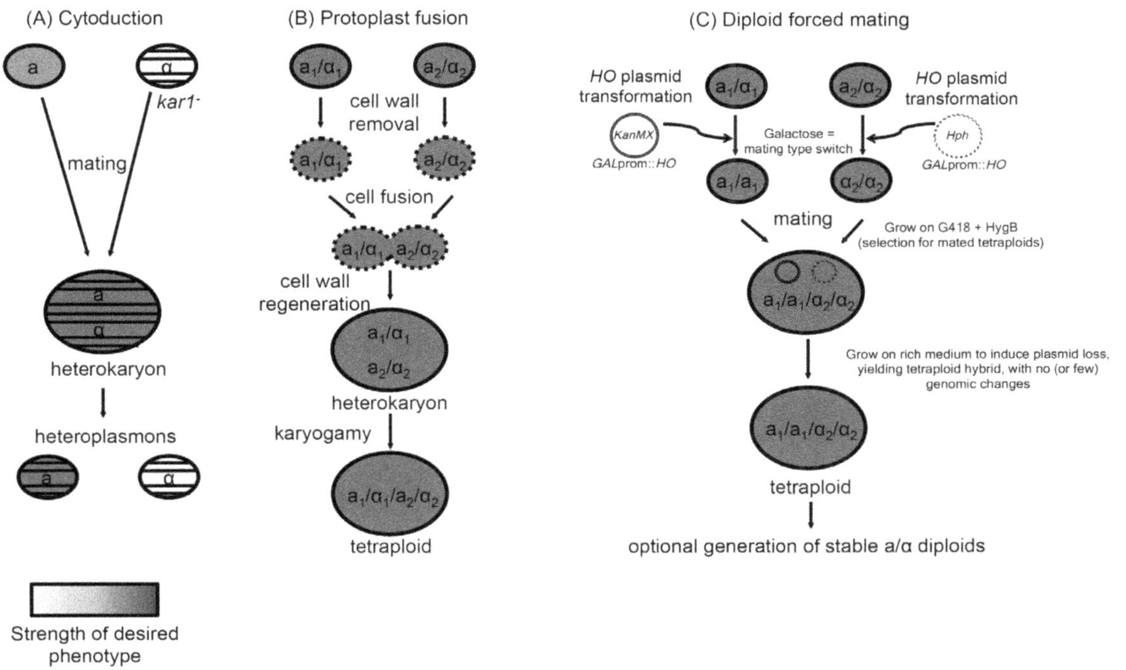

Figure 5.4 Other strain improvement techniques. As in Fig 5.3, parental strains are shown in the top row, and the greyscale bar at lower left indicates the strength of the phenotype. (A) Cytoduction. Cytoduction (cell fusion without nuclear fusion) can be used to transfer cytoplasmically inherited traits, such as mitochondria or other organelles between cells. It can also result in transfer of single chromosomes between nuclei. The parental strain containing the desired cytoplasmic trait first needs its *KAR1* gene inactivated (shown as kar1⁻). Next, both parental strains are mated, if capable (if not, they can be fused by protoplast fusion). Because nuclear fusion (karyogamy) is blocked due to the kar1⁻ defect, the heterokaryon (cell containing two unfused nuclei) subsequently divides into cells containing a nucleus of only one parent but the cytoplasmic components of both parents (=heteroplasmons). With proper genetic selection, this technique can also yield so-called disomic strains that contain the full chromosome complement of one parent plus one chromosome from the other parent. (B) Protoplast fusion. For protoplast fusion, cell walls are first removed (usually by enzymatic means), after which the cells are asexually merged by incubation in osmotically supportive medium and in high concentration so cells are in close proximity. After fusion, the cell wall regenerates and the heterokaryons may undergo karyogamy to form asexually stable hybrids. (C) HO-induced mating and hybridization. This technique allows genetic analysis of a sterile or otherwise intractable strain via a tetraploid intermediate, as well as efficient production of *Saccharomyces* interspecific hybrids. Each parent strain is transformed with a plasmid containing an inducible HO gene, and because each plasmid has a different dominant selectable marker, this allows selection of successful mating events on double selective media. In the example shown here, parental diploids are mated to form a tetraploid, but other starting ploidies may be used.

breeding techniques; this must be established on a strain-by-strain basis, which can be a laborious process. Strains also vary in their sporulation efficiency, and sporulation conditions developed for laboratory strains might not be optimal for all brewing yeasts. For example, lager yeasts prefer lower sporulation temperatures than do laboratory *S. cerevisiae* strains (Bilinski *et al.*, 1987), and sporulation conditions may thus need to be optimized on a strain-by-strain basis, again a laborious endeavour. However, once viable spores have been isolated, breeding lines for future mating experiments can be developed.

The direct mating approach has been carried out with brewing yeasts, either with strains that have high spore viability, or with rare surviving spores, as shown by the following examples. Gjermansen and Sigsgaard (1981) isolated rare viable spores of lager yeast and established breeding lines to develop strains with fermentation performances similar to the parental strains. Bilinski and Casey (1989) performed mating of rare viable *S. pastorianus* spores with *S. cerevisiae* ale strain spores, observing altered (and in some cases desired) fermentation characteristics. Similarly, Sanchez *et al.* (2012) improved the thermotolerance, osmotolerance, and ethanol tolerance of lager yeast by mating rare viable *S. pastorianus* spores with spores from *S. cerevisiae* strains derived from multiple sources. Steensels *et al.* (2014a) used direct mating to improve the aroma characteristics in ale yeast, but first screened out yeasts that were incapable of forming stable haploid breeding lines. Krogerus *et al.* (2015) mated spores from a natural auxotrophic mutant (ura-) *S. cerevisiae* ale yeast strain to spores from a natural auxotrophic mutant (lys-) *S. eubayanus* yeast strain, and selected for prototrophy, thus isolating novel lager yeast-type interspecific hybrids. Similarly, Mertens *et al.* (2015) also generated novel lager-type hybrids without underlying auxotrophic mutations, by directly manipulating spores from unaltered ale yeast strains to be adjacent to spores from unaltered *S. eubayanus* yeast strains, thus allowing fusion and mating without any selection steps. In both studies many of the novel lager-type hybrids showed promising unique fermentation properties and aroma profiles not seen in production lager strains.

A very recent study has characterized the whole genome sequences of over 100 *S. cerevisiae* ale and other brewing yeast strains (Gallone *et al.*, 2016), yielding a rich trove of genomic data, including polymorphisms that are known or suspected to contribute to industrially relevant phenotypes. Such knowledge can allow the use of PCR or other DNA-based assays, on a fairly large scale, to select only those segregants or cells carrying the desired genetic variants for further breeding; this has been called 'marker-assisted breeding' and has been used for many years in crop and livestock breeding. As a proof-of-concept for brewing yeasts, Gallone *et al.* (2016) created new *S. cerevisiae* intraspecific hybrids with altered aromatic properties using marker-assisted breeding.

In general, however, direct mating can be laborious and time-consuming, especially for strains, such as many brewing strains, that sporulate poorly. Furthermore, the direct mating strategy is a gamble: because brewing strains are generally heterozygous, meiotic segregation results in spores that can be very diverse genetically, so that some of the spores used in matings may carry inferior traits; also, it is extremely challenging to select for genes that are closely physically linked to loci that cause spore inviability. Thus, other hybridization methods have been developed that do not rely on the ability to generate viable spores, as follows.

Rare mating within or between *Saccharomyces* species

Diploid (and higher ploidy) yeast cells generally do not express individual mating types and therefore do not participate in mating. However, at very low frequencies these cells can undergo mating type switching to become homozygous for either the 'a' or 'α' mating type (Gunge and Nakatomi, 1972; Spencer and Spencer, 1996); they are then able to mate to a spore or cell of the opposite mating type (Fig. 5.3C). Since yeast populations can be quite large, these rare events do occur in large populations. If the resulting mated cells are able to be isolated by selection or screening, hybrid strains that are triploid, tetraploid, or higher ploidy, can be obtained (de Barros Lopes *et al.*, 2002). While rare mating has been successful in producing commercially used novel wine yeasts (Bellon *et al.*, 2011, 2013, 2015), few have utilized this approach in the breeding of brewing yeasts; an example is the rare mating of *S. cerevisiae* ale yeast to *S. bayanus* to increase fermentation performance at low temperatures (Sato *et al.*, 2002).

Cytoduction

Cytoduction (Fig. 5.4A) is the transfer of subcellular organelles between cells without the transfer of nuclear genes (Spencer and Spencer, 1996); this process can be used to specifically transfer mitochondria and mitochondrial genes, or the virus-like-particles that cause the 'killer' phenotype (reviewed by Bussey, 1991; Wickner and Edskes, 2015). The use of cytoduction to introduce the 'killer' phenotype, as well as mitochondria, into ale and lager yeast strains has been reported (Hammond and Eckersley, 1984) with the eventual goal(s) of creating brewing yeasts that could kill contaminating yeasts, or display altered fermentation characteristics, respectively. The method works by fusing or mating a 'recipient' strain carrying a mutation (in the *KAR1* gene) that makes its nucleus unable to fuse with the 'donor' cell's nucleus after mating between the two cells has occurred. Cytoplasmic contents from the donor cell can then be transferred into the cytoplasm of the recipient cell; additionally, single chromosomes can 'leak out' from the nucleus of the donor cell and get taken up by the nucleus of the recipient strain. Nilsson-Tillgren *et al.* (1981) performed early experiments using this method with lager yeasts, transferring single lager yeast chromosomes into *S. cerevisiae* lab strains. This allowed some genetic characterization of the transferred lager chromosomes and also helped deduce the interspecific hybrid nature of lager yeasts; however this method is laborious and restricted to genetic characterization, and is thus not very useful in the development of new and improved brewing yeast strains.

Protoplast fusion

Yeast, like all fungi, have a cell wall; the *S. cerevisiae* cell wall is primarily made of mannoproteins, polysaccharides, and chitin (Klis *et al.*, 2006). When the cell wall is removed from a yeast cell, the result is a membrane-exposed protoplast (also called a spheroplast). Under electrical current, protoplasts can fuse together to form hybrid cells, which combine the genomes of each parental strain, resulting in a higher ploidy cell (Fig. 5.4B). The fusion product cell is capable of unlimited asexual growth, and may be capable of sexual reproduction if the parental strains are genetically compatible. Studies that have used protoplast fusion to create novel yeast hybrids for brewing applications are plentiful and span several decades (Russell and Stewart, 1979; Stewart, 1981), and this approach has specifically been used to transfer the killer phenotype (Young, 1981), or to optimize characteristics such as dextrin utilization (Barney *et al.*, 1980; de Figueroa and de van Broock, 1985), flocculation (Urano *et al.*, 1993), and ester production (Mukai *et al.*, 2001). Protoplast fusion also works between *Saccharomyces* yeasts and various other fungal genera, since the normal reproductive barriers are overcome by this process (Spencer *et al.*, 1988). This idea of generating intergeneric hybrids (i.e. hybrids between two organisms where each belongs not only to a different species, but to a different genus) by protoplast fusion has not been explored extensively in brewing yeast breeding, although it has been discussed (Morales and Dujon, 2012; Steensels *et al.*, 2014b), and recent examples in cider (Ye *et al.*, 2013) and wine (Carrau *et al.*, 1994) yeast exist.

HO-induced switching and hybridization

Another method for creating hybrids that remains unexplored for brewing yeast breeding is HO-induced hybridization (Fig. 5.4C), which was originally reported as a method to dissect the genetic basis of pentose utilization in a naturally occurring sterile hybrid yeast (Schwartz *et al.*, 2012). This technique involves transforming each parent strain with a plasmid that can transiently induce expression of the site-specific HO endonuclease, which facilitates mating type switching in higher ploidy cells (normally the HO endonuclease enzyme is active only in haploid cells as described in 'Life cycle of *Saccharomyces* yeasts; genomes, ploidy, aneuploidy, and interspecific hybrids' above). Each parent strain is transformed with a plasmid containing an inducible HO gene; each plasmid has a different dominant selectable marker, enabling easy selection of successful mating events on double selective media. The HO-induced hybridization method works on a principle similar to rare mating, where normally mating-incompetent higher ploidy cells switch to mating type 'a' or 'α'; however, in this case the switch occurs at a higher rate due to HO enzyme induction. Once mating has occurred, the resulting hybrid yeast can be grown non-selectively for several generations. This causes the yeast to eject the plasmids, leaving no trace of genetic modification in the hybrid (Fig. 5.4C).

This method has been successfully used to create a tetraploid saison ale yeast hybrid (D.J. Kvitek, unpublished), formed between two parental saison yeasts with very different characteristics: parent one is a classic saison yeast with an earthy phenolic aroma profile but poor fermentation characteristics, while parent two is a very strong fermenting saison yeast with a subdued ester-dominant aroma profile. The hybrid between these yeasts produced a beer that combined the earthy aroma of parent one with the strong fermenting characteristics of parent two. The hybridization of these two strains thus eliminated the need to pitch multiple strains of yeast into the same wort to get the desired combination of aromas and fermentation behaviour; instead, the same results were obtained from a single hybrid strain, making both brewing and yeast strain maintenance easier and more streamlined. Recently, a similar HO-induced hybridization method has been reported and used to create synthetic interspecific hybrid lager-, Belgian ale- and cider-type yeasts (Alexander *et al.*, 2016). These studies suggest that this method of generating both interspecific and intraspecfic hybrids, especially for cases where the desired parental strains do not produce viable spores, could be very useful in generating brewing yeasts with novel flavour and aroma profiles that can be further combined with desired fermentation behaviours.

Future directions for breeding and genetic manipulation of brewing yeasts

Recent advances in DNA sequencing technologies, as well as novel applications of older genetic techniques – especially when such technologies are combined – hold much promise for the future development of brewing-related yeast strains. These methods, discussed below, include directed construction of hybrid yeasts, genome shuffling, 'classic' genetic modification techniques, and recent techniques for minimally invasive genetic modifications of genomes.

The future of hybrid yeasts in brewing

Novel yeast strains created through both interspecific (mating between two strains of different species) and intraspecific (mating between two strains of the same species) yeast hybridizations have been commercialized in industries such as biofuels and wine; it does not appear that any such 'laboratory-made' hybrids are currently used commercially in the brewing industry. Whether this is due to the potential perception (warranted or not) of laboratory–made hybrids as 'genetically modified organisms' (GMOs), a previous failure to obtain successful beer strains through hybridization, or for other intangible reasons, it is clear that recent discoveries from whole-genome sequencing, along with advances in genetic techniques, make it easier and possibly more rewarding than ever to carry out creative hybridizations using brewing yeasts. For example, in the wine arena, interspecific hybrids between many of the different members of the *Saccharomyces* genus have been constructed via the rare mating method, using a diploid wine yeast as the *S. cerevisiae* parent; the resulting hybrids often provide unique and desirable sensory characteristics to wine, and some of the strains are now being used commercially (Bellon *et al.*, 2011, 2013, 2015).

There is some movement towards these types of studies in the beer arena. With the recent discovery of free-living *S. eubayanus* as the non-*S. cerevisiae* component of the lager yeast genome (Libkind *et al.*, 2011), novel lager-like yeast hybrids are beginning to be deliberately constructed to create strains with unique fermentation, flavour, and aroma characteristics, as discussed above in the 'Direct mating' section (Krogerus *et al.*, 2015; Mertens *et al.*, 2016). Phenotypic screens of ale yeasts (see, for example, Steensels *et al.*, 2014a), including 'heirloom' strains (see, for example, Parker *et al.*, 2014), has highlighted the diversity that exists for aroma profiles and other traits among these yeasts (Fig. 5.2A). The wide spectrum of different ale yeasts that can serve as the *S. cerevisiae* parent, along with the very recent isolations of several genetically diverse and geographically far-flung strains of *S. eubayanus* (Bing *et al.*, 2014; Peris *et al.*, 2014, 2015; Baker *et al.*, 2015), points to an expanding universe of possible 'lager-like' hybrid combinations. Deliberate intraspecific hybridization (e.g. between different strains of *S. cerevisiae* yeasts; see discussion above of Steensels *et al.*, 2014a, crosses of ale and sake yeasts, and Kvitek, unpublished results of crossing two different saison yeasts) appears to have been more rarely studied among brewing strains than interspecific hybridizations. This is possibly because lager beer is far and away the dominant beer style in the world, and thus experiments to optimize lager yeasts may be more common and/or able to be funded. Overall, experimental hybridization studies have not been thoroughly explored in a brewing context, and may provide increased genetic diversity that can subsequently be screened/selected for novel and desirable features (Figs. 5.3 and 5.4). Additionally, intergeneric yeast hybrids – mainly obtained through protoplast fusion – have been explored in other industrial applications (Morales and Dujon, 2012; Steensels *et al.*, 2014b) but not in brewing yeasts; this approach may provide a wealth of strains with even more traits of interest.

Genome shuffling to combine traits and/or to discover the genetic basis of phenotypic traits in brewing yeasts

Other genetic techniques, based on combining and/or mixing the genomes of strains, offer new possibilities for brewing yeast strain construction. Because virtually all brewing strains are non-mating (due to being of diploid or even higher ploidy, or aneuploid), it is not possible to just merely mate two such strains together to mix their genomes (as one can do with stably haploid, *HO*-mutant laboratory strains). However, it is often possible to perform mass mating of spores, or mass cell fusion, to rapidly combine desired traits from two or more brewing yeast strains into a single strain. These techniques can also work for starting strains that are heterogeneous, such as mixed populations (for example, a mutagenized population of cells, or a pool of meiotically recombined cells). First, a very large number of cells (or spores, if the strains can produce viable spores) from each starting 'parental' population are generated (note that there can be one, two, or even more starting populations); they are then all mixed together, allowing random mating (if haploid cells or spores) or random cell fusion (if asexual cells are made into protoplasts) to occur. This leads to a mixing of the genotypes of the various starting populations into single cells, often resulting in cells bearing the combined traits of interest. If selection or enrichment for cells carrying the desired combination of traits is possible, several rounds of mass mating (or mass fusion) followed by selection can be performed iteratively to give further refinement or stronger expression of the desired phenotypic traits. In evolutionary terms, this process consists of genetic recombination followed by natural selection of genotypes displaying advantageous phenotypes in the imposed environmental condition, repeated over several generations.

The goal in these typical mass-mating or mass-fusion methods has usually been to achieve a stable hybrid line that expresses the desired traits. Related techniques that result in 'genome shuffling' may prove to be a valuable research tool to discover the genetic bases of phenotypic traits important in beer production, and may also be of use in generating novel strains that combine beneficial traits. Genome shuffling experiments are similar to the mass-mating or mass-fusion methods, but are performed such that genome recombination occurs at every round, eventually mixing the two starting genomes together into single cells, but in a very patchwork manner that varies from cell to cell, thus allowing different phenotypic combinations to be observed. Again, this can be used to bring together and enhance many positive traits from different backgrounds into one superior strain. But the recombined populations can also be used to map where the locations of genes controlling the traits reside. Brief descriptions of techniques and examples from the brewing industry (if available) are discussed below.

Mass mating

The traditional mass-mating protocol (i.e. mixing pools of spores; Fig. 5.3D) was employed by Wang and Hou (2010) using a diploid ale strain that sporulated well and produced viable spores. The starting diploid strain was subjected to both chemical and UV irradiation mutagenesis, then sporulated, and the haploid spores allowed to mass-mate in a random manner. The resulting mated diploid cells were then subjected to growth under increased sugar concentrations and ethanol concentrations, and the best-growing strains were chosen to repeat the process; a total of 3 rounds were performed. This resulted in a strain that showed increased tolerance to high sugar and ethanol concentrations, faster fermentation times in high sugar and ethanol, higher ethanol yields, and better flavour profiles. However, this method has seen limited use for brewing yeasts due to the frequent difficulty in sporulating and mating most such strains.

Mass mating and genome shuffling in asexual strains via transient HO induction

Traditional mass mating can be performed only with sexually competent yeast strains. However, it is possible that forced hybridization of asexual yeasts, via transient HO induction as discussed above (Fig. 5.4C), could yield sexually competent cells (e.g. by doubling an aneuploid or hybrid chromosome complement to make a functionally diploid cell) that are able to proceed through sporulation and produce meiotically recombined offspring. Performing several rounds of this regime could serve as a potent genome shuffling technique for genetically intractable strains, and allow screening or selection of strains with novel combinations of beneficial traits.

Mass protoplast fusion

Another technique for performing genome shuffling on genetically intractable strains is based on recursive protoplast fusion: cells from the two starting strains (which can first be mutagenized) are stripped of their cells walls, then mixed and allowed to fuse, with the process then repeated iteratively. This technique has been used for non-brewing industrial strains of various yeasts and bacteria (reviewed by Gong *et al.*, 2009) and has recently been performed with a diploid *S. cerevisiae* brewing yeast to yield strains with increased production of the natural antioxidant glutathione (Yin *et al.*, 2016).

RTG/LOH-based shuffling

A very recently described technique that leads to genome shuffling, based on a 'return to growth' (RTG) strategy, essentially interrupts the process of meiosis, which is initiated when diploid cells are starved for certain nutrients. This interruption is achieved by adding back nutrients at a point after meiosis has been initiated but before any commitment to meiosis (and therefore before meiotic cell division) has occurred (Dayani *et al.*, 2011; see also review by Simchen, 2009). When RTG is performed, the cells return to the mitotic growth pattern, but with an important change – their chromosomes have undergone meiotic recombination, which causes double-stranded breaks in the chromosomes; the repair of these breaks leads to regions of the chromosomes that have experienced a loss of heterozygosity. Loss of heterozygosity (LOH) refers to a chromosomal region that was originally heterozygous (where DNA sequences differ at various sites along each parental chromosome), but has become homozygous, i.e. both chromosomes now share the identical DNA sequence across that region, and thus, for that 'LOH' region, the cell has lost the genetic information from one of the parents. These LOH patches occur in different places for each different cell in the population that underwent RTG, so that within the population as a whole, most or all of each parental genome is 'uncovered' as an LOH patch (Laureau *et al.*, 2016). Successive rounds of RTG leads to progressively smaller LOH patches, so that eventually a population of diploid cells is obtained in which the chromosomes are essentially entirely homozygous but contain a patchwork of contributions from each parent across a single chromosome, i.e. the genome has both been shuffled and homozygosed. Importantly, this technique works even with 'sterile' strains that do not sporulate and/or only produce inviable spores, and furthermore, it requires no genetic modifications ('GM'; see below section 3.5) including the introduction of plasmids (even transient), making it an attractive option for brewing yeasts and other industrial strains. An even more recent study describes the use of the CRISPR-Cas9 genome editing system (described below under 'Precise genetic modifications') in a heterozygous *S. cerevisiae* diploid to produce targeted mitotic recombination events at any desired location with high frequency (Sadhu *et al.*, 2016). Similar to RTG, this generates a population of cells with shuffled genomes of LOH patches; however, the use of plasmids carrying the editing system components may be considered as 'genetic modification'.

Genome shuffling to discover the genetic basis of phenotypic traits

Once a genome-shuffled population is derived from the desired starting strains, it can be used to perform phenotypic screens to identify the gene(s) responsible for the trait; this is especially important for QTL that have many genes contributing to the trait (see Liti and Louis, 2012, for review of different QTL screening methods). Such shuffling may also reveal novel beneficial genetic combinations, leading to new traits of interest. If a selection is available for the phenotype of interest, the population can be subjected to the selection, such that only those cells with the phenotype survive and can be harvested; this group of cells (the genetic 'segregants' showing the trait of interest) can be analysed by DNA sequencing or other whole-genome assays and compared to the un-selected population to determine which mutations or gene alleles are causing the new phenotype; this type of study is called 'bulk segregant analysis'. When used with extremely large populations, segregating for genetic and trait variation, the technique can be very powerful and specific in detecting the various genes contributing to the trait (Wenger *et al.*, 2010; Ehrenreich *et al.*, 2011; Parts *et al.*, 2011). If high-throughput screens are available for the phenotype(s) of interest, thousands of single colonies can be picked and assayed robotically for the phenotype; those colonies displaying the two extremes of the phenotype are then separately pooled (one pool for each phenotype extreme) and each pool sequenced. Those shared genetic variants found among all the clones in a given pool that are rarely or never present in the other pool can be explored as possible causal variants of the phenotype. The larger the number of clones per pool, and the more extreme the phenotypic difference between the two pools, the more sensitive bulk segregant analysis becomes for detecting the underlying genetic loci responsible for the trait (e.g. Snoek *et al.*, 2015). The insights gleaned from these types of genome shuffling studies, such as the genes and pathways contributing to desired traits, can be used to construct novel brewing yeast strains either directly through genetic modifications or metabolic engineering/synthetic biology, or indirectly through adaptive evolution, as described below.

Adaptive (directed) evolution

Adaptive evolution (also called directed or experimental evolution, or evolutionary engineering) refers to methods of selecting among a cell population those individuals that contain beneficial genetic mutations for a particular trait; this can be done by growing the population of cells for long periods of time (i.e. many generations of cell division) under conditions that favour the beneficial mutations (Fig 5.2B). This can be performed either in repetitive steps ('batch' evolution), or in a long continuous manner (e.g. chemostat or other continuous culturing methods), under a growth condition that favours cells with any beneficial genetic variation (i.e. that allows them to grow better in that condition). The end-point of the adaptive evolution regime is a new population that consists mainly of cells with the beneficial mutation(s). In these experiments, the genetic variation that contains such beneficial mutations either already exists in the population (e.g. the experiment starts with a mixture of genetically diverse strains), or arises by spontaneous mutation during the evolution; the population can also be deliberately enriched for genetic variation (such as a mutagenized or genome-shuffled population) prior to the evolution. Adaptive evolution can also be used on a single strain, to incrementally enhance an already existing phenotype. In these types of experiments, the selection is not just a one-step process, but instead occurs over hundreds of generations. The strategies employed in adaptive evolution include serial transfers of batch cultures, continuous culturing methods, plate selection, and other methods. Chambers *et al.* (2009) give detailed examples and descriptions of various types and techniques of adaptive evolution, especially as pertaining to yeasts used in the food and beverage industries, while a recent review by Fisher and Lang (2016) describes methods, results and future directions for experimental evolution studies in various fungal model systems, including *Saccharomyces*.

Whole-genome and high-throughput sequencing

Recent advances in DNA sequencing technology have led to the ability to rapidly and very inexpensively (< $100; and probably < $10 in the immediate future) sequence the whole genome of a yeast strain (see Reuter *et al.*, 2015, for an up-to-date review of high-throughput sequencing technologies); in turn, this has led to an explosion in the number of yeast genomes that have been fully sequenced – at current count, the number of *S. cerevisiae* whole-genome sequences in the literature and public databases is on the order of a few hundred, representing strains isolated from all manner of environments, including brewing-related yeasts; but this number is likely to grow into the thousands as sequencing costs fall even further. Other chapters in this book (see Chapters 4 and 6) will delve in much more detail into the insights about lager, ale, and other brewing yeasts that have come from the analyses of their genome sequences; we merely note here that the availability of high-throughput sequencing will be invaluable for all genetic work with brewing yeasts going forward, for example performing the types of genome-shuffling QTL analyses described above, and for possible genetic modifications as described below.

Genetic modification strategies

Beer-making has been performed over many millennia and has long been an occupation of much importance, both socially and economically. The brewing industry has often served as an impetus for scientific progress, especially in the areas of chemistry and microbiology, and has even been cited as the foundation for the field of biotechnology (Bud,

1994). This is probably due to many aspects: e.g. the need for pre-processing and sterilization of wort and the ensuing need for cleaning and sterilization of equipment; the necessity of recognizing and transferring only those yeast with desired qualities into the sterile wort; the eventual isolation of pure strains of brewing yeasts and the understanding of their genetics; the monitoring and understanding of the resulting fermentation process; and the understanding of chemical and microbiological changes that can occur upon storage of beers. However, despite the historic scientific underpinnings of beer-making, it is currently generally perceived by the brewing industry that most customers/consumers of beers would be unwilling to purchase or consume beers that are produced using genetically modified organisms (GMOs); there is thus much trepidation about modifying brewing yeasts using any type of technique that could be construed as 'genetic modification' (GM). However, the precise characteristics that comprise a GMO can be difficult to define universally. On the one hand, most of the directed breeding techniques described above, specifically directed mating, rare mating, traditional mass mating, and cytoduction, do not require any introduction of foreign DNA into the yeasts and thus most likely would not be considered by scientists, regulatory agencies, and (hopefully) the beer-drinking public as resulting from GMOs. Likewise, brewing yeasts subjected to mutagenic agents and then put through genetic screens (as discussed below) are also not considered to be GMOs. On the other end of the spectrum; however, any technique yielding a yeast strain that permanently contains within its genome any 'exogenous' DNA – i.e. DNA that does not originate from that species of yeast, especially DNA coming from organisms not closely related to yeast – probably would be considered by scientists, regulatory agencies, and the lay public alike as a GMO (Fig. 5.2B). This type of 'permanent exogenous DNA' genetic modification of the yeast genome is very easy to perform due to the long history of yeast being used as a model organism in molecular biology studies; in many cases, such GMO yeasts are used in biotechnology applications to produce drugs or for bioethanol production (Steensels *et al.*, 2014b). However, both the brewing and wine industries currently avoid the use of GM yeasts.

There are, however, genetic manipulation methods that straddle the middle grounds in the concept of what comprises a GMO. For example, hybrid organisms created through protoplast fusion are considered in some countries to be GMOs, but not in other countries (Pérez-Través *et al.*, 2012; Steensels *et al.*, 2014b). Also, some plasmid-based methods, such as the HO-induction based techniques described above, are interesting because they introduce plasmids containing bacterial and yeast DNA into a yeast strain, but only transiently; the plasmids do not interact with the nuclear genome of the cell, and after the plasmids are ejected from the cells, there is no change to the cell's genome other than the mating-type switch, which could have occurred in a normal haploid yeast cell. Other genomic techniques can result in precise changes to the genome of a brewing yeast strain, but are changes that merely represent DNA regions from other yeasts of the same species. Finally, new techniques allow the precise deletion of genes from yeast and/or the precise addition of suites of genes, coming either from yeast or other species, that can encode entire metabolic pathways, allowing the synthesis of new molecules in the cell, a type of 'synthetic biology'. Whether any these types of manipulations would be perceived by the public and/or regulatory agencies as representing the application of a GMO is an open question. We briefly describe some of these types of genetic modification methods below; most are described in greater detail in the review by Steensels *et al.* (2014b). Note that in contrast to the mating and shuffling techniques described above, which aim to generate an increase in, and mixing of, genetic diversity, the GM techniques (except mutagenesis) described below offer the possibility to start with just one starting yeast strain and alter only a desired number of pre-selected genes while retaining the remainder of the strain's genome, and thus hopefully retaining its beneficial brewing characteristics (Fig. 5.2B).

Genome-wide mutagenesis

Brewing yeasts have been subjected to growth in the presence of mutagens, such as irradiation with UV light or treatment with ethane methylsulfonate (EMS), resulting in a population of cells carrying different spontaneous mutations in the genome that could lead to the acquisition of desired traits that can be screened or selected for (Fig. 5.2B). This approach has been used quite widely in brewing yeasts, as the resulting strains are not considered to be GMOs. Mutagenesis has been used in combination with mass mating and genome shuffling (e.g. by Wang and Hou, 2010, as discussed above), but has also been used as a stand-alone method as a way to increase genetic diversity and screen for beneficial phenotypes. For example, a lager yeast strain was mutagenized by EMS; the resulting mutants were screened for growth in high-gravity and high-ethanol conditions, and some clones were obtained that show better growth in these conditions (Yu *et al.*, 2012). In another study, sulfur pathway mutants of a *S. cerevisiae* brewing strain were obtained by performing two rounds of UV mutagenesis, combined with plating for cadmium sulfate resistance (to increase glutathione levels). The best candidate clone produced higher levels of sulfur dioxide and glutathione (antioxidants), but lower levels of hydrogen sulfide (off-flavours), all of which should improve beer flavour stability (Chen *et al.*, 2012).

Plasmid-based genetic modification

Plasmids, circular pieces of self-replicating DNA usually bearing both bacterial and yeast-encoded genes and regulatory regions, represented the first method used to introduce exogenous DNA into yeast cells in the laboratory (Beggs, 1978; Hinnen *et al.*, 1978). Such plasmids can easily be introduced ('transformed') into brewing yeast strains, provided that there is a dominant selectable marker on the plasmid, such as a gene that provides resistance to a drug that would normally be toxic to the brewing yeast. These are the types of plasmids that are used in the HO-induction methods to promote mating, as described above. In some early applications of plasmid technology to brewing applications, libraries of plasmids – each carrying a different overexpressed yeast gene (or in some cases, genes from other organisms) – were inserted into brewing strains; the strains were then screened for beneficial traits. An interesting recent autobiographical review includes the saga of how UK government approval was pursued and achieved for a plasmid-modified brewing strain (Hammond, 2016). However, since these circular self-replicating plasmids can easily be lost from the cells upon the lack of the selective agent (e.g. a selective drug), most types of plasmid-based genetic modification are unstable and thus more often used for research purposes, as in the example above to identify novel genes of interest, and not generally used to introduce new functions or attributes into brewing (or other biotechnology-related) yeasts destined for production use.

Chromosomal integration of DNA

Yeast cells, due to their robust homologous recombination capacity, have an extraordinary ability to take up linear pieces of DNA into their cells and then precisely integrate those DNA pieces into their chromosomes, provided that the linear DNA has stretches (greater than ≈30 base pairs) at either end that exactly match the DNA on one of the yeast chromosomes. In this way, novel exogenous DNA (usually as linearized plasmids, as PCR products, or as synthetic oligonucleotides), carrying genes or other genetic regions from yeast or non-yeast sources, can easily be introduced into yeast cells and become inserted between the two flanking matching regions into the DNA of the yeast chromosome (see Botstein and Fink, 2011, for diagrams of different types of plasmid insertions). Note that there must generally be a selection to enrich for cells that have actually taken up and inserted the DNA into their chromosome; again, as described above, for brewing yeasts the selectable marker would probably need to be a dominant drug-resistance gene. Because this novel exogenous DNA is inserted directly into the chromosome, it is permanently embedded in the yeast genome and will be transmitted to all clonal descendants of that yeast cell, resulting in a strain that stably retains the DNA even in the absence of further selection for the inserted DNA. Note that many industrial yeasts have less ability, compared to laboratory yeasts, to take up and chromosomally insert DNA, so plasmids designed to insert into repeated regions of the genome – such as ribosomal DNA repeats or transposon-related sequences – have been utilized to help obtain integrants and boost expression of the inserted genes by increasing the number of possible integration sites (Lopes *et al.*, 1989; Kudla and Nicolas, 1992). Recently, a system that enables rapid modular construction of yeast integration plasmids, designed for integration and subsequent 'pop-out' using a series of selectable and counter-selectable genes, was shown to yield stable integration of multiple genes into multiple loci of yeast with precise excision of the selection marker (Siddiqui *et al.*, 2014); this may represent a plasmid-based approach to multilocus yeast genetic modifications in addition to the techniques described in the next section, and may be desirable in the brewing industry (should plasmid-exposed strains become acceptable) since the final strains do not contain drug-resistance genes.

Precise genetic modifications

Methods have been developed recently that allow the direct modification of chromosomal DNA in yeast, such that there is no foreign or unwanted DNA (such as selectable gene markers) remaining in the cell's genome. In general, the small number of dominant, counter-selectable and other auxotrophic markers still poses a bottleneck for making modifications of multiple genes in industrial yeast strains, so any method that removes the selectable marker (and thus allows the marker to be re-used in a subsequent step) is very attractive. Most of these methods employ selectable/counter-selectable marker genes, such as the 'delitto perfetto' technique, in which a DNA fragment ('CORE' cassette) is inserted into the genome of the starting yeast strain; the cassette contains gene(s) that are both selectable (for initially detecting cells with the insertion) and counter-selectable (for subsequent removal of the reporter) and is designed with flanking homologous regions so that it will insert into the region of interest (Storici *et al.*, 2001). Note that, for brewing and industrial yeasts, the selection and counter-selection markers should preferably act in a dominant fashion and not require pre-existing auxotrophic mutations (Kutyna *et al.*, 2014; Siewers, 2014). Next, an oligonucleotide or PCR fragment that 'covers' the regions flanking the inserted CORE cassette and contains the desired genetic modification(s) is transformed into the cell; 'counter-selection' then results in the precise insertion, via homologous recombination, of the desired genetic modification, with concomitant deletion of the reporter gene, resulting in a 'scarless' genetic modification. Many types of genetic modifications can be generated with this technique, including gene deletions, promoter replacement, one- or multigene insertion, and nucleotide substitutions (i.e. targeted point mutations; see Stuckey and Storici, 2013). A modification of the 'delitto perfetto' method, where the CORE cassette now also carries both the gene for an inducible endonuclease and the restriction site for the nuclease, generates an induced double-strand break (DSB) in the cassette, which greatly increases the efficiency of the process (up to 20% of cells carry the modification) by recruiting the homology-directed repair machinery to the site; this makes it feasible to carry out these modifications in diploid cells (Storici *et al.*, 2003) and also allows large deletions and targeted gross chromosomal rearrangements to be obtained (Storici and Resnick, 2006). As mentioned above, a very important aspect of this technique is that it allows one to 'rescue' (or more accurately 'recycle') the reporter gene with each round; it can therefore be used in a series of selection/counter-selection steps to generate many different genetic modifications into a single yeast strain. While there apparently have been no reports of this method having been used for modifying brewing yeasts, a series of genetic modification steps using the 'delitto perfetto' technique has recently been employed to generate wine strains producing lower ethanol than the starting strain; however, this was done in a haploid strain (Varela *et al.*, 2012). Overall, though, the many sequential transformations needed for 'delitto perfetto'-based multigene modifications are time-consuming, and induction of DSBs can cause unwanted mutations or genomic rearrangements (Solis-Escalante *et al.*, 2014); it thus remains to be seen how well this type of technique can apply to industrial strains.

A particularly exciting new methodology for precise 'scarless' genome modification, which works in virtually all organisms (even those without robust homologous recombination capacity), is the CRISPR/Cas9 system (Gasiunas *et al.*, 2012; Jinek *et al.*, 2012; also recently reviewed by Hsu *et al.*, 2014; Sander and Joung, 2014). The basis for this technique lies in the harnessing of a bacterial RNA–enzyme interaction, originally used by the bacteria as a primitive immune response to evade viral infection. To briefly summarize this technique, the Cas9 protein, a DNA nuclease, is expressed from a self-replicating plasmid in a cell, along with a pre-designed 'guide RNA' (gRNA). The Cas9 protein aligns the gRNA (which needs to contain ≈20 bases matching the desired region of genome modification) to the target DNA sequence in the yeast genome, and then creates a DSB in the yeast chromosome at the target site. Oligonucleotides that 'cover' this chromosomal break can be co-transformed into the cell, and just as in the 'delitto perfetto' method, homologous recombination (for yeast and other cells with robust homologous recombination systems) will repair the break, but will incorporate the genetic modifications present in the 'covering' oligonucleotide (note that for organisms without robust homologous recombination systems, non-homologous-end-joining is the usual repair method). Methods based on this genetic modification technique are proving to be successful not just in mammalian and other previously intractable cells, but also in yeast (e.g. DiCarlo *et al.*, 2013), with CRISPR/Cas9 mediated genetic modifications approaching 100% efficiency. This means that CRISPR/Cas9 system allows for the directed modification of the yeast genome without the use of any integrated selectable markers (note that dominant drug-resistant markers are used for the Cas9 and gRNA plasmids, but cells lacking the plasmids can easily be obtained after brief growth on non-selective medium); this is truly an important advance for the modification of industrial yeasts. Additionally, many gRNAs, each targeting a different genomic site, can be introduced at the same time, meaning that multiple genetic modifications can be made at once; this vastly speeds up the process of obtaining multilocus changes in a yeast genome, by avoiding the cycles of marker use and removal needed in the 'delitto perfetto' and related methods. The past 2 years have seen a flurry of research papers describing the use of the CRISPR/Cas9 system to modify yeast genomes, with reports of up to five or six genes or chromosomal regions being modified in a single transformation (Bao *et al.*, 2015; Jakočiūnas *et al.*, 2015a,b; Mans *et al.*, 2015; Stovicek *et al.*, 2015), and even one report of a successful high-efficiency CRISPR/Cas9-induced genetic modification that was seen to occur at the desired locus on all of the homologous chromosomes in a polyploid industrial yeast strain (Ryan *et al.*, 2014). The latter paper also speculates that their modified CRISPR/Cas9 method (which rapidly generates marker-less barcoded gene mutations) could be used to link genes to phenotypes in industrial yeasts via large-scale pooled fitness studies or QTL mapping. A more recent study has directly demonstrated use of the CRISPR/Cas9 system to map traits in a heterozygous diploid *S. cerevisiae* strain, first performing targeted LOH genome shuffling (discussed in RTG/LOH shuffling section above), and then using the LOH panel to very accurately and quickly map a manganese sensitivity trait (Sadhu *et al.*, 2016); this implies that such a system could possibly be used for QTL mapping in industrial yeasts.

The recent explosion of genetic modification techniques that are both high efficiency and precise, and that can be used for diploid (or higher ploidy) prototrophic yeast cells, means that there are ample opportunities for directed modification of brewing yeasts, should the brewing industry wish to pursue this path. Two recent papers (David and Siewers, 2014; Sander and Joung, 2014) review some of the newest advances in genomic engineering techniques applicable for industrial yeasts, especially CRISPR/Cas9-based techniques, as well as other methods that employ endonucleases and targeted DSBs to generate targeted genome modifications.

Metabolic engineering and synthetic biology

Many yeast strains currently used in biotechnology applications (e.g. for production of fuel or drug molecules) were originally constructed by a series of plasmid integrations – often with many exogenous genes and genetic regulatory regions from non-related organisms, comprising an entirely new metabolic pathway, introduced into these yeast genomes. This 'rewiring' of a metabolic pathway in order to produce novel traits, or new or higher levels of a desired molecule, is often referred to as metabolic engineering (Krivoruchko *et al.*, 2011; Steensels *et al.*, 2014b). The term 'synthetic biology' is used somewhat similarly in terms of creating new metabolic pathways in a cell, except that it often implies that the component genes and regulatory regions are synthetic (e.g. assembled from oligonucleotides and not necessarily naturally occurring) and/or are part of a system that integrates all the pieces easily (Krivoruchko *et al.*, 2011; Steensels *et al.*, 2014b), even to the level of whole chromosomes (Koszul, 2016). Aside from the sheer difficulty of obtaining yeast strains with precise multilocus genetic modifications (until recently, at least), one of the most difficult aspects of metabolic engineering is predicting how the cell will respond to the presence of novel molecules and how the altered flux of metabolites within the cell will affect cell growth; computer modelling algorithms have been designed to predict such 'flux analyses' (e.g. Bordel, 2014; Borodina and Nielsen, 2014). With the availability of such algorithms, and also with the increasing number of more precise and efficient techniques for genetic modification in industrial yeast strains, especially the marker-free rapid multilocus genetic modifications that can be made by the CRISPR/Cas9 system, there are many recent reports of yeast strains designed as microbial 'bio-factories' that are producing molecules of interest at industrial scales (reviewed by Borodina and Nielsen, 2014). Finally, taking synthetic biology to its extreme, a large ongoing international research project, called the Synthetic Yeast Genome (Sc2.0) Project, has the goal of creating a yeast cell bearing an entirely synthetic genome (i.e. all chromosomes constructed by designing DNA sequences that are then synthesized in a laboratory) by the year 2018 (Dymond *et al.*, 2011; Pérez-Través *et al.*, 2012; Pretorius, 2016; Steensels *et al.*, 2014). As part of this project, several synthetic chromosome arms and one entire synthetic chromosome have already been constructed and introduced successfully into yeast strains (Annaluru *et al.*, 2014; Dymond *et al.*, 2011). An interesting and useful twist has also been incorporated into the Sc2.0 synthetic chromosomes: the ability to easily and densely shuffle the genome at will, using a technique called SCRaMbLE which employs an inducible recombination system, thus generating genomic diversity rapidly and easily for further selection and trait mapping purposes (Dymond *et al.*, 2011; Shen *et al.*, 2016). Obviously, the ability to insert any number of desired genes, in an effort to express entirely novel metabolic pathways, can be easily designed into such synthetic yeast chromosomes, and this idea has been optimistically championed for the case of wine yeast (Pretorius, 2016). At the same time, the Sc2.0 project is well aware of the ethical and safety concerns associated with the large-scale employment of synthetic chromosomes and synthetic organisms, and has taken care to address these issues publicly and propose a policy of self-regulation, where synthetic biology researchers follow a common set of principles including institutional and governmental oversight regulations where applicable (Pretorius, 2016; Sliva *et al.*, 2015).

In principle, the introduction of novel metabolic pathways by the above-described concerted genetic modifications or even by the introduction of entire synthetic chromosomes could very well be helpful in improving brewing yeasts' performance in fermentation and/or in altering or improving beer sensory characteristics; however, as we have mentioned repeatedly, there is resistance to using such GM yeasts, especially those bearing 'permanent' integration of foreign (non-yeast) or synthetic DNA, in the brewing industry and for now it appears that full-scale metabolic engineering is not being attempted with brewing yeasts.

Summary

We have briefly described the challenges and opportunities of performing different types of genetic manipulations – both traditional and 'next-generation' – on brewing yeasts, as a way to introduce desired fermentation, flavour, aroma, or other, (possibly novel) attributes into these yeasts. Many of these manipulations, namely traditional genetic breeding or mating techniques, require no introduction of exogenous DNA into the yeasts, and therefore the resulting yeast strains would not be considered as GMOs. Other techniques – for example, those utilizing HO induction to induce mating – do introduce exogenous DNA into the strain, but in a transient manner such that no trace of genetic modification is left in the final strain. Yet other techniques, such as the CRISPR/Cas9 system, introduce very precise mutations into the endogenous yeast genome, such that there may be no exogenous non-yeast DNA residing in the genome of the final strain, an example of a 'minimally invasive' genetic modification.

Finally, more traditional genetic modifications and synthetic biology approaches result in strains with non-yeast DNA permanently embedded in the genome of the final strain. Whether any or all of these latter types of techniques ('transient DNA introduction', 'minimally invasive modification' or 'permanent exogenous DNA modification') might eventually result in production-level brewing yeast strains that are 'palatable' to the beer-drinking public remains to be seen. However, it is clear that even if the genetic toolkit for producing novel brewing yeast strains remains restricted to those techniques that require absolutely no introduction of exogenous DNA into the strains at any point in the process, there are still methods (e.g. direct mating either within or between species, rare mating, cytoduction, spheroplast fusion,

RTG-based genome-shuffling, and adaptive evolution) that are available to brewers interested in creating brewing yeast strains with novel desired traits and/or precise combinations of desired traits. The combination (with or without reiterative cycling) of various different techniques – e.g. rounds of mass mating alternating with mutagenesis and/or adaptive evolution, or other combinations that cycle between generating genetic variation and screening or selection – may also prove to be very useful in strain development. In general, however, these types of manipulations have not yet been widely explored in the brewing yeast field, and thus the door for genetic experimentation in brewing yeasts is wide open and promises to yield fruitful results.

References

Alexander, W.G., Peris, D., Pfannenstiel, B.T., Opulente, D.A., Kuang, M., and Hittinger, C.T. (2016). Efficient engineering of marker-free synthetic allotetraploids of *Saccharomyces*. Fungal. Genet. Biol. *89*, 10–17. http://dx.doi.org/10.1016/j.fgb.2015.11.002

Annaluru, N., Muller, H., Mitchell, L.A., Ramalingam, S., Stracquadanio, G., Richardson, S.M., Dymond, J.S., Kuang, Z., Scheifele, L.Z., Cooper, E.M., *et al.* (2014). Total synthesis of a functional designer eukaryotic chromosome. Science *344*, 55–58. http://dx.doi.org/10.1126/science.1249252

Baker, E., Wang, B., Bellora, N., Peris, D., Hulfachor, A.B., Koshalek, J.A., Adams, M., Libkind, D., and Hittinger, C.T. (2015). The Genome Sequence of *Saccharomyces eubayanus* and the Domestication of Lager-Brewing Yeasts. Mol. Biol. Evol. 32, 2818–2831. http://dx.doi.org/10.1093/molbev/msv168

Bao, Z., Xiao, H., Liang, J., Zhang, L., Xiong, X., Sun, N., Si, T., and Zhao, H. (2015). Homology-integrated CRISPR-Cas (HI-CRISPR) system for one-step multigene disruption in *Saccharomyces cerevisiae*. ACS Synth. Biol. *4*, 585–594. http://dx.doi.org/10.1021/sb500255k

Barney, M.C., Jansen, G.P., and Helbert, J.R. (1980). Use of spheroplast fusion and genetic transformation to introduce dextrin utilization into *Saccharomyces uvarum*. J. Am. Soc. Brew. Chem. (USA) *38*, 1–5. http://dx.doi.org/10.1094/ASBCJ-38-0001

Barrio, E., González, S., Arias, A., Belloch, C., and Querol, A. (2006). Molecular mechanisms involved in the adaptive evolution of industrial yeasts. In Yeasts in Food and Beverages (The Yeast Handbook), A. Querol, and G.H. Fleet, eds. (Berlin: Springer-Verlag), pp. 153–174.

Beggs, J.D. (1978). Transformation of yeast by a replicating hybrid plasmid. Nature *275*, 104–109.

Bellon, J.R., Eglinton, J.M., Siebert, T.E., Pollnitz, A.P., Rose, L., de Barros Lopes, M., and Chambers, P.J. (2011). Newly generated interspecific wine yeast hybrids introduce flavour and aroma diversity to wines. Appl. Microbiol. Biotechnol. *91*, 603–612. http://dx.doi.org/10.1007/s00253-011-3294-3

Bellon, J.R., Schmid, F., Capone, D.L., Dunn, B.L., and Chambers, P.J. (2013). Introducing a new breed of wine yeast: Interspecific hybridisation between a commercial *Saccharomyces cerevisiae* wine yeast and *Saccharomyces mikatae*. PLOS ONE *8*, e62053. http://dx.doi.org/10.1371/journal.pone.0062053

Bellon, J.R., Yang, F., Day, M.P., Inglis, D.L., and Chambers, P.J. (2015). Designing and creating *Saccharomyces* interspecific hybrids for improved, industry relevant, phenotypes. Appl. Microbiol. Biotechnol. *99*, 8597-8609. http://dx.doi.org/10.1007/s00253-015-6737-4

Bilinski, C.A., Russell, I., and Stewart, G.G. (1987). Physiological requirements for induction of sporulation in lager yeast. J. Inst. Brew. *93*, 216–219. http://dx.doi.org/10.1002/j.2050-0416.1987.tb04502.x

Bilinski, C.A., and Casey, G.P. (1989). Developments in sporulation and breeding of brewer's yeast. Yeast *5*, 429–438. http://dx.doi.org/10.1002/yea.320050603

Bing, J., Han, P.J., Liu, W.Q., Wang, Q.M., and Bai, F.Y. (2014). Evidence for a Far East Asian origin of lager beer yeast. Curr. Biol. *24*, R380–381. http://dx.doi.org/10.1016/j.cub.2014.04.031

Bordel, S. (2014). Genome-scale metabolic models of yeast, methods for their reconstruction, and other applications. Methods Mol. Biol. *1152*, 269–279. http://dx.doi.org/10.1007/978-1-4939-0563-8_16

Borneman, A.R., and Pretorius, I.S. (2015). Genomic insights into the *Saccharomyces sensu stricto* complex. Genetics *199*, 281–291. http://dx.doi.org/10.1534/genetics.114.173633

Borneman, A.R., Desany, B.A., Riches, D., Affourtit, J.P., Forgan, A.H., Pretorius, I.S., Egholm, M., and Chambers, P.J. (2011). Whole-genome comparison reveals novel genetic elements that characterize the genome of industrial strains of *Saccharomyces cerevisiae*. PLOS Genet. *7*, e1001287. http://dx.doi.org/10.1371/journal.pgen.1001287

Borodina, I., and Nielsen, J. (2014). Advances in metabolic engineering of yeast *Saccharomyces cerevisiae* for production of chemicals. Biotechnol. J. *9*, 609–620. http://dx.doi.org/10.1002/biot.201300445

Botstein, D., and Fink, G.R. (2011). Yeast: an experimental organism for 21st Century biology. Genetics *189*, 695–704. http://dx.doi.org/10.1534/genetics.111.130765

Bud, R. (1994). The Uses of Life: A History of Biotechnology (Cambridge, UK: Cambridge University Press). http://dx.doi.org/10.1002/jctb.280590129

Bussey, H. (1991). K1 killer toxin, a pore-forming protein from yeast. Mol. Microbiol. *5*, 2339–2343.

Carrau, J.L., Dillon, A.J., Serafini, L.A., and Pazqual, M.S. (1994). L-Malic acid degrading yeast for wine making. United States Patent 5,330,774 1–4.

Chambers, P.J., Bellon, J.R., Schmidt, S.A., Varela, C., and Pretorius, I.S. (2009). Non-Genetic Engineering Approaches for Isolating and Generating Novel Yeasts for Industrial Applications. (Dordrecht: Springer Netherlands), pp. 433–457.

Chen, Y., Yang, X., Zhang, S., Wang, X., Guo, C., Guo, X., and Xiao, D. (2012). Development of *Saccharomyces cerevisiae* producing higher levels of sulfur dioxide and glutathione to improve beer flavor stability. Appl. Biochem. Biotechnol. *166*, 402–413. http://dx.doi.org/10.1007/s12010-011-9436-3

Cosma, M.P. (2004). Daughter-specific repression of *Saccharomyces cerevisiae* HO: Ash1 is the commander. EMBO Rep. *5*, 953–957. http://dx.doi.org/10.1038/sj.embor.7400251

Darwin, C. (1859). On the Origin of Species by Means of Natural Selection, or the Preservation of Favoured Races in the Struggle for Life (London, UK: John Murray).

Darwin, C. (1868). The Variation of Animals and Plants under Domestication, Vols I and II (London, UK: John Murray).

Darwin, C., and Wallace, A. (1858). On the tendency of species to form varieties; and on the perpetuation of varieties and species by natural means of selection. J. Proc. Linnean Soc. London. Zoology 3, 45–62.

David, F., and Siewers, V. (2015). Advances in yeast genome engineering. FEMS Yeast Res. *15*, 1-14. http://dx.doi.org/10.1111/1567-1364.12200

Dawson, G. (2013). Beer Domesticated Man. http://nautil.us/issue/8/home/beer-domesticated-man?

Dayani, Y., Simchen, G., and Lichten, M. (2011). Meiotic recombination intermediates are resolved with minimal crossover formation during Return-to-Growth, an analogue of the mitotic cell cycle. PLOS Genet. *7*, e1002083. http://dx.doi.org/10.1371/journal.pgen.1002083

de Barros Lopes, M., Bellon, J.R., Shirley, N.J., and Ganter, P.F. (2002). Evidence for multiple interspecific hybridization in *Saccharomyces sensu stricto* species. FEMS Yeast Res. *1*, 323–331.

de Figueroa, L.I., and de van Broock, M.R. (1985). Spheroplast fusion of a brewing yeast strain with *Saccharomyces diastaticus*. Appl. Microbiol. Biotechnol. *21*, 206–209.

DiCarlo, J.E., Norville, J.E., Mali, P., Rios, X., Aach, J., and Church, G.M. (2013). Genome engineering in *Saccharomyces cerevisiae* using CRISPR-Cas systems. Nucleic Acids Res. *41*, 4336–4343. http://dx.doi.org/10.1093/nar/gkt135

Dunham, M.J., Badrane, H., Ferea, T., Adams, J., Brown, P.O., Rosenzweig, F., and Botstein, D. (2002). Characteristic genome rearrangements in experimental evolution of *Saccharomyces cerevisiae*. Proc. Natl. Acad. Sci. U.S.A. *99*, 16144–16149. http://dx.doi.org/10.1073/pnas.242624799

Dunn, B., and Sherlock, G. (2008). Reconstruction of the genome origins and evolution of the hybrid lager yeast *Saccharomyces pastorianus*. Genome. Res. *18*, 1610–1623. http://dx.doi.org/10.1101/gr.076075.108

Dymond, J.S., Richardson, S.M., Coombes, C.E., Babatz, T., Muller, H., Annaluru, N., Blake, W.J., Schwerzmann, J.W., Dai, J., Lindstrom, D.L., *et al.* (2011). Synthetic chromosome arms function in yeast and generate phenotypic diversity by design. Nature *477*, 471–476. http://dx.doi.org/10.1038/nature10403

Ehrenreich, I.M., Torabi, N., Jia, Y., Kent, J., Martis, S., Shapiro, J.A., Gresham, D., Caudy, A.A., and Kruglyak, L. (2010). Dissection of genetically complex traits with extremely large pools of yeast segregants. Nature *464*, 1039–1042. http://dx.doi.org/10.1038/nature08923

Fisher, K.J., and Lang, G.I. (2016). Experimental evolution in fungi: An untapped resource. Fungal. Genet. Biol. *94*, 88–94. http://dx.doi.org/10.1016/j.fgb.2016.06.007

Gallone, B., Steensels, J., Prahl, T., Soriaga, L., Saels, V., Herrera-Malaver, B., Merlevede, A., Roncoroni, M., Voordeckers, K., Miraglia, L., *et al.* (2016). Domestication and Divergence of *Saccharomyces cerevisiae* Beer Yeasts. Cell *166*, 1397–1410.e16. http://dx.doi.org/10.1016/j.cell.2016.08.020

Gasiunas, G., Barrangou, R., Horvath, P., and Siksnys, V. (2012). Cas9-crRNA ribonucleoprotein complex mediates specific DNA cleavage for adaptive immunity in bacteria. Proc. Natl. Acad. Sci. U.S.A. *109*, E2579–86.

Gibson, B., and Liti, G. (2014). *Saccharomyces pastorianus*: Genomic insights inspiring innovation for industry. Yeast *32*, 17–27. http://dx.doi.org/10.1002/yea.3033

Gjermansen, C., and Sigsgaard, P. (1981). Construction of a hybrid brewing strain of *Saccharomyces carlsbergensis* by mating of meiotic segregants. Carlsberg Res. Comm. *46*, 1–11.

Gonçalves, M., Pontes, A., Almeida, P., Barbosa, R., Serra, M., Libkind, D., Hutzler, M., Gonçalves, P., and Sampaio, J.P. (2016). Distinct Domestication Trajectories in Top-Fermenting Beer Yeasts and Wine Yeasts. Curr. Biol. *26*, 2750–2761.

Gong, J., Zheng, H., Wu, Z., Chen, T., and Zhao, X. (2009). Genome shuffling: Progress and applications for phenotype improvement. Biotechnol. Adv. *27*, 996–1005. http://dx.doi.org/10.1016/j.biotechadv.2009.05.016

Greig, D. (2007). A screen for recessive speciation genes expressed in the gametes of F1 hybrid yeast. PLOS Genet. *3*, e21.

Greig, D., Louis, E.J., Borts, R.H., and Travisano, M. (2002). Hybrid speciation in experimental populations of yeast. Science *298*, 1773–1775. http://dx.doi.org/10.1126/science.1076374

Gunge, N., and Nakatomi, Y. (1972). Genetic Mechanisms of Rare Matings of the Yeast *Saccharomyces cerevisiae* Heterozygous for Mating Type. Genetics *70*, 41–58.

Haber, J.E. (2012). Mating-type genes and MAT switching in *Saccharomyces cerevisiae*. Genetics *191*, 33–64. http://dx.doi.org/10.1534/genetics.111.134577

Hammond, J.R.M., and Eckersley, K.W. (1984). Fermentation properties of brewing yeast with killer character. J. Inst. Brewing *90*, 167-177.

Hanson, S.J., Byrne, K.P., and Wolfe, K.H. (2014). Mating-type switching by chromosomal inversion in methylotrophic yeasts suggests an origin for the three-locus *Saccharomyces cerevisiae* system. Proc. Natl. Acad. Sci. USA *111*, E4851–E4858. http://dx.doi.org/10.1073/pnas.1416014111

Herskowitz, I. (1988). Life cycle of the budding yeast *Saccharomyces cerevisiae*. Microbiol. Rev. *52*, 536–553.

Hewitt, S.K., Donaldson, I.J., Lovell, S.C., and Delneri, D. (2014). Sequencing and characterisation of rearrangements in three *S. pastorianus* strains reveals the presence of chimeric genes and gives evidence of breakpoint reuse. PLOS ONE 9, e92203. http://dx.doi.org/10.1371/journal.pone.0092203

Hinnen, A., Hicks, J.B., and Fink, G.R. (1978). Transformation of yeast. Proc. Natl. Acad. Sci. U.S.A. *75*, 1929–1933.

Hittinger, C.T. (2013). *Saccharomyces* diversity and evolution: a budding model genus. Trends Genet. *29*, 309–317. http://dx.doi.org/10.1016/j.tig.2013.01.002

Hou, J., Friedrich, A., Gounot, J.-S., and Schacherer, J. (2015). Comprehensive survey of condition-specific reproductive isolation reveals genetic incompatibility in yeast. Nat. Commun. *6*, 7214. http://dx.doi.org/10.1038/ncomms8214

Hsu, P.D., Lander, E.S., and Zhang, F. (2014). Development and applications of CRISPR-Cas9 for genome engineering. Cell *157*, 1262–1278. http://dx.doi.org/10.1016/j.cell.2014.05.010

Jakočiūnas, T., Bonde, I., Herrgård, M., Harrison, S.J., Kristensen, M., Pedersen, L.E., Jensen, M.K., and Keasling, J.D. (2015a). Multiplex metabolic pathway engineering using CRISPR/Cas9 in *Saccharomyces cerevisiae*. Metab. Eng. *28*, 213–222. http://dx.doi.org/10.1016/j.ymben.2015.01.008

Jakočiūnas, T., Rajkumar, A.S., Zhang, J., Arsovska, D., Rodriguez, A., Jendresen, C.B., Skjødt, M.L., Nielsen, A.T., Borodina, I., Jensen, M.K., *et al.* (2015b). CasEMBLR: Cas9-facilitated multiloci genomic integration of in vivo assembled DNA parts in *Saccharomyces cerevisiae*. ACS Synth. Biol. *4*, 1226–1234. http://dx.doi.org/10.1021/acssynbio.5b00007

Jinek, M., Chylinski, K., Fonfara, I., Hauer, M., Doudna, J.A., and Charpentier, E. (2012). A programmable dual-RNA-guided DNA endonuclease in adaptive bacterial immunity. Science *337*, 816–821. http://dx.doi.org/10.1126/science.1225829

Kahn, J.P. (2015). How Beer Gave Us Civilization. http://www.nytimes.com/2013/03/17/opinion/sunday/how-beer-gave-us-civilization.html

Kao, K.C., Schwartz, K., and Sherlock, G. (2010). A genome-wide analysis reveals no nuclear dobzhansky-muller pairs of determinants of speciation between *S. cerevisiae* and *S. paradoxus*, but suggests more complex incompatibilities. PLOS Genet. *6*, e1001038. http://dx.doi.org/10.1371/journal.pgen.1001038

Katz, S.E. (2011). Fermentation as a co-evolutionary force. In Cured, Fermented and Smoked Foods. Proceedings of the Oxford Symposium on Food and Cookery 2010, H. Saberi, ed. (Totnes and London, UK: Prospect Books), pp. 165-174.

Klis, F.M., Boorsma, A., and De Groot, P.W.J. (2006). Cell wall construction in *Saccharomyces cerevisiae*. Yeast *23*, 185–202. http://dx.doi.org/10.1002/yea.1349

Knop, M. (2006). Evolution of the hemiascomycete yeasts: on life styles and the importance of inbreeding. Bioessays *28*, 696–708. http://dx.doi.org/10.1002/bies.20435

Koszul, R. (2016). Beyond the bounds of evolution: Synthetic chromosomes… How and what for? C.R. Biol. *339*, 324–328. http://dx.doi.org/10.1016/j.crvi.2016.05.007

Krivoruchko, A., Siewers, V., and Nielsen, J. (2011). Opportunities for yeast metabolic engineering: Lessons from synthetic biology. Biotechnol. J. *6*, 262–276. http://dx.doi.org/10.1002/biot.201000308

Krogerus, K., Magalhães, F., Vidgren, V., and Gibson, B. (2015). New lager yeast strains generated by interspecific hybridization. J. Ind. Microbiol. Biotechnol. *42*, 769–778. http://dx.doi.org/10.1007/s10295-015-1597-6

Kudla, B., and Nicolas, A. (1992). A multisite integrative cassette for the yeast *Saccharomyces cerevisiae*. Gene *119*, 49–56.

Kumaran, R., Yang, S.Y., and Leu, J.Y. (2013). Characterization of chromosome stability in diploid, polyploid and hybrid yeast cells. PLOS ONE *8*, e68094. http://dx.doi.org/10.1371/journal.pone.0068094

Kutyna, D.R., Cordente, A.G., and Varela, C. (2014). Genetic engineering of industrial *Saccharomyces cerevisiae* strains using a selection/counter-selection approach. In Methods in Molecular Biology, (New York: Springer New York), pp. 157–168. http://dx.doi.org/10.1007/978-1-4939-0563-8_9

Laureau, R., Loeillet, S., Salinas, F., Bergström, A., Legoix-Né, P., Liti, G., and Nicolas, A. (2016). Extensive Recombination of a Yeast Diploid Hybrid through Meiotic Reversion. PLOS Genet. *12*, e1005781. http://dx.doi.org/10.1371/journal.pgen.1005781

Lee, H.Y., Chou, J.Y., Cheong, L., Chang, N.H., Yang, S.Y., and Leu, J.Y. (2008). Incompatibility of nuclear and mitochondrial genomes causes hybrid sterility between two yeast species. Cell *135*, 1065–1073. http://dx.doi.org/10.1016/j.cell.2008.10.047

Legras, J.L., Merdinoglu, D., Cornuet, J.M., and Karst, F. (2007). Bread, beer and wine: *Saccharomyces cerevisiae* diversity reflects human history. Mol. Ecol. *16*, 2091–2102.

Libkind, D., Hittinger, C.T., Valério, E., Gonçalves, C., Dover, J., Johnston, M., Gonçalves, P., and Sampaio, J.P. (2011). Microbe domestication and the identification of the wild genetic stock of lager-brewing yeast. Proc. Natl. Acad. Sci. U.S.A. *108*, 14539–14544. http://dx.doi.org/10.1073/pnas.1105430108

Liti, G., and Louis, E.J. (2012). Advances in quantitative trait analysis in yeast. PLOS Genet. *8*, e1002912. http://dx.doi.org/10.1371/journal.pgen.1002912

Lopes, T.S., Klootwijk, J., Veenstra, A.E., van der Aar, P.C., van Heerikhuizen, H., Raúe, H.A., and Planta, R.J. (1989). High-copy-number integration into the ribosomal DNA of *Saccharomyces cerevisiae*: a new vector for high-level expression. Gene *79*, 199–206.

Ludlow, C.L., Cromie, G.A., Garmendia-Torres, C., Sirr, A., Hays, M., Field, C., Jeffery, E.W., Fay, J.C., and Dudley, A.M. (2016). Independent Origins of Yeast Associated with Coffee and Cacao Fermentation. Curr. Biol. *26*, 965–971. http://dx.doi.org/10.1016/j.cub.2016.02.012

Mans, R., van Rossum, H.M., Wijsman, M., Backx, A., Kuijpers, N.G.A., van den Broek, M., Daran-Lapujade, P., Pronk, J.T., van Maris, A.J.A., and Daran, J.M.G. (2015). CRISPR/Cas9: a molecular Swiss army knife for simultaneous introduction of multiple genetic modifications in *Saccharomyces cerevisiae*. FEMS Yeast Res. *15*, fov004. http://dx.doi.org/10.1093/femsyr/fov004

Marinoni, G., Manuel, M., Petersen, R.F., Hvidtfeldt, J., Sulo, P., and Piskur, J. (1999). Horizontal transfer of genetic material among *Saccharomyces* yeasts. J. Bacteriol. *181*, 6488–6496.

McGovern, P.E. (2009). Uncorking the Past: The Quest for Wine, Beer, and Other Alcoholic Beverages (Berkeley and Los Angeles, CA, USA: University of California Press).

McGovern, P.E. (2003). Ancient Wine: The Search for the Origins of Viniculture (Princeton, NJ, USA: Princeton University Press).

Mendel, G. (1865). Versuche über Pflanzen-Hybriden. Verhandl naturf Verein Brünn *4* (Abhandl.), 3-47.

Merlini, L., Dudin, O., and Martin, S.G. (2013). Mate and fuse: how yeast cells do it. Open. Biol. 3, 130008. http://dx.doi.org/10.1098/rsob.130008

Mertens, S., Steensels, J., Saels, V., De Rouck, G., Aerts, G., and Verstrepen, K.J. (2015). A Large Set of Newly Created Interspecific *Saccharomyces* Hybrids Increases Aromatic Diversity in Lager Beers. Appl. Env. Microbiol. *81*, 8202–8214. http://dx.doi.org/10.1128/AEM.02464-15

Monerawela, C., James, T.C., Wolfe, K.H., and Bond, U. (2015). Loss of lager specific genes and subtelomeric regions define two different *Saccharomyces cerevisiae* lineages for *Saccharomyces pastorianus* Group I and II strains. FEMS Yeast Res. *15*, fou008. http://dx.doi.org/10.1093/femsyr/fou008

Morales, L., and Dujon, B. (2012). Evolutionary role of interspecies hybridization and genetic exchanges in yeasts. Microbiol. Mol. Biol. Rev. *76*, 721–739. http://dx.doi.org/10.1128/MMBR.00022-12

Mukai, N., Nishimori, C., Fujishige, I.W., Mizuno, A., Takahashi, T., and Sato, K. (2001). Beer brewing using a fusant between a sake yeast and a brewer's yeast. J. Biosci. Bioeng. *91*, 482–486.

Nguyen, H.V., Legras, J.L., Neuvéglise, C., and Gaillardin, C. (2011). Deciphering the hybridisation history leading to the Lager lineage based on the mosaic genomes of *Saccharomyces bayanus* strains NBRC1948 and CBS380. PLOS ONE *6*, e25821. http://dx.doi.org/10.1371/journal.pone.0025821

Nilsson-Tillgren, T., Gjermansen, C., Kielland-Brandt, M.C., Petersen, J.G.L., and Holmberg, S. (1981). Genetic differences between *Saccharomyces carlsbergensis* and *S. cerevisiae*. Analysis of chromosome III by single chromosome transfer. Carlsberg Res. Comm. *46*, 65–76.

Okuno, M., Kajitani, R., Ryusui, R., Morimoto, H., Kodama, Y., and Itoh, T. (2016). Next-generation sequencing analysis of lager brewing yeast strains reveals the evolutionary history of interspecies hybridization. DNA Res. *23*, 67-80. http://dx.doi.org/10.1093/dnares/dsv037

Parker, N., James, S., Dicks, J., Bond, C., Nueno-Palop, C., White, C., and Roberts, I.N. (2014). Investigating flavour characteristics of British ale yeasts: techniques, resources and opportunities for innovation. Yeast 32, 281-7. http://dx.doi.org/10.1002/yea.3052

Parts, L., Cubillos, F.A., Warringer, J., Jain, K., Salinas, F., Bumpstead, S.J., Molin, M., Zia, A., Simpson, J.T., Quail, M.A., *et al.* (2011). Revealing the genetic structure of a trait by sequencing a population under selection. Genome. Res. *21*, 1131–1138. http://dx.doi.org/10.1101/gr.116731.110

Peris, D., Langdon, Q.K., Moriarty, R.V., Sylvester, K., Bontrager, M., Charron, G., Leducq, J.-B., Landry, C.R., Libkind, D., and Hittinger, C.T. (2016). Complex Ancestries of Lager-Brewing Hybrids Were Shaped by Standing Variation in the Wild Yeast *Saccharomyces eubayanus*. PLOS Genetics *12*, e1006155. http://dx.doi.org/10.1371/journal.pgen.1006155

Peris, D., Sylvester, K., Libkind, D., Gonçalves, P., Sampaio, J.P., Alexander, W.G., and Hittinger, C.T. (2014). Population structure and reticulate evolution of *Saccharomyces eubayanus* and its lager-brewing hybrids. Mol. Ecol. *23*, 2031–2045. http://dx.doi.org/10.1111/mec.12702

Pérez-Través, L., Lopes, C.A., Barrio, E., and Querol, A. (2012). Evaluation of different genetic procedures for the generation of artificial hybrids in *Saccharomyces* genus for winemaking. Int. J. Food Microbiol. *156*, 102–111. http://dx.doi.org/10.1016/j.ijfoodmicro.2012.03.008

Pérez-Través, L., Lopes, C.A., Querol, A., and Barrio, E. (2014). On the complexity of the *Saccharomyces bayanus* taxon: hybridization and potential hybrid speciation. PLOS ONE *9*, e93729. http://dx.doi.org/10.1371/journal.pone.0093729

Plech, M., de Visser, J.A., and Korona, R. (2014). Heterosis is prevalent among domesticated but not wild strains of *Saccharomyces cerevisiae*. G3 *4*, 315–323. http://dx.doi.org/10.1534/g3.113.009381

Pretorius, I.S. (2016). Synthetic genome engineering forging new frontiers for wine yeast. Crit. Rev. Biotechnol. *18*, 1–25. http://dx.doi.org/10.1080/07388551.2016.1214945

Rainieri, S., Kodama, Y., Kaneko, Y., Mikata, K., Nakao, Y., and Ashikari, T. (2006). Pure and mixed genetic lines of *Saccharomyces bayanus* and *Saccharomyces pastorianus* and their contribution to the lager brewing strain genome. Appl. Environ. Microbiol. *72*, 3968–3974. http://dx.doi.org/10.1128/AEM.02769-05

Robinson, H.A., Pinharanda, A., and Bensasson, D. (2016). Summer temperature can predict the distribution of wild yeast populations. Ecol. Evol. *6*, 1236–1250. http://dx.doi.org/10.1002/ece3.1919

Reuter, J.A., Spacek, D.V., and Snyder, M.P. (2015). High-throughput sequencing technologies. Mol. Cell *58*, 586–597. http://dx.doi.org/10.1016/j.molcel.2015.05.004

Russell, I., and Stewart, G.G. (1979). Spheroplast fusion of brewer's yeast strains. J. Inst. Brew. *85*, 95–98.

Ryan, O.W., Skerker, J.M., Maurer, M.J., Li, X., Tsai, J.C., Poddar, S., Lee, M.E., DeLoache, W., Dueber, J.E., Arkin, A.P., *et al.* (2014). Selection of chromosomal DNA libraries using a multiplex CRISPR system. Elife 3, e03703. http://dx.doi.org/10.7554/eLife.03703

Sadhu, M.J., Bloom, J.S., Day, L., and Kruglyak, L. (2016). CRISPR-directed mitotic recombination enables genetic mapping without crosses. Science *352*, 1113–1116. http://dx.doi.org/10.1126/science.aaf5124

Sanchez, R.G., Solodovnikova, N., and Wendland, J. (2012). Breeding of lager yeast with *Saccharomyces cerevisiae* improves stress resistance and fermentation performance. Yeast *29*, 343–355. http://dx.doi.org/10.1002/yea.2914

Sander, J.D., and Joung, J.K. (2014). CRISPR-Cas systems for editing, regulating and targeting genomes. Nat. Biotechnol. *32*, 347–355. http://dx.doi.org/10.1038/nbt.2842

Sato, M., Kishimoto, M., Watari, J., and Takashio, M. (2002). Breeding of brewer's yeast by hybridization between a top-fermenting yeast *Saccharomyces cerevisiae* and a cryophilic yeast *Saccharomyces bayanus*. J. Biosci. Bioeng. *93*, 509–511. http://dx.doi.org/10.1016/S1389-1723(02)80101-3

Schwartz, K., Wenger, J.W., Dunn, B., and Sherlock, G. (2012). APJ1 and GRE3 homologs work in concert to allow growth in xylose in a natural *Saccharomyces sensu stricto* hybrid yeast. Genetics *191*, 621–632. http://dx.doi.org/10.1534/genetics.112.140053

Selmecki, A.M., Maruvka, Y.E., Richmond, P.A., Guillet, M., Shoresh, N., Sorenson, A.L., De, S., Kishony, R., Michor, F., Dowell, R., *et al.* (2015). Polyploidy can drive rapid adaptation in yeast. Nature *519*, 349–352. http://dx.doi.org/10.1038/nature14187

Shen, Y., Stracquadanio, G., Wang, Y., Yang, K., Mitchell, L.A., Xue, Y., Cai, Y., Chen, T., Dymond, J.S., Kang, K., et al. (2016). SCRaMbLE generates designed combinatorial stochastic diversity in synthetic chromosomes. Genome Res. *26*, 36–49. http://dx.doi.org/10.1101/gr.193433.115

Siddiqui, M.S., Choksi, A., and Smolke, C.D. (2014). A system for multilocus chromosomal integration and transformation-free selection marker rescue. FEMS Yeast Res. *14*, 1171–1185. http://dx.doi.org/10.1111/1567-1364.12210

Siewers, V. (2014). An overview on selection marker genes for transformation of *Saccharomyces cerevisiae*. Methods Mol. Biol. *1152*, 3–15. http://dx.doi.org/10.1007/978-1-4939-0563-8_1

Simchen, G. (2009). Commitment to meiosis: what determines the mode of division in budding yeast? Bioessays *31*, 169–177. http://dx.doi.org/10.1002/bies.200800124

Sliva, A., Yang, H., Boeke, J.D., and Mathews, D.J. (2015). Freedom and Responsibility in Synthetic Genomics: The Synthetic Yeast Project. Genetics *200*, 1021–1028. http://dx.doi.org/10.1534/genetics.115.176370

Snoek, T., Picca Nicolino, M., Van den Bremt, S., Mertens, S., Saels, V., Verplaetse, A., Steensels, J., and Verstrepen, K.J. (2015). Large-scale robot-assisted genome shuffling yields industrial *Saccharomyces cerevisiae* yeasts with increased ethanol tolerance. Biotechnol. Biofuels. *8*, 32. http://dx.doi.org/10.1186/s13068-015-0216-0

Solis-Escalante, D., Kuijpers, N.G.A., van der Linden, F.H., Pronk, J.T., Daran, J.-M., and Daran-Lapujade, P. (2014). Efficient simultaneous excision of multiple selectable marker cassettes using I-SceI-induced double-strand DNA breaks in *Saccharomyces cerevisiae*. FEMS Yeast Res. *14*, 741–754. http://dx.doi.org/10.1111/1567-1364.12162

Spencer, J.F., and Spencer, D.M. (1996). Rare-mating and cytoduction in *Saccharomyces cerevisiae*. Methods Mol. Biol. *53*, 39–44. http://dx.doi.org/10.1385/0-89603-319-8:39

Spencer, J.F., Spencer, D.M., and Reynolds, N. (1988). Genetic manipulation of non-conventional yeasts by conventional and non-conventional methods. J. Basic Microbiol. *28*, 321–333.

Steensels, J., Meersman, E., Snoek, T., Saels, V., and Verstrepen, K.J. (2014a). Large-scale selection and breeding to generate industrial yeasts with superior aroma production. Appl. Environ. Microbiol. *80*, 6965–6975. http://dx.doi.org/10.1128/AEM.02235-14

Steensels, J., Snoek, T., Meersman, E., Picca Nicolino, M., Voordeckers, K., and Verstrepen, K.J. (2014b). Improving industrial yeast strains: exploiting natural and artificial diversity. FEMS Microbiol. Rev. *38*, 947–995. http://dx.doi.org/10.1111/1574-6976.12073

Steensels, J., and Verstrepen, K.J. (2014). Taming wild yeast: potential of conventional and nonconventional yeasts in industrial fermentations. Annu. Rev. Microbiol. *68*, 61–80. http://dx.doi.org/10.1146/annurev-micro-091213-113025

Stewart, G.G. (1981). The genetic manipulation of industrial yeast strains. Can. J. Microbiol. *27*, 973–990.

Storchová, Z. (2014). Ploidy changes and genome stability in yeast. Yeast *31*, 421–430. http://dx.doi.org/10.1002/yea.3037

Storici, F., Durham, C.L., Gordenin, D.A., and Resnick, M.A. (2003). Chromosomal site-specific double-strand breaks are efficiently targeted for repair by oligonucleotides in yeast. Proc. Natl. Acad. Sci. U.S.A. *100*, 14994–14999. http://dx.doi.org/10.1073/pnas.2036296100

Storici, F., Lewis, L.K., and Resnick, M.A. (2001). In vivo site-directed mutagenesis using oligonucleotides. Nat. Biotechnol. *19*, 773–776. http://dx.doi.org/10.1038/90837

Stovicek, V., Borodina, I., and Forster, J. (2015). CRISPR–Cas system enables fast and simple genome editing of industrial *Saccharomyces cerevisiae* strains. Metab. Eng. Commun. *2*, 13–22. http://dx.doi.org/10.1016/j.meteno.2015.03.001

Storici, F., and Resnick, M.A. (2006). The delitto perfetto approach to in vivo site-directed mutagenesis and chromosome rearrangements with synthetic oligonucleotides in yeast. In Methods in Enzymology, (Elsevier), pp. 329–345. http://dx.doi.org/10.1016/S0076-6879(05)09019-1

Stuckey, S., and Storici, F. (2013). Gene Knockouts, in vivo Site-Directed Mutagenesis and Other Modifications Using the Delitto Perfetto System in *Saccharomyces cerevisiae*. Methods Enzymol. *533*, 103-131. http://dx.doi.org/10.1016/B978-0-12-420067-8.00008-8

Sunshine, A.B., Payen, C., Ong, G.T., Liachko, I., Tan, K.M., and Dunham, M.J. (2015). The fitness consequences of aneuploidy are driven by condition-dependent gene effects. PLOS Biol. *13*, e1002155. http://dx.doi.org/10.1371/journal.pbio.1002155

Urano, N., Sato, M., Sahara, H., and Koshino, S. (1993). Conversion of a non-flocculent brewer's yeast to flocculent ones by electrofusion. 2. Small-scale brewing by fusants. J. Biotechnol. *28*, 249–261.

van den Broek, M., Bolat, I., Nijkamp, J.F., Ramos, E., Luttik, M.A.H., Koopman, F., Geertman, J.M., de Ridder, D., Pronk, J.T., and Daran, J.M. (2015). Chromosomal Copy Number Variation in *Saccharomyces pastorianus* Is Evidence for Extensive Genome Dynamics in Industrial Lager Brewing Strains. Appl. Env. Microbiol. *81*, 6253–6267. http://dx.doi.org/10.1128/AEM.01263-15

Varela, C., Kutyna, D.R., Solomon, M.R., Black, C.A., Borneman, A., Henschke, P.A., Pretorius, I.S., and Chambers, P.J. (2012). Evaluation of gene modification strategies for the development of low-alcohol-wine yeasts. Appl. Environ. Microbiol. *78*, 6068–6077. http://dx.doi.org/10.1128/AEM.01279-12

Wang, H., and Hou, L. (2010). Genome shuffling to improve fermentation properties of top-fermenting yeast by the improvement of stress tolerance. Food Sci. Biotechnol. *19*, 145–150. http://dx.doi.org/10.1007/s10068-010-0020-3

Wendland, J. (2014). Lager yeast comes of age. Eukaryotic. Cell *13*, 1256–1265. http://dx.doi.org/10.1128/EC.00134-14

Wenger, J.W., Schwartz, K., and Sherlock, G. (2010). Bulk segregant analysis by high-throughput sequencing reveals a novel xylose utilization gene from *Saccharomyces cerevisiae*. PLOS Genet. *6*, e1000942. http://dx.doi.org/10.1371/journal.pgen.1000942

Wickner, R.B., and Edskes, H.K. (2015). Yeast killer elements hold their hosts hostage. PLOS Genet. *11*, e1005139. http://dx.doi.org/10.1371/journal.pgen.1005139

Ye, M., Yue, T., Yuan, Y., and Wang, L. (2013). Production of yeast hybrids for improvement of cider by protoplast electrofusion. Biochem. Eng. J. *81*, 162–169. http://dx.doi.org/10.1016/j.bej.2013.10.016

Yin, H., Ma, Y., Deng, Y., Xu, Z., Liu, J., Zhao, J., Dong, J., Yu, J., and Chang, Z. (2016). Genome shuffling of *Saccharomyces cerevisiae* for enhanced glutathione yield and relative gene expression analysis using fluorescent quantitation reverse transcription polymerase chain reaction. J. Microbiol. Methods *127*, 188–192. http://dx.doi.org/10.1016/j.mimet.2016.06.012

Young, T.W. (1981). The genetic manipulation of killer character into brewing yeast. J. Inst. Brew. *87*, 292–295.

Yu, Z., Zhao, H., Li, H., Zhang, Q., Lei, H., and Zhao, M. (2012). Selection of *Saccharomyces pastorianus* variants with improved fermentation performance under very high gravity wort conditions. Biotechnol. Lett. *34*, 365–370. http://dx.doi.org/10.1007/s10529-011-0780-8

Genomics and Evolution of Beer Yeasts

6

Brigida Gallone[1,2,3,4,5], Stijn Mertens[1,2,5], Sam Crauwels[5,6], Bart Lievens[5,6], Kevin J. Verstrepen[1,2,5] and Jan Steensels[1,2,5*]

[1]Department of Microbial and Molecular Systems (M2S), Centre of Microbial and Plant Genetics (CMPG), Laboratory for Genetics and Genomics, KU Leuven, Leuven, Belgium..
[2]Systems Biology Laboratory, VIB Center for Microbiology, Bio-Incubator, Leuven, Belgium.
[3]Department of Plant Systems Biology, VIB, Ghent, Belgium.
[4]Department of Plant Biotechnology and Bioinformatics, Ghent University, Gent, Belgium.
[5]Leuven Institute for Beer Research, KU Leuven, Bio-Incubator, Leuven, Belgium.
[6]Department of Microbial and Molecular Systems (M2S), Cluster for Bioengineering Technology (CBeT), Laboratory for Process Microbial Ecology and Bioinspirational Management (PME&BIM), KU Leuven, Campus De Nayer, Sint-Katelijne Waver, Belgium.

*Correspondence: jan.steensels@biw.vib-kuleuven.be

https://doi.org/10.21775/9781910190616.06

Abstract

The evolutionary adaptation of organisms to a specific niche is one of the most fascinating processes in biology. Classic Darwinian theory explains how the interplay between (genetic and phenotypic) variation on one hand, and selection on the other, drives evolution. For certain traits, evolutionary adaptation is forced by man-mediated selection, which results in 'domestication', the adaptation of organisms to man-made niches, as is commonly observed in crops, livestock, and pet animals. Yeasts serve as a very interesting model organism for adaptive evolution, since their small and compact genomes provide a very attractive and powerful model for comparative genomics and genome-evolution studies. Moreover, several species, such as the traditional baker's or brewer's yeast *Saccharomyces cerevisiae*, have been subjected to natural as well as human selection, both of which shaped their genotypes and phenotypes. The emergence of whole-genome sequencing technologies resulted in an overwhelming amount of high-quality and highly detailed yeast genome sequences, allowing researchers to investigate how these yeasts interacted with (and adapted to) their environment. In this chapter, we describe how evolutionary processes shaped the genome and phenome of three intriguing yeast species associated with beer brewing: *S. cerevisiae*, *Saccharomyces pastorianus* and *Brettanomyces bruxellensis*.

Introduction

Few people realize that yeast serves as an important eukaryotic model organism for basic research. Between 2001 and 2016, five Nobel prizes (in Physiology or Medicine and Chemistry) have been awarded to researchers who used yeast to understand fundamental aspects of (human) biology, including cell cycle regulation, gene expression, cell ageing, autophagy and protein secretion (www.nobelprize.org). Moreover, in 1996, the baker's yeast *S. cerevisiae* (strain S288c), was the first eukaryotic organism to have its genome fully sequenced. This multimillion-dollar project was driven by an international consortium of researchers from 19 countries working in 94 laboratories, using several different sequencing methods and technologies (Goffeau, 2000; Goffeau *et al.*, 1996). Now, two decades later, as whole-genome sequencing (WGS) becomes an almost trivial procedure, we are faced with a burst of high-quality and highly detailed yeast genome sequences; in-depth sequencing of a yeast genome now takes only a few days and costs a few hundred dollars (Engel and Cherry, 2013; Engel *et al.*, 2014). Consequently, to date, the whole genome sequence of hundreds of yeast species has been published, with thousands of additional genomes in the pipeline. This rich dataset makes it possible to systematically explore the interplay between genetic variation (i.e. changes in DNA sequences) and phenotypic variation (i.e. changes in properties and behaviour).

In the last 20 years, our basic knowledge of the physiology and evolution of yeasts has increased drastically. This growing body of information is largely rooted in the recent development of high-throughput techniques and analytical tools, often called 'Omic' methodologies, that allow simultaneous investigation of thousands of genes (genomics), transcripts (transcriptomics), proteins (proteomics), and metabolites (metabolomics), providing an in-depth analysis of cellular processes at a global and systems level. Using these groundbreaking techniques, several research groups have invested in large-scale characterization of diverse yeast strains (Bergström *et al.*, 2014; Borneman *et al.*, 2011, 2014, 2016; Dunn and Sherlock, 2008; Gallone *et al.*, 2016; Gonçalves *et al.*, 2016; Liti *et al.*, 2009; Schacherer *et al.*, 2009; Strope *et al.*, 2015). Moreover, while initial genomic studies merely focused on a small number of laboratory isolates [e.g. S288c, a *S. cerevisiae* lab strain bred for easy laboratory use and therefore often not representative of natural or industrial yeasts (Warringer *et al.*, 2011)], there is a growing interest in 'natural' or 'wild' isolates (i.e. isolated from environments not associated with human activity, such as soil or tree bark) and isolates originating from industrial niches (e.g. beer or wine fermentations). Although these studies emerged only in the past couple of years, they already yielded valuable information on genetic diversity, genome evolution, and population history and structure of industrially relevant yeast species, such as *S. cerevisiae*, *S. pastorianus*, and *B. bruxellensis*, which are heavily influenced by their association with man-made fermentation processes.

This chapter will start with a chronological overview of how yeast characterization, identification, taxonomy, and phylogeny evolved from basic phenotyping to advanced sequencing approaches. Next, we summarize the current knowledge of the evolution and genomic features of the three main beer-related yeast species (*S. cerevisiae*, *S. pastorianus*, and *B. bruxellensis*), and illustrate how these observations are related to their behaviour in a brewing environment.

From phenotyping to whole-genome sequencing

This section describes how techniques used for beer yeast typing and characterization evolved from low-resolution, error-prone, and labour-intensive phenotype-based genotyping to WGS. We focus on three species closely associated with beer brewing: the common brewer's yeast *S. cerevisiae* (used in ale production), the lager yeast *S. pastorianus*, and *Brettanomyces* (*Dekkera*) *bruxellensis*, an organism involved in spontaneous beer fermentation.

Before the emergence of molecular typing techniques, early yeast classification and speciation was based on the yeast's morphology and physiological properties (Boulton and Quain, 2009; Kurtzman and Fell, 1997). This is at the least remarkable, as the definition of species relies on the basic principle of genetic isolation, and not on the subjective appraisal and weighting of phenotypic properties (Tornai-Lehoczki and Dlauchy, 2000). Morphological differences between beer yeasts are rather limited; they are all unicellular fungi (Querol and Bond, 2009), the cellular morphology of which can vary from spheroidal to ovoidal, with multilateral budding observed when clonally reproducing. Cells occur isolated, in pairs, or sometimes as short chains or clusters, and in some conditions even as pseudohyphae (Boulton and Quain, 2009; Deak, 2008; Kurtzman and Fell, 1997; Voordeckers *et al.*, 2012a). Besides highly similar cell morphology, brewing yeasts also generally form highly similar white, smooth colonies when plated on agar medium.

The shortcoming of brewing yeast classification based solely on morphological properties was already described by Pasteur in his 'Études sur la bière' (Pasteur, 1876) and he clearly stressed the need for other classification criteria. Therefore, several new and more detailed typing methods were developed for the identification and classification of yeasts in the early twentieth century. These tests were based on phenotypic identification procedures in which several physiological and biochemical tests were used, typically targeting the yeast's ability to ferment different sugars, assimilate (=grow aerobically on) carbon and nitrogen compounds and grow in different stressful conditions (e.g. high or low temperatures or vitamin-free medium) (reviewed in Boulton and Quain, 2009; and Kurtzman and Fell, 1997).

Since there is often a need in the brewing industry to quickly and effectively distinguish ale (*S. cerevisiae*) and lager (*S. pastorianus*) yeasts, specific and rapid phenotypic tests to discriminate these two species were developed (Table 6.1) (see Chapter 4). First, these species were shown to differ in their flocculation behaviour. Flocculation is the ability of yeasts to forms flocs (clumps of cells) after the fermentation process. The flocculation process involves flocculins, which are lectin-like proteins that are associated with the cell wall of flocculating cells. These flocculins selectively bind mannose residues present in the cell wall of adjacent yeast cells and are activated by calcium ions present in the medium. Interestingly, *S. cerevisiae* ale yeast tends to adhere to the ascending CO_2 bubbles towards the surface of the fermenting wort, whereas *S. pastorianus* lager yeast will sediment to the bottom of the fermentation vessel. This property has been ascribed to the different properties of the *FLO1*/Lg-*FLO1* gene. The lager yeast-specific Lg-*FLO1* gene can not only bind mannose residues, but also has a high affinity towards glucose residues, and is therefore responsible for the 'NewFlo' phenotype of lager yeasts, and makes them to floc out to the bottom of the fermentor towards the end of the fermentation (Jin and Speers, 1998; Verstrepen and Klis, 2006; Verstrepen *et al.*, 2003) (see Chapter 1). For this reason, *S. cerevisiae* yeast was dubbed top-fermenting yeast and *S. pastorianus* is generally known as the bottom-fermenting yeast. Second, they differ in their carbon metabolism. Most notably, *S. pastorianus* yeasts produce the extracellular enzyme α-galactosidase (melibiase, encoded by the *MEL1* gene), which enables the hydrolysis of melibiose into the readily assimilable sugars galactose and glucose, while *S. cerevisiae* is unable to do so (Boulton and Quain, 2009; Deak, 2008; Gibson and Liti, 2014; Gibson *et al.*, 2013a). Moreover, it was also shown that *S. cerevisiae* yeast only can partly ferment the trisaccharide raffinose (Deak, 2008). Third, ale and lager yeasts show differences in temperature tolerance. Whereas the optimal fermentation temperature of lager yeasts was shown to be below 15°C, it is generally higher for ale yeasts, typically between 20–30°C (Querol and Bond, 2009). Additionally, ale yeasts are more tolerant towards high temperatures, and can grow at up to 41 or even 42°C, whereas lager yeasts can grow only at temperatures up to 32–34°C (Deak, 2008; Hebly *et al.*, 2015; Meersman, 2011; Mertens *et al.*, 2015).

Since *Brettanomyces* and *Saccharomyces* yeasts are genetically so distinct and their characteristics often differ widely, development of quick screening tests to distinguish between the two was relatively straightforward. Especially in the wine industry, where *Brettanomyces* is a vicious spoilage organism, much research has been dedicated to develop reliable methods for detection and identification of *Brettanomyces* in the fermentation environment. One of the most widely used methods involves a plating assay on semi-selective media with ethanol as carbon source, combined with bromocresol

Table 6.1 Overview of phenotypic differences between the lager yeast *S. pastorianus* and the ale yeast *S. cerevisiae* (adapted from Deak, 2008)

Characteristics	Lager strains	Ale strains
Mode of flocculation	Bottom	Top
Fermentation temperature	4–15°C	15–24°C
Maximum growth temperature	32–34°C	38–42°C
Utilization of maltotriose	More complete	Less efficient
Utilization of melibiose	Yes	No
SO_2 production	>4 mg/l	<2 mg/l
Fructose transport	Active proton symport	Facilitated diffusion
Sporulation	No	Yes

green and/or phenolic precursors such as hydrocinnamic acids to distinguish the genus *Brettanomyces* from other yeasts after a long period of cultivation (Rodríguez *et al.*, 2014).

However, it soon became apparent that yeast classification solely based on morphologic and phenotypic characteristics was insufficient to deal with the wide variety of yeast used and found in the fermentation industry. Moreover, diverse process developments and changes in the beer-brewing industry have undermined some of the earlier mentioned physiologic classification markers. For example, the use of large cylindroconical fermentation vessels induces ale yeast to sediment after the main fermentation to the cone, a property characteristic of lager yeast (Boulton and Quain, 2009). Therefore, new techniques, based on the genomic rather than phenotypic features of the yeast, were developed for the detection, identification, and classification of yeasts (Campbell, 1972; Tornai-Lehoczki and Dlauchy, 2000). Moreover, these techniques led to a new area for ecological surveys, and enabled researchers to have a closer look into the population dynamics of fermentative yeasts (Legras and Karst, 2003).

The first developed molecular method for the differentiation between lager and ale beer yeast was based on mitochondrial DNA (mtDNA) restriction profiling (Aigle *et al.*, 1984). Gel electrophoresis of the resulting genomic fragments revealed that there were some clear and consistent differences between ale and lager yeasts. Moreover, the obtained patterns of beer yeasts (both ale and lager) were in turn very different to the patterns of non-beer yeasts, suggesting that it also is a good technique to detect possible contaminations.

In the following decades, the portfolio of molecular typing techniques for beer yeast differentiation was further expanded: DNA–DNA homology (Tornai-Lehoczki *et al.*, 1996; Vaughan Martini and Kurtzman, 1985; Martini and Martini, 1987), electrophoretic karyotyping (Tornai-Lehoczki *et al.*, 1996; Vezinhet *et al.*, 1990), random amplified polymorphic DNA analysis (RAPD) (Baleiras Couto *et al.*, 1994), amplification of interdelta regions (Ness *et al.*, 1993), and ribosomal RNA coding DNA restriction fragment length polymorphism (RFLP) (Baleiras Couto *et al.*, 1996; Messner and Prillinger, 1995; Smole Mozina *et al.*, 1997), or a combination of different techniques (Tornai-Lehoczki and Dlauchy, 2000). Later, more advanced DNA-based techniques with higher resolution, such as microsatellite comparison (Goddard *et al.*, 2010; Katz Ezov *et al.*, 2006; Legras *et al.*, 2005, 2007b); restriction site-associated sequencing (RAD-seq) (Cromie *et al.*, 2013); multilocus sequence typing (Bing *et al.*, 2014; Fay and Benavides, 2005; Ramazzotti *et al.*, 2012; Wang *et al.*, 2012); tiling array hybridization (Schacherer *et al.*, 2009) and ultimately WGS (Liti *et al.*, 2009) were developed and established for yeast characterization. The advent of WGS revolutionized the way to investigate and characterize genetic and phenotypic diversity in yeast. The analysis of whole genomes rapidly progressed from the study of one or a handful of yeast isolates to simultaneous investigation of tens or even hundreds of individuals, enabling the development of a population genetic perspective. Examining genome-wide patterns of sequence variation within and between closely related species is providing the first comprehensive view of the evolutionary history of *S. cerevisiae, S. pastorianus* and *B. bruxellensis*.

Saccharomyces cerevisiae – an alcohol producer tuned to perfection

In order to thrive in a fermentation environment, *Saccharomyces* spp. in general (and *S. cerevisiae* in particular) possess several phenotypic features that make them the ultimate fermentation specialists they are today. These features arose gradually during evolution, both inside and outside man-made fermentation environments.

Natural selection shaped the *Saccharomyces* genome

One of the most striking attributes of *Saccharomyces* spp. is their perfect adaptation to sugar-rich, oxygen-limited environments. It was hypothesized that the emergence of fruit-bearing plants (a new niche that provided a rich, but highly competitive source of ready-to-use sugars) approximately 80–150 million years ago triggered the selection of a cascade of specific genetic adaptations, all targeted towards colonization of these new niches. Therefore, it seems obvious that these adaptations happened by natural selection, and no man-mediated (artificial) selection was involved.

One of the most striking examples is the emergence of the so-called 'Crabtree effect' in *Saccharomyces*. Crabtree-positive yeasts show a metabolism in which glucose (above a certain threshold concentration) represses respiration, so that even when oxygen is still available, cells will favour fermentation. This persistent fermentative behaviour has several advantages over respiration. First, fermentation enables a higher carbon flux and faster production of energy. Second, the main end-product of a fermentative metabolism is ethanol, which can serve as an effective antimicrobial agent, to which *Saccharomyces* itself is highly tolerant. Therefore, the Crabtree effect fits a make-accumulate-consume strategy, an ecological strategy in which ethanol is first produced and accumulated to high concentrations to inhibit the growth of other microbes and later consumed again when all fermentable sugars have been converted (Thomson *et al.*, 2005). Apart from the Crabtree effect, there are several other physiological features that provide *Saccharomyces* spp. with a competitive advantage in fermentation-like environments (e.g. rotting fruit). They have evolved a high tolerance to several environmental stresses (such as high temperatures and a high concentration of osmolytes), a very high glycolytic flux, and the ability to grow in both aerobic and anaerobic conditions (Conant and Wolfe, 2007; Goddard, 2008; Piškur *et al.*, 2006). It is interesting to note that these individual properties are also present in various other yeasts, but they are only uniquely combined in a few species, including *S. cerevisiae* and its closest relatives, providing a strong competitive advantage over other wild yeasts (and bacteria) in many fermentation environments (Piškur *et al.*, 2006).

While most of these properties are now common knowledge, the underlying genetics were only investigated recently. While many questions remain unsolved, these studies led to the first hints towards the evolutionary pathways *Saccharomyces* spp. went through in response to these newly emerged niches. For example, it was shown that the duplication of several key genes, such as those encoding alcohol dehydrogenase (Hagman *et al.*, 2013; Thomson *et al.*, 2005), hexose transporters (Lin and Li, 2011), and enzymes linked to glycolysis (Conant and Wolfe, 2007), as well as global rewiring of the transcriptional network after whole-genome duplication (Ihmels *et al.*, 2005), played a major role in the evolution of *Saccharomyces*.

Saccharomyces cerevisiae as the first domesticated microbe

While these adaptations, presumably to the presence of fruit-bearing plants occurred millions of years prior, as described above, new opportunities arose for *Saccharomyces* yeasts relatively more recently (approximately 8,000–10,000 years ago). Humankind abandoned its hunter–gatherer lifestyle and introduced a horticultural tradition during the Neolithic evolution. After that, it did not take long before people realized that exposing fruits and grains to the environment (sometimes) positively changed the characteristics of these products significantly, and additionally prolonged their shelf life. From then on, beer and other fermented beverages such as wine, sake, cider or mead were an inherent part of the human diet, as they served as a source of nutrition, as medicine, and as a vital supply of uncontaminated water (Gibson and Liti, 2014; Hornsey, 2003; Kodama *et al.*, 2006; Libkind *et al.*, 2011). Later, skilled artisans found out that it paid to keep a small sample of fermented dough or beer sediment and mix this sample with a new, unfermented batch. In this way, without realizing it, they transferred specific, well-adapted microbes from one fermentation cycle to the next, thereby introducing the concept of starter cultures. During these consecutive fermentation steps, several novel superior yeast mutants and variants emerged through (mainly unintentional) artificial selection by breeding and directed evolution. It is now hypothesized that this practice induced the gradual adaptation of *Saccharomyces* yeasts to man-made conditions, resulting in organisms specialized in specific fermentation environments but behaving suboptimal in most other, more 'natural' habitats, thereby making *S. cerevisiae* one of the oldest domesticated organisms on the planet (Fay and Benavides, 2005; Liti *et al.*, 2009; Sicard and Legras, 2011).

Although these initial steps in microbial domestication happened haphazardly and are rarely documented, genetic analysis of current biodiversity enables the identification of specific adaptation. Not unexpectedly, these characteristics are often unique for specific types of industrial fermentation processes, e.g. brewing, baking, and winemaking (Borneman *et al.*, 2011; Sicard and Legras, 2011; Spor *et al.*, 2009; Will *et al.*, 2010) (Table 6.2). However, it is important to note that some of these trait dissimilarities might be due rather to genetic drift and in fact pre-date or coincide with the emergence of synthetic fermentation environments (as suggested in Warringer *et al.*, 2011).

Beer fermentations are a highly selective niche

Yeasts used in the beer industry provide an excellent model to investigate the effects of artificial selection. For example, they can be recovered and reused after the fermentation process (unlike, e.g. bread yeasts) and are employed continuously throughout the year (unlike, for example, wine yeasts, where the fermentation scheme is tightly linked to the grape harvest season). Therefore, many current beer yeast strains can be considered to be the result of a centuries-long evolution experiment, performed by brewers in a highly selective niche, the brewing environment. It is therefore not surprising that multiple genetic adaptations to the beer-making process have been described.

First and foremost, beer yeasts show a remarkable efficiency in utilization of maltose, the prime carbon source in beer wort. Through extensive duplication and subsequent functional divergence of subtelomeric genes involved in maltose metabolism, present-day brewer's yeast is capable of hydrolysing a variety of sugars much more efficiently than its ancestors (Brown *et al.*, 2010; Charron and Michels, 1988; Dunn and Sherlock, 2008; Gallone *et al.*, 2016; Gonçalves *et al.*, 2016; Pougach *et al.*, 2014; Voordeckers *et al.*, 2012b).

Table 6.2 Examples of (suspected) domestication traits of *S. cerevisiae*. Domestication traits are characteristics that have diverged between the domesticated strains and their wild ancestors

Trait	Industry	Responsible gene(s)	Reference
Stress tolerance			
Copper tolerance	Wine	*CUP1*	Fay *et al.* (2004), Liti *et al.* (2009), Warringer *et al.* (2011)
Molasses toxin tolerance	Beer, distillery	*RTM1*	Borneman *et al.* (2011), Ness and Aigle (1995)
Sulfite tolerance	Wine	*SSU1*	Pérez-Ortín *et al.* (2002)
Nutrient utilization			
Fructose utilization	Wine	*FSY1*, *HXT3*	Galeote *et al.* (2010), Novo *et al.* (2009)
Malto(trio)se utilization	Beer	*AGT1*, *MAL*	Brown *et al.* (2010), Gallone *et al.* (2016), Gonçalvez *et al.* (2016), Charron and Michels (1988), Dunn and Sherlock (2008), Stambuk *et al.* (2009), Steensels *et al.* (2014), Voordeckers *et al.* (2012b)
Xylose utilization	Wine	*XDH1*	Wenger *et al.* (2010)
Sensory quality			
General wine aroma	Wine	Unknown	Hyma *et al.* (2011)
Acetate ester production	Fermented beverages	ND	Steensels *et al.* (2014)
Phenolic off-flavour (POF) production	Beer	*PAD1*, *FDC1*	Dunn and Sherlock (2008), Gallone *et al.* (2016), Gonçalves *et al.* (2016), Mukai *et al.* (2010, 2014)
Other			
Flocculation/flor-formation	Sherry, beer	*FLO* genes	Fidalgo *et al.* (2006), Christiaens *et al.* (2012)
Lag phase	Wine, bakery	*ARO8*, *ADE5,7*, *VBA3*	Carmona-Gutierrez *et al.* (2013)
Mesophilic behaviour	Lager beer	ND	Dunn and Sherlock (2008)
Vitamin biosynthesis	Biofuel	*SNO*, *SNZ*	Stambuk *et al.* (2009)

Another trait in which beer yeasts often excel is flocculation, and more specifically the timing of the onset of flocculation. Flocculation is the ability of cells to stick to each other and form aggregates that rapidly sediment to the bottom, or rise to the top, of the fermentation medium (see Chapter 1). This is an important trait in the beer industry, since it provides an easy and cheap way to separate the yeast cells from the finished beverage (Verstrepen *et al.*, 2003). However, early flocculation leads to inefficient or even stuck fermentation processes. Therefore, many brewer's yeasts have been selected to flocculate at the exact moment when all fermentable sugars have been converted into carbon dioxide and ethanol. Moreover, some reports suggest that brewers have fine-tuned the flocculation behaviour of their yeast strain by selecting specific layers of yeast sediment for re-inoculation of a subsequent fermentation batch (Powell *et al.*, 2004). The genetic basis of this phenotype, and the remarkable speed at which yeasts are able to switch their flocculation behaviour, has been studied intensively (Christiaens *et al.*, 2012; Verstrepen *et al.*, 2003). It was shown that flocculation behaviour is controlled by the *FLO* genes, encoding flocculins, and that the instability of the tandem repeats present in these genes enables relatively rapid expansions and contractions in the gene size, thereby allowing for the fast isolation of spontaneous mutants with altered flocculation characteristics (Verstrepen *et al.*, 2005). Lastly, recent evidence also pinpoints 4-vinylguaiacol (4-VG) production (or rather the absence thereof) as an important feature of brewing yeasts. Indeed, genetic analysis shows that disruptive genetic mutations in the causative genes (*PAD1* and *FDC1*) were heavily selected for in brewing yeasts, while this was never encountered in wild strains (Gallone *et al.*, 2016; Gonçalves *et al.*, 2016). Interestingly, this strong adaptation to the beer environment came with a cost. General stress resistance (temperature tolerance, ethanol tolerance, salt tolerance), which is vital for survival in nature, is often impaired in brewing yeast. This loss of 'survival skills' is typical for domesticated organisms (imagine releasing a Chihuahua in the wilderness), and is one of the clearest signs of human interference with the organism's evolution.

In conclusion, *S. cerevisiae* combines several natural features that allow it to thrive in industrial fermentation processes. Moreover, these features were further enhanced or specialized during domestication.

Saccharomyces cerevisiae phylogenetics and population structure

With their small and compact genomes, *Saccharomyces* yeasts represent a very attractive and powerful model for comparative genomics and genome evolution. Despite the current knowledge on the life history of wild and industrial strains being limited, recent population genetic studies provide the first steps in understanding how the *Saccharomyces* population is structured and stratified, how this evolved over time, and to what extent these processes are influenced by human interactions.

In 2005, Fay and coworkers analysed the genetic diversity of 81 strains of *S. cerevisiae*, isolated from a variety of human and natural fermentation environments, as well as sources unrelated to alcoholic beverage production (such as tree exudates and immunocompromised patients), at five unlinked genetic loci. This study provided for the first time genetic evidence for domestication events, and established that diverged populations of wild *S. cerevisiae* exist independently of domesticated isolates. This latter observation was particularly interesting, since it was commonly believed that *S. cerevisiae* is a domesticated species with no truly natural strains existing, and that isolates from the wild were simply escaped industrial strains (Martini, 1993; Naumov, 1996). The authors further concluded that all biodiversity of industrial *S. cerevisiae* strains could be traced back to (at least) two independent domestication events; one leading to the current wine, beer, and bread yeasts (most likely in Europe or the Middle East), and one leading to the current sake yeasts (in Asia). In a later study, Legras and coworkers confirmed this theory, but additionally hypothesized that since up to 28% of the genetic diversity within yeasts from the same industry was associated with geographical origin, strains were further domesticated locally in the past millennia (Legras *et al.*, 2007).

In 2009, two seminal papers showed for the first time genome-wide maps at the nucleotide level of large collections of *S. cerevisiae* isolates, sampled from diverse ecological (wine, beer, spirits, bread, immunocompromised individuals, soil, etc.) and geographical sources (Liti *et al.*, 2009; Schacherer *et al.*, 2009). The first study employed a high-density Affimetrix Yeast Tilling Microarray (YTM) to genetically characterize 63 diverse *S. cerevisiae* strains and compare them to the genome of the lab strain S288c (Schacherer *et al.*, 2009). The second study, that gave rise to the *Saccharomyces* Genome Resequencing Project (SGRP), used low-coverage (1- to 4-fold) WGS to study the haploid derivatives of 36 *S. cerevisiae* and 35 *S. paradoxus* strains. The inclusion of *S. paradoxus* in this study is especially interesting, since this species is the closest known relative of *S. cerevisiae,* but has never been associated with industrial fermentation processes, and can thus be regarded as a completely non-domesticated species. In both studies, the authors could identify several clean subpopulations ('lineages') in *S. cerevisiae.* Liti *et al.* (2009) identified five (specific to geographic location or ecological niches; 'Malaysian', 'West African', 'Sake', 'North America', and 'Wine/European') (Fig. 6.1A), while Schacherer *et al.* (2009) identified three ('Sake', 'Wine, and 'Laboratory') (Fig. 6.1B). Each clean lineage was monomorphic for the majority of segregating sites that were private within the lineage. However, it has to be noted that the strain collections used in these studies are not exhaustive, since intensive sampling revealed that at least eight more wild lineages are present in China alone (Wang *et al.*, 2012). Nevertheless, these studies provided valuable insight into *S. cerevisiae* genetic diversity, population history, and evolution. First, it was established that most wine strains, although they were collected from dispersed geographical locations, were members of a single subpopulation in both studies. Recently, a new non-industrial sublineage has been identified within the wine group that includes Mediterranean oak isolates. Comparative genomics and demographic analysis suggested that this wild population included the ancestral genetic stock of today's wine yeasts (Almeida *et al.*, 2015). Furthermore, the wine subpopulation showed in all studies a remarkably low level of genetic polymorphism and harboured an excess of low-frequency single-nucleotide polymorphisms (SNPs), all features that can be explained by a population bottleneck associated to a single domestication event in wine strains, followed by human direct or accidental dispersal around the world. In 2016, WGS of 212 *S. cerevisiae* strains, including 106 commercial wine starter isolates, corroborated previous findings and confirmed the presence of high level of inbreeding and substantial strain redundancy across the full catalogue of commercial wine strains (Borneman *et al.*, 2016). As a result, the genetic diversity within these industrial strains is rather limited compared to the full spectrum of *S. cerevisiae* biodiversity. It was calculated that while the nucleotide diversity (π, the average number of nucleotide differences per site between any two DNA sequences chosen randomly from the sample population) was 0.56×10^{-3} in 14 representative wine yeasts, a more elaborate study of 138 strains (including both industrial and wild strains) revealed a sequence diversity that was more than one order of magnitude larger (7.27×10^{-3}) (Wang *et al.*, 2012). Interestingly, this diversity is much broader than observed in the human population, which is estimated to be 'only' $1 \text{x} 10^{-3}$ (Jorde and Wooding, 2004). Second, it became clear that there has been extensive transition of strains from one industry to the other, as well as 'back to nature' events after domestication (Schacherer *et al.*, 2009). For instance, the wine yeast lineage also included strains originating from distilleries, immunocompromised patients, and natural environments. Third, it became evident that in both studies the majority of strains (including many beer, bread, and clinical isolates) did not harbour the genome of one clean lineage, but rather included genomic fractions of various lineages, and can thus be considered as 'recombinant' (mosaic) strains (Fig. 6.1C and D). Interestingly, this peculiar population structure was not observed in the wild species *S. paradoxus,* supporting the idea that close association of *S. cerevisiae* with human activity facilitated crossbreeding of geographically isolated lineages and thus generation of new combinations of pre-existing variations. This in turn resulted in phenotypically divergent variants that were in turn spread across the globe. Whereas in theory these results support the hypothesis about two independent domestication events (as raised by Fay *et al.*, 2005), Liti *et al.* (2009) also proposed an alternative scenario. They argue that it is equally possible that human activity may have used existing natural strains with specific fermentation capabilities, providing opportunities for outcrossing and recombination between isolates originating from disparate environments (Goddard and Greig, 2015; Liti *et al.*, 2009). Unfortunately, beer yeasts

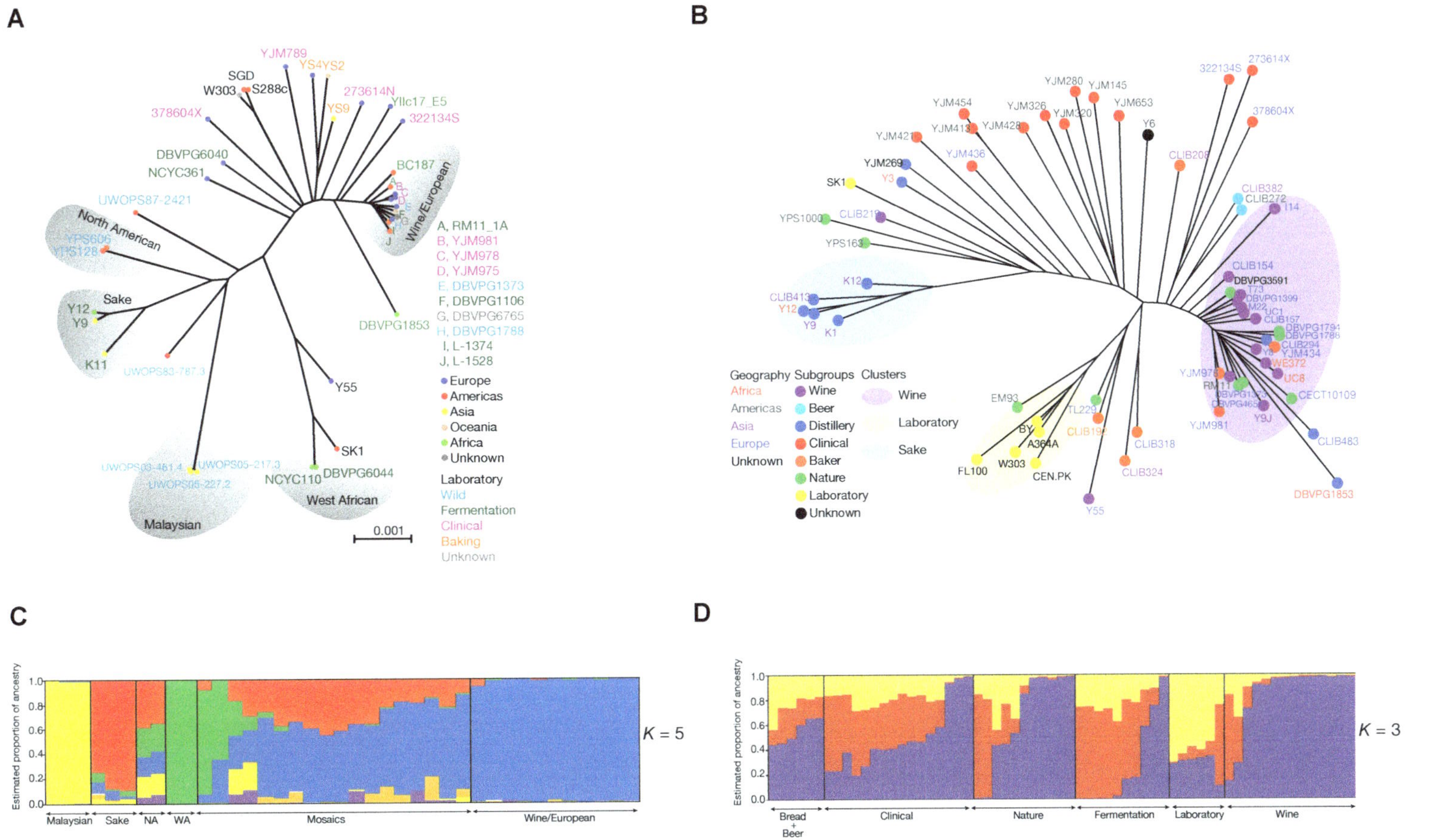

Figure 6.1 *Saccharomyces cerevisiae* phylogenetics and population structure. (A) Neighbour-joining tree based on SNP differences of the *S. cerevisiae* strains sequenced in Liti *et al*. (2009). The five clean lineages are highlighted in grey, colour denotes ecological niche (font) or geographic origin (dots); scale bar indicate frequencies of base-pair differences; (B) Neighbour-joining tree based on SNP differences of *S. cerevisiae* strains sequenced in Schacherer *et al*. (2009). The three main lineages are highlighted with coloured ellipses. Colour denotes geographic origin (font) or ecological niche (dots). (C) Inferred population structure of *S. cerevisiae* strains sequenced in Liti *et al*. (2009), based on SNP data. The most likely number of inferred subpopulations is five (estimated ancestry $K=5$). Each strain is represented by a single vertical bar which is partitioned into coloured segments indicating the estimated coefficient of membership to each subpopulation. (D) Inferred population structure of *S. cerevisiae* strains sequenced in Schacherer *et al*. (2009), based on SNP data. All panels are adapted from Liti *et al*. (2009) and Schacherer *et al*. (2009).

are poorly represented in these studies, which are often biased towards natural and/or wine strains. Moreover, these studies also suffer from several technical constraints, which results in the loss of valuable genetic information. For example, both studies used S288c as a reference strain, either for genome mapping (Liti *et al.*, 2009) or array design (Schacherer *et al.*, 2009). Therefore, in 2011, Borneman and coworkers documented the first *de novo* WGS of six industrial *S. cerevisiae* strains in their natural ploidy (Borneman *et al.*, 2011). While this paper also mainly focused on wine strains (Lalvin QA23, AWRI769, Vin13, and VL3), two Australian ale-brewing strains (FostersO and FostersB) were included. These six strains were sequenced at high coverage (average 20-fold, compared with 1- to 4-fold in Liti *et al.*, 2009), with a combination of shotgun and paired-end sequencing, which resulted in high-quality genomic assemblies. Phylogenetic analysis revealed that the wine and ale strains clustered in two separate populations; the wine strains clustered with the previously established 'Wine/European' lineage, while the beer strains were not part of one of the previously defined clean lineages. However, since these clusters were shown to be much more closely related to each other than to strains with origin outside Europe, it was hypothesized that two distantly related, but both European, *S. cerevisiae* lineages are the ancestors of the wine and beer subgroups.

In 2016, Gallone and coworkers reported WGS and phenotyping of a collection of 157 industrial yeasts including 102 commercial beer *S. cerevisiae* strains, providing comprehensive insight into the evolution and diversification of brewing isolates (Gallone *et al.*, 2016). Surprisingly, the pool of commercial *S. cerevisiae* beer isolates was not exclusively composed by ale-brewing strains, but also included 10 *S. cerevisiae* strains used for the industrial production of lager beers that were assumed to have been brewed using only strains of *S. pastorianus* (see section below, '*Saccharomyces pastorianus* – the odd one out of the *Saccharomyces* genus'). Phylogenetic analysis revealed the presence of five distinct lineages including only industrial strains, which were genetically and phenotypically separated from wild strains. Wine and sake strains clustered in the well-established 'Wine' and 'Asia' lineages (Liti *et al.*, 2009). The majority of beer yeasts clustered in two only distantly related lineages, 'Beer 1' and 'Beer 2'. The last industrial lineage, dubbed 'Mixed' lineage, included all bread strains and a large portion of the 'atypical' beer yeasts used for bottle refermentation of strong Belgian ales. These findings have the exciting implication that multiple domestication events are responsible for the present-day ale yeast biodiversity. Moreover, the large set of sequencing data allowed the authors to estimate the origin of these Beer yeast lineages. They found that the Beer 1 lineage originated around 1573–1604 AD, while the Beer 2 lineage is more recent and originated about 50 years later. This means that beer yeast domestication started well after the first reported beer production (3000–4000 BC), but before the first yeast isolation by Emil Christian Hansen, or even the discovery of microbes by Louis Pasteur, in the nineteenth century. Interestingly, the calculated dates coincide with the gradual switch from home-centred beer brewing, in which every family produced their own beer, to more professional large-scale brewing, first in pubs and monasteries, and later also in breweries (Hornsey, 2003).

Gallone and coworkers further describe that whereas the specific industrial niche (wine, beer, sake, bread, etc.) was shown to have a major influence in the diversification of industrial yeasts, the influence of geographical isolation was also observed. Most notably, the Beer 1 lineage consisted of three geographically distinct groups: 'Belgium/Germany', 'Britain' and 'US'. Moreover, the data point towards an English origin of US beer yeasts. It was hypothesized that English settlers introduced UK beer yeasts not long after their colonization of the New World (around 1600 AD), after which the yeasts started diverging locally, resulting in the current biodiversity. The Beer 2 lineage, on the other hand, showed absence of geographical substructure and included strains originating from Eastern Europe, Belgium, Germany, UK, and US. Perhaps not surprisingly, subclustering based on beer-styles was generally not observed. This is in line with brewers' common practice of reusing the same yeast strain for a wide variety of beers. Remarkable exceptions to this trend were identified in strains used for the production of specialty beers that are traditional to specific geographic areas, such as Hefeweizen, English stout, and Belgian Saison. These results are in line with the study of Gonçalvez *et al.* (2016), who performed comparative genomics on a set of 28 *S. cerevisiae* ale beer yeasts.

In conclusion, the overall picture gathered by recent studies shows how the genetic diversity of *S. cerevisiae* has been influenced by complex dynamics related to its close association to human technology through history, as well as by genetic drift and migration, leading to progressively genetically differentiated populations. While the history of commercial beer strains was obscure for a very long time, various recent papers (Borneman *et al.*, 2011; Gallone *et al.*, 2016; Gonçalves *et al.*, 2016) unravelled the interesting population structure and evolution of these industrial yeasts.

The complex genome structure of *S. cerevisiae* ale yeasts

The recent genetic analysis of ale beer yeasts in their natural ploidy and the unbiased, *de novo* mapping approach exposed various genetic elements of industrial yeast strains that remained concealed in previous investigations. Major alterations of ploidy, large chromosomal copy number variations (CNVs), intra-strain variation between homologous chromosomes, gene content variation across strains, and novel strain-specific gene clusters were identified, further emphasizing the complexity of industrial *S. cerevisiae* genomes (Borneman *et al.*, 2011; Gallone *et al.*, 2016; Gonçalves *et al.*, 2016). Many of these observations are probably a direct consequence of the lack of a sexual life cycle in ale yeasts: continuous clonal (vegetative) reproduction, especially in a stressful environment, is known to lead to heterozygosity, gross chromosomal rearrangements and poly- and aneuploidies over time (Dunham *et al.*, 2002; Masel and Lyttle, 2011; Selmecki *et al.*, 2008). Wine and natural strains usually possess a fully functional sexual lifecycle and are mostly clean diploids. In contrast, ale strains are often obligate asexual or show little or no spore viability. For instance, Gallone and coworkers (2016) observed that almost half of the strains from the Beer 1 lineage were unable to sporulate, while this trait was rarely detected in the other lineages. Furthermore, strains from the Beer 1 lineage that sporulated, produced in 80% of the cases only unviable spores.

Partial and whole chromosome copy number modifications are frequently observed in laboratory strains of *S. cerevisiae* (Torres *et al.*, 2007) as well as in industrial (Dunn *et al.*, 2012) and wild isolates (Hose *et al.*, 2015; Liti *et al.*, 2013). The extent of these modifications can vary greatly according to the origin of the strains. Ale strains typically possess a more irregular genomic structure and many of them are polyploid or aneuploid (Gallone *et al.*, 2016; Legras *et al.*, 2007; Mortimer, 2000). Interestingly, laboratory strains often show growth deficiencies when forced to acquire extra copies of chromosomes, while non-laboratory (e.g. industrial) yeast strains are more tolerant to aneuploidy and display for some gene classes gene-dosage compensation effect at the transcriptional level (Hose *et al.*, 2015). Indeed, Borneman and coworkers observed large chromosomal amplifications (a 600 kb region chrII, a 200 kb region of chrX, and the whole chrIII of FostersO, and the whole chrIII, chrV, and chrXV of FostersB) and deletions (a 400-kb region in chrVII and the whole chrXIV of FostersO) in beer strains, but to a lesser extent in wine and wild strains.

Copy number variations (gene duplications or deletions) can have substantial phenotypic effects and emerging evidence describes aneuploidy/polyploidy as an evolutionary capacitor with a crucial role in the process of rapid adaptation to harsh environments and genetic perturbations (Ames *et al.*, 2010; Dunn *et al.*, 2012; Selmecki *et al.*, 2015; Yona *et al.*, 2012). Furthermore, in many industrial strains, such variations are more frequent in specific regions such as subtelomeric sites. For instance, subtelomeric gene families regularly subjected to contraction or expansion include sugar utilization genes, vitamins metabolism, ion transport, stress resistance, DNA recombination, and regulation of meiotic cell cycle and reproduction (see 'Beer fermentations are a highly selective niche', above). This suggests that in many industrial strains aneuploidies and CNVs might be adaptive and/or linked to desirable industrial features (resistance to stress, high sugar content, extreme pH, nutrient limitations, etc.). One clear example of adaptive CNVs is the strongly increased number of *MAL* genes, responsible for the utilization of maltose, in ale beer yeasts (Gallone *et al.*, 2016; Gonçalves *et al.*, 2016).

Many industrial *S. cerevisiae* strains contain a high level of heterozygous SNPs compared to their wild counterparts. However, the proportion of these sites was shown to vary immensely. Beer yeasts typically exhibited a higher degree of heterozygosity compared to wine yeasts, with a

5.10-fold and 2.04-fold increase of heterozygous sites detected in strains from Beer 1 and Beer 2 lineages, respectively (Gallone *et al.*, 2016). The level of intra-strain allelic differences (i.e. heterozygosity) is an important indicator of the life history and provides insights into the role of asexual *versus* sexual reproduction, mitotic recombination, and outcrossing (Magwene *et al.*, 2011). The high level of (heterozygous) SNPs in beer yeasts compared to wine yeasts is in line with the 'genome renewal' theory posed by Mortimer, in which cells with a fully functional sexual lifecycle (such as the majority of wine and wild strains) can eliminate deleterious mutations (accumulated during clonal reproduction) by one round of sporulation and self-mating (Mortimer, 2000; Mortimer *et al.*, 1994). In this context, highly heterozygous strains could represent isolates that are less likely to undergo sexual cycle, preserving high degree of intra-specific variability. However, an alternative hypothesis suggests that heterozygosity can also be generated through (recent) outcrossing, and human activities might have favoured recombination between strains with different genetic backgrounds. Indeed, recent population genomic studies demonstrated the presence of high degrees of admixture between lineages within the global population of *S. cerevisiae*, with many mosaic strains identified, specifically among industrial and clinical isolates (Liti *et al.*, 2009; Gallone *et al.*, 2016; Gonçalves *et al.*, 2016; Schacherer *et al.*, 2009; Strope *et al.*, 2015).

Ale yeasts: probably the oldest microbial pets known to mankind

Taken together, these findings shed light on how the combination of life history and niche adaptation shaped the genome of *S. cerevisiae* ale yeasts. It becomes increasingly clear that today's industrial *S. cerevisiae* yeasts are genetically and phenotypically separated from wild stocks due to human selection and trafficking. Maybe even more interesting is that the thousands of industrial yeasts that are available today seem to stem from only a few ancestral strains that made their way into food fermentations and subsequently evolved into separate lineages, each used for specific industrial applications. Within each cluster, strains are sometimes further subdivided along geographical boundaries, as is the case for the Beer 1 clade (Gallone *et al.*, 2016), which is divided into three main subgroups.

However, it is important to note that strains from different fermentation environments experienced fundamentally different evolutionary paths. Continuous growth in man-made, rich beer medium led to large changes in the genome (e.g. aneuploidies, major chromosomal rearrangements) and the loss of survival skills outside this specialized niche. This is in sharp contrast to wine yeasts, for example, which experience the grape must environment only for a short period during the year and persist the rest of the time in and around vineyards or in gut of insects. Therefore, wine strains are exposed more often to natural, nutrient-poor environments, and consequently have characteristics that are more similar to the strains encountered in the wild. Moreover, these frequent 'back to nature' events probably induce sporulation (sexual reproduction) and thus favour hybridization with wild yeasts. In addition, the different common practices for wine and brewing industries have a strong influence on the effective population size of yeast populations and as a consequence, on the patterns of molecular evolution and variation. Because beer is produced throughout the year and beer yeasts are recycled for a few batches of fermentation each time, trillions of cells are transferred when a new batch is inoculated. By contrast, for wine yeasts only a relatively small amount of cells will contribute to the next harvest season grape must. This has resulted in a high genetic diversity within beer yeasts compared to the more uniform wine yeast population. Together, this makes ale strains the most domesticated *S. cerevisiae* strains around, and explains why they are such a perfect fit for beer fermentations. However, this also implies that these yeasts fail to perform well in more innovative brewing conditions (e.g. very high gravity brewing), as they encounter new stresses for which they are usually not adapted, while other strains from different lineages often are.

Saccharomyces pastorianus – the odd one out of the *Saccharomyces* genus

Lagers currently account for more than 90% of the global beer market (www.statista.com/statistics/270275/worldwide-beer-production/). They are typically fermented at a lower temperature (8°C to 15°C), after which a period of cold storage (i.e. lagering, a traditional practice vital for sensorial quality) is performed. It is believed that the lager beer production process originally was introduced in the fifteenth century in Bavaria (Germany), when brewing became legally restricted to wintertime (at colder temperatures) to minimize the microbial spoilage of Bavarian beers. Later, the advent of refrigeration in the nineteenth century enabled lager brewing throughout the whole year (Gibson and Liti, 2014; Kodama *et al.*, 2006; Querol and Bond, 2009). Because of the high appreciation of this type of beer, it quickly spread around the globe. The yeasts used in this practice typically sink to the bottom (and do not rise to the top) of the fermentation vessel towards the end of the fermentation, and are therefore often called 'bottom-fermenting yeasts'. Interestingly, this bottom-fermenting phenotype was described very quickly after the dawn of lager beer brewing in Nuremberg, a town in the state of Bavaria (Bond, 2009; Gibson and Liti, 2014; Hornsey, 2003; Kodama *et al.*, 2006; Smart, 2007). In the nineteenth century, groundbreaking work by Louis Pasteur established that this phenomenon was caused by yeasts (Barnett, 2000; Meussdoerffer, 2009; Pasteur, 1873, 1876; Rees, 1870), and the species name *S. pastorianus* was first coined by the German scientist Max Rees in 1870 as a tribute to Pasteur's work in the field. Emil Christian Hansen isolated shortly thereafter the first pure yeast cultures from lager beer fermentation during his work in the Carlsberg laboratory in Copenhagen. He classified the three isolated pure yeast lineages as separate species; one as *S. pastorianus*, one as *Saccharomyces carlsbergenesis* ('Unterhefe Nr. 1'), and one as *Saccharomyces monacensis* ('Unterhefe Nr. 2') (Barnett, 2000; Meussdoerffer, 2009; Regenberg and Hansen, 2001). After Hansen's findings, starter cultures became general practice in the breweries and bottom-fermenting yeast was classified as *S. carlsbergensis*, disregarding the earlier classification of bottom fermenting yeast as *S. pastorianus* by Max Rees. However, genetic analysis of the various isolates in 1985 showed that the type strains of *S. carlsbergensis*, *S. monacensis*, and *S. pastorianus* (the original isolates from Hansen) were almost identical, resulting in the reclassification of all bottom-fermenting yeasts to *S. pastorianus* (Bond, 2009; Gibson and Liti, 2014; Kodama *et al.*, 2006; Polaina, 2002; Smart, 2007; Wendland, 2014). Furthermore, recent research has revealed that *S. pastorianus* is not a true species at all, but instead an interspecific hybrid of *S. cerevisiae x Saccharomyces eubayanus* (see following section and Chapter 4); nevertheless, in practice the species name *S. pastorianus* is still used to denote this lineage of interspecific hybrids. Because of its industrial importance, much research has been dedicated to the characterization of the lager yeast genome. This led to novel insights in the peculiar genome of this species, provided clues about its origin and shed light on some evolutionary processes that enabled this species to thrive in lager beer fermentation.

Hybrid nature of lager yeast

The physiology of lager yeasts differs fundamentally from the physiology of other brewing yeasts (Barnett, 2000; Pasteur, 1876; Rees, 1870). One of the most peculiar differences is the inability of *S. pastorianus* to sporulate and form viable spores, a property that is still present in many ale-type (and other *S. cerevisiae*) yeast strains (Anderson and Martin, 1975; Kodama *et al.*, 2006; Snoek *et al.*, 2015; Steensels *et al.*, 2014). This inability to form viable offspring is a trait typically encountered in interspecific cross-breeding, e.g. mules (horse × donkey) or ligers (lion × tiger). Indeed, early genetic analysis showed that *S. pastorianus* harboured genetic material of (at least) two different species, and was thus not a clean yeast lineage, but rather the result of a hybridization event between *S. cerevisiae* and another (non-*cerevisiae*) *Saccharomyces* species. The first molecular evidence of the hybrid nature of lager yeast was obtained by a technique called kar-mediated single chromosome transfer (Nilsson-Tillgren *et al.*, 1981), and revealed that the chromosomes of lager yeasts could be divided into three types: (i) homologous (*cerevisiae*-like) chromosomes, (ii) homeologous (non-*cerevisiae* like) chromosomes, and (iii) mosaic chromosomes, i.e. chromosomes composed of both homologous and homeologous segments. Later, these findings were confirmed by several DNA hybridization experiments (e.g. by using southern blot or *S. cerevisiae*-specific gene arrays) (Dunn and Sherlock, 2008; Paul Casey, 1986; Tamai *et al.*, 1998; Yamagishi and Ogata, 1999).

By comparing the DNA sequence of 11 independent loci, Dunn and Sherlock further suggested that the *S. cerevisiae* parent of *S. pastorianus* was closely related to ale-type *S. cerevisiae* strains, and not to wild isolates or strains used in other fermentation industries (Dunn and Sherlock, 2008). This result was in line with the previous work in which allelic variation in 12 microsatellite loci of 651 diverse *S. cerevisiae* and 15 *S. pastorianus* strains was investigated (Legras *et al.*, 2007).

The first research suggesting a potential origin of the non-*cerevisiae* part of the lager genome was published in 1985 (Vaughan Martini and Kurtzman, 1985). Using DNA–DNA hybridization methods, the authors revealed a similarity of 72% between the non-*cerevisiae* moiety of the lager yeast CBS1513 (at that time classified as *S. carlsbergensis*) and *S. bayanus*, a cold tolerant species commonly encountered in wine fermentations. This finding was later confirmed by PCR/RFLP analysis of 48 genes of the same lager yeast strain (Rainieri *et al.*, 2006). Interestingly, this hypothesis was later questioned when it was established that part of the *S. pastorianus* *MET2* gene showed a significant sequence difference with the corresponding *S. bayanus* sequence, suggesting a closely related *Saccharomyces* species, rather than *S. bayanus*, as the non-*cerevisiae* parent (Hansen and Kielland-Brandt, 1994). In 2009, the analysis of the first whole-genome sequence of a lager strain (Weihenstephan 34/70) confirmed this hypothesis. By aligning annotated open reading frames (ORFs) of the lager brewing strain Weihenstephan 34/70 to annotated ORFs of *S. cerevisiae* S288c and *S. bayanus* CBS7001 reference genomes, the *cerevisiae*-type subgenome showed very high similarity to *S. cerevisiae* S288c (> 99%) as opposed to the *bayanus*-type subgenome, exhibiting lower sequence identity with *S. bayanus* CBS7001 (average of 92.7%) (Nakao *et al.*, 2009). Moreover, the authors identified eight genes in the lager yeast genome that were not present in the genomes of the *S. cerevisiae* yeast S288c nor in the genome of the *S. bayanus* yeast CBS7001, further indicating that probably a different, yet closely related species is the second lager yeast parent.

In 2011, Argentinean researchers sampling for cryotolerant *Saccharomyces* yeasts in the Patagonian forest stumbled upon a new *Saccharomyces* species (Libkind *et al.*, 2011). The draft genome obtained through WGS of this species (dubbed *Saccharomyces eubayanus*), showed a remarkable high degree of similarity (99.56%) to the non-*cerevisiae* portion of the lager yeast genome, indicating that this species is very likely the missing link in the *S. pastorianus* origin. The authors further suggest a possible scenario where the initial hybridization event between a diploid *S. cerevisiae* cell and a diploid *S. eubayanus* cell gave rise to an allotetraploid hybrid (the original *S. pastorianus* strain), which was subsequently subjected to extensive genome rearrangement and mitotic recombination, resulting in loss of heterozygosity and recombinant chimeric chromosomes. Since these adaptations occurred in the highly selective and man-made environment of (lager) beer fermentation, they considered this the 'domestication' of lager yeasts (Fig. 6.2). However, several questions still remain unanswered. While *S. eubayanus* was originally discovered in Argentina, it is rather unlikely that lager yeasts originated in South America. Initially, Libkind and coworkers hypothesized that the South American *S. eubayanus* strain was introduced in Europe via the transatlantic travel between Europe and America (Libkind *et al.*, 2011; Peris *et al.*, 2014). However, while transatlantic travel was only established after Columbus' first voyage to the new world and the first reports of the bottom-fermenting phenotype predate Columbus' travels, lager brewing yeasts most likely originated earlier (probably in the early 1400s in Bavaria) (Bond, 2009; Gibson and Liti, 2014; Hornsey, 2003; Kodama *et al.*, 2006; Smart, 2007). More recent discoveries of genetically distinct lineages of *S. eubayanus* in other parts of the world (North America and China) suggest that *S. eubayanus* is not unique to South America (Bing *et al.*, 2014; Peris *et al.*, 2014). Moreover, genetic evidence suggests that the non-*cerevisiae* moiety of the Weihenstephan 34/70 is more closely related to a *S. eubayanus* lineage isolated in Tibet (sequencing of 12 loci indicated a 99.82% similarity to the non-*cerevisiae* moiety of lager yeast of the Asian *S. eubayanus* isolate, compared to the 99.56% similarity obtained by WGS with the Argentinean isolate described by Libkind *et al.*, 2011), suggesting that this lineage is more likely the direct ancestor of this lager yeast (Bing *et al.*, 2014). Therefore, it is now hypothesized that *S. eubayanus* made his way to Europe via the 2000-year-old Silk Road. Nevertheless, *S. eubayanus* is up till now not yet discovered in Europe and it could be that this yeast species occupies a highly specific niche in Europe and still awaits discovery (Gibson and Liti, 2014).

However, the question of when and how *S.*

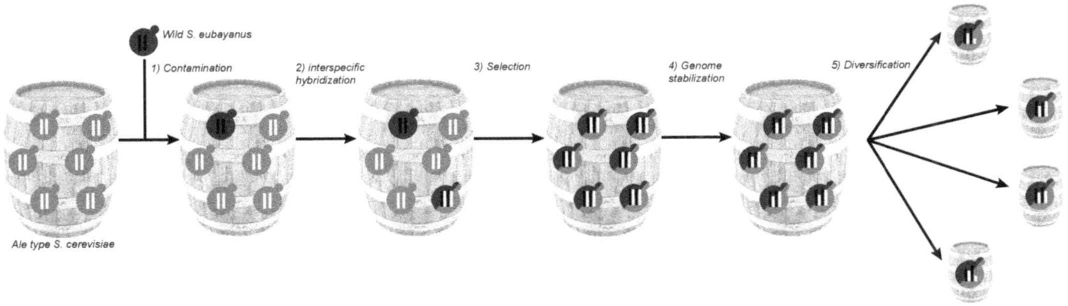

Figure 6.2 The origin of lager yeast. Current hypothesis about the origin of lager yeast involves (1) a contamination of ancient Bavarian fermentation (suggested to be originally conducted by ale yeast *S. cerevisiae*) by a wild *S. eubayanus* yeast contaminant; (2) a rare interspecific hybridization event between both yeast species; (3) selection of the interspecific hybrid due to its likely ability to combine the fermentation capacity and ethanol tolerance of its *S. cerevisiae* parent with the cold tolerance of its *S. eubayanus* parent; (4–5) followed by genome stabilization and diversification of the ancient *S. pastorianus* yeast to its current conformation. This whole process happened at least twice in history, giving rise to the two groups of lager yeasts (Saaz and Frohberg types).

eubayanus got into contact with the lager-brewing environment remains open. It is hypothesized that this event took place 500–600 years ago, triggered by a law enforcing brewing at cold temperatures in Bavaria. The non-*cerevisiae* parent (*S. eubayanus*) is thought to occur as a wild yeast contaminant around the brewing environment, and being better equipped to withstand the fermentation conditions in cold temperatures compared to the native ale yeasts. However, phenotypic analysis of the first two *S. eubayanus* strains isolated showed an inferior fermentation profile of *S. eubayanus* compared to *S. cerevisiae* (e.g. it is unable to ferment maltotriose, shows a lower ethanol tolerance, and produces an inferior aroma profile), prohibiting its use as starter cultures for lager beer fermentation. An interspecific hybridization event (between the *S. eubayanus* contaminant and the ale-type *S. cerevisiae*), that probably happened within the brewing tank, resolved the shortcomings of both species, and resulted in a hybrid species (*S. pastorianus*) that possessed the combined advantage of cold tolerance and fermentation capacity. This species was therefore able to outcompete its parental strains in lager beer fermentation, and was in this way (unintentionally) selected by the brewers, as traditionally part of the fermented beer was used to inoculate the next batch (Gibson and Liti, 2014).

Lager yeasts can be divided into two genetically and phenotypically distinct lineages

Over the years, the interspecific hybridization between *S. cerevisiae* and *S. eubayanus* yeasts probably occurred numerous times. However, to date, only two lineages of lager yeast remain present in industry. These two archetypes are referred to as 'Saaz' or 'group I' and 'Frohberg' or 'group II' type *S. pastorianus* yeasts (Dunn and Sherlock, 2008; Fabian and McCullough, 1933; Gibson and Liti, 2014; Gibson *et al.*, 2013b; Liti *et al.*, 2005; Walther *et al.*, 2014). These names refer to the work of Paul Lindner (1909), who reported the isolation of two individual *S. pastorianus* yeast lineages, which he named 'Saaz' and 'Frohberg' after the locations in Bohemia and Germany in which these strains were originally used (Gibson and Liti, 2014; Gibson *et al.*, 2013b; Lindner, 1909).

Today, a clear trend exists between the lager yeast archetype and the country in which these yeasts are used. Group I (Saaz type), which are not used frequently any more in industry, are mainly used in Czech breweries, as well as the Carlsberg brewery in Denmark, whereas Group II (Frohberg type) are more widespread in other European and North-American breweries (Dunn and Sherlock, 2008). While these archetypes show many similarities, there are some interesting genetic and phenotypic differences as well.

Genetic differences

The separation of *S. pastorianus* yeasts into two genetically distinct groups was firstly suggested based on RFLP analysis and the analysis of transposon sequence distribution among different *S. pastorianus* yeasts (Liti *et al.*, 2005; de Barros Lopes *et al.*, 2002). Later, this hypothesis was confirmed by the use of array Comparative Genomic Hybridization (array CGH) and DNA sequencing of the adjacent and intronic regions of 11 intron-containing genes (Dunn and Sherlock, 2008). In the same experiments, it also became clear that both archetypes did not share a common ancestry. Initially, it was predicted that Saaz-type yeasts originated from a haploid–haploid hybridization event, whereas Frohberg-type yeasts derived from a diploid–haploid hybridization event, where the diploid moiety originates from the *S. cerevisiae* parent (Bond, 2009; Dunn and Sherlock, 2008). However, recent WGS analysis revealed that both Weihenstephan 34/70 (a Frohberg-type yeast) and CBS1513 (Unterhefe I, a Saaz-type yeast) are allotetraploid strains, both originating from a diploid-diploid rare mating event between *S. cerevisiae* and *S. eubayanus* (Nakao *et al.*, 2009; Walther *et al.*, 2014). Interestingly, the two strains differed remarkably in the ratio in which both parental strains were retained in the genome: while Weihenstephan 34/70 harbours a tetraploid genome of 23.6Mb, with an approx. 1 : 1 ratio of both parental strains (36 different chromosome structures, 64 chromosomes in total), the genome of CBS1513 is much smaller (19.5Mb), and the *S. cerevisiae* parental strain is underrepresented compared to *S. eubayanus* (allotriploid (3n-1) with 29 different chromosome structures, 47 chromosomes in total). This interesting parental imbalance in the CBS1513 genome was shown to be caused by a large loss of the *S. cerevisiae* genome, including three complete *S. cerevisiae* chromosomes (VI, XI and XII). This is in line with previous findings based on PCR-RFLP (Rainieri *et al.*, 2006) and array CGH (Dunn and Sherlock, 2008). Additionally, the CBS1513 genome was characterized by numerous large regions of loss of heterozygosity in chromosomes originating from *S. cerevisiae*, resulting in homozygous sequences derived from the *S. eubayanus* parent. Indeed, it was shown that approximately 1.44 Mb of *S. cerevisiae*-derived DNA got replaced by its *S. eubayanus* complement on four different chromosomes (chromosome IV, XIII, XV and XVI), whereas the opposite was only true for 0.22 Mb.

More recently, Okuno *et al.* (2016) analysed in more detail the genetic composition of ten lager yeasts (five Saaz type: CBS1503, CBS1513, CBS1538, CBS1174, CBS2440 and five Frohberg type W34/70, CBS1483, CBS1484, CBS2156, CBS5832) using Illumina next generation-sequencing. The ploidy of the strains was estimated by mapping the obtained paired-end reads onto the genomes of both parental species. *S. cerevisiae* derived chromosomes in Saaz-type yeasts were haploid or missing, whereas most *S. eubayanus*-derived chromosomes appeared to be diploid or triploid. Moreover, partial or whole deletions of *S. cerevisiae* derived chromosomes were observed frequently (e.g. deletions of the right arm of chromosome IV, left arm of chromosome XIII and the entire chromosome XII were common in all five sequenced Saaz-type lager yeasts). In contrast, the genome of Frohberg type lager yeasts was composed of a haploid or diploid set of *S. cerevisiae* derived chromosomes, whereas the ploidy of *S. eubayanus* derived chromosomes ranged from haploid to triploid. The observed chromosomal imbalance implied that the five Saaz type *S. pastorianus* strains, together with Frohberg type lager yeast CBS2156 can be regarded as being triploids, with a total genome size ranging from 14.4 Mb to 19.2 Mb and that Frohberg type lager yeasts (except for CBS2156) are tetraploids (Okuno *et al.*, 2016). Analysis of the genome structure of *S. pastorianus* indicates a complex evolutionary history and allows inference on the origin of the new species.

Throughout the years, there has been some uncertainty about whether both *S. pastorianus* archetypes originate from a single hybridization between *S. cerevisiae* and *S. eubayanus*, or if the Saaz-type and Frohberg-type lager yeasts are derived from two independent hybridization events. Based on the recent identification of three shared translocation sites (*HSP82*, *XRN1/KEM1*, and *MAT*) between the Weihenstephan 34/70 and the CBS1513 yeast, it was suggested that both types of lager yeasts share a joint history and a common ancestor, backing up previous work in which the close relationship between Weihenstephan 34/70 and CBS1513 was appointed based on single gene analysis (Nguyen *et al.*, 2011; Walther *et al.*, 2014). More recent work rejected this hypothesis of a shared ancestor for both types of lager yeasts, and restored the hypothesis that the two lager yeast lineages originate from two separate hybridization events (Monerawela *et al.*, 2015). This study showed that the Frohberg strains originated from a stout-type yeast, while the *S. cerevisiae* moiety of the Saaz strains had the highest similarity with Foster O-like ale strains (an Australian ale yeast with European roots).

However, the sequencing of ten lager yeasts by Okuno and coworkers identified novel shared interchromosomal translocations between *S. cerevisiae*-type and *S. eubayanus*-type genomes in both groups of lager yeasts, favouring the presence of at least one shared hybridization event (Okuno *et al.*, 2016). For instance, this hypothesis was supported by the presence of two translocations in *HSP82* and *KEM1* loci, occurring in eight out of the 10 tested lager yeasts (for yeasts CBS1538 and CBS1174 the HSP82 locus got lost via a chromosomal deletion).

In addition, the mtDNA of *S. pastorianus* showed some interesting trends. In a study on 22 different lager yeasts by RFLP analysis (using four different restriction enzymes), all lager yeasts tested showed a similar uniparental inheritance of the mtDNA of their non-*cerevisiae* parent (Rainieri *et al.*, 2008). These findings were later confirmed in the work of Dunn and Sherlock (where none of the 17 tested lager yeasts seemed to contain *S. cerevisiae*-derived mtDNA; Dunn and Sherlock, 2008) and the WGS of Weihenstephan 34/70 (Nakao *et al.*, 2009). This suggests that *S. eubayanus* mtDNA might harbour one or more genes that provide a competitive advantage for *S. pastorianus* in a lager beer environment. This theory finds further support in the work of González and coworkers, who also observed a similar uniparental inheritance of mtDNA in natural interspecific hybrids between *S. cerevisiae* and the cold-tolerant species *S. bayanus* and *S. kudriavzevii*, discovered in European wine fermentation environments (González *et al.*, 2006). However, the genes causing this trend are yet to be identified.

Phenotypic differences

The two lager yeast archetypes were originally described in 1909 (Lindner, 1909), but the first systematic phenotypic screening of these groups was only published recently (Gibson and Liti, 2014; de Barros Lopes *et al.*, 2002; Mertens *et al.*, 2015; Walther *et al.*, 2014). In these studies, several remarkable differences were revealed.

First, Saaz-type yeasts showed a higher growth capacity at 10°C than Frohberg-type yeasts, suggesting that Saaz-type yeasts harbour a higher tolerance towards cold temperatures (Gibson *et al.*, 2013b; Walther *et al.*, 2014). Second, Frohberg-type yeasts showed a faster fermentation profile and higher degree of attenuation in 22°P fermentation at 15°C (Gibson *et al.*, 2013b), 14°P fermentation at 14°C (Walther *et al.*, 2014), and 12°P fermentation at 16°C (Mertens *et al.*, 2015). This remarkable difference was explained by the incapability of Saaz-type lager yeasts to efficiently metabolize maltotriose (Gibson *et al.*, 2013b). Third, Saaz-type yeasts showed lower cell viability and formed more respiration-deficient 'petite' cells at the end of fermentation. These phenotypes are generally undesired in lager yeasts, since lager yeasts are traditionally reused for seven to twenty-one consecutive fermentation batches, and thus require high phenotypic stability. Finally, the aroma profile of both lager yeast archetypes also differed significantly. In general, Frohberg-type yeasts produced higher concentrations of ethyl acetate, isoamyl acetate, and isoamyl alcohol, but less acetaldehyde (Gibson *et al.*, 2013b; Mertens *et al.*, 2015; Walther *et al.*, 2014). These phenotypic differences may partly explain why Frohberg-type lager yeasts are generally preferred over Saaz-type lager yeasts in today's beer industry.

Future prospects of lager yeasts

The earlier described limited genetic and phenotypic diversity of lager yeasts [especially when compared to the immense genetic and aromatic diversity of ale *S. cerevisiae* yeast strains (Gallone *et al.*, 2016; Steensels *et al.*, 2014a)] has inspired several researchers to develop new and more diverse lager yeasts (for the most recent review, see Gibson and Liti, 2014). Moreover, the recent discovery and isolation of *S. eubayanus* lent the opportunity to develop new interspecific hybrids between *S. cerevisiae* and *S. eubayanus* in the lab, providing a powerful tool to generate strains with superior brewing properties (Hebly *et al.*, 2015; Krogerus *et al.*, 2015; Mertens *et al.*, 2015). Indeed, the development of interspecific hybrids has proven to be an efficient approach to develop novel yeast variants with enhanced characteristics for wine- and beer-making. Typically, an industrial strain of *S. cerevisiae* yeast is crossed with a wild, non-*cerevisiae* member of the *Saccharomyces sensu stricto* complex, such as *S. bayanus* (Marinoni *et al.*, 1999; Sato *et al.*, 2002; Sebastiani *et al.*, 2002; Serra *et al.*, 2005), *S. kudriavzevii* (Bellon *et al.*, 2011; Pérez-Través *et al.*, 2012), *S. uvarum* (Dunn *et al.*, 2013; Piotrowski *et al.*, 2012), *S. mikatae* (Bellon *et al.*, 2013), or *S. eubayanus* (Hebly *et al.*, 2015; Krogerus *et al.*, 2015; Mertens *et al.*, 2015). The first attempt to generate new lager yeasts was established through pair-wise mass mating of ale type *S. cerevisiae* yeasts and a *S. bayanus* yeast strain (for further description of mass mating and other breeding techniques, see Chapter 5). Resulting hybrids showed a higher fermentation capacity than both their respective parental strains, and some showed a similar fermentation capacity to their control bottom-fermenting yeast, already highlighting the potential of newly formed interspecific yeast hybrids for lager beer brewing (Sato *et al.*, 2002). More recently, two papers were published in which a similar, auxotrophic marker-assisted mass-mating strategy was used to generate four interspecific yeast hybrids between *S. eubayanus* type strain CBS12357 (wild isolate isolated in Patagonia; Libkind *et al.*, 2011) and *S. cerevisiae* strains IMK439 (Hebly *et al.*, 2015) or VTT-A81062 (Krogerus *et al.*, 2015). The hybrids inherited interesting properties of both parental strains (cold tolerance, maltotriose utilization, and strong flocculation) and showed hybrid vigour for several traits, such as fermenting speed and fermentation capacity in lab-scale lager beer fermentation tests (Hebly *et al.*, 2015; Krogerus *et al.*, 2015). In a more elaborate study, a set of 31 novel interspecific yeast hybrids were developed, resulting from large-scale robot-assisted selection and breeding between six different *S. cerevisiae* and two different *S. eubayanus* strains, with the aim to increase the aromatic diversity in lager beers (Mertens *et al.*, 2015). Many of these new hybrids produced an aromatic profile significantly different from those produced by currently available lager yeast in both lab and pilot scale fermentation tests, therefore providing a source of lager yeasts able to produce new, aromatically diverse lager beers.

Brettanomyces bruxellensis – the locomotive of spontaneous beer fermentation

A yeast that received considerably less scientific attention compared to *Saccharomyces*, but which is strongly linked with the production process of specific beer styles, is *B. bruxellensis*. While the vast majority of beers are brewed by pure starter cultures of *S. cerevisiae* or *S. pastorianus* yeasts (see above), this species plays a key role in beer fermentation processes relying on a natural inoculum, such as the American Coolship Ales (Bokulich *et al.*, 2012) and the lambic and gueuze beers produced in the surroundings of Brussels, Belgium (Bokulich *et al.*, 2013; Martens, 1997). In these complex microbial matrices, *Brettanomyces* lives in perfect harmony with various other microbial groups, such as lactic acid bacteria, and accounts for many of the typical organoleptic characteristics of the beer. Moreover, the incredible potential and unique aromatic properties of this species are increasingly recognized, with more and more artisans adding it deliberately to their fermentations, either as a pure culture or in combination with more traditional brewing strains (see Chapter 7).

Interestingly, *Brettanomyces* yeasts were amongst the first yeasts ever to be isolated and described in detail. The etymological origin of the *Brettanomyces* genus lies in Great Britain, where it was first isolated in 1904 by Niels Hjelte Claussen, a younger colleague of the famous Emile Christian Hansen at the Carlsberg brewery. He retrieved the yeasts from strong English stock ales, and was inspired by this niche when naming his new discovery (*Brettano* = British, *Myces* = fungus). Claussen quickly recognized the peculiar flavour characteristics of these yeasts compared to more conventional *Saccharomyces* strains, and therefore protected the application of *Brettanomyces* in beer fermentations with a patent (UK patent GB190328184). Interestingly, this is the first patented microorganism in history.

One century later, however, the role of *Brettanomyces* in the food industry is confounded and ambiguous. While their presence is still imperative in some (mainly spontaneously fermented) beers, they are considered to be some of the worst spoilage microbes in wine, mostly due to their typical aroma profile, which can be described as 'burnt plastic', 'barnyard', 'horse sweat', and 'leather' [also referred to as the co-called 'Brett character' (Licker *et al.*, 1999; Wedral *et al.*, 2010)]. Mainly driven by their role as a villain in wine production, several research groups adopted this yeast for a more in-depth study. The advent of next-generation sequencing (see above) recently provided these researchers with the tools necessary for the in-depth analysis of *Brettanomyces* (population) genomics, yielding fascinating data on the genomic build-up of this evolutionarily intriguing yeast. In this section, we will provide a concise overview of the current knowledge of the *Brettanomyces* genomics, evolution, and population structure. More elaborate studies on this yeast are given elsewhere (Crauwels *et al.*, 2015a; Curtin and Pretorius, 2014; Schifferdecker *et al.*, 2014; Steensels *et al.*, 2015).

Brettanomyces taxonomy and phylogeny

Over the years, a plethora of different *Brettanomyces* species were suggested and the names of these species were freely used in scientific publications. However, *Brettanomyces* taxonomy has changed regularly in the past decades and there have been many reclassifications over the years, making direct comparisons between old and more recent papers often challenging. A first attempt to describe the *Brettanomyces* genus comprehensively was performed by Mathieu Custers in 1940. He defined a classification based on a few asexually reproducing (anamorphic) variants (Custers, 1940; Wijsman *et al.*, 1984). A few decades later, in 1960, the formation of ascospores was observed in some strains and the genus *Dekkera* was introduced in the taxonomy as the teleomorphic (sexual) counterpart of *Brettanomyces*. In the first edition of their manual on yeast characteristics and identification, Barnett and co-workers described the following 9 *Brettanomyces* and *Dekkera* species: *Brettanomyces abstinens*, *Brettanomyces anomalus*, *Brettanomyces claussenii*, *Brettanomyces custersianus*, *Brettanomyces custersii*, *Brettanomyces lambicus*, *Brettanomyces naardenensis*, *Dekkera bruxellensis* and *Dekkera intermedia* (Barnett *et al.*, 1983). Today, five species are formally described, based on molecular analysis of the genera: the anamorphs *B. bruxellensis*, *B. anomalus*, *B. custersianus*, *B. naardenensis*, and *Brettanomyces nanus*, with teleomorphs existing for the first two species, *D. bruxellensis* and *D. anomala*. However, according to the recent guidelines of the International Code of Nomenclature for algae, fungi, and plants (the Melbourne Code), fungal species should be assigned only a single valid name. Since the name *Brettanomyces* is well-known and used more commonly in the food and beverage industries, it will likely be prioritized over *Dekkera* (Daniel *et al.*, 2014). Moreover, after the first description of spore formation, spores have not been reported again (Schifferdecker *et al.*, 2014), suggesting the teleomorphic state is very rare (see further).

Despite that *Brettanomyces* spp. have several traits in common with *S. cerevisiae*, these two genera are genetically only distantly related (Kurtzman and Robnett, 2013). Research papers focusing on the phylogeny of *B. bruxellensis* and related species remain scarce, but it is shown that *Brettanomyces* is part of a phylogenetic clade comprising methylotrophic yeast species of the genus *Ogataea*, such as *Ogataea angusta*, *Ogataea glucozyma*, *Ogataea parapolymorpha*, and *Ogataea polymorpha*, as well as other methylotrophs such as *Kuraishia capsulata*, *Candida boidinii*, and *Komagetaella pastoris* (previously *Pichia pastoris*), although the inclusion of the latter species in this clade is uncertain. This clade seems to have diverged from the 'CTG-clade' (containing, for example, *Candida albicans*, *Debaryomyces hansenii*, and *Scheffersomyces stipitis*) progenitor after sharing a common ancestor with *S. cerevisiae*. Using these phylogenetic relationships, it was calculated that *B. bruxellensis* and *S. cerevisiae* separated roughly estimated 200 million years ago (Rozpędowska *et al.*, 2011). Nevertheless, both species share several interesting traits. For example, they have independently acquired resistance to high ethanol concentrations, as well as the ability to produce ethanol even in the presence of oxygen (the Crabtree effect, see earlier section, 'Natural selection shaped the *Saccharomyces* genome'), which enables both species to thrive in alcoholic fermentation processes (see further). This is a clear example of parallel evolution, since these traits arose independently in different lineages that share a similar niche (Rozpędowska *et al.*, 2011).

A first glimpse at the peculiar genome and population structure of *Brettanomyces bruxellensis*

To date, genome assemblies of six *B. bruxellensis* strains have been published, with several more on the way. Five of these six strains originate from the wine industry (Borneman *et al.*, 2014; Curtin *et al.*, 2012; Piškur *et al.*, 2012; Valdes *et al.*, 2014; Woolfit *et al.*, 2007), while one was isolated from lambic beer (Crauwels *et al.*, 2014). In 2007, Woolfit and coworkers published the first exploratory genome survey of the French wine spoilage strain CBS2499, providing a first glimpse of the surprising features of the *B. bruxellensis* genome (Woolfit *et al.*, 2007). Initially, CBS2499 was believed to be haploid, but this was quickly debunked by a follow-up study in which a more in-depth de-novo assembly was performed, showing that the strain was actually diploid (Piškur *et al.*, 2012). This latter study described a total assembly size of 13.4 Mb and was able to identify 5600 genes, from which 75% could be functionally annotated. Around the same time, a second strain was sequenced (AWRI 1499), yielding a 12.7-Mb assembly and 4969 predicted genes (Curtin *et al.*, 2012). ST05.12/22, the only non-wine isolate sequenced to date, was shown to be 13.0 Mbp and 5255 genes were predicted (Crauwels *et al.*, 2014). Recently, the first South American isolate was sequenced (Valdes *et al.*, 2014), and Borneman and coworkers published an in-depth genomic comparison of four wine isolates: the previously sequenced AWRI 1499 and CBS2499 and two newly sequenced strains, AWRI 1608 and AWRI 1613 (Borneman *et al.*, 2014). Analysis of the sequences revealed some interesting similarities of, but also peculiar differences between, the different strains.

First, it was shown that the ploidy of *B. bruxellensis* is variable: while CBS2499, AWRI 1613, and ST05.12/22 were shown to be diploid, haplotyping analysis revealed that AWRI 1499 and AWRI 1608 consist of two moderately heterozygous sets of chromosomes (a core, diploid fraction) and a third haploid set that is divergent (Borneman *et al.*, 2014; Crauwels *et al.*, 2014). Fascinatingly, the variable haploid fraction was shown to be phylogenetically distant, suggesting that multiple independent hybridization events were involved in the emergence of these allotriploids. It is not yet clear if the divergent haploid set originates from a separate species or distantly related *B. bruxellensis* strains.

Second, it was shown that the karyotype varies drastically from strain to strain (Hellborg and Piškur, 2009). *B. bruxellensis* strains can contain between four and nine chromosomes, and the size of these chromosomes can range from 1 to 6 Mbp. This is peculiar, since chromosome configuration is usually well preserved among populations belonging to the same species; for example, different strains of *S. cerevisiae* are collinear and consist of 16 chromosomes. However, since newly formed hybrid genomes tend to be very unstable (as often shown for *Saccharomyces* hybrids; Antunovics *et al.*, 2005; Dunn *et al.*, 2013; Piotrowski *et al.*, 2012), mechanisms that drive genome stabilization could at least explain some of the extreme karyotype variability observed. Additionally, the karyotype variability could suggest that *B. bruxellensis* might employ frequent variations in chromosome structure to increase their genome variability and competitiveness. However, although genomic mutability is beneficial for the adaptability of the species, it can impede sexual reproduction and drive speciation (Fischer *et al.*, 2000). This could explain the absence (or very low frequency) of sexual reproduction in *B. bruxellensis* (see further).

Third, even though *Brettanomyces* did not undergo a whole genome duplication (WGD) event, many localized duplications (CNVs)] and even opposing copy number changes in the same region between strains, were observed (Borneman *et al.*, 2014; Crauwels *et al.*, 2014, 2015b). Since gene duplication is a major source of new genes, it is a central factor influencing genome evolution (Ohno, 1970; Wolfe and Li, 2003). The observed CNVs were frequent in subtelomeric regions and often included genes involved in nutrient (sugar) metabolism, indicating that they might affect utilization of specific carbon sources (Borneman *et al.*, 2014). However, a more in-depth study on the paralogues is required to draw strong conclusion.

Brettanomyces is a well-adapted fermentation scavenger

The occurrence of *Brettanomyces* in natural habitats is only sporadically described (Renouf and Lonvaud-Funel, 2007), but it has been shown that they frequently colonize man-made ecological niches, such as alcoholic fermentation processes (wine, beer, bioethanol, cider, etc.), soft drinks, dairy products, kombucha, and sourdough (Crauwels *et al.*, 2015a; Steensels *et al.*, 2015). A common thread in these niches is the presence of harsh environmental conditions that are lethal for many microbes: high ethanol concentrations, low pH, the absence of readily fermentable nitrogen and carbon sources, low oxygen, etc. While resistance to these stressors is not uncommon in microbes, there are few species that combine all of these traits, thus withstanding such challenging environments. In this section, we will further elaborate on how *Brettanomyces* evolved to become such a highly specialized fermentation organism and how it has acquired different mechanisms to outcompete, or rather outlive, its main fungal competitor in alcoholic fermentations, *S. cerevisiae*.

Comparative analysis of the *B. bruxellensis* genomes revealed some interesting properties that could be linked to their behaviour and ecological niches. First, similar to *S. cerevisiae*, *B. bruxellensis* seems to have evolved a mechanism that allows them to accumulate and be highly tolerant to high concentrations of a toxic metabolite (ethanol), even in the presence of oxygen, a strategy that heavily contributes to their dominance over ethanol-sensitive microbes in sugar-rich environments, and is called the 'make-accumulate-consume' strategy (Piškur *et al.*, 2006). Recently, a study performed by Rozpedowska and coworkers (2011) deciphered how *B. bruxellensis* has evolved this phenotype similarly to, but independently of, *Saccharomyces* yeasts: both lineages used the same strategy relying on global promoter rewiring to change the expression pattern of respiration-associated genes. Interestingly, *B. bruxellensis* seems to have evolved an additional strategy to outcompete other microbes. Besides ethanol, they are also capable of producing, accumulating, and later consuming acetic acid in aerobic conditions, and withstand the resulting low-pH environment. The (lineage-specific) duplication of oxidoreductase genes might explain the capacity to produce acetate under aerobic conditions (Curtin *et al.*, 2012; Piškur *et al.*, 2012). It is important to note, however, that not all *Brettanomyces* species share this trait. For example, *B. naardenensis*, which separated approximately 100 million years ago from *B. bruxellensis*, is unable to grow in the absence of oxygen and their metabolism is completely respiratory (and thus no ethanol or acetic acid is formed), indicating that they are Crabtree-negative (Rozpędowska *et al.*, 2011).

Second, it was shown that *B. bruxellensis* is well-equipped to withstand nutrient-poor environments. For example, many *B. bruxellensis* strains are equipped with a gene cluster containing a nitrate transporter, nitrate reductase, nitrite reductase, and two Zn(II)2 Cys6-type transcription factors, which enables the utilization of nitrate as a sole nitrogen source (Borneman *et al.*, 2014; Crauwels *et al.*, 2014, 2015b; Woolfit *et al.*, 2007). This trait might provide an important fitness advantage over other species, such as *S. cerevisiae* (typically unable to utilize nitrate), in low-nitrogen environments like molasses, soft drinks, or late stages of beer and wine fermentations. Additionally, *B. bruxellensis* lacks (or only shows very limited) glycerol 3-phosphate phosphatase activity, resulting in no (or very low) glycerol production (Tiukova *et al.*, 2013; Wijsman *et al.*, 1984). This trait can be disadvantageous in certain conditions (e.g. it reduces growth speed in environments deprived from oxygen, since glycerol is an important factor in maintaining the redox balance in anaerobiosis), but it provides a competitive advantage over *S. cerevisiae* in nutrient-limiting environments (where energy efficiency is pivotal), since glycerol production is an energy-consuming process. All these adaptations to harsh, nutrient-limiting environments support the scavenging lifestyle of *B. bruxellensis*.

Third, gene content analysis revealed a relative enrichment in cell membrane-related genes compared with two genetically closely related species (*Pichia angusta* and *Komagetaella pastoris*) and *S. cerevisiae* (Curtin et *al.*, 2012). Interestingly, many of the enriched membrane-related genes (e.g. *FIG2*, *FLO1*, *FLO5*, *FLO9*, *HKR1*, and *MUC1*) might be advantageous for survival in wine or beer matured in oak barrels, where they could mediate the adhesion of the cells to the internal wall of the barrel and protect them from washing out during high-pressure cleaning (Christiaens *et al.*, 2012; Verstrepen and Klis, 2006).

Finally, *B. bruxellensis* shows some interesting trends in carbon-source utilization, which can sometimes even be variable amongst different isolates (see further). Compared to *S. cerevisiae*, the genome of *B. bruxellensis* shows a significant enrichment for membrane-associated nutrient transporters and genes involved in the metabolism of alternative carbon sources (such as chitin, N-acetylglucosamine, galactose, mannose, and lactose) (Curtin *et al.*, 2012; Woolfit *et al.*, 2007). Moreover, it shows very efficient sucrose utilization, accounted for by the expression of a high-efficiency sucrose transporter for which no homologues exist in *S. cerevisiae*. This trait might be key for the high competitiveness of *B. bruxellensis* in sucrose-based fermentations such as certain types of molasses (De Barros Pita *et al.*, 2011; Tiukova *et al.*, 2013). Additionally, *B. bruxellensis* shows a higher affinity for glucose in carbon-limiting conditions, possibly mediated by an orthologue of the *Candida albicans HGT1* gene, encoding a high-affinity H^+-symport glucose transporter (Leite *et al.*, 2013).

In short, the behaviour of *B. bruxellensis* shows strong similarities with *Saccharomyces* species, which is reflected in the numerous niches that they share. However, *B. bruxellensis* seems to have evolved an extensive specialization in more challenging conditions, which are for example created during the course of beverage fermentation processes.

Intraspecific *Brettanomyces bruxellensis* variability: adaptations to wine and beer environment?

Genetic diversity studies using classic DNA fingerprinting techniques have revealed significant genotypic intraspecific variability of *B. bruxellensis* (Fig. 6.3) (Conterno *et al.*, 2006; Crauwels *et al.*, 2014; Martorell *et al.*, 2006; Miot-Sertier and Lonvaud-Funel, 2007; Mitrakul *et al.*, 1999; Vigentini *et al.*, 2012). Moreover, much like in *S. cerevisiae*, some of these studies reported a correlation between genotype groups of *B. bruxellensis* and their source of isolation (e.g. beer or wine) (Conterno *et al.*, 2006; Crauwels *et al.*, 2014; Vigentini *et al.*, 2012), suggesting niche adaptation. Furthermore, recently, this correlation was also established for *B. bruxellensis* phenotypes (Crauwels *et al.*, 2015b). In this latter

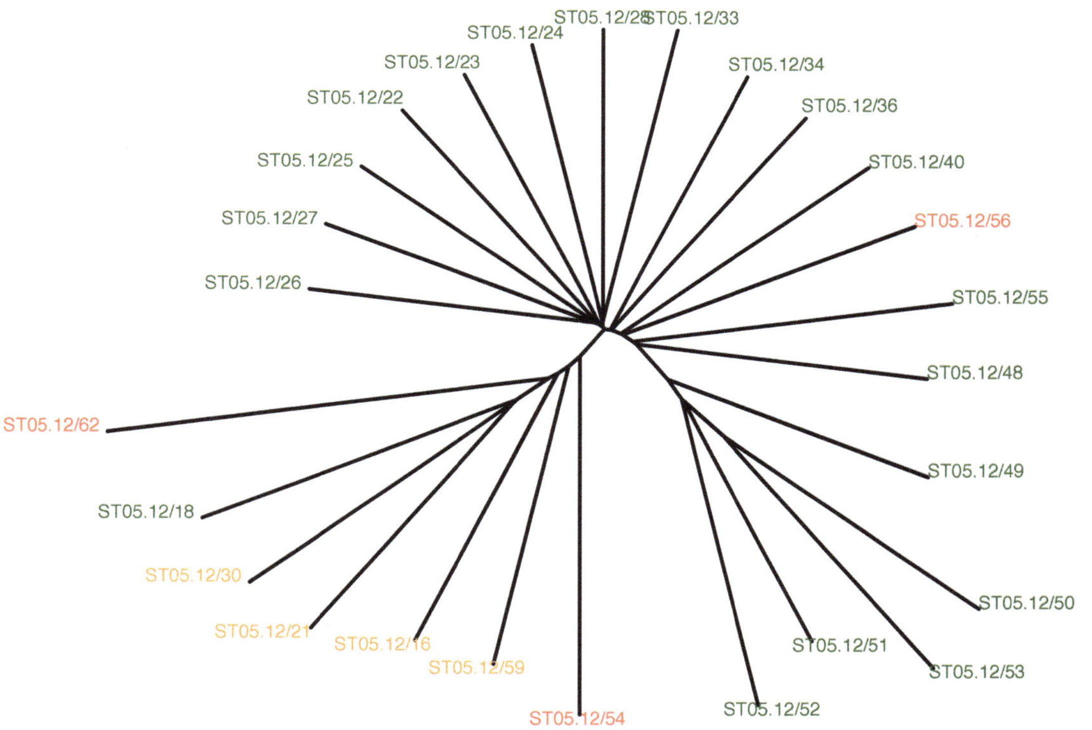

Figure 6.3 *Brettanomyces bruxellensis* phylogenetics. Neighbour-joining tree of *B. bruxellensis* strains from diverse ecological niches, including beer (indicated in green), soft drinks (orange), and wine (red). Studied strains were genotyped using seven established DNA fingerprinting techniques. Data analysis was performed on the combined dataset as described in Crauwels *et al.* (2014).

study, the authors performed high-throughput phenotyping experiments (using Biolog Phenotype Microarrays), and identified several consistent differences between strains from different origins, such as wine and beer fermentations. For example, the ability to assimilate particular α- and β-glycosides as well as α- and β-substituted monosaccharides was shown to be highly variable amongst isolates, but consistent within strains from the same origin. While strains isolated from wine were able to utilize D-galactose, this is not the case for beer isolates. Coinciding with these observations, strains unable to grow on galactose were found to lack at least one of the genes involved in the Leloir pathway of galactose metabolism. Further, brewing strains are, in contrast to wine strains, not capable of using the β-glycoside disaccharides cellobiose and gentiobiose as well as the β-substituted monosaccharides β-methyl-glucoside and arbutine, suggesting that these strains lack the enzyme(s) responsible for the breakage of specific β-bonded sugars. Indeed, WGS revealed that while wine strains contain two (distinct) β-glucosidase genes, beer strains lack one of these genes (Crauwels *et al.*, 2015b). Interestingly, these β-glucosidases are industrially very relevant, since they also play an important role in flavour development of beer and wine (Daenen *et al.*, 2004). They enable the hydrolysis of locked, glycosidically bound flavour compounds and can thus enrich the flavour profile of these fermented beverages (Pogorzelski and Wilkowska, 2007). However, further research is needed to investigate the exact role of these β-glucosidases in flavouring capability of *B. bruxellensis* strains.

Interestingly, the genes involved in the Leloir pathway as well as the above-mentioned β-glucosidase genes are clustered in a ≈36 kb region encompassing 13 genes, the majority of which are involved in carbon metabolism. This region was found to be completely absent in the beer strain ST05/12.22 (Crauwels *et al.*, 2014). Moreover, a more thorough investigation revealed that this cluster has been gradually lost over time in beer strains: some lack only a few genes, other lack all 13 genes, but all beer strains lack the β-glucosidase gene. In contrast, this cluster of genes was entirely present in wine strains (at least in eight of the nine studied strains) (Table 6.3) (Crauwels *et al.*, 2015b). Interestingly, this genomic region is also prone to CNVs and loss-of-heterozygosity (Crauwels *et al.*, 2015b). Based on these findings, it may be speculated that this gene cluster carries a fitness cost (e.g. a metabolic burden) for *B. bruxellensis* in certain fermentation systems such as beer brewing, thereby providing a selective pressure for its loss. These observations are reminiscent of the concerted loss of the galactose catabolism cluster in Japanese *Saccharomyces kudriavzevii* isolates compared to European isolates (Hittinger *et al.*, 2010). However, further research is required to draw firm conclusions.

Despite the increasing knowledge about these genetic and phenotypic differences between *B. bruxellensis* strains, only little is known about the behavior of different *B. bruxellensis* strains in different ecological niches. Recent research by Crauwels *et al.* (2016) has shown that sugar consumption and aroma production are determined by both the yeast strain and the composition of the medium. Furthermore, this study reinforces the hypothesis of niche adaptation of *B. bruxellensis*, most clearly for wine strains. For example, only strains originally isolated from wine were able to thrive well and produce the typical *Brettanomyces*-related phenolic off-flavors 4-ethylguaiacol and 4-ethylphenol when inoculated in red wine. Sulfite tolerance was found as a key factor explaining the observed differences in fermentation performance and off-flavor production.

It is also interesting to note the possibility that the ploidy level may be linked to the yeast's ecological niche. More particularly, triploidy seems to be predominant in the Australian *B. bruxellensis* population, since it is observed in 92% of all isolates from Australian wines (Borneman *et al.*, 2014; Curtin *et al.*, 2007). Moreover, microsatellite typing hints towards the existence of similar populations in French and South African wine isolates (Albertin *et al.*, 2014). In contrast, the majority of *B. bruxellensis* beer strains are found to be diploid (Crauwels *et al.*, 2015b). This interesting allotriploid genome structure is not rare in fungi and is evocative of the interspecific hybrids identified in the *Saccharomyces sensu stricto* clade, such as the lager yeast *S. pastorianus* and the *S. cerevisiae*/*S. kudriavzevii* hybrids isolated from wine and ale beer fermentations (González *et al.*, 2006, 2008). In the case of *Saccharomyces* hybrids, it was suggested that the additional set of chromosomes confers a selective advantage in an industrial environment (see above), but it

Table 6.3 Distribution of 13 clustered genes[a] across different *Brettanomyces bruxellensis* strains originating from beer and wine[b,c]

GenBank Accession No.	Function	Beer strains ST05.12/																	Wine strains ST05.12/								
		22[c]	33	25	27	34	23	26	48	49	50	51	52	53	24	28	40	18	70	54	56	62	63	66	67	68	69
EIF45404	Mfs drug transporter																										
EIF45405	Putative mfs-mdr transporter																										
EIF45406	Mfs multidrug																										
EIF45407	High-affinity glucose transporter																										
EIF45408	Galactose-1-phosphate uridylyltransferase																										
EIF45409	Galactokinase																										
EIF45410	Gal10 bifunctional protein																										
EIF45411	Dtdp-glucose-dehydratase																										
EIF45412	Hexose transporter																										
EIF45413	Maltase																										
EIF45414	Multidrug resistance regulator 1																										
EIF45415	β-Glucosidase																										
EIF45416	Hexose transporter																										

[a]Clustered genes in a region of ≈36 kb, corresponding to a part of *B. bruxellensis* AWRI 1499 (ST05.12/62) contig AHIQ01000280.

[b]Determined by PCR amplification using primers targeting the almost complete ORF, followed by agarose gel electrophoresis. Grey, target band detected; black, target band not detected.

[c]Adapted from Crauwels *et al*. (2015b).

remains to be determined whether similar evolutionary driving forces are at play in the allotriploid *B. bruxellensis* strains. It was suggested that the ability of wine strains to withstand high levels of sulfite, the main antimicrobial agent in wine fermentations, might be at least partially explained by the triploidy state (Borneman *et al.*, 2014; Curtin *et al.*, 2012), but further research is required.

Brettanomyces genomics: catching up with *Saccharomyces*

All in all, our knowledge of *Brettanomyces* is still very limited compared with the vast number of data available for *S. cerevisiae*, the model fungus par excellence. However, several research groups currently are trying to bridge this gap and unravel the peculiar genomic properties of this industrially highly relevant yeast. However, several important challenges remain.

First, despite the fact that some research groups reported ascospore formation (albeit at low frequency) of *B. bruxellensis* (e.g. CBS74 and CBS4914; Van der Walt *et al.*, 1964) and *B. anomalus* (e.g. CBS8139; Th. Smith and van Grinsven, 1984) strains, it still remains unclear to what extent *Brettanomyces* can reproduce sexually, how often this occurs, and what the influence of these events is on the evolution of the species. The extreme karyotypic variability and frequently observed allotriploidy might hint towards the lack (or at least the rarity) of sexual reproduction. In line with this theory, an extensive genetic investigation of the *Brettanomyces* population within a winery (spanning several vintages) revealed that the same persistent clonal *Brettanomyces* population was responsible for wine spoilage in this winery for over half a century (Albertin *et al.*, 2014).

Second, the genetic and molecular toolbox for *Brettanomyces* should be expanded in order to facilitate more in-depth genetic analysis. For example, an optimal transformation protocol is still lacking, auxotrophic mutants are scarce, and only a limited amount of dedicated commercial tools, such as plasmid vectors or gene arrays, are available (Schifferdecker *et al.*, 2014). However, some initial attempts to develop molecular tools and protocols were conducted. Miklenić and coworkers developed various transformation protocols (based on spheroplast transformation, electroporation, and the LiAc/PEG method) and described the first (reported) genetic transformation of *B. bruxellensis*, using a kanMX4 marker cassette that allowed for selection on agar plates containing geneticin (Miklenić *et al.*, 2013). Additionally, Schifferdecker and colleagues developed a *URA3*-deficient mutant strain and a plasmid (named P892, derived from pUC57) containing a functional *URA3* gene of CBS2499 (Schifferdecker *et al.*, 2014). Using these tools, an auxotrophic transformation system and an expression vector was developed, allowing overexpression of individual genes with a constitutive *TEF1* promoter (Schifferdecker *et al.*, 2016). While commercial microarray kits for *Brettanomyces* are still lacking, a first large-scale transcriptomic analysis of *B. bruxellensis* was recently performed (Tiukova *et al.*, 2013). Using whole transcriptome sequencing, the authors managed to detect the expression of 3715 out of the 4861 annotated genes of CBS11270, and the results elucidated survival strategies of this yeast in harsh environments. For example, they observed a low expression of genes involved in glycerol production and a high number of expressed sugar transporter genes, two mechanisms possibly accounting for the high competitiveness of *B. bruxellensis* in oxygen-limited and nutrient-deprived environments (Tiukova *et al.*, 2013).

Third, to fully grasp the *Brettanomyces* population structure and evolutionary history, analysis of a broader collection of ecologically and geographically diverse strains is necessary. While the main focus thus far is put on wine spoilers (mainly originating from Australia), preliminary investigation of strains from different niches and geographical origins already revealed a large degree of intraspecific (phenotypic and genetic) diversity, and sometimes revealed links between phenotypic behaviour and source of isolation (Crauwels *et al.*, 2014, 2015b). While some of these correlations might be due to genetic drift and shared population histories of strains originating from the same environment, some might reflect adaptive mechanisms developed in industrial niches. However, more elaborate studies, targeting the genetic and phenotypic characterization of large collections of diverse strains, will yield more information on some fundamental questions, and shed light on the currently obscure *Brettanomyces* evolutionary history. For example, it will be interesting to see if (and to what extent) *Brettanomyces* is domesticated, and how the *Brettanomyces* population is structured.

Conclusions

With their small and compact genome, short replication time and easy handling in the lab, yeasts pose the ultimate model organism for genetics and population genomics. However, surprisingly, the number of sequenced yeast strains currently available is still limited, especially compared to other eukaryotic model organisms, such as the fruit fly, *Arabidopsis thaliana* and humans, which have more complex and much larger genomes (Peter and Schacherer, 2016). However, current technological developments provide the opportunity to obtain whole-genome sequences for just a few 100 dollars, resulting in a recent burst of yeast genomics data (Almeida *et al.*, 2014, 2015; Borneman *et al.*, 2011, 2014, 2016; Gallone *et al.*, 2016; Jeffares *et al.*, 2015; Okuno *et al.*, 2016; Strope *et al.*, 2015). While these studies reveal crucial aspects of yeast evolution and domestication, they only cover the tip of the iceberg of yeast biodiversity. Therefore, future studies that include a larger number of yeast strains (including non-*cerevisiae* species such as *B. bruxellensis*), will further elucidate the evolutionary path these yeasts went through during their association with human-made fermentation environments. Furthermore, these studies will give valuable insights in genotype–phenotype associations, providing opportunities for targeted strain improvement strategies, such as marker-assisted breeding.

References

Akao, T., Yashiro, I., Hosoyama, A., Kitagaki, H., Horikawa, H., Watanabe, D., Akada, R., Ando, Y., Harashima, S., Inoue, T., *et al.* (2011). Whole-genome sequencing of sake yeast *Saccharomyces cerevisiae* Kyokai no. 7. DNA Res. *18*, 423–434. http://dx.doi.org/10.1093/dnares/dsr029

Albertin, W., Panfili, A., Miot-Sertier, C., Goulielmakis, A., Delcamp, A., Salin, F., Lonvaud-Funel, A., Curtin, C., and Masneuf-Pomarede, I. (2014). Development of microsatellite markers for the rapid and reliable genotyping of *Brettanomyces bruxellensis* at strain level. Food Microbiol. *42*, 188–195. http://dx.doi.org/10.1016/j.fm.2014.03.012

Almeida, P., Gonçalves, C., Teixeira, S., Libkind, D., Bontrager, M., Masneuf-Pomarède, I., Albertin, W., Durrens, P., Sherman, D.J., Marullo, P., *et al.* (2014). A Gondwanan imprint on global diversity and domestication of wine and cider yeast *Saccharomyces uvarum*. Nat. Commun. *5*, 4044. http://dx.doi.org/10.1038/ncomms5044

Almeida, P., Barbosa, R., Zalar, P., Imanishi, Y., Shimizu, K., Turchetti, B., Legras, J.L., Serra, M., Dequin, S., Couloux, A., *et al.* (2015). A population genomics insight into the Mediterranean origins of wine yeast domestication. Mol. Ecol. *24*, 5412–5427. http://dx.doi.org/10.1111/mec.13341

Ames, R.M., Rash, B.M., Hentges, K.E., Robertson, D.L., Delneri, D., and Lovell, S.C. (2010). Gene duplication and environmental adaptation within yeast populations. Genome Biol. Evol. *2*, 591–601. http://dx.doi.org/10.1093/gbe/evq043

Anderson, B.E., and Martin, P.A. (1975). The sporulation and mating of brewing yeasts. J. Inst. Brew.. *81*, 242–247. http://dx.doi.org/10.1002/j.2050-0416.1975.tb03685.x

Antunovics, Z., Nguyen, H.V., Gaillardin, C., and Sipiczki, M. (2005). Gradual genome stabilisation by progressive reduction of the *Saccharomyces uvarum* genome in an interspecific hybrid with *Saccharomyces cerevisiae*. FEMS Yeast Res. *5*, 1141–1150. http://dx.doi.org/10.1016/j.femsyr.2005.04.008

Baleiras Couto, M.M., van der Vossen, J.M., Hofstra, H., and Huis in 't Veld, J.H. (1994). RAPD analysis: a rapid technique for differentiation of spoilage yeasts. Int. J. Food Microbiol. *24*, 249–260. http://dx.doi.org/10.1016/0168-1605(94)90123-6

Baleiras Couto, M.M., Eijsma, B., Hofstra, H., Huis in't Veld, J.H., and van der Vossen, J.M. (1996). Evaluation of molecular typing techniques to assign genetic diversity among *Saccharomyces cerevisiae* strains. Appl. Environ. Microbiol. *62*, 41–46.

Barnett, J.A. (2000). A history of research on yeasts 2: Louis Pasteur and his contemporaries, 1850-1880. Yeast. *16*, 755–771. http://dx.doi.org/10.1002/1097-0061(20000615)16:8<755::AID-YEA587>3.0.CO;2-4

Barnett, J.A., and Lichtenthaler, F.W. (2001). A history of research on yeasts 3: Emil Fischer Eduard Buchner and their contemporaries, 1880-1900. Yeast. *18*, 363–388. http://dx.doi.org/10.1002/1097-0061(20010315)18:4<363::AID-YEA677>3.0.CO;2-R

Barnett, J.A., Payne, R.W., and Yarrow, D. (1983). Yeasts: characteristics and identification. 1st edition (Cambridge: Cambridge University Press).

Bellon, J.R., Eglinton, J.M., Siebert, T.E., Pollnitz, A.P., Rose, L., De Barros Lopes, M., and Chambers, P.J. (2011). Newly generated interspecific wine yeast hybrids introduce flavour and aroma diversity to wines. Appl. Microbiol. Biotechnol. *91*, 603–612. http://dx.doi.org/10.1007/s00253-011-3294-3

Bellon, J.R., Schmid, F., Capone, D.L., Dunn, B.L., and Chambers, P.J. (2013). Introducing a new breed of wine yeast: interspecific hybridisation between a commercial *Saccharomyces cerevisiae* wine yeast and *Saccharomyces mikatae*. PLoS ONE *8*, e62053. http://dx.doi.org/10.1371/journal.pone.0062053

Bergström, A., Simpson, J.T., Salinas, F., Barré, B., Parts, L., Zia, A., Nguyen Ba, A.N., Moses, A.M., Louis, E.J., Mustonen, V., *et al.* (2014). A high-definition view of functional genetic variation from natural yeast genomes. Mol. Biol. Evol. *31*, 872–888. http://dx.doi.org/10.1093/molbev/msu037

Bing, J., Han, P.J., Liu, W.Q., Wang, Q.M., and Bai, F.Y. (2014). Evidence for a Far East Asian origin of lager beer yeast. Curr. Biol. *24*, R380–381. http://dx.doi.org/10.1016/j.cub.2014.04.031

Bokulich, N.A., and Bamforth, C.W. (2013). The microbiology of malting and brewing. Microbiol. Mol. Biol. Rev. *77*, 157–172. http://dx.doi.org/10.1128/MMBR.00060-12

Bokulich, N.A., Bamforth, C.W., and Mills, D.A. (2012). Brewhouse-resident microbiota are responsible for multi-stage fermentation of American coolship ale. PLoS ONE *7*, e35507. http://dx.doi.org/10.1371/journal.pone.0035507

Bond, U. (2009). Chapter 6: The genomes of lager yeasts. Adv. Appl. Microbiol. *69*, 159–182. http://dx.doi.org/10.1016/S0065-2164(09)69006-7

Borneman, A.R., Desany, B.A., Riches, D., Affourtit, J.P., Forgan, A.H., Pretorius, I.S., Egholm, M., and Chambers, P.J. (2011). Whole-genome comparison reveals novel genetic elements that characterize the genome of industrial strains of *Saccharomyces cerevisiae*. PLoS Genet. *7*, e1001287. http://dx.doi.org/10.1371/journal.pgen.1001287

Borneman, A.R., Zeppel, R., Chambers, P.J., and Curtin, C.D. (2014). Insights into the *Dekkera bruxellensis* genomic landscape: comparative genomics reveals variations in ploidy and nutrient utilisation potential amongst wine isolates. PLoS Genet. *10*, e1004161. http://dx.doi.org/10.1371/journal.pgen.1004161

Borneman, A.R., Forgan, A.H., Kolouchova, R., Fraser, J.A., and Schmidt, S.A. (2016). Whole Genome comparison reveals high levels of inbreeding and strain redundancy across the spectrum of commercial wine strains of *Saccharomyces cerevisiae*. G3 *6*, 957–971. http://dx.doi.org/10.1534/g3.115.025692

Boulton, C., and Quain, D. (2009). Brewing Yeast and Fermentation. (Oxford: Blackwell Science).

Brown, C.A., Murray, A.W., and Verstrepen, K.J. (2010). Rapid expansion and functional divergence of subtelomeric gene families in yeasts. Curr. Biol. *20*, 895–903. http://dx.doi.org/10.1016/j.cub.2010.04.027

Campbell, I. (1972). Numerical analysis of the genera *Saccharomyces* and *Kluyveromyces*. J. Gen. Microbiol. *73*, 279–301. http://dx.doi.org/10.1099/00221287-73-2-279

Carmona-Gutierrez, D., Alavian-Ghavanini, A., Habernig, L., Bauer, M.A., Hammer, A., Rossmann, C., Zimmermann, A.S., Ruckenstuhl, C., Büttner, S., Eisenberg, T., *et al.* (2013). The cell death protease Kex1p is essential for hypochlorite-induced apoptosis in yeast. Cell Cycle *12*, 1704–1712. http://dx.doi.org/10.4161/cc.24801

Charron, M.J., and Michels, C.A. (1988). The naturally occurring alleles of *MAL1* in *Saccharomyces* species evolved by various mutagenic processes including chromosomal rearrangement. Genetics *120*, 83–93.

Christiaens, J.F., Van Mulders, S.E., Duitama, J., Brown, C.A., Ghequire, M.G., De Meester, L., Michiels, J., Wenseleers, T., Voordeckers, K., and Verstrepen, K.J. (2012). Functional divergence of gene duplicates through ectopic recombination. EMBO Rep. *13*, 1145–1151. http://dx.doi.org/10.1038/embor.2012.157

Conant, G.C., and Wolfe, K.H. (2007). Increased glycolytic flux as an outcome of whole-genome duplication in yeast. Mol. Syst. Biol. *3*, 129. http://dx.doi.org/10.1038/msb4100170

Conterno, L., Joseph, C.M.L., Arvik, T.J., Henick-Kling, T., and Bisson, L.F. (2006). Genetic and physiological characterization of *Brettanomyces bruxellensis* strains isolated from wines. Am. J. Enol. Vitic. *57*, 139–147.

Crauwels, S., Zhu, B., Steensels, J., Busschaert, P., De Samblanx, G., Marchal, K., Willems, K. a., Verstrepen, K.J., and Lievens, B. (2014). Assessing genetic diversity among *Brettanomyces* yeasts by DNA fingerprinting and whole-genome sequencing. Appl. Environ. Microbiol. *80*, 4398–4413. http://dx.doi.org/10.1128/AEM.00601-14

Crauwels, S., Steensels, J., Aerts, G., Willems, K.A., Verstrepen, K., and Lievens, B. (2015a). Brettanomyces bruxellensis, essential contributor in spontaneous beer fermentations providing novel opportunities for the brewing industry. BrewingScience *68*, 110–121.

Crauwels, S., Van Assche, A., de Jonge, R., Borneman, A.R., Verreth, C., Prahl, T., De Samblanx, G., Marchal, K., Van de Peer, Y., Willems, K.A., *et al.* (2015b). Comparative phenomics and targeted use of genomics reveals variation in carbon and nitrogen assimilation amongst different *Brettanomyces bruxellensis* strains. Appl. Microbiol. Biotechnol. *99*, 9123-34. http://dx.doi.org/10.1007/s00253-015-6769-9

Crauwels, S., Van Opstaele, F., Jaskula-Goiris, B., Steensels, J., Verreth, C., Bosmans, L., Paulussen, C., Herrera-Malaver, B., de Jonge, R., De Clippeleer, J., *et al.* (2016). Fermentation assays reveal differences in sugar and (off-) flavor metabolism across different *Brettanomyces bruxellensis* strains. FEMS Yeast Res. *17*, fow105. http://dx.doi.org/10.1093/femsyr/fow105

Cromie, G.A., Hyma, K.E., Ludlow, C.L., Garmendia-Torres, C., Gilbert, T.L., May, P., Huang, A. a, Dudley, A.M., and Fay, J.C. (2013). Genomic sequence diversity and population structure of *Saccharomyces cerevisiae* assessed by RAD-seq. G3 *3*, 2163–2171. http://dx.doi.org/10.1534/g3.113.007492

Curtin, C.D., and Pretorius, I.S. (2014). Genomic insights into the evolution of industrial yeast species *Brettanomyces bruxellensis*. FEMS Yeast Res. *14*, 997–1005. http://dx.doi.org/10.1111/1567-1364.12198

Curtin, C.D., Bellon, J.R., Henschke, P.A., Godden, P.W., and de Barros Lopes, M.A. (2007). Genetic diversity of *Dekkera bruxellensis* yeasts isolated from Australian wineries. FEMS Yeast Res. *7*, 471–481. http://dx.doi.org/10.1111/j.1567-1364.2006.00183.x

Curtin, C.D., Borneman, A.R., Chambers, P.J., and Pretorius, I.S. (2012). De-novo assembly and analysis of the heterozygous triploid genome of the wine spoilage yeast *Dekkera bruxellensis* AWRI1499. PLoS ONE *7*, e33840. http://dx.doi.org/10.1371/journal.pone.0033840

Custers, M.T.J. (1940). Onderzoekingen over het gistgeslacht *Brettanomyces*. (Delft, NL: Dissertation).

Daenen, L., Vanderhaegen, B., Verachtert, H., and Derdelinckx, G. (2004). Flavour enhancement in beer by yeast beta-glucosidase activity. Commun. Agric. Appl. Biol. Sci. *69*, 73–76. http://dx.doi.org/10.1111/j.1567-1364.2008.00421.x

Daniel, H.M., Lachance, M.A., and Kurtzman, C.P. (2014). On the reclassification of species assigned to *Candida* and other anamorphic ascomycetous yeast genera based on phylogenetic circumscription. Antonie van Leeuwenhoek, Int. J. Gen. Mol. Microbiol. *106*, 67–84. http://dx.doi.org/10.1007/s10482-014-0170-z

de Barros Lopes, M., Bellon, J.R., Shirley, N.J., and Ganter, P.F. (2002). Evidence for multiple interspecific hybridization in *Saccharomyces sensu stricto* species. FEMS Yeast Res. *1*, 323–331. http://dx.doi.org/10.1111/j.1567-1364.2002.tb00051.x

De Barros Pita, W., Leite, F.C.B., De Souza Liberal, A.T., Simões, D.A., and De Morais, M.A. (2011). The ability to use nitrate confers advantage to *Dekkera bruxellensis* over S. *cerevisiae* and can explain its adaptation to industrial fermentation processes. Antonie van Leeuwenhoek, Int. J. Gen. Mol. Microbiol. *100*, 99–107. http://dx.doi.org/10.1007/s10482-011-9568-z

Deak, T. (2008). Handbook of Food Spoilage Yeasts. 2nd edn. (Boca Raton, FL: CRC Press).

Dunham, M.J., Badrane, H., Ferea, T., Adams, J., Brown, P.O., Rosenzweig, F., and Botstein, D. (2002). Characteristic genome rearrangements in experimental evolution of *Saccharomyces cerevisiae*. Proc. Natl. Acad. Sci. U.S.A. *99*, 16144–16149. http://dx.doi.org/10.1073/pnas.242624799

Dunn, B., and Sherlock, G. (2008). Reconstruction of the genome origins and evolution of the hybrid lager yeast *Saccharomyces pastorianus*. Genome Res. *18*, 1610–1623. http://dx.doi.org/10.1101/gr.076075.108

Dunn, B., Richter, C., Kvitek, D.J., Pugh, T., and Sherlock, G. (2012). Analysis of the *Saccharomyces cerevisiae* pan-genome reveals a pool of copy number variants distributed in diverse yeast strains from differing industrial environments. Genome Res. *22*, 908–924. http://dx.doi.org/10.1101/gr.130310.111

Dunn, B., Paulish, T., Stanbery, A., Piotrowski, J., Koniges, G., Kroll, E., Louis, E.J., Liti, G., Sherlock, G., and Rosenzweig, F. (2013). Recurrent rearrangement during adaptive evolution in an interspecific yeast hybrid suggests a model for rapid introgression. PLoS Genet. *9*, e1003366. http://dx.doi.org/10.1371/journal.pgen.1003366

Engel, S.R., and Cherry, J.M. (2013). The new modern era of yeast genomics: community sequencing and the resulting annotation of multiple *Saccharomyces cerevisiae* strains at the *Saccharomyces* Genome Database. Database (Oxford) *2013*,bat012. http://dx.doi.org/10.1093/database/bat012

Engel, S.R., Dietrich, F.S., Fisk, D.G., Binkley, G., Balakrishnan, R., Costanzo, M.C., Dwight, S.S., Hitz, B.C., Karra, K., Nash, R.S., *et al.* (2014). The reference genome sequence of *Saccharomyces cerevisiae*: then and now. G3 *20*, 389–398. http://dx.doi.org/10.1534/g3.113.008995

Fabian, F.W., and McCullough, N.B. (1933). Dissociation in yeasts. (Michigan State College of Agriculture and Applied Science).

Fay, J.C., and Benavides, J.A. (2005). Evidence for domesticated and wild populations of *Saccharomyces cerevisiae*. PLoS Genet. *1*, 66–71. http://dx.doi.org/10.1371/journal.pgen.0010005

Fay, J.C., McCullough, H.L., Sniegowski, P.D., and Eisen, M.B. (2004). Population genetic variation in gene expression is associated with phenotypic variation in *Saccharomyces cerevisiae*. Genome Biol. 5, R26. http://dx.doi.org/10.1186/gb-2004-5-4-r26

Fidalgo, M., Barrales, R.R., Ibeas, J.I., and Jimenez, J. (2006). Adaptive evolution by mutations in the FLO11 gene. Proc. Natl. Acad. Sci. U.S.A. *103*, 11228–11233. http://dx.doi.org/10.1073/pnas.0601713103

Fischer, G., James, S.A., Roberts, I.N., Oliver, S.G., and Louis, E.J. (2000). Chromosomal evolution in *Saccharomyces*. Nature *405*, 451–454. http://dx.doi.org/10.1038/35013058

Galeote, V., Novo, M., Salema-Oom, M., Brion, C., Valério, E., Gonçalves, P., and Dequin, S. (2010). FSY1, a horizontally transferred gene in the *Saccharomyces cerevisiae* EC1118 wine yeast strain, encodes a high-affinity fructose/H+ symporter. Microbiology *156*, 3754–3761. http://dx.doi.org/10.1099/mic.0.041673-0

Gallone, B., Steensels, J., Prahl, T., Soriaga, L., Saels, V., Herrera-Malaver, B., Merlevede, A., Roncoroni, M., Voordeckers, K., Miraglia, L., *et al.* (2016). Domestication and divergence of *Saccharomyces cerevisiae* beer yeasts. Cell *166*, 1397–1410.e16. http://dx.doi.org/10.1016/j.cell.2016.08.020

Gibson, B., and Liti, G. (2014). *Saccharomyces pastorianus*: genomic insights inspiring innovation for industry. Yeast *32*, 17-27. http://dx.doi.org/10.1002/yea.3033

Gibson, B.R., Londesborough, J., Rautio, J., Mattinen, L., and Vidgren, V. (2013). Transcription of α-glucoside transport and metabolism genes in the hybrid brewing yeast *Saccharomyces pastorianus* with respect to gene provenance and fermentation temperature. J. Inst. Brew. *119*, 23–31. http://dx.doi.org/10.1002/jib.66

Gibson, B.R., Storgårds, E., Krogerus, K., and Vidgren, V. (2013). Comparative physiology and fermentation performance of Saaz and Frohberg lager yeast strains and the parental species *Saccharomyces eubayanus*. Yeast *30*, 255–266. http://dx.doi.org/10.1002/yea.2960

Goddard, M.R. (2008). Quantifying the complexities of *Saccharomyces cerevisiae's* ecosystem engineering via fermentation. Ecology *89*, 2077–2082. http://dx.doi.org/10.1890/07-2060.1

Goddard, M.R., and Greig, D. (2015). *Saccharomyces cerevisiae*: a nomadic yeast with no niche? FEMS Yeast Res. *15*, fov009. http://dx.doi.org/10.1093/femsyr/fov009

Goddard, M.R., Anfang, N., Tang, R., Gardner, R.C., and Jun, C. (2010). A distinct population of *Saccharomyces cerevisiae* in New Zealand: Evidence for local dispersal by insects and human-aided global dispersal in oak barrels. Environ. Microbiol. *12*, 63–73. http://dx.doi.org/10.1111/j.1462-2920.2009.02035.x

Goffeau, A. (2000). Four years of post-genomic life with 6,000 yeast genes. FEBS Lett. *480*, 37–41. http://dx.doi.org/10.1016/S0014-5793(00)01775-0

Goffeau, A., Barrell, B.G., Bussey, H., Davis, R.W., Dujon, B., Feldmann, H., Galibert, F., Hoheisel, J.D., Jacq, C., Johnston, M., *et al.* (1996). Life with 6000 genes. Science *274*, 546, 563–567. http://dx.doi.org/10.1126/science.274.5287.546

Gonçalves, M., Pontes, A., Almeida, P., Barbosa, R., Serra, M., Libkind, D., Hutzler, M., Gonçalves, P., and Sampaio, J.P. (2016). Distinct domestication trajectories in top-fermenting beer yeasts and wine yeasts. Curr. Biol. *26*, 2750–2761. http://dx.doi.org/10.1016/j.cub.2016.08.040

González, S.S., Barrio, E., Gafner, J., and Querol, A. (2006). Natural hybrids from *Saccharomyces cerevisiae*,

Saccharomyces bayanus and *Saccharomyces kudriavzevii* in wine fermentations. FEMS Yeast Res. *6*, 1221–1234. http://dx.doi.org/10.1111/j.1567-1364.2006.00126.x

González, S.S., Barrio, E., and Querol, A. (2008). Molecular characterization of new natural hybrids of *Saccharomyces cerevisiae* and *S. kudriavzevii* in brewing. Appl. Environ. Microbiol. *74*, 2314–2320. http://dx.doi.org/10.1128/AEM.01867-07

Hagman, A., Säll, T., Compagno, C., and Piskur, J. (2013). Yeast 'make-accumulate-consume' life strategy evolved as a multi-step process that predates the whole genome duplication. PLoS ONE *8*, e68734. http://dx.doi.org/10.1371/journal.pone.0068734

Hansen, J., and Kielland-Brandt, M.C. (1994). *Saccharomyces carlsbergensis* contains two functional MET2 alleles similar to homologues from *S. cerevisiae* and *S. monacensis*. Gene *140*, 33–40. http://dx.doi.org/10.1016/0378-1119(94)90727-7

Hebly, M., Brickwedde, A., Bolat, I., Driessen, M.R., de Hulster, E.A., van den Broek, M., Pronk, J.T., Geertman, J.M., Daran, J.M., and Daran-Lapujade, P. (2015). S. *cerevisiae* × S. *eubayanus* interspecific hybrid, the best of both worlds and beyond. FEMS Yeast Res. *15*, fov005. http://dx.doi.org/10.1093/femsyr/fov005

Hellborg, L., and Piškur, J. (2009). Complex nature of the genome in a wine spoilage yeast, *Dekkera bruxellensis*. Eukaryot. Cell *8*, 1739–1749. http://dx.doi.org/10.1128/EC.00115-09

Hittinger, C.T., Gonçalves, P., Sampaio, J.P., Dover, J., Johnston, M., and Rokas, A. (2010). Remarkably ancient balanced polymorphisms in a multi-locus gene network. Nature *464*, 54–58. http://dx.doi.org/10.1038/nature08791

Hornsey, I.S. (2003). A History of Beer and Brewing. (Cambridge: Royal Society of Chemistry).

Hose, J., Yong, C.M., Sardi, M., Wang, Z., Newton, M.A., and Gasch, A.P. (2015). Dosage compensation can buffer copy- number variation in wild yeast. Elife *4*, 1–27. http://dx.doi.org/10.7554/eLife.05462

Hyma, K.E., Saerens, S.M., Verstrepen, K.J., and Fay, J.C. (2011). Divergence in wine characteristics produced by wild and domesticated strains of *Saccharomyces cerevisiae*. FEMS Yeast Res. *11*, 540–551. http://dx.doi.org/10.1111/j.1567-1364.2011.00746.x

Ihmels, J., Bergmann, S., Gerami-Nejad, M., Yanai, I., McClellan, M., Berman, J., and Barkai, N. (2005). Rewiring of the yeast transcriptional network through the evolution of motif usage. Science *309*, 938–940. http://dx.doi.org/10.1126/science.1113833

Jeffares, D.C., Rallis, C., Rieux, A., Speed, D., Převorovský, M., Mourier, T., Marsellach, F.X., Iqbal, Z., Lau, W., Cheng, T.M., *et al.* (2015). The genomic and phenotypic diversity of *Schizosaccharomyces pombe*. Nat. Genet. *47*, 235–241. http://dx.doi.org/10.1038/ng.3215

Jin, Y.L., and Alex Speers, R. (1998). Flocculation of *Saccharomyces cerevisiae*. Food Res. Int. *31*, 421–440. http://dx.doi.org/10.1016/S0963-9969(99)00021-6

Jorde, L.B., and Wooding, S.P. (2004). Genetic variation, classification and 'race'. Nat. Genet. *36* (Suppl. 11), S28–33. http://dx.doi.org/10.1038/ng1435

Katz Ezov, T., Boger-Nadjar, E., Frenkel, Z., Katsperovski, I., Kemeny, S., Nevo, E., Korol, A., and Kashi, Y. (2006). Molecular-genetic biodiversity in a natural population of the yeast *Saccharomyces cerevisiae* from 'Evolution Canyon': Microsatellite polymorphism, ploidy and controversial sexual status. Genetics *174*, 1455–1468. http://dx.doi.org/10.1534/genetics.106.062745

Kodama, Y., Kielland-Brandt, M.C., and Hansen, J. (2006). Lager brewing yeast. Top. Curr. Genet. *15*, 145–164.

Krogerus, K., Magalhães, F., Vidgren, V., and Gibson, B. (2015). New lager yeast strains generated by interspecific hybridization. J. Ind. Microbiol. Biotechnol. *42*, 769–778. http://dx.doi.org/10.1007/s10295-015-1597-6

Kurtzman, C.P., and Fell, J.W. (1997). The Yeasts: a taxonomic study. (Amsterdam: Elsevier B.V.).

Kurtzman, C.P., and Robnett, C.J. (2013). Relationships among genera of the *Saccharomycotina* (Ascomycota) from multigene phylogenetic analysis of type species. FEMS Yeast Res. *13*, 23–33. http://dx.doi.org/10.1111/1567-1364.12006

Legras, J.L., and Karst, F. (2003). Optimisation of interdelta analysis for *Saccharomyces cerevisiae* strain characterisation. FEMS Microbiol. Lett. *221*, 249–255. http://dx.doi.org/10.1016/S0378-1097(03)00205-2

Legras, J.L., Ruh, O., Merdinoglu, D., and Karst, F. (2005). Selection of hypervariable microsatellite loci for the characterization of *Saccharomyces cerevisiae* strains. Int. J. Food Microbiol. *102*, 73–83. http://dx.doi.org/10.1016/j.ijfoodmicro.2004.12.007

Legras, J.L., Merdinoglu, D., Cornuet, J.M., and Karst, F. (2007). Bread, beer and wine: *Saccharomyces cerevisiae* diversity reflects human history. Mol. Ecol. *16*, 2091–2102. http://dx.doi.org/10.1111/j.1365-294X.2007.03266.x

Leite, F.C.B., Basso, T.O., Pita, W.D.B., Gombert, A.K., Simões, D.A., and De Morais, M.A. (2013). Quantitative aerobic physiology of the yeast *Dekkera bruxellensis*, a major contaminant in bioethanol production plants. FEMS Yeast Res. *13*, 34–43. http://dx.doi.org/10.1111/1567-1364.12007

Libkind, D., Hittinger, C.T., Valério, E., Gonçalves, C., Dover, J., Johnston, M., Gonçalves, P., and Sampaio, J.P. (2011). Microbe domestication and the identification of the wild genetic stock of lager-brewing yeast. Proc. Natl. Acad. Sci. U.S.A. *108*, 14539–14544. http://dx.doi.org/10.1073/pnas.1105430108

Lin, Z., and Li, W.H. (2011). Expansion of hexose transporter genes was associated with the evolution of aerobic fermentation in yeasts. Mol. Biol. Evol. *28*, 131–142. http://dx.doi.org/10.1093/molbev/msq184

Lindner, P. (1909). Mikroskopische Betriebskontrolle in den Garungswerben. 6th edn. (Berlin: Paul Parey).

Liti, G., Peruffo, A., James, S.A., Roberts, I.N., and Louis, E.J. (2005). Inferences of evolutionary relationships from a population survey of LTR-retrotransposons and telomeric-associated sequences in the *Saccharomyces sensu stricto* complex. Yeast *22*, 177–192. http://dx.doi.org/10.1002/yea.1200

Liti, G., Carter, D.M., Moses, A.M., Warringer, J., Parts, L., James, S.A., Davey, R.P., Roberts, I.N., Burt, A., Koufopanou, V., *et al.* (2009). Population genomics of domestic and wild yeasts. Nature *458*, 337–341. http://dx.doi.org/10.1038/nature07743

Liti, G., Nguyen Ba, A.N., Blythe, M., Müller, C.A., Bergström, A., Cubillos, F.A., Dafhnis-Calas, F., Khoshraftar, S., Malla, S., Mehta, N., *et al.* (2013).

High quality de novo sequencing and assembly of the *Saccharomyces arboricolus* genome. BMC Genomics *14*, 69. http://dx.doi.org/10.1186/1471-2164-14-69

Magwene, P.M., Kayıkçı, Ö., Granek, J.A., Reininga, J.M., Scholl, Z., and Murray, D. (2011). Outcrossing, mitotic recombination, and life-history trade-offs shape genome evolution in *Saccharomyces cerevisiae*. Proc. Natl. Acad. Sci. U.S.A. *108*, 1987–1992. http://dx.doi.org/10.1073/pnas.1012544108

Marinoni, G., Manuel, M., Petersen, R.F., Hvidtfeldt, J., Sulo, P., and Piskur, J. (1999). Horizontal transfer of genetic material among *Saccharomyces* yeasts. J. Bacteriol. *181*, 6488–6496.

Martens, H. (1997). Microbiological aspects of a mixed yeast-bacterial fermentation in the production of a special Belgian acidic ale. J. Inst. Brew. *103*, 85–91.

Martini, A. (1993). Origin and domestication of the wine yeast *Saccharomyces cerevisiae*. J. Wine Res. *4*, 165–176.

Martini, A.V., and Martini, A. (1987). Three newly delimited species of *Saccharomyces sensu stricto*. Antonie Van Leeuwenhoek *53*, 77–84.

Martorell, P., Barata, A., Malfeito-Ferreira, M., Fernández-Espinar, M.T., Loureiro, V., and Querol, A. (2006). Molecular typing of the yeast species *Dekkera bruxellensis* and *Pichia guilliermondii* recovered from wine related sources. Int. J. Food Microbiol. *106*, 79–84. http://dx.doi.org/10.1016/j.ijfoodmicro.2005.05.014

Masel, J., and Lyttle, D.N. (2011). The consequences of rare sexual reproduction by means of selfing in an otherwise clonally reproducing species. Theor. Popul. Biol. *80*, 317–322. http://dx.doi.org/10.1016/j.tpb.2011.08.004

Meersman, E. (2011). Selectie van specifieke gisten voor een optimale cacaofermentatie. (Leuven, Belgium: Dissertation).

Mertens, S., Steensels, J., Saels, V., De Rouck, G., Aerts, G., and Verstrepen, K.J. (2015). A large set of newly created interspecific yeast hybrids increases aromatic diversity in lager beers. Appl. Environ. Microbiol. *81*, 8202–8214. http://dx.doi.org/10.1128/AEM.02464-15

Messner, R., and Prillinger, H. (1995). *Saccharomyces* species assignment by long range ribotyping. Antonie Van Leeuwenhoek *67*, 363–370.

Meussdoerffer, F.G. (2009). A comprehensive history of beer brewing. In: Handbook of Brewing: Processes, Technology, Markets. (Hoboken: Wiley).

Miklenić, Marina, Štafa, A., Baji, A., Žunar, B., Lisni, B., and Svetec, I.K. (2013). Genetic transformation of the yeast *Dekkera/Brettanomyces bruxellensis* with non-homologous DNA. J. Microbiol. Biotechnol. *23*, 674–680. http://dx.doi.org/JMB023-05-12

Miot-Sertier, C., and Lonvaud-Funel, A. (2007). Development of a molecular method for the typing of *Brettanomyces bruxellensis* (*Dekkera bruxellensis*) at the strain level. J. Appl. Microbiol. *102*, 555–562. http://dx.doi.org/10.1111/j.1365-2672.2006.03069.x

Mitrakul, C.M., Henick-Kling, T., and Egli, C.M. (1999). Discrimination of *Brettanomyces/Dekkera* yeast isolates from wine by using various DNA finger-printing methods. Food Microbiol. *16*, 3–14. http://dx.doi.org/10.1006/fmic.1998.0217

Monerawela, C., James, T.C., Wolfe, K.H., and Bond, U. (2015). Loss of lager specific genes and subtelomeric regions define two different *Saccharomyces cerevisiae* lineages for *Saccharomyces pastorianus* Group I and II strains. FEMS Yeast Res. *15*, 1–11. http://dx.doi.org/10.1093/femsyr/fou008

Mortimer, R.K. (2000). Evolution and variation of the yeast (*Saccharomyces*) genome. Genome Res. *10*, 403–409. http://dx.doi.org/10.1101/gr.10.4.403

Mortimer, R.K., Romano, P., Suzzi, G., and Polsinelli, M. (1994). Genome renewal: A new phenomenon revealed from a genetic study of 43 strains of *Saccharomyces cerevisiae* derived from natural fermentation of grape musts. Yeast. *10*, 1543–1552. http://dx.doi.org/10.1002/yea.320101203

Mukai, N., Masaki, K., Fujii, T., Kawamukai, M., and Iefuji, H. (2010). PAD1 and FDC1 are essential for the decarboxylation of phenylacrylic acids in *Saccharomyces cerevisiae*. J. Biosci. Bioeng. *109*, 564–569. http://dx.doi.org/10.1016/j.jbiosc.2009.11.011

Mukai, N., Masaki, K., Fujii, T., and Iefuji, H. (2014). Single nucleotide polymorphisms of *PAD1* and *FDC1* show a positive relationship with ferulic acid decarboxylation ability among industrial yeasts used in alcoholic beverage production. J. Biosci. Bioeng. *118*, 50–55. http://dx.doi.org/10.1016/j.jbiosc.2013.12.017

Nakao, Y., Kanamori, T., Itoh, T., Kodama, Y., Rainieri, S., Nakamura, N., Shimonaga, T., Hattori, M., and Ashikari, T. (2009). Genome sequence of the lager brewing yeast, an interspecies hybrid. DNA Res. *16*, 115–129. http://dx.doi.org/10.1093/dnares/dsp003

Naumov, G. (1996). Genetic identification of biological species in the *Saccharomyces sensu stricto* complex. J. Ind. Microbiol. Biotechnol. *17*, 295–302. http://dx.doi.org/10.1007/BF01574704

Ness, F., and Aigle, M. (1995). RTM1: a member of a new family of telomeric repeated genes in yeast. Genetics *140*, 945–956.

Ness, F., Lavallee, F., Dubourdieu, D., Aigle, M., and Dulau, L. (1993). Identification of yeast strains using the polymerase chain-reaction. S. J. Sci. Food Agric. *62*, 89–94.

Nguyen, H.V., Legras, J.L., Neuvéglise, C., and Gaillardin, C. (2011). Deciphering the hybridisation history leading to the Lager lineage based on the mosaic genomes of *Saccharomyces bayanus* strains NBRC1948 and CBS380. PLoS ONE *6*, e25821. http://dx.doi.org/10.1371/journal.pone.0025821

Nilsson-Tillgren, T., Gjermansen, C., Kielland-Brandt, M.C., Petersen, J.G.L., and Holmberg, S. (1981). Genetic differences between *Saccharomyces carlsbergensis* and *S. cerevisiae*. Analysis of chromosome III by single chromosome transfer. Carlsberg Res. Commun. *46*, 65–76. http://dx.doi.org/10.1007/BF02906199

Novo, M., Bigey, F., Beyne, E., Galeote, V., Gavory, F., Mallet, S., Cambon, B., Legras, J.L., Wincker, P., Casaregola, S., *et al.* (2009). Eukaryote-to-eukaryote gene transfer events revealed by the genome sequence of the wine yeast *Saccharomyces cerevisiae* EC1118. Proc. Natl. Acad. Sci. U.S.A. *106*, 16333–16338. http://dx.doi.org/10.1073/pnas.0904673106

Ohno, S. (1970). Evolution by Gene Duplication. (Berlin and New York: Springer-Verlag).

Okuno, M., Kajitani, R., Ryusui, R., Morimoto, H., Kodama, Y., and Itoh, T. (2016). Next-generation sequencing analysis of lager brewing yeast strains reveals the evolutionary history of interspecies hybridization. DNA Res. *23*, 67-80. http://dx.doi.org/10.1093/dnares/dsv037

Pasteur, L. (1873). Etudes sur le vin, ses maladies, causes qui les provoquent, Procédés Nouveaux, pour le conserver et pour le vieillir. (Paris).

Pasteur, L. (1876). Etudes sur la bière ses maladies, causes qui les provoquent, procédé pour la rendre inaltérable, avec une théorie nouvelle de la fermentation.

Paul Casey, G. (1986). Molecular and genetic analysis of chromosomes X in *Saccharomyces carlsbergensis*. Carlsberg Res. Commun. *51*, 343–362.

Pérez-Ortín, J.E., Querol, A., Puig, S., and Barrio, E. (2002). Molecular characterization of a chromosomal rearrangement involved in the adaptive evolution of yeast strains. Genome Res. *12*, 1533–1539. http://dx.doi.org/10.1101/gr.436602

Pérez-Través, L., Lopes, C.A., Barrio, E., and Querol, A. (2012). Evaluation of different genetic procedures for the generation of artificial hybrids in *Saccharomyces* genus for winemaking. Int. J. Food Microbiol. *156*, 102–111. http://dx.doi.org/10.1016/j.ijfoodmicro.2012.03.008

Peris, D., Sylvester, K., Libkind, D., Gonçalves, P., Sampaio, J.P., Alexander, W.G., and Hittinger, C.T. (2014). Population structure and reticulate evolution of *Saccharomyces eubayanus* and its lager-brewing hybrids. Mol. Ecol. *23*, 2031–2045. http://dx.doi.org/10.1111/mec.12702

Peter, J., and Schacherer, J. (2016). Population genomics of yeasts: towards a comprehensive view across a broad evolutionary scale. Yeast 33:73-81. http://dx.doi.org/10.1002/yea.3142

Piotrowski, J.S., Nagarajan, S., Kroll, E., Stanbery, A., Chiotti, K.E., Kruckeberg, A.L., Dunn, B., Sherlock, G., and Rosenzweig, F. (2012). Different selective pressures lead to different genomic outcomes as newly-formed hybrid yeasts evolve. BMC Evol. Biol. *12*, 46. http://dx.doi.org/10.1186/1471-2148-12-46

Piškur, J., Rozpedowska, E., Polakova, S., Merico, A., and Compagno, C. (2006). How did *Saccharomyces* evolve to become a good brewer? Trends Genet. *22*, 183–186. http://dx.doi.org/10.1016/j.tig.2006.02.002

Piškur, J., Ling, Z., Marcet-Houben, M., Ishchuk, O.P., Aerts, A., LaButti, K., Copeland, A., Lindquist, E., Barry, K., Compagno, C., *et al.* (2012). The genome of wine yeast *Dekkera bruxellensis* provides a tool to explore its food-related properties. Int. J. Food Microbiol. *157*, 202–209. http://dx.doi.org/10.1016/j.ijfoodmicro.2012.05.008

Pogorzelski, E., and Wilkowska, A. (2007). Flavour enhancement through the enzymatic hydrolysis of glycosidic aroma precursors in juices and wine beverages: a review. Flavour Fragr. J. *22*, 251–254. http://dx.doi.org/10.1002/ffj.1784

Polaina, J. (2002). Brewer's Yeast: Genetics and Biotechnology. Applied Mycology and Biotechnology *38*, 1133-1144.

Pougach, K., Voet, A., Kondrashov, F.A., Voordeckers, K., Christiaens, J.F., Baying, B., Benes, V., Sakai, R., Aerts, J., Zhu, B., *et al.* (2014). Duplication of a promiscuous transcription factor drives the emergence of a new regulatory network. Nat. Commun. *5*, 4868. http://dx.doi.org/10.1038/ncomms7543

Powell, C.D., Quain, D.E., and Smart, K.A. (2004). The impact of sedimentation on cone yeast heterogeneity. J. Am. Soc. Brew. Chem. *62*, 8–17. http://dx.doi.org/10.1094/ASBCJ-62-0008

Querol, A., and Bond, U. (2009). The complex and dynamic genomes of industrial yeasts. FEMS Microbiol. Lett. *293*, 1–10. http://dx.doi.org/10.1111/j.1574-6968.2008.01480.x

Rainieri, S., Kodama, Y., Kaneko, Y., Mikata, K., Nakao, Y., and Ashikari, T. (2006). Pure and mixed genetic lines of *Saccharomyces bayanus* and *Saccharomyces pastorianus* and their contribution to the lager brewing strain genome. Appl. Environ. Microbiol. *72*, 3968–3974. http://dx.doi.org/10.1128/AEM.02769-05

Rainieri, S., Kodama, Y., Nakao, Y., Pulvirenti, A., and Giudici, P. (2008). The inheritance of mtDNA in lager brewing strains. FEMS Yeast Res. *8*, 586–596. http://dx.doi.org/10.1111/j.1567-1364.2008.00363.x

Ramazzotti, M., Berná, L., Stefanini, I., and Cavalieri, D. (2012). A computational pipeline to discover highly phylogenetically informative genes in sequenced genomes: application to *Saccharomyces cerevisiae* natural strains. Nucleic Acids Res. *40*, 3834–3848. http://dx.doi.org/10.1093/nar/gks005

Rees, M. (1870). Botanische Untersuchungen über die Alkoholgährungspilze. (Leipzig).

Renouf, V., and Lonvaud-Funel, A. (2007). Development of an enrichment medium to detect *Dekkera/Brettanomyces bruxellensis*, a spoilage wine yeast, on the surface of grape berries. Microbiol. Res. *162*, 154–167. http://dx.doi.org/10.1016/j.micres.2006.02.006

Rodríguez, M.E., Pérez-Través, L., Sangorrín, M.P., Barrio, E., and Lopes, C.A. (2014). *Saccharomyces eubayanus* and *Saccharomyces uvarum* associated with the fermentation of Araucaria araucana seeds in Patagonia. FEMS Yeast Res. *14*, 948–965. http://dx.doi.org/10.1111/1567-1364.12183

Rozpędowska, E., Hellborg, L., Ishchuk, O.P., Orhan, F., Galafassi, S., Merico, A., Woolfit, M., Compagno, C., and Piskur, J. (2011). Parallel evolution of the make-accumulate-consume strategy in *Saccharomyces* and *Dekkera* yeasts. Nat. Commun. *2*, 302. http://dx.doi.org/10.1038/ncomms1305

Sato, M., Kishimoto, M., Watari, J., and Takashio, M. (2002). Breeding of brewer's yeast by hybridization between a top-fermenting yeast *Saccharomyces cerevisiae* and a cryophilic yeast *Saccharomyces bayanus*. J. Biosci. Bioeng. *93*, 509–511. http://dx.doi.org/10.1016/S1389-1723(02)80101-3

Schacherer, J., Shapiro, J.A., Ruderfer, D.M., and Kruglyak, L. (2009). Comprehensive polymorphism survey elucidates population structure of *Saccharomyces cerevisiae*. Nature *458*, 342–345. http://dx.doi.org/10.1038/nature07670

Schifferdecker, A.J., Dashko, S., Ishchuk, O.P., and Piškur, J. (2014). The wine and beer yeast *Dekkera bruxellensis*. Yeast. *31*, 323-332. http://dx.doi.org/10.1002/yea.3023

Schifferdecker, A.J., Siurkas, J., Andersen, M.R., Joerck-Ramberg, D., Ling, Z., Zhou, N., *et al.* (2016). Alcohol dehydrogenase gene ADH3 activates glucose alcoholic fermentation in genetically engineered *Dekkera bruxellensis* yeast. Applied Microbiology and Biotechnology *100*, 3219–3231. http://dx.doi.org/10.1007/s00253-015-7266-x

Sebastiani, F., Barberio, C., Casalone, E., Cavalieri, D., and Polsinelli, M. (2002). Crosses between *Saccharomyces cerevisiae* and *Saccharomyces bayanus* generate fertile hybrids. Res. Microbiol. *153*, 53–58. http://dx.doi.org/10.1016/S0923-2508(01)01286-4

Selmecki, A., Gerami-Nejad, M., Paulson, C., Forche, A., and Berman, J. (2008). An isochromosome confers drug resistance in vivo by amplification of two genes, ERG11 and TAC1. Mol. Microbiol. *68*, 624–641. http://dx.doi.org/10.1111/j.1365-2958.2008.06176.x

Selmecki, A.M., Maruvka, Y.E., Richmond, P.A., Guillet, M., Shoresh, N., Sorenson, A.L., De, S., Kishony, R., Michor, F., Dowell, R., *et al.* (2015). Polyploidy can drive rapid adaptation in yeast. Nature *519*, 349–352. http://dx.doi.org/10.1038/nature14187

Serra, A., Strehaiano, P., and Taillandier, P. (2005). Influence of temperature and pH on *Saccharomyces bayanus* var. *uvarum* growth; impact of a wine yeast interspecific hybridization on these parameters. Int. J. Food Microbiol. *104*, 257–265. http://dx.doi.org/10.1016/j.ijfoodmicro.2005.03.006

Sicard, D., and Legras, J.L. (2011). Bread, beer and wine: yeast domestication in the *Saccharomyces sensu stricto* complex. C.R. Biol. *334*, 229–236. http://dx.doi.org/10.1016/j.crvi.2010.12.016

Smart, K.A. (2007). Brewing yeast genomes and genome wide expression and proteome profiling during fermentation. Yeast *24*, 993–1013. http://dx.doi.org/10.1002/yea.1553

Smole Mozina, S., Dlauchy, D., Deak, T., and Raspor, P. (1997). Identification of *Saccharomyces sensu stricto* and *Torulaspora* yeasts by PCR ribotyping. Lett. Appl. Microbiol. *24*, 311–315. http://dx.doi.org/10.1046/j.1472-765X.1997.00068.x

Snoek, T., Picca Nicolino, M., Van den Bremt, S., Mertens, S., Saels, V., Verplaetse, A., Steensels, J., and Verstrepen, K.J. (2015). Large-scale robot-assisted genome shuffling yields industrial *Saccharomyces cerevisiae* yeasts with increased ethanol tolerance. Biotechnol. Biofuels *8*, 1–32. http://dx.doi.org/10.1186/s13068-015-0216-0

Spor, A., Nidelet, T., Simon, J., Bourgais, A., de Vienne, D., and Sicard, D. (2009). Niche-driven evolution of metabolic and life-history strategies in natural and domesticated populations of *Saccharomyces cerevisiae*. BMC Evol. Biol. *9*, 296. http://dx.doi.org/10.1186/1471-2148-9-296

Stambuk, B.U., Dunn, B., Alves, S.L., Duval, E.H., and Sherlock, G. (2009). Industrial fuel ethanol yeasts contain adaptive copy number changes in genes involved in vitamin B1 and B6 biosynthesis. Genome Res. *19*, 2271–2278. http://dx.doi.org/10.1101/gr.094276.109

Steensels, J., Snoek, T., Meersman, E., Picca Nicolino, M., Voordeckers, K., and Verstrepen, K.J. (2014). Improving industrial yeast strains: exploiting natural and artificial diversity. FEMS Microbiol. Rev. *38*, 947–995. http://dx.doi.org/10.1111/1574-6976.12073

Steensels, J., Meersman, E., Snoek, T., Saels, V., and Verstrepen, K.J. (2014). Large-scale selection and breeding to generate industrial yeasts with superior aroma production. Appl. Environ. Microbiol. *80*, 6965–6975. http://dx.doi.org/10.1128/AEM.02235-14

Steensels, J., Daenen, L., Malcorps, P., Derdelinckx, G., Verachtert, H., and Verstrepen, K.J. (2015). *Brettanomyces* yeasts – From spoilage organisms to valuable contributors to industrial fermentations. Int. J. Food Microbiol. *206*, 24–38. http://dx.doi.org/10.1016/j.ijfoodmicro.2015.04.005

Strope, P.K., Skelly, D.A., Kozmin, S.G., Mahadevan, G., Stone, E.A., Magwene, P.M., Dietrich, F.S., and Mccusker, J.H.. (2015). The 100-genomes strains, an S. *cerevisiae* resource that illuminates its natural phenotypic and genotypic variation and emergence as an opportunistic pathogen. Genome Res. *25*, 762–774. http://dx.doi.org/10.1101/gr.185538

Tamai, Y., Momma, T., Yoshimoto, H., and Kaneko, Y. (1998). Co-existence of two types of chromosome in the bottom fermenting yeast, *Saccharomyces pastorianus*. Yeast *14*, 923–933. http://dx.doi.org/10.1002/(SICI)1097-0061(199807)14:10<923::AID-YEA298>3.0.CO;2-I

Smith, M.T., and van Grinsven, A.M. (1984). *Dekkera anomala* sp. nov., the teleomorph of *Brettanomyces* anomalus, recovered from spoiled soft drinks. Antonie Van Leeuwenhoek *50*, 143–148.

Thomson, J.M., Gaucher, E.A., Burgan, M.F., De Kee, D.W., Li, T., Aris, J.P., and Benner, S.A. (2005). Resurrecting ancestral alcohol dehydrogenases from yeast. Nat. Genet. *37*, 630–635. http://dx.doi.org/10.1038/ng1553

Tiukova, I.A., Petterson, M.E., Tellgren-Roth, C., Bunikis, I., Eberhard, T., Pettersson, O.V., and Passoth, V. (2013). Transcriptome of the alternative ethanol production strain *Dekkera bruxellensis* CBS 11270 in sugar limited, low oxygen cultivation. PLoS ONE *8*, e58455. http://dx.doi.org/10.1371/journal.pone.0058455

Tornai-Lehoczki, J., and Dlauchy, D. (2000). Delimination of brewing yeast strains using different molecular techniques. Int. J. Food Microbiol. *62*, 37–45. http://dx.doi.org/10.1016/S0168-1605(00)00356-1

Tornai-Lehoczki, J., Péter, G., Dlauchy, D., and Deák, T. (1996). Some remarks on 'a taxonomic key for the genus *Saccharomyces*' (Vaughan Martini and Martini 1993). Antonie Van Leeuwenhoek *69*, 229–233. http://dx.doi.org/10.1007/BF00399611

Torres, E.M., Sokolsky, T., Tucker, C.M., Chan, L.Y., Boselli, M., Dunham, M.J., and Amon, A. (2007). Effects of aneuploidy on cellular physiology and cell division in haploid yeast. Science *317*, 916–924. http://dx.doi.org/10.1126/science.1142210

Valdes, J., Tapia, P., Cepeda, V., Varela, J., Godoy, L., Cubillos, F.A., Silva, E., Martinez, C., and Ganga, M.A. (2014). Draft genome sequence and transcriptome analysis of the wine spoilage yeast *Dekkera bruxellensis* LAMAP2480 provides insights into genetic diversity, metabolism and survival. FEMS Microbiol. Lett. *361*, 104–106. http://dx.doi.org/10.1111/1574-6968.12630

Vaughan Martini, A., and Kurtzman, C.P. (1985). Deoxyribonucleic acid relatedness among species of the genus *Saccharomyces sensu stricto*. Int. J. Syst. Bacteriol. *35*, 508–511.

Verachtert, H. (1992). Lambic and gueze brewing: mixed cultures in action. (COMETT). Course on Microbial Contaminants, Helsinki.

Verstrepen, K.J., and Klis, F.M. (2006). Flocculation, adhesion and biofilm formation in yeasts. Mol.

Microbiol. *60*, 5–15. http://dx.doi.org/10.1111/j.1365-2958.2006.05072.x

Verstrepen, K.J., Derdelinckx, G., Verachtert, H., and Delvaux, F.R. (2003). Yeast flocculation: what brewers should know. Appl. Microbiol. Biotechnol. *61*, 197–205. http://dx.doi.org/10.1007/s00253-002-1200-8

Verstrepen, K.J., Jansen, A., Lewitter, F., and Fink, G.R. (2005). Intragenic tandem repeats generate functional variability. Nat. Genet. *37*, 986–990. http://dx.doi.org/10.1038/ng1618

Vezinhet, F., Blondin, B., and Hallet, J.N. (1990). Chromosomal DNA patterns and mitochondrial DNA polymorphism as tools for identification of enological strains of *Saccharomyces cerevisiae*. Appl. Microbiol. Biotechnol. *32*, 568–571. http://dx.doi.org/10.1007/BF00173729

Vigentini, I., De Lorenzis, G., Picozzi, C., Imazio, S., Merico, A., Galafassi, S., Piškur, J., and Foschino, R. (2012). Intraspecific variations of *Dekkera/Brettanomyces bruxellensis* genome studied by capillary electrophoresis separation of the intron splice site profiles. Int. J. Food Microbiol. *157*, 6–15. http://dx.doi.org/10.1016/j.ijfoodmicro.2012.02.017

Voordeckers, K., De Maeyer, D., van der Zande, E., Vinces, M.D., Meert, W., Cloots, L., Ryan, O., Marchal, K., and Verstrepen, K.J. (2012a). Identification of a complex genetic network underlying *Saccharomyces cerevisiae* colony morphology. Mol. Microbiol. *86*, 225–239. http://dx.doi.org/10.1111/j.1365-2958.2012.08192.x

Voordeckers, K., Brown, C.A., Vanneste, K., van der Zande, E., Voet, A., Maere, S., and Verstrepen, K.J. (2012b). Reconstruction of ancestral metabolic enzymes reveals molecular mechanisms underlying evolutionary innovation through gene duplication. PLoS Biol. *10*, e1001446. http://dx.doi.org/10.1371/journal.pbio.1001446

Walther, A., Hesselbart, A., and Wendland, J. (2014). Genome sequence of *Saccharomyces carlsbergensis*, the world's first pure culture lager yeast. G3 *4*, 783–793. http://dx.doi.org/10.1534/g3.113.010090

Wang, Q.M., Liu, W.Q., Liti, G., Wang, S.A., and Bai, F.Y. (2012). Surprisingly diverged populations of *Saccharomyces cerevisiae* in natural environments remote from human activity. Mol. Ecol. *21*, 5404–5417. http://dx.doi.org/10.1111/j.1365-294X.2012.05732.x

Warringer, J., Zörgö, E., Cubillos, F.A., Zia, A., Gjuvsland, A., Simpson, J.T., Forsmark, A., Durbin, R., Omholt, S.W., Louis, E.J., *et al.* (2011). Trait variation in yeast is defined by population history. PLoS Genet. *7*, e1002111. http://dx.doi.org/10.1371/journal.pgen.1002111

Wedral, D., Shewfelt, R., and Frank, J. (2010). The challenge of *Brettanomyces* in wine. LWT– Food Sci. Technol. *43*, 1474–1479. http://dx.doi.org/10.1016/j.lwt.2010.06.010

Wendland, J. (2014). Lager yeast comes of age. Eukaryot. Cell *13*, 1256–1265. http://dx.doi.org/10.1128/EC.00134-14

Wenger, J.W., Schwartz, K., and Sherlock, G. (2010). Bulk segregant analysis by high-throughput sequencing reveals a novel xylose utilization gene from *Saccharomyces cerevisiae*. PLoS Genet. *6*, e1000942. http://dx.doi.org/10.1371/journal.pgen.1000942

Wijsman, M.R., van Dijken, J.P., van Kleeff, B.H., and Scheffers, W.A. (1984). Inhibition of fermentation and growth in batch cultures of the yeast Brettanomyces intermedius upon a shift from aerobic to anaerobic conditions (Custers effect). Antonie Van Leeuwenhoek *50*, 183–192.

Will, J.L., Kim, H.S., Clarke, J., Painter, J.C., Fay, J.C., and Gasch, A.P. (2010). Incipient balancing selection through adaptive loss of aquaporins in natural *Saccharomyces cerevisiae* populations. PLoS Genet. *6*, e1000893. http://dx.doi.org/10.1371/journal.pgen.1000893

Wolfe, K.H., and Li, W.H. (2003). Molecular evolution meets the genomics revolution. Nat. Genet. *33*, 255–265. http://dx.doi.org/10.1038/ng1088

Woolfit, M., Rozpedowska, E., Piskur, J., and Wolfe, K.H. (2007). Genome survey sequencing of the wine spoilage yeast *Dekkera* (Brettanomyces) *bruxellensis*. Eukaryot. Cell 6, 721–733. http://dx.doi.org/10.1128/EC.00338-06

Yamagishi, H., and Ogata, T. (1999). Chromosomal structures of bottom fermenting yeasts. Syst. Appl. Microbiol. *22*, 341–353. http://dx.doi.org/10.1016/S0723-2020(99)80041-1

Yona, A.H., Manor, Y.S., Herbst, R.H., Romano, G.H., Mitchell, A., Kupiec, M., Pilpel, Y., and Dahan, O. (2012). Chromosomal duplication is a transient evolutionary solution to stress. Proc. Natl. Acad. Sci. U.S.A. *109*, 21010–21015. http://dx.doi.org/10.1073/pnas.1211150109

Microbial Ecology of Traditional Beer Fermentations

7

Freek Spitaels[1], Anneleen D. Wieme[1,2], Isabel Snauwaert[1], Luc De Vuyst[3] and Peter Vandamme[1*]

[1]Laboratory of Microbiology, Faculty of Sciences, Ghent University, Ghent, Belgium.
[2]Laboratory of Biochemistry and Brewing, Faculty of Bioscience Engineering, Ghent University, Ghent, Belgium.
[3]Research Group of Industrial Microbiology and Food Biotechnology (IMDO), Faculty of Sciences and Bioengineering Sciences, Vrije Universiteit Brussel, Brussels, Belgium.

*Correspondence: peter.vandamme@ugent.be

https://doi.org/10.21775/9781910190616.07

Abstract

Traditional beer fermentations do not apply inoculation of pure cultures to initiate the fermentation process. In contrast, they rely on spontaneous inoculation, 'backslopping', or inoculation with an undetermined mixture of microorganisms for the production of the beer. Knowledge about the microbiota responsible for these special fermentation processes does not only enable a closer quality management, but the isolation of the microorganisms involved enables their application as starter cultures. Very commonly, the same microorganisms are involved in these processes, i.e. the yeasts *Saccharomyces cerevisiae* and/or *Saccharomyces pastorianus* plus *Brettanomyces* (*Dekkera*) *bruxellensis*, lactic acid bacteria and/or acetic acid bacteria. All traditional beers share a desirable tartness and their production processes take from several days to several years to complete. This chapter reviews notable types of traditional beer fermentations, their microbial ecology, and methods used to investigate their composition.

Introduction to traditional and mixed-culture beer fermentations

Beer is among the oldest fermented beverages and is the product of a fermented sugar extract from grains. Historically, beers were the result of spontaneous fermentations, involving several yeasts and bacteria. Conversely, pure yeast cultures have been commonly used in most modern breweries ever since they were first proposed by Hansen (Claussen, 1904). Most commercial beer brands are produced using proprietary *Saccharomyces cerevisiae* or *Saccharomyces pastorianus* and *Saccharomyces bayanus* strains for the production of ales and lagers, respectively. The resulting beers have a standardized batch-to-batch flavour and are appreciated by beer lovers around the world. The presence of wild-type yeasts and/or bacteria during the fermentation process or in the final beer can alter the flavour and physical characteristics of the beers, turning them unsellable. The economic losses related to these problems are a threat in commercial breweries worldwide, emphasizing the need for quality control of the final beers prior to their shipping to the customers.

In contrast to modern commercial ale and lager brands, some traditional beers do not require axenic yeast cultures to pitch their worts and some breweries do not even inoculate the wort to induce the fermentation of their beers. These products are thought to resemble ancient beers, since they involve both bacteria and yeasts and have a vinous acidic character. Two major types of such beers are brewed in Belgium. For the production of lambic and gueuze beers, no yeast or bacteria are pitched to start the fermentation, whereas red-brown acidic ales re-pitch a lactic acid bacteria (LAB)-harbouring yeast suspension, that is harvested after every fermentation. Nowadays, there are traditional beers brewed outside Belgium that also apply mixed fermentations, either by spontaneous inoculation or through the use of mixed starter cultures. Lambic beers in particular and spontaneously fermented beers in general are currently highly appreciated all over the world, since they are trademarks of traditional craftsmanship. Traditionally, Belgian lambic beers were produced in the Senne river valley (south-west of Brussels) and in the southeast of Brussels, but due to their growing popularity lambic-type beers are now produced elsewhere in Belgium and internationally.

The use of culture-dependent and culture-independent techniques in microbial biodiversity studies

While the microbiota of some traditional beers had been characterized with classical methods several decades ago, more recent microbial biodiversity studies have been performed using culture-independent community fingerprinting techniques in combination with traditional cultivation methods (Dolci *et al.*, 2010; Laureys and De Vuyst, 2014; Scheirlinck *et al.*, 2008; Van der Meulen *et al.*, 2007; Wouters *et al.*, 2013). Some of these studies have focused on specific microbial groups of those communities, e.g. LAB or yeasts, while others address the entire community (Martens *et al.*, 1991; Shanta Kumara and Verachtert, 1991; Van Oevelen *et al.*, 1977; Verachtert and Iserentant, 1995). In recent years, more advanced and high-throughput culture-independent methods, such as bar-coded amplicon sequencing (BAS) and shotgun metagenomics, are rapidly replacing community fingerprinting methods such as denaturing gradient gel electrophoresis (DGGE) of polymerase chain reaction (PCR) gene amplicons. The data obtained are considered superior to culture-dependent data. Indeed, methods that involve cultivation of microorganisms are often considered less informative, as some microorganisms can be present in a viable but non-culturable (VBNC) state, while isolation media favour the cultivation of specific microorganisms only (Gorski, 2012; Millet and Lonvaud-Funel, 2000). Moreover, metagenomic techniques are also superior to classical fingerprint-based culture-independent techniques such as PCR-DGGE, because of their ability to detect low-abundance species in the communities (Bokulich and Bamforth, 2013).

However, culture-independent techniques also introduce biases, for instance through differences in DNA extraction efficacy or in PCR-based amplification of target sequences (Hong *et al.*, 2009; Yuan *et al.*, 2012), and potentially detect not only metabolically inactive but also dead cells. Therefore, although these modern culture-independent approaches provide a more in-depth analysis of the microbial communities, they are not without limitations and also do not reveal which species are metabolically most active. Other tools, not only transcriptomics or meta-metabolomics but also the availability of pure cultures of community members will be required to reveal a more complete image of the microbial diversity present in an ecosystem.

Ironically, the potential to examine microbial ecosystems by means of modern metagenomics approaches triggered a renewed interest in the development of new approaches to cultivate a larger fraction of microorganisms that are known through the detection of their DNA only (*Nature Reviews: Microbiology* Editorial, 2013; Rappé, 2013; Teske, 2010). Based on metagenomics information about the genes and metabolic potential present, new culture media have been developed that target the isolation of specific microbial groups (Bomar *et al.*, 2011). An approach referred to as culturomics (Lagier *et al.*, 2012) is currently gaining momentum in the field of gut microbiome research. Culturomics refers to the high-throughput and miniaturized application of numerous classical media for the isolation of microorganisms and is limited only by the rate of identification of the isolates obtained. For this purpose, sequence-based identification methods are too slow and expensive but matrix-assisted laser desorption/ionization time-of-flight mass spectrometry (MALDI-TOF MS) has been advocated as an ideal identification technique in culturomics approaches to study microbial diversity (Lagier *et al.*, 2012).

Ecology of lambic beers produced in traditional and industrial lambic beer breweries

The traditional lambic beer fermentation process

Acidic lambic beers, obtained by spontaneous fermentation, are probably the oldest known beers (De Keersmaecker, 1996). They are the products of a mixed fermentation that can last up to 3 years and are traditionally fermented in wooden casks (Verachtert and Iserentant, 1995). Notwithstanding the increasing popularity of lambic beers, the fermentation process was dealt with only in some detail in studies performed between 1976 and 1995 by the research group of Professor-Emeritus Hubert Verachtert (Martens *et al.*, 1991, 1992; Shanta Kumara and Verachtert, 1991; Spaepen *et al.*, 1978, 1979; Van Oevelen *et al.*, 1976, 1977; Verachtert and Iserentant, 1995). These studies were performed using biochemical methods solely and were limited in the number of isolates examined and in the taxonomical information obtained (Van Oevelen *et al.*, 1977; Verachtert and Iserentant, 1995). Since the publication of these early studies, the taxonomy of bacteria and yeasts involved in the lambic beer fermentation process underwent several changes and biochemical identification methods were shown inadequate to reliably identify these microorganisms (Cleenwerck *et al.*, 2008; De Bruyne *et al.*, 2008; Kämpfer and Glaeser, 2012; Kurtzman and Robnett, 1998; Nhung *et al.*, 2007). More recently, the microbial composition of these beers was updated with the current taxonomic knowledge and by the use of more advanced identification techniques (Spitaels *et al.*, 2014c, 2015b).

Lambic beer is traditionally brewed during the cold winter months only, i.e. from October until March, because the lambic wort has to be cooled to approximately 20°C within the timeframe of one night. Lambic beer is traditionally produced using about 66% malted barley and 33% unmalted wheat. The use of at least 30% unmalted wheat in the mash is regulated by Belgian law. Traditionally, the lambic wort production starts with a turbid mash method, which is a combination of the English infusion and German decoction processes (Fig. 7.1). Hot water is added during the English infusion process to increase the temperature of the mash. During decoction, the brewer boils a part of the mash separately to rupture the starch granules and subsequently reintroduces it into the mash tun to increase the total mash temperature, ensuring the rests at the enzymes' optimal temperatures (Briggs *et al.*, 2004). During turbid mashing, the brewer does not reintroduce the separately boiled wort (called slime) into the mash tun, so that not all of the wort passes through all temperature rests (Fig. 7.1). The use of unmalted wheat and the turbid mashing step with separate slime cooking results in a wort that is rich in malto-oligosaccharides or dextrins. These dextrins are non-fermentable by conventional *Saccharomyces* brewing yeasts (Shanta Kumara and Verachtert, 1991), but they can be fermented by *Brettanomyces* (the sporulating form of this yeast is named *Dekkera*) yeasts that are also present during the maturation of red-brown acidic ales of south-west Flanders (Martens *et al.*, 1997). The wort is boiled for 3 h, which is a long period compared with other beer types, and a large amount of aged hops is added to enhance the microbiological stability of the beer without resulting in a bitter hop flavour (Verachtert and Derdelinckx, 2005; Vriesekoop *et al.*, 2012). After wort cooking, the wort is cooled in an open vessel, called the cooling tun or coolship, which is mostly located in the attic of the brewery (Fig. 7.1). After overnight cooling in the coolship, the wort becomes exposed to the environmental microbiota that initiate fermentation. As lambic beers were originally only produced in the Senne river valley (southwest of Brussels) and in the southeast of Brussels, it was believed that the responsible microbiota were present in the air of this region (Verachtert and Iserentant, 1995).

The first reports of the microbiota and their metabolites divided the lambic beer fermentation process into four phases: the *Enterobacteriaceae* phase, the main fermentation phase, the acidification phase, and the maturation phase; more recent studies considered the acidification phase and maturation stage as a single period (Spitaels *et al.*, 2014c; Van Oevelen *et al.*, 1976, 1977; Verachtert and Iserentant, 1995). Each phase was characterized by the presence of specific microorganisms and metabolites (Van Oevelen *et al.*, 1977; Verachtert and Iserentant, 1995). The culture media used were selected based on previous studies and the observation of increased concentrations of acetic acid and lactic acid, indicating the presence of acetic acid bacteria (AAB) and LAB (Van Oevelen *et al.*, 1976).

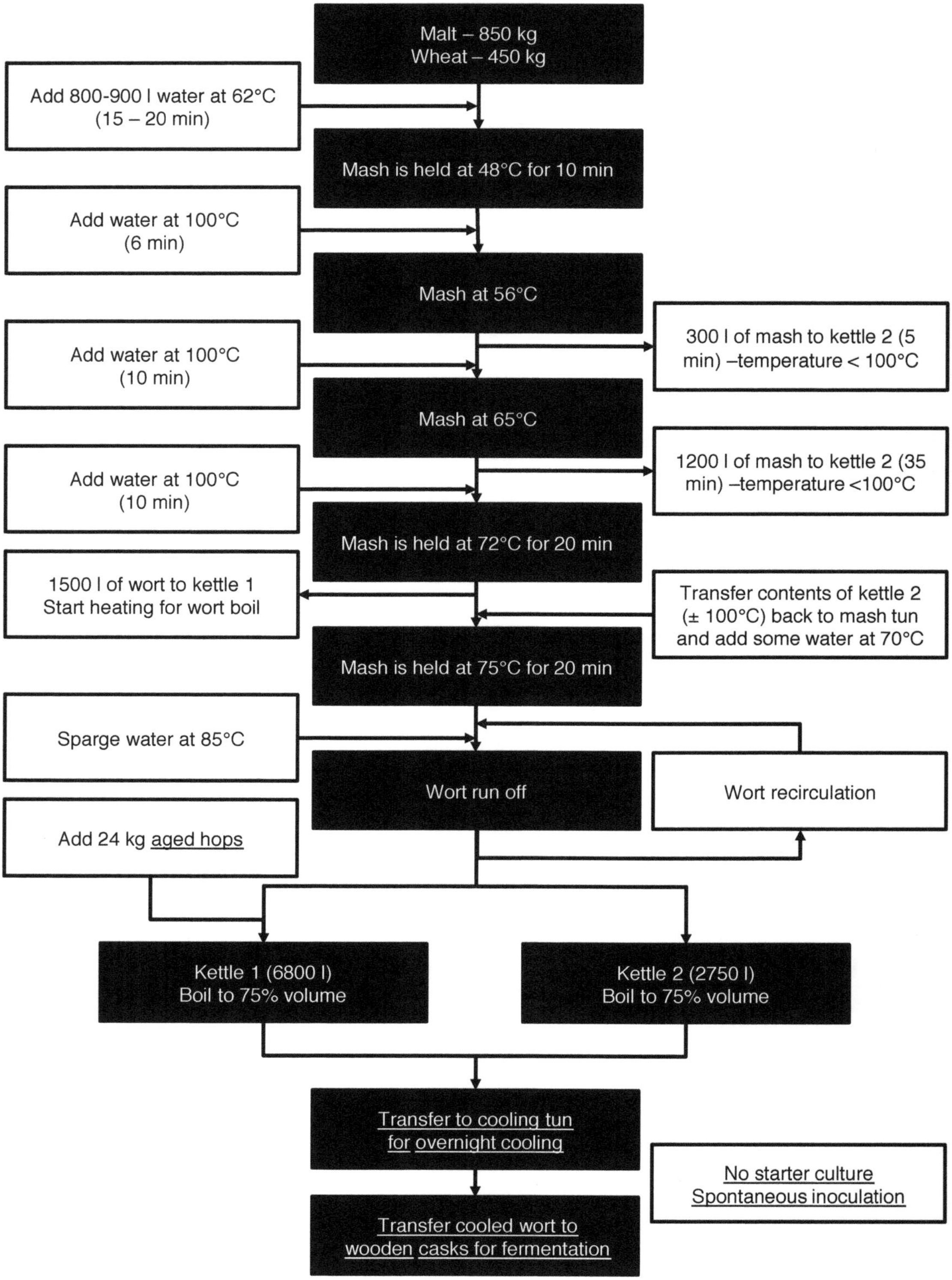

Figure 7.1 Example of a brewing scheme in a traditional lambic beer brewery, making use of turbid mashing and two boiling kettles. The typical lambic beer fermentation process characteristics, next to the unusual mashing scheme, are underlined.

Enterobacteriaceae were isolated from the cooled wort in the cooling tun and the cask (Martens *et al.*, 1991; Van Oevelen *et al.*, 1977; Verachtert and Iserentant, 1995). The *Enterobacteriaceae* phase was reported to start after 3–7 days of fermentation, when bacterial counts reached up to 10^8 CFU/ml, to proceed for 30 to 40 days, and to be characterized by *Enterobacter cloacae* and *Klebsiella pneumoniae* (*Klebsiella aerogenes*) (Brisse *et al.*, 2006) as the predominantly isolated *Enterobacteriaceae* species (Martens *et al.*, 1991). *Enterobacter aerogenes, Citrobacter freundii, Escherichia coli* and *Hafnia alvei* were additionally isolated (Martens *et al.*, 1991). The number of *Enterobacteriaceae* cells present in brewery air is however low, and it has been hypothesized that wort inoculation during cooling is not homogeneous and bacteria are probably adsorbed to particles present in the air (Martens *et al.*, 1991).

Oxidative yeasts, such as the cycloheximide-resistant *Hanseniaspora uvarum* [its asexual form is named *Kloeckera apiculata* (Meyer *et al.*, 1978)] and *Naumovia dairensis* (previously known as *Saccharomyces dairensis*) (Kurtzman, 2003) as well as *Saccharomyces uvarum* [previously known as *S. globosus* (Nguyen and Gaillardin, 2005)] are the main yeast species present during the enterobacterial phase of the lambic beer fermentation process (Van Oevelen *et al.*, 1977; Verachtert and Iserentant, 1995). *H. uvarum* has a low fermentative capacity and is commonly found during the spontaneous fermentation of wines and ciders, where its contribution to flavour complexity is increasingly appreciated (Bezerra-Bussoli *et al.*, 2013). Similar to *B. bruxellensis, H. uvarum* is capable of producing ethyl esters, and was long considered as a wine spoilage yeast (Moreira *et al.*, 2011; Romano *et al.*, 2003). During this phase, the pH dropped one value, and considerable levels of butanediol and dimethyl sulfide were formed, along with formic acid, acetic acid, lactic acid, and ethanol (Verachtert and Iserentant, 1995). The disappearance of the *Enterobacteriaceae* after about one month of fermentation is explained by the depletion of glucose, the increase in ethanol concentration and the decreased pH of the wort (Martens *et al.*, 1991).

A similar start of the lambic beer fermentation was found during a more recent study of a traditional lambic beer brewing process (Spitaels *et al.*, 2014c). A fast dereplication technique based on MALDI-TOF MS enabled the processing of more isolates from several phases during lambic beer production. Overnight cooled wort already contained high counts of *Enterobacteriaceae* in the cooling tun (10^6–10^7 CFU/ml). No yeasts were found in the lambic beer samples from the cooling tun (Spitaels *et al.*, 2014c). The counts of the *Enterobacteriaceae* were the highest (10^8 CFU/ml) after 1–2 weeks. *E. coli* was again isolated, whereas *Hafnia paralvei* [an opportunistic human and animal pathogen (Huys *et al.*, 2010)] was isolated instead of *H. alvei*. It is likely that isolates from previous studies were in fact also *H. paralvei*, as this species was separated from *H. alvei* only very recently (Huys *et al.*, 2010). Other isolates in the *Enterobacteriaceae* phase were identified as *Enterobacter hormaechei, Enterobacter kobei, Klebsiella oxytoca, Citrobacter gillenii* and *Raoultella terrigena* (Spitaels *et al.*, 2014c). All these species are coliform bacteria and thus indicator microorganisms for faecal contamination of surface waters and foods. Although these species are considered to be opportunistic pathogens, they are commonly found in various spontaneously fermented foods and beverages, and some were previously isolated from lambic beer as well (Chao *et al.*, 2013; Martens *et al.*, 1991). Remarkably, most of these microorganisms have also been reported as spoilage microorganisms in sweet unfermented wort and pitching yeast (Bokulich and Bamforth, 2013; Van Vuuren and Priest, 2003; Vriesekoop *et al.*, 2012) (see Chapter 10).

Debaryomyces hansenii and *S. cerevisiae* were two yeast species isolated immediately after the transfer of the wort into the cask. *S. pastorianus* and *Naumovia castelii* were subsequently isolated from a 1-week-old wort sample. *Debaryomyces hansenii* is a known beer spoilage yeast (Bokulich and Bamforth, 2013), whereas *N. castellii* was previously known as *Saccharomyces castellii* and part of the *Saccharomyces sensu stricto* group (Kurtzman, 2003). DNA originating from *H. uvarum* was detected in the community profiles of wort samples of traditionally produced lambic beer, but this species was not isolated (Spitaels *et al.*, 2014c). *Hanseniaspora uvarum* originates from the fruit surfaces when thriving in wine and cider fermentation processes (Beltran *et al.*, 2002), but it is known as an environmental contaminant in ales and lagers (Bokulich and Bamforth, 2013; Manzano *et al.*, 2011).

The *Enterobacteriaceae* phase was followed by the main fermentation phase or alcoholic fermentation phase, which started after 3 to 4 weeks. *Saccharomyces* spp. dominated the fermentation process from month 1 until month 4 (Spitaels *et al.*, 2014c; Van Oevelen *et al.*, 1977; Verachtert and Iserentant, 1995). *S. cerevisiae, S. bayanus/pastorianus* and *S. uvarum* were identified as the main actors during this stage (Van Oevelen *et al.*, 1977; Verachtert and Iserentant, 1995). High counts of *S. cerevisiae* and *S. pastorianus* were present at the start of the main fermentation phase, but after 3 months of fermentation most isolates were identified as *S. pastorianus* (Spitaels *et al.*, 2014c). Previous studies only reported the presence of *S. cerevisiae, S. bayanus/S. pastorianus* and *S. uvarum* in the main fermentation phase and did not provide detailed information for different sampling moments. It is however not clear why *S. pastorianus* can outlive *S. cerevisiae* in the lambic beer fermentation process of a traditional brewery. The genomic background of the hybrid species *S. pastorianus* was recently elucidated (Libkind *et al.*, 2011). In *Saccharomyces*, hybridization events between cryotolerant and non-cryotolerant *Saccharomyces* species offered a benefit for the resulting hybrids, because of the capacity of these hybrids to ferment at lower temperature (Peris *et al.*, 2012). Consequently, all commonly used lager-type yeasts are domesticated strains of the initial pastorianus and bayanus hybrids (Libkind *et al.*, 2011). The ambient temperature of the rooms where lambic beers are fermenting is rarely 20°C or higher during the first fermentation months, which may explain the predominance of *S. pastorianus* in a traditional lambic beer brewery process. Vidgren *et al.* (2010) reported that ale (generally *S. cerevisiae*) and lager (generally *S. pastorianus* or *S. bayanus*) strains exhibit a similar maltose transport activity at 20°C, but at 0°C the activity of lager strains is higher by the expression of cryotolerant maltose and maltotriose transporters. The different temperature sensitivity of the maltose and maltotriose transporters could have an influence on the survival of different *Saccharomyces* hybrids, since the transport of these molecules is assumed to be the rate-limiting step in the utilization of these saccharides (Cousseau *et al.*, 2013).

After the main fermentation phase, oxidative yeasts, i.e. *Cryptococcus* spp., *Candida* spp., *Pichia* spp. and *Torulopsis* spp. form a pellicle at the top of the liquid and serve as an oxygen barrier (Van Oevelen *et al.*, 1977; Verachtert and Iserentant, 1995). *Pediococcus damnosus* (*Pediococcus cerevisiae*) was commonly isolated during and after this main fermentation phase, in addition to low counts of AAB [*Acetobacter* spp. and *Gluconobacter* (*Acetomonas*) spp.], which were isolated irregularly (Spitaels *et al.*, 2014c; Van Oevelen *et al.*, 1977; Verachtert and Iserentant, 1995). *Acetobacter lambici* (Spitaels *et al.*, 2014a) and *Gluconobacter cerevisiae* (Spitaels *et al.*, 2014b), two newly described AAB species, were occasionally isolated during the lambic beer fermentation process. From 2 months onwards, *P. damnosus* was consistently present. During this phase, the majority of the ethanol present in the lambic beers was produced, and the level of dimethyl sulfide (produced during the *Enterobacteriaceae* phase) decreased, driven off by CO_2 bubbles produced by the yeasts still present from the fermenting lambic beer (Verachtert and Iserentant, 1995). Simultaneously, higher alcohols, fatty acids and esters, including hexanoate, octanoate, decanoate and their ethyl esters, respectively, were formed as well (Spaepen *et al.*, 1978; Van Oevelen *et al.*, 1976; Verachtert and Iserentant, 1995).

After 2–3 months of main fermentation and the depletion of the carbon sources that can be fermented by *Saccharomyces* spp. (simple saccharides up to maltotriose), an acidification phase has been reported that was characterized by the increasing isolation of *Pediococcus* and occasionally *Lactobacillus* strains (only in breweries with large casks), while *Brettanomyces* strains became prevalent after 4 to 8 months of fermentation. Simultaneously, the number of *Saccharomyces* yeasts decreased (Van Oevelen *et al.*, 1977; Verachtert and Iserentant, 1995). The acidification was characterized by a strong increase in lactic acid and ethyl lactate concentrations, which are typical metabolites of lambic beers (Van Oevelen *et al.*, 1976; Verachtert and Iserentant, 1995). During the warm summer months, LAB can also cause slime in the fermenting lambic beer, which is undesirable (Van Oevelen *et al.*, 1977; Van Oevelen and Verachtert, 1979).

Brettanomyces and LAB species have a synergistic effect on beer attenuation (Andrews and Gilliland, 1952; Shanta Kumara and Verachtert, 1991). *Brettanomyces* spp. in combination with LAB degrade the residual dextrins that are not fermented by *Saccharomyces* spp. (Shanta Kumara and Verachtert, 1991). Lambic beers reach a high attenuation during the maturation phase, resulting in a residual gravity that may be below 1°P (Shanta Kumara and Verachtert, 1991; Verachtert and Iserentant, 1995). Super-attenuation or over-attenuation was already described by Andrews and Gilliland (1952), who demonstrated that a primary attenuation limit, typical for an axenic *S. cerevisiae* culture, and a secondary attenuation limit, typical for an axenic *B. bruxellensis* culture, can still be overcome by the use of a mixed culture of yeasts and bacteria (Andrews and Gilliland, 1952). Hence, there is probably a synergistic effect of the yeast and bacterial cultures during the degradation of dextrins and starch (Andrews and Gilliland, 1952). A similar finding was reported by Shanta Kumara and Verachtert (1991), who demonstrated that *Brettanomyces* is the main contributor to the super-attenuation of lambic beers, but its effect is more pronounced in a mixed culture with *Pediococcus* (Shanta Kumara and Verachtert, 1991). *Brettanomyces* produces an α-glucosidase, an enzyme capable of dextrin degradation (Shanta Kumara and Verachtert, 1991). This α-glucosidase shows intracellular as well as extracellular activities and acts by removing a single glucose molecule from the dextrin polymer (De Cort *et al.*, 1994; Shanta Kumara *et al.*, 1993). The enzyme is fast-acting, as under optimal conditions malto-oligosaccharides shorter than maltotetraose are not found in the presence of the enzyme (Shanta Kumara *et al.*, 1993). The low pH of lambic beers, however, may explain the slow process of over-attenuation *in situ* in lambic beers (Shanta Kumara *et al.*, 1993).

After 10 months, the bacterial counts decrease and a new phase in the lambic beer fermentation process is initiated by the increase of *Brettanomyces* spp. During this phase, cell-bound esterases of *Brettanomyces* yeasts can form and degrade several esters in the fermenting lambic beer (Spaepen and Verachtert, 1982) and several metabolites and flavour compounds are produced by the synergistic action of LAB and *Brettanomyces* yeasts (Shanta Kumara and Verachtert, 1991; Van Oevelen *et al.*, 1976, 1977; Verachtert and Iserentant, 1995). These include the esters ethyl acetate and ethyl lactate as well as the long-chain fatty acids and their esters such as ethyl caprylate and ethyl caprate (Spaepen *et al.*, 1978). Only minimal concentrations of ethyl caprate are present in most other beers and this can thus be considered as a typical aroma compound of lambic beers (Spaepen *et al.*, 1978). However, a beer produced by the mixed fermentation of a LAB-harbouring pitching yeast with a secondary cask fermentation contains comparable concentrations of these long-chain fatty acids and their esters (Spaepen *et al.*, 1979). At the end of this phase, after about 2 years, the number of LAB and *Brettanomyces* yeasts was reported to decrease (Van Oevelen *et al.*, 1977; Verachtert and Iserentant, 1995). AAB were also isolated during these phases.

However, as mentioned above, a detailed analysis of the microbiota at 3 and 6 months of fermentation in a more recent study could not discriminate between the acidification and maturation phases (Spitaels *et al.*, 2014c). A gradual decrease of *Saccharomyces* yeasts and a consecutive increase of *Brettanomyces* yeasts was not found, which is characteristic for the acidification phase. The number of LAB was elevated in the 6-month-old sample compared to the 3-month old sample, and reached counts that were comparable to those of *Brettanomyces* yeasts. Consequently, the acidification probably occurred very rapidly between the sampling at 3 and 6 months and it was therefore considered as a part of the long maturation phase. Indeed, yeast isolates from the 3-month-old sample were identified as *Saccharomyces* spp., while those of the 6-month-old sample were identified as *Brettanomyces* spp.

The ambient temperature of the cask storage room can influence the pace of fermentation and the ability of *B. bruxellensis* to dominate the fermentation, as shown by a batch-dependent species distribution during the fermentation phase (Spitaels *et al.*, 2014c). This phase can be dominated by *B. bruxellensis* or can have a more complex microbiota and include several other yeast species, such as *Candida patagonica, Brettanomyces anomalus, Pichia membranifaciens, Priceomyces carsonii,* and *Wickerhamomyces anomalus* (Spitaels *et al.*, 2014c).

LAB were isolated from all samples after the *Enterobacteriaceae* phase. In contrast to previous studies, *P. damnosus* was the only LAB species found (Spitaels *et al.*, 2014c). It is unclear why *P. damnosus* was the only LAB species isolated, while other studies reported the general presence of *Lactobacillus* spp., including *Lactobacillus brevis,* a well-known beer spoilage bacterium (De Cort *et al.*, 1994; Shanta Kumara and Verachtert, 1991; Vriesekoop *et al.*, 2012).

The microbial communities present in 3-year-old

lambic beer were highly similar to those present in the 2-year-old lambic beer and consisted of *P. damnosus*, *A. lambici*, *P. membranifaciens*, *B. bruxellensis*, *B. anomalus*, *C. patagonica* and *W. anomalus* (Spitaels *et al.*, 2015a). This contrasted with the results of Verachtert and Iserentant (1995), who reported a decrease in the counts of LAB and yeasts towards the end of the fermentation process and suggested that this microbiota is highly adapted to growth and survival in lambic beer.

Besides lambic beers, lambic beer brewers produce gueuze and fruit lambic beers, while gueuze blenders buy lambic beers from lambic beer brewers to produce their own beers. Gueuze beers are produced by the re-fermentation of a mixture of young lambic beer that contains a lot of dextrins and old lambic beer that contains dextrin-hydrolysing microorganisms (Verachtert and Iserentant, 1995). The pellicle yeasts survive in the initial stages of the re-fermentation process, although they do not multiply (Verachtert and Iserentant, 1995). Their presence can be explained by breaking of the pellicle during emptying of the casks.

The industrial lambic beer fermentation process

Nowadays, lambic beers are also produced on an industrial scale in several breweries in Belgium and the brewing processes have even been adopted in the USA, where the resulting beers are called American coolship ales (ACAs; see below). The technical characteristics of both traditional and industrial production processes differ, which might influence the microbiota and thus the fermentation process. In an industrial process, lambic wort is made using an infusion mashing rather than a turbid mashing scheme, the wort is acidified at the end of the wort boiling to pH 4.0 using lactic acid and, finally, the wort is pre-chilled after boiling, before being transferred to the cooling tun. Together, this enables an industrial brewery to produce lambic beers throughout the year. The data presented below were taken from Spitaels *et al.* (2015b).

In contrast to a traditional lambic beer fermentation process, none of the cooling tun samples of an industrial brewing process yielded DNA or isolates. Furthermore, members of the *Enterobacteriaceae* could not be isolated, nor could their DNA be detected through PCR-DGGE experiments. Most likely, the acidification of the boiled wort before chilling prevented the growth of *Enterobacteriaceae*, which is known to be inhibited below pH 4.0 (Priest and Stewart, 2006). Bacteria and yeasts were isolated as soon as the cooled wort was transferred into the casks. These early isolates were identified as *Pichia kudriavzevii*, *D. hansenii* and *Acetobacter orientalis* and AAB were isolated from the start of fermentation up to 6 months, which again contrasted with a traditional fermentation process. *S. cerevisiae* was already dominant after 1 week of fermentation, but also *B. bruxellensis* was isolated from that sample. *Pediococcus damnosus* was present from 3 weeks onwards. If the dominance of *B. bruxellensis* and *P. damnosus* is again used to demarcate the maturation phase, such as in a traditional lambic beer fermentation process (Spitaels *et al.*, 2014c; Van Oevelen *et al.*, 1977; Verachtert and Iserentant, 1995), then the main fermentation phase lasted for one month only. *Hanseniaspora uvarum* was not characteristic, as it was only found in the initial samples of a sluggishly starting fermentation batch, which confirmed that this species has a low fermentative capacity, as discussed above.

Overall, the microbiota present during the maturation phase was the same as the one in a traditional lambic beer fermentation process, although the species diversity was simpler. Similar to a traditional lambic beer fermentation process, the main fermentation phase of an industrial lambic beer fermentation process was dominated by *Saccharomyces* spp. The dominant isolation of *S. cerevisiae* from one batch in the warm summer months supported the hypothesis that the dominance of *S. pastorianus* at the end of the lambic beer fermentation process in a traditional brewery was due to the tolerance of the maltose and maltotriose transporter of this species towards low temperatures.

The inoculation source of the spontaneous lambic beer fermentation process

Little is known about the inoculation sources that introduce the microbiota necessary for a spontaneous lambic beer fermentation process. Previously, it has been claimed that a specific air microbiota is introduced into the wort in the cooling tun, and that lambic beer only could be produced in the Brussels area where the necessary microbiota were assumed to be indigenous. Air sampling revealed that some species, such as *Brettanomyces* spp., that are characteristic of lambic beer, were present in the air of lambic beer breweries (Spitaels *et al.*, 2014c; Verachtert and Iserentant, 1995). It is however unclear if this presence is a consequence of the production of lambic beers at that location, or a cause of the inoculation of the cooling wort. Cooling wort samples in the cooling tun of a traditional lambic beer brewery yielded various *Enterobacteriaceae* species (both through cultivation and PCR-DGGE), but no LAB or other microorganisms (Spitaels *et al.*, 2014c). LAB and yeasts might have been present in very low numbers in the cooling tun sample, compared to *Enterobacteriaceae*, which may have inhibited their detection by cultivation or via PCR-DGGE.

Besides the brewery air, the wood used in the brewery is another potential source of inoculation. Construction wood that is not covered or treated with paint is present in the truss of the cooling tun room and can be used as ceiling of the cask storage room of traditional breweries. The Belgian Federal Agency for the Safety of the Food Chain published a guide for auto-control of hygienic conditions in breweries in 2007 (Belgian Federal Agency for the Safety of the Food Chain, 2007), which outlines that breweries should adopt easy-to-maintain surfaces in the production area and prohibit the use of untreated wood. However, several exceptions were included for the production of spontaneously fermented beers. Ceilings of a brewery should be free of moisture, except for the cooling tun room when lambic beer is brewed. Additionally, the presence of untreated wood was allowed in the production areas of lambic beer and red-brown ale breweries (Belgian Federal Agency for the Safety of the Food Chain, 2007). These guidelines therefore acknowledge the importance of untreated wooden surfaces in the brewery as a potential source of microbiota that could be introduced into the wort.

Most likely, however, the reuse of non-sterile wooden barrels enhances the success of the fermentation, as is the case in natural cider fermentation processes (del Campo *et al.*, 2003). Wooden tools and casks are known as safe harbours for bacteria and yeasts that are present during spontaneous fermentation of wines and ciders (Swaffield and Scott, 1995; Swaffield *et al.*, 1997). These microorganisms can penetrate the wood in a short period of time, where they are protected from cleaning procedures (Barata *et al.*, 2013; Guzzon *et al.*, 2011). Additionally, these microorganisms can survive for a prolonged time in the pores via micro-oxygenation (De Rosso *et al.*, 2008; Hidalgo *et al.*, 2010; Torija *et al.*, 2009). Clearly, it is conceivable also that the lambic beer microbiota will persist in cask wood after the cleaning procedure, which consists of washing the inside of the cask with cold water and a treatment with low-pressure steam. Yet spontaneous fermentation processes were reported successful even when new, unused casks and stainless steel fermenters were used for the production of lambic beers (Verachtert and Derdelinckx, 2005). The microbiota and metabolites of the lambic beers in the latter study were monitored over a period of 18 months and the typical characteristics of lambic beers were found (Verachtert and Derdelinckx, 2005). These characteristics include the presence of *Enterobacteriaceae*, *Saccharomyces* spp., *Brettanomyces* spp. and LAB, together with the presence of ethanol, acetic acid and lactic acid, and ethyl acetate and ethyl lactate (Verachtert and Derdelinckx, 2005). Furthermore, the wort was highly attenuated and there was a clear drop in pH (Verachtert and Derdelinckx, 2005). The authors also stated that all lambic worts will reach the expected characteristics of a lambic beer after a one-year fermentation, irrespective of the fermentation profile or initial microbial load of the individual worts (Verachtert and Derdelinckx, 2005). Moreover, lambic beer fermentation was reported successful and yielded similar products when wort, brewed and cooled in a lambic brewery, was fermented in other lambic breweries (Verachtert and Derdelinckx, 2005). Additionally, it should be noted that carbon dioxide produced during the main fermentation phase in traditional lambic beer breweries causes an overflow of beer from the cask, which is sealed with a loose bung. The brewer tops the casks off with fermenting wort from other casks to decrease the cask headspace and only replaces this temporary plug with a permanent wooden plug or rubber stopper after the main fermentation phase. These practices will influence the microbiota composition and the fermentation process. Together, these observations demonstrate that the sources of inoculation in a traditional lambic beer brewery may be diverse and brewery-dependent and/or that the impact of individual microorganisms on the resulting lambic beers may be overestimated. These data also demonstrated that lambic beer fermentation is robust in the Senne river valley (Verachtert and Derdelinckx, 2005).

Traditionally, the production of lambic beers was assumed to be only possible in the Senne river valley and the use of the cooling tun enabled the inoculation of the wort with the microbiota that were uniquely present in the air of the Senne river valley (Verachtert and Iserentant, 1995). However, two breweries in west Flanders and therefore located outside the Senne river valley, and several American craft breweries have successfully adopted spontaneous fermentation processes for the production of lambic beers and ACAs, respectively (Bokulich *et al.*, 2012). It was found that overnight cooled wort samples from the cooling tun of an industrial lambic brewing process were not inoculated (Spitaels *et al.*, 2015b). In contrast, as soon as wort was transferred into the casks, there was a detectable microbiota through cultivation. This unambiguously indicated that in such a brewing process, the microbiota responsible for the fermentation did not originate from the air and, hence, the Senne river valley is not a *conditio sine qua non*. In addition, the industrial lambic beer fermentation process lacked an *Enterobacteriaceae* phase, while such a phase was present in a traditional lambic beer fermentation process (Spitaels *et al.*, 2014c, 2015b; Van Oevelen *et al.*, 1977; Verachtert and Iserentant, 1995) and in the ACA fermentation process (Bokulich *et al.*, 2012). In an industrial brewery all surfaces are typically covered or treated with antifungal paint, which renders the surfaces smooth and easy to clean. Hence, there are no untreated wooden surfaces in such a brewery, except for the cask wood. Consequently, the influence of the cask wood on the successful inoculation and subsequent lambic beer fermentation is likely to be higher in an industrial brewing process. The wort in an industrial lambic brewing process is chilled and the air flow in and near the cooling tun is upwards instead of sideways or downwards. The chilling temperature may influence the microbial growth in the cooling tun. Indeed, the wort of a sluggishly fermenting batch in such a brewery (Spitaels *et al.*, 2015b) was chilled to a lower temperature than applied normally. Another batch was correctly chilled and isolates were obtained from the freshly transferred cask wort sample and the fermentation was initiated within one week (Spitaels *et al.*, 2015b). Hence, the wort did not contain bacteria when it was transferred to the cask and only the low fermentative *H. uvarum* yeasts grew, rather than *Saccharomyces* spp. (Spitaels *et al.*, 2015b). 'Rebooting' such sluggishly fermenting lambic beer batches by mixing it with another batch is a commonly applied procedure in industrial brewing processes and new barrels are filled with fermenting lambic wort prior to their first use. Most likely, the production of lambic beers in such a brewery is facilitated through a lambic beer core microbiota that is enriched in the cask wood. The close monitoring and mixing of aberrantly fermenting batches enables the brewery to control the fermentation process outcome. The lambic beer fermentation in an industrial brewing process is therefore successful without inoculation in the cooling tun and the transfer of chilled wort into a cask used previously for lambic beer production is sufficient to obtain a normal lambic beer fermentation process. However, when the wort would be cooled and transferred directly into a clean stainless steel tank, the fermentation probably will not take place, as described in the production of cider (del Campo *et al.*, 2003).

The microbiota and metabolites of ageing gueuze beers

Young and old lambic beers are blended by the brewer to make gueuze beers, which spontaneously re-ferment after bottling, a process that is referred to as 'ageing'. Gueuze beers bottled between 5 months and 17 years prior to the sampling have been examined to study the changes in microbiota and metabolites that can be used as a proxy of the processes that occur during gueuze maturation (Spitaels *et al.*, 2015a). All gueuze beers showed the characteristic presence of *B. bruxellensis* and comprised acetic acid, lactic acid, ethyl acetate, and ethyl lactate as the most abundant metabolites. While pediococci were readily isolated from 1-, 2- and 3-year-old lambic beers, and from a gueuze beer bottled a few months before sampling, LAB or any other bacteria were no longer isolated from gueuze beers bottled more than 3 years prior to sampling (Spitaels *et al.*, 2015a). Different yeast species including *B. bruxellensis, B. anomalus, P. membranifaciens*, and *S. cerevisiae* were isolated from recently bottled gueuze, but this diversity decreased with age until only *B. bruxellensis* was isolated via enrichment culturing, even from the 17-year-old gueuze beer. The low nutritional demand of this yeast species probably enabled its long-term survival in this environment (Aguilar Uscanga *et al.*, 2000; Renouf *et al.*, 2007). The latter is also supported by the versatile metabolism of *Brettanomyces* yeasts, which can both produce and assimilate carbon sources such as acetic acid and ethanol (Renouf *et al.*, 2007). The yeast cells in these old gueuze beers most likely occurred in a VBNC state that could be reversed by the enrichment procedure. The VBNC state allows yeast (and bacterial) cells to withstand several stress conditions (Millet and Lonvaud-Funel, 2000).

Further, the metabolite analyses revealed that the ageing of gueuze beer is probably limited in time by the depletion of the available malto-oligosaccharides (Spitaels *et al.*, 2015a). Malto-oligosaccharide concentrations were very low in 9-year-old and 17-year-old gueuze beer samples and no further increase in the concentrations of lactic acid was found in these gueuze beer samples (Spitaels *et al.*, 2015a). Furthermore, the acetic acid concentrations were the lowest in the 17-year-old gueuze beers, suggesting that acetic acid was further metabolized and that no new acetic acid was produced. Lactic acid concentrations increased steadily for beers aged up to 10 years, but no further increase was noticed in the beers that were aged for 17 years.

Additionally, the typical fruitiness was no longer perceived in the sensory analysis in the oldest gueuze beers examined (Spitaels *et al.*, 2015a). This was probably caused by the degradation of the fatty acid ethyl esters, which are known to add fruitiness in beer. The increasing concentrations of ethyl lactate and decreasing concentrations of ethyl decanoate could be considered as positive and negative gueuze beer-ageing metabolite biomarkers, respectively.

Ecology of lambic-style ACA beers

Acidic beers are currently attracting interest worldwide, especially in the USA (Bokulich *et al.*, 2012). In the American craft brewing industry, which is the collective name for small- to mid-scale breweries, the production of lambic beers is mimicked and the resulting beers are called ACAs (Bokulich *et al.*, 2012). Breweries adopt the open cooling vessels and fermentations are performed in wooden casks or stainless steel fermentation tanks (Bokulich *et al.*, 2012). The microbiota of an ACA fermentation process has been studied using primarily culture-independent techniques (Bokulich *et al.*, 2012). Culture-dependent techniques were limited to the use of two aerobically incubated bacterial isolation media, the collection of two bacterial isolates per colony morphotype and their identification using 16S rRNA gene sequence analysis (Bokulich *et al.*, 2012), a technique not sufficiently discriminatory for accurate species level identification (Cleenwerck *et al.*, 2010; De Bruyne *et al.*, 2007, 2008; Mollet *et al.*, 1997; Naser *et al.*, 2007). The community diversity of multiple barrels of multiple fermentation batches was studied through bar-coded 16S rRNA gene amplicon sequencing and terminal restriction fragment length polymorphism (T-RFLP), a technique with a sensitivity similar to that of DGGE but with a higher automation capacity (Bokulich and Mills, 2012). Bacterial numbers were studied using quantitative PCR assays (Bokulich *et al.*, 2012).

Bokulich *et al.* (2012) reported fermentation phases in the production of ACAs similar to those of lambic beer fermentation processes. *Enterobacteriaceae* and some oxidative yeasts dominated the first phase of the fermentation, but *S. cerevisiae* was immediately the most dominant yeast (Bokulich *et al.*, 2012). Members of the *Enterobacteriaceae* family were dominant up to 1 month, but some species were isolated up to 12 weeks into the fermentation process (Bokulich *et al.*, 2012). ACAs are seasonal products of these breweries and it is likely that *S. cerevisiae* is enriched in the brewery environment by its use in other types of beers produced in these breweries, probably explaining their early dominance during ACA fermentation (Bokulich *et al.*, 2012). From week 4 onwards, LAB were the most dominant bacteria; *Lb. brevis* was the only bacterial species isolated during the whole fermentation process, but *Pediococcus* was the most dominant LAB from week 4 onwards based on T-RFLP and BAS analyses (Bokulich *et al.*, 2012). *Brettanomyces* was detected from week 11 onwards in minor numbers, but was dominant after 1 year. Low numbers of AAB were found during the whole fermentation process (Bokulich *et al.*, 2012). Interestingly, the fermentation profiles were very similar between batches and between barrels, even when barrels had a different origin, were new or reused (Bokulich *et al.*, 2012).

Ecology of red-brown acidic beers

West of the Scheldt in Belgium, non-spontaneous mixed acid beer fermentation was originally applied for beer production where hops were replaced by herbs (called 'gruyt'). Non-spontaneous mixed fermentation is used in two types of Belgian acidic ales, namely the red acidic ales of south-west Flanders (Roeselare) and the red-brown acidic ales that are produced in south-west and south-east Flanders. The red acidic ales of south-west Flanders were traditionally produced using an in-house starter culture that contains yeasts and LAB, by the reuse of the starter in every fermentation batch (Martens *et al.*, 1997). The level of bacteria is kept low by acid washing (mainly with phosphoric acid) of the yeast suspension (Martens *et al.*, 1997). These ales have a vinous acidic character and their production starts with mashing of malted barley and cooked unmalted maize (Martens *et al.*, 1997). The main ethanol fermentation phase proceeds for about seven days and is followed by a secondary lactic acid fermentation phase that proceeds for another 4–5 weeks (Martens *et al.*, 1997). Finally, a long maturation phase of 20 to 24 months occurs in large oak casks and *P. damnosus* and *Pediococcus parvulus*, together with *Brettanomyces* spp. and AAB (supported by natural microoxygenation through the wood) are an active part of the microbiota during this phase (Martens *et al.*, 1997). These results were confirmed by a recent study using BAS on red-brown finished beer samples, which revealed that the most abundant microbial species present in these beers were *P. damnosus*, *B. bruxellensis* and *A. pasteurianus* (Snauwaert *et al.*, 2016). The draft genome representative strain of *P. damnosus*, isolated from a maturation phase sample of these beers, was recently sequenced (Snauwaert *et al.*, 2015). Furthermore, additional operational taxonomic units were assigned to *Candida*, *Lactobacillus* and an unclassified fungal community member. This 'tandem' fermentation process (the main fermentation phase followed by cask maturation) was imported into Belgium around 1860 from northern England, where it was used for the production of old English Porter beer (Claussen, 1904; Martens *et al.*, 1997). *Brettanomyces* spp. were present in old English Porter beer until the production process was altered and stainless steel fermentation vessels were used instead of wooden casks, indicating that contact with wooden casks is an important feature for supporting the growth of this yeast (Martens *et al.*, 1997). The production of red-brown acidic ales is very similar, with the fermentation being initiated by re-pitching of LAB-harbouring yeast starter cultures and the use of open fermentation vessels from which the yeast is harvested at the end of the fermentation phase, followed by maturation in oak vessels (Martens, 1996; Martens *et al.*, 1997). Red-brown acidic ales differ from old brown ales in that the latter beers are not oak-aged, but they are also produced in south-east Flanders.

Other mixed beer fermentations

Berliner Weisse

Berliner Weisse is a trademarked beer that may be brewed only in Berlin, following European Union regulations (Burberg and Zarnkow, 2009). It is minimally characterized microbiologically. The mash is made with a 2:1 to 3:1 ratio of wheat malt and barley malt, has a low initial gravity of around 7°P to 9°P and the level of carbonation in the finished product is high (Burberg and Zarnkow, 2009; Verachtert and Derdelinckx, 2005). Traditionally, the wort is not boiled, but rather cooled directly after lautering, with the hops being added during the mashing, although in modern Berliner Weisse production a heating step is incorporated (Burberg and Zarnkow, 2009; Verachtert and Derdelinckx, 2005). The fermentation is traditionally carried out in an open fermenter by the reuse of a yeast culture that harbours LAB and generally has a 4:1 to 6:1 yeast: LAB ratio (Burberg and Zarnkow, 2009). The secondary fermentation is carried out in bottles by the addition of Kräusen (a yeast layer that is formed on top of the fermenting beer) to the green beer in the bottles, after which the bottles are stored for from 3 weeks to 2 years (Burberg and Zarnkow, 2009). The resulting beer is 95% attenuated and has a pH of 3.0 (Burberg and Zarnkow, 2009). More recently, these beers are also produced by dividing the wort into two parts, after which one half is fermented with a homofermentative *Lactobacillus* and the other half with ale yeast (Verachtert and Derdelinckx, 2005).

Sorghum beers

On the African continent, tropical cereals such as maize and sorghum are used for the production of beers, since barley (a cool-season, temperate cereal) cultivation is not viable (Taylor, 2003). Moreover, sorghum is the only viable food grain in regions with semi-arid and subtropical climatic conditions (Taylor, 2003). Sorghum beers are widely produced in sub-Saharan Africa and are well-known under their local names, such as burukutu (Nigeria), tchapalo (Ivory Coast), dolo (Burkina Faso), pito (Ghana), munkoyo (Zambia) and bili bili (Chad) (Abegaz, 2007; Faparusi *et al.*, 1973; Lyumugabe *et al.*, 2010, 2013; Marcellin *et al.*, 2009; N'guessan *et al.*, 2011; Nanadoum and Pourquie, 2009; Sawadogo-Lingani *et al.*, 2007; Schoustra *et al.*, 2013; Taylor, 2003; van der Aa Kühle *et al.*, 2001; Zulu *et al.*, 1997). Sorghum beers are traditionally opaque, but some commercial clear versions exist as well (Hibbett and Taylor, 2013; Nanadoum and Pourquie, 2009). These beers are mostly produced by the women of agropastoral families (which perform agriculture by growing crops and keeping livestock) on a weekly basis and are often sold (Dancause *et al.*, 2010). Production methods differ between countries and recipes are often household specific (Taylor, 2003).

As an example, the production of bili bili starts with a malting of the sorghum grains; steeping, germination, and drying of the grains takes about 1 week (Nanadoum and Pourquie, 2009). After the milling of the sorghum malt, the flour is steeped for at least 2 hours, after which the supernatant is removed from the residue (Nanadoum and Pourquie, 2009). The residue is cooked for an average of 2 hours to ensure gelatinization of the starch (Nanadoum and Pourquie, 2009). The thick mash of the residue is mixed with the supernatant at a temperature of 65°C to 70°C. Subsequently, this mixture is left to cool overnight in open air (Nanadoum and Pourquie, 2009). During overnight cooling, the wort acidifies through the action of LAB, which are spontaneously inoculated from either the sorghum malt or from the surrounding air (Nanadoum and Pourquie, 2009). Alternatively, in some sorghum beers, the LAB are introduced by backslopping (Taylor, 2003). In dolo and pito beers, *Lactobacillus fermentum* is reported as the predominant LAB (Sawadogo-Lingani *et al.*, 2007), whereas in burukutu beer the predominant LAB are identified as *Leuconostoc mesenteroides*, *Lactobacillus plantarum*, *Lb. brevis*, and *Lactobacillus delbrueckii* (Faparusi *et al.*, 1973). After boiling the acidified wort, either dried yeast obtained from a previous fermentation or backslopping from the previous beer is added to start the fermentation, in which *S. cerevisiae* dominates (Nanadoum and Pourquie, 2009). This yeast species is dominant in all sorghum beer main fermentation phases, which take place overnight (Faparusi *et al.*, 1973; N'guessan *et al.*, 2011; Nanadoum and Pourquie, 2009; van der Aa Kühle *et al.*, 2001). The next morning, the beer is ready to be sold and has a shelf-life of about one day (Nanadoum and Pourquie, 2009). In burukutu beer, also a high number of AAB is found, which are now all classified as *Gluconobacter oxydans* (Faparusi *et al.*, 1973).

Other cereal-based beverages

Chicha

Chicha is a traditional beverage produced in South America and was already produced by the Incas (Vallejo *et al.*, 2013). The production starts by steeping and germination of maize grains to get a maize malt (Vallejo *et al.*, 2013). Alternatively, the maize is chewed to convert the starch into fermentable sugars by the action of the amylase in saliva (Gomes *et al.*, 2009). Besides maize, also cassava and cane sugar can be used in the production of chicha (Gomes *et al.*, 2009). After cooking, the mixture is poured into clay pots, which are buried, and the liquid is left to ferment for 1 up to 6 days (Gomes *et al.*, 2009; Vallejo *et al.*, 2013). The end-product of the fermentation primarily contains *S. cerevisiae* yeasts (Vallejo *et al.*, 2013). As these beers are produced locally and the recipe can differ between two producers, the microorganisms present in the beers can differ greatly. Several LAB are described to be involved in chicha fermentation, including *Lb. plantarum*, *Weissella viridescens*, *Enterococcus faecium* and *Leuc. mesenteroides* (Elizaquivel *et al.*, 2015). Since the clay fermentation pots are reused for every fermentation and no bacteria or yeasts are pitched to start the fermentation, the microbiota involved in the fermentation of chicha probably penetrate into the clay surface; a new yeast species, *Candida theae*, was recently isolated from chicha clay fermentation pots found in a tomb (Chang *et al.*, 2012).

Boza

Boza is a fermented beverage that is produced in Turkey and other Balkan countries (Kabak and Dobson, 2011). Its production starts with the boiling of a mixed flour of millet, rice, and wheat, and water (Kabak and Dobson, 2011). After filtering, the supernatant is inoculated with a part of a previous fermentation batch of boza, sourdough, or yoghurt (Altay *et al.*, 2013; Kabak and Dobson, 2011). The mixture ferments at 30°C for 24 hours (Altay *et al.*, 2013; Botes *et al.*, 2007; Kabak and Dobson, 2011). The microbiota present during the fermentation can vary significantly, depending on the inoculum and region of production. Generally, a range of LAB is found during these fermentations, including *Lactobacillus* and *Leuconostoc* spp. (Altay *et al.*, 2013; Botes *et al.*, 2007; Kabak and Dobson, 2011). Several yeast species are found, but in contrast to other fermented cereal-based beverages, *Saccharomyces* spp. are not always present (Botes *et al.*, 2007). Instead, *Candida* spp. and *Pichia* spp. can be the dominant yeasts in boza fermentation (Altay *et al.*, 2013; Botes *et al.*, 2007; Kabak and Dobson, 2011). Opportunistic pathogenic yeasts have been isolated from Bulgarian boza, highlighting the need for starter cultures (Botes *et al.*, 2007). The shelf-life of boza is about 15 days and it is acceptable for consumption until the pH drops below 3.5 (Altay *et al.*, 2013).

Future perspectives

Further research of the lambic beer fermentation process and other traditional beer fermentation processes in other breweries (traditional, industrial, or those with intermediate characteristics) will reveal the extent to which the overall microbiota described is generic, or if novel species will be discovered. The isolates obtained in these studies can be used for the production of these traditional beers in a controlled way, but they can also be applied in other beer types, since they are thought to be more persistent in the fermentation process, as they dominate the fermentation without the need of being pitched in large amounts. Consequently, brewing with these isolates would make the beer less prone to microbial spoilage. The yeasts present in traditional beers are also known to survive a long time during bottle re-fermentation, which is a very interesting property of yeast strains for ale bottle re-fermentation as well.

The role of yeast species that are not commonly isolated from fermented beverages also deserves further attention. The isolation of *Yarrowia lipolytica* at late phases of the lambic beer fermentation from both traditional and industrial lambic brewing processes suggests an advantage of this species to survive and prevail in lambic beers. Cheese is the main food product from which this yeast is isolated, next to fermented and raw meat products, but it has also been isolated occasionally from soft drinks, wines, and ciders (Groenewald *et al.*, 2014).

Finally, although the source of the microbiota in an industrial lambic brewing process seems to be known, the sources of lambic microbiota are unclear in traditional lambic beer brewing processes. Several findings indicate a role for both brewery air and untreated construction wood and cask wood, but this may differ between breweries (Verachtert and Derdelinckx, 2005; Verachtert and Iserentant, 1995). All these breweries have produced lambic beers for a long time and the entire environment is probably enriched with the lambic beer microbiota. The precise mode of inoculation might therefore remain part of the mystery and tradition of lambic beer and traditional beers in general.

References

Abegaz, K. (2007). Isolation, characterization and identification of lactic acid bacteria involved in traditional fermentation of borde, an Ethiopian cereal beverage. Afr. J. Biotechnol. *6*, 1469-1478.

Aguilar Uscanga, M.G., Delia, M.L., and Strehaiano, P. (2000). Nutritional requirements of *Brettanomyces bruxellensis*: growth and physiology in batch and chemostat cultures. Can. J. Microbiol. *46*, 1046–1050.

Altay, F., Karbancioglu-Guler, F., Daskaya-Dikmen, C., and Heperkan, D. (2013). A review on traditional Turkish fermented non-alcoholic beverages: microbiota, fermentation process and quality characteristics. Int. J. Food Microbiol. *167*, 44-56.

Andrews, B.J., and Gilliland, R. (1952). Superattenuation of beer: A study of three organisms capable of causing abnormal attenuations. J. Inst. Brew. *58*, 189-196.

Barata, A., Laureano, P., D'Antuono, I., Martorell, P., Stender, H., Malfeito-Ferreira, M., Querol, A., and Loureiro, V. (2013). Enumeration and identification of 4-ethylphenol producing yeasts recovered from the wood of wine ageing barriques after different sanitation treatments. J. Food Res. 2, 140-149.

Belgian Federal Agency for the Safety of the Food Chain (2007). Leidraad voor de brouwerijen. PB 03 - LD 07 - REV 00 – 2007.

Beltran, G., Torija, M.J., Novo, M., Ferrer, N., Poblet, M., Guillamón, J.M., Rozès, N., and Mas, A. (2002). Analysis of yeast populations during alcoholic fermentation: a six year follow-up study. Syst. Appl. Microbiol. *25*, 287–293.

Bezerra-Bussoli, C., Baffi, M.A., Gomes, E., and Da-Silva, R. (2013). Yeast diversity isolated from grape musts during spontaneous fermentation from a Brazilian winery. Curr. Microbiol. *67*, 356-361.

Bokulich, N.A., and Bamforth, C.W. (2013). The microbiology of malting and brewing. Microbiol. Mol. Biol. Rev. *77*, 157–172. http://dx.doi.org/10.1128/MMBR.00060-12

Bokulich, N.A., Bamforth, C.W., and Mills, D.A. (2012). Brewhouse-resident microbiota are responsible for multi-stage fermentation of American coolship ale. PLOS ONE *7*, e35507. http://dx.doi.org/10.1371/journal.pone.0035507

Bokulich, N.A., and Mills, D.A. (2012). Next-generation approaches to the microbial ecology of food fermentations. BMB Rep. *45*, 377–389.

Bomar, L., Maltz, M., Colston, S., and Graf, J. (2011). Directed culturing of microorganisms using metatranscriptomics. MBio *2*, e00012-11. http://dx.doi.org/10.1128/mBio.00012-11

Botes, A., Todorov, S.D., von Mollendorff, J.W., Botha, A., and Dicks, L.M.T. (2007). Identification of lactic acid bacteria and yeast from boza. Process Biochemistry *42*, 267-270.

Briggs, D.E., Boulton, C., Brookes, P., and Stevens, R. (2004). Brewing Science and Practice (Cambridge, England: Woodhead Publishing Limited).

Brisse, S., Grimont, F., and Grimont, P.A.D. (2006). The genus *Klebsiella*. In The Prokaryotes, M. Dworkin, S. Falkow, E. Rosenberg, K.-H. Schleifer, and E. Stackebrandt, eds. (New York, NY, USA: Springer), pp. 159-196.

Burberg, F., and Zarnkow, M. (2009). Special production methods. In Handbook of Brewing, H.M. Eßlinger, ed. (Weinheim, Germany: Wiley-VCH Verlag GmbH & Co. KGaA), pp. 235-256.

Chang, C.F., Lin, Y.C., Chen, S.F., Carvaja Barriga, E.J., Barahona, P.P., James, S.A., Bond, C.J., Roberts, I.N., and Lee, C.F. (2012). *Candida theae* sp. nov., a new anamorphic beverage-associated member of the *Lodderomyces* clade. Int. J. Food Microbiol. *153*, 10–14. http://dx.doi.org/10.1016/j.ijfoodmicro.2011.09.012

Chao, S.-H., Huang, H.-Y., Kang, Y.-H., Watanabe, K., and Tsai, Y.-C. (2013). The diversity of lactic acid bacteria in a traditional Taiwanese millet alcoholic beverage during fermentation. LWT - Food Sci. Technol. *51*, 135-142.

Claussen, N.H. (1904). On a method for the application of Hansen's pure yeast system in the manufacturing of well-conditioned English stock beers. J. Inst. Brew. *10*, 308-331.

Cleenwerck, I., De Vos, P., and De Vuyst, L. (2010). Phylogeny and differentiation of species of the genus *Gluconacetobacter* and related taxa based on multilocus sequence analyses of housekeeping genes and reclassification of *Acetobacter xylinus* subsp. *sucrofermentans* as *Gluconacetobacter sucrofermentans* (Toyosaki et al. 1996) sp. nov., comb. nov. Int. J. Syst. Evol. Microbiol. *60*, 2277-2283.

Cleenwerck, I., Gonzalez, A., Camu, N., Engelbeen, K., De Vos, P., and De Vuyst, L. (2008). *Acetobacter fabarum* sp. nov., an acetic acid bacterium from a Ghanaian cocoa bean heap fermentation. Int. J. Syst. Evol. Microbiol. *58*, 2180–2185. http://dx.doi.org/10.1099/ijs.0.65778-0

Cousseau, F.E., Alves, S.L., Jr., Trichez, D., and Stambuk, B.U. (2013). Characterization of maltotriose transporters from the *Saccharomyces eubayanus* subgenome of the hybrid *Saccharomyces pastorianus* lager brewing yeast strain Weihenstephan 34/70. Lett. Appl. Microbiol. *56*, 21-29.

Dancause, K.N., Akol, H.A., and Gray, S.J. (2010). Beer is the cattle of women: sorghum beer commercialization and dietary intake of agropastoral families in Karamoja, Uganda. Soc. Sci. Med. *70*, 1123–1130. http://dx.doi.org/10.1016/j.socscimed.2009.12.008

De Bruyne, K., Franz, C.M., Vancanneyt, M., Schillinger, U., Mozzi, F., de Valdez, G.F., De Vuyst, L., and Vandamme, P. (2008). *Pediococcus argentinicus* sp. nov. from Argentinean fermented wheat flour and identification of *Pediococcus* species by *pheS*, *rpoA* and *atpA* sequence analysis. Int. J. Syst. Evol. Microbiol. *58*, 2909–2916. http://dx.doi.org/10.1099/ijs.0.65833-0

De Bruyne, K., Schillinger, U., Caroline, L., Boehringer, B., Cleenwerck, I., Vancanneyt, M., De Vuyst, L., Franz, C.M., and Vandamme, P. (2007). *Leuconostoc holzapfelii* sp. nov., isolated from Ethiopian coffee fermentation and assessment of sequence analysis of housekeeping genes for delineation of *Leuconostoc* species. Int. J. Syst. Evol. Microbiol. *57*, 2952-2959.

De Cort, S., Shantha Kumara, H.M.C., and Verachtert, H. (1994). Localization and characterization of alpha-glucosidase activity in *Lactobacillus brevis*. Appl. Environ. Microbiol. *60*, 3074–3078.

De Keersmaecker, J. (1996). The mystery of lambic beer. Sci. Am. *275*, 74-81.

De Rosso, M., Cancian, D., Panighel, A., Dalla Vedova, A., and Flamini, R. (2008). Chemical compounds released from five different woods used to make barrels for aging wines and spirits: volatile compounds and polyphenols. Wood Sci. Technol. *43*, 375-385.

del Campo, G., Santos, J., Berregi, I., Velasco, S., Ibarburu, I., Duenas, M., and Irastorza, A. (2003). Ciders produced by two types of presses and fermented in stainless steel and wooden vats. J. Inst. Brew. *109*, 342-348.

Dolci, P., Alessandria, V., Rantsiou, K., Bertolino, M., and Cocolin, L. (2010). Microbial diversity, dynamics and activity throughout manufacturing and ripening of Castelmagno PDO cheese. Int. J. Food Microbiol. *143*, 71–75. http://dx.doi.org/10.1016/j.ijfoodmicro.2010.07.007

Elizaquivel, P., Perez-Cataluna, A., Yepez, A., Aristimuno, C., Jimenez, E., Cocconcelli, P.S., Vignolo, G., and Aznar, R. (2015). Pyrosequencing vs. culture-dependent approaches to analyze lactic acid bacteria associated to chicha, a traditional maize-based fermented beverage from Northwestern Argentina. Int. J. Food Microbiol. *198*, 9-18.

Faparusi, S.I., Olofinboba, M.O., and Ekundayo, J.A. (1973). The microbiology of burukutu beer. Z. Allg. Mikrobiol. *13*, 563–568.

Gomes, F., Lacerda, I., Libkind, D., Lopes, C., Carvajal, J., and Rosa, C. (2009). Traditional foods and beverages from South America: microbial communities and production strategies. In Industrial Fermentation: Food Processes, Nutrient Sources and Production Strategies, J. Krause, and O. Fleischer, eds. (New York, NY, USA: Nova Science Publishers), pp. 79-114.

Gorski, L. (2012). Selective enrichment media bias the types of *Salmonella enterica* strains isolated from mixed strain cultures and complex enrichment broths. PLOS ONE 7, e34722.

Groenewald, M., Boekhout, T., Neuvéglise, C., Gaillardin, C., van Dijck, P.W., and Wyss, M. (2014). *Yarrowia lipolytica*: safety assessment of an oleaginous yeast with a great industrial potential. Crit. Rev. Microbiol. *40*, 187–206. http://dx.doi.org/10.3109/1040841X.2013.770386

Guzzon, R., Widmann, G., Malacarne, M., Nardin, T., Nicolini, G., and Larcher, R. (2011). Survey of the yeast population inside wine barrels and the effects of certain techniques in preventing microbiological spoilage. Eur. Food Res. Technol. *233*, 285-291.

Hibbett, D.S., and Taylor, J.W. (2013). Fungal systematics: is a new age of enlightenment at hand? Nat. Rev. Microbiol. *11*, 129–133. http://dx.doi.org/10.1038/nrmicro2963

Hidalgo, C., Vegas, C., Mateo, E., Tesfaye, W., Cerezo, A.B., Callejón, R.M., Poblet, M., Guillamón, J.M., Mas, A., and Torija, M.J. (2010). Effect of barrel design and the inoculation of *Acetobacter pasteurianus* in wine vinegar production. Int. J. Food Microbiol. *141*, 56–62. http://dx.doi.org/10.1016/j.ijfoodmicro.2010.04.018

Hong, S., Bunge, J., Leslin, C., Jeon, S., and Epstein, S.S. (2009). Polymerase chain reaction primers miss half of rRNA microbial diversity. ISME J. *3*, 1365–1373. http://dx.doi.org/10.1038/ismej.2009.89

Huys, G., Cnockaert, M., Abbott, S.L., Janda, J.M., and Vandamme, P. (2010). *Hafnia paralvei* sp. nov., formerly known as *Hafnia alvei* hybridization group 2. Int. J. Syst. Evol. Microbiol. *60*, 1725–1728. http://dx.doi.org/10.1099/ijs.0.018606-0

Kabak, B., and Dobson, A.D. (2011). An introduction to the traditional fermented foods and beverages of Turkey. Crit. Rev. Food Sci. Nutr. *51*, 248–260. http://dx.doi.org/10.1080/10408390903569640

Kämpfer, P., and Glaeser, S.P. (2012). Prokaryotic taxonomy in the sequencing era – the polyphasic approach revisited. Environ. Microbiol. *14*, 291–317. http://dx.doi.org/10.1111/j.1462-2920.2011.02615.x

Kurtzman, C.P. (2003). Phylogenetic circumscription of *Saccharomyces*, *Kluyveromyces* and other members of the *Saccharomycetaceae*, and the proposal of the new genera *Lachancea*, *Nakaseomyces*, *Naumovia*, *Vanderwaltozyma* and *Zygotorulaspora*. FEMS Yeast Research 4, 233-245.

Kurtzman, C.P., and Robnett, C.J. (1998). Identification and phylogeny of ascomycetous yeasts from analysis of nuclear large subunit (26S) ribosomal DNA partial sequences. Antonie Van Leeuwenhoek 73, 331–371.

Lagier, J.C., Armougom, F., Million, M., Hugon, P., Pagnier, I., Robert, C., Bittar, F., Fournous, G., Gimenez, G., Maraninchi, M., *et al.* (2012). Microbial culturomics: paradigm shift in the human gut microbiome study. Clin. Microbiol. Infect. *18*, 1185–1193. http://dx.doi.org/10.1111/1469-0691.12023

Laureys, D., and De Vuyst, L. (2014). Microbial species diversity, community dynamics, and metabolite kinetics of water kefir fermentation. Appl. Environ. Microbiol. *80*, 2564–2572. http://dx.doi.org/10.1128/AEM.03978-13

Libkind, D., Hittinger, C.T., Valério, E., Gonçalves, C., Dover, J., Johnston, M., Gonçalves, P., and Sampaio, J.P. (2011). Microbe domestication and the identification of the wild genetic stock of lager-brewing yeast. Proc. Natl. Acad. Sci. U.S.A. *108*, 14539–14544. http://dx.doi.org/10.1073/pnas.1105430108

Lyumugabe, F., Bajyana Songa, E., Wathelet, J.P., and Thonart, P. (2013). Volatile compounds of the traditional sorghum beers 'ikigage' brewed with *Vernonia amygdalina* 'umubirizi'. Cerevisia, Bel. J. Brew. Biotechnol. *37*, 89-96.

Lyumugabe, F., Kamaliza, G., Bajyana, E., and Thonart, P. (2010). Microbiological and physico-chemical characteristic of Rwandese traditional beer 'ikigage'. Afr. J. Biotechnol. *9*, 4241-4246.

Manzano, M., Iacumin, L., Vendrames, M., Cecchini, F., Comi, G., and Buiatti, S. (2011). Craft beer microflora identification before and after a cleaning process. J. Inst. Brew. *117*, 343-351.

Marcellin, D.K., Celestin, Y.K., and Guillaume, L.Y. (2009). Predominant lactic acid bacteria involved in the spontaneous fermentation step of tchapalo process, a traditional sorghum beer of Côte d'Ivoire. Res. J. Biol. Sci. *4*, 789-795.

Martens, H. (1996). Microbiology and biochemistry of the acid ales of Roeselare. In Dissertationes de Agricultura (Belgium) no 294 (Leuven: Katholieke Universiteit Leuven).

Martens, H., Dawoud, E., and Verachtert, H. (1991). Wort enterobacteria and other microbial-populations involved during the 1st month of lambic fermentation. J. Inst. Brew. *97*, 435-439.

Martens, H., Dawoud, E., and Verachtert, H. (1992). Synthesis of aroma compounds by wort enterobacteria during the 1st stage of lambic fermentation. J. Inst. Brew. *98*, 421-425.

Martens, H., Iserentant, D., and Verachtert, H. (1997). Microbiological aspects of a mixed yeast-bacterial fermentation in the production of a special Belgian acidic ale. J. Inst. Brew. *103*, 85-91.

Meyer, S.A., Smith, M.T., and Simione, F.P. (1978). Systematics of *Hanseniaspora* Zikes and *Kloeckera* Janke. Antonie Van Leeuwenhoek 44, 79–96.

Millet, V., and Lonvaud-Funel, A. (2000). The viable but non-culturable state of wine micro-organisms during storage. Lett. Appl. Microbiol. *30*, 136–141.

Mollet, C., Drancourt, M., and Raoult, D. (1997). *rpoB* sequence analysis as a novel basis for bacterial identification. Mol. Microbiol. *26*, 1005–1011.

Moreira, N., Pina, C., Mendes, F., Couto, J.A., Hogg, T., and Vasconcelos, I. (2011). Volatile compounds contribution of *Hanseniaspora guilliermondii* and *Hanseniaspora uvarum* during red wine vinifications. Food Control *22*, 662-667.

N'guessan, K.F., Brou, K., Jacques, N., Casaregola, S., and Dje, K.M. (2011). Identification of yeasts during alcoholic fermentation of tchapalo, a traditional sorghum beer from Côte d'Ivoire. Antonie Van Leeuwenhoek *99*, 855–864. http://dx.doi.org/10.1007/s10482-011-9560-7

Nanadoum, M., and Pourquie, J. (2009). Sorghum beer: Production, nutritional value and impact upon human health. In Beer in Health and Disease Prevention, V.R. Preedy, ed. (San Diego, CA, USA: Academic Press), pp. 53-60.

Naser, S.M., Dawyndt, P., Hoste, B., Gevers, D., Vandemeulebroecke, K., Cleenwerck, I., Vancanneyt, M., and Swings, J. (2007). Identification of lactobacilli by *pheS* and *rpoA* gene sequence analyses. Int. J. Syst. Evol. Microbiol. *57*, 2777–2789.

Nature Reviews: Microbiology Editorial (2013). The cultural revolution. Nature Rev: Microbiol. *11*, 1-1.

Nguyen, H.-V., and Gaillardin, C. (2005). Evolutionary relationships between the former species *Saccharomyces uvarum* and the hybrids *Saccharomyces bayanus* and *Saccharomyces pastorianus*; reinstatement of *Saccharomyces uvarum* (Beijerinck) as a distinct species. FEMS Yeast Research *5*, 471-483.

Nhung, P.H., Ohkusu, K., Mishima, N., Noda, M., Monir Shah, M., Sun, X., Hayashi, M., and Ezaki, T. (2007). Phylogeny and species identification of the family *Enterobacteriaceae* based on *dnaJ* sequences. Diagn. Microbiol. Infect. Dis. *58*, 153–161.

Peris, D., Lopes, C., Belloch, C., Querol, A., and Barrio, E. (2012). Comparative genomics among *Saccharomyces cerevisiae* × *Saccharomyces kudriavzevii* natural hybrid strains isolated from wine and beer reveals different origins. BMC Genomics *13*, doi:10.1186/1471-2164-1113-1407.

Priest, F.G., and Stewart, G.G. (2006). Microbiology and microbiological control in the brewery. In Handbook of Brewing; Second edition (Boca Raton, FL: CRC Press), pp. 607-629.

Rappé, M.S. (2013). Stabilizing the foundation of the house that 'omics builds: the evolving value of cultured isolates to marine microbiology. Curr. Opin. Microbiol. *16*, 618–624. http://dx.doi.org/10.1016/j.mib.2013.09.009

Renouf, V., Lonvaud-Funel, A., and Coulon, J. (2007). The origin of *Brettanomyces bruxellensis* in wines: A review. JISVV *41*, 161-173.

Romano, P., Fiore, C., Paraggio, M., Caruso, M., and Capece, A. (2003). Function of yeast species and strains in wine flavour. Int. J. Food Microbiol. *86*, 169–180.

Sawadogo-Lingani, H., Lei, V., Diawara, B., Nielsen, D.S., Møller, P.L., Traoré, A.S., and Jakobsen, M. (2007). The biodiversity of predominant lactic acid bacteria in dolo and pito wort for the production of sorghum beer. J. Appl. Microbiol. *103*, 765–777.

Scheirlinck, I., Van der Meulen, R., Van Schoor, A., Vancanneyt, M., De Vuyst, L., Vandamme, P., and Huys, G. (2008). Taxonomic structure and stability of the bacterial community in Belgian sourdough ecosystems as assessed by culture and population fingerprinting. Appl. Environ. Microbiol. *74*, 2414-2423.

Schoustra, S.E., Kasase, C., Toarta, C., Kassen, R., and Poulain, A.J. (2013). Microbial community structure of three traditional zambian fermented products: mabisi, chibwantu and munkoyo. PLOS ONE *8*, e63948. http://dx.doi.org/10.1371/journal.pone.0063948

Shantha Kumara, H.M.C., De Cort, S., and Verachtert, H. (1993). Localization and characterization of alpha-glucosidase activity in *Brettanomyces lambicus*. Appl. Environ. Microbiol. *59*, 2352–2358.

Shanta Kumara, H.M.C., and Verachtert, H. (1991). Identification of lambic superattenuating micro-organisms by the use of selective antibiotics. J. Inst. Brew. *97*, 181-185.

Snauwaert, I., Stragier, P., De Vuyst, L., and Vandamme, P. (2015). Comparative genome analysis of *Pediococcus damnosus* LMG 28219, a strain well-adapted to the beer environment. BMC Genomics *16*, 267. http://dx.doi.org/10.1186/s12864-015-1438-z

Snauwaert, I., Roels, S.P., Van Nieuwerburg, F., Van Landschoot, A., De Vuyst, L., and Vandamme, P. (2016). Microbial diversity and metabolite composition of Belgian red-brown acidic ales. Int. J. Food Microbiol. *221*, 1–11. http://dx.doi.org/10.1016/j.ijfoodmicro.2015.12.009

Spaepen, M., Van Oevelen, D., and Verachtert, H. (1978). Fatty acids and esters produced during the spontaneous fermentation of lambic and gueuze. J Inst Brew *84*, 278-282.

Spaepen, M., Van Oevelen, D., and Verachtert, H. (1979). Higher fatty acid (HFA) and HFA-ester content of spontaneously fermented Belgian beers and evaluation of their analytical determination. Brauwissenschaft *32*, S1-S6.

Spaepen, M., and Verachtert, H. (1982). Esterase activity in the genus *Brettanomyces*. J Inst Brew *88*, 11-17.

Spitaels, F., Li, L., Wieme, A., Balzarini, T., Cleenwerck, I., Van Landschoot, A., De Vuyst, L., and Vandamme, P. (2014a). *Acetobacter lambici* sp. nov., isolated from fermenting lambic beer. Int. J. Syst. Evol. Microbiol. *64*, 1083–1089. http://dx.doi.org/10.1099/ijs.0.057315-0

Spitaels, F., Van Kerrebroeck, S., Wieme, A.D., Snauwaert, I., Aerts, M., Van Landschoot, A., De Vuyst, L., and Vandamme, P. (2015a). Microbiota and metabolites of aged bottled gueuze beers converge to the same composition. Food Microbiol. *47*, 1–11. http://dx.doi.org/10.1016/j.fm.2014.10.004

Spitaels, F., Wieme, A., Balzarini, T., Cleenwerck, I., Van Landschoot, A., De Vuyst, L., and Vandamme, P. (2014b). *Gluconobacter cerevisiae* sp. nov., isolated from the brewery environment. Int. J. Syst. Evol. Microbiol. *64*, 1134–1141. http://dx.doi.org/10.1099/ijs.0.059311-0

Spitaels, F., Wieme, A.D., Janssens, M., Aerts, M., Daniel, H.M., Van Landschoot, A., De Vuyst, L., and Vandamme, P. (2014c). The microbial diversity of traditional spontaneously fermented lambic beer. PLOS ONE *9*, e95384. http://dx.doi.org/10.1371/journal.pone.0095384

Spitaels, F., Wieme, A.D., Janssens, M., Aerts, M., Van Landschoot, A., De Vuyst, L., and Vandamme, P. (2015b). The microbial diversity of an industrially produced lambic beer shares members of a traditionally produced one and reveals a core microbiota for lambic beer fermentation. Food Microbiol. *49*, 23–32. http://dx.doi.org/10.1016/j.fm.2015.01.008

Swaffield, C.H., and Scott, J.A. (1995). Existence and development of natural microbial populations in wooden storage vats used for alcoholic cider maturation. J. Am. Soc. Brew. Chem. 53, 117-120.

Swaffield, C.H., Scott, J.A., and Jarvis, B. (1997). Observations on the microbial ecology of traditional alcoholic cider storage vats. Food Microbiol. *14*, 353-361.

Taylor, J.R.N. (2003). Overview: Importance of sorghum in Africa. In AFRIPRO, Workshop on the proteins of sorghum and millets: Enhancing nutritional and functional properties for Africa, P.S. Belton, and J.R.N. Taylor, eds. (Pretoria, South Africa: http://www.afripro.org.uk/papers/Paper01Taylor.pdf).

Teske, A. (2010). Grand challenges in extreme microbiology. Front. Microbiol. *1*, 111. http://dx.doi.org/10.3389/fmicb.2010.00111

Torija, M.-J., Mateo, E., Vegas, C.-A., Jara, C., González, A., Poblet, M., Reguant, C., Guillamon, J., and Mas, A. (2009). Effect of wood type and thickness on acetification kinetics in traditional vinegar production. Int. J. Wine Res. *1*, 155-160.

Vallejo, J.A., Miranda, P., Flores-Felix, J.D., Sanchez-Juanes, F., Ageitos, J.M., Gonzalez-Buitrago, J.M., Velazquez, E., and Villa, T.G. (2013). Atypical yeasts identified as *Saccharomyces cerevisiae* by MALDI-TOF MS and gene sequencing are the main responsible of fermentation of chicha, a traditional beverage from Peru. Syst. Appl. Microbiol. *36*, 560-564.

van der Aa Kühle, A., Jesperen, L., Glover, R.L., Diawara, B., and Jakobsen, M. (2001). Identification and characterization of *Saccharomyces cerevisiae* strains isolated from West African sorghum beer. Yeast *18*, 1069-1079.

Van der Meulen, R., Scheirlinck, I., Van Schoor, A., Huys, G., Vancanneyt, M., Vandamme, P., and De Vuyst, L. (2007). Population dynamics and metabolite target analysis of lactic acid bacteria during laboratory fermentations of wheat and spelt sourdoughs. Appl. Environ. Microbiol. 73, 4741-4750.

Van Oevelen, D., L'Escaille, F., and Verachtert, H. (1976). Synthesis of aroma components during the spontaneous fermentation of lambic and gueuze. J. Inst. Brew. *82*, 322-326.

Van Oevelen, D., Spaepen, M., Timmermans, P., and Verachtert, H. (1977). Microbiological aspects of spontaneous wort fermentation in the production of lambic and gueuze. J. Inst. Brew. *83*, 356-360.

Van Oevelen, D., and Verachtert, H. (1979). Slime production by brewery strains of *Pediococcus cerevisiae*. J. Am. Soc. Brew. Chem. *37*, 34-37.

Van Vuuren, H.J., and Priest, F.G. (2003). Gram-negative brewery bacteria. In Brewing Microbiology, F.G. Priest, and I. Campbell, eds. (New York, NY, USA: Springer), pp. 219-245.

Verachtert, H., and Derdelinckx, G. (2005). Acidic beers: enjoyable reminiscences of the past. Cerevisia, Bel. J. Brew. Biotechnol. *30*, 38-47.

Verachtert, H., and Iserentant, D. (1995). Properties of Belgian acid beers and their microflora. Part I. The production of gueuze and related refreshing acid beers. Cerevisia, Bel. J. Brew. Biotechnol. *20*, 37–41.

Vidgren, V., Multanen, J.P., Ruohonen, L., and Londesborough, J. (2010). The temperature dependence of maltose transport in ale and lager strains of brewer's yeast. FEMS Yeast Res. *10*, 402–411. http://dx.doi.org/10.1111/j.1567-1364.2010.00627.x

Vriesekoop, F., Krahl, M., Hucker, B., and Menz, G. (2012). 125th Anniversary review: Bacteria in brewing: The good, the bad and the ugly. J. Inst. Brew. *118*, 335-345.

Wouters, D., Bernaert, N., Conjaerts, W., Van Droogenbroeck, B., De Loose, M., and De Vuyst, L. (2013). Species diversity, community dynamics, and metabolite kinetics of spontaneous leek fermentations. Food Microbiol. *33*, 185–196. http://dx.doi.org/10.1016/j.fm.2012.09.016

Yuan, S., Cohen, D.B., Ravel, J., Abdo, Z., and Forney, L.J. (2012). Evaluation of methods for the extraction and purification of DNA from the human microbiome. PLOS ONE *7*, e33865. http://dx.doi.org/10.1371/journal.pone.0033865

Zulu, R.M., Dillon, V.M., and Owens, J.D. (1997). Munkoyo beverage, a traditional Zambian fermented maize gruel using *Rhynchosia* root as amylase source. Int. J. Food Microbiol. *34*, 249–258.

Fungal Contamination of Barley and Malt

8

Ludwig Niessen

Technical University of Munich, TUM School of Life Sciences Weihenstephan, Chair for Technical Microbiology, Freising, Germany

Correspondence: niessen@wzw.tum.de

https://doi.org/10.21775/9781910190616.08

Abstract

Fungal contamination of barley and malt results in a high economic burden from malt yield loss, quality failure, and costs connected to the presence of toxic fungal secondary metabolites, i.e. mycotoxins. This chapter describes the routes and factors that influence fungal contamination of cereals and malt, both qualitatively and quantitatively, as well as the consequences which fungal growth may have in terms of malt and beer quality. Focus is given to the role of mycotoxins and their fate during malting and brewing as well as to recent research in the field of beer gushing. Since analysis is a strong tool to prevent fungi and their metabolites from entering the malt-beer chain, recent developments in the analytical toolbox are discussed, including chemical and molecular biology-based approaches. Finally, possible ways for the prevention of fungal growth in the field and during malting are discussed.

Introduction

Fungal contamination of agricultural commodities results in worldwide economic burden from yield loss, quality failure, and costs connected to the presence of toxic fungal secondary metabolites, i.e. mycotoxins. Some rough calculations may help to illustrate the magnitude of the problems induced by fungal contamination of barley and it can be concluded that they may be transferable to other crops such as wheat and sorghum as crops of minor worldwide importance for the brewing industry. Based on data from 17 different regions of the world for the period 1996–1998, Oerke and Dehne (2004) estimated loss potentials due to four different pest groups (weeds, animal pests, fungal and bacterial pathogens, viruses) for major crops such as wheat, rice, maize, barley, potatoes, soybeans, sugar beet and cotton. They estimated loss potential to be about 50% for barley with actual losses of up to 26–30% occurring under unfavourable conditions. Of the actual losses, about 15% are accounted for by fungal and bacterial pathogens with no information given about the proportion of fungal pathogens in this estimate. We would tend to give it a 50% share in this calculation. According to the UN Food and Agricultural Organization, (FAO, 2009) barley is the second most important coarse grain crop with a worldwide production of 136 million metric tons in 2007 harvested on 56.6 million hectares of acreage. Taking the above figures into consideration, an estimated loss of 10.2 million metric tons in worldwide barley production may occur due to fungal pathogens. Based on an average price of 180 $US per metric ton (feed/malt barley), this amounts to a worldwide annual loss of $1.8 billion (US) in agribusiness. Of the 136 million metric tons of barley harvested in 2007, an estimated 21.36 million metric tons of malt have been produced (FAO, 2009) meaning a raw material consumption of approximately 27 million metric tons of screened barley (conversion factor from raw material to malt 1.267). Thus there is a total world production of approximately 30 million metric tons of malting barley or $6.2 billion of turnover in global agribusiness.

The FAO have previously published a 25% estimate for the proportion of total food supply contaminated by the major mycotoxins in precariously high concentrations (Mannon and Johnson, 1985). Given the fact that this estimate is also true for malting barley, five million metric tons of that commodity can be supposed to be either unfit for processing in the malt house and have to go to alternative uses (with reduction in producer benefit) or can be sold only to the malt producer at a reduced price, both measures adding up to further losses. Costs for mycotoxin management can only be roughly estimated since they are influenced by several different cost factors such as research, testing, detoxification, discarding, health costs and insurance. For the situation in the USA, Vardon (2003) estimated an annual range of cumulated losses from $0.5 million to over $1.5 billion only from the major mycotoxins aflatoxin (maize and peanuts), fumonisin (maize, wheat, barley), and deoxynivalenol (DON) (wheat, barley). Uncertainties were built into the cost model based on commodity outputs, prices, and contamination levels based on surveillance samples and compliance with FDA regulatory limits.

Wheat and barley scab (head blight) induced by *Fusarium graminearum* and *F. culmorum* is an excellent example of the high economic and social impact of fungal contamination of cereals. According to Windels (2000), the losses in the United States agricultural production of wheat and barley caused by several severe outbreaks of the disease in the 1990s and early 2000s were estimated to be close to $3 billion including losses of $400 million in barley production. Costs were due to reduction both of yield and quality and led to a considerable change both in agricultural practice and cropping systems and also in the social structure of rural communities in affected regions due to farm exits and depopulation. As a consequence, malt producers and breweries suffered from problems with mycotoxins (DON, fumonisin) and gushing due to fungal contamination as well as from a shortage in barley supply which had to be compensated for by import of (more expensive) raw materials.

However, fungal contamination of brewing cereals means more than the potential contamination of commodities with mycotoxins. Fungi influence the quality of barley, wheat and sorghum and the malts produced from them by altering their composition and structure thereby influencing parameters such as malt colour, wort filterability, fermentation performance, beer colour and aroma. They may influence the brewing process by interfering with the metabolism of the barley grain during malting or with the yeast metabolism during brewing. They often leave tiny amounts of highly potent proteins causing the gushing of beer.

Current research on the taxonomy, physiology, and toxicology of barley- and malt-related fungi has provided a deeper insight into the way these organisms interact with their environment and how this interaction influences the quality of brewing malt and beer. Results from this research may help to cope with some of the problems related to fungal contamination of brewing raw materials and thus help to find solutions to the benefit of the brewing industry.

The following chapter provides an overview of published research about the influences fungal organisms have on barley and malt and how they interfere with the brewing process and with the final product. Moreover, the following chapter will discuss gaps in our knowledge regarding fungal contamination of barley and malt as well as its management.

Ecological considerations (microbial ecology)

The associated mycobiota describes the community of fungi (filamentous fungi and yeasts) present in a particular type of habitat such as barley (Andersen *et al.*, 1996). From a brewer's perspective the habitat of major interest is the seed produced by the cereal plant at the end of the vegetation period. However, there are numerous other habitats with their own typical associated mycobiota which may interfere and influence the mycobiota associated with the cereal seeds used as raw material in the processes of malt and beer production. These mutual interactions between associated mycobiotas of, for example, soil, air, rain, insects, weed plants or agricultural equipment can be supposed to have an immediate influence on the composition of the spectrum of species present on or in the grain after harvest. However, this influence is quite small as compared to the influences acting selectively on a fungal propagule in a particular habitat. Once a fungal propagule has been deposited on a cereal plant there are several environmental factors that determine whether the association is temporary or becomes permanent. Physical factors such as water potential, temperature, pH, light and radiation, redox potential and surface texture as well as physiological factors such as water availability, presence and availability of micro and macro nutrients, presence of antimicrobials produced by the plant or by other microorganisms, presence of predators, parasites, competitors or pathogens will decide the survival of a propagule in its microenvironment. All these factors have a selective action that stabilizes a typical mycobiota associated with a particular type of habitat. Moreover, the physiological properties of the propagule and the host influence the type of association with the host. Interactions can be neutral, saprophytic or parasitic.

In order to survive under environmentally stressful conditions fungi may follow three fundamentally different strategies of competition with other microorganisms (Pugh and Boddy, 1988; Widden, 1997; Magan *et al.*, 2003). The first of those can be described as outcompeting competitors by fast rates of growth and reproduction (ruderal type or r-selected type). The ruderal strategy is regularly followed by many species in the *Chytridiomycota* and in the *Zygomycota*. The second strategy is to outcompete other microorganisms by tolerance to stress against environmental conditions such as low temperature, low water activity, high salt content or use of antifungal compounds (stress type or s-selected type). The extreme xerophilic fungus *Xeromyces bisporus* and some species in *Eurotium* would be examples for that group. The third group of species uses a strategy of confrontation by competing with other species for nutrients and space by producing a variety of (anti-microbial) secondary metabolites (competitive type or c-selected type). Some of the c-selected species even combine production of metabolites with high reproduction rates, e.g. species of *Penicillium* and *Aspergillus*. Because of their ability to produce secondary metabolites which eventually are harmful for animals and humans, i.e. mycotoxins, the latter category of species is the one giving the highest concern in regard to the safety and hygiene of brewing cereals and malt.

Routes of fungal colonization of cereal grains and malt

Colonization of cereal grain used for malt production by fungi can basically occur via several different routes. Unspecific superficial colonization occurs by fungal and yeast propagules that are transported through air movement or are washed out from the atmosphere with rainfall. Such propagules are deposited directly on any part of the plant, including the ears. Fungi in that category mostly have small dry and mostly pigmented spores enabling them to survive long periods of dryness and irradiation by sunlight without suffering any physiological or genetic damage. Typically the propagules are unable to germinate and grow under the conditions in the field. During harvesting, threshing and transport manipulations, propagules resting on parts of the plant other than the grain are homogeneously distributed to the harvested grains. Under the conditions of storage they can survive for longer periods of time and will germinate and grow as the water content of barley exceeds 14–15% and CO_2 accumulates at elevated temperatures (Magan and Lacey, 1984). Hence the name storage fungi.

Fungal species collectively summarized as the field fungi have evolved mechanisms enabling them to colonize the cereal plant by growing on its surface or even by penetrating its tissues. Deposition by airflow over greater distances is rather rare in such species since they are mostly unfit for survival in dryness and under sunlight irradiation. Species of field fungi are rather distributed by short-range deposition of ascospores by wind flow (Osborne and Stein, 2007; Xu and Nicholson, 2009), by rain splash (Ooka and Kommedahl, 1977; Madden, 1997) or by using insect vectors (Hajek and St. Leger, 1994) with a distribution radius of viable inoculum within a short distance ranging between centimetres and a few metres. Cereal grains are colonized by field species either by propagules deposited directly on the ear or indirectly by mycelia that have penetrated tissues remote from the ear and grow towards it subcutaneously (Kang and Buchenauer, 2000). True plant pathogens such as *Septoria tritici* or *Septoria nodorum* can actively penetrate tissues (Cohen and Eyal, 1993; Eyal, 1999) and infect cereal grains and other parts of the plant. Other species such as *Fusarium culmorum* and *F. graminearum* can only enter through injuries (mechanical damage, insect bite), preformed entrances (stomata) or very soft tissues (stamina, pistil and stigma of the gynoecium) and use production of mycotoxins such as DON and nivalenol (NIV) as virulence factors to undergo cellular defence mechanisms (Bai *et al.*, 2002; Wagacha and Muthomi, 2007). Several pathogenic species of field fungi such as *F. culmorum*, *Microdochium nivale* or *Bipolaris sorokiniana* penetrate the plant during seedling germination either from inoculum present in the surrounding soil or from inoculum present as a seed borne infection (Knudsen *et al.*, 1995; Bonde *et al.*, 1997; Al-Sadi and Deadman, 2010). For some species symptomless growth within infected plants has been demonstrated where infection eventually may reach the developing grain.

Once the malting process has started by steeping the cereal grains in water, fungal propagules present on the surface or within the grain tissue will be activated and start growing and multiplying according to their ecological and physiological preferences. Initial inoculum of r-selected species will be distributed very quickly and evenly over the complete batch to grow and multiply immediately. An inoculum of s-selected species will be distributed quite evenly over the batch but since conditions will not be optimal for them, they will be unable to compete. Inoculum of c-selected species such as *Fusarium* spp., *Alternaria* spp., *Epicoccum nigrum* or *Bipolaris sorokiniana* will further develop on grains in which the species are present with some superficial spread first to neighbouring grains and later to be wider distributed as the malting process proceeds with mechanical turning of the green malt.

The microbial community of brewing barley and malt

In general, the microbial community found on or in barley seeds may contain numerous species from five principle groups, i.e. viruses, bacteria, fungi, slime moulds and protozoa. The presence of viruses on the surface of barley seeds or within their interior parts is a generally unexplored field of research. Because of its role as a seed borne pathogen of barley, the barley stripe mosaic virus and its association with barley grains have been studied more intensively (Mink, 1993). However, the presence and impact of symptomless barley viruses on the microbiota present on or in the barley seed has not been studied to date. Also the impact that viruses associated with fungi (mycoviruses) or bacteria (bacteriophages) may have on individual species or on the community of microorganisms present on or in barley seeds is as yet unknown. However, an influence can be assumed since many plant pathogenic fungi and endophytic fungi of grasses have been found to carry mycoviruses (see comprehensive literature compilations by Pearson *et al.* (2009) and by Herrero *et al.* (2009). According to the literature, many fungal species undergo modifications of their phenotype leading to increased or decreased virulence of virus infected strains of plant pathogens. This effect has been observed in species which are important to the quality and yield of barley, e.g. *F. graminearum* or *F. culmorum* (Chu *et al.*, 2002), *Alternaria alternata* (Aoki *et al.*, 2009) including potential production of mycotoxins by *Aspergillus* species (Varga *et al.*, 1994). On the other hand, the link between mycoviruses and the production of fungal secondary metabolites has been observed as well (Detroy and Worden, 1979). Where knowledge about specific interaction of fungal viruses with their host and of the impact of this interaction on the barley microbiota is scarce, knowledge of this relation in bacteria and their respective phages is even more negligible. However, this relation has been intensively studied in areas such as medicine (Merril *et al.*, 2003), biotechnology or food fermentation, where bacteriophages play an important role (Emond and Moineau, 2007). It can therefore be speculated that bacteria present on or in barley seeds will interact with bacterial viruses and that their numbers and activity will also be influenced, either positively or negatively.

Occurrence of slime moulds (*Mycetozoa*) and protozoans on cereal seeds has been reported by some authors (Pepper and Kiesling, 1963; Mills and Frydman, 1980). However, no representative data about their presence or about the species and groups (plasmodial or cellular slime moulds, amoeboid or ciliate protozoa) prevailing have been presented or published elsewhere. The fact that slime moulds and protozoa feed on living and dead bacteria, yeasts, and fungi might point to a certain yet unspecified effect of slime moulds and protozoans on the microbial community when present on barley seeds (Hohl and Raper, 1993).

Bacteria together with fungal organisms can be supposed to be the groups with the greatest influence on the properties of barley grain since they occur regularly in higher numbers and many of them are physiologically able to use grain components as nutrient source. According to Roberts *et al.* (2005), the bacterial consortium consists of aerobic mesophilic bacteria (no growth at 10°C or less, optimum growth between 20°C and 40°C, colony count from 10^2 to 10^6 cfu/g), psychrotrophic bacteria (growth occurs at 7°C or less, growth optimum >20°C, colony count from 10^4 to $>10^5$ cfu/g), actinomycetes (up to 10^6 cfu/g), aerobic spore-forming bacteria (colony counts from 10^0 to 10^5 cfu/g) and coliform bacteria (colony counts from 10^2 to 10^4 cfu/g). An overview of bacterial species that have been isolated from barley grain was compiled by Pepper and Kiesling (1963) and, more recently, by Noots *et al.* (1999). The spectrum shows a variety of different species. Many of them belong to Gram-negative genera such as *Alcaligenes, Clavibacterium, Enterobacter, Erwinia, Escherichia, Flavibacterium, Pseudomonas* or *Xanthomonas.* Gram positive species represent the genera *Arthrobacter, Bacillus, Corynebacterium, Lactobacillus, Leuconostoc, Micrococcus* and *Thermoactinomyces.* Some genera are represented by only one species, e.g. *Arthrobacter, Aplanobacter, Brevibacterium, Corynebacterium, Kurthia* and *Pediococcus.* Moreover, groups of unidentified bacteria were summarized as white and yellow bacteria. Filamentous bacteria were summarized as actinomycetes with no further differentiation into species.

The community of fungal species on or in barley seeds and malt grain is subject to change through the production process, starting from the developing ear and grain, through harvested grain and ending with the kilned malt. As depicted in Fig. 8.1, the mycobiota is subject to dynamic change in biomass and species spectrum over time. Typical field fungi such as *Alternaria* spp., *Cladosporium* spp., *Curvularia* spp., *Drechslera* spp., *Epicoccum nigrum, Fusarium* spp., *Microdochium* spp., *Nigrospora* spp., *Septoria* spp., *Trichoderma* spp., dominate the spectrum of fungal species because they are able to use the developing grain as substrate without necessarily damaging or killing the embryo. However, many other fungal species representing all major taxonomic groups can be found upon plating of whole barley grains or dilutions of barley meal. The spectrum of fungal species identified from a barley sample varies greatly with time, especially during storage since many species in the genera *Aspergillus* and *Penicillium* but also typical xerophiles such as *Eurotium* spp. or *Wallemia sebi* only start to develop and multiply after harvest when the water activity of grains decreases to low values during drying and storage. During malt production considerable changes in the prevailing growth conditions mark another fundamental change in the fungal community of the barley grain. High water activities during steeping and high carbon dioxide concentrations during subsequent germination are ideally suited for growth of quite selective species such as *Alternaria* spp., *Epicoccum nigrum, Fusarium avenaceum, F. graminearum, F. culmorum* and *F. tricinctum* from inside the grain or *Geotrichum candidum, Mucor mucedo, Rhizopus oryzae* and *Rhizopus stolonifer* as well as various white and red yeast species on the surface of the grain. Kilning of malt results in heat denaturation of most of the microbiota present on germinated barley. Accordingly, the fungal community undergoes another change in which fast-growing and strongly sporulating Zygomycetes such as *Rhizopus* ssp. and *Mucor* spp. as well as yeasts and yeast-like fungi such as *Geotrichum candidum* and *Ramichloridium schulzeri* strongly develop under the conditions whereas other members of the typical malting flora are inactivated and eventually killed.

Table 8.1 shows an overview of the spectrum of species frequently encountered on raw barley seeds from the author's own research using culture-based microbiological analysis (mycological status). For the analysis, grains are surface disinfected by immersing them in a sodium hypochlorite solution (1% active chlorine) for 10 minutes before washing them twice with sterile tap water. One hundred seeds (five per plate) are plated to SNA medium (Nirenberg, 1976; Nirenberg, 1981) containing streptomycin and aureomycin for inhibition of bacterial growth and 2,4-dichlorophenoxyacetic acid (2,4-D) for repression of seed germination. Fungal identification is performed after 14 days of incubation at 17°C in a 12 hours dark/light rhythm with a mixture of white light and $UV_{360\,nm}$ light. Light microscopy of the mycelia growing from the seeds is performed at 100-fold magnification directly into the open agar plates. Results show that only a small number of species occur regularly and with relatively high numbers of contaminated seeds per sample. Examples for this group of fungi

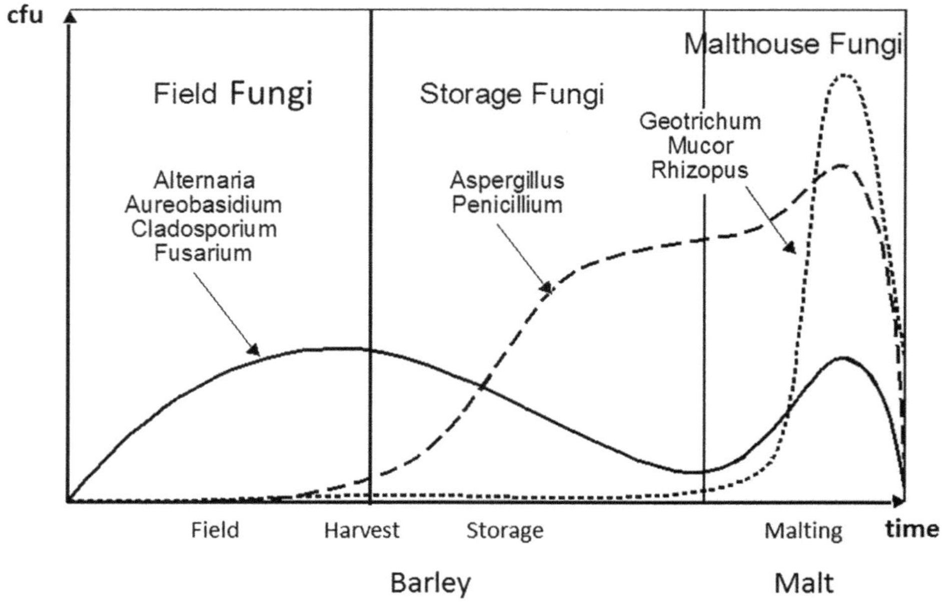

Figure 8.1 Dynamics of the fungal community on barely grain from the field to kilned malt. Redrawn from Müller (1995), with kind permission of Fachverlag Hans Carl GmbH, Nürnberg, Germany.

are *Alternaria* spp., *Epicoccum nigrum*, and *Fusarium tricinctum* which can be found in nearly all samples with relatively low variation in their individual contamination rates. Other species such as *Botrytis cinerea, Cladosporium herbarum, Bipolaris sorokiniana, Fusarium avenaceum, F. equiseti, F. culmorum, F. graminearum, F. poae, Microdochium majus* and *M. nivale* and red yeasts do regularly occur but their numbers show a much higher variation between samples and between years analysed. Others are found unregularly in some years on a varying percentage of samples but fail to be found in others. The spectrum of species detected with the method described above is much more restricted as compared to the spectrum published by Noots *et al.* (1999). However, it has to be realized that the list of species given in that publication was compiled from the literature and has to be interpreted as the maximum of contamination potentially occurring on a sample of barley. The list of species provided by Flannigan (1969) as a compilation of species occurring on barley seeds with and without surface disinfection and after incubation under various conditions of media and temperature shows much more similarity with the spectrum of species given in Table 8.1. The routine use of surface disinfection and the incubation under fairly low temperatures may provide an explanation for the low numbers of *Aspergillus* and *Penicillium* species as well as the low counts and incidence of other typical surface contaminants. In fact, the method used to set up Table 8.1 favours the detection of typical species of field fungi such as *Alternaria, Bipolaris, Cladosporium, Epicoccum, Fusarium* or *Microdochium*, which can invade deeper layers of the cereal grain and can thus escape being killed during surface disinfection. The reason for applying a method that strongly selects for field fungi rather than a broader species spectrum by this author is the connection between *Fusarium* contamination of barley and wheat and the occurrence of gushing in beers produced from such contaminated malt. Niessen *et al.* (1992) demonstrated a correlation between the percentage of grains in a sample that are contaminated with either *Fusarium graminearum, F. cerealis* or *F. culmorum* and the potential of a sample of barley or wheat to cause gushing. Authors set up a maximum level at 3% of grains contaminated by the sum of any of the three species above. It appeared that samples in which this contamination level was exceeded almost always caused gushing in the gushing test described by Donhauser *et al.* (1990).

Table 8.1 Spectrum of fungal species occurring on selected brewing barley samples from Germany between 2009 and 2014 ($n = \pm 30$ per year) and PCR-based assays for detection and identification

Fungal genus/species	2009	2010	2011	2012	2013	2014	Average (%)	STD (%)	PCR-based detection assays
Acremonium strictum	0.08	0.2		0.14	0.4	0.36	0.20	±35.30	Doss and Welty (1995), Haugland and Vesper (2000), Meklin *et al*. (2004)
Alternaria spp.	57.80	73.3	46.6	52.71	59.1	69.1	59.77	±8.35	Zur *et al*. (1999, Haugland and Vesper (2000), Johnson *et al*. (2000)
Aspergillus spp.			0.23				0.04	±122.47	Mukherjee *et al*. (2006), Suanthie *et al*. (2009)
Aureobasidium pullulans	0.17						0.03	±122.47	Schena *et al*. (2000a), Schena *et al*. (2000b), Martini *et al*. (2009)
Botrytis cinerea	1.17	0.33	0.23	0.86	0.05	0.07	0.45	±50.82	Rigotti *et al*. (2002), Brouwer *et al*. (2003), Gachon and Saindreman (2004)
Chaetomium globosum			0.07		0.6		0.11	±107.85	Haugland and Vesper (2000)
Cladosporium cladosporioides	1.17			0.21		0.43	0.30	±76.02	Haugland and Vesper (2000), Zeng *et al*. (2006)
Cladosporium herbarum	0.92	0.73	1.3	0.79	5.25	4.14	2.19	±45.31	Haugland and Vesper (2000), Zeng *et al*. (2006)
Bipolaris sorokiniana	3.30	7	10.1	6.14	5.45	1	5.50	±28.44	Matusinsky *et al*. (2010), Aggarwal *et al*. (2011)
Epicoccum nigrum	16.10	9.38	20.9	23.9	8.85	15.8	15.82	±122.47	Haugland and Vesper (2000), Martini *et al*. (2009)
Fusarium acuminatum	0.40			0.14		0.28	0.14	±19.026	Williams *et al*. (2002)
Fusarium anthophilum			0.07				0.01	±62.50	–
Fusarium avenaceum	2.20	1.1	1.4	1.36	0.3	0.86	1.20	±134.16	Schilling *et al*. (1996), Turner *et al*. (1998), Waalwijk *et al*. (2004)
Fusarium camptoceras		0.07					0.01	±37.78	–
Fusarium crookwellense				0.07			0.01	±40.62	Yoder and Christianson (1998)
Fusarium culmorum		0.07	0.38	0.57	0.2	0.5	0.29	±122.47	Schilling *et al*. (1996), Nicholson *et al*. (1998), Mishra *et al*. (2003), Leisova *et al*. (2005)
Fusarium equiseti	0.33	0.33	0.31	0.43	0.05	0.64	0.35	±122.47	Mishra *et al*. (2003), Nicholson *et al*. (2004), Jurado *et al*. (2005), Nicolaisen *et al*. (2009)
Fusarium graminearum	3.60	2.1	0.77	2.64	0.85	4	2.33	±27.45	Niessen and Vogel (1997), Nicholson *et al*. (1998), Doohan *et al*. (1998), Yang *et al*. (2008), Yin *et al*. (2009)
Fusarium langsethiae		0.9	0.07	0.43	1.3	0.64	0.56	±29.11	Niessen *et al*. (2004), Wilson *et al*. (2004), Riazantsev *et al*. (2008), Nicolaisen *et al*. (2009)
Fusarium poae	0.30	0.6	2.1	0.93	1.95	2.36	1.37	±31.66	Parry and Nicholson (1996), Yli-Mattila *et al*. (2004), Niessen *et al*. (2004), Kulik *et al*. (2008), Stakheev *et al*. (2011)
Fusarium sacchari					1.15	0.14	0.22	±107.32	–

Table 8.1 Continued

Fungal genus/species	2009	2010	2011	2012	2013	2014	Average (%)	STD (%)	PCR-based detection assays
Fusarium sporotrichioides				0.07		0.21	0.05	±90.83	Kulik *et al.* (2004), Niessen *et al.* (2004), Yli-Mattila *et al.* (2004), Konstantinova and Yli-Mattila (2004), Demeke *et al.* (2005)
Fusarium solani		0.07		0.07			0.02	±77.46	Alexandrakis *et al.* (1998), Jaeger *et al.* (2000), Lievens *et al.* (2006)
Fusarium subglutinans	0.08	0.13					0.04	±80.69	Möller *et al.* (1999), Mulé *et al.* (2004), Nicolaisen *et al.* (2009)
Fusarium tricinctum	6.20	2.4	7.2	6.93	4.35	6.36	5.57	±16.57	Kulik (2008), Nicolaisen *et al.* (2009), Riazantsev *et al.* (2008)
Fusarium verticillioides			0.07			0.07	0.02	±77.46	Beck and Barnett (2003), Mulé *et al.* (2004), Patiño *et al.* (2004), Sanchez-Rangel *et al.* (2005), Nicolaisen *et al.* (2009)
Geotrichum candidum		0.2	0.77	0.29	0.75		0.34	±51.97	Nakamura *et al.* (2007)
Gonatobotrys simplex		0.73	0.07	0.93	0.05	0.14	0.32	±62.90	–
Harzia acremonioides			0.14				0.02	±122.47	–
Microdochium majus	3.00	0.33	1.4	1.29	0.6	0.14	1.13	±46.53	Nicholson *et al.* (1996), Nicholson and Parry (1996), Glynn *et al.* (2005)
Microdochium nivale	4.30	0.2	0.23	2.5	0.55	0.21	1.33	±64.027	Nicholson *et al.* (1996), Glynn *et al.* (2005)
Mucor spp.		0.07	0.62		0.1	0.07	0.14	±82.69	Voigt *et al.* (1999)
Nigrospora sphaerica		0.27	0.38	0.07	0.05	0.14	0.15	±48.06	–
Penicillium spp.		0.07	0.62		0.1	0.07	0.14	±82.69	Pedersen *et al.* (1997), Mukherjee *et al.* (2006), Suanthie *et al.* (2009)
Phoma spp.				0.07			0.01	±122.47	Keinath *et al.* (2001)
Ramichloridium schulzeri					1.7	0.29	0.33	±102.56	–
Rhizopus stolonifer			0.23		0.05		0.05	±98.59	Nagao *et al.* (2005)
Red yeast	0.25	0.07	1.3		4.25	0.86	1.12	±75.59	García *et al.* (2004), Hierro *et al.* (2006)
Trichoderma spp.			0.07		0.05		0.02	±79.06	Hagn *et al.* (2007)
Ulocladium atrum	0.70			0.14	0.15	0.57	0.26	±57.78	Haugland and Vesper (2000), Meklin *et al.* (2004)
White yeast		0.07					0.01	±122.47	García *et al.* (2004), Hierro *et al.* (2006)
No fungal contamination	1.00	3.7	4	1.5	4.2	1.5	2.65	±27.59	

Problems related to fungal contamination of brewing cereals and malt

Barley grains provide a rich source of nutrients for all kinds of microorganisms. According to Gupta *et al.* (2010), whole barley grain contains about 65% (w/w) to 68% starch, 10% to 17% protein, 4% to 9% β-glucan, 2% to 3% free lipids, and 1.5% to 2.5% minerals with a percentage of total dietary fibre (DF) ranging from 11% to 34% and soluble DF within a range of 3% to 20%. However, most of this richness can only be explored if an organism's toolbox is ready to provide appropriate enzymes and transporters necessary to break down polymeric substrates into monomers and to transport these monomers properly into the cell for further metabolic processing. Also, the presence of catabolic and anabolic pathways must be present in a fungus in order to generate energy and cellular matter from the substrate. This is the reason why the microbial community of raw barley grains is strongly dominated by those fungal species that are able to exploit the developing seed with its starchy endosperm and its protein-rich aleuronic layers. Fungi of lesser dominance as well as yeasts and bacteria are characterized by their ability to use only deposited organic matter and the few exudations of the hull or the husk that will not allow for excessive growth. Moreover, the decreasing water activity occurring during maturation of the seeds has a greatly adverse effect on all microorganisms present. Many species will expire from drying out and only those species will survive which have developed mechanisms to counter the conditions, e.g. formation of survival structures such as spores and chlamydospores or accumulation of compatible solutes in vegetative structures. Organisms such as *Aspergillus glaucus*, *A. penicillioides*, *Eurotium amstelodami*, *E. chevalieri*, *E. repens*, *E. rubrum*, *Penicillium aurantiogriseum*, *P. brevicompactum*, *P. chrysogenum*, *P. citrinum*, *P. crustosum*, *P. glabrum* and *Wallemia sebi* are ubiquitous species that grow particularly well under reduced water activities between a_w 0.72 and a_w 0.80.

The following three sections deal with the three principal problems related to the fungal contamination of brewing cereals and derived malt.

Yield reduction

Yield reduction due to fungal contamination of barley grains (see introduction) can result via different mechanisms. Fungi such *as Microdochium nivale*, *Fusarium graminearum* and other seed-borne and soil-borne fungal plant diseases can result in conditions such as seedling blight or damping off of seedlings, leading to retarded germination of the seedling, which is eventually killed prematurely. This effect will decrease the number of plants growing per unit area, leading to a yield reduction as compared to a reference area with healthy plants. Aside from being seed-borne, the above-mentioned pathogens, as well as a variety of others, can also have a devastating effect when present in the soil to which sound barley has been sown (soil-borne).

Once germinated without microbial attack, the young barley plant develops into the flowering state where it becomes vulnerable to fungal attacking mechanisms that aim at either the glumes or anthers and gynoecium of individual spikelets or at the spike's rachis. Glume infection is mainly by entering of the germination tubes of airborne spores into the stomata with subsequent dispersal of the fungus into the glume tissue and, later on, into the developing endosperm (Pritsch *et al.*, 2000; Xu, 2003). Fungal infection via anthers and the gynoecium apparatus depends strongly on the presence of appropriate climatic conditions during flowering, i.e. a period of high humidity and medium temperatures in order to establish infection. Provided the prevalence of optimum conditions over 24–48 hours post infection, growth of cereal pathogens through the anther tissue into the ovaries of the developing seed can lead to deep seated infection of the endosperm, eventually killing the young embryo or establishing seed-borne infection. Growth in species such as *F. graminearum* was found to be highly stimulated by the presence of choline and betaine, which are typically present in cereal anthers in high concentrations (Strange *et al.*, 1974). Deep-seated infections will strongly reduce grain development and grain filling, resulting in small and often shrivelled grain with a strong diminution in grain yield. Spread of fungal infection from single spikelets to the rachis may result in cutting off the water supply for spikelets above the point of the primary infection. Affected spikelets will ripen early and show symptoms of head blight with grains showing strong reduction in size and filling. All symptoms described cause a reduction in thousand-kernel weight, i.e. reduced yield.

A third mechanism responsible for yield

reduction due to fungal infection is the lowering of assimilative leaf area. Fungal pathogens such as *Erysiphe graminis* (powdery mildew), *Puccinia hordei* (brown rust, Scholes and Farrar, 1986), *Septoria tritici* and *Rhynchosporium secalis* (leaf blotch of wheat and barley, respectively, Fowler and Owen, 1971) or *Pyrenophora teres* (net blotch of barley, Evans, 1969) strongly reduce the assimilative capacity of cereal plants, resulting in reduction of grain filling and overall yield.

Fungal enzymatic activity

Fungi are an important source of enzymes for the brewing industry, both as producers of the generic enzymes (Østergaard and Olsen, 2011) and as a host for the transgenic production of non-fungal industrial enzymes (Olempska-Beer *et al.*, 2006). Fungal enzymes are useful as additives to enhance hydrolysis during mashes involving partial or sole addition of raw cereals (Bajomo and Young, 1993) or cereal adjuncts (Linko *et al.*, 1998) and to enhance filterability of wort for mashes of less than fully modified malt. By contrast, fungal enzymatic activity can have quite negative effects if developed as natural contaminants through the endogenous fungal community of barley and malt. Fungi were estimated to make up approximately 0.1% of the microbial biomass present on or in naturally contaminated cereal grains (Van Nierop *et al.*, 2006). Fungal enzymatic activity has repeatedly been linked to poor quality of cereals and malt e.g. insufficient malt modification, malt yield reduction, or reduction of diastatic power of malts (Van Nierop *et al.*, 2006). Such mycobiota-associated modification of the malt parameters may, among other things, result in abnormal fermentation, intensification of wort and beer colour, gushing of beer, off taste and off flavour (Spicher, 1989). Information available on fungal enzymes produced and excreted by fungi in or on cereal grains is limited to the few species showing the most deleterious effects on the commodity, i.e. *Fusarium graminearum* and *F. culmorum*, *Alternaria* spp., *Drechslera* spp. It can, however, be assumed that excretion of any amylolytic, proteolytic or lipolytic fungal enzyme has a potential to modify the composition of cereal grains and, in turn, may potentially influence the quality of the cereal raw material used in malt production and beer fermentation. Changes that occur range from complete decomposition of components to various degrees of chemical modification. Also, the colour and odour of mould-contaminated barley and malt may be negatively influenced by enzymatic activity, especially when the grain is stored under suboptimum conditions (Christensen and Kaufmann, 1965). Fungi have been demonstrated to produce cellulases (Hoy *et al.*, 1981), xylanases, β-glucanases, and proteases during infection of barley and other cereals (Schwarz *et al.*, 2002). These authors detected an increase in β-glucanase, xylanase, and protease activities in barley grain and in barley malt upon inoculation with *Fusarium graminearum* and *F. poae*, respectively. It was concluded that the enzymes were of fungal origin and that their activity levels appear to be such that they may affect the quality of the malt and the wort produced from it. The quality of beer produced from such worts is also affected. Fungal proteases play an important role here since they have considerable influence on the protein concentration and its composition in the grain and in products made therefrom (Nightingale *et al.*, 1999). Fungal proteases produced by *Fusarium* spp., which belong to the so-called *Fusarium* head blight complex of species, have been particularly well characterized. Pekkarinen *et al.* (2003) detected several alkaline proteases in barley samples after field infection with *Fusarium* species. The presence of these proteases correlated well with the degradation of barley grain storage proteins (C- and D-hordeins) in the infested grains. Moreover, Hecht and Hippeli (2007) postulated the presence of a heat-stable protease and its degradation of the beer foam protein nsLtp1 as a major cause for beer gushing since they observed considerably lower concentrations of the foam protein in gushing beers as compared to non-gushing beers.

Activity of fungal xylanases and glucanases has been suspected as the major reason for the presence of factors leading to premature yeast flocculation (PYF) during beer fermentation. Van Nierop *et al.* (2004) postulated that degradation of the malt husk arabinoxylan by fungal enzymes produces polysaccharides of sufficient size to cross-link yeast cell lectins resulting in flocculation and precipitation of fermenting yeast at still high sugar concentrations. Resultant beers are of low quality since they have high sugar contents and low end-of-fermentation cell counts (Verstrepen *et al.*, 2003).

Fungal lipases play a major role as pathogenicity

factors in many plant pathogenic fungi but have also great importance in biotechnology (Subramoni *et al.*, 2010). Secretion of lipases by *F. graminearum* has been demonstrated as a virulence factor during the infection of cereals (Voigt *et al.*, 2005).

In response to the secretion of fungal lytic enzymes, the plant reacts by changing its metabolic and gene expression profile (Geddes *et al.*, 2008) and by producing antimicrobial substances to defend itself against attack (van Nierop, 2006). Besides a battery of enzymes that may directly attack microbial cells (β-1,3-glucanases, chitinases, proteases), the products of some enzymes (peroxidases, oxalate oxidase, ammonia lyase) may have adverse effects on microorganisms as well (van Loon *et al.*, 2006). Moreover, pathogenesis-related (PR) proteins (Sels *et al.*, 2008), which are antimicrobial peptides including the thionins, plant defensins, hevein- and knottin-like proteins and non-specific lipid transfer proteins (nsLTP), have been identified as having antifungal properties (Selitrennikoff, 2001). Interestingly, the heat-denatured and glycosylated form of the latter protein is supposed to be one of the major foam-stabilizing proteins in beer (Douliez *et al.*, 2000) and its degradation by fungal proteases has been discussed as a causative agent of beer gushing (Hippeli and Elstner, 2002).

Fungal secondary metabolites

Secondary metabolites are distinguished from primary metabolites that are produced and distributed almost universally by the intermediary metabolism of living organisms. Secondary metabolites are often bioactive, usually of low molecular weight, and are produced as families of related compounds at restricted stages in the life cycle, with production often being correlated with a specific stage in the fungal development or morphological differentiation (Calvo *et al.*, 2002). Secondary metabolites are dispensable for the producing fungus and have restricted taxonomic distribution with only a small group of organisms producing each metabolite (Bennett and Bentley, 1989; Keller *et al.*, 2005). Filamentous fungi produce an enormous variety of secondary metabolites in pure culture but also when growing on natural substrates such as brewing cereals and malt. Nielsen and Smedsgaard (2003) provided a list of 474 individual and structurally well-characterized substances isolated from extracts of pure liquid cultures of filamentous fungi. To date, the most recent version (2012) of the AntiBase database (Wiley-VCH, Weinheim, Germany) used for the LC-MS based identification of microbial secondary compounds contains 3000 fungal secondary metabolites (Klitgaard *et al.*, 2014). However, this number is likely to represent only a fraction of the total secondary metabolites produced by filamentous fungi. Fungal secondary metabolites can have a diversity of physiological and ecological functions (Vining 1990) but in the majority of cases their functions are obscure. However, it can be assumed that they are bound to play an important role in the fungal life cycle and in the interaction with the environment since their production is highly regulated and complex in most cases. Regulatory links between secondary metabolism, light and sexual/asexual reproduction have been established, which might explain their function to a wider extent (Fox and Howlet, 2008).

Antimicrobials

Antimicrobials such as the antibacterial penicillins, cephalosporins and other β-lactam antibiotics, the antiprotozoal trypacidin, the antibacterial enniatins (fusafungin), the bacteriostatic fusidic acid (fucidin, ramycin) or the antifungals cerulenin, cordycepin, siccanin and viridian, just to name a few, are produced by filamentous fungi to cope with other microorganisms competing for nutrients and space. Many of them have been used in modern medicine but some can also be harmful to humans and animals. Similar observations can be made regarding fungal pigments. The red discoloration often observed in *Fusarium*-infected barley and malt results from bikaverin and norbikaverin production by many species in the genus (Wieman *et al.*, 2009). However, those pigments also have antimicrobial properties against certain protozoa and fungi (Limón *et al.*, 2010). Also, compounds such as the naphtho-γ-pyrones aurofusarin and rubrofusarin, which are responsible for the intense red to violet pigmentation of many *Fusarium* spp., have mycotoxin properties but are nonetheless an interesting subject for pharmacological research due to their anti-oxidant, antimicrobial, anti-cancer, anti-HIV, anti-hyperuricuric, anti-tubercular, or mammalian triacylglycerol synthesis inhibitive activities (Choque *et al.*, 2015).

Pigments

Melanins, the pigments mainly responsible for black discoloration of weathered barley and malt after infection with dematiaceous fungi such as *Alternaria* spp. and *Drechslera* spp., are anti-stress compounds with antioxidant abilities but also function as virulence factors during invasion of the host plant (Henson *et al.*, 1999). Other fungal pigments seem to have no known adverse effects and their exploitation as natural colorants in the food industry is hence intensively studied (Durán *et al.*, 2002; Mapari *et al.*, 2005, 2010).

Fungal hormones

Fungal hormones are a group of secondary metabolites to which a function can clearly be attributed. They are produced by cells as very specific molecules that have morphogenetic effects on other cells of the same or closely related species and are aimed at regulating and coordinating the temporal and spatial sequence of events leading to the pairing and fusion of nuclei of two cells during sexual reproduction (see reviews by Gooday, 1974 and by Gooday and Davis, 1993). The mechanism of selective attraction of mating partners may involve the mutual production of compounds binding specifically to hormone receptors on or in the respective partner and triggering the morphogenetic differentiations leading to sexual reproduction like in *Saccharomyces cerevisiae* (Herskowitz, 1988). In other species, one of the two mating partners produces a hormone that triggers morphogenetic differentiation in the other. Production of antheridiol in the water mould *Achlya* spp. would be an example for this type (Gooday, 1974). The third type of mechanism is the mutual production of two different precursors by each mating partner, which mutually supplements a biosynthetic pathway in the respective mating partners resulting in the production of a compound that triggers the formation of mating organs in both partners. This type of mechanism can be found in the *Mucorales* such as in *Mucor* spp. or *Rhizopus* spp. (Werkman and van den Ende, 1974).

Mycotoxins

Several overviews of the history of mycotoxins in food and feed, including cereals and malt, as well as coverage of their impact on human and animal health and on the economy have been published in the past (Kampelmacher, 1973; Charmley *et al.*, 1995; D'Mello and MacDonald, 1997; Hussein and Brasel, 2001; Richard, 2007; Bhat *et al.*, 2010; Zain, 2011). Despite the wide variety of substances that fall into the category of mycotoxins (about 400), the number of compounds present in substrates for human consumption and consumption by livestock is relatively limited (≈30). This is due to the fact that the majority of compounds are intermediates of known mycotoxins so far described only in lab cultures. Under natural conditions, however, they are converted to one of the major components found regularly in food products. Most of the known mycotoxins produced on cereals and malt come from species within the ascomyceteous genera *Alternaria, Aspergillus, Penicillium,* and *Fusarium.* However, also species in *Acremonium* (crotocin), *Chaetomium* (chaetoglobosin, cochliodinol), *Eurotium* (xanthocillin), *Stachybotrys* (roridin, satratoxins, trichoverrins and trichoverrols, verrucarin), *Trichoderma* (trichodermin) *or Trichothecium* (trichothecin) may infrequently grow in cereal seeds and can produce the mycotoxins shown in parentheses. The following sections give an overview of the mycotoxins that have been reported to occur in brewing cereals, malt or beer. Information on legal limits for the mycotoxins in different countries of the world were taken from a survey published by the FAO (2004). More background information on legal limits for mycotoxins in the EU and in the USA can be found on the homepage of the German Society for Mycotoxin Research www.mycotoxin.de/.

Toxins of *Aspergillus* and *Penicillium* in beer production

Aflatoxins

Aflatoxins are produced by a group of 14 different species in three sections of the fungal genus Aspergillus, i.e. sections *Flavi, Nidulantes,* and *Ochraceorosei* (Varga *et al.*, 2009). Aflatoxins are known as the most potent carcinogens among natural products. They act as acute liver toxicants and trigger liver carcinomas in addition to impairing respiration, renal and gastrointestinal function, and the nervous and immune systems in humans and animals (Coulombe, 1994). Owing to its high toxicological potential, the compound and its derivatives are regulated in most countries worldwide (van Egmond and Jonker, 2005; van Egmond *et al.*, 2007). *Aspergillus flavus* and *A. parasiticus* are the species that occur on brewing cereals and brewing adjuncts and aflatoxins have been isolated from barley and malt in different parts of the world (Park *et al.*, 2002; Mateo *et al.*, 2011). However, contamination of beer in most cases resulted from the use of contaminated maize grits rather than from cereal malt (Pietri *et al.*, 2010). Due to its low water solubility, only a small proportion of the initial aflatoxin concentration is transmitted into the final beer so that its occurrence in the product is rather an exception, at least in Europe and in Northern America (Mably *et al.*, 2005; Bertuzzi *et al.*, 2011; Burdaspal and Legarda, 2013).

Citrinin

Citrinin is an isocoumarin (pentaketide) mycotoxin produced by some *Penicillium* species, i.e. *P. citrinum*, *P. expansum*, *P. radicicola* and *P. verrucosum* as well as *Monascus ruber* (Samson *et al.* 2010; Ostry *et al.*, 2013). *P. verrucosum* has been identified as the major producer of the toxin in barley where it mostly co-occurs together with the structurally related mycotoxin ochratoxin A (see below) (Krogh *et al.*, 1973; Hökby *et al.*,1979; Domoglou *et al.* 1984; Vrabcheva *et al.*, 2000). Biosynthesis of both mycotoxins uses a common isocoumarin moiety precursor (Larsen *et al.*, 2001) and it has recently been established that one toxin will be produced at the expense of the other depending on the prevailing environmental conditions (Schmidt-Heydt *et al.*, 2015).

Citrinin is nephrotoxic (Krogh *et al.*, 1970) and teratogenic (Reddy *et al.*, 1982) in mammals and has been found to inhibit cholesterol biosynthesis (Endo and Kuroda, 1976). Since it usually occurs together with ochratoxin A, a participation in the development of porcine nephropathy and human Balkan endemic nephropathy (BEN) has been discussed (Vrabcheva *et al.*, 2000). Krogh *et al.* (1974) established that citrinin is unstable under the conditions prevailing during mashing and will therefore not be transmitted into the final beer, though Ikalafeng *et al.* (2013) reported the presence of the toxin at elevated levels in traditionally brewed indigenous beers in South Africa.

Cyclopiazonic acid

Production of cyclopiazonic acid has been found in cultures of *Aspergillus flavus*, *A. oryzae*, *Penicillium camemberti*, *P. commune*, *P. dipodomyicola*, *P. griseofulvum*, and *P. palitans* (Frisvad *et al.*, 2004; Samson *et al.*, 2010). The toxin is uncommon in brewing cereals and malt but has been detected in maize after infection with *P. commune*, *A. flavus* and *A. orzyzae* as a co-contaminant, together with aflatoxins (Gqaleni *et al.*, 1996). No published reports on the compound as a contaminant of beer are available. Owing to structural similarities, symptoms of toxicity encountered in laboratory animals after administration of cyclopiazonic acid were very similar to aflatoxin toxicosis, i.e. hepatotoxic, nephrotoxic, neurotoxic and carcinogenic effects. So far no legal limits have been adopted for cyclopiazonic acid.

Ochratoxin A (OTA)

Ochratoxin A is produced in cultures of *Petromyces alliaceus*, *Aspergillus carbonarius*, *A. niger* group species, *A. ochraceus*, *A. steynii*, *A. westerdijkeae*, *P. nordicum* and *P. verrucosum* (Samson *et al.*, 2010). *P. verrucosum* has been identified as the major producer of the toxin in brewing barley where it mostly co-occurs together with the structurally related isocoumarine mycotoxin citrinin (see above) (Krogh *et al.*, 1973; Hökby *et al.*, 1979; Vrabcheva *et al.*, 2000). Contamination of maize kernels is regularly due to infection with *A. carbonarius*, *A. niger*, and *A. ochraceus*, depending on the origin of the tested materials (Magnoli *et al.*, 2006, 2007). The toxin has been detected in various food products such as cereal products, nuts, coffee, cocoa, wine, grape juice, vegetables, dried fruits and spices. It has been demonstrated to contaminate a high percentage of beer samples although in low concentrations (Visconti *et al.*, 2000; Tangni *et al.*, 2002; Medina *et al.*, 2005; Varga *et al.*, 2005; Bellver Soto *et al.*, 2014). Concentration of the compound has been shown to decrease considerably during mashing and during the brewing process so that 2% to 25% of the original OTA amounts were recovered from the final beer (Krogh *et al.*, 1974; Chu *et al.*, 1975). Ochratoxin A is nephrotoxic, cytotoxic, immunotoxic, teratogenic and genotoxic (Petzinger and Ziegler, 2000) and has been classified a B2 cancer compound by the IARC (1993). Classification into the group of probably carcinogenic compounds (2A) has been proposed (Kuiper-Goodman, 1996). While the induction of cancer is not proven in humans there are clear indications of a connection to BEN (Vrabcheva *et al.*, 2000). Legal limits for the contamination of food and food raw materials with OTA are in effect in many countries, including the EU, with maximum values ranging from 2.0 to 20 µg/kg in different food sources (van Egmond, 2007). A limit of 0.2 µg/kg has been set up for beer by the European Commission.

Patulin and penicillic acid

Patulin is a lactone compound that can be produced by cultures of *A. clavatus, Byssochlamys nivea, P. carneum, P. expansum, P. griseofulvum, P. paneum* and *P. sclerotigenum* whereas the structurally related compound penicillic acid is produced by strains of *A. ochraceus, A. steynii, A. westerdijkiae, P. aurantiogriseum, P. cyclopium, P. freii, P. melanoconidium, P. polonicum* and *P. viridicatum*. Among the species listed above, *A. clavatus* has been frequently isolated from malted barley and from malt dust. Patulin and penicillic acid have cytotoxic properties (Schaeffer *et al.*, 1975). Both compounds induce DNA strand breaks and inhibition of DNA synthesis (Stětina and Votava, 1986). Inhibition of the aminoacyl-tRNA synthetase plays a significant role in the observed toxicological effects (Arafat *et al.*, 1985). The carcinogenicity of patulin is currently under investigation and induction of skin tumours after superficial application of patulin have been discussed (Saxena *et al.*, 2011). Penicillic acid often occurs together with OTA in food and feed materials. Neither patulin nor penicillic acid has been found in beer yet. However, this is not due to their absence from the brewing process but rather to their denaturation during the fermentation process (Inoue *et al.*, 2013). Patulin is actively decomposed to non-toxic ascladiol E and ascladiol Z by brewing yeast under fermentative conditions (Moss and Long, 2002). Because of its suspected carcinogenicity, legal regulations for patulin exist in various countries (FAO, 2004), including the EU. Maximum legal values vary between 20 and 50 µg/kg and are almost always related to consumption of apple products rather than cereals.

Sterigmatocystin

Production of sterigmatocystin has recently been found to be a widely distributed trait in various genera within the *Pezizomycotina*. Sterigmatocystin can be produced by strains of species within genera *Emericella, Aspergillus, Chaetomium, Botryotrichum* and *Humicola* (Rank *et al.*, 2011) with *A. versicolor* being the main producer of the toxin in cereals. The toxin can be found in a variety of food sources such as cheese, fruits, vegetables, nuts, spices, green coffee, rice, maize, cereals and cereal products. Veršilovskis *et al.* (2008a) detected sterigmatocystin in about 25% of grain samples from Latvia and were able to show that it occurs also in a low percentage of local beers (Veršilovskis *et al.*, 2008b). It can be anticipated that the toxin will occur in grain from countries other than Latvia as well. The fact that several advanced techniques have been established recently for the sensitive detection and quantification of sterigmatocystin in cereals may point to elevated awareness towards the problem (see Hossain and Goto, 2015; Sasaki *et al.*, 2014; Li *et al.*, 2014). The furofuran (dekaketide) toxin occurs as an intermediate during aflatoxin biosynthesis in laboratory cultures of typical aflatoxin producers (*A. flavus. A. parasiticus*) but is formed as a metabolic end product by these species. According to its structural similarity with aflatoxins, sterigmatocystin is also supposed to be closely related in regard to its toxicology. No legal limits have been set up yet (FAO, 2004) but the substance should be treated in a way similar to that of aflatoxins.

Fusarium toxins in beer production

Trichothecenes

Trichothecenes are the biggest group of chemically related compounds among the mycotoxins. These sesquiterpenes are produced in cultures of several different species belonging to the *Fusarium* sections *Discolor, Gibbosum, Sporotrichiella* and *Arthrosporiella*. Moreover, production of trichothecene derivatives (macrocyclic trichothecenes) was found in species from genera *Trichoderma, Myrothecium, Cephalosporium, Stachybotrys, Cladosporium, Trichothecium, Verticimonosporium* and *Spicellum* as well as in the leaves of the Brazilian shrub *Baccharis megapotamica* (Jarvis *et al.*, 1991). Trichothecenes occur in all kinds of cereals and food product made therefrom, including malt and beer as well as in vegetables and fruits but also in materials such as building materials, textiles or in the air (spores of *Stachybotrys chartarum*). According to their chemical structure, all trichothecenes can be categorized into four basic types, i.e. 8-keto-trichothecenes (type A), 8-hydroxy-trichothecenes (type B), di-epoxythrichothecenes (C) and macrocyclic trichothecenes (type D) (see McCormick *et al.*, 2011). Compounds belonging to type A and type B are the most common in cereals and cereal products (Rodriguez-Carrasco *et al.*, 2013). The trichothecenes most commonly detected in wheat and barley grown in different geographic regions are the type B trichothecenes deoxynivalenol (DON) and its acetylated derivatives 3-acetyl DON and 15-acetyl DON, nivalenol (NIV), and fusarenone-X (FX) as well as the type A trichothecenes T2-toxin, HT2-toxin, and diacetoxyscirpenol (DAS) (Barthel *et al.*, 2012; Ibáñez-Vea *et al.*, 2012; Rasmussen *et al.*, 2012; Tittlemier *et al.*, 2013).

Several of the trichothecenes have been found to conjugate with sugars to form glycosides, masking their detection during routine analytical procedures (Berthiller *et al.*, 2013). However, the toxins can readily be released from their glycosidic partner by microbial activity in the human or animal gut to display their full toxicity (Dall'Erta *et al.*, 2013). Assessment of the toxicological potential of regulated trichothecenes such as DON in a sample may therefore be underestimated if only concentrations of the parent toxins are measured during routine analysis. Transfer of deoxynivalenol-3-glucoside (DON-3-Glc) from barley through malting and the brewing process was observed with a steady increase of DON-3-Glc resulting in an excess of the glucoside over the free form of the toxin in the final beer (Lancova *et al.*, 2008; Kostelanska *et al.*, 2009). Recently, Zachariasova *et al.* (2012) observed the presence of DON-3-oligoglucosides of varying chain length in addition to the monoglucoside and the unmodified DON in malt and beer, showing that trichothecene analysis may become a more complex task in the future.

The fate of DON as the most widely distributed *Fusarium* mycotoxin has been studied using immunochemical (Niessen and Donhauser, 1993) or HPLC-based analysis (Kostelanska *et al.*, 2011). Studies revealed that a considerable proportion of DON is washed away from a field contaminated sample during steeping (Beattie *et al.*, 1998). However, levels of the toxin increased during the malting process due to *de novo* synthesis, adding up to the concentrations already present inside barley grains from field contamination. Most of the toxin was transferred to the wort and no reduction was found during wort boiling, fermentation and ripening so that the finished beer had similarly high concentrations as compared to the corresponding malt (Lancova *et al.*, 2008). It was most interesting to see that final concentrations of 'masked' DON glycosides were higher than the unmasked derivative.

Rocha *et al.* (2005) reviewed the toxicological effects and underlying cellular mechanisms of the most common trichothecene mycotoxins in mammals and in plants. Acute symptoms of trichothecene uptake are diarrhoea (Matsuoka and Kubota, 1987), vomiting (Ishii *et al.*, 1975), impairment of the immune system, reduced performance of the heart muscle, disorder of the nervous system and skin irritations (see Beasly, 1989). Acute toxic aleukia (ATA, Joffe, 1971), the Kashin-Beck disease (Joffe, 1986; Kolsteren, 1992) and the red mould disease (Yoshizawa, 1983) have been described as human conditions that are associated with trichothecenes.

Legal limits have been set for the trichothecenes deoxynivalenol and T2-toxin in several countries worldwide, including the EU, Canada, Russia, and the USA (FAO, 2004). Moreover, a few countries have additional regulations for HT-2 toxin and diacetoxyscirpenol in certain food and feed commodities. Although more toxic than DON, NIV concentrations have undergone no legal regulation yet. However, the toxin might need to be given more attention from a legal perspective in the future (European Commission, 2002).

Fumonisins

Fumonisins comprise a group of structurally similar polyketide mycotoxins. They have been found to be produced by several of the >34 species within the *Gibberella fujikuroi* complex (GFC) of *Fuarium* species (Kvas *et al.*, 2009), e.g. *F. fujikuroi*, *F. globosum*, *F. proliferatum*, *F. nygamai*, *F. subglutinans* and *F. verticillioides* (Proctor *et al.*, 2004). Outside this group of closely related species, fumonisins have been found to be produced in some strains of *F. oxysporum* and, surprisingly, in *Aspergillus niger* (Frisvad *et al.*, 2007) and in biotechnologically important species within the fungal genus *Tolypocladium* (Mogensen *et al.*, 2011). Whilst the most common fumonisin in *Fusarium* spp. is fumonisin B_1, *A. niger* strains produce fumonisins B_2, B_4 and B_6 (Månsson *et al.*, 2010; Mogensen *et al.*, 2010). Fumonisins have rarely been detected as natural contaminants of barley or wheat (Stanković *et al.*, 2012). However, they can be detected in beers on a regular basis (Soriano and Dragacci, 2004; Bertuzzi *et al.*, 2011). Similar to aflatoxins, it can therefore be assumed that maize-based brewing adjuncts will be the major source of fumonisin contamination in beer. Beside fumonisin B_1, which is typically produced by *Fusarium* spp., fumonisin B_2 has been found in beer also, obviously produced by *A. niger* contamination. Other fumonisins from that source, FB_4, and FB_6, have not been found in beer yet, but can be anticipated in the product. Moreover, fumonisins can be regularly found in traditional African beers brewed from sorghum and maize (Nkwe *et al.*, 2005; Shephard *et al.*, 2005). Pietri *et al.* (2010) demonstrated that about half of the fumonisin present in contaminated brewing raw materials can be transferred through the brewing process. This is a much higher percentage as compared to aflatoxin, most of which is bound to spent grains and eliminated from the brewing process.

Fumonsin B_1 has the highest toxicity among this group of compounds (Musser and Plattner, 1997). It is neurotoxic, hepatotoxic, and nephrotoxic in animals, and it has been classified as a possible carcinogen to humans (Stockmann-Juvala and Savolainen, 2008). Major mechanisms of toxicity are related to the structural similarities between sphingolipids and fumonisins, which interfere with ceramide synthase and lead to accumulation of sphinganine in cells and tissues. A correlation between high fumonisin concentrations in the diets of people in parts of China (Yoshizawa *et al.*, 1994) and south African Transkei (Sydenham *et al.*, 1990) with the occurrence of oesophageal cancer has been suspected.

The European Commission has set legal limits for fumonisins in certain food stuffs ranging from 200 μg/kg to 4000 μg/kg for the sum of FB_1 and FB_2 (European Commission, 2006). Many countries, which take reference to the EU legislation, such as Turkey, Bosnia and Herzegovina, Norway and Switzerland, have established similar maximum limits for fumonisins. In the international markets, however, no specific maximum limits for fumonisins are established in major markets including China, Japan, India, Gulf Cooperation Council (GCC), Russia, Canada and many Latin American countries.

Emerging mycotoxins: moniliformin, fusaproliferin, enniatins and beauvericin

Emerging mycotoxins are considered as less important in comparison to 'classical' mycotoxins since they are probably not of acute toxicity. However, their high prevalence in foodstuffs in occasionally high concentrations (up to mg/kg) warrants an assessment of their importance for food safety. No legal limits have been set up so far for any of the emerging mycotoxins.

Moniliformin is a 1-hydroxycyclobutene 3,4-dione mycotoxin produced by various *Fusarium* spp. (*F. acuminatum*, *F. avenaceum*, *F. oxysporum*, *F. subglutinans*, *F. tricinctum*, *F. verticillioides*) on substrates such as maize, rice, cereals and millet. Natural contamination has been observed in low concentrations in wheat and barley mainly grown in Scandinavian countries (Jestoi *et al.*, 2004; Uhlig *et al.*, 2004) but was found to be at quite high levels in maize (Sharman *et al.*, 1991; Lew *et al.*, 1996). Whilst the toxin may be present in barley and malt and in even higher concentrations in maize-based brewing adjuncts, it has never been detected in beer. However, according to experiments performed by Pineda-Valdes and Bullerman (2010) moniliformin is fairly stable under the pH and temperature conditions prevailing during mashing. As a consequence, the fate of the toxin during the brewing process should be elucidated in future studies. Data on the toxicity of moniliformin is scarce and has been reviewed by Jestoi (2008). It was suspected to be involved in the development of human oesophageal cancer, possibly interacting with fumonisins. The relationship between moniliformin and a human cardiomyopathy known as the Keshan disease has been suggested (Zhao, 1993). Inhibition of glutathione peroxidase and glutathione reductase by the toxin in the myocardium seems to play an important role in the condition (Chen *et al.*, 1990).

Beauvericin and enniatins are cyclohexadepsipeptides produced by several species in the genus *Fusarium* (see Kulik *et al.*, 2007 for list of species) as well as by the entomopathogenic fungal species *Beauveria bassiana* and *Verticillium hemipterigenum* and by the mangrove fungus *Halosarpheia* sp. Several chemically different enniatins (A, A1, B, B1, B2, B3, B4, D, E, F, G, H and I) and one beauvericin derivative have been described so far. Enniatins A and B were frequently isolated from wheat and barley samples marketed in Spain (Meca *et al.*, 2010), Morocco (Mahnine *et al.*, 2011), Tunisia (Oueslati *et al.*, 2011), and Norway (Uhlig *et al.*, 2006). Recently, Bolechová *et al.* (2015) detected enniatins in all of 52 analysed barley and malt samples from the Czech Republic. Beauvericin has frequently been detected together with enniatins but its frequency of occurrence and the concentrations found are generally lower. Both types of cyclodepsipeptides are produced via non-ribosomal peptide synthesis using unique multifunctional enzymes, enniatin synthetase (Hornbogen *et al.*, 2002) and beauvericin synthetase (Peeters *et al.*, 1988). Recently, Hu *et al.* (2014) followed the fate of beauvericin and enniatins A, A1, B and B1 through the malting and brewing processes. Considerable amounts of all toxins were produced during green malt production with a similarly strong decrease after kilning. During brewing, considerable amounts of the toxins were eliminated from the process with the spent grains and after adsorption to yeast cells (Vaclavikova *et al.*, 2013). Whilst most of the toxins were eliminated, enniatins and beauvericin were present in the final beer in detectable concentrations. Meca *et al.* (2013) report degradation rates between 23% and 82% for beauvericin during the brewing process, which means that considerable levels will be present in the final beer. Moreover, since spent grain and yeast are often used as animal feed, problems with the toxins may arise in downstream areas of beer production.

Knowledge of the toxicology of enniatins and beauvercin is fragmentary and needs further elucidation. Both compounds were found to be cytotoxic (Calò *et al.*, 2004) and antibiotic (Dobler *et al.*, 1969). They act as ionophores in cell cultures leading to an inward calcium flux (Kamyar *et al.*, 2004, 2006). Beside their toxic effects at higher concentrations, enniatins showed profound apoptosis-inducing effects especially against various human cancer cell types at low micromolar concentrations (Dornetshuber *et al.*, 2007; Hyun *et al.*, 2009).

Fusaproliferin is a sesterterpene mycotoxin, production of which has been reported by strains of different *Fusarium* spp. within the *Gibberella fujikuroi* complex, with *F. anthophilum*, *F. guttiforme*, *F. proliferatum* and *F. subglutinans* being the major producers (Fotso *et al.*, 2002; Moretti *et al.*, 2007). The compound has been frequently observed to contaminate barley, wheat, maize and rice from different climatic regions and to be co-occurring with beauvericin and fumonisin B_1, however with lower frequency and concentrations (Ritieni *et al.*, 1997; Meca *et al.*, 2010; Zinedine *et al.*, 2011). Rubert *et al.* (2011) detected no fusaproliferin in any of 25 commercial beers analysed for the toxin. Recently, Ezekiel *et al.* (2015) observed a 99.4% reduction of fusaproliferin concentrations during the production of *kunu-zaki*, a traditional maize-based fermented alcoholic beverage produced in rural Nigeria. The result may indicate a behaviour similar to that of beauvericin and enniatins, which are widely eliminated from the brewing process with spent grains and yeast (Vaclavicova *et al.*, 2013).

Fusaproliferin is a teratogenic fungal metabolite that causes cephalic dichotomy, acrocephaly, and limb asymmetry in chicken embryos (Ritieni *et al.*, 1997) and is toxic to IARC/LCL 171 human B lymphocytes (Logrieco *et al.*, 1996). No human diseases have been connected with the toxin and no legal limits have been set up for it so far.

Alternaria toxins in beer production

Alternaria toxins [alternariol (AOH), alternariol monomethyl ether (AME), altenuene (ALT), altertoxins I, II, and III (ATX-I, -II, and -III), L-tenuazonic acid (TeA), *Alternaria alternata* lycopersici toxin (AAL)] have been found to be produced by cultures of a wide variety of *Alternaria* spp., including *A. alternata*, *A. solani*, and *A. tenuissima* as well as *Phoma sorghina* and *Pyricularia oryzae* (Ostry, 2008). Producing species occur on a wide variety of host plants, including wheat and barley, under different climatic conditions. *Alternaria* toxins represent four different groups of chemical structures: dibenzopyrone derivatives (AOH, AME, ALT), perylene derivatives (ATX-I, -II, -III), tetramic acid derivatives (TeA) and polyketides (AAL) (Bottalico and Logrieco, 1998). *Alternaria* species are regularly found to contaminate barley and wheat grains with incidences of up to 90% in a sample. Several surveys have elucidated the spectrum of *Alternaria* toxins produced by pure cultures isolated from wheat and barley. However, there seems to be an absence of investigations regarding the incidence of *Alternaria* toxins in wheat, barley and malts made thereof. Mycotoxins from *Alternaria* spp. have been detected in weathered wheat from China (Li and Yoshizawa, 2000). Only recently Müller and Korn

(2013) analysed German wheat over a period of 10 years and detected TeA in 30% of the samples with other *Alternaria* toxins being of lower frequency. When present in the malt, a considerable proportion of TeA seems to be transferred into the final beer as shown by Siegel *et al.* (2010a) who detected the toxin in 37 of 43 commercial beer samples in concentrations up to 175 μg/kg. The same authors observed only minor degradation of AOH, AME, and ALT during bread baking which would hint that degradation might be rather low during mashing (Siegel *et al.*, 2010b).

TeA has been shown to exert an inhibitory effect on the growth of mammalian cells. It has been used *in vitro* for the inhibition of human tumour cells (Kaczka *et al.* 1964). AME showed weak mutagenicity in the Ames test (Scott and Stoltz, 1980) and is teratogenic in tests with hamsters (Pollock *et al.*, 1982). Alternariol is activated by exposure to light. Under the action of UV light, cross-linking of double-stranded DNA was observed (DiCosmo and Straus, 1985). Mutagenic and carcinogenic properties of Alternariol and AME and participation in the development of laryngeal cancer in areas of China were suspected (Liu *et al.*, 1992). No specific limits or guidelines have been adopted for any of the *Alternaria* toxins.

Gushing

Gushing describes the spontaneous over foaming of a carbonated beverage upon opening of a bottle without previous agitation. The phenomenon has been observed in carbonated beverages such as beer (Christian *et al.*, 2011; Fischer, 2001), champagne (Kemp *et al.*, 2015) or sparkling juice drinks (Schuhmacher, 2002). The phenomenon is caused by the presence of high concentrations of hydrophobic condensation nuclei at which dissolved carbon dioxide will instantly turn into the gas phase upon pressure release during opening of the bottle, forming gas-filled bubbles that grow fast and rise upwards, leading to overfoaming of the liquid (Pellaud, 2002; Casey, 1996).

Gushing is a multicausal phenomenon and two types are commonly distinguished in regard to the causative factors involved (Amaha and Kitabatake 1981; Casey 1996). The term 'secondary gushing' is used for all technological factors, e.g. dust or other particulate matter, particles leaking from filter materials, or particulate calcium oxalate crystals (Carrington *et al.*, 1972), that cause introduction of nucleation particles into the bottled beverage. Secondary factors can typically be handled by modifying the brewing and filling process to result in exclusion of such particles or surfactants. Primary type gushing is exclusively related to the use of malt or unmalted cereals that have been infected by certain fungi during growth in the field or in the malt house (Gjertsen and Trolle, 1963). Based on assumptions about possible structure/effect relationships, Hippeli and Elstner (2002) were the first to publish speculations on a possible role of hydrophobins as gushing inducers in beer. Authors were obviously unaware of the fact that Haikara *et al.* (1999) had filed a PCT patent (WO 99/54725) already in 1999 (with priority to a national Finnish patent from 1998) in which hydrophobins were used as indicators for gushing in carbonated beverages (Haikara *et al.*, 1999). Today it has become a generally accepted doctrine that these extremely amphiphilic fungal proteins are responsible for the induction of primary gushing in beer (Sarlin *et al.*, 2005a; Garbe *et al.*, 2011; Specker, 2014).

Hydrophobins have been shown to be produced by a great variety of species within the filamentous fungi so that their production seems to be a general principle in that group (Talbot, 1997). One of their obvious natural functions is to decrease the surface tension of water by self-assembling at water/air interfaces, thus enabling transition of this barrier during the production of aerial mycelia (Wösten *et al.*, 1999; Cox *et al.*, 2007). Another role may be to establish proper contact between fungal cells and host tissue during plant infection (Wösten *et al.*, 1994; Kim *et al.*, 2005). All known proteins of that type share a common pattern of eight cysteines at conserved positions. Sequences between cysteines are however highly variable but the four-domain secondary structure always results in the formation of extremely amphiphilic proteins. Two subgroups of hydrophobins, class 1 and class 2, have been identified according to differences in spacing and sequence between cysteines, hydrophobicity patterns and solubility in organic solvents (Wessels *et al.*, 1994). Among the few hydrophobins available in purified form from transgenic *Pichia pastoris* cultures only those belonging to class 2 have been shown to induce gushing (Stübner *et al.*, 2010; Lutterschmid *et al.*, 2011; Niu *et al.*, 2012; Sarlin *et al.*, 2012). Hydrophobins enter the barley-to-beer chain upon use of fungal-contaminated brewing malt (Sarlin *et al.*, 2007). Contaminations with *Fusarium* spp. such as *F. graminearum*, *F. culmorum* or *F. poae* have been found to be highly correlated with gushing induction in the beer produced (Gjertsen *et al.*, 1965; Niessen *et al.*, 1992; Schwarz *et al.*, 1996; Sarlin *et al.*, 2005b).

The mechanism of action of hydrophobin-induced gushing as well as the way they interact with other promoting or inhibiting factors is still a matter of debate. Currently, a mechanism that is in accordance with the thermodynamic approach of the 'nano-bomb' theory described by Shokribousjein *et al.* (2011) and refined by Deckers *et al.* (2013) seems to provide many explanations for the phenomenon observed during primary gushing. The theory is in line with observations about interactions found between hydrophobins and the regular beer foam proteins nsLtp1 and Z4 (Stübner *et al.*, 2010; Specker *et al.*, 2014) as well as their interaction with lipophilic hop components (Gardner *et al.*, 1973; Lutterschmid *et al.*, 2010; Müller *et al.*, 2010; Shokribousjein *et al.*, 2014). Hydrophobin layers fulfil the basic assumptions made in the varying permeability model (for a review see Pellaud, 2002). Simulation of molecular dynamics of carbon dioxide condensation resulted in evidence for a clustering of CO_2 molecules at the hydrophobin's hydrophobic patch, thus supporting the interaction of CO_2 and hydrophobins (Deckers *et al.*, 2012a). According to the nano-bomb theory, small particles of 5–10 nm in diameter represent hydrophobin-coated CO_2 micro-bubbles. These develop during yeast fermentation, filling and shaking of bottles. At a critical diameter, the hydrophobic/hydrophilic monolayer hydrophobin film surrounding the bubble becomes impermeable and further shrinkage is prevented according to the varying permeability model. The resulting nano-bubbles were calculated to possess an internal pressure of about 4 bar (Deckers *et al.*, 2010, 2012b). During opening of the bottle, the gas–liquid equilibrium between beer and the atmosphere is abruptly misbalanced and nano-bubbles present in the beverage will expand explosively and CO_2 from the surrounding liquid phase will diffuse into the bubble leading to uncontrolled bubble growth (Pellaud, 2002). The rapid expansion of micro bubbles provides the energy which is used to break bonds between CO_2 and water molecules in the vicinity of an expanding bubble, hence the name 'nano bomb'. This eventually forces CO_2 molecules to transit from the water-soluble state into the gas phase by free diffusion and formation of unstabilized secondary gas bubbles that rise to the surface in masses resulting in gushing (Deckers *et al.*, 2010).

The model recently presented by Specker (2014) is in broad agreement with the nano-bomb model. However, this author demonstrated that addition of purified transgenic nsLtp1 to beer or carbonated water previously mixed with a gushing-inducing concentration of purified transgenic class 2 hydrophobin FcHyd5p from *F. culmorum* resulted in a significant decrease of gushing volumes compared with a nsLtp1-free control. Results obtained from atomic force microscopy analysis of mixed hydrophobin/nsLtp1 surface films suggested that nano-bubbles present in gushing beer may in fact be surrounded by mixed layers of amphiphilic proteins which tend to be more prone to disruption than films of either pure protein. Moreover, the author deduced that formation of secondary bubbles during gushing events comes from CO_2 nucleation at the hydrophobic inner layers of now exposed fragments of the disrupted bubble skin.

Detection and identification of fungi and fungal metabolites in cereals, malt, and beer

Physical, chemical, and affinity-based methods

Visual and olfactory examination of ingredients for signs of fungal contamination has been the principal method of quality assessment since humans first made beer. Even today with the availability of advanced instrumentation, experienced brewers and maltsters can recognize a low-quality barley or malt by its colour, smell and hand feel. Musty smells are an indication of attack by typical storage fungi, which in turn can be related to improper storage of barley or malt. Also, changes in grain colour may be indicative for darker beer colours resulting from a malt lot or even for the potential to induce primary gushing. Many German brewers use a method in which the number of red-discoloured kernels is counted in a malt lot. The 'relevant' grains show a discoloration typical for *Fusarium* contamination and malt lots with more than five such grains in

200 g will lead to rejection by the brewery (Niessen *et al.*, 1991; Niessen *et al.*, 1992; Engelmann *et al.*, 2012). The method is rapid and requires little effort to be implemented. However, it requires some experience to differentiate 'relevant' grains from 'non-relevant' red discolorations. Kernels may be mistakenly stained in shades of red by growth of other moulds, e.g. *Epicoccum nigrum*, or red yeast belonging to genera *Rhodotorula*, *Rhodosporidium*, *Sporidiobolus*, *Sporobolomyces*, and *Phaffia*.

In order to assess fungal contamination and fungal secondary metabolites in brewing cereals and malt, physical and chemical sensors are increasingly applied in modern quality control because results are more objective and reliable as compared to visual and olfactory analysis (Logrieco *et al.*, 2005). Visual and acoustic sensors make use of differences between contaminated and sound grains in regard to absorption or reflection of light or acoustic waves. Both types of analysis provide non-destructive measurement of fungal biomass and secondary metabolites in cereal samples. Acoustic wave sensors have been applied to the detection of trichothecene mycotoxins in wheat using either transmission of acoustic waves at frequencies of 5–36 kHz or reflection of an acoustic impulse in a frequency range 0 – 125 kHz (Juodeikiene *et al.*, 2004). Sensors have been developed for the detection of DON (Juodeikiene *et al.*, 2004, 2008) and DON in co-occurrence with T2-toxin/HT2-toxin (Juodeikiene *et al.*, 2011) in wheat samples. Available visual techniques use different parts of the light spectrum from UV/visible to near- and far-infrared. Levasseur-Garcia (2012) provides an overview of the application of infrared spectroscopy for the identification and detection of fungi and fungal metabolites on cereal grain. The method makes use of the fact that infrared (IR, 2500 nm to 25 μm) and near infrared (NIR, 760–2500 nm) light induces molecular vibration in organic molecules, which can be measured very sensitively. The analysis of resulting spectra both of acoustic wave-based analysis or IR- and NIR-based analysis uses multivariate, chemometric methods in which statistical correlations between the spectra and certain sample parameters such as mycotoxin concentrations or fungal biomass are established. In order to address such correlations specifically, the range of wavelengths has to be determined for each parameter at which a maximum correlation between deviation from a non-infected sample and the quantity of the assessed parameter, e.g. a mycotoxin or fungal biomass, can be observed. NIR spectroscopy has been applied to the assessment of DON concentrations concurrently with ergosterol and numbers of scabby grains as an indicator for fungal biomass in wheat (Dowell *et al.*, 1999) and barley (Roberts *et al.*, 1991; Börjessen *et al.*, 2007). Also chitin is a compound typically produced by fungi and yeasts that has been applied to the detection of mould contamination in barley and other food sources (Roberts *et al.*, 1991; Cousin, 1996). Since both parameters are prevalent in all fungal organisms, no species-specific detection of fungi in contaminated materials is possible with this method. Fourier-transformed infrared microscopy (FTIR microscopy) is a method using the same principles as described above but it is used to analyse pure cultures of microorganisms including moulds and yeasts and identify them at the species level by comparing sample spectra with reference spectra from a database (Santos *et al.*, 2010; Wenning and Scherer, 2013). In hyperspectral imaging the reflectance of a sample is analysed at various wavelength bands ranging from UV to NIR. For analysis, colour and light intensity of each pixel of the image is analysed for each wavelength band and differences between colour and light distribution in images of contaminated and non-contaminated reference samples are compared. Systems have been commercialized for brewing applications such as detection of *Fusarium* contamination in wheat (Delwiche *et al.*, 2011).

Chemical detection of fungal contamination and fungal secondary metabolites is a field that has been extensively studied for several decades and is widely used for the analysis of brewing raw materials and beer (Lattanzio *et al.*, 2009). Analytical protocols for all known mycotoxins have been extensively reviewed by several authors (Jarvis, 2003; Shephard, 2008; Rahmani *et al.*, 2009; Turner *et al.*, 2009). Protocols usually comprise sampling, sample preparation, extraction, cleanup, separation, detection and quantification of mycotoxins. Extraction from contaminated samples is performed using organic solvents or solvent mixtures of optimized polarity in order to separate the analytical target compound from the matrix and other compounds interfering with the analysis. Further cleanup by liquid-liquid extraction or solid phase extraction (reversed phase, ion exchange, immunoaffinity) can be applied to remove non-target compounds and to concentrate analytes. Thin-layer chromatography (TLC), high performance liquid chromatography (HPLC), gas chromatography (GC) and electrophoresis have been used as technical platforms to separate extracted analytes. Absorption of UV and visible light, fluorescence with or without derivatization as well as mass spectrometry are applied to detect the previously separated analytes. Identification of analytes has been accomplished by comparison of retention times with reference materials. HPLC and GC separation have been combined with mass spectrometry (LC-MS, GC-MS) for identification of individual compound peaks. Currently, the most sophisticated analytical systems use tandem mass spectrometry (LC-MS/MS) in which the first MS (MS1) is used to separate different compounds present in an LC-peak after electrospray ionization (ESI) and the second MS (MS2) provides further analysis of selected mass fragments from MS1 for identification. Analysis of each sample in positive and negative ionization mode enables detection and quantification of >130 different secondary metabolites including frequently occurring mycotoxins and their glycosylated derivatives (Vishwanath *et al.*, 2009; Streit *et al.*, 2013). Using an LC-MS/MS-based method that could detect 15 different mycotoxins in parallel analysis, Tamura *et al.* (2011) analysed samples of beer and beer-based drinks from the Japanese market and found NIV, DON and fumonisins in low concentrations. Romero-Gonzalez *et al.* (2009) used an LC-MS/MS-based system to detect 12 different mycotoxins in beer. Analysis of a small set of commercial samples revealed occurrence of T2- and HT-2 toxins, aflatoxin B_1, and fumonisin B_2 in low concentrations in some samples. Zachariasova *et al.* (2010) developed a multimycotoxin method for the screening of 32 different compounds in beer but did not show results of sample analyses. Quantification of analytes in all the chemical detection methods described above is possible by calibration with external standards. Calibration with internal standards has been demonstrated for several mycotoxins with isotopically labelled derivatives in stable isotope dilution assays (SIDA) (Rychlik and Asam, 2008).

Affinity-based detection of mycotoxins has relied mainly on the use of antibodies or fragments thereof. However, recently DNA-based aptamers and molecularly imprinted polymers (MIP) have been developed and applied for the specific analysis of mycotoxins (Maragos *et al.*, 2009b). Aflatoxin B_1 (Ma *et al.*, 2014), aflatoxin M_1 (Malhotra *et al.*, 2014), DON (Eifler, 2014), Fumonisin B_1 (McKeague *et al.*, 2010), OTA (Cruz-Aguado and Penner., 2008), T2-toxin (Chen *et al.*, 2014) and zearalenone (Chen *et al.*, 2013) can now be detected on the basis of single-stranded DNA oligonucleotide aptamers of 50–100 nucleotides in length which form a three-dimensional structure specifically upon contact with the analyte molecule (Sampson, 2003). MIP form hollow three-dimensional structures in which the target molecule can be trapped specifically by mutual electrostatic interaction of side groups (Haupt and Mosbach, 2000). They provide solid-phase materials with a molecular memory for the target compounds and have been utilized in solid-phase extraction of and sensors for the detection of fumonisin analogues (de Smet *et al.*, 2009), DON and zearalenone (Weiss *et al.*, 2003), OTA (Baggiani *et al.*, 2001) and moniliformin (Appell *et al.*, 2007).

Antibody-based immunochemical detection methods are based on the stereo-specific binding reaction between the variable domain of an immunoglobulin (antibody) and its antigen. Extensive information about development and use of immunoglobulins is available in specialized literature and in textbooks (Owen *et al.*, 2013; Murphy, 2012; Delves *et al.*, 2011; Subramanian, 2004; Campbell, 2000).

Various immunochemical assays have been described for the detection of fungal mycelia. However, the largest part of the literature covers detection of clinically relevant species. A smaller number of publications deals with the analysis of phytopathogenic fungi and other plant-associated species. The following literature review describes immunochemical assays that have been set up for the analysis of fungi that have been described to grow on cereals or cereal-based foods. The monoclonal antibodies and polyclonal antisera described have not necessarily been described for analysis of barley and wheat or malt produced therefrom but the methods used can in principle be used for that purpose after proper modification of the sample preparation protocol. Notermans and Heuvelman (1985) were the first to develop specific antisera for the detection of foodborne moulds. Their antisera were raised by immunization of rabbits with freeze-dried preparations of protein precipitates from cultures of *Mucor racemosus, Fusarium oxysporum,* and *Penicillium verrucosum.* The enzyme-linked immunosorbent assay (ELISA) assay set up with the *P. verrucosum* specific antiserum was shown to detect most *Penicillium* species (Notermans *et al.*, 1986). Kamphuis *et al.* (1989) developed an immunochemical latex agglutination assay for the rapid detection of a broad spectrum of foodborne *Aspergillus* and *Penicillium* species. Later, Dewey *et al.* (1990) used antibodies specific for *Penicillium islandicum* to set up an immunological test strip format for the analysis of rice samples. Tsai and Cousin (1990) published an ELISA for the simultaneous detection of different moulds such as *Aspergillus versicolor, Cladosporium herbarum, Geotrichum candidum, Mucor circinelloides* and *Penicillium chrysogenum* in yoghurt and cheese. An even wider range of grain-associated fungi, including *Penicillium* spp. and *Aspergillus* spp. as well as typical field fungi such as *Septoria* spp. and *Fusarium* spp., can be detected with the assays described by Banks and colleagues (Banks *et al.*, 1994, 1996). *Penicillium aurantiogriseum* was detected in cereals by Lu *et al.* (1994) in a highly specific manner using a monoclonal antiserum against the fungus. Chang and Yu (1997) developed a rapid immunochemical detection method for *Aspergillus parasiticus* and *Penicillium citrinum* and applied it to the analysis of rice and maize. Tsai and Yu (1999) used the same antiserum for the analysis of cereals. The detection of mycelia of aflatoxin-producing fungi in cereals and foodstuffs using an immunochemical method was described by several authors (Shapira *et al.*, 1997; Yong and Cousin, 2001). Immunochemical methods have also been developed for the detection of *Fusarium* contaminations in cereals. A polyclonal serum for the collective determination of *F. avenaceum, F. culmorum* and *F. graminearum* contamination was obtained after immunization with the supernatant of a still culture of *F. culmorum* by Beyer *et al.* (1993). An even broader spectrum of detected *Fusarium* species was reported by Iyer and Cousin (2003), who applied the assay to the analysis of grains and food. Rohde and Rabenstein (2005) reported similar results with polyclonal antisera produced in rabbits against mycelia of *F. graminearum* and *F. culmorum*, respectively, which they applied for the analysis of wheat grains. Banks *et al.* (1996) described the use of a monoclonal antibody for the specific detection of *F. avenaceum* in cereals. The use of polyclonal antibodies raised in chicken eggs for the detection of *F. poae* was described by Gan *et al.* (1997). In an attempt to detect infection of maize with fumonisin-producing *Fusarium* spp., Meirelles *et al.* (2006) set up an ELISA-based immunoassay using a polyclonal antiserum raised against an unknown peptide from a *F. verticillioides* culture. Finally, Meyer and Dewey (2000) used a culture supernatant to raise monoclonal antibodies for the detection of *Botrytis cinerea* in a wide variety of plants.

Even more importance has been given to the immunochemical analysis of mycotoxins as harmful fungal secondary metabolites in brewing cereals and malt. Zheng *et al.* (2006), Maragos (2006) and Maragos and Busman (2010) provide comprehensive reviews of the literature describing classical and modern methods of immunochemical mycotoxin analysis. Assays based on ELISA have been described for the detection of a great variety of different mycotoxins, including those described previously in section, 'Mycotoxins'. Those assays are commercially available for application in brewing cereals, malt, and beer after appropriate sample preparation (Rahmani *et al.*, 2009). This type of assay is based on a competitive reaction between free and solid-phase-bound mycotoxin for binding to an enzyme-labelled specific antibody. Usually reactions are performed in 96-well microtitre plates as the solid phase. Following specific binding, unbound analyte and antibodies are removed by washing and detection of the binding event is performed after addition of a chromogenic enzyme substrate (Turner *et al.*, 2009). The developed colour is measured spectrophotometrically and light absorption is inversely proportional to analyte concentration. The major advantages of ELISA are speed, low-cost and user-friendliness, because it is portable and easy to perform even under on-site conditions in the brewery (Pleadin *et al.*, 2012). Numerous variations of the ELISA format have been described, differing mainly in enzyme labels and substrates used and in the choice of the solid phase-bound reaction partner.

Other variations of immunoassays involve the use of fluorochrome-labelled mycotoxins or antibodies enabling direct signal detection.

Fluorescence polarization immunoassays do not involve solid phase binding of components. They have been set up for many important mycotoxins. The method measures the change in fluorescence intensity in a solution upon binding of a fluorescently labelled antigen to a specific antibody in relation to the concentration of free antigen in a sample, thus allowing its direct quantification with high sensitivity within minutes (Maragos, 2009a). All major mycotoxins potentially occurring in raw materials and beer can be detected with this method even though no application has been described so far for that specific purpose.

Another variation of the immunosorbent assay is a technology in which the specific antibody or the antigen is immobilized on the surface of a membrane and brought into contact with the free antigen in a sample solution. The method is marketed as a lateral flow device (LFD) assay and is available for the detection of all important mycotoxins, including those occurring in the barley-to-beer chain (Anfossi *et al.*, 2013). Indirect and direct assay formats provide the carrier-bound mycotoxin or the mycotoxin-specific antibody, respectively, immobilized on a nylon membrane. Upon sample application, the free mycotoxin binds to a specific antibody bound to gold nano particles (GNP) in the indirect format or is just mixed with a GNP-bound conjugate of the toxin in direct format. As the sample fluid moves along the LFD membrane by capillary force, hitherto unreacted antibodies are retained by binding to the immobilized antigen in the indirect assay or free and GNP-bound toxin molecules compete for binding to the immobilized antibody in direct assays. In both cases the immobilized GNP result in formation of a red-coloured line, the intensity of which is inversely proportional to the concentration of free mycotoxin in a sample. The same principle has recently been used for the parallel analysis of aflatoxins, DON, and zearalenone in a multiplex assay (Song *et al.*, 2014).

Apart from the aforementioned mycotoxin assays, in which the affinity-based binding event is transduced visually, mostly as a colour change, several other principles of biosensoric signal transduction have been applied to the study of mycotoxins (Pohanka *et al.*, 2007). Electrochemical transducers measure electron movement (potentiometric), current change (amperometric) or changes in conductivity (conductometric) occurring due to the binding event or due to activity of a label enzyme. Optical biosensors use optical phenomena occurring due to binding of affinity molecules such as antibodies or aptamers to a glass surface or a gold-coated glass surface, the light reflective properties of which are influenced by the binding event. Surface plasmon resonance (SPR)-based sensors have been used to detect and quantify the most important mycotoxins occurring in food (Li *et al.*, 2012). Fibreoptic or optical waveguide biosensors make use of the induction of an evanescent light wave when light is totally reflected at the inner surface of a glass fibre. Provided the evanescent wave has got the right wavelength, it can be absorbed by fluorophore molecules in close proximity to the surface of the fibre and induce fluorescence, which can be measured by coupling back into the fibre (Maragos and Thompson, 1999). Biosenors based on the technology have been developed for aflatoxins and fumonisins (Thompson and Maragos, 1996; Maragos and Thompson, 1999).

Microbiological methods

Cultural methods for the detection and quantification of mould propagules in food and food raw materials have long been used (Jarvis *et al.*, 1983; Beuchat, 1987; Gourama and Bullerman, 1995; Pitt and Hocking, 2009; Samson *et al.*, 2010). They make use of the fact that viable moulds and yeasts can be detected and counted after cultivation of a sample on microbiological culture media, either as numbers of contaminated individual particles or as colony-forming units. In addition to mere detection and enumeration of fungal contamination, exact identification of the species is of prime importance because species identification may provide an indication of possible quality problems associated with the investigated sample. This is particularly important when assessing a possible mycotoxin contamination. Microbiological testing methods have the advantage that they can be easily handled in the laboratory without much equipment, other than a microscope. They can be applied to a variety of different materials to be tested.

The microbiological growth media applied in fungal analysis of food can be divided into general growth media for detection of a wide range of fungi and yeasts and selective media allowing growth of a restricted number of species. Besides the choice of the medium, selective growth of certain species or groups of species can be achieved by the choice of proper incubation conditions. As an example, an incubation temperature of 37°C is applied for the selective cultivation of human pathogenic yeasts from food samples using a non-selective culture medium such as malt extract agar or YPG where the high incubation temperature leads to exclusion of mesophilic yeasts, most of which are non-pathogenic to humans. Also for the investigation of the presence of heat-resistant moulds in pasteurized foods, the selection is done through a pre-treatment of the sample at 70°C before incubation of cultures at elevated temperatures rather than by the use of a selective growth medium (Pitt and Hocking, 2009). Today, non-selective detection and enumeration of a broad spectrum of moulds and yeasts from food sources and from the air is routinely done on Dichloran Rose Bengal Chloramphenicol agar (DRBC agar, King *et al.*, 1979). The medium is selective for ascomyceteous and basidiomycetous filamentous and yeast fungi but growth of bacteria and Zygomycetes is largely suppressed by additives. For the selective investigation of xerophilic fungi as an important group of food-borne and airborne fungi, Dichloran 18% Glycerol agar is often used (DG18 agar, Hocking and Pitt, 1980). DRBC and DG18 have been certified by the International Organization for Standardization (ISO) for enumeration of yeasts and moulds in high water activity food and animal feedstuffs (a_w > 0.95, ISO 21527-1) and low water activity foods and animal feedstuffs (a_w < 0.95, ISO 21527–2), respectively. However, both culture media are better suited for general growth and enumeration than for identification of fungi because they do not properly develop the micro- and macro-morphological features that are needed for identification. As a consequence, subcultures have to be prepared on optimal media in order to identify an isolate to the species level. In regard to morphological identification of species, it is particularly important that the morphological features of the moulds are properly expressed. Therefore, different media are used for enumeration, isolation, or identification. Complex media such as malt extract agar, oatmeal agar, maize meal agar, potato dextrose agar are often used for cultivation but also synthetic media, such as SNA have been used and are superior for some groups of fungi. Some species or genera may also need specific media such as clove leaf agar, Czapek yeast autolysate agar or cherry decoction to form the characteristic structures that are important for their identification. Many species have to be incubated at different growth conditions or incubated under UV light or in darkness in order to bring about the typical morphological structures needed for their identification.

In studies focusing on detection and enumeration of single fungal species or groups of physiologically similar species, selective media may provide a tool to circumvent extensive isolation for species identification. Selective growth conditions are created both by specifying certain growth parameters such as a_w, pH, salinity or sugar content. Moreover, the selection of certain fungi on the growth medium can also be affected by addition of substances that kill unwanted organisms, or at least strongly inhibit their development. An example for this is addition of substances such as iprodione (Abildgren *et al.*, 1987), dichloran (Andrews and Pitt, 1986; Conner, 1992), rose Bengal (Newhouse and Hunter, 1983) or pentachloro-nitorbenzole (PCNB, Nash and Snyder, 1962; Nishikawa and Kohgo, 1975; Gyllang *et al.*, 1981; Burgess *et al.*, 1988) to general growth media such as Szapek Dox agar in order to create selective conditions for the examination of *Fusarium* infestation in cereals. *Fusarium* spp. are highly resistant to both compounds and can therefore be selectively enumerated in a genus-specific manner. However, isolations still have to be made for species identification. Based on PCNB agar, further development led to species-selective media that can be applied to investigate either *F. graminearum* (mannitol-PCNB agar, Böhm-Schraml *et al.*, 1993) or *F. culmorum* (malachite green agar, Böhm-Schraml, 1995) in brewing cereals and malt. The use of dichloran rose bengal yeast extract sucrose agar (DRYS, Frisvad, 1983) or dichloran yeast-extract sucrose 18% glycerol agar (DYSG, Elmholt *et al.*, 1999) proved to be a valuable tool in the microbiological analysis of food commodities for mycotoxin-producing *Penicillium* spp. The latter medium is even useful in the selective differentiation of *Penicillium verrucosum* from *P. nordicum*, both producers of ochratoxin A. AFPA agar is useful for the selective identification of aflatoxin-producing species such as *Aspergillus flavus* and *A. parasiticus*. Aflatoxinogenic species can be recognized by their typical orange-red colour when colonies are observed from their reverse (Pitt *et al.*, 1983).

Molecular biological methods

A disadvantage of using microbiological methods routinely for detection, enumeration and identification of foodborne moulds and yeasts is the long period of 5–14 days necessary for the investigation. Moreover, as extensive knowledge of fungal morphology and systematics is necessary for a qualitative assessment, a high degree of specialization is needed for the analysis. Also, the fact that microbiological analysis can only detect living fungal mycelia and yeasts may appear as a problem. Sample materials in which fungal propagules are not viable or not able to be cultivated on the media used cannot be studied. However, many of the materials tested in food mycology have undergone longer periods of storage or have been processed before analysis and will therefore contain no or only few living propagules. Especially for processed materials, the knowledge of the microbiological history of a sample is of great importance for the assessment of potential consumer risks associated with a sample. Methods based on the detection and analysis of DNA or RNA circumvent many of the mentioned problems because they have the advantage of being much faster and highly specific. Moreover they and can be applied for the detection of both living and dead organisms.

PCR-based methods

The right column in Table 8.1 gives a list of publications describing PCR-based detection for fungal species frequently encountered on raw and malted seeds of barley and wheat. The list of publications does not reflect the complete literature available but has been restricted to cite the original descriptions of a primer pair or, in cases where several different primer pairs have been described for the same gene in a species, the first publication for each of the different genes is shown. The list of publications clearly shows that almost all species are covered by the availability of at least one pair of species-specific primers. *Fusarium anthophilum*, *F. camptoceras*, *F. sacchari*, *Gonatobotrys simplex*, *Harzia acremonioides*, *Nigrospora sphaerica* and *Ramichloridium schulzeri* are the only species among the 41 species and species groups listed in Table 8.1 for which no records of a species-specific PCR-based detection assay were found in the literature. The reason for this failure of a specific PCR-based detection system may be either their highly infrequent occurrence, the fact that they have very low importance as pathogens or food spoilage organisms, or the unavailability of a suitable sequence source. It is interesting to observe that many of the species can be specifically detected by primers covered by a patent (Haugland and Vesper, 2000) that was granted to the US Environmental Protection Agency (US-EPA) in 2002. The patent documentation contains all the primer and probe sequences needed to set up TaqMan®-based detection assays for more than 130 fungal species and some important groups of fungal species. Although licenses for the commercialization of primers have been granted to companies in the USA, Germany and United Kingdom the sequences are available on the US-EPO homepage under www.epa.gov/microbes/moldtech.htm#primers for non-commercial use. All primers and probes listed by that source have been set up to bind to sequences of genes coding for genomic or mitochondrial ribosomal RNA in the target organisms, including the ribosomal RNA genes, the internal transcribed spacers (ITS), the intergenic spacers (IGS) and the non-transcribed spacers (NTS) present in such genes. Similarly, various of the listed references have used that sequence source for primer design. However, other sequence sources, such as universal genes coding for proteins with cellular housekeeping functions (calmodulin, β-tubulin, elongation factor 1α, chitin synthase, topoisomerase), functions in fungal reproduction (mating type genes *MAT1*, *MAT2*, *MAT3*) or universal mitochondrial genes (*cytb*, *cox1*, *cox2*) have been applied for primer design. Several authors used genes coding for enzymes or regulatory proteins involved in the production of secondary metabolites or other enzymes and structural proteins, which turn out to be characteristic for the target fungus to develop species-specific PCR primers, e.g. mycotoxin biosynthetic pathway genes, alkaline protease, chitinase, actin or histone 3. Niessen *et al.* (2008) reviewed the PCR-based methods available for the diagnosis of mycotoxin-producing fungi as the most important group in terms of hygiene in brewing cereals and malt. Authors showed that PCR-based diagnostic assays have been developed for the vast majority of mycotoxigenic fungi potentially occurring on brewing cereals and malt.

LAMP-based methods

Apart from enzyme-free self-templated amplification systems currently under development (Dong *et al.*, 2012; Michaelis *et al.*, 2014; Jung and Ellington, 2014), a variety of different methods for isothermal enzymatic *in vitro* DNA amplification have been developed over the past 20 years (see reviews by Gill and Ghaemi, 2008; Fakruddin *et al.*, 2013; Yan *et al.*, 2014; Li and Macdonald, 2015). Most isothermal amplification systems have the advantage of being easily operated with simple equipment since no thermal cycling is necessary as in PCR. Moreover, they are useful tools in point-of-care (POC) applications in clinical settings making them highly attractive for the diagnostic industry. Loop-mediated isothermal amplification (LAMP) is an approach to nucleic acid amplification that is especially suitable due to its high specificity, rapidness, user friendliness and low price. Niessen (2015) reviewed its application for the diagnosis of filamentous fungi and yeasts. The method has been applied for the species-specific diagnosis of *Fusarium graminearum* (Niessen and Vogel, 2010) and *F. tricinctum* (Niessen *et al.*, 2012) in wheat and barley as well as for the group-specific detection of gushing-inducing *Fusarium* spp. (Denschlag *et al.*, 2012, 2013) and producers of trichothecene mycotoxins (Denschlag *et al.*, 2014) in cereals and malt. Moreover, is has been applied to the detection and identification of *Saccharomyces* brewing yeasts and wild yeasts in beer and other sources (Hayashi *et al.*, 2007, 2009).

The method relies on auto-cycling strand displacement DNA synthesis performed by thermophilic DNA polymerases under isothermal conditions with a set of four specifically designed primers. These hybridize to six different parts of the target DNA sequence (Notomi *et al.*, 2000). A comprehensive explanation of the reaction mechanisms involved can be found in the literature (Notomi *et al.*, 2000; Tomita *et al.*, 2008; Niessen, 2015). Fig. 8.2 shows a schematic representation of the different reaction steps leading to DNA synthesis during LAMP. The method basically makes use of the large fragment of the *Bst* DNA polymerase from *Geobacillus stearothermophilus*. The large fragment of the enzyme contains the 5′→3′ polymerase activity but lacks 5′→3′ exonuclease activity. Similar enzymes (Bsm, GspM, GsM 2.0, GspSSP) from other bacterial hosts as well as optimized versions of the original *Bst* polymerase (*Bst* 2.0, Bst warmstart) are now commercially available (Chander *et al.*, 2014; Woźniakowski and Samorek-Salamonowicz, 2014; Kang *et al.*, 2014). The *Bst* DNA polymerase large fragment displaces third-strand DNA with high efficiency during primer-initiated polymerization of new DNA, leaving a double-stranded product and a single-stranded DNA strand, which can act as the matrix for further primer annealing and DNA polymerization.

Since *Bst* DNA polymerase has a very high activity, vast amounts of high-molecular-weight DNA are produced within a short time. The exceptionally high specificity of LAMP is because a set of four primers with six binding sites must hybridize correctly to their target sequence before DNA biosynthesis occurs. A third pair of primers (loop primers) can be added optionally to the reaction in order to further amplify the amount of DNA produced during LAMP (Nagamine *et al.*, 2002). One of the primer pairs is constructed in such a way that the reverse complement of a binding site downstream of the F2c/B2c binding site (F1c/B1c) is attached to the 5′-end of a primer binding to that site. These composite primers are essential for the specificity of the amplification reaction and thus have to be chosen very carefully. Both parts of each FIP/BIP primer should be checked for cross-reactivity by *in silico* analysis (e.g. BLAST) prior to application in LAMP reactions. A pair of outer primers (F3/B3) anneals upstream of the F2c/B2c binding site to displace the initial LAMP product strand from the DNA matrix. Specificity of outer primers can be regarded as being of lower importance since they are not involved in any of the following amplification reactions and a low number of base mismatches will not prevent amplification. The process is initiated by attachment of primers to the DNA target. Primers are elongated and the second matrix strand is displaced from the target DNA. The newly synthesized product itself is displaced from the matrix strand by the F3/B3 product strand. Primers F3 and B3 have no further function once the amplification process has been initiated. As the final product of the amplification initiation step, a dumbbell-structured, single-stranded DNA is formed by hybridization of both

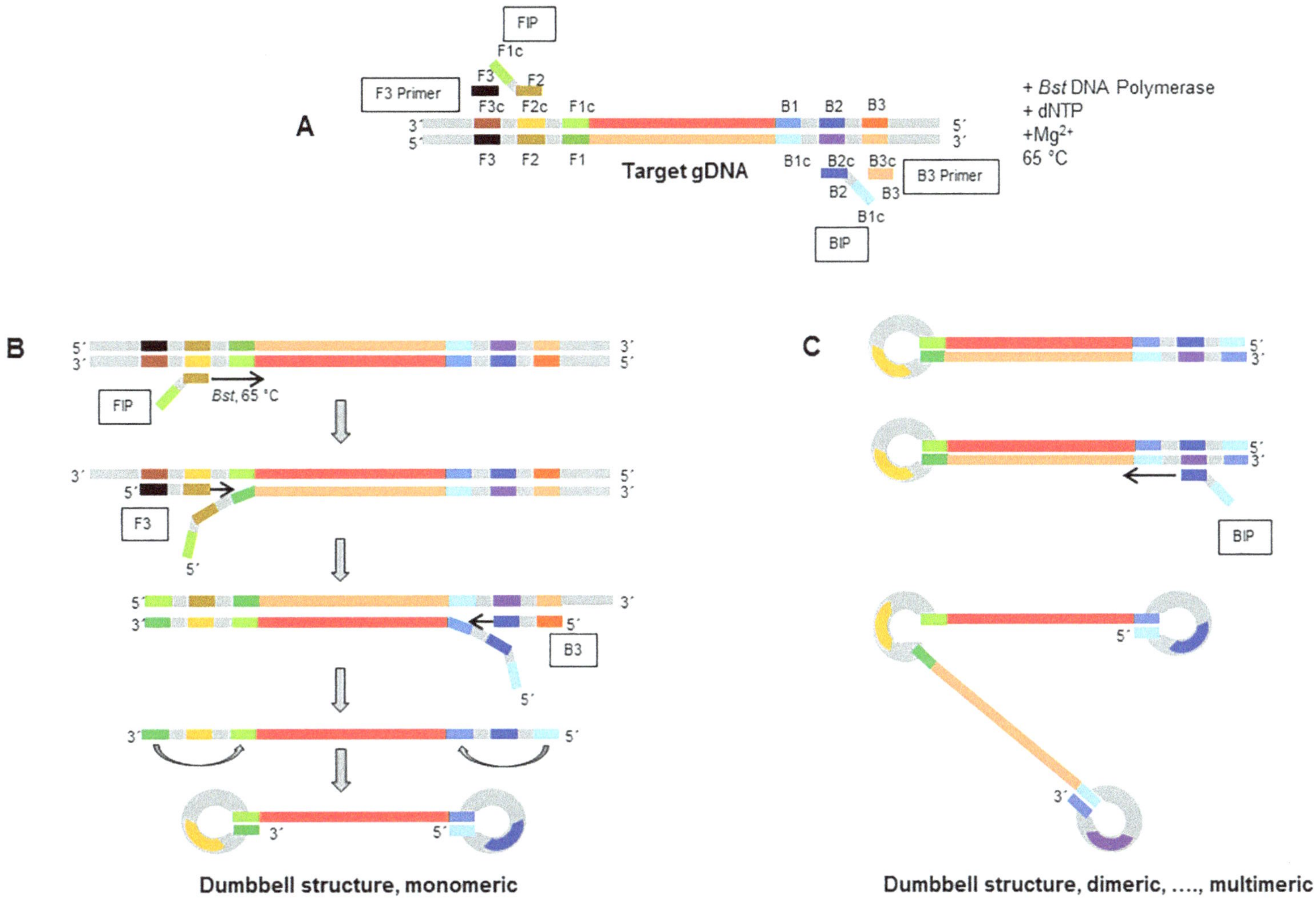

Figure 8.2 Schematic representation of the LAMP reaction. A, Primers, binding sites, and reaction conditions. B, Initiation of the LAMP reaction resulting in production of a double-loop stem structure (dumbbell structure). C, Autocycling enzymatic DNA amplification during LAMP resulting in multimers of different size of the monomeric double-loop stem structure. Redrawn from Niessen (2015), with kind permission of Springer-Verlag, Heidelberg, Germany.

ends of the molecule to complementary downstream sequences, forming two loops. Starting from this structure, primers FIP and BIP continuously hybridize to newly generated binding sites and are elongated, displaced and refolded while forming ever-longer multimers of the basic dumbbell structure. Loop primers are designed to hybridize to the single-stranded loop structures present in the dumbbell structures as well as in the multimeric DNA formed during autocycling DNA amplification. They prime the production of novel template DNA to which FIP/BIP primers can bind to initiate synthesis of even higher concentrations of DNA. Addition of loop primers therefore does not increase the sensitivity of amplification but rather enables earlier detection of a LAMP signal as compared with a reaction run without loop primers.

Direct detection of amplification in LAMP can be done by addition of DNA intercalating dyes (SYTO 9, SYBR Green 1, ethidium bromide) or fluorescent hybridization probes. Indirect detection is accomplished via Mg-pyrophosphate turbidity or calcein fluorescence (see reviews by Niessen *et al.*, 2013; Niessen, 2015). Quantification of template DNA concentrations is possible but not very accurate due to the autocycling nature of the amplification reaction (Denschlag *et al.*, 2013; Niessen, 2015). Beside its speed and ease of use, robustness is another major advantage of LAMP assays over PCR. It has been demonstrated that the reaction is quite insensitive against inhibitors from the sample matrix (Kaneko *et al.*, 2007; Francois *et al.*, 2011). Simple procedures for sample preparation are therefore sufficient in many cases to obtain a signal after addition of mycelia or fungal spores just washed off cereal or malt grains directly to the LAMP reaction mix (Luo *et al.*, 2012, 2014; Denschlag *et al.*, 2014). Application of the LAMP method for the analysis of brewing cereals and malt was demonstrated by Denschlag *et al.* (2012, 2013) who designed primers that detected the *hyd5* gene coding for the class 2 hydrophobin Hyd5p in *Fusarium* spp., which have been associated with gushing in beer. The assay detected *F. cerealis*, *F. culmorum*, and *F. graminearum* and had a detection limit of three *Fusarium*-contaminated grains in 200 g. Analysis of gushing-positive and gushing-negative malts revealed good correlation with gushing-test results (modified Carlsberg's test) and showed that the latter test seems to overestimate gushing potential. The same authors developed another LAMP-based assay detecting *tox5* and *tox6*, two genes involved in the production of trichothecene mycotoxins in *Fusarium* spp. (Denschlag *et al.*, 2014). The assay was applied to the analysis of wheat. LAMP results corresponded well with the presence of DON in respective samples at threshold values of 163 ppb and 1000 ppb when DNA extraction or a simple lavage of the samples was used for sample preparation, respectively. Niessen *et al.* (2012) demonstrated the usefulness of LAMP for the detection of *F. tricinctum* in barley samples and in single barley grains by immersing single seeds into the LAMP master mix prior to the reaction.

Control of fungal contamination in brewing cereals

The previous sections have discussed problems arising from fungal contamination of brewing cereals and malt, and the opportunities for detection and identification of fungi and their metabolic products. However, for maltsters and brewers the prevention of contamination or at least the prevention of fungal growth and release of secondary metabolites would be a preferred goal. Basically, two different strategies can be applied to reach the goal of minimizing the adverse effects of filamentous fungi and yeasts on the quality of malt and beer: prevention of fungal contamination of raw materials and prevention of fungal growth during storage and malt production.

Prevention of fungal contamination of raw materials

As described earlier in this chapter, the interaction between fungal contaminants and the cereal plant is highly complex, influenced by a variety of physiological and environmental factors. Plant responses to different stresses are highly complex and involve changes at the transcriptome, cellular, and physiological levels both in plant and fungus (Atkinson and Urwin, 2012). Infection of plant tissues by a fungus is only possible if the plant has no mechanisms to defend itself against the attack, i.e. a compatible interaction. In incompatible interactions the plant has an appropriate mechanism in place that prevents it from being attacked by the fungus. Such mechanisms can be directed to a particular pathogen in a highly specific way, e.g. sacrificing infected tissue areas in which cells die shortly after the attack, depriving the fungus of nutrients and water (hypersensitive reaction, suicide mechanism) or more generalized to fungal attackers by forming mechanical barriers such as the cuticle (Martin, 1964), production of low-molecular-weight antifungal compounds (phytoanticipins, phytoalexins) (Morrissey and Osbourne, 1999), antifungal proteins and peptides (De Lucca *et al.*, 2005), protease inhibitors (Ryan, 1990), production of chitinases, β-1,3-glucanases and other hydrolytic enzymes (Li *et al.*, 2001), just to name a few (see review by Heitefuss, 2001). Establishing or improving specific or unspecific defence mechanisms against fungal attack and spreading of fungal mycelia in plants is the aim of cereal plant breeding. Since the devastating head blight epidemics of the 1990s and ensuing years, resistance to *Fusarium graminearum* and *F. culmorum* as well as reduction of DON accumulation has become a major goal in barley and wheat breeding (Mesterhazy, 2014; McMullen *et al.*, 2012) since infections have a high economical as well as human and animal health impact (Windels, 2000).

In addition to plant breeding, agrochemicals have long been used to combat fungal growth in the field. Compounds are applied either as a seed treatment to protect the growing plant from attack by soil borne fungi or by spraying plants during the vegetation period. No chemical treatment is available to date providing general protection from fungal attack or general reduction of fungal growth and spreading. However, specific treatments for particular pathogens are available and widely used to prevent yield loss. Some fungi of specific relevance to the brewing industry such as *Fusarium graminearum* are difficult to control in the field. However, recent research has allowed some improvements. Salgado *et al.* (2014) demonstrated that integration of growing moderately resistant varieties, application of a mixed treatment with tebuconazole and prothioconazole (also metconazole, Tateishi *et al.*, 2014) as well as the use of appropriate harvesting equipment and settings resulted in a significant reduction of DON concentrations at increased grain yields and thus in reduction of discounts from wheat prices. Similar effects can be anticipated for the barley-to-malthouse chain. Crop rotation is an additional technical measure that can be taken in order to reduce the presence of pathogens and quality-reducing fungi in cereal crops (Marburger *et al.*, 2015). It is well established now that planting maize prior to wheat or barley, or wheat prior to barley, will accumulate *Fusarium graminearum* inoculum in the soil for infection of the cereals during germination, resulting in an increase of DON concentrations and gushing potential of malts (Bilikova and Hudec, 2014). Microorganisms have been investigated as control agents for cereal diseases caused by *Fusarium* species in the field, including FHB (Khan and Doohan, 2009) and seedling blight (Khan *et al.*, 2006). The application of antagonistic microbial cultures in the field has proven to be efficient in reducing *Fusarium* spp. and other fungal contaminations (Weller, 1988; Kiss, 2003; Junaid *et al.*, 2013).

Prevention of fungal growth during storage and malt production

The environmental conditions prevailing during long-term storage of brewing cereals are essential for growth of fungi and for their metabolic activity. Availability of water and oxygen are the most important growth-limiting factors alongside temperature, pH, and availability of nutrients (Christensen and Kaufmann, 1965; Lacey, 1989; Doohan *et al.*, 2003; Magan and Aldred, 2007). Also, the production of mycotoxins during storage of cereals is influenced and limited by environmental factors (Magan *et al.*, 1984, 2010; Miller, 1995; Schrödter *et al.*, 2004). Reduction of mycotoxin levels during storage of barley and other cereals has been achieved by controlling environmental conditions, including water activity (Magan and Aldred, 2007), and by addition of natural substances such as essential oils (Paster *et al.*, 1995; Juglal *et al.*, 2002; Chulze, 2010) or microbial starter cultures. Yeast strains (Björnberg and Schnürer, 1993; Petersson and Schnürer, 1995; Ädel-Druvefors and Schnürer, 2005), bacteria (Frändberg and Schnürer, 1995) or even fungi (Jensen *et al.*, 2000) have been added to stored grain and were demonstrated to have a reducing or at least preserving effect on the fungal community.

Malting is a process in which a complex ecosystem evolves due to the prevailing moisture and temperature conditions, thus allowing contaminating microorganisms to develop and to have a negative influence on the quality of malt (Laitila, 2007; Laitila *et al.*, 2007; Noots *et al.*, 1999; Wolf-Hall, 2007; Raulio *et al.*, 2009). The fungal community and other microorganisms growing during malting compete with grain metabolism for oxygen and may therefore considerably reduce grain germination during malting (Doran and Briggs, 1993; Noots *et al.*, 1999). The steeping step as part of the malting process strongly promotes growth of bacteria, yeasts, and fungi with the establishment of stable biofilms on the grain surface (Laitila *et al.*, 2007; Raulio *et al.*, 2009). Beside fungal growth, significant increases in levels of several mycotoxins can occur during malting (Vegi *et al.*, 2011; Oliveira *et al.*, 2012). As shown in Fig. 8.1, several *Fusarium* species were demonstrated to proliferate from steeping through germination until early stages of kilning (Oliveira *et al.*, 2012; Sarlin *et al.*, 2005b; Vegi *et al.*, 2011). *Fusarium* mould depletes grain nutrients, such as starch and protein, and colonizes its interior via exo-proteolytic and cellulolytic enzymes (Kang and Buchenauer, 2000; Oliveira *et al.*, 2012, 2013), which can result in malting losses. Prevention of fungal growth during malting can be accomplished with chemical, physical, and biological methods. Addition of antifungal chemicals or general microbiocides is possible and has some inhibitive effect on the malt fungi. Presence of residues from the treatment will compromise the quality and safety of the malt produced and are therefore not recommend (Wolf-Hall, 2007). However, methods using gaseous ozone or hydrogen peroxide may be promising since they do not leave any residues in the malt and have been shown to have some effect in reducing fungal growth during malt production (Kottapalli *et al.*, 2005). Also, non-degrading physical methods such as electron-beam irradiation (Kottapalli *et al.*, 2006) or microwave treatment (Akaranuchat *et al.*, 2008) have been demonstrated to show some reductive effect on fungi during malt production. Growth reduction by addition of appropriate starter cultures has been achieved in analogy to grain storage by addition of starter cultures during malt production (Boivin *et al.*, 1997; Linko *et al.*, 1998; Laitila *et al.*, 2002; Wolf-Hall, 2007). Addition of starters was either done during steeping or by spraying during soaking or germination. Special regulations may exist in national law of individual states or the EU. According to German national law any additives to the malting process are prohibited unless the malt is intended for export.

Conclusions

At the first glance fungal contamination of cereal crops is a rather two-faced problem, depending on the viewpoint of either the agricultural or brewing industry. Farmers are primarily interested in high yield quantity whereas malt producers and brewers are interested in high quality of their raw materials. A second glance, however, may reveal that also maltsters and brewers should be interested in high quantity since low barley prices always depend on high yield. Most fungal species prevailing on cereal grain do not seem to directly influence quantity and quality of either grain or malt, and much research has been given to just the few species involved in quantitative or qualitative traits. However, not much is known about the mutual influence that fungal species and other microbes have in terms of grain and malt quality. It has been shown in previous work that partners in microbial communities interact with each other in a positive or negative way (Laitila *et al.*, 2007). Most interesting from the maltster's or brewer's perspective are interactions in which quality traits are positively influenced by competition, modification or antagonism between microbes (O'Mahoni *et al.*, 2000). However, the details about these interactions still have to be elucidated in order to utilize the mechanisms involved to the advantage of grain and malt quality. Preliminary results utilizing antagonistic bacteria, yeasts and fungi to reduce the growth of mainly *Fusarium* species during malt production showed promising results and have led to the development of modified malting processes (Rouse and van Sinderen, 2008; Laitila *et al.*, 2006). Aside from growth reduction, deterioration of mycotoxins produced in the field during the malting or even the brewing process by otherwise harmless microbes will be a challenge for future research (Karlovsky, 1999).

Meanwhile, the malting and brewing industry is forced to take precautions in order to deal with the threat of quality problems posed by fungal contamination of their raw materials, in order to produce safe high-quality products. Measures must be taken to prevent mould- and mycotoxin-contaminated product from being processed and marketed. A variety of tools and protocols are now available that can be used to detect mycotoxins at every step along the line of the production processes in malting and brewing. There is a choice between highly sophisticated and specialized methods of high specificity and sensitivity or more or less simple and rapid assays, many of which can be applied on-site during processing. Also, methods for the detection of fungal contamination have evolved greatly during recent years from very time-consuming microbiological techniques to very rapid immunochemical or molecular biological assay systems. Even the presence of certain genes in a sample can be detected and used to predict the probability of mycotoxin contaminations or quality failures such as gushing. Although the spectrum of analytical methods is wide and much of it has been applied in research and in model industrial processes, it will still take time to transfer them into procedures that will be accepted as standards by the brewing and malting industry.

References

Abildgren, M.P., Lund, F., Thrane, U., and Elmholt, S. (1987). Czapek-Dox agar containing iprodione and dicloran as a selective medium for the isolation of *Fusarium* species. Lett. Appl. Microbiol. *5*, 83–87.

Ädel Druvefors, U., and Schnürer, J. (2005). Mold-inhibitory activity of different yeast species during airtight storage of wheat grain. FEMS Yeast Res. *5*, 373–378. http://dx.doi.org/10.1016/j.femsyr.2004.10.006

Aggarwal, R., Gupta, S., Banerjee, S., and Singh, V.B. (2011). Development of a SCAR marker for detection of *Bipolaris sorokiniana* causing spot blotch of wheat. Can. J. Microbiol. *57*, 934–942. http://dx.doi.org/10.1139/w11-089

Akaranuchat, P., Noimanee, P., Krittigamas, N., von Hoersten, D., and Vearasilp, S. (2008). Control of seed-borne fungi by radio frequency heat treatment as alternative seed treatment in barley (Hordeum vulgare). J. Plant Dis. Protect. *115* 45–45.

Alexandrakis, G., Jalali, S., and Gloor, P. (1998). Diagnosis of *Fusarium* keratitis in an animal model using the polymerase chain reaction. Br. J. Ophthalmol. *82*, 306–311. http://dx.doi.org/10.1136/bjo.82.3.306

Al-Sadi, A.M., and Deadman, M.L. (2010). Influence of seed-borne *Cochliobolus sativus* (Anamorph *Bipolaris sorokiniana*) on crown rot and root rot of barley and wheat. J. Phytopathol. *158*, 683–690. http://dx.doi.org/10.1111/j.1439-0434.2010.01684.x

Amaha, M., and Kitabatake, K. (1981). Gushing in beer. In Brewing Science vol. 2, J.R.C. Pollock, ed. (London, UK: Academic Press), pp. 457–489.

Andersen, B., Thrane, U., Svendsen, A., and Rasmussen, I.A. (1996). Associated field mycobiota on malt barley. Can. J. Bot. *74*, 854–858.

Andrews, S., and Pitt, J.I. (1986). Selective medium for isolation of *Fusarium* species and dematiaceous hyphomycetes from cereals. Appl. Environ. Microbiol. *51*, 1235–1238.

Anfossi, L., Baggiani, C., Giovannoli, C., D'Arco, G., and Giraudi, G. (2013). Lateral-flow immunoassays for mycotoxins and phycotoxins: a review. Anal. Bioanal. Chem. *405*, 467–480. http://dx.doi.org/10.1007/s00216-012-6033-4

Aoki, N., Moriyama, H., Kodama, M., Arie, T., Teraoka, T., and Fukuhara, T. (2009). A novel mycovirus associated with four double-stranded RNAs affects host fungal growth in *Alternaria alternata*. Virus Res. *140*, 179–187. http://dx.doi.org/10.1016/j.virusres.2008.12.003

Appell, M., Kendra, D.F., Kim, E.K., and Maragos, C.M. (2007). Synthesis and evaluation of molecularly imprinted polymers as sorbents of moniliformin. Food Addit. Contam. *24*, 43–52. http://dx.doi.org/10.1080/02652030600887586

Arafat, W., Kern, D., and Dirheimer, G. (1985). Inhibition of aminoacyl-tRNA synthetases by the mycotoxin patulin. Chem. Biol. Interact. *56*, 333–349.

Atkinson, N.J., and Urwin, P.E. (2012). The interaction of plant biotic and abiotic stresses: from genes to the field. J. Exp. Bot. *63*, 3523–3543. http://dx.doi.org/10.1093/jxb/ers100

Baggiani, C., Giraudi, G., and Vanni, A. (2001). A molecular imprinted polymer with recognition properties towards the carcinogenic mycotoxin ochratoxin A. Bioseparation *10*, 389–394. http://dx.doi.org/10.1023/A:1021506223136

Bai, G.H., Desjardins, A.E., and Plattner, R.D. (2002). Deoxynivalenol-nonproducing *Fusarium graminearum* causes initial infection, but does not cause disease spread in wheat spikes. Mycopathologia *153*, 91–98. http://dx.doi.org/10.1023/A:1014419323550

Bajomo, M.F., and Young, T.W. (1993). The properties, composition and fermantability of worts made from 100 % raw sorghum and commercial enzymes. J. Inst. Brew. *99*, 153–158.

Banks, J.N., Cox, S.J., Northway, B.J., and Rizvi, R.H. (1994). Monoclonal antibodies to fungi of significance to the quality of foods and feeds. Food Agric. Immunol. *6*, 321–327.

Banks, J.N., Rivzi, R.H., Barker, I., Turner, J.A., Rahman, S., and Northway, B.J. (1996). Specific monoclonal antibodies to *Fusarium* species and *Microdochium nivale*. Food Agric. Immunol. *8*, 249–268.

Barthel, J., Gottschalk, C., Rapp, M., Berger, M., Bauer, J., and Meyer, K. (2012). Occurrence of type A, B and D trichothecenes in barley and barley products from the Bavarian market. Mycotoxin Res. *28*, 97–106. http://dx.doi.org/10.1007/s12550-012-0123-1

Beasly, V.R. (1989). Trichothecene Mycotoxins: Pathophysiological Effects, vol. 1 (Boca Raton, USA: CRC Press).

Beattie, S., Schwarz, P.B., Horsley, R., Barr, J., and Casper, H.H. (1998). The effect of grain storage conditions on the viability of *Fusarium* and deoxynivalenol production in infested malting barley. J. Food Prot. *61*, 103–106.

Beck, J.J., and Barnett, C.J. (2003). Detection of *Fusarium* species infecting corn using the polymerase chain reaction. U.S. Patent Application #200330113722, june 19, 2003, 23 pp.

Bellver Soto, J., Fernández-Franzón, M., Ruiz, M.J., and Juan-García, A. (2014). Presence of ochratoxin A (OTA) mycotoxin in alcoholic drinks from southern European countries: wine and beer. J. Agric. Food Chem. *62*, 7643–7651. http://dx.doi.org/10.1021/jf501737h

Bennett, J.W., and Bentley, R. (1989). What's in a name? Microbial secondary metabolism. Adv. Appl. Microbiol. *34,* 1–28.

Berthiller, F., Crews, C., Dall'Asta, C., De Saeger, S., Haesaert, G., Karlovsky, P., Oswald, I.P., Seefelder, W., Speijers, G., and Stroka, J. (2013). Masked mycotoxins: a review. Mol. Nutr. Food Res. *57,* 165–186. http://dx.doi.org/10.1002/mnfr.201100764

Bertuzzi, T., Rastelli, S., Mulazzi,, A., Donadini, G., and Pietri, A. (2011). Mycotoxin occurrence in beer produced in several European countries. Food Contr. *22,* 2059–2064. http://dx.doi.org/10.1016/j.foodcont.2011.06.002

Beuchat, L.R. ed. (1987). Food and beverage mycology (New York, USA: Springer Science & Business Media).

Beyer, W., Höxter, H., Miedaner, T., Sander, E., and Geiger, H.H.. (1993). Indirect ELISA for quantitative assessment of *Fusarium* spp. in rye. Z. Pflanzenkrankh. Pflanzensch. *100,* 278–284.

Bhat, R., Rai, R.V., and Karim, A.A. (2010). Mycotoxins in food and feed: present status and future concerns. Comp. Rev. Food Sci. Food Safety *9,* 57–81. http://dx.doi.org/10.1111/j.1541-4337.2009.00094.x

Bilikova, J., and Hudec, K. (2015). Occurrence of *Fusarium* head blight of barley in Slovakia. J. Cent. Eur. Agric. *15,* 175–184.

Björnberg, A., and Schnürer, J. (1993). Inhibition of the growth of grain-storage molds in vitro by the yeast *Pichia anomala* (Hansen) Kurtzman. Can. J. Microbiol. *39,* 623–628.

Böhm-Schraml, M., Niessen, L., Donhauser, S., Engelhard, G., and Wallnöfer, P. (1993). Medium zur selektiven Abgrenzung von *Fusarium graminearum.* Deutsche Lebensmittel-Rundschau *89,* 152–154.

Böhm-Schraml, M. (1995). Fusarienwachstum und Toxinbildung während der Mälzung — Entwicklung von Selektivnährböden zum Nachweis toxinbildender Fusarien. Doctoral thesis, Technische Universität München, München, Germany,180 pp.

Börjesson, T., Stenberg, B., and Schnürer, J. (2007). Near-infrared spectroscopy for estimation of ergosterol content in barley: A comparison between reflectance and transmittance techniques. Cereal Chem. *84,* 231–236. http://dx.doi.org/10.1094/CCHEM-84-3-0231

Boivin, P., and Malanda, M. (1997). Improvement of malt quality and safety by adding starter culture during the malting process. Tech. Quart. MBAA *34,* 96–101.

Bolechová, M., Benešová, K., Běláková, S., Čáslavský, J., Pospíchalová, M., and Mikulíková, R. (2015). Determination of seventeen mycotoxins in barley and malt in the Czech Republic. Food Contr. *47,* 108–113. http://dx.doi.org/10.1016/j.foodcont.2014.06.045

Bonde, M.R., Peterson, G.L., Schaad, N.W., and Smilanick, J.L. (1997). Karnal bunt of wheat. Plant Dis. *81,* 1370–1377.

Bottalico, A., and Logrieco, A. (1998). Toxigenic *Alternaria* species of economic importance. In Mycotoxins in Agriculture and Food Safety, K.K. Sinha, and D. Bhatnagar, eds. (New York, USA: Marcel Dekker), pp. 65–108.

Brouwer, M., Lievens, B., Van Hemelrijck, W., Van den Ackerveken, G., Cammue, B.P., and Thomma, B.P. (2003). Quantification of disease progression of several microbial pathogens on *Arabidopsis thaliana* using real-time fluorescence PCR. FEMS Microbiol. Lett. *228,* 241–248. http://dx.doi.org/10.1016/S0378-1097(03)00759-6

Burdaspal, P.A., and Legarda, T.M. (2013). Survey on aflatoxin in beer sold in Spain and other European countries. World Mycotox. J. *6,* 93–101. http://dx.doi.org/10.3920/WMJ2012.1465

Burgess, L.W., Liddell, C.M., and Summerell, B.A. (1988). Laboratory Manual for *Fusarium* Research, 2nd edition (Sydney, AUS: University of Sydney).

Calò, L., Fornelli, F., Ramires, R., Nenna, S., Tursi, A., Caiaffa, M.F., and Macchia, L. (2004). Cytotoxic effects of the mycotoxin beauvericin to human cell lines of myeloid origin. Pharmacol. Res. *49,* 73–77. http://dx.doi.org/10.1016/j.phrs.2003.07.002

Calvo, A.M., Wilson, R.A., Bok, J.W., and Keller, N.P. (2002). Relationship between secondary metabolism and fungal development. Microbiol. Mol. Biol. Rev. *66,* 447–459. http://dx.doi.org/10.1128/MMBR.66.3.447-459.2002

Campbell, A.M. (2000). Monoclonal Antibody Technology: The Production and Characterization of Rodent and Human Hybridomas (Amsterdam, NL: Elsevier).

Carrington, R., Collett, R.C., Dunkin, I.R., and Halek, G. (1972). Gushing promoters and suppressants in beer and hops. J. Inst. Brew. *78,* 243–254. http://dx.doi.org/10.1002/j.2050-0416.1972.tb03443.x

Casey, G.P. (1996). Primary versus secondary gushing and assay procedures used to assess malt/beer gushing potential. Tech. Quart.– Master Brew. Assoc. Am. *33,* 229–235.

Chander, Y., Koelbl, J., Puckett, J., Moser, M.J., Klingele, A.J., Liles, M.R., Carrias, A., Mead, D.A., and Schoenfeld, T.W. (2014). A novel thermostable polymerase for RNA and DNA loop-mediated isothermal amplification (LAMP). Front. Microbiol. *5,* 136–146. http://dx.doi.org/10.3389/fmicb.2014.00395

Chang, G.H., and Yu, R.C. (1997). Rapid immunoassay of fungal mycelia in rice and corn. J. Chin. Agric. Chem. Soc. 35, 533–539 (English abstract of the Chinese original text).

Charmley, L.L., Trenholm, H.L., Prelusky, D.B., and Rosenberg, A. (1995). Economic losses and decontamination. Nat. Toxins 3, 199–203.

Chen, L.Y., Tian, X.L., and Yang, B. (1990). A study on the inhibition of rat myocardium glutathione peroxidase and glutathione reductase by moniliformin. Mycopathologia *110,* 119–124.

Chen, X., Huang, Y., Duan, N., Wu, S., Ma, X., Xia, Y., Zhu, C., Jiang, Y., and Wang, Z. (2013). Selection and identification of ssDNA aptamers recognizing zearalenone. Anal. Bioanal. Chem. *405,* 6573–6581. http://dx.doi.org/10.1007/s00216-013-7085-9

Chen, X., Huang, Y., Duan, N., Wu, S., Xia, Y., Ma, X., Zhu, C., Jiang, Y., and Wang, Z. (2014). Screening and identification of DNA aptamers against T-2 toxin assisted by graphene oxide. J. Agric. Food. Chem. *62,* 10368–10374. http://dx.doi.org/10.1021/jf5032058

Choque, E., El Rayess, Y., Raynal, J., and Mathieu, F. (2015). Fungal naphtho-γ-pyrones – secondary metabolites of industrial interest. Appl. Microbiol. Biotechnol. *99,*

1081–1096. http://dx.doi.org/10.1007/s00253-014-6295-1

Christensen, C.M., and Kaufmann, H.H. (1965). Deterioration of stored grains by fungi. Ann. Rev. Phytopath. *3*, 69–84.

Christian, M., Titze, J., Ilberg, V., and Jacob, F. (2011). Novel perspectives in gushing analysis: A review. J. Inst. Brew. *117*, 295–313. http://dx.doi.org/10.1002/j.2050-0416.2011.tb00474.x

Chu, F.S., Chang, C.C., Ashoor, S.H., and Prentice, N. (1975). Stability of aflatoxin B-1 and ochratoxin A in brewing. Appl. Microbiol. *29*, 313–316.

Chu, Y.M., Jeon, J.J., Yea, S.J., Kim, Y.H., Yun, S.H., Lee, Y.W., and Kim, K.H. (2002). Double-stranded RNA mycovirus from *Fusarium graminearum*. Appl. Environ. Microbiol. *68*, 2529–2534. http://dx.doi.org/10.1128/AEM.68.5.2529-2534.2002

Chulze, S.N. (2010). Strategies to reduce mycotoxin levels in maize during storage: a review. Food Addit. Contam. Part A Chem. Anal. Control Expo. Risk Assess. *27*, 651–657. http://dx.doi.org/10.1080/19440040903573032

Cohen, L., and Eyal, Z. (1993). The histology of processes associated with the infection of resistant and susceptible wheat cultivars with *Septoria tritici*. Plant Pathol. *42*, 737–743.

Conner, D.E. (1992). Evaluation of methods for selective enumeration of *Fusarium* species in feedstuffs. In Modern Methods in Food Mycology, R.A. Samson, A.D. Hocking, J.I. Pitt, and A.D King, eds. (Amsterdam, NL: Elsevier), pp. 299–302.

Coulombe, R.A. (1994). Nonhepatic disposition and effects of aflatoxin B1. In The Toxicology of Aflatoxins – Human Health, Veterinary, and Agricultural Significance, D.L. Eaton, and J.D. Groopman, eds. (San Diego, USA: Academic Press), pp. 89–98.

Cousin, M.A. (1996). Chitin as a measure of mold contamination of agricultural commodities and foods. J. Food Prot. *59*, 73–81.

Cox, A.R., Cagnol, F., Russell, A.B., and Izzard, M.J. (2007). Surface properties of class ii hydrophobins from *Trichoderma reesei* and influence on bubble stability. Langmuir *23*, 7995–8002. http://dx.doi.org/10.1021/la700451g

Cruz-Aguado, J.A., and Penner, G. (2008). Determination of ochratoxin a with a DNA aptamer. J. Agric. Food Chem. *56*, 10456–10461. http://dx.doi.org/10.1021/jf801957h

Dall'Erta, A., Cirlini, M., Dall'Asta, M., Del Rio, D., Galaverna, G., and Dall'Asta, C. (2013). Masked mycotoxins are efficiently hydrolyzed by human colonic microbiota releasing their aglycones. Chem. Res. Toxicol. *26*, 305–312. http://dx.doi.org/10.1021/tx300438c

Damoglou, A.P., Downey, G.A., and Shannon, W. (1984). The production of ochratoxin A and citrinin in barley. J. Sci. Food Agric. *35*, 395–400.

Deckers, S., Gebruers, K., Baggerman, G., Lorgouilloux, Y., Delcour, J., Michiels, C., Derdelinckx, G., Martens, J., and Neven, H. (2010). CO^2-Hydrophobin structurtes acting as nanobombs in beer. Part 1: a critical review of hypotheses and mechanisms. Monatsschrift für Brauwissenschaft, 63(3/4), 54–61.

Deckers, S.M., Venken, T., Khalesi, M., Gebruers, K., Baggerman, G., Lorgouilloux, Y., Shokribousjein, Z., Ilberg, V., Schönberger, C., Titze, J., Verachtert, H., Michiels, C., Neven, H., Delcour, J., Martens, J., Dederlinckx, G., and de Maeyer, M. (2012a). Combined modeling and biophysical characterisation of CO_2 interaction with class II hydrophobins: New insight into the mechanism underpinning primary gushing. J. Am. Soc. Brew. Chem. *70*, 249–256. http://dx.doi.org/10.1094/ASBCJ-2012-0905-01

Deckers, S.M., Vissers, L., Gebruers, K., Shokribousjein, Z., Khalesi, M., Riveros-Galan, D., Schönberger, C., Verachtert, H., Neven, H., Delcour, J., Michiels, C., Ilberg, V., Derdelinckx, G., Titze, J., and Martens, J. (2012b). Doubly modified Carlsberg test combined with dynamic light scattering allows prediction of the primary gushing potential of harvested barley and malt. Cerevisia *37*, 77–81. http://dx.doi.org/10.1016/j.cervis.2012.10.001

Deckers, S.M., Vissers, L., Khalesi, M., Shokribousjein, Z., Verachtert, H., Gebruers, K., Pirlot, X., Rock, J.M., Ilberg, V., Titze, J., Neven, H., and Derdelinckx, G. (2013). Thermodynamic view of primary gushing. J. Am. Soc. Brew. Chem. *71*, 149–152.

De Lucca, A.J., Cleveland, T.E., and Wedge, D.E. (2005). Plant-derived antifungal proteins and peptides. Can. J. Microbiol. *51*, 1001–1014.

Delves, P.J., Martin, S.J., Burton, D.R., and Roitt, I.M. (2011). Roitt's Essential Immunology (West Sussex, UK: Wiley-Blackwell).

Delwiche, S.R., Kim, M.S., and Dong, Y. (2011). *Fusarium* damage assessment in wheat kernelsby Vis/NIR hyperspectral imaging. Sens. Instrum. Food Qual. *5*, 63–71. http://dx.doi.org/10.1007/s11694-011-9112-x

Demeke, T., Clear, R.M., Patrick, S.K., and Gaba, D. (2005). Species-specific PCR-based assays for the detection of *Fusarium* species and a comparison with the whole seed agar plate method and trichothecene analysis. Int. J. Food Microbiol. *103*, 271–284. http://dx.doi.org/10.1016/j.ijfoodmicro.2004.12.026

Denschlag, C., Vogel, R.F., and Niessen, L. (2012). Hyd5 gene-based detection of the major gushing-inducing *Fusarium* spp. in a loop-mediated isothermal amplification (LAMP) assay. Int. J. Food Microbiol. *156*, 189–196. http://dx.doi.org/10.1016/j.ijfoodmicro.2012.03.009

Denschlag, C., Vogel, R.F., and Niessen, L. (2013). Hyd5 gene based analysis of cereals and malt for gushing-inducing *Fusarium* spp. by real-time LAMP using fluorescence and turbidity measurements. Int. J. Food Microbiol. *162*, 245–251. http://dx.doi.org/10.1016/j.ijfoodmicro.2013.01.019

Denschlag, C., Rieder, J., Vogel, R.F., and Niessen, L. (2014). Real-time loop-mediated isothermal amplification (LAMP) assay for group specific detection of important trichothecene producing Fusarium species in wheat. Int. J. Food Microbiol. *177*, 117–127. http://dx.doi.org/10.1016/j.ijfoodmicro.2014.02.010

De Smet, D., Dubruel, P., Van Peteghem, C., Schacht, E., and De Saeger, S. (2009). Molecularly imprinted solid-phase extraction of fumonisin B analogues in bell pepper, rice and corn flakes. Food Addit. Contam. Part A Chem.

Anal. Control Expo. Risk Assess. *26*, 874–884. http://dx.doi.org/10.1080/02652030902788920

Detroy, R.W., and Worden, K.A. (1979). Interactions of fungal viruses and secondary metabolites, In Fungal Viruses, H.P Molitoris, M. Hollings and H.A. Wood, eds. (Berlin, DE, Springer), pp. 94–107.

Dewey, F.M., MacDonald, M.M., Phillips, S.I., and Priestley, R.A. (1990). Development of monoclonal-antibody-ELISA and -DIP-STICK immunoassays for *Penicillium islandicum* in rice grains. J. Gen. Microbiol. *136*, 753–760. http://dx.doi.org/10.1099/00221287-136-4-753

DiCosmo, F., and Straus, N.A. (1985). Alternariol, a dibenzopyrone mycotoxin of *Alternaria* spp., is a new photosensitizing and DNA cross-linking agent. Experientia *41*, 1188–1190.

D'Mello, J.F., Macdonald, A.M., Postel, D., Dijksma, W.T., Dujardin, A., and Placinta, C.M. (1998). Pesticide use and mycotoxin production in *Fusarium* and *Aspergillus* phytopathogens. Eur. J. Plant Pathol. *104*, 741–751. http://dx.doi.org/10.1023/A:1008621505708

Dobler, M., Dunitz, J.D., and Krajewski, J. (1969). Structure of the K+ complex with enniatin B, a macrocyclic antibiotic with K+ transport properties. J. Mol. Biol. *42*, 603–606.

Dong, J., Cui, X., Deng, Y., and Tang, Z. (2012). Amplified detection of nucleic acid by G-quadruplex based hybridization chain reaction. Biosens. Bioelectron. *38*, 258–263. http://dx.doi.org/10.1016/j.bios.2012.05.042

Donhauser, S., Weideneder, A., Winnewisser, W., and Geiger, E. (1990). Test zur Ermittlung der Gushingneigung von Rohfrucht, Malz, Würze und Bier. Brauwelt *130*, 1317–1320.

Doohan, F.M., Parry, D.W., Jenkinson, P., and Nicholson, P. (1998). The use of species-specific PCR-based assays to analyse *Fusarium* ear blight of wheat. Plant Pathol. (Oxford) *47*, 197–205. http://dx.doi.org/10.1046/j.1365-3059.1998.00218.x

Doohan, F.M., Brennan, J., and Cooke, B.M. (2003). Influence of climatic factors on *Fusarium* species pathogenic to cereals. In Epidemiology of Mycotoxin Producing Fungi, X. Xu, J.A. Bailey, and B.M. Cooke, eds. (Amsterdam, NL: Springer), pp. 755–768.

Doran, P., and Briggs, D. (1993). Microbes and grain germination. J. Inst. Brew. *99*, 165–170.

Dornetshuber, R., Heffeter, P., Kamyar, M.R., Peterbauer, T., Berger, W., and Lemmens-Gruber, R. (2007). Enniatin exerts p53-dependent cytostatic and p53-independent cytotoxic activities against human cancer cells. Chem. Res. Toxicol. *20*, 465–473. http://dx.doi.org/10.1021/tx600259t

Doss, R.P., and Welty, R.E. (1995). A polymerase chain reaction-based procedure for detection of *Acremonium coenophialum* in tall fescue. Phytopathology *85*, 913–917.

Douliez, J.P., Michon, T., Elmorjani, K., and Marion, D. (2000). Structure, biological and technological functions of lipid transfer proteins and indolines, the major lipid binding proteins from cereal kernels. J. Cereal Sci. *32*, 1–20. http://dx.doi.org/10.1006/jcrs.2000.0315

Dowell, F.E., Ram, M.S., and Seitz, L.M. (1999). Predicting scab, vomitoxin, and ergosterol in single wheat kernels using near-infrared spectroscopy. Cereal Chem. *76*, 573–576. http://dx.doi.org/10.1094/CCHEM.1999.76.4.573

Durán, N., Teixeira, M.F., De Conti, R., and Esposito, E. (2002). Ecological-friendly pigments from fungi. Crit. Rev. Food Sci. Nutr. *42*, 53–66. http://dx.doi.org/10.1080/10408690290825457

Eifler, J. (2014). Electronic Nose-Based *Fusarium* Detection and Deoxynivalenol Aptamer Development. Dotctoral thesis, Universität Göttingen, Göttingen, Germany.

Elmholt, S., Labouriau, R., Hestbjerg, H., and Nielsen, J.M. (1999). Detection and estimation of conidial abundance of *Penicillium verrucosum* in soil by dilution plating on a selective and diagnostic agar medium (DYSG). Mycol. Res. *103*, 887–895. http://dx.doi.org/10.1017/S0953756298007965

Emond, E., and Moineau, S. (2007). Bacteriophages and food fermentations. In Interactions of fungal viruses and secondary metabolites, S. McGrath and D. van Sinderen, eds. (Norfolk, UK, Caister Academic Press), pp. 93–123.

Endo, A., and Kuroda, M. (1976). Citrinin, an inhibitor of cholesterol synthesis. J. Antibiot. *29*, 841–843. http://doi.org/10.7164/antibiotics.29.841

Engelmann, J., Huber, W., Illing, S., and Jacob, F. (2012). Das Phänomen Gushing und rote Körner im Malz (Teil 1). Brauwelt *32/2012*, 912–915.

European Commission (2002). Opinion of the Scientific Committee on Food on *Fusarium* toxins. Part 6: Group evaluation of T-2 toxin, HT-2 toxin, nivalenol and deoxynivalenol. European Commission, Health & Consumer Protection Directorate-General, Scientific Committee on Food, Document SCF/CS/CNTM/MYC/27 Final. Brussels, 27 February 2002.

European Commissionn (2006). Commission Regulation (EC) No 1881/2006 of 19 December 2006 setting maximum levels for certain contaminants in foodstuffs. European Commission, Brussels, 35 pp.

Evans, S.G. (1969). Observations on the development of leaf blotch and net blotch of barley from barley debris. Plant Pathol. *18*, 116–118.

Eyal, Z. (1999). The *Septoria tritici* and *Stagonospora nodorum* blotch diseases of wheat. Eur. J. Plant Pathol. *105*, 629–641. http://dx.doi.org/10.1023/A:1008716812259

Ezekiel, C.N., Abia, W.A., Ogara, I.M., Sulyok, M., Warth, B., and Krska, R. (2015). Fate of mycotoxins in two popular traditional cereal-based beverages (kunu-zaki and pito) from rural Nigeria. LWT Food Sci. Technol. *60*, 137–141. http://dx.doi.org/10.1016/j.lwt.2014.08.018

Fakruddin, M., Mannan, K.S., Chowdhury, A., Mazumdar, R.M., Hossain, M.N., Islam, S., and Chowdhury, M.A. (2013). Nucleic acid amplification: Alternative methods of polymerase chain reaction. J. Pharm. Bioallied Sci. *5*, 245–252. http://dx.doi.org/10.4103/0975-7406.120066

FAO (2004). Worldwide regulations for mycotoxins in food and feed in 2003. FAO Food and Nutrition Paper 81 (Rome, IT: Food and Agriculture Organization of the United Nations).

FAO (2009). Agribusiness Handbook – Barley, Malt, Beer (Rome, IT: Food and Agriculture Organization of the United Nations).

Fischer, S. (2001). Blasenbildung von in Flüssigkeiten gelösten Gasen. Doctoral thesis, Technische Universität München, Munich, Germany.

Flannigan, B. (1969). Microflora of dreid barley grain. Trans. Br. Mycol. Soc. *53*, 371–379. http://dx.doi.org/10.1016/S0007-1536(69)80095-1

Fotso, J., Leslie, J.F., and Smith, J.S. (2002). Production of beauvericin, moniliformin, fusaproliferin, and fumonisins B_1, B_2, and B_3 by fifteen ex-type strains of fusarium species. Appl. Environ. Microbiol. *68*, 5195–5197. http://dx.doi.org/10.1128/AEM.68.10.5195-5197.2002

Fowler, A.M., and Owen, H. (1971). Studies on leaf blotch of barley (*Rhynchosporium secalis*). Transact. Brit. Mycol. Soc. *56*, 137–152.

Fox, E.M., and Howlett, B.J. (2008). Secondary metabolism: regulation and role in fungal biology. Curr. Opin. Microbiol. *11*, 481–487. http://dx.doi.org/10.1016/j.mib.2008.10.007

Frändberg, E., and Schnürer, J. (1998). Antifungal activity of chitinolytic bacteria isolated from airtight stored cereal grain. Can. J. Microbiol. *44*, 121–127.

Francois, P., Tangomo, M., Hibbs, J., Bonetti, E.J., Boehme, C.C., Notomi, T., Perkins, M.D., and Schrenzel, J. (2011). Robustness of a loop-mediated isothermal amplification reaction for diagnostic applications. FEMS Immunol. Med. Microbiol. *62*, 41–48. http://dx.doi.org/10.1111/j.1574-695X.2011.00785.x

Frisvad, J.C. (1983). A selective and indicative medium for groups of *Penicillium viridicatum* producing different mycotoxins in cereals. J. Appl. Bacteriol. *54*, 409–416. http://dx.doi.org/10.1111/j.1365-2672.1983.tb02636.x

Frisvad, J.C., Smedsgaard, J., Ostenfeld Larsen, T., and Samson, R.A. (2004). Mycotoxins, drugs and other extrolites produced by species in *Penicillium* subgenus *Penicillium*. Stud. Mycol. *49*, 201–241.

Frisvad, J.C., Smedsgaard, J., Samson, R.A., Larsen, T.O., and Thrane, U. (2007). Fumonisin B2 production by *Aspergillus niger*. J. Agric. Food Chem. *55*, 9727–9732. http://dx.doi.org/10.1021/jf0718906

Gachon, C., and Saindrenan, P. (2004). Real-time PCR monitoring of fungal development in *Arabidopsis thaliana* infected by *Alternaria brassicicola* and *Botrytis cinerea*. Plant Physiol. Biochem. *42*, 367–371. http://dx.doi.org/10.1016/j.plaphy.2004.04.001

Gan, Z., Marquardt, R.R., Abramson, D., and Clear, R.M. (1997). The characterization of chicken antibodies raised against *Fusarium* spp. by enzyme-linked immunosorbent assay and immunoblotting. Int. J. Food Microbiol. *38*, 191–200. http://dx.doi.org/10.1016/S0168-1605(97)00108-6

Garbe, L.A., Schwarz, P., and Ehmer, A. (2011). Beer gushing. In Beer: A Quality Perspective – Handbook of Alcoholic Beverages Series, C.W. Bamforth, ed. (San Diego, USA: Academic Press), pp. 185–212.

García, T., Mayoral, B., González, I., López-Calleja, I., Sanz, A., Hernández, P.E., and Martín, R. (2004). Enumeration of yeasts in dairy products: a comparison of immunological and genetic techniques. J. Food Prot. *67*, 357–364.

Gardner, R.J., Laws, D.R.J., and McGuinness, J.D. (1973). The suppression of gushing by the use of hop oil. J. Inst. Brew. *79*, 209–211. http://dx.doi.org/10.1002/j.2050-0416.1973.tb03525.x

Geddes, J., Eudes, F., Laroche, A., and Selinger, L.B. (2008). Differential expression of proteins in response to the interaction between the pathogen *Fusarium graminearum* and its host, *Hordeum vulgare*. Proteomics *8*, 545–554. http://dx.doi.org/10.1002/pmic.200700115

Gill, P., and Ghaemi, A. (2008). Nucleic acid isothermal amplification technologies: a review. Nucleosides Nucleotides Nucleic Acids *27*, 224–243. http://dx.doi.org/10.1080/15257770701845204

Gjertsen, P., Trolle, B., and Andersen, K. (1963). Weathered barley as a contributory cause of gushing in beer. Proc. Eur. Brew. Conv. *9*, 320–341.

Gjertsen, P., Trolle, B., and Andersen, K. (1965). Studies on gushing II. Gushing caused by microorganisms, specially *Fusarium* species. Proc. Eur. Brew. Conv. *10*, 428–438.

Glynn, N.C., Hare, M.C., Parry, D.W., and Edwards, S.G. (2005). Phylogenetic analysis of EF-1 alpha gene sequences from isolates of *Microdochium nivale* leads to elevation of varieties *majus* and *nivale* to species status. Mycol. Res. *109*, 872–880. https://doi.org/10.1017/S0953756205003370

Gooday, G.W. (1974). Fungal sex hormones. Annu. Rev. Biochem. *43*, 35–87. http://dx.doi.org/10.1146/annurev.bi.43.070174.000343

Gooday, G.W., and Adams, D.J. (1993). Sex hormones and fungi. Adv. Microb. Physiol. *34*, 69–145. http://dx.doi.org/10.1016/S0065-2911(08)60028-4

Gourama, H., and Bullerman, L.B. (1995). Detection of molds in foods and feeds: potential rapid and selective methods. J. Food Protect. *58*, 1389-1394.

Gqaleni, N., Smith, J.E., Lacey, J., and Gettinby, G. (1996). The production of cyclopiazonic acid by *Penicillium commune* and cyclopiazonic acid and aflatoxins by *Aspergillus flavus* as affected by water activity and temperature on maize grains. Mycopathologia *136*, 103–108. http://dx.doi.org/10.1007/BF00437503

Gupta, M., Abu-Ghannam N., and Gallaghar, E. (2010). Barley for brewing: Characteristic changes during malting, brewing and applications of its by-products. Comp. Rev. Food Sci. Food Safety *9*, 318–328. http://dx.doi.org/10.1111/j.1541-4337.2010.00112.x

Gyllang, H., Kjellen, K., Haikara, A., and Sigsgaard, P. (1981). Evaluation of fungal contamination on barley and malt. J. Inst. Brew. *87*, 248–251. http://dx.doi.org/10.1002/j.2050-0416.1981.tb04026.x

Hagn, A., Wallisch, S., Radl, V., Munch, J.C., and Schloter, M. (2007). A new cultivation independent approach to detect and monitor common *Trichoderma* species in soils. J. Microbiol. Methods *69*, 86–92.

Haikara, A., Sarlin, T., Nakari-Setälä, T., and Penttila, M. (1999). Method for determining a gushing factor in a berverage. Patent WO 99/54725, International Publication Date 28.10.99, 13 pp.

Hajek, A.E., and St. Leger, R.J. (1994). Interactions between fungal pathogens and insect hosts. Annu. Rev. Entomol. *39*, 293–322.

Heitefuss, R. (2001). Defence reactions of plants to fungal pathogens: principles and perspectives, using powdery mildew on cereals as an example. Naturwissenschaften *88*, 273–283. http://dx.doi.org/10.1007/s001140100226

Haugland, R.A., and Vesper, S.J. (2000). Method for identifying and quantifying specific fungi and bacteria.

US Patent and Trademark Office, Patent 6,387,652, May 14, 2002, 29 pp.

Haupt, K., and Mosbach, K. (2000). Molecularly imprinted polymers and their use in biomimetic sensors. Chem. Rev. *100,* 2495–2504. http://dx.doi.org/10.1021/cr990099w

Hayashi, N., Arai, R., Tada, S., Taguchi, H., and Ogawa, Y. (2007). Detection and identification of *Brettanomyces/Dekkera* sp. yeasts with a loop-mediated isothermal amplification method. Food Microbiol. *24,* 778–785.

Hayashi, N., Minato, T., Kanai, K., Ikushima, S., Yoshida, S., Tada, S., Taguchi, H., and Ogawa, Y. (2009). Differentiation of species belonging to *Saccharomyces sensu stricto* using a loop-mediated isothermal amplification method. J. Am. Soc. Brew. Chem. *67,* 118–126.

Henson, J.M., Butler, M.J., and Day, A.W. (1999). The dark side of the mycelium: melanins of phytopathogenic fungi. Annu. Rev. Phytopathol. *37,* 447–471.

Herrero, N., Sánchez Márquez, S., and Zabalgogeazcoa, I. (2009). Mycoviruses are common among different species of endophytic fungi of grasses. Arch. Virol. *154,* 327–330. http://dx.doi.org/10.1007/s00705-008-0293-5

Herskowitz, I. (1988). Life cycle of the budding yeast *Saccharomyces cerevisiae.* Microbiol. Rev. *52,* 536–553.

Hierro, N., Esteve-Zarzoso, B., González, A., Mas, A., and Guillamón, J.M. (2006). Real-time quantitative PCR (QPCR) and reverse transcription-QPCR for detection and enumeration of total yeasts in wine. Appl. Environ. Microbiol. *72,* 7148–7155. http://dx.doi.org/10.1128/AEM.00388-06

Hippeli, S., and Elstner, E.F. (2002). Are hydrophobins and/or non-specific lipid transfer proteins responsible for gushing in beer? New hypotheses on the chemical nature of gushing inducing factors. Z. Naturforsch. C. J. Biosci. *57,* 1–8. https://doi.org/10.1515/znc-2002-1-201

Hecht, D., and Hippeli, S. (2007). Role of ns-LTP1 in the development of primary gushing. Monatsschr. Brauwiss. *60,* 1–9.

Hocking, A.D., and Pitt, J.I. (1980). Dichloran-glycerol medium for enumeration of xerophilic fungi from low-moisture foods. Appl. Environ. Microbiol. *39,* 488–492.

Hohl, H.R., and Raper, K.B. (1963). Nutrition of cellular slime molds. I. Growth on living and dead bacteria. J. Bacteriol. *85,* 191–198.

Hökby, E., Hult, K., Gatenbeck, S., and Rutqvist, L. (1979). Ochratoxin A and citrinin in 1976 crop of barley stored on farms in Sweden. Acta Agric. Scand. *29,* 174–178.

Hornbogen, T., Glinski, M., and Zocher, R. (2002). Biosynthesis of depsipeptide mycotoxins in *Fusarium.* Eur. J. Plant Pathol. *108,* 713–718. http:/dx.doi.org/10.1023/A:1020687231810

Hossain, M.Z., and Goto, T. (2015). Determination of sterigmatocystin in grain using gas chromatography-mass spectrometry with an on-column injector. Mycotoxin Res. *31,* 17–22. http://dx.doi.org/10.1007/s12550-014-0214-2

Hoy, J.L., McCauley, B.J., and Fincher, B. (1981). Cellulases of plant and microbial origin in germinating barley. J. Inst. Brew. *87,* 77–80. http://dx.doi.org/10.1002/j.2050-0416.1981.tb03990.x

Hu, L., Gastl, M., Linkmeyer, A., Hess, M., and Rychlik, M. (2014). Fate of enniatins and beauvericin during the malting and brewing process determined by stable isotope dilution assays. LWT– Food Sci. Technol. *56,* 469–477. http://dx.doi.org/10.1016/j.lwt.2013.11.004

Hussein, H.S., and Brasel, J.M. (2001). Toxicity, metabolism, and impact of mycotoxins on humans and animals. Toxicology *167,* 101–134. http://dx.doi.org/10.1016/S0300-483X(01)00471-1

Hyun, U., Lee, D.H., Lee, C., and Shin, C.G. (2009). Apoptosis induced by enniatins H and MK1688 isolated from Fusarium oxysporum FB1501. Toxicon *53,* 723–728. http://dx.doi.org/10.1016/j.toxicon.2009.02.012

IARC (1993). Ochratoxin A. In IARC Monographs on the Evaluation of Carcinogenic Risks to Humans. Vol. 56. Some Naturally Occurring Substances: Food Items and Constituents, Heterocyclic Aromatic Amines and Mycotoxins. Anonyous (Lyon, FR: International Agency for Research on Cancer), pp. 489–521.

Ibáñez-Vea, M., Lizarraga, E., González-Peñasa, E., and López de Cerain, A. (2012). Co-occurrence of type-A and type-B trichothecenes in barley from a northern region of Spain. Food Contr. *25,* 81–88.

Ikalafeng, B.K., Pool, E.J., Lues, R., and Groenewald, W.H. (2013). Seasonal variation of citrinin in traditionally brewed African beer. Afric. J. Microbiol. Res. *7,* 5126–5129.

Inoue, T., Nagatomi, Y., Uyama, A., and Mochizuki, N. (2013). Fate of mycotoxins during beer brewing and fermentation. Biosci. Biotechnol. Biochem. *77,* 1410–1415.

Ishii, K., Ando, Y., and Ueno, Y. (1975). Toxicological approaches to the metabolites of Fusaria. IX. Isolation of vomiting factor from moldy corn infected with *Fusarium* species. Chem. Pharm. Bull. *23,* 2162–2164. http://doi.org/10.1248/cpb.23.2162

Iyer, M.S., and Cousin, M.A. (2003). Immunological detection of *Fusarium* species in cornmeal. J. Food Prot. *66,* 451–456.

Jaeger, E.E., Carroll, N.M., Choudhury, S., Dunlop, A.A., Towler, H.M., Matheson, M.M., Adamson, P., Okhravi, N., and Lightman, S. (2000). Rapid detection and identification of *Candida, Aspergillus,* and *Fusarium* species in ocular samples using nested PCR. J. Clin. Microbiol. *38,* 2902–2908.

Jarvis, B., Seiler, D.A., Ould, A.J., and Williams, A.P. (1983). Observations on the enumeration of moulds in food and feedingstuffs. J. Appl. Bacteriol. *55,* 325–336. http://dx.doi.org/10.1111/j.1365-2672.1983.tb01329.x

Jarvis, B.B., Mokhtari-Rejali, N., Schenkel, E.P., Barros, C.S., and Matzenbacher, N.I. (1991). Trichothecene mycotoxins from Brazilian *Baccharis* species. Phytochemistry *30,* 789–797. http://dx.doi.org/10.1016/0031-9422(91)85253-V

Jarvis, B.B. (2003). Analysis for mycotoxins: the chemist's perspective. Arch. Environ. Health *58,* 479–483. http://dx.doi.org/10.3200/AEOH.58.8.479-483

Jensen, B., Knudsen, I.M., and Jensen, D.F. (2000). Biological seed treatment of cereals with fresh and long-term stored formulations of *Clonostachys rosea*: Biocontrol efficacy against *Fusarium culmorum.* Eur. J. Plant Pathol. *106,* 233–242. http://dx.doi.org/10.1023/A:1008794626600

Jestoi, M., Rokka, M., Yli-Mattila, T., Parikka, P., Rizzo, A., and Peltonen, K. (2004). Presence and concentrations of the *Fusarium*-related mycotoxins beauvericin, enniatins and moniliformin in Finnish grain samples. Food Addit. Contam. *21*, 794–802. http://dx.doi.org/10.1080/02652030410001713906

Jestoi, M. (2008). Emerging fusarium-mycotoxins fusaproliferin, beauvericin, enniatins, and moniliformin: a review. Crit. Rev. Food Sci. Nutr. *48*, 21–49. http://dx.doi.org/10.1080/10408390601062021

Joffe, A.Z. (1971). Alimentary toxic Aleucia. In Microbial Toxins, vol. 7: Algal and Fungal Toxins, S.J. Ajl, A. Ciegler, S. Kadis, T.C. Montie, and G. Weinbaum, eds. (New York, USA: Academic Press), pp. 139–189.

Joffe, A.Z. (1986). *Fusarium* Species: Their Biology and Toxicology (New York, USA: Wiley & Sons).

Johnson, R.D., Johnson, L., Kohmoto, K., Otani, H., Lane, C.R., and Kodama, M. (2000). A polymerase chain reaction-based method to specifically detect *Alternaria alternata* apple pathotype (*A. mali*), the causal agent of Alternaria Blotch of apple. Phytopathology *90*, 973–976. http://dx.doi.org/10.1094/PHYTO.2000.90.9.973

Juglal, S., Govinden, R., and Odhav, B. (2002). Spice oils for the control of co-occurring mycotoxin-producing fungi. J. Food Prot. *65*, 683–687.

Junaid, J.M., Dar, N.A., Bhat, T.A., Bhat, A.H., and Bhat, M.A. (2013). Commercial biocontrol agents and their mechanism of action in the management of plant pathogens. Int. J. Mod. Plant Anim. Sci. *1*, 39–57.

Jung, C., and Ellington, A.D. (2014). Diagnostic applications of nucleic acid circuits. Acc. Chem. Res. *47*, 1825–1835. http://dx.doi.org/10.1021/ar500059c

Juodeikiene, G., Kunigėlis, V., Vidmantienė, D., and de Koe, W.J. (2004). Acoustic screening for the determination of deoxynivalenol (DON) in wheat. Vet. Ir Zootech. *25*, 52–59.

Juodeikiene, G., Basinskiene, L., Vidmantiene, D., Bartkiene, E., Kunigelis, V., and de Koe, W.J. (2008). Rapid acoustic screening of deoxynivalenol (DON) in grain. World Mycotox. J. *1*, 267–274.

Juodeikiene, G., Basinskiene, L., Vidmantiene, D., Bartkiene, E., Bakutis, B., and Baliukoniene, V. (2011). Acoustic sensing of deoxynivalenol in co-occurrence with zearalenone and T-2/HT-2 toxin in winter wheat cultivar Sirvinta from Lithuania. World Mycotox. J. *4*, 395–404.

Jurado, M., Vazquez, C., Patino, B., and Gonzalez-Jaen, M.T. (2005). PCR detection assays for the trichothecene-producing species *Fusarium graminearum*, *Fusarium culmorum*, *Fusarium poae*, *Fusarium equiseti* and *Fusarium sporotrichioides*. Syst. Appl. Microbiol. *28*, 562–568. http://dx.doi.org/10.1016/j.syapm.2005.02.003

Kaczka, E.A., Gitterman, C.O., Dulaney, E.L., Smith, M.C., Hendlin, D., Woodruff, H.B., and Folkers, K. (1964). Discovery of inhibitory activity of tenuazonic acid for growth of human adenocarcinoma-1. Biochem. Biophys. Res. Commun. *14*, 54–57.

Kampelmacher, E.H. (1973). 'Since Eve ate apples'. Can. J. Public Health *64*, 231–237.

Kamyar, M., Rawnduzi, P., Studenik, C.R., Kouri, K., and Lemmens-Gruber, R. (2004). Investigation of the electrophysiological properties of enniatins. Arch. Biochem. Biophys. *429*, 215–223. http://dx.doi.org/10.1016/j.abb.2004.06.013

Kamyar, M.R., Kouri, K., Rawnduzi, P., Studenik, C., and Lemmens-Gruber, R. (2006). Effects of moniliformin in presence of cyclohexadepsipeptides on isolated mammalian tissue and cells. Toxicol. In Vitro *20*, 1284–1291. http://dx.doi.org/10.1016/j.tiv.2006.03.001

Kamphuis, H.J., Notermans, S., Veeneman, G.H., van Boom, J.H., and Rombouts, F.M. (1989). A rapid and reliable method for the detection of molds in foods: Using the latex agglutination assay. J. Food Prot. *52*, 244–247.

Kaneko, H., Kawana, T., Fukushima, E., and Suzutani, T. (2007). Tolerance of loop-mediated isothermal amplification to a culture medium and biological substances. J. Biochem. Bioph. Methods *70*, 499–501. http://dx.doi.org/10.1016/j.jbbm.2006.08.008

Kang, Z., and Buchenauer, H. (2000). Cytology and ultrastructure of the infection of wheat spikes by *Fusarium culmorum*. Mycol. Res. *104*, 1083–1093.

Kang, J.S., Kim, A.Y., Han, H.R., Moon, Y.S., and Koh, Y.H. (2014). Development of two alternative loop-mediated isothermal amplification tools for detecting pathogenic pine wood nematodes. For. Path. http://dx.doi.org/10.1111/efp.12147

Karlovsky, P. (1999). Biological detoxification of fungal toxins and its use in plant breeding, feed and food production. Nat. Toxins 7, 1–23.

Kottapalli, B., Wolf-Hall, C.E., and Schwarz, P. (2005). Evaluation of gaseous ozone and hydrogen peroxide treatments for reducing *Fusarium* survival in malting barley. J. Food Prot. *68*, 1236–1240.

Kottapalli, B., Wolf-Hall, C.E., and Schwarz, P. (2006). Effect of electron-beam irradiation on the safety and quality of *Fusarium*-infected malting barley. Int. J. Food Microbiol. *110*, 224–231. http://dx.doi.org/10.1016/j.ijfoodmicro.2006.04.007

Keinath, A.P., Somai, B.M., and Dean, R.A. (2001). Method for diagnosis of gummy stem blight in plants using a polymerase chain reaction assay. US Patent and Trademark Office, United States Patent Application 20010024785 A1, 16pp.

Keller, N.P., Turner, G., and Bennett, J.W. (2005). Fungal secondary metabolism – from biochemistry to genomics. Nat. Rev. Microbiol. 3, 937–947. http://dx.doi.org/10.1038/nrmicro1286

Kemp, B., Alexandre, H., Robillard, B., and Marchal, R. (2015). Effect of production phase on bottle-fermented sparkling wine quality. J. Agric. Food Chem. *63*, 19–38. http://dx.doi.org/10.1021/jf504268u

Khan, M.R., Fischer, S., Egan, D., and Doohan, F.M. (2006). Biological control of fusarium seedling blight disease of wheat and barley. Phytopathology *96*, 386–394. http://dx.doi.org/10.1094/PHYTO-96-0386

Khan, M.R., and Doohan, F.M. (2009). Bacterium-mediated control of *Fusarium* head blight disease of wheat and barley and associated mycotoxin contamination of grain. Biol. Contr. *48*, 42–47. http://dx.doi.org/10.1016/j.biocontrol.2008.08.015

Kim, S., Ahn, I.P., Rho, H.S., and Lee, Y.H. (2005). MHP1, a *Magnaporthe grisea* hydrophobin gene, is required for fungal development and plant colonization. Mol. Microbiol. *57*, 1224–1237. http://dx.doi.org/10.1111/j.1365-2958.2005.04750.x

King, A.D., Hocking, A.D., and Pitt, J.I. (1979). Dichloran-rose bengal medium for enumeration and isolation of molds from foods. Appl. Environ. Microbiol. *37*, 959–964.

Kiss, L. (2003). A review of fungal antagonists of powdery mildews and their potential as biocontrol agents. Pest. Manag. Sci. *59*, 475–483. http://dx.doi.org/10.1002/ps.689

Klitgaard, A., Iversen, A., Andersen, M.R., Larsen, T.O., Frisvad, J.C., and Nielsen, K.F. (2014). Aggressive dereplication using UHPLC-DAD-QTOF: screening extracts for up to 3000 fungal secondary metabolites. Anal. Bioanal. Chem. *406*, 1933–1943. http://dx.doi.org/10.1007/s00216-013-7582-x

Knudsen, I.M., Hockenhull, J., and Jensen, D.F. (1995). Biocontrol of seedling diseases of barley and wheat caused by *Fusarium culmorum* and *Bipolaris sorokiniana*: effects of selected fungal antagonists on growth and yield components. Plant Pathol. *44*, 467–477. http://dx.doi.org/10.1111/j.1365-3059.1995.tb01669.x

Kolsteren, P. (1992). Kashin-Beck disease. Ann. Soc. Belg. Med. Trop. *72*, 81–91.

Konstantinova, P., and Yli-Mattila, T. (2004). IGS-RFLP analysis and development of molecular markers for identification of Fusarium *poae*, Fusarium *langsethiae*, Fusarium *sporotrichioides* and Fusarium *kyushuense*. Int. J. Food Microbiol. *95*, 321–331. http://dx.doi.org/10.1016/j.ijfoodmicro.2003.12.010

Kostelanska, M., Hajslova, J., Zachariasova, M., Malachova, A., Kalachova, K., Poustka, J., Fiala, J., Scott, P.M., Berthiller, F., and Krska, R. (2009). Occurrence of deoxynivalenol and its major conjugate, deoxynivalenol-3-glucoside, in beer and some brewing intermediates. J. Agric. Food Chem. *57*, 3187–3194. http://dx.doi.org/10.1021/jf803749u

Kostelanska, M., Zachariasova, M., Lacina, O., Fenclova, M., Kollos, A.L., and Hajslova, J. (2011). The study of deoxynivalenol and its masked metabolites fate during the brewing process realised by UPLC-TOFMS method. Food Chem. *126*, 1870–1876. http://dx.doi.org/10.1016/j.foodchem.2010.12.008

Krogh, P., Hasselager, E., and Friis, P. (1970). Studies on fungal nephrotoxicity. 2. Isolation of two nephrotoxic compounds from *Penicillium viridicatum* Westling: citrinin and oxalic acid. Acta Pathol. Microbiol. Scand. B Microbiol. Immunol. *78*, 401–413.

Krogh, P., Hald, B., and Pedersen, E.J. (1973). Occurrence of ochratoxin A and citrinin in cereals associated with mycotoxic porcine nephropathy. Acta Pathol. Microbiol. Scand. B Microbiol. Immunol. *81*, 689–695.

Krogh, P., Hald, B., Gjertsen, P., and Myken, F. (1974). Fate of ochratoxin A and citrinin during malting and brewing experiments. Appl. Microbiol. *28*, 31–34.

Kulik, T., Fordoński, G., Pszczółkowska, A., Płodzień, K., and Łapiński, M. (2004). Development of PCR assay based on ITS2 rDNA polymorphism for the detection and differentiation of *Fusarium sporotrichioides*. FEMS Microbiol. Lett. *239*, 181–186. http://dx.doi.org/10.1016/j.femsle.2004.08.037

Kulik, T., Pszczółkowska, A., Fordoński, G., and Olszewski, J. (2007). PCR approach based on the esyn1 gene for the detection of potential enniatin-producing *Fusarium* species. Int. J. Food Microbiol. *116*, 319–324. http://dx.doi.org/10.1016/j.ijfoodmicro.2007.02.003

Kulik, T. (2008). Detection of *Fusarium tricinctum* from cereal grain using PCR assay. J. Appl. Genet. *49*, 305–311. http://dx.doi.org/10.1007/BF03195628

Kuiper-Goodman, T. (1996). Risk assessment of ochratoxin A: an update. Food Addit. Contam. *13*, 53–57.

Kvas, M., Marasas, W.F.O., Wingfield, B.D., Wingfield, M.J., and Steenkamp, E.T. (2009). Diversity and evolution of *Fusarium* species in the *Gibberella fujikuroi* complex. Fung. Div. *34*, 1–21.

Lacey, J. (1989). Pre- and post-harvest ecology of fungi causing spoilage of foods and other stored products. Soc. Appl. Bacteriol. Symp. Ser. *18*, 11S–25S.

Laitila, A., Alakomi, H.L., Raaska, L., Mattila-Sandholm, T., and Haikara, A. (2002). Antifungal activities of two *Lactobacillus plantarum* strains against *Fusarium* moulds in vitro and in malting of barley. J. Appl. Microbiol. *93*, 566–576. http://dx.doi.org/10.1046/j.1365-2672.2002.01731.x

Laitila, A., Sweins, H., Vilpola, A., Kotaviita, E., Olkku, J., Home, S., and Haikara, A. (2006). *Lactobacillus plantarum* and *Pediococcus pentosaceus* starter cultures as a tool for microflora management in malting and for enhancement of malt processability. J. Agric. Food Chem. *54*, 3840–3851. http://dx.doi.org/10.1021/jf052979j

Laitila, A. (2007). Microbes in the Tailoring of Barley Malt Properties. Doctoral thesis, University of Helsinki, Helsinki, Finland (VTT Publications 645), 112 pp.

Laitila, A., Kotaviita, E., Peltola, P., Home, S., and Wilhelmson, A. (2007). Indigenous microbial community of barley greatly influences grain germination and malt quality. J. Inst. Brew. *113*, 9–20. http://dx.doi.org/10.1002/j.2050-0416.2007.tb00250.x

Lancova, K., Hajslova, J., Poustka, J., Krplova, A., Zachariasova, M., Dostalek, P., and Sachambula, L. (2008). Transfer of *Fusarium* mycotoxins and 'masked' deoxynivalenol (deoxynivalenol-3-glucoside) from field barley through malt to beer. Food Addit. Contam. Part A Chem. Anal. Control Expo. Risk Assess. *25*, 732–744. http://dx.doi.org/10.1080/02652030701779625

Larsen, T.O., Svendsen, A., and Smedsgaard, J. (2001). Biochemical characterization of ochratoxin A-producing strains of the genus *Penicillium*. Appl. Environ. Microbiol. *67*, 3630–3635. http://dx.doi.org/10.1128/AEM.67.8.3630-3635.2001

Lattanzio, V.M.T., Pascale, M., and Visconti, A. (2009). Current analytical methods for trichothecene mycotoxins in cereals. Trends Anal. Chem. *28*, 758–768. http://dx.doi.org/10.1016/j.trac.2009.04.012

Leisova, L., Kucera, L., Minarikova, V., and Ovesna, J. (2005). AFLP-based PCR markers that differentiate spot and net forms of Pyrenophora teres. Plant Pathol. *54*, 66–73. http://dx.doi.org/10.1111/j.1365-3059.2005.01117.x

Lew, H., Chelkowski, J., Pronczuk, P., and Edinger, W. (1996). Occurrence of the mycotoxin moniliformin in maize (*Zea mays* L.) ears infected by *Fusarium subglutinans* (Wollenw. & Reinking) Nelson *et al.* Food Addit. Contam. *13*, 321–324. http://dx.doi.org/10.1080/02652039609374414

Li, F.q., and Yoshizawa, T. (2000). *Alternaria* mycotoxins in weathered wheat from China. J. Agric. Food Chem. *48,* 2920–2924. http://dx.doi.org/10.1021/jf0000171

Li, W.L., Faris, J.D., Muthukrishnan, S., Liu, D.J., Chen, P.D., and Gill, B.S. (2001). Isolation and characterization of novel cDNA clones of acidic chitinases and β-1,3-glucanases from wheat spikes infected by *Fusarium graminearum*. Theor. Appl. Genet. *102,* 353–362. http://dx.doi.org/10.1007/s001220051653

Li, Y., Liu, X., and Lin, Z. (2012). Recent developments and applications of surface plasmon resonance biosensors for the detection of mycotoxins in foodstuffs. Food Chem. *132,* 1549–1554. http://dx.doi.org/10.1016/j.foodchem.2011.10.109

Li, M., Li, P., Wu, H., Zhang, Q., Ma, F., Zhang, Z., Ding, X., and Wang, H. (2014). An ultra-sensitive monoclonal antibody-based competitive enzyme immunoassay for sterigmatocystin in cereal and oil products. PLoS ONE *9,* e106415. http://dx.doi.org/10.1371/journal.pone.0106415

Li, J., and Macdonald, J. (2015). Advances in isothermal amplification: novel strategies inspired by biological processes. Biosens. Bioelectron. *64,* 196–211. http://dx.doi.org/10.1016/j.bios.2014.08.069

Lievens, B., Brouwer, M., Vanachter, A.C.R.C., Cammune, B.P.A., and Thomma, B.P.H.J. (2006). Real-time PCR for detection and quantification of fungal and oomycete tomato pathogens in plant and soil samples. Plant Sci. *171,* 155–165. http://dx.doi.org/10.1016/j.plantsci.2006.03.009

Levasseur-Garcia, C. (2012). Infrared spectroscopy applied to identification and detection of microorganisms and their metabolites on cereals (corn, wheat, and barley). In Agricultural Science, G. Aflakpui, ed. (Shanghai, CN: InTech), pp. 185–196.

Limón, M.C., Rodríguez-Ortiz, R., and Avalos, J. (2010). Bikaverin production and applications. Appl. Microbiol. Biotechnol. *87,* 21–29. http://dx.doi.org/10.1007/s00253-010-2551-1

Linko, M., Haikara, A., Ritala, A., and Penttila, M. (1998). Recent advances in the malting and brewing industry. J. Biotechnol. *65,* 85–98. http://dx.doi.org/10.1016/S0168-1656(98)00135-7

Liu, G.T., Qian, Y.Z., Zhang, P., Dong, W.H., Qi, Y.M., and Guo, H.T. (1992). Etiological role of *Alternaria alternata* in human esophageal cancer. Chin. Med. J. *105,* 394–400.

Logrieco, A., Moretti, A., Fornelli, F., Fogliano, V., Ritieni, A., Caiaffa, M.F., Randazzo, G., Bottalico, A., and Macchia, L. (1996). Fusaproliferin production by *Fusarium subglutinans* and its toxicity to *Artemia salina,* SF-9 insect cells, and IARC/LCL 171 human B lymphocytes. Appl. Environ. Microbiol. *62,* 3378–3384.

Logrieco, A., Arrigan, D.W., Brengel-Pesce, K., Siciliano, P., and Tothill, I. (2005). DNA arrays, electronic noses and tongues, biosensors and receptors for rapid detection of toxigenic fungi and mycotoxins: a review. Food Addit. Contam. *22,* 335–344. http://dx.doi.org/10.1080/02652030500070176

Lu, P., Marquardt, R.R., Fröhlich, A.A., and Mills, J.T. (1994). Detection of *Penicillium aurantiogriseum* by ELISA utilizing antibodies produced against exoantigens. Microbiol. (Reading) *140,* 3267-3276.

Luo, J., Vogel, R.F., and Niessen, L. (2012). Development and application of a loop-mediated isothermal amplification assay for rapid identification of aflatoxigenic molds and their detection in food samples. Int. J. Food Microbiol. *159,* 214–224. http://dx.doi.org/10.1016/j.ijfoodmicro.2012.08.018

Luo, J., Taniwaki, M.H., Iamanaka, B.T., Vogel, R.F., and Niessen, L. (2014). Application of loop-mediated isothermal amplification assays for direct identification of pure cultures of *Aspergillus flavus, A nomius,* and *A caelatus* and for their rapid detection in shelled Brazil nuts. Int. J. Food Microbiol. *172,* 5–12. http://dx.doi.org/10.1016/j.ijfoodmicro.2013.12.001

Lutterschmid, G., Stübner, M., Vogel, R.F., and Niessen, L. (2010). Induction of gushing with recombinant class II hydrophobin FcHyd5p from *Fusarium culmorum* and the impact of hop compounds on its gushing potential. J. Inst. Brew. *116,* 339–347. http://dx.doi.org/10.1002/j.2050-0416.2010.tb00784.x

Lutterschmid, G., Muranyi, M., Stübner, M., Vogel, R.F., and Niessen, L. (2011). Heterologous expression of surface-active proteins from barley and filamentous fungi in *Pichia pastoris* and characterization of their contribution to beer gushing. Int. J. Food Microbiol. *147,* 17–25. http://dx.doi.org/10.1016/j.ijfoodmicro.2011.02.030

Ma, X., Wang, W., Chen, X., Xia, Y., Wu, S., Duan, N., and Wang, Z. (2014). Selection, identification, and application of Aflatoxin B_1 aptamer. Eur. Food Res. Technol. *238* 919–925. http//dx.doi.org/10.1007/s00217-014-2176-1

Mably, M., Mankotia, M., Cavlovic, P., Tam, J., Wong, L., Pantazopoulos, P., Calway, P., and Scott, P.M. (2005). Survey of aflatoxins in beer sold in Canada. Food Addit. Contam. *22,* 1252–1257. http://dx.doi.org/10.1080/02652030500241884

Madden, L.V. (1997). Effects of rain on splash dispersal of fungal pathogens. Can. J. Plant Pathol. *19,* 225–230. http://dx.doi.org/10.1080/07060669709500557

Magan, N., and Lacey, J. (1984). Effects of gas composition and water activity on growth of field and storage fungi and their interactions. Transact. Brit. Mycol. Soc. *82,* 305–314.

Magan, N., Cayley, G.R., and Lacey, J. (1984). Effect of water activity and temperature on mycotoxin production by *Alternaria alternata* in culture and on wheat grain. Appl. Environ. Microbiol. *47,* 1113–1117.

Magan, N., Hope, R., Cairns, V., and Aldred, D. (2003). Post-harvest fungal ecology: impact of fungal growth and mycotoxin accumulation in stored grain. Eur. J. Plant Pathol. *109,* 723–730. http://dx.doi.org/10.1023/A:1026082425177

Magan, N., and Aldred, D. (2007). Post-harvest control strategies: minimizing mycotoxins in the food chain. Int. J. Food Microbiol. *119,* 131–139. http://dx.doi.org/10.1016/j.ijfoodmicro.2007.07.034

Magan, N., Aldred, D., Mylona, K., and Lambert, R.J. (2010). Limiting mycotoxins in stored wheat. Food Addit. Contam. Part A Chem. Anal. Control Expo. Risk Assess. *27,* 644–650. http://dx.doi.org/10.1080/19440040903514523

Magnoli, C., Hallak, C., Astoreca, A., Ponsone, L., Chiacchiera, S., and Dalcero, A.M. (2006). Occurrence of ochratoxin A-producing fungi in commercial corn kernels in Argentina. Mycopathologia *161*, 53–58. http://dx.doi.org/10.1007/s11046-005-0237-5

Magnoli, C.E., Astoreca, A.L., Chiacchiera, S.M., and Dalcero, A.M. (2007). Occurrence of ochratoxin A and ochratoxigenic mycoflora in corn and corn based foods and feeds in some South American countries. Mycopathologia *163*, 249–260. http://dx.doi.org/10.1007/s11046-007-9005-z

Mahnine, N., Meca, G., Elabidi, A., Fekhaoui, M., Saoiabi, A., Font, G., Mañes, J., and Zinedine, A. (2011). Further data on the levels of emerging Fusarium mycotoxins enniatins (A, A1, B, B1), beauvericin and fusaproliferin in breakfast and infant cereals from Morocco. Food Chem. *124*, 481–485. http://dx.doi.org/10.1016/j.foodchem.2010.06.058

Malhotra, S., Pandey, A.K., Rajput, Y.S., and Sharma, R. (2014). Selection of aptamers for aflatoxin M1 and their characterization. J. Mol. Recognit. *27*, 493–500. http://dx.doi.org/10.1002/jmr.2370

Mannon, J., and Johnson, E. (1985). Fungi down on the farm. New Scientist *105*, 12–16.

Månsson, M., Klejnstrup, M.L., Phipps, R.K., Nielsen, K.F., Frisvad, J.C., Gotfredsen, C.H., and Larsen, T.O. (2010). Isolation and NMR characterization of fumonisin B2 and a new fumonisin B6 from *Aspergillus niger*. J. Agric. Food Chem. *58*, 949–953. http://dx.doi.org/10.1021/jf902834g

Mapari, S.A., Nielsen, K.F., Larsen, T.O., Frisvad, J.C., Meyer, A.S., and Thrane, U. (2005). Exploring fungal biodiversity for the production of water-soluble pigments as potential natural food colorants. Curr. Opin. Biotechnol. *16*, 231–238. http://dx.doi.org/10.1016/j.copbio.2005.03.004

Mapari, S.A., Thrane, U., and Meyer, A.S. (2010). Fungal polyketide azaphilone pigments as future natural food colorants? Trends Biotechnol. *28*, 300–307. http://dx.doi.org/10.1016/j.tibtech.2010.03.004

Maragos, C.M., and Thompson, V.S. (1999). Fiber-optic immunosensor for mycotoxins. Nat. Toxins *7*, 371–376.

Maragos, C. (2006). Emerging technologies for mycotoxin detection. J. Toxicol. *23*, 317–344. http://dx.doi.org/10.1081/TXR-200027859

Maragos, C. (2009a). Fluorescence polarization immunoassay of mycotoxins: a review. Toxins *1*, 196–207. http://dx.doi.org/10.3390/toxins1020196

Maragos, C.M. (2009b). Recent advances in the development of novel materials for mycotoxin analysis. Anal. Bioanal. Chem. *395*, 1205–1213. http://dx.doi.org/10.1007/s00216-009-2728-6

Maragos, C.M., and Busman, M. (2010). Rapid and advanced tools for mycotoxin analysis: a review. Food Addit. Contam. Part. A. Chem. Anal. Control Expo. Risk Assess. *27*, 688–700. http://dx.doi.org/10.1080/19440040903515934

Marburger, D.A., Venkateshwaran, M., Conley, S.P., Esker, P.D., Lauer, J.G., and Ané, J.M. (2015). Crop rotation and management effect on *Fusarium* spp. populations. Crop Sci. *55*, 365–376. http://dx.doi.org/10.2135/cropsci2014.03.0199

Martin, J.T. (1964). Role of cuticule in the defense against plant disease. Annu. Rev. Phytopathol. *2*, 81–100.

Martini, M., Musetti, R., Grisan, S., Polizzotto, R., Borselli, S., Pavan, F., and Osler, R. (2009). DNA-dependent detection of the grapevine fungal endophytes *Aureobasidium pullulans* and *Epicoccum nigrum*. Plant Dis. *93*, 993–998. http://dx.doi.org/10.1094/PDIS-93-10-0993

Mateo, E.M., Gil-Serna, J., Patiño, B., and Jiménez, M. (2011). Aflatoxins and ochratoxin A in stored barley grain in Spain and impact of PCR-based strategies to assess the occurrence of aflatoxigenic and ochratoxigenic *Aspergillus* spp. Int. J. Food Microbiol. *149*, 118–126. http://dx.doi.org/10.1016/j.ijfoodmicro.2011.06.006

Matsuoka, Y., and Kubota, K. (1987). Studies on mechanisms of diarrhea induced by fusarenon-X, a trichothecene mycotoxin from *Fusarium* species: fusarenon-X-induced diarrhea is not mediated by cyclic nucleotides. Toxicol. Appl. Pharmacol. *91*, 326–332.

Matusinsky, P., Frei, P., Mikolasov, R., Svacinov, I., Tvaruzek, L., and Spitzer, T. (2010). Species specific detection of *Bipolaris sorokiniana* from wheat and barley tissues. Crop Prot. *29*, 1325–1330. http://dx.doi.org/10.1016/j.cropro.2010.07.013

McCormick, S.P., Stanley, A.M., Stover, N.A., and Alexander, N.J. (2011). Trichothecenes: from simple to complex mycotoxins. Toxins *3*, 802–814. http://dx.doi.org/10.3390/toxins3070802

McKeague, M., Bradley, C.R., De Girolamo, A., Visconti, A., Miller, J.D., and Derosa, M.C. (2010). Screening and initial binding assessment of fumonisin B_1 aptamers. Int. J. Mol. Sci. *11*, 4864–4881. http://dx.doi.org/10.3390/ijms11124864

McMullen, M., Bergstrom, G., De Wolf, E., Dill-Macky, R., Hershman, D., Shaner, G., and Van Sanford, D. (2012). A unified effort to fight an enemy of wheat and barley: Fusarium head blight. Plant Dis. *96*, 1712–1728. http://dx.doi.org/10.1094/PDIS-03-12-0291-FE

Meca, G., Zinedine, A., Blesa, J., Font, G., and Mañes, J. (2010). Further data on the presence of *Fusarium* emerging mycotoxins enniatins, fusaproliferin and beauvericin in cereals available on the Spanish markets. Food Chem. Toxicol. *48*, 1412–1416. http://dx.doi.org/10.1016/j.fct.2010.03.010

Meca, G., Zhou, T., Li, X.Z., and Mañes, J. (2013). Beauvericin degradation during bread and beer making. Food Contr. *34*, 1–8. http://dx.doi.org/10.1016/j.foodcont.2013.03.032

Medina, A., Jiménez, M., Gimeno-Adelantado, J.V., Valle-Algarra, F.M., and Mateo, R. (2005). Determination of ochratoxin A in beer marketed in Spain by liquid chromatography with fluorescence detection using lead hydroxyacetate as a clean-up agent. J. Chromatogr. A. *1083*, 7–13. http://dx.doi.org/10.1016/j.chroma.2005.05.089

Meirelles, P.G., Ono, M.A., Ohe, M.C.T., Maroneze, D.M., Itano, E.N., Garcia, G.T., Sugiura, Y., Ueno, Y., Hirooka, E.Y., and Ono, E.Y.S. (2006). Detection of *Fusarium* sp. contamination in corn by enzyme-linked immunosorbent assay. Food Agric. Immunol. *17*, 79-89. http://dx.doi.org/10.1080/09540100600688754

Meklin, T., Haugland, R.A., Reponen, T., Varma, M., Lummus, Z., Bernstein, D., Wymer, L.J., and Vesper, S.J. (2004). Quantitative PCR analysis of house dust can reveal abnormal mold conditions. J. Environ. Monit. *6*, 615–620. http://dx.doi.org/10.1039/b400250d

Merril, C.R., Scholl, D., and Adhya, S.L. (2003). The prospect for bacteriophage therapy in Western medicine. Nat. Rev. Drug Discov. *2,* 489–497. http://dx.doi.org/10.1038/nrd1111

Mesterhazy, A. (2014). Breeding for resistance to *Fusarium* head blight in wheat. In Mycotoxin Reduction on Grain Chains, J.F. Leslie, and A. Logrieco, eds. (Oxford, UK: John Wiley and Sons), pp. 189–208.

Meyer, U.M., and Dewey, F.M. (2000). Efficacy of different immunogens for raising monoclonal antibodies to *Botrytis cinerea*. Mycol. Res. *104,* 979–987.

Michaelis, J., Roloff, A., and Seitz, O. (2014). Amplification by nucleic acid-templated reactions. Org. Biomol. Chem. *12,* 2821–2833. http://dx.doi.org/10.1039/c4ob00096j

Miller, J.D. (1995). Fungi and mycotoxins in grain: implications for stored product research. J. Stored Prod. Res. *31,* 1–16.

Mills, J.T., and Frydman, C. (1980). Mycoflora and condition of grains from overwintered fields in Manitoba, 1977–78. Can. Plant Dis. Surv. *60,* 1–7.

Mink, G.I. (1993). Pollen and seed-transmitted viruses and viroids. Annu. Rev. Phytopathol. *31,* 375–402. http://dx.doi.org/10.1146/annurev.py.31.090193.002111

Mishra, P.K., Fox, R.T., and Culham, A. (2003). Development of a PCR-based assay for rapid and reliable identification of pathogenic Fusaria. FEMS Microbiol. Lett. *218,* 329–332. http://dx.doi.org/10.1111/j.1574-6968.2003.tb11537.x

Möller, E.M., Chelkowski, J., and Geiger, H.H. (1999). Species-specific PCR assays for the fungal pathogens *Fusarium moniliforme* and *Fusarium subglutinans* and their application to diagnose maize ear rot disease. J. Phytopath. (Berlin) *147,* 497–508. http://dx.doi.org/10.1046/j.1439-0434.1999.00380.x

Mogensen, J.M., Frisvad, J.C., Thrane, U., and Nielsen, K.F. (2010). Production of Fumonisin B2 and B4 by *Aspergillus niger* on grapes and raisins. J. Agric. Food Chem. *58,* 954–958. http://dx.doi.org/10.1021/jf903116q

Mogensen, J.M., Møller, K.A., von Freiesleben, P., Labuda, R., Varga, E., Sulyok, M., Kubátová, A., Thrane, U., Andersen, B., and Nielsen, K.F. (2011). Production of fumonisins B2 and B4 in *Tolypocladium* species. J. Ind. Microbiol. Biotechnol. *38,* 1329–1335. http://dx.doi.org/10.1007/s10295-010-0916-1

Moretti, A., Mulè, G., Ritieni, A., and Logrieco, A. (2007). Further data on the production of beauvericin, enniatins and fusaproliferin and toxicity to *Artemia salina* by *Fusarium* species of *Gibberella fujikuroi* species complex. Int. J. Food Microbiol. *118,* 158–163. http://dx.doi.org/10.1016/j.ijfoodmicro.2007.07.004

Morrissey, J.P., and Osbourn, A.E. (1999). Fungal resistance to plant antibiotics as a mechanism of pathogenesis. Microbiol. Mol. Biol. Rev. *63,* 708–724.

Moss, M.O., and Long, M.T. (2002). Fate of patulin in the presence of the yeast *Saccharomyces cerevisiae*. Food Addit. Contam. *19,* 387–399. http://dx.doi.org/10.1080/02652030110091163

Müller, C. (1995). Möglichkeiten zur Bewertung des Gesundheitszustandes von Braugerste und Malz. Brauwelt *134,* 1036–1054.

Müller, M.P., Schmid, F., Becker, T., and Gastl, M. (2010). Impact of different hop compounds on the overfoaming volume of beer caused by primary gushing. J. Inst. Brew. *116,* 459–463. http://dx.doi.org/10.1002/j.2050-0416.2010.tb00797.x

Müller, M.E.H., and Korn, U. (2013). *Alternaria* mycotoxins in wheat – A 10 years survey in the Northeast of Germany. Food Contr. *34,* 191–197. http://dx.doi.org/10.1016/j.foodcont.2013.04.018

Mukherjee, M., Watt, D.A., and Berjak, P. (2006). Molecular detection and diagnosis of fungal contaminants of recalcitrant seeds: *Quercus robur* L. acorns as a model system. Seed Sci. Technol. *34,* 415–427.

Mulé, G., Susca, A., Stea, G., and Moretti, A. (2004). A species-specific PCR assay based on the calmodulin partial gene for identification of *Fusarium verticillioides, F. proliferatum* and *F. subglutinans*. Eur. J. Plant Pathol. *110,* 495–502. http://dx.doi.org/10.1023/B:EJPP.0000032389.84048.71

Murphy, K.M. (2012). Janeway's Immunobiology. 8th edn (New York, USA: Taylor & Francis).

Musser, S.M., and Plattner, R.D. (1997). Fumonisin composition in cultures of *Fusarium moniliforme, Fusarium proliferatum,* and *Fusarium nygamai*. J. Agric. Food Chem. *45,* 1169–1173. http://dx.doi.org/10.1021/jf960663t

Nagamine, K., Hase, T., and Notomi, T. (2002). Accelerated reaction by loop-mediated isothermal amplification using loop primers. Mol. Cell. Probes *16,* 223–229. http://dx.doi.org/10.1006/mcpr.2002.0415

Nagao, K., Ota, T., Tanikawa, A., Takae, Y., Mori, T., Udagawa, S., and Nishikawa, T. (2005). Genetic identification and detection of human pathogenic *Rhizopus* species, a major mucormycosis agent, by multiplex PCR based on internal transcribed spacer region of rRNA gene. J. Dermatol. Sci. *39,* 23–31. http://dx.doi.org/10.1016/j.jdermsci.2005.01.010

Nakamura, M., Suprata, D.N., and Iwai, H. (2007). Differentiation of pathogenic and nonpathogenic isolates of *Geotrichum candidum* sensu Suprapta *et al.* (1995) on citrus fruit based on PCR-RFLP analysis of rDNA ITS and PCR using specific primers designed in polygalacturonase genes. Mycoscience *49,* 155–158. http://dx.doi.org/10.1007/s10267-007-0397-x

Nash, S.M., and Snyder, W.C. (1962). Quantitative estimations by plate counts of propagules of the bean root rot of *Fusarium* species in field soils. Phytopathology *52,* 567–572.

Newhouse, J.R., and Hunter, B.B. (1983). Selective media for recovery of *Cylindrocladium* and *Fusarium* species from roots and stems of tree seedlings. Mycologia *75,* 228–233.

Nicholson, P., Lees, A.K., Maurin, N., Parry, D.W., and Rezanoor, H.N. (1996). Development of a PCR assay to identify and quantify *Microdochium nivale var. nivale* and *Microdochium nivale var. majus* in wheat. Phys. Mol. Plant Pathol. *48,* 257–271. http://dx.doi.org/10.1006/pmpp.1996.0022

Nicholson, P., and Parry, D.W. (1996). Development and use of a PCR assay to detect *Rhizoctonia cerealis,* the cause of sharp eyespot in wheat. Plant Path.(Oxford) *45,* 872–883. http://dx.doi.org/10.1111/j.1365-3059.1996.tb02898.x

Nicholson, P., Simpson, D.R., Weston, G., Rezanoor, H.N., Lees, A.K., Parry, D.W., and Joyce, D. (1998). Detection and quantification of *Fusarium culmorum* and *Fusarium graminearum* in cereals using PCR assays. Phys. Mol. Plant Pathol. *53,* 17–37. http://dx.doi.org/10.1006/pmpp.1998.0170

Nicholson, P., Simpson, D.R., Wilson, A.H., Chandler, E., and Thomsett, M. (2004). Detection and differentiation of trichothecene and enniatin-producing *Fusarium* species on small-grain cereals. Eur. J. Plant Pathol. *110,* 503–514.

Nicolaisen, M., Suproniene, S., Nielsen, L.K., Lazzaro, I., Spliid, N.H., and Justesen, A.F. (2009). Real-time PCR for quantification of eleven individual *Fusarium* species in cereals. J. Microbiol. Methods *76,* 234–240. http://dx.doi.org/10.1016/j.mimet.2008.10.016

Nielsen, K.F., and Smedsgaard, J. (2003). Fungal metabolite screening: database of 474 mycotoxins and fungal metabolites for dereplication by standardised liquid chromatography–UV–mass spectrometry methodology. J. Chrom. A *1002,* 111–136. http://dx.doi.org/10.1016/S0021-9673(03)00490-4

Niessen, L., Donhauser, S., Weideneder, A., Geiger, E., and Vogel, H. (1991). Möglichkeiten einer verbesserten visuellen Beurteilung des mikrobiologischen Status von Malzen. Brauwelt *131,* 1556–1562.

Niessen, L., Donhauser, S., Weideneder, A., Geiger, E., and Vogel, H. (1992). Mykologische Untersuchungen an Cerealien und Malzen im Zusammenhang mit dem Wildwerden (Gushing) des Bieres. Brauwelt *132,* 702–714.

Niessen, L., and Donhauser, S. (1993). Fate of deoxynivalenol in the process of brewing and its prevalence in commercial beer. In Occurrence and Significance of Mycotoxins (Slough, UK: Central Science Laboratory), pp. 203–207.

Niessen, M.L., and Vogel, R.F. (1997). Specific identification of *Fusarium graminearum* by PCR with *gaoA* targeted primers. Syst. Appl. Microbiol. *20,* 111–123. http://dx.doi.org/10.1016/S0723-2020(97)80055-0

Niessen, L., Schmidt, H., and Vogel, R.F. (2004). The use of tri5 gene sequences for PCR detection and taxonomy of trichothecene-producing species in the *Fusarium* section *Sporotrichiella*. Int. J. Food Microbiol. *95,* 305–319. http://dx.doi.org/10.1016/j.ijfoodmicro.2003.12.009

Niessen, L. (2008). PCR-based diagnosis and quantification of mycotoxin-producing fungi. Adv. Food Nutr. Res. *54,* 81–138. http://dx.doi.org/10.1016/S1043-4526(07)00003-4

Niessen, L., and Vogel, R.F. (2010). Detection of *Fusarium graminearum* DNA using a loop-mediated isothermal amplification (LAMP) assay. Int. J. Food Microbiol. *140,* 183–191. http://dx.doi.org/10.1016/j.ijfoodmicro.2010.03.036

Niessen, L., Gräfenhan, T., and Vogel, R.F. (2012). ATP citrate lyase 1 (acl1) gene-based loop-mediated amplification assay for the detection of the *Fusarium tricinctum* species complex in pure cultures and in cereal samples. Int. J. Food Microbiol. *158,* 171–185. http://dx.doi.org/10.1016/j.ijfoodmicro.2012.06.021

Niessen, L., Luo, J., Denschlag, C., and Vogel, R.F. (2013). The application of loop-mediated isothermal amplification (LAMP) in food testing for bacterial pathogens and fungal contaminants. Food Microbiol. *36,* 191–206. http://dx.doi.org/10.1016/j.fm.2013.04.017

Niessen, L. (2015). Current state and future perspectives of loop-mediated isothermal amplification (LAMP)-based diagnosis of filamentous fungi and yeasts. Appl. Microbiol. Biotechnol. *99,* 553–574. http://dx.doi.org/10.1007/s00253-014-6196-3

Nightingale, M.J., Marchylo, B.A., Clear, R.M., Dexter, J.E., and Preston, K.R. (1999). Fusarium head blight: Effect of fungal proteases on wheat storage proteins. Cereal Chem. *76,* 150 158. http://dx.doi.org/10.1094/CCHEM.1999.76.1.150

Nirenberg, H.I. (1976). Untersuchungen über die morphologische und biologische Differenzierung in der *Fusarium*-Sektion Sporotrichiella. Mitt. Biol. Bundesanst. Land Forstwirtsch. Berlin Dahlem *127,* 1–93.

Nirenberg, H.I. (1981). A simplified method for identifying *Fusarium* spp. occurring on wheat. Can. J. Bot. *59,* 1599–1609.

Nishikawa, N., and Kohgo, M. (1975). A selective medium for genera *Fusarium* and *Gibberella* contaminating on barley and malt grains. Bull. Brew. Sci. *21,* 9–17.

Niu, B., Wang, D., Yang, Y., Xu, H., and Qiao, M. (2012). Heterologous expression and characterization of the hydrophobin HFBI in *Pichia pastoris* and evaluation of its contribution to the food industry. Amino Acids *43,* 763–771. http://dx.doi.org/10.1007/s00726-011-1126-5

Nkwe, D.O., Taylor, J.E., and Siame, B.A. (2005). Fungi, aflatoxins, fumonisin Bl and zearalenone contaminating sorghum-based traditional malt, wort and beer in Botswana. Mycopathologia *160,* 177–186. http://dx.doi.org/10.1007/s11046-005-6867-9

Noots, I., Delcour, J.A., and Michiels, C.W. (1999). From field barley to malt: detection and specification of microbial activity for quality aspects. Crit. Rev. Microbiol. *25,* 121–153. http://dx.doi.org/10.1080/10408419991299257

Notermans, S., and Heuvelman, C.J. (1985). Immunological detection of molds in food by using the ELISA: Preparation of antigens. Int. J. Food Microbiol. *2,* 247–258. http://dx.doi.org/10.1016/0168-1605(85)90015-7

Notermans, S., Heuvelman, C.J., Beumer, R.R., and Maas, R. (1986). Immunological detection of moulds in food: relation between antigen production and growth. Int. J. Food Microbiol. *3,* 253–261. http://dx.doi.org/10.1016/0168-1605(86)90002-4

Notomi, T., Okayama, H., Masubuchi, H., Yonekawa, T., Watanabe, K., Amino, N., and Hase, T. (2000). Loop-mediated isothermal amplification of DNA. Nucleic Acids Res. *28,* E63. http://dx.doi.org/10.1093/nar/28.12.e63

Oerke, E.C., and Dehne, H.W. (2004). Safeguarding production-losses in major crops and the role of crop protection. Crop Protect. *23,* 275–285. http://dx.doi.org/10.1016/j.cropro.2003.10.001

Olempska-Beer, Z.S., Merker, R.I., Ditto, M.D., and DiNovi, M.J. (2006). Food-processing enzymes from recombinant microorganisms – a review. Regul. Toxicol. Pharmacol. *45,* 144–158. http://dx.doi.org/10.1016/j.yrtph.2006.05.001

Oliveira, P.M., Mauch, A., Jacob, F., Waters, D.M., and Arendt, E.K. (2012). Fundamental study on the influence of *Fusarium* infection on quality and ultrastructure of barley malt. Int. J. Food Microbiol. *156*, 32–43. http://dx.doi.org/10.1016/j.ijfoodmicro.2012.02.019

Oliveira, P.M., Waters, D.M., and Arendt, E.K. (2013). The impact of *Fusarium culmorum* infection on the protein fractions of raw barley and malted grains. Appl. Microbiol. Biotechnol. *97*, 2053–2065. http://dx.doi.org/10.1007/s00253-013-4696-1

O'Mahony, A., O'Sullivan, T., Walsh, Y., Vaughan, A., Maher, M., Fitzgerald, G.F., and Sinderen, D. (2000). Characterisation of antimicrobial producing lactic acid bacteria from malted barley. J. Inst. Brew. *106*, 403–410. http://dx.doi.org/10.1002/j.2050-0416.2000.tb00531.x

Ooka, J.J., and Kommedahl, T. (1977). Wind and rain dispersal of *Fusarium moniliforme* in corn fields. Phytopathology *67*, 1023–1026.

Osborne, L.E., and Stein, J.M. (2007). Epidemiology of *Fusarium* head blight on small-grain cereals. Int. J. Food Microbiol. *119*, 103–108. http://dx.doi.org/10.1016/j.ijfoodmicro.2007.07.032

Østergaard, L.H., and Olsen, H.S. (2011). Industrial applications of fungal enzymes. In The Mycota, vol. 10 Industrial Applications, M. Hofrichter, ed. (New York, USA: Springer), pp. 269–290.

Ostry, V. (2008). *Alternaria* mycotoxins: an overview of chemical characterization, producers, toxicity, analysis and occurrence in foodstuffs. World Mycotox. J. *1*, 175–188. http://dx.doi.org/10.3920/WMJ2008.x013

Ostry, V., Malir, F., and Ruprich, J. (2013). Producers and important dietary sources of ochratoxin A and citrinin. Toxins *5*, 1574–1586. http://dx.doi.org/10.3390/toxins5091574

Oueslati, S., Meca, G., Mliki, A., Ghorbel, A., and Mañes, J. (2011). Determination of *Fusarium* mycotoxins enniatins, beauvericin and fusaproliferin in cereals and derived products from Tunisia. Food Contr. *22*, 1373–1377. http://dx.doi.org/10.1016/j.foodcont.2011.02.015

Owen, J., Punt, J., and Stranford, S. (2013). Kuby Immunology, 7th edn (New York, USA: W.H. Freeman and Company).

Park, J.W., Kim, E.K., Shon, D.H., and Kim, Y.B. (2002). Natural co-occurrence of aflatoxin B1, fumonisin B1 and ochratoxin A in barley and corn foods from Korea. Food Addit. Contam. *19*, 1073–1080. http://dx.doi.org/10.1080/02652030210151840

Parry, D.W., and Nicholson, P. (1996). Development of a PCR assay to detect *Fusarium poae* in wheat. Plant Pathol. (Oxford) *45*, 383–391. http://dx.doi.org/10.1046/j.1365-3059.1996.d01-133.x

Paster, N., Menasherov, M., Ravid, U., and Juven, B. (1995). Antifungal activity of oregano and thyme essential oils applied as fumigants against fungi attacking stored grain. J. Food Protect. *58*, 81–85.

Patiño, B., Mirete, S., González-Jaén, M.T., Mulé, G., Rodríguez, M.T., and Vázquez, C. (2004). PCR detection assay of fumonisin-producing *Fusarium verticillioides* strains. J. Food Prot. *67*, 1278–1283.

Pearson, M.N., Beever, R.E., Boine, B., and Arthur, K. (2009). Mycoviruses of filamentous fungi and their relevance to plant pathology. Mol. Plant Pathol. *10*, 115–128. http://dx.doi.org/10.1111/j.1364-3703.2008.00503.x

Pedersen, L.H., Skouboe, P., Boysen, M., Soule, J., and Rossen, L. (1997). Detection of *Penicillium* species in complex food samples using the polymerase chain reaction. Int. J. Food Microbiol. *35*, 169–177. http://dx.doi.org/10.1016/S0168-1605(96)01216-0

Peeters, H., Zocher, R., and Kleinkauf, H. (1988). Synthesis of beauvericin by a multifunctional enzyme. J. Antibiot. *41*, 352–359. http://doi.org/10.7164/antibiotics.41.352

Pekkarinen, A.I., Sarlin, T.H., Laitila, A.T., Haikara, A.I., and Jones, B.L. (2003). *Fusarium* species synthesize alkaline proteinases in infested barley. J. Cereal Sci. *37*, 349–356. http://dx.doi.org/10.1006/jcrs.2002.0512

Pellaud, J. (2002). Gushing: State of the art. Cervisia *27*, 189–205.

Pepper, E.H., and Kiesling, R.L. (1963). A list of bacteria, fungi, yeasts, nematodes, and viruses occurring on and within barley kernels. Proc. Assoc. Off. Seed Anal. *53*, 199–207.

Petersson, S., and Schnürer, J. (1995). Biocontrol of mold growth in high-moisture wheat stored under airtight conditions by *Pichia anomala, Pichia guilliermondii,* and *Saccharomyces cerevisiae*. Appl. Environ. Microbiol. *61*, 1027–1032.

Petzinger, E., and Ziegler, K. (2000). Ochratoxin A from a toxicological perspective. J. Vet. Pharmacol. Ther. *23*, 91–98. http://dx.doi.org/10.1046/j.1365-2885.2000.00244.x

Pietri, A., Bertuzzi, T., Agosti, B., and Donadini, G. (2010). Transfer of aflatoxin B1 and fumonisin B1 from naturally contaminated raw materials to beer during an industrial brewing process. Food Addit. Contam. Part A. Chem. Anal. Control Expo. Risk Assess. 27, 1431–1439. http://dx.doi.org/10.1080/19440049.2010.489912

Pineda-Valdes, G., and Bullerman, L.B. (2010). Thermal stability of moniliformin at varying temperature, pH, and time in an aqueous environment. J. Food Prot. *11*, 1467–1609.

Pitt, J.I., Hocking, A.D., and Glenn, D.R. (1983). An improved medium for the detection of *Aspergillus flavus* and *A. parasiticus*. J. Appl. Bacteriol. *54*, 109–114. http://dx.doi.org/10.1111/j.1365-2672.1983.tb01307.x

Pitt, J.I., and Hocking, A.D. (2009). Methods for isolation, enumeration and identification. In Fungi and Food Spoilage, 2nd. edition, J.I. Pitt, and A.D. Hocking, eds. (New York, USA: Springer), pp. 19–52.

Pleadin, J., Perši, N., Zadravec, M., Sokolović, M., Vulić, A., Jaki, V., and Mitak, M. (2012). Correlation of deoxynivalenol and fumonisin concentration determined in maize by ELISA methods. J. Immunoassay Immunochem. 33, 414–421. http://dx.doi.org/10.1080/15321819.2012.662569

Pohanka, M., Jun, D., and Kuca, K. (2007). Mycotoxin assays using biosensor technology: a review. Drug Chem. Toxicol. *30*, 253–261. http://dx.doi.org/10.1080/01480540701375232

Pollock, G.A., DiSabatino, C.E., Heimsch, R.C., and Hilbelink, D.R. (1982). The subchronic toxicity and teratogenicity of alternariol monomethyl ether produced by *Alternaria solani*. Food Chem. Toxicol. *20*, 899–902.

Pritsch, C., Muehlbauer, G.J., Bushnell, W.R., Somers, D.A., and Vance, C.P. (2000). Fungal development and induction of defense response genes during early infection of wheat spikes by *Fusarium graminearum*. Mol. Plant Microbe Interact. *13*, 159–169. http://dx.doi.org/10.1094/MPMI.2000.13.2.159

Proctor, R.H., Plattner, R.D., Brown, D.W., Seo, J.A., and Lee, Y.W. (2004). Discontinuous distribution of fumonisin biosynthetic genes in the *Gibberella fujikuroi* species complex. Mycol. Res. *108*, 815–822. https://doi.org/10.1017/S0953756204000577

Pugh, G.J.F., and Boddy, L. (1988). A view of disturbance and life strategies in fungi. Proc. Royal Soc. Edinb. Sect. B. Biol. Sci. *94*, 3–11.

Rahmani, A., Jinap, S., and Soleimany, F. (2009). Qualitative and quantitative analysis of Mycotoxins. Comp. Rev. Food Sci. Food Safety *8*, 202–251. http://dx.doi.org/10.1111/j.1541-4337.2009.00079.x

Rank, C., Nielsen, K.F., Larsen, T.O., Varga, J., Samson, R.A., and Frisvad, J.C. (2011). Distribution of sterigmatocystin in filamentous fungi. Fungal Biol. *115*, 406–420. http://dx.doi.org/10.1016/j.funbio.2011.02.013

Rasmussen, P.H., Nielsen, K.F., Ghorbani, F., Spliid, N.H., Nielsen, G.C., and Jørgensen, L.N. (2012). Occurrence of different trichothecenes and deoxynivalenol-3-β-D-glucoside in naturally and artificially contaminated Danish cereal grains and whole maize plants. Mycotoxin Res. *28*, 181–190. http://dx.doi.org/10.1007/s12550-012-0133-z

Raulio, M., Wilhelmson, A., Salkinoja-Salonen, M., and Laitila, A. (2009). Ultrastructure of biofilms formed on barley kernels during malting with and without starter culture. Food Microbiol. *26*, 437–443. http://dx.doi.org/10.1016/j.fm.2009.02.004

Reddy, R.V., Mayura, K., Hayes, A.W., and Berndt, W.O. (1982). Embryocidal, teratogenic and fetotoxic effects of citrinin in rats. Toxicology *25*, 151–160.

Richard, J.L. (2007). Some major mycotoxins and their mycotoxicoses – an overview. Int. J. Food Microbiol. *119*, 3–10. http://dx.doi.org/10.1016/j.ijfoodmicro.2007.07.019

Rigotti, S., Gindro, K., Richter, H., and Viret, O. (2002). Characterization of molecular markers for specific and sensitive detection of *Botrytis cinerea* Pers.: Fr. in strawberry (*Fragaria × ananassa* Duch.) using PCR. FEMS Microbiol. Lett. *209*, 169–174. http://dx.doi.org/10.1111/j.1574-6968.2002.tb11127.x

Ritieni, A., Moretti, A., Logrieco, A., Bottalico, A., Randazzo, G., Monti, S.M., Ferracane, R., and Fogliano, V. (1997). Occurrence of fusaproliferin, fumonisin B1, and beauvericin in maize from Italy. J. Agric. Food Chem. *45*, 4011–4016. http://dx.doi.org/10.1021/jf9702151

Roberts, C.A., Marquardt, R.R.. Frohlich, A.A., McGraw, R.L., Rotter, R.G., and Hening, J.C. (1991). Chemical and spectral quantification of mold in contaminated barley. Cereal Chem. *68*, 272–275.

Roberts, T.A., Cordier, J.L., Gram, L., Tompkin, R.B., Pitt, J.I., Gorris, L.G.M., and Swanson, K.M.J. (2005). Microorganisms in Food. In Microbial Ecology of Food Commodities, vol. 6, 2nd. edition (New York, USA: Springer), pp. 392–439.

Rocha, O., Ansari, K., and Doohan, F.M. (2005). Effects of trichothecene mycotoxins on eukaryotic cells: a review. Food Addit. Contam. *22*, 369–378. http://dx.doi.org/10.1080/02652030500058403

Rodríguez-Carrasco, Y., Ruiz, M.J., Font, G., and Berrada, H. (2013). Exposure estimates to Fusarium mycotoxins through cereals intake. Chemosphere *93*, 2297–2303. http://dx.doi.org/10.1016/j.chemosphere.2013.07.086

Rohde, S., and Rabenstein, F. (2005). Standardization of an indirect PTA-ELISA for detection of *Fusarium* spp. in infected grains. Mycotoxin Res. *21*, 100–104. http://dx.doi.org/10.1007/BF02954429

Romero-Gonzalez, R., Martinez Vidal, J.L., Aguilera-Luiz, M.M., and Garrido Frenich, A. (2009). Application of conventional solid-phase extraction for multimycotoxin analysis in beers by ultrahigh-performance liquid chromatography-tandem mass spectrometry. J. Agric. Food Chem. *57*, 9385–9392. http://dx.doi.org/10.1021/jf903154a

Rouse, S., and van Sinderen, D. (2008). Bioprotective potential of lactic acid bacteria in malting and brewing. J. Food Prot. *71*, 1724–1733.

Rubert, J., Mañes, J., James, K.J., and Soler, C. (2011). Application of hybrid linear ion trap-high resolution mass spectrometry to the analysis of mycotoxins in beer. Food Addit. Contam. Part A Chem. Anal. Control Expo. Risk Assess. *28*, 1438–1446. http://dx.doi.org/10.1080/19440049.2011.595015

Ryan, C.A. (1990). Protease inhibitors in plants: Genes for improving defenses against insects and pathogens. Annu. Rev. Phytopathol. *28*, 425–449.

Riazantsev, D.Iu., Abramova, S.L., Evstratova, S.V., Gagkaeva, T.Iu., and Zavriev, S.K. (2008). [FLASH-PCR diagnostics of toxigenic fungi of the genus *Fusarium*.] Bioorg. Khim. *34*, 799–807.

Rychlik, M., and Asam, S. (2008). Stable isotope dilution assays in mycotoxin analysis. Anal. Bioanal. Chem. *390*, 617–628. http://dx.doi.org/10.1007/s00216-007-1717-x

Salgado, J.D., Madden, L.V., and Paul, P.A. (2014). Efficacy and economics of integrating in-field and harvesting strategies to manage *Fusarium* head blight of wheat. Plant Dis. *98*, 1407–1421. http://dx.doi.org/10.1094/PDIS-01-14-0093-RE

Sampson, T. (2003). Aptamers and SELEX: the technology. World Pat. Inform. *25*, 123–129.

Samson, R.A., Houbraken, J., Thrane, U., Frisvad, J.C., and Anderson, B. (2010). Food and Indoor Fungi (Utrecht, NL: Centraalbureau voor Schimmelcultures).

Sanchez-Rangel, D., SanJuan-Badillo, A., and Plasencia, J. (2005). Fumonsin production by *Fusarium verticillioides* strains isolated from maize in Mexico and development of a polymerase chain reaction to detect potential toxigenic strains in grains. Agric. Food Chem. *53*, 8565-8571. http://dx.doi.org/10.1021/jf0514827

Santos, C., Fraga, M.E., Kozakiewicz, Z., and Lima, N. (2010). Fourier transform infrared as a powerful technique for the identification and characterization of filamentous fungi and yeasts. Res. Microbiol. *161*, 168–175. http://dx.doi.org/10.1016/j.resmic.2009.12.007

Sarlin, T., Nakari-Setälä, T., Linder, M., Penttilä, M., and Haikara, A. (2005a). Fungal hydrophobins as predictors of the gushing activity of malt. J. Inst. Brew. *111*, 105–111. http://dx.doi.org/10.1002/j.2050-0416.2005.tb00655.x

Sarlin, T., Laitila, A., Pekkarinen, A., and Haikara, A. (2005b). Effects of three Fusarium species on the quality of barley and malt. J. Am. Soc. Brew. Chem. *63*, 43–49.

Sarlin, T., Vilpola, A., Kotaviita, E., Olkku, J., and Haikara, A. (2007). Fungal hydrophobins in the barley-to-beer chain. J. Inst. Brew. *113*, 147–153. http://dx.doi.org/10.1002/j.2050-0416.2007.tb00271.x

Sarlin, T., Kivioja, T., Kalkkinen, N., Linder, M.B., and Nakari-Setälä, T. (2012). Identification and characterization of gushing-active hydrophobins from *Fusarium graminearum* and related species. J. Basic Microbiol. *52*, 184–194. http://dx.doi.org/10.1002/jobm.201100053

Sasaki, R., Hossain, M.Z., Abe, N., Uchigashima, M., and Goto, T. (2014). Development of an analytical method for the determination of sterigmatocystin in grains using LCMS after immunoaffinity column purification. Mycotoxin Res. *30*, 123–129. http://dx.doi.org/10.1007/s12550-014-0196-0

Saxena, N., Ansari, K.M., Kumar, R., Chaudhari, B.P., Dwivedi, P.D., and Das, M. (2011). Role of mitogen activated protein kinases in skin tumorigenicity of patulin. Toxicol. Appl. Pharmacol. *257*, 264–271. http://dx.doi.org/10.1016/j.taap.2011.09.012

Schaeffer, W.I., Smith, N.E., and Payne, P.A. (1975). Physiological and biochemical effects of the mycotoxin patulin on Chang liver cell cultures. In Vitro *11*, 69–77.

Schena, L., Finetti Sialer, M., and Gallitelli, D. (2000a). Specific identification of *Aureobasidium pullulans* strain L47 using Scorpion PCR. EPPO Bull. *30*, 559–562. http://dx.doi.org/10.1111/j.1365-2338.2000.tb00947.x

Schena, L., Finetti Sialer, M., and Gallitelli, D. (2002b). Molecular detection of strain L47 of *Aureobasidium pullulans*, a biocontrol agent of postharvest diseases. Plant Dis. *86*, 54–60. http://dx.doi.org/10.1094/PDIS.2002.86.1.54

Schilling, A.G., Moeller, E.M., and Geiger, H.H. (1996). Polymerase chain reaction-based assays for species-specific detection of *Fusarium culmorum*, *F. graminearum* and *F. avenaceum*. Phytopathology *86*, 515–522.

Schmidt-Heydt, M., Stoll, D., Schütz, P., and Geisen, R. (2015). Oxidative stress induces the biosynthesis of citrinin by *Penicillium verrucosum* at the expense of ochratoxin. Int. J. Food Microbiol. *192*, 1–6. http://dx.doi.org/10.1016/j.ijfoodmicro.2014.09.008

Scott, P.M., and Stoltz, D.R. (1980). Mutagens produced by *Alternaria alternata*. Mutat. Res. *78*, 33–40.

Shephard, G.S. (2008). Determination of mycotoxins in human foods. Chem. Soc. Rev. *37*, 2468–2477. http://dx.doi.org/10.1039/b713084h

Shokribousjein, Z., Deckers, S.M., Gebruers, K., Lorgouilloux, Y., Baggerman, G., Verachtert, H., Delcour, J.A., Etienne, P., Rock, J.M., Michiels, C., and Derdelinckx, G. (2011). Hydrophobins, beer foaming and gushing. Cerevisia *35*, 85–101. http://dx.doi.org/10.1016/j.cervis.2010.12.001

Shokribousjein, Z., Philippaerts, A., Riveros, D., Titze, J., Ford, Y., Deckers, S.M., Khalesi, M., Delcour, J.A., Gebruers, K., Verachtert, H., Ilberg, V., Derdelinckx, G., and Sels, B. (2014). A curative method for primary gGushing of beer and carbonated beverages: Characterization and application of antifoam based on hop oils. J. Am. Soc. Brew. Chem. *72*, 12–21.

Scholes, J.D., and Farrar, J.F. (1986). Increased rates of photosynthesis in localized regions of a barley leaf infected with brown rust. New Phytol. *104*, 601–612.

Schrödter, R. (2004). Influence of harvest and storage conditions on trichothecenes levels in various cereals. Toxicol. Lett. *153*, 47–49. http://dx.doi.org/10.1016/j.toxlet.2004.04.043

Schumacher, T. (2002). Gushing in Fruchtsaftschorlen. Teil 1– Ursachen für die starke Schaumbildung. Getränkeindustrie *7*, 8–10.

Schwarz, P.B., Beattie, S., and Casper, H.H. (1996). Relationship between *Fusarium* infestation of barley and the gushing potential of malt. J. Inst. Brew. *102*, 93–96. http://dx.doi.org/10.1002/j.2050-0416.1996.tb00899.x

Schwarz, P.B., Jones, B.L., and Steffenson, B.J. (2002). Enzymes associated with *Fusarium* infection of barley. J. Am. Soc. Brew. Chem. *60*, 130–134.

Selitrennikoff, C.P. (2001). Antifungal proteins. Appl. Environ. Microbiol. *67*, 2883–2894. http://dx.doi.org/10.1128/AEM.67.7.2883-2894.2001

Sels, J., Mathys, J., De Coninck, B.M., Cammue, B.P., and De Bolle, M.F. (2008). Plant pathogenesis-related (PR) proteins: a focus on PR peptides. Plant Physiol. Biochem. *46*, 941–950. http://dx.doi.org/10.1016/j.plaphy.2008.06.011

Shapira, R., Paster, N., Menasherov, M., Eyal, O., Mett, A., Meiron, T., Kuttin, E., and Salomon, R. (1997). Development of polyclonal antibodies for detection of aflatoxigenic molds involving culture filtrate and chimeric proteins expressed in *Escherichia coli*. Appl. Environ. Microbiol. *63*, 990 995.

Sharman, M., Gilbert, J., and Chelkowski, J. (1991). A survey of the occurrence of the mycotoxin moniliformin in cereal samples from sources worldwide. Food Addit. Contam. *8*, 459–466. http://dx.doi.org/10.1080/02652039109373996

Shephard, G.S., van der Westhuizen, L., Gatyeni, P.M., Somdyala, N.I., Burger, H.M., and Marasas, W.F. (2005). Fumonisin mycotoxins in traditional Xhosa maize beer in South Africa. J. Agric. Food Chem. *53*, 9634–9637. http://dx.doi.org/10.1021/jf0516080

Shokribousjein, Z., Deckers, S.M., Gebruers, K., Lorgouilloux, Y., Baggerman, G., Verachtert, H., Delcour, J.A., Etienne, P., Rock, J.M., Michiels, C., and Derdelinckx, G. (2011). Hydrophobins, beer foaming and gushing. Cerevisiae *35*, 85–101. http://dx.doi.org/10.1016/j.cervis.2010.12.001

Siegel, D., Merkel, S., Koch, M., and Nehls, I. (2010a). Quantification of the *Alternaria* mycotoxin tenuazonic acid in beer. Food Chem. *120*, 902–906. http://dx.doi.org/10.1016/j.foodchem.2009.10.070

Siegel, D., Feist, M., Proske, M., Koch, M., and Nehls, I. (2010b). Degradation of the *Alternaria* mycotoxins alternariol, alternariol monomethyl ether, and altenuene upon bread baking. J. Agric. Food Chem. *58*, 9622–9630. http://dx.doi.org/10.1021/jf102156w

Song, S., Liu, N., Zhao, Z., Njumbe Ediage, E., Wu, S., Sun, C., De Saeger, S., and Wu, A. (2014). Multiplex lateral flow immunoassay for mycotoxin determination. Anal.

Chem. *86*, 4995–5001. http://dx.doi.org/10.1021/ac500540z

Soriano, J.M., and Dragacci, S. (2004). Occurrence of fumonisins in foods. Food Res. Int. *37*, 985–1000. http://dx.doi.org/10.1016/j.foodres.2004.06.009

Specker, C., Niessen, L., and Vogel, R.F. (2014). In vitro studies on the main beer protein Z4 of *Hordeum vulgare* concerning heat stability, protease inhibition and gushing. J. Inst. Brew. *120*, 85–92. http://dx.doi.org/10.1002/jib.118

Specker, C. (2014). Analysis of the interaction of gushing inducing hydrophobins with beer foam proteins. Doctoral thesis, Technische Universität München, Munich, Germany 174 pp.

Spicher, G. (1989). Fungi on cereals and their effect on quality with special consideration of malting barley. Monatsschr. Brauwiss. *42*, 68–77.

Stakheev, A.A., Ryazantsev, D.Y., Gagkaeva, T.Y., and Zavriev, S.K. (2011). PCR detection of *Fusarium* fungi with similar profiles of the produced mycotoxins. Food Control *22*, 462–468. http://dx.doi.org/10.1016/j.foodcont.2010.09.028

Stĕtina, R., and Votava, M. (1986). Induction of DNA single-strand breaks and DNA synthesis inhibition by patulin, ochratoxin A, citrinin, and aflatoxin B1 in cell lines CHO and AWRF. Folia Biol. *32*, 128–144.

Stockmann-Juvala, H., and Savolainen, K. (2008). A review of the toxic effects and mechanisms of action of fumonisin B1. Hum. Exp. Toxicol. *27*, 799–809. http://dx.doi.org/10.1177/0960327108099525

Strange, R.N., Majer, J.R., and Smith, H. (1974). The isolation and identification of choline and betaine as the two major components in anthers and wheat germ that stimulate *Fusarium graminearum* in vitro. Physiol. Plant Pathol. *4*, 277–290.

Stanković, S., Lević, J., Ivanović, D., Krnjaja, V., Stanković, G., and Tančić, S. (2012). Fumonisin B1 and its co-occurrence with other fusariotoxins in naturally-contaminated wheat grain. Food Contr. *23*, 384–388. http://dx.doi.org/10.1016/j.foodcont.2011.08.003

Streit, E., Schwab, C., Sulyok, M., Naehrer, K., Krska, R., and Schatzmayr, G. (2013). Multi-mycotoxin screening reveals the occurrence of 139 different secondary metabolites in feed and feed ingredients. Toxins *5*, 504–523. http://dx.doi.org/10.3390/toxins5030504

Stübner, M., Lutterschmid, G., Vogel, R.F., and Niessen, L. (2010). Heterologous expression of the hydrophobin FcHyd5p from *Fusarium culmorum* in *Pichia pastoris* and evaluation of its surface activity contributing to gushing of carbonated beverages. Int. J. Food Microbiol. *141*, 110–115. http://dx.doi.org/10.1016/j.ijfoodmicro.2010.03.003

Suanthie, Y., Cousin, M.A., and Woloshuk, C.P. (2009). Multiplex real-time PCR for detection and quantification of mycotoxigenic *Aspergillus*, *Penicillium* and *Fusarium*. J. Stored Prod. Res. *45*, 139–145. http://dx.doi.org/10.1016/j.jspr.2008.12.001

Subramanian, G. (2004). Antibodies: Volume 1–Production and Purification (New York, USA: Kluwer Academic).

Subramoni, S., Suárez-Moreno, Z.R., and Venturi, V. (2010). Lipases as pathogenicity factors of plant pathogens, In Handbook of Hydrocarbon and Lipid Microbiology, K.N. Timmis, ed. (New York, USA: Springer), pp 3269–3277.

Sydenham, E.W., Thiel, P.G., Marasas, W.F., Shephard, G.S., Van Schalkwyk, D.J., and Koch, K.R. (1990). Natural occurrence of some *Fusarium* mycotoxins in corn from low and high esophageal cancer prevalence areas of the Transkei, Southern Africa. J. Agric. Food Chem. *38*, 1900–1903.

Tateishi, H., Miyake, T., Mori, M., Sakuma, Y., and Saishoji, T. (2014). Effect of application timing of metconazole on *Fusarium* head blight development and mycotoxin contamination in wheat and barley. J. Pest. Sci. *39*, 1–6. http://doi.org/10.1584/jpestics.D12-077

Tamura, M., Uyama, A., and Mochizuki, N. (2011). Development of a multi-mycotoxin analysis in beer-based drinks by a modified QuEChERS method and ultra-high-performance liquid chromatography coupled with tandem mass spectrometry. Anal. Sci. *27*, 629–635. http://doi.org/10.2116/analsci.27.629

Tangni, E.K., Ponchaut, S., Maudoux, M., Rozenberg, R., and Larondelle, Y. (2002). Ochratoxin A in domestic and imported beers in Belgium: occurrence and exposure assessment. Food Addit. Contam. *19*, 1169–1179. http://dx.doi.org/10.1080/02652030210007859

Talbot, N.J. (1997). Fungal Biology: Growing into the air. Curr. Biol. *7*, R78–81. http://dx.doi.org/10.1016/S0960-9822(06)00041-8

Thompson, V.S., and Maragos, C.M. (1996). Fiber-optic immunosensor for the detection of fumonisin B_1. J. Agric. Food Chem. *44*, 1041–1046. http://dx.doi.org/10.1021/jf950661t

Tittlemier, S.A., Gaba, D., and Chan, J.M. (2013). Monitoring of *Fusarium* trichothecenes in Canadian cereal grain shipments from 2010 to 2012. J. Agric. Food Chem. *61*, 7412–7418. http://dx.doi.org/10.1021/jf4019257

Tomita, N., Mori, Y., Kanda, H., and Notomi, T. (2008). Loop-mediated isothermal amplification (LAMP) of gene sequences and simple visual detection of products. Nat. Protoc. *3*, 877–882. http://dx.doi.org/10.1038/nprot.2008.57

Tsai, G.J., and Cousin, M.A. (1990). Enzyme-linked immunosorbent assay for detection of molds in cheese and yogurt. J. Dairy Sci. *73*, 3366–3378.

Tsai, G.J., and Yu, S.C. (1999). Detecting *Aspergillus parasiticus* in cereals by an enzyme-linked immunosorbent assay. Int. J. Food Microbiol. *50*, 181–189. http://dx.doi.org/10.1016/S0168-1605(99)00084-7

Turner, A.S., Lees, A.K., Rezanoor, H.N., and Nicholson, P. (1998). Refinement of PCR-detection of *Fusarium avenaceum* and evidence from DNA marker studies for phenetic relatedness to *Fusarium tricinctum*. Plant Pathol. (Oxford) *47*, 278–288. http://dx.doi.org/10.1046/j.1365-3059.1998.00250.x

Turner, N.W., Subrahmanyam, S., and Piletsky, S.A. (2009). Analytical methods for determination of mycotoxins: a review. Anal. Chim. Acta *632*, 168–180. http://dx.doi.org/10.1016/j.aca.2008.11.010

Uhlig, S., Torp, M., Jarp, J., Parich, A., Gutleb, A.C., and Krska, R. (2004). Moniliformin in Norwegian grain. Food Addit. Contam. *21*, 598–606. http://dx.doi.org/10.1080/02652030410001704258

Uhlig, S., Torp, M., and Heier, B.T. (2006). Beauvericin and enniatins A, A1, B and B1 in Norwegian grain: a survey. Food Chem. *94*, 193–201. http://dx.doi.org/10.1016/j.foodchem.2004.11.004

Vaclavikova, M., Malachova, A., Veprikova, Z., Dzuman, Z., Zachariasova, M., and Hajslova, J. (2013). 'Emerging' mycotoxins in cereals processing chains: changes of enniatins during beer and bread making. Food Chem. *136*, 750–757. http://dx.doi.org/10.1016/j.foodchem.2012.08.031

van Egmond, H.P., and Jonker, M.A. (2005). Worldwide regulations on aflatoxins. In Aflatoxin and Food Safety–Food Science and Technology, H.K. Abbas, ed. (Boca Raton, USA: CRC Press), pp. 77–93.

van Egmond, H.P., Schothorst, R.C., and Jonker, M.A. (2007). Regulations relating to mycotoxins in food: perspectives in a global and European context. Anal. Bioanal. Chem. *389*, 147–157. http://dx.doi.org/10.1007/s00216-007-1317-9

van Loon, L.C., Rep, M., and Pieterse, C.M. (2006). Significance of inducible defense-related proteins in infected plants. Annu. Rev. Phytopathol. *44*, 135–162. http://dx.doi.org/10.1146/annurev.phyto.44.070505.143425

Van Nierop, S.N., Evans, D.E., Axcell, B.C., Cantrell, I.C., and Rautenbach, M. (2004). Impact of different wort boiling temperatures on the beer foam stabilizing properties of lipid transfer protein 1. J. Agric. Food Chem. *52*, 3120–3129. http://dx.doi.org/10.1021/jf035125c

Van Nierop, S.N.E., Rautenbach, M., Axcell, B.C., and Cantrell, I.C. (2006). The Impact of microorganisms on barley and malt quality—A review. J. Am. Soc. Brew. Chem. *64*, 69–78.

Vardon, P.J. (2003). Mycotoxins: risks in plant, animal and human systems. In Potential Economic Costs of Mycotoxins in the United States. CAST Task Force Report No. 139, January, pp. 136–142.

Varga, J., Kevei, F., Vágvölgyi, C., Vriesema, A., and Croft, J.H. (1994). Double-stranded RNA mycoviruses in section *Nigri* of the *Aspergillus* genus. Can. J. Microbiol. *40*, 325–329.

Varga, J., Kiss, R., Matrai, T., Matrai, T., and Teren, J. (2005). Detection of ochratoxin A in hungarian wines and beers. Act. Aliment. *34*, 381–392. http://dx.doi.org/10.1556/AAlim.34.2005.4.6

Varga, J., Frisvad, J.C., and Samson, R.A. (2011). Two new aflatoxin producing species, and an overview of *Aspergillus* section *Flavi*. Stud. Mycol. *69*, 57–80. http://dx.doi.org/10.3114/sim.2011.69.05

Vegi, A., Schwarz, P., and Wolf-Hall, C.E. (2011). Quantification of Tri5 gene, expression, and deoxynivalenol production during the malting of barley. Int. J. Food Microbiol. *150*, 150–156. http://dx.doi.org/10.1016/j.ijfoodmicro.2011.07.032

Veršilovskis, A., Bartkevičs, V., and Miķelsone, V. (2008a). Sterigmatocystin presence in typical Latvian grains. Food Chem. *109*, 243–248. http://dx.doi.org/10.1016/j.foodchem.2007.12.066

Veršilovskis, A., De Saeger, S., and Mikelsone, V. (2008b). Determination of sterigmatocystin in beer by high performance liquid chromatography with ultraviolet detection. World Mycotox. J. *1*, 161–166. http://dx.doi.org/10.3920/WMJ2008.x012

Verstrepen, K.J., Derdelinckx, G., Verachtert, H., and Delvaux, F.R. (2003). Yeast flocculation: what brewers should know. Appl. Microbiol. Biotechnol. *61*, 197–205. http://dx.doi.org/10.1007/s00253-002-1200-8

Vining, L.C. (1990). Functions of secondary metabolites. Annu. Rev. Microbiol. *44*, 395–427. http://dx.doi.org/10.1146/annurev.mi.44.100190.002143

Visconti, A., Pascale, M., and Centonze, G. (2000). Determination of ochratoxin A in domestic and imported beers in italy by immunoaffinity clean-up and liquid chromatography. J. Chromatogr. A. *888*, 321–326. http://dx.doi.org/10.1016/S0021-9673(00)00549-5

Vishwanath, V., Sulyok, M., Labuda, R., Bicker, W., and Krska, R. (2009). Simultaneous determination of 186 fungal and bacterial metabolites in indoor matrices by liquid chromatography/tandem mass spectrometry. Anal. Bioanal. Chem. *395*, 1355–1372. http://dx.doi.org/10.1007/s00216-009-2995-2

Voigt, K., Cigelnik, E., and O'donnell, K. (1999). Phylogeny and PCR identification of clinically important Zygomycetes based on nuclear ribosomal-DNA sequence data. J. Clin. Microbiol. *37*, 3957–3964.

Voigt, C.A., Schäfer, W., and Salomon, S. (2005). A secreted lipase of *Fusarium graminearum* is a virulence factor required for infection of cereals. Plant J. *42*, 364–375. http://dx.doi.org/10.1111/j.1365-313X.2005.02377.x

Vrabcheva, T., Usleber, E., Dietrich, R., and Märtlbauer, E. (2000). Co-occurrence of ochratoxin A and citrinin in cereals from Bulgarian villages with a history of Balkan endemic nephropathy. J. Agric. Food Chem. *48*, 2483–2488. http://dx.doi.org/10.1021/jf990891y

Waalwijk, C., van der Heide, R., de Vries, I., van der Lee, T., Schoen, C., Costrel-de Corainville, G., Häuser-Hahn, I., Kastelein, P., Köhl, J., Lonnet, P., Demarquet, T., and Kema, G.H.. (2004). Quantitative detection of *Fusarium* species in wheat using TaqMan. Eur. J. Plant Pathol. *110*, 481–494.

Wagacha, J.M., and Muthomi, J.W. (2007). *Fusarium culmorum*: Infection process, mechanisms of mycotoxin production and their role in pathogenesis in wheat. Crop Protect. *26*, 877–885. http://dx.doi.org/10.1016/j.cropro.2006.09.003

Weiss, R., Freudenschuss, M., Krska, R., and Mizaikoff, B. (2003). Improving methods of analysis for mycotoxins: molecularly imprinted polymers for deoxynivalenol and zearalenone. Food Addit. Contam. *20*, 386–395. http://dx.doi.org/10.1080/0265203031000065827

Weller, D.M. (1988). Biological control of soilborne plant pathogens in the rhizosphere with bacteria. Annu. Rev. Phytopath. *26*, 379–407.

Wenning, M., and Scherer, S. (2013). Identification of microorganisms by FTIR spectroscopy: perspectives and limitations of the method. Appl. Microbiol. Biotechnol. *97*, 7111–7120. http://dx.doi.org/10.1007/s00253-013-5087-3

Werkman, B.A., and Van den Ende, H. (1974). Trisporic acid synthesis in homothallic and heterothallic *Mucorales*. J. Gen. Microbiol. *82*, 273–278.

Wessels, J.G.H. (1994). Developmental regulation of fungal cell wall formation. Ann. Rev. Phytopathol. *32*, 413–437.

Widden, P. (1997). Competition and the fungal community. Environmental and microbial Relationships. In The

Mycota IV., D.T. Wicklow, and B. Söderström, eds. (Berlin, DE: Springer-Verlag), pp. 135–147.

Wiemann, P., Willmann, A., Straeten, M., Kleigrewe, K., Beyer, M., Humpf, H.U., and Tudzynski, B. (2009). Biosynthesis of the red pigment bikaverin in *Fusarium fujikuroi*: genes, their function and regulation. Mol. Microbiol. *72,* 931–946. http://dx.doi.org/10.1111/j.1365-2958.2009.06695.x

Williams, K.J., Dennis, J.I., Smyl, C., and Wallwork, H. (2002). The application of species-specific assays based on the polymerase chain reaction to analyse *Fusarium* crown rot of durum wheat. Austral. Plant Pathol. *31,* 119–127. http://dx.doi.org/10.1071/AP01079

Wilson, A., Simpson, D., Chandler, E., Jennings, P., and Nicholson, P. (2004). Development of PCR assays for the detection and differentiation of *Fusarium sporotrichioides* and *Fusarium langsethiae.* FEMS Microbiol. Lett. *233,* 69–76. http://dx.doi.org/10.1016/j.femsle.2004.01.040

Windels, C.E. (2000). Economic and social impacts of fusarium head blight: changing farms and rural communities in the northern great plains. Phytopathology *90,* 17–21. http://dx.doi.org/10.1094/PHYTO.2000.90.1.17

Wösten, H.A., Schuren, F.H., and Wessels, J.G. (1994). Interfacial self-assembly of a hydrophobin into an amphipathic protein membrane mediates fungal attachment to hydrophobic surfaces. EMBO J. *13,* 5848–5854.

Wösten, H.A., van Wetter, M.A., Lugones, L.G., van der Mei, H.C., Busscher, H.J., and Wessels, J.G. (1999). How a fungus escapes the water to grow into the air. Curr. Biol. *9,* 85–88. http://dx.doi.org/10.1016/S0960-9822(99)80019-0

Wolf-Hall, C.E. (2007). Mold and mycotoxin problems encountered during malting and brewing. Int. J. Food Microbiol. *119,* 89–94. http://dx.doi.org/10.1016/j.ijfoodmicro.2007.07.030

Woźniakowski, G., and Samorek-Salamonowicz, E. (2014). Direct detection of Marek's disease virus in poultry dust by loop-mediated isothermal amplification. Arch. Virol *159,* 3083–3087. http://dx.doi.org/10.1007/s00705-014-2157-5

Xu, X. (2003). Effects of environmental conditions on the development of *Fusarium* ear blight. Eur. J. Plant Pathol. *109,* 683–689. http://dx.doi.org/10.1023/A:1026022223359

Xu, X., and Nicholson, P. (2009). Community ecology of fungal pathogens causing wheat head blight. Annu. Rev. Phytopathol. *47,* 83–103. http://dx.doi.org/10.1146/annurev-phyto-080508-081737

Yan, L., Zhou, J., Zheng, Y., Gamson, A.S., Roembke, B.T., Nakayama, S., and Sintim, H.O. (2014). Isothermal amplified detection of DNA and RNA. Mol. Biosyst. *10,* 970–1003. http://dx.doi.org/10.1039/c3mb70304e

Yang, L., van der Lee, T., Yang, X., Yu, D., and Waalwijk, C. (2008). *Fusarium* populations on Chinese barley show a dramatic gradient in mycotoxin profiles. Phytopathology *98,* 719–727. http://dx.doi.org/10.1094/PHYTO-98-6-0719

Yin, Y., Liu, X., and Ma, Z. (2009). Simultaneous detection of *Fusarium asiaticum* and *Fusarium graminearum* in wheat seeds using a real-time PCR method. Lett. Appl. Microbiol. *48,* 680–686. http://dx.doi.org/10.1111/j.1472-765X.2009.02595.x

Yli-Mattila, T., Paavaen-Huhtala, S., Parikka, P., Konstantinova, P., and Gagkaeva, T.Y. (2004). Molecular and morphological diversity of *Fusarium* species in Finland and north western Russia. Eur. J. Plant Pathol. *110,* 573–585. http://dx.doi.org/10.1023/B:EJPP.0000032397.65710.69

Yoder, W.T., and Christianson, L.M. (1998). Species-specific primers resolve members of *Fusarium* section *Fusarium*: Taxonomic status of the edible 'Quorn' fungus reevaluated. Fung. Gen. Biol. *23,* 68–80. http://dx.doi.org/10.1006/fgbi.1997.1027

Yong, R.K., and Cousin, M.A. (2001). Detection of moulds producing aflatoxins in maize and peanuts by an immunoassay. Int. J. Food Microbiol. *65,* 27–38. http://dx.doi.org/10.1016/S0168-1605(00)00505-5

Yoshizawa, T. (1983). Red-mold disease and natural occurrence in Japan. In Trichothecenes – Chemical, Biological and Toxicological Aspects, Y. Ueno, ed. (Amsterdam, NL: Elsevier), pp. 195–209.

Yoshizawa, T., Yamashita, A., and Luo, Y. (1994). Fumonisin occurrence in corn from high- and low-risk areas for human esophageal cancer in China. Appl. Environ. Microbiol. *60,* 1626–1629.

Zachariasova, M., Cajka, T., Godula, M., Malachova, A., Veprikova, Z., and Hajslova, J. (2010). Analysis of multiple mycotoxins in beer employing (ultra)-high-resolution mass spectrometry. Rapid. Commun. Mass Spectrom. *24,* 3357–3367. http://dx.doi.org/10.1002/rcm.4746

Zachariasova, M., Vaclavikova, M., Lacina, O., Vaclavik, L., and Hajslova, J. (2012). Deoxynivalenol oligoglycosides: new 'masked' fusarium toxins occurring in malt, beer, and breadstuff. J. Agric. Food. Chem. *60,* 9280–9291. http://dx.doi.org/10.1021/jf302069z

Zain, M.E. (2011). Impact of mycotoxins on humans and animals. J. Saudi Chem. Soc. *15,* 129–144. http://dx.doi.org/10.1016/j.jscs.2010.06.006

Zeng, Q.Y., Westermark, S.O., Rasmuson-Lestander, A., and Wang, X.R. (2006). Detection and quantification of *Cladosporium* in aerosols by real-time PCR. J. Environ. Monit. *8,* 153–160. http://dx.doi.org/10.1039/b509515h

Zhao, D., Feng, Q., Yan, X., Li, C., Pan, Y., and Cui, Q. (1993). Ultrastructural study of moniliformin induced lesions of myocardium in rats and mice. Biomed. Environ. Sci. *6,* 37–44.

Zheng, M.Z., Richard, J.L., and Binder, J. (2006). A review of rapid methods for the analysis of mycotoxins. Mycopathologia *161,* 261–273. http://dx.doi.org/10.1007/s11046-006-0215-6

Zinedine, A., Meca, G., Mañes, J., and Font, G. (2011). Further data on the occurrence of Fusarium emerging mycotoxins enniatins (A, A1, B, B1), fusaproliferin and beauvericin in raw cereals commercialized in Morocco. Food Contr. *22,* 1–5. http://dx.doi.org/10.1016/j.foodcont.2010.05.002

Zur, G., Hallerman, E.M., Sharf, R., and Kashi, Y. (1999). Development of a polymerase chain reaction-based assay for the detection of *Alternaria* fungal contamination in food products. J. Food Prot. *62,* 1191–1197.

Investigation of Beer Spoilage Lactic Acid Bacteria Using Omic Approaches

9

Jordyn Bergsveinson and Barry Ziola*

Department of Pathology and Laboratory Medicine, College of Medicine, University of Saskatchewan, Royal University Hospital, Saskatoon, SK, Canada.

*Correspondence: b.ziola@usask.ca

https://doi.org/10.21775/9781910190616.09

Abstract

Consistent production of quality beer requires brewers be concerned with not only the health of fermenting yeast and optimizing brewing conditions, but also with the potential for bacterial contamination at every stage of production. This perpetual threat has driven investigation into mechanisms of bacterial beer spoilage, with significant emphasis placed on isolates belonging to the group broadly known as lactic acid bacteria (LAB). These organisms are problematic for the global brewing industry, as they are not only able to grow in and spoil the harsh, niche environment of beer, but also frequently elude even current means of detection and control due to the lack of genetic uniformity among diverse brewing-related LAB. This chapter summarizes relevant background in beer spoilage LAB characterization, detection, and control within the context of beer spoilage LAB genetic variability. While traditional methods of analysis remain more accessible to brewers for quality control, the advantages of incorporating powerful omics-based methods within the industry are presented. Lastly, current omics methods are discussed in terms of their notable ability to help solve developing issues related to the use of LAB as controlled flavouring and/or fermentation agents by popular craft or specialty brewing operations.

Introduction to lactic acid bacteria in beer: '*The good, the bad and the ugly*'

One only has to perform a cursory literature search of lactic acid bacteria (LAB) to be overwhelmed by available information extolling their industrial importance for fields ranging from human health to food production. Ultimately, LAB are a collection of Gram-positive, catalase-negative, non-sporulating, non-motile and acid-tolerant organisms that share the capacity to produce lactic acid as a primary product of sugar fermentation while being incredibly heterogeneous in terms of physiological attributes, metabolic and fermentation capabilities, and ability to inhabit diverse niches (Pfeiler and Klaenhammer, 2007). These diverse attributes greatly increase both their presence and utility in the production of multiple food and beverage products, from wine and beer to cheese, dairy, meat and vegetable products (Makarova *et al.*, 2006). Further, members of the LAB group that naturally occupy food and beverage niches have been ascribed the generally regarded as safe (GRAS) designation, allowing them to be exploited for improvement and preservation of a wide range of food and beverage products, and for the production of probiotics (Klaenhammer *et al.*, 2005). Unfortunately, the unwanted presence or uncontrolled over-growth of these organisms during food or beverage production can occur, posing challenges, as well as opportunities, especially to the brewing industry.

Beer is an unexpected environment to support microbial growth given that beer-spoiling bacteria must simultaneously overcome several physiological hurdles, including the antimicrobial action of ethanol and hop-derived bitter acids, low pH, limited available nutrients, and low O_2, with concurrent high CO_2 levels (Fernandez and Simpson, 1993; Sakamoto and Konings, 2003; Simpson, 1993b). Nonetheless, LAB probably have always been associated with and/or involved in the production of beer, either as a naturally occurring agent for traditional spontaneous-fermentation styles such as lambic or Weisse beers, producing characteristic 'sour' flavours via production of lactic or acetic acid (Tonsmeire, 2014) (see Chapter 7), or as an unseen source of spoilage that results in an undesired or poor quality product. Though the participation of LAB in beer-production and spoilage was not appreciated until Louis Pasteur began to isolate these bacterial cells from beer (Suzuki, 2011), their diverse historical role(s) in brewing allows for the general characterization of 'the good, the bad, and the ugly' outcomes of LAB involvement.

LAB isolates can indeed be helpful, if not necessary components, of specialized fermentations that produce specific beer styles, both traditional and new (Tonsmeire, 2014). This is increasingly important to consider in discussions of beer spoilage-related (BSR) LAB, as the current global beer market is experiencing a significant expansion in the numbers of 'local' craft beers. The influx of new companies necessitates that breweries distinguish themselves with consumers through unique products and this need has led to innovative use of raw materials and production processes, with the inclusion of both fermenting LAB and non-traditional yeasts. These 'helpful' fermenting LAB, however, must have several important attributes, chief among them being the inability to overgrow in beer and to not inhibit normal yeast function. Thus, current industry trends highlight an important context in which to consider brewing-associated LAB, and open up interesting avenues for how best to investigate and further the role of LAB in the spectrum of modern beer production.

Putting aside the expansion in modern brew styles, the fact remains that since the industrial revolution, global brewing practices increasingly have focused on producing 'clean' and consistent brew products, free of bacteria and their metabolites (Tonsmeire, 2014). With the advent of pasteurization and appreciation for hygienic practices during food and beverage production, the average global beer consumer today is probably accustomed to 'conventional' or non-sour products, beers that are free from characteristic signs of LAB overgrowth. This means the beer should have no cloudy 'haze', no 'sour' taste or other unappealing off-flavours such as 'buttery' diacetyl, and be free of bacterial sedimentation or exopolysaccharide 'slime' (Back, 2005). Such occurrences in most beer products are unexpected, and encapsulate both the bad and the ugly results of unwanted LAB being present. The outcome of these spoilage events causes a loss of consumer and brand confidence when compromised beer is consumed, or significant revenue and time loss to the brewery in the event of batch contamination. As LAB are attributed with causing 60 – 90% of the brewing spoilage events worldwide (Asano *et al.*, 2009; Back, 1994), significant interest has gone into ascertaining how they spoil beer and how this is best controlled. Despite this interest, incidence of BSR LAB contamination remains difficult to delineate due to under-appreciation of how diverse a group they are, even though relevant research constantly highlights this diversity.

The promise of omics for BSR LAB research

Recent review articles cover the evolution and current state of understanding of BSR LAB prevalence, genetics (Bokulich and Bamforth, 2013; Sakamoto and Konings, 2003; Suzuki, 2011b) and research techniques used during investigation (Ben-Amor *et al.*, 2007; Bokulich and Mills, 2012b; Bokulich *et al.*, 2012a,b). Though such background knowledge is of critical importance to understanding the current issues facing the brewing field, this chapter is not meant to be exhaustive of all relevant literature history to BSR LAB. Rather, this information is used to highlight the apparent gaps in knowledge and need for the expansion of research methods into omics applications.

The ability for transcriptomics or proteomics to profile, in a rapid and high-throughput manner, how a specific microbe grows under defined conditions and/or provide information on a microbial community's genetics, activities and ecology means that these omic approaches can effectively balance the interests of academia and industry, and overcome the problem of understanding BSR LAB variability. To date, research into BSR LAB has often failed to provide data of equal value to research investigators and brewers tasked with carrying out detection of contaminating BSR LAB. For example, detailed study of genetic or physiological stress response mechanisms of BSR LAB is of great value to LAB and brewing research writ large; however, these data alone present little utility to individual brewers. Further, the targeted analysis of just a few genes, or one physiological stressor in few specific isolates, has provided only minimal and incremental expansion to our current knowledge regarding LAB. Most importantly, findings from targeted-analysis experiments are frequently inconsistent for all BSR LAB, thus curtailing the value of these data from both academic and industry perspectives.

Omics approaches have proven to be a powerful way to investigate LAB genetic and metabolic diversity (Claesson *et al.*, 2007; Horvath *et al.*, 2009; Marakova and Koonin, 2007), and when applied broadly, produce large amounts of data that can be mined to give statistically relevant genetic or metabolic markers for beer spoilage that could be effectively screened within breweries. Secondly, these approaches help distinguish potentially helpful LAB from BSR LAB for use in specialty brews, by correlating limited beer-growth ability with desirable genetic or metabolic traits, without having to develop optimal strains through the use of laborious genetic modification techniques. The *meta* data that is produced from omics approaches thus allows for the conversion of information obtained by broad-scale or community-analysis of BSR LAB to specific application required for application in the brewery.

Diversity, relatedness and maintenance of BSR LAB

General LAB characteristics

Problematic BSR LAB and sour-beer fermenting LAB alike traditionally belong to the *Firmicutes* phylum, order *Lactobacillales,* in the genera *Lactobacillus* and *Pediococcus* (Priest, 2003), with *Lactobacillus brevis, Lactobacillus linderni* and *Pediococcus damnosus* being the most commonly encountered bacteria that spoil beer (Back, 2005; Menz *et al.*, 2010; Priest, 2003; Suzuki, 2008; Thelen *et al.*, 2006). Additional LAB species also have been detected with varying frequencies in brewing environments, including *Lactobacillus amylolyticus* (Bohak *et al.*, 1998), *Lactobacillus backii* (Bohak *et al.*, 2006), *Lactobacillus brevisimilis* (Back, 1987), *Lactobacillus buchneri, Lactobacillus casei, Lactobacillus coryneformis, Lactobacillus curvatus, Lactobacillus delbrueckii* (Priest, 2003), *Lactobacillus dextrinicus* (Haakensen *et al.*, 2009a), *Lactobacillus fermentum, Lactobacillus fructivorans, Lactobacillus malefermentans* (Russell and Walker, 1953), *Lactobacillus parabuchneri,* (Priest, 2003), *Lactobacillus paracollinoides* (Suzuki, 2004), *Lactobacillus plantarum* (Priest, 2003; Thelen *et al.*, 2006), *Lactobacillus paraplantarum* (Curk *et al.*, 1996), *Lactobacillus paucivorans* (Ehrmann *et al.*, 2010), *Lactobacillus rossiae* (Corsetti *et al.*, 2005), *Pediococcus claussenii* (Dobson *et al.*, 2002), *Pediococcus inopinatus, Pediococcus parvulus* (Martens *et al.*, 1997) and *Pediococcus pentosaceous* (Jespersen and Jakobsen, 1996).

Despite the reported incidence of certain species within breweries, it must be emphasized that the ability for BSR LAB to grow in and spoil beer is *not a species attribute,* but is an isolate-specific capability, as is the case for most spoilage lactobacilli (Sanders *et al.*, 2015). As this phenomenon indicates that there must be a level of genetic specialization in a BSR LAB isolate that allows for the beer spoilage phenotype, the search for a small number of 'detectable' genetic markers has long been the focus of analysis. However, this narrow scope of investigation fails to appreciate the degree of genetic dissimilarity inherent among BSR LAB given that these isolates belong to the LAB group as a result of shared functional characteristics (i.e. particular metabolic capacities) and not necessarily genetic relatedness among LAB (Sun *et al.*, 2014). Additionally, the use of 'common' BSR LAB isolates for the study of genetic differences probably produces results that are non-universal for BSR LAB. Common isolates are those that can grow in routine culture media, and are thus probably consistently over-represented during detection procedures (Suzuki *et al.*, 2008). Given that not all *Lactobacillus* and *Pediococcus* isolates can grow in these media, those that can likely skew incidence reports of beer spoilage, and therefore the amount of research interest and available information on BSR LAB.

BSR LAB diversity

To better understand the genetic adaptations that separate BSR LAB from non-spoiling isolates of the same species, and their origins, we must first examine the diversity of species involved. Both *Lactobacillus* and *Pediococcus* genera are comprised of Gram-positive, catalase-negative isolates and share overlapping DNA G+C content (*Lactobacillus*: 32–55% mol and *Pediococcus*: 35–44 mol%). Although these two genera are closely related to each other and to the genus *Leuconostoc*, as demonstrated by 16S rRNA gene sequence analysis, they have several distinctive features (Schleifer and Ludwig, 1995). *Pediococcus* isolates grow under a range of facultatively aerobic to microaerophilic conditions and are homofermentative in that they do not generate CO_2 when they produce lactic acid from fermentation of glucose (Holzapfel *et al.*, 2009). Further, pediococci are not capable of reducing nitrate, while some lactobacilli isolates can (Hammes and Hertel, 2006; Hammes and Vogel, 1995). *Lactobacillus* species are generally anaerobic, although some are aerotolerant and may be either homofermentative like *Pediococcus*, or heterofermentative and produce lactic acid, CO_2, and ethanol and/or acetic acid as primary end products of fermentation.

Lactobacillus spp. are currently organized into three distinct metabolic or fermentative groups, prior to further phylogenetic arrangement based on genetic relatedness (Holzapfel and Wood, 2014; Sun *et al.*, 2014). The first fermentation group is that of the obligate homofermentative (OHO) species, which can only ferment hexoses and do so via the Embden–Meyerhof–Parnas (EMP) pathway, largely producing lactic acid as a by-product (Hammes and Vogel, 1995). Those species that are capable of homofermentation, but during starvation or glucose limitation can degrade pentoses and gluconate via the pentose phosphate pathway (PPP) to produce acetic acid, ethanol and formic acid as by-products, are referred to as facultative heterofermentative (FHE). Finally, the obligate heterofermentative (OHE) group will metabolize pentoses and hexoses solely through the first part of the PPP via the phosphogluconate pathway and produce lactic acid, CO_2, and ethanol or acetic acid (Holzapfel and Wood, 2014; Sun *et al.*, 2014; Zheng *et al.*, 2015). In the context of brewing, common BSR LAB belong to all three groups; for example, *L. brevis* and *L. lindneri* are OHE and *L. plantarum* is FHE. The different metabolic capacities of these isolates therefore may influence not only the style of beer or brewery location they are able to grow in as a result of available nutrients, but also the severity and type of spoilage they cause as a result of their metabolic by-products.

To further illustrate lactobacilli diversity, Sun *et al.* (2014) characterized eight 'niche type' environments where lactobacilli are commonly found, including plant or plant-associated fermentation products, sourdough, meat products, dairy products, wine products, human or animal gastrointestinal (GI) tracts, human or animal non-GI sources, and the general environment. Notably, breweries or beer products are not included likely because LAB are not necessarily an essential component of beer fermentation or production. Further, many BSR LAB species can be isolated from different environments; for example, *L. brevis* has been isolated from the human GI tract, and *L. lindneri* and *L. plantarum* can be recovered from plant materials and dairy products (Salvetti *et al.*, 2012), as well as from beer. The ability of different isolates of the same species to occupy multiple niches and exhibit different fermentation types is common for *Lactobacillus* species (Douillard *et al.*, 2013). Thus, it is not surprising that BSR LAB isolates occupying the same niche have different genomic features, underscoring the idea that different genetic mechanisms allow for adaptation to a given environment and/or stress (Sun *et al.*, 2014).

As food production industries are principally concerned with LAB adaptation to their specific application (i.e. unique environment), LAB genomics and phylogenetic relationships have received considerable attention (Hammes and Vogel 1995; Salvetti *et al.*, 2012; Zhang *et al.*, 2011). Whole genome sequencing, phylogenomics and other bioinformatic approaches to compare LAB species have resolved questions of group diversity, evolutionary relatedness and provided a wealth of information concerning general genetic composition. Multiple LAB genomes are available publicly through the National Center for Biotechnology Information database (www.ncbi.nlm.nih.gov/

genome), with over a hundred of these being *Lactobacillus* isolates and a dozen being *Pediococcus* isolates, and an estimated 80 unreleased, ongoing projects worldwide (Sun *et al.*, 2014). These genomic data are of great general utility, however, with respect to brewing-microbiology, only a small percentage of these genomes or projects belong to BSR isolates (Bergsveinson *et al.*, 2015c; Kelly *et al.*, 2012; Pittet *et al.*, 2012a,b). The continued sequencing of LAB genomes is essential, as general analysis of genomic content will be more robust and less inclined towards bias if LAB isolates from a variety of different sources are included (Pfeiler and Klaenhammer, 2007; Sun *et al.*, 2014). As there is assumed genetic variation between BSR isolates from beer and non-BSR isolates, and BSR LAB species isolated from any source, more data must be made available for both BSR and non-BSR LAB from multiple isolation sources in order to effectively determine the evolution and distinguishing characteristics of BSR LAB.

Traditional and emerging methods for BSR LAB detection and identification

Culture-based methods

Culture methods are still the most commonly used approach for routine detection and identification of BSR LAB in the brewery, for reason of their ease of use, limited need for specialized training, relatively low monetary and space cost, and proven utility. However, culture methods have inherent disadvantages, owing to the variable nature of BSR LAB isolates, e.g. differences in their fastidious aerotolerance or nutritional requirements, and the different adaptive states they may exist in when isolated from beer. These factors make the primary isolation of some LAB contaminants via growth quite difficult (Deng *et al.*, 2014; Suzuki *et al.*, 2008). More importantly, there is no single medium that effectively screens and supports growth for all possible beer spoilage LAB (Taskila *et al.*, 2011).

de Man Rogosa Sharpe (MRS) medium (de Man *et al.*, 1960), which was designed for the cultivation of LAB, remains the most relied-upon medium in brewery settings (Sakamoto and Konings, 2003). There are several descriptions of supplementing MRS medium with varying concentrations of beer, expecting that the added beer enables cultivation of beer-adapted (hard-to-culture) organisms and that the nutrients provided by the MRS medium allow for more rapid growth (Haakensen *et al.*, 2009b; Holzapfel, 1992; Suzuki *et al.*, 2008b). Further modifications to beer-supplemented MRS include adding reducing agents to remove oxygen tension in the medium to facilitate the growth of a wide range of BSR LAB in addition to microaerophilic strains (Nishikawa and Kohgo, 1985; Taskila *et al.*, 2010). Similarly, other developments such as the Advanced Beer Detection (ABD) medium, developed by Suzuki *et al.* (2008b), seek to reduce medium osmolarity with the goal of isolating hard-to-culture BSR LAB.

Often there is a need to exclude other non-LAB brewing microorganisms from growing while concurrently enriching the medium to cultivate specific or hard-to-culture LAB isolates. Enrichment culturing prior to plating is a common technique to influence the number and identity of isolates grown and is often critical for the efficiency of downstream molecular detection techniques. Thus, contaminating yeast or Gram-negative bacteria are excluded from growing in detection media by the inclusion of cycloheximide and 2-phenylethanol, respectively (Taskila *et al.*, 2011). To select for specific or hard to cultivate BSR LAB, enrichment media are typically differentiated based on carbon sources present to exploit differences in substrate utilization between species (Endo *et al.*, 2011). The most general substitution that can be made is removal of glucose in favour of another carbohydrate, so as to limit the growth of very fast-growing LAB, thereby 'levelling the playing field', giving hard-to-culture isolates that are out-competed in most standard growth media a chance to grow (Endo *et al.*, 2011). In addition, some metabolites produced by LAB, such as lactic acid or bacteriocins that have antimicrobial action, may also add to the selectivity of the enrichment cultivation (Moneke *et al.*, 2009).

Ultimately, primary cultivation and even use of specialized culture media to detect and identify BSR LAB are not fully effective for the accurate detection of BSR LAB (Suzuki, 2011; Taskila *et al.*, 2011). Nonetheless, culture methods remain an important area of investigation for the reason that culturing is often a preliminary step for molecular analysis and because culture-based tests traditionally have provided the most information for spoilage incidence reports, which has greatly influenced our current understanding concerning relevant BSR LAB.

Molecular techniques

Molecular methods typically have higher associated cost and the need for specialized training, and thus have different utility for research or industry interests. Further, molecular methods can often be labour-intensive despite their ascribed benefit of being 'rapid', since pre-enrichment culture or isolation is often required before the molecular detection limit can be achieved or to remove inhibitory molecules found in beer (Back, 2005; Taskila *et al.*, 2011). However, the allure of molecular techniques for the brewing industry is centred on their increased specificity and sensitivity in detecting and identifying BSR LAB. Recent reviews (Bokulich *et al.*, 2012a; Bokulich and Mills 2012b) provide an extensive comparison of methodology concerning microbial community profiling in the brewing industry, and thus only a general overview of current community- or microbe-targeted molecular methods for BSR LAB analysis is presented here.

rDNA and RNA detection

PCR and qPCR

Many molecular techniques that seek to profile the microbes within a community specifically target ribosomal genes or rDNA (i.e. 16S rDNA, 23S rDNA, rDNA inter-space regions) given the ubiquitous presence of rRNA (i.e. in both viable and non-viable cells) and the conserved nature of these sequences, enabling the ability to distinguish between species and isolates (Ben-Amor *et al.*, 2007). As brewers are often solely interested in the presence of viable cells, other genes that increase the discriminatory power of these nucleic acid techniques, such as elongation-factor genes or other hop tolerance genes can also be targeted (Juvonen *et al.*, 2010).

The polymerase chain reaction (PCR) and various adaptations thereof are arguably the most frequently used means of performing targeted-interrogations of the 16S rDNA and target genes of interest (i.e. hop tolerance genes) for BSR LAB (Haakensen *et al.*, 2007; Pittet *et al.*, 2011; Suzuki *et al.*, 2004a,b). Perhaps the most important variation of traditional end-point PCR for beer spoilage LAB detection is that of the multiplex PCR assay, which is used to interrogate multiple targets at one time. Distinct primers targeting separate genes or regions of interest have been used within both the brewing and wine industry to profile hop tolerance genes and/or to rapidly identify LAB genera and species (Haakensen *et al.*, 2008; Petri *et al.*, 2013; Pfannebecker and Fröhlich, 2008). These tests give same-day results, require relatively low expertise to run, and have sensitive detection limits, thus presenting an attractive method for brewery use.

Quantitative PCR (qPCR) usage has increased in brewery application because it allows the rapid quantitation of target DNA at an extremely sensitive level (such as from a single cell) (Bokulich *et al.*, 2012a). The notable drawbacks include the high cost of the initial instrument and software, increased expertise required over conventional PCR, as well as concerns for quantitation accuracy given that the signal output does not discriminate living and non-living cells, and that results are influenced by the target gene copy number. Thus, an appropriately controlled and validated system is critical for drawing accurate conclusions. Since qPCR is not a community-profiling technique, it has limited use in interrogating mixed-culture fermentations or unknown isolates. Nonetheless, qPCR allows for the accurate monitoring of both the presence and quantity of specific populations in the brewing environment, and has notable advantages over other methods in terms of analysis speed and achievable sensitivity. Reverse transcription qPCR (RT-qPCR) assays have also been developed which analyse actively transcribed mRNA content, and thus viable cells, such as for the detection of the BSR LAB hop tolerance genes, though these assays must still be stringently and appropriately controlled (Bergsveinson *et al.*, 2012; Sami *et al.*, 1997b; Haakensen *et al.*, 2007).

PCR assays continue to be optimized as the methodology itself evolves. A recent application is droplet digital PCR (ddPCR), which operates on the principle of absolute target quantification without need for internal control genes or excessive reaction replicates (Hindson *et al.*, 2011; Pinheiro *et al.*, 2012). ddPCR was recently used to investigate the copy number of hop tolerance genes within a brewery setting (Bokulich *et al.*, 2015). Though ddPCR is limited in the number of targets it can interrogate, and by its cost, it most certainly can be further developed and applied to investigate gene target distribution and abundance within a brewery or contaminated sample (Hindson *et al.*, 2011).

Molecular techniques

Several techniques apart from PCR have been developed to detect specific DNA and rDNA targets. Fluorescence *in situ* hybridization (FISH) is a method that targets singular or groups of isolates in a community and operates on the basis of using different fluorescently labelled probes that hybridize to specific target regions (e.g. 16S rRNA) in a chemically fixed cell sample, followed by fluorescent microscopy (Bokulich *et al.*, 2012a; Bottari *et al.*, 2006). Based on the fluorescent signals observed, populations of different cells can be assessed. However, this method is limited as to how many unique cell populations it can identify and is thus more suited for targeted-analysis. FISH can also be coupled to Flow Cytometry (FCM), which performs automated cell sorting based on fluorescent signals, which allows for the quantitation of different cells within a population. Together, these methods provide a semi-automated means of acquiring quantitative data within a few days for isolates of interest without the need for excessive pre-processing (e.g. DNA extraction). However, expensive equipment is required and the probes needed are often not commercially available (Bokulich *et al.*, 2012a). Therefore, FISH and FISH-FCM do not lend themselves well to in-house application for the brewery, yet provide interesting data when performed as an out-sourced procedure or in a research setting. Thus far, this methodology has only been well developed for characterization of yeast populations, with few reported studies of application to BSR LAB (Meng *et al.*, 2012; Thelen *et al.*, 2002, 2004). However, recent work involving cider fermentations showed that FCM could distinguish and separate mixed yeast and bacterial cultures based on membrane integrity and esterase activity, and could identify different physiological states resulting from differences in fermentation conditions, thus having interesting implications for beer fermentations (Herrero *et al.*, 2006).

While FISH (and/or FISH-FCM) are targeted-analysis methods, given they are limited by the number of probes that can be used during experimentation, denaturing gradient gel electrophoresis (DGGE) allows for a more robust identification of microbial community members through the 16S rRNA gene, and has been applied to beer-related LAB (Bokulich *et al.*, 2012a; Manzano *et al.*, 2005; Tsuchiya *et al.*, 1994). This method uses universal PCR primers to amplify specific DNA sequences in a community, then separates the amplicons in a polyacrylamide gels containing a gradient of urea and formamide on the basis of differences in GC content (melting temperature), thereby allowing detection of DNA sequence heterogeneity in microbial communities (Bokulich *et al.*, 2012a; Muyzer *et al.*, 1993). Again, this method has limited use within the brewery in that it is technically difficult and requires DNA extraction, and has a detection threshold that is often above the cell concentration found in beer samples (Cocolin *et al.*, 2001). Further, it requires subsequent processing and sequencing steps following the gel separation to produce accurate identification of the bacteria yielding the resolved bands, making it a laborious process fraught with the inherent errors and biases related to PCR amplification and DNA extraction (Bokulich *et al.*, 2012a; Cocolin *et al.*, 2001; de Lipthay *et al.*, 2004).

Another very useful method for assaying microbial community diversity is Terminal Restriction Fragment Length Polymorphism (TRFLP). Universal PCR primers targeting the 16S rRNA gene that have been fluorescently labelled are used to amplify this DNA region from a mixed culture. Amplicons are then purified and in separate reactions, digested by one or more restriction enzymes, followed by capillary electrophoresis. The separation of the fluorescently labelled DNA fragments allows for unique patterns to emerge for a given organism (Bokulich *et al.*, 2012a; Liu *et al.*, 1997). This method is flexible in terms of its ability to provide either high throughput data or more targeted analysis of mixed microbial communities, and is relative easy to use with low cost, making it a more attractive option for routine use in contaminant surveillance within breweries (Bokulich and Mills, 2012a; Bokulich *et al.*, 2012a). Further, this method can be adapted to provide greater resolution for specific BSR LAB targets through modification of the target sequences and restriction enzymes used (Bokulich *et al.*, 2015).

Multilocus sequence typing (MLST)

The use of multilocus sequence typing (MLST) has increased in tandem with whole genome sequencing in order to answer many questions of LAB relatedness and evolution (Enright and Spratt, 1999; Maiden *et al.*, 1998; Sun *et al.*, 2014). MLST relies on DNA sequence analysis of conserved housekeeping genes (or other protein-coding sequences) to type bacteria (Enright and Spratt, 1999; Maiden, 2008) and reveal insight into the overall diversity of a species. MLST has direct appeal to the brewing industry not only because of lower cost and required time compared to whole genome sequencing, but also due to the potential of distinguishing same-species isolates recovered from different sources and thereby the potential influence of the beer niche on genetic adaptations. However, in order to effectively develop MLST into a rapid means of screening for BSR versus non-BSR LAB, whole genome data provided by deep sequencing needs to be available to inform on specific assay targets.

Omics

Deep sequencing of DNA and mRNA

Genome sequencing, or in the case of microbial community profiling, meta-genome sequencing, provides the entire genome or identity of each organism in the sample under analysis (Table 9.1). The amount of genetic information obtained by this technique is exponentially greater than provided by targeted-sequence analysis (i.e. of housekeeping genes or 16S rDNA). Given that a small handful of genes have not yet proven adequate to distinguish between BSR and non-BSR LAB, the wealth of data from deep DNA and mRNA sequencing is critical for better understanding of the total genetic character and higher-level metabolic regulation that differentiates these two groups of organisms and those LAB able to provide helpful fermentation for craft beers. Further, emerging patterns of species- or genus-level genetic content may be identified and then incorporated into routine brewery-level diagnostic approaches.

Applications of transcriptomics or metatranscriptomics, or the profiling of the genetic expression (mRNA) within an organism or community, is by far the most accurate means of studying genetic pathways required for growth in a given condition, the interactions between members of a community such as quorum signalling, and overall process and stress regulation mechanisms (Bokulich *et al.*, 2012a,b; Simon and Daniel, 2011; Warnecke and Hess, 2009). (Meta)transcriptomics builds upon genomics or metagenomics to reveal what genetic content is specifically active and therefore important for growth and/or activity of a BSR LAB isolate on its own or in a microbial community (Bron *et al.*, 2012; Table 9.1). Only three transcriptomic studies of BSR LAB isolates have been completed to date: on *L. brevis* BSO 464 in degassed and gassed beer, and on *P. claussenii* ATCC BAA-344^T during growth in beer with undetermined dissolved CO_2 content and on both organisms grown in the presence of growth-limiting hop concentrations (Bergsveinson *et al.*, 2016a,b; Pittet *et al.*, 2013). These studies have revealed insights into not only the complexity of BSR LAB adaptation to the beer environment, but have also confirmed the importance of plasmids for the beer-growth phenotype. Further, these two studies indicate that cell membrane modification and nutrient scavenging (and general membrane transport) are critical responses to the beer environment, and further confirm the importance of biogenic amines production and metabolism as a common hallmark of LAB beer spoilage (Bergsveinson *et al.*, 2016b; Geissler *et al.*, 2016; Izquierdo-Oulido *et al.*, 1996; Kalač *et al.*, 2002). Overall, transcriptomic studies are beginning to reveal genetic adaptations shared by BSR LAB and indicate important next-step investigation efforts.

Deep-sequencing applications represent the current interface of academic research and industrial interests in the brewing field because though they are readily applied in a research setting, they do not presently lend themselves to routine use within the brewery. However, these technologies continually decrease in cost, and are currently being used in clinical settings, making it reasonable to predict that these methods will become part of routine practice in a variety of fields, including the brewing industry. Until such time, support of current academic research by the brewing industry is important, as omics data has the power to delineate specific markers for LAB beer spoilage ability, allowing for development of better detection methodology for brewery use.

Table 9.1 Omics[a]

Technology and purpose	Input	Targets	Output
Genomics Provides the genetic profile of an organism	DNA extracted from single organism	The entire DNA content is sequenced	Genomic DNA sequence of an organism. Basis for comparing gene content between and among organisms
Metagenomics Details organisms present in a bacterial community and the relative abundance of community members	DNA extracted from a community – a sample with multiple organisms present	One genetic marker (e.g. 16S rDNA) that defines a species/genus. These sequences are often termed operational taxonomic unit (OTU)	Relative abundance and identity of each OTU that comprises that community
Transcriptomics Determines which genetic pathways are important for growth under *X* condition(s). Changes in expression over time can be detailed (many other questions can be answered)	Extracted messenger RNA (mRNA) from single organism growing under condition(s) of interest	All mRNA sequences extracted	mRNA transcripts are mapped to the genes they originate from and quantified. Thus, it is possible to know what genes or pathways are being highly expressed. *This methodology integrates with the simultaneous or previous application of genomics*
Metatranscriptomics Tracks how a bacterial community respond to changes in *X* condition(s). Determines if specific genetic pathways are expressed in specific conditions	Extracted messenger RNA (mRNA) from community/environmental sample	All mRNA extracted (expression) and total rRNA (abundance)	mRNA sequences are mapped to specific pathways and pathway function is classified. Similar rRNA sequences (OTUs) are grouped to determine relative abundance of community members and how these abundances might shift with treatment. *This methodology integrates with the simultaneous or previous application of (meta) genomics*
Proteomics Characterizes the structure/function/identity/interaction of proteins in a sample at *X* time, under *X* condition(s)	Purified proteins from sample of interest at a given time point	Depends on the intention of study – e.g. studying protein–protein interaction or need to characterize the type of proteins present	Detailed information on the nature of proteins produced in a sample. *This methodology can be integrated with the application of transcriptomics or metabolomics*
Metaproteomics Discovery of all proteins present in a community/environmental sample	Purified proteins from community at given time point	Structure or specific molecular signatures of protein(s)	Discovery based-approach: catalogues all proteins present. Potential protein 'biomarkers' – a protein that is indicative of a specific physiological state or growth ability. *This methodology can be integrated with application of metatranscriptomics or metabolomics*
Metabolomics Characterization of specific chemical signals/small molecules that characterize a specific metabolic or chemical process	Total isolated metabolites from a sample (intermediate molecules of metabolism)	Structure, mass/size or polarity of each metabolite to determine identity and function	Total metabolite characterization and quantitation gives a 'snapshot' of cell physiology at a given time. *This methodology can be integrated with (meta)genomics, (meta)transcriptomics, and (meta) proteomics*
Lipidomics Characterization of the total lipid content (lipidomics) within an organism or community sample	Extraction of specific lipid classes (i.e. glycerophospholipids, fatty acids, cholesteryl esters, glycerolipids, sterols)	Structure, mass/size, polarity of lipid samples	Determination of lipid profile of a cell in response to a given sample. *This methodology can be integrated with (meta)genomics, (meta)transcriptomics, and (meta) proteomics*

[a]Omics is used denote a study of the totality of something (e.g. genomics, the *total* genetic content of an organism).

Proteomics and metabolomics

The use of gas or liquid chromatography (GC or LC) and mass spectrometry (MS) to analyse the total protein, metabolite, or volatile compounds in a beer sample or microbial community can reveal insight into the complex process of microbial energy metabolism, quorum sensing, and protein production during fermentation and spoilage (di Cagno *et al.*, 2011; Picariello *et al.*, 2012). Further, these techniques can also be applied to resolve community proteomes, and investigate probiotic and bacteriocin production (Baugher and Klaenhammer, 2011). These methods are beginning to be applied with greater frequency to BSR LAB (Behr *et al.*, 2007; de Bok *et al.*, 2011, Wieme *et al.*, 2014), with notable recent application of matrix-assisted laser desorption/ionization time-of-flight mass spectrometry (MALDI-TOF MS) to help characterize the microbial community of lambic beer and to distinguish between *L. brevis* isolates of different beer-spoiling virulence (Kern *et al.*, 2014; Spitaels *et al.*, 2014). Very recently, a comprehensive analysis and comparison of metabolites present in beer during the growth of 26 LAB strains of varying beer spoilage strength was conducted with high-performance liquid chromatography (HPLC) (Geissler *et al.*, 2016). This study found that carbohydrate and amino acids metabolism did not correlate with spoilage potential of an isolate, but with fermentation type (heterofermentative/homofermentative) and LAB species (Geissler *et al.*, 2016). Further, a distinct amino acid metabolism profile was observed for heterofermentive LAB, resulting in biogenic amine production, which is presumed to assist in maintaining energy supply and pH homeostasis (Geissler *et al.*, 2016). These advances aside, the brewing industry lags behind other LAB-related fields in applying high-throughput protein analysis or sequencing techniques to solve the problem of LAB beer spoilage as well as help characterize useful LAB (Baugher and Klaenhammer, 2011; Mozzi *et al.*, 2013).

Methods of control

The use of omics approaches has the potential to reveal not only the nature and ecology of brewing-related LAB, thus leading to better diagnostic approaches, but to also provide information that can be exploited as potential means of controlling both BSR LAB and non-BSR LAB.

Physical means

Control or intervention strategies for BSR LAB traditionally have included the use of pasteurization, filtration, and chemical sanitation or detergents, although all these techniques can have undesirable outcomes or effects. For example, the use of high heat in pasteurization can cause denaturation or isomerization of important flavour compounds (e.g. hop-derived compounds) and decrease the 'fresh' taste of beer (Asano *et al.*, 2007; Franchi *et al.*, 2013). Further, filtration is not always efficacious given that small LAB often escape (Asano *et al.*, 2007; Back *et al.*, 1992) and chemical means of sterilization presents a selective pressure for resistant contaminating bacteria and, based on toxicity, can have varying implications for safe use on equipment and for human contact (Deasy *et al.*, 2011).

High-pressure treatments or homogenization (HPH) represents a promising physical means of controlling contaminating microorganisms in beer and is used in other LAB-associated fields (Kheadr *et al.*, 2002; Tribst *et al.*, 2013). HPH forces molecular conformational changes in proteins, enzymes and polysaccharides, and damages the overall integrity of cell membranes, and has specifically been shown to inactivate the hop tolerance protein HorA in BSR LAB (Tribst *et al.*, 2008, 2013; Ulmer *et al.*, 2002). HPH treatment efficacy is influenced by not only the Gram character of the cell wall, but the cell size (Wuytack *et al.*, 2002); however, Franchi *et al.* (2011, 2013) further characterized the conditions that facilitate effective HPH treatment for BSR LAB such that decreased pressure levels may be used for inactivation, thus increasing the utility of this approach for the brewer.

Bacteriocins

In many food-production fields, LAB-produced bacteriocins (antimicrobial peptides) are exploited for their ability to biologically preserve food through elimination of other microorganisms (Cintas *et al.*, 2001; Cotter *et al.*, 2005). Bacteriocins are a diverse collection of peptides and small proteins that have a wide range of activity against LAB and other Gram-positive bacteria (Cintas *et al.*, 2001; Klaenhammer, 1993). In the brewing industry, bacteriocins may eliminate other LAB or non-LAB beer spoilage organisms either through direct action against the cell or indirectly by contributing to the acidification of wort, making it less hospitable for bacterial growth (Lewis, 1998; Vaughan *et al.*, 2005; Rouse and van Sinderen, 2008). The most studied bacteriocin in the context of brewing is nisin (Delves-Broughton *et al.*, 1996; Vaughan *et al.*, 2005); however, chitosan (Gil *et al.*, 2004) and the bacteriocinogenic activity of the barley isolate *Lactococcus lactic* M30 have also been evaluated for use (Basanta *et al.*, 2007).

Bacteriocins are an attractive means for controlling BSR LAB growth, as they could be used in place of potentially harmful sanitizers or acid-washing which may affect yeast viability (Delves-Broughton *et al.*, 1996; Ogden *et al.*, 1988; Vaughan *et al.*, 2005). Further, bacteriocins theoretically could be added at most stages of the brewing production process in order to target either specific contaminated sites or introduce an inherent method of control (Delves-Broughton *et al.*, 1996; Ogden *et al.*, 1988; Vaughan *et al.*, 2004, 2005). Interest in bacteriocins has increased to the point that there is potential to develop and use yeast starter cultures genetically modified to produce bacteriocins or develop wort bioacidifying LAB (Dequin, 2001; James *et al.*, 2013; Van Reenen *et al.*, 2003). This approach would potentially allow for the selection or development of LAB starter cultures suitable for use in craft-fermentations, such that the LAB involved eventually limit their own growth by the production of, or exposure to, bacteriocins.

Phage therapy

Bacteriophages and the products they produce have also been used for antibacterial action in a variety of food-industries and are gaining interest for application in the brewing industry (Kelly *et al.*, 2011, 2012; Mahony *et al.*, 2011, 2012). Bacteriophages are an excellent candidate for exploitation given that they are safe for human consumption, as they are specific to their bacterial host and are ubiquitous in nature (Park *et al.*, 2011). Deasy *et al.* (2011) recently described *L. brevis* bacteriophage SAC12 with infectious activity against three *L. brevis* beer spoilage strains, which could effectively control the number of viable *L. brevis* cells present at a relatively high contaminant level. In light of the apparent success of bacteriophage therapy, further exploration of this treatment for beer via either a singular agent or a cocktail of phage agents appears worthwhile. These investigations can be aided by metagenomics and metatranscriptomics data of BSR LAB cells, which in turn will increase our understanding of both bacteriophage-host LAB relationships and LAB bacteriophage resistance.

BSR LAB and the brewing environment

Niche adaptation and horizontal gene transfer

Distinction between differently adapted LAB isolates lies not only with the analysis of the LAB core genomes, but also in the investigation of chromosomal sequences that appear to have originated in another species and mobile genetic elements (MGEs) such as plasmids (Broadbent *et al.*, 2012). The latter two genetic features are frequently acquired through horizontal gene transfer (HGT) between isolates of the same or different species. By comparing recently divergent as well as ecologically distinct genomes, it is revealed that HGT is important for the transfer of sequences or clusters of sequences, and drives the existence of diversification (Heuer and Smalla, 2007; Wiedenbeck and Cohan 2011). In fact, HGT events are promoted by environmental stress, resulting in faster adaptation or 'short-term' evolution in challenging environments (Dziewit and Bartosik, 2014).

For LAB, HGT events mediated by plasmids are important to a variety of industries (de Angelis and Gobetti, 2011; Cai *et al.*, 2009). In the brewing industry, conventional genetic markers of beer spoilage such as the exopolysaccharide gene *gtf*, and the hop tolerance genes *hitA*, *horA*, and *horC* are all plasmid-encoded and exhibit a very high degree of sequence identity in many different species (Suzuki, 2011; Walling *et al.*, 2005). The existence of these markers suggests not only the occurrence and support of HGT in and by the brewery, but also the importance of investigating other plasmid-harboured genes that demarcate BSR from non-BSR LAB.

Given that the ecological diversity among LAB appears to be driven in general by genome reduction mechanisms, the acquisition of niche-specific genes through the transfer of plasmids is an important area of investigation (Schroeter and Klaenhammer, 2009). Indeed, recent omics-based studies support the notion that plasmids are important for conferring beer spoilage ability. New genomic data for several *L. brevis* isolates has revealed that an increased number of plasmids may correlate with the ability of isolates to withstand increasingly harsh and specific environments. For example, *L. brevis* KB290 originally isolated from a traditional Japanese fermented vegetable and also able to grow in simulated gastric and intestinal juices, has nine plasmids ranging in size from 5.8 to 42 kb (Fukao *et al.*, 2013). Similarly, the rapid beer-spoiling isolate *L. brevis* BSO 464 has eight plasmids ranging from 2.3 to 85 kb (Bergsveinson *et al.*, 2015a). These two isolates are incapable of growth in the other isolate's niche-environment (J. Bergsveinson, unpublished), indicating that each possesses specific genes that do not confer immediate cross-resistance to another stressful environment; as such, these isolates have niche-specific tolerance genes. In contrast, the type strain *L. brevis* ATCC 367^{T} only harbours two plasmids (13 and 35 Kb) (Makarova *et al.*, 2006) and is unable to spoil beer and cannot grow in gastric juices (J. Bergsveinson, unpublished; Fukao *et al.*, 2013). This further suggests that increased plasmid-coding capacity likely supports the ability of *L. brevis* strains to infiltrate diverse environments. This idea is supported by a recent study showing that the sequential loss of plasmids from *L. brevis* BSO 464 results in loss of its original beer spoilage phenotype, indicating that beer spoilage is mediated by specific plasmid-encoded functions (Bergsveinson *et al.*, 2015a). Similarly, transcriptomic analysis performed on BSR LAB *L. brevis* BSO 464 (Lb464) (Bergsveinson *et al.*, 2016b) and *P. claussenii* ATCC BAA-344^{T} (Pc344) (Pittet *et al.* 2013) reveals that several significant plasmid-based transcripts were active across their respective eight plasmids when in the beer environment, notably on the plasmids that already harbour hop tolerance genes (*horC* for Lb464 and *horA* for Pc344) (Bergsveinson *et al.*, 2015a, 2016b; Pittet *et al.*, 2010, 2013). Collectively, these results strongly suggest that specific plasmids encode previously undescribed beer spoilage-related functions and that detailed investigation of plasmid genes in relation to growth in niche environments, beer, brewery, or otherwise, will prove useful.

Increased transcriptomic studies, in conjunction with comparative genomics, will most accurately and fully reveal the importance of plasmid-mediated functions for BSR LAB. Once more it is emphasized, that for these data to be of utility to the brewing industry, this analysis must be performed with more frequency on BSR LAB of both same and different species. As the cost of this analysis decreases and bioinformatics tools become more sensitive (Thayer, 2014; Mardis, 2011), it will be possible to investigate the broad importance of widely conserved plasmid sequences in BSR LAB, as has been done for other niche-adapted organisms (Dziewit and Bartosik, 2014; Papadimitriou *et al.*, 2015). Such analysis is reasonably expected to increase the number of species-independent, but beer spoilage-specific genes (and/or their transcripts) that can be screened for during quality control routines in the brewery.

Origin of BSR LAB

Phylogenetics and comparative genomics can help answer questions on the evolutionary development of BSR LAB; however, the answer to how and when these isolates emerged likely lies within the brewery itself. BSR LAB likely emerged with inclusion of hops in beer between the fifth and ninth centuries (Suzuki, 2011; Tonsmeire, 2014). Following genetic adaptation to this specific stress, BSR LAB then adapted further and have since remained tightly linked with the brewing environment (Suzuki *et al.*, 2008a; Suzuki, 2011). Indeed, BSR LAB isolates are rarely isolated elsewhere than breweries or beer, though non-BSR LAB isolates of the same species are (Suzuki *et al.*, 2008a; Suzuki, 2011). Breweries thus are both the selective environment and the reservoir for their own contaminants.

A recent study by Bokulich *et al.* (2015) investigated the distribution pattern of LAB species and putative hop tolerance genes in a brewery producing several different kinds of beer, using LAB-specific TRFLP (LAB-TRFLP) and ddPCR, respectively. The brewery involved produces 'conventional' beer (potential BSR LAB contaminants), sour beer (potentially helpful LAB fermenters and/or BSR LAB) and coolship beer (BSR LAB and environmental microflora). The LAB-TRFLP applied in this study was found to more sensitively discriminate between species of the *Lactobacillales* order and most genera of the *Bacillales* order present in mixed culture (Bokulich and Mills, 2012a).

The LAB-TRFLP also identified organisms from other phyla not previously reported as recovered from beer, likely as a result of the fact the organisms in question are present at low abundance and are never actively selected for during detection (Bokulich *et al.*, 2015). By applying this technique to analyse the LAB community profile throughout a brewery, Bokulich *et al.* (2015) were able to conclude that the brewery microbiota is probably driven by contact with raw substrates (grains, hops, yeast and beer), with this contact resulting in the profile of LAB present within a given brewery. For example, they found that wort samples contained a mixture of *L. delbrueckii*, *L. hilgardii*, *L. sakei*, *Lactococcus lactis*, *Leuconostoc mesenteroides*, *Streptococcus* spp., as well as a *Bacillus* spp., most of which were only rarely detected in other fermenting and bottled beer samples (Bokulich *et al.*, 2015). Many of these species, while not necessarily found in finished beer, are apparently associated with grain and therefore their detection in wort is unsurprising (Bokulich and Bamforth, 2013).

Interestingly, distinct LAB profiles from specific brewery samples were detected at different sites, presumably as result of potential contact with the beer sample. For example, sour and coolship beers were dominated by *Pediococcus* spp. and *L. lindneri*, though fermenters and barrel surfaces that contacted these sour fermentations around the time of sampling exhibited similar community composition; however, *L. brevis* and *Lactobacillus* sp. were found to be more common on these surfaces then on other surfaces or in the beer. Floor and packaging area surfaces contained a more diverse composition of LAB, with the predominant organisms being *L. brevis*, *L. delbrueckii*, and *L. lindneri*, which were also detected in the sour wort and beer. Perhaps most interesting was the finding that only *Pediococcus* spp. were detected on grain samples, while *L. brevis*, *L. lindneri* and *Pediococcus* spp. were recovered from hop pellets. This is noted as to be potentially due to the weak amplification from grain samples as a result of either inhibition of PCR by grain polyphenols or as a function of low LAB populations (Bokulich *et al.*, 2015). Though the data gathered are of exceptionally high detail, ultimately this work cautions against ascribing raw substrates as causing contamination of all areas or equipment that share similar microbial community compositions, as there are alternative means for microbial transfer within the environment such as fruit flies, or more likely, human activity (Bokulich *et al.*, 2015).

Given the ubiquitous presence of LAB in and on natural sources such as plants and humans, it is likely that the introduction of specific LAB species into the brewing environment, and their prevalence and distribution throughout, is an outcome of the specific raw materials (grain, hop, water, yeast) used and is a further function of a given brewery's specific geographical location; facility history; recipe, processing, and production lines; and personnel hygiene. The individual nature of a brewery has been underscored by the analysis of LAB-contamination in Australian breweries wherein specific contamination was found to be associated more with the particular brewery, rather than with specific antimicrobial challenges present by the beer sample that they were isolated from (ethanol, pH, hops) (Menz *et al.*, 2010). The microbiological quality and hygiene of a brewery thus is apparently dependent more on production practices and sanitation regimes than it is on the beer characteristics (e.g. highly hopped or alcoholic beers) (Menz *et al.*, 2010).

The work presented in Bokulich *et al.* (2015) is a foundational study from which to model further analysis of other breweries. Though it can be restated that the presence of LAB isolates and prevalence/distribution of them in a brewery will probably be brewery specific, ultimately an understanding of where bacterial (LAB) contamination is taking place within a given brewery should allow for the identification of specific contamination sources (i.e. raw materials vs. personnel) and help to strategize how best to prevent, or treat and recover contaminated product.

Hop resistance

Antimicrobial effect of hops

Beer and the brewery are stressful environments *in toto*, though hops are certainly considered the defining stress for microorganism growth. When hops are added to wort and boiled, alpha acids are extracted and transformed into iso-alpha-acids, the principal bitter components in beer (Steenackers *et al.*, 2015). These bitter compounds have a demonstrated antimicrobial effect on Gram-positive bacteria (e.g. LAB), which was originally recognized to be through increasing the permeability of the bacterial cell wall, causing leakage of the cytoplasmic membrane, inhibiting normal cellular processes, and causing changes in the cellular proton ionophore activity and uptake of leucine (Simpson, 1993a,b). Ultimately, bitter compounds act as ionophores, which sequester protons within the cell and dissipate the pH gradient across the cellular membrane, thereby reducing the proton motive force (PMF) and all PMF-dependent cellular activities, such as nutrient uptake (Sakamoto and Konings, 2003; Simpson, 1993a). Further, the strength of the inhibitory effect of hops was found to be dependent on pH and mediated by a cation (K^+, Mn^{2+}) exchange across the bacterial membrane (Simpson, 1993b). Importantly, these same mechanisms were not found to be active against Gram-negative isolates, probably as a result of the protection afforded by their outer membrane. Further investigation found that a transmembrane redox reaction of hop compounds occurred at low pH (such as in beer) and in the presence of Mn^{2+}, and that this redox activity causes cellular oxidative damage (Behr and Vogel, 2010). Therefore, hop resistance is a multifactorial process where at least two distinct levels of resistance mediate the stress of hops, namely proton ionophore-induced and oxidative stress mechanisms (Behr and Vogel, 2010).

Hop tolerance mechanisms

To combat the intrusion of hop bitter acids into the cell, it has been suggested that hop-tolerant LAB isolates produce higher molecular weight lipoteichoic acids (LTAs) in the cell wall as a response to hop presence (Behr *et al.*, 2006; Suzuki, 2011; Yasui and Yoda, 1997). The change in LTA composition increases the natural barrier function of the cell wall, fortifying it against the damaging invasion of hop bitter acids (Suzuki, 2011). Further, these LTAs are purported to act as reservoirs of divalent cations (Mn^{2+}) that can complex with hop bitter compounds as they move across the cell. The competitive binding of LTAs and Mn^{2+} limit the extent to which Mn^{2+}–hop bitter compound complexes are formed, further reducing the potential effects of hops against the cell (Behr *et al.*, 2006; Suzuki, 2011; Vogel *et al.*, 2010). This layer of defence or resistance is probably passive and of very little energy burden to the cell if established (Suzuki, 2011). Additionally, in a *L. brevis* model, it has been found that intracellular Mn^{2+}-dependent enzymes are induced by the presence of hop bitter compounds and this induction may help to maintain redox homeostasis and generate energy in response to a loss of PMF and depletion of Mn^{2+} reserves (Behr *et al.*, 2007). As these types of enzymes are responsible for maintaining cellular redox homeostasis, this cellular response is targeted at ameliorating the oxidative stress induced by hop bitter acids (Behr and Vogel, 2010; Vogel *et al.*, 2010).

Proposed hop tolerance genes

Potential genetic elements responsible for conferring some level of hop resistance to isolates have historically received a great deal of attention, given these elements would have utility for rapidly screening and distinguishing BSR LAB from non-BSR LAB. Hop resistance genes described to date all share the characteristics of being plasmid-located and having gene products associated with the cytoplasmic membrane, working to either remove hop-compounds from the intracellular space or maintain cellular homeostasis (Suzuki, 2011).

The first gene described was *horA*, recovered from a *L. brevis* isolate (Sami *et al.*, 1997a) and its product found to act as an ATP-binding cassette (ABC) transporter that purportedly exports hop bitter acids out of the cell (Sakamoto *et al.*, 2001). The *horC* gene, also originally described in *L. brevis*, is predicted to function as a PMF-dependent multidrug transporter belonging to the resistance–nodulation–cell division (RND) superfamily that can export bitter compounds from the cell (Ijima *et al.*, 2006; Suzuki *et al.*, 2005). Interestingly, HorA and HorC can both confer resistance to multiple structurally unrelated drugs (Sakamoto *et al.*, 2001; Suzuki *et al.*, 2005). HorB is often included in some lists of hop tolerance genes, though its proposed function as the transcriptional regulator of HorC has not been convincingly confirmed (Bergsveinson *et al.*, 2012; Iijima *et al.*, 2006). The third major hop tolerance gene described, *hitA*, is suggested to function in the uptake of divalent cations (Mn^{2+}) following reduction of the intracellular concentration of these cations by hop bitter compounds and thereby helping maintain Mn^{2+}-dependent cellular functions (Hayashi *et al.*, 2001). There have also been reports of two genes specific to *Pediococcus* spp., *bsrA* and *bsrB*, which presumably also function as multidrug ABC transporters against the action of hop compounds (Haakensen *et al.*, 2009c).

Although the mechanisms of action and prevalence of these genes in LAB have been well documented, questions remain as to ability of these hop tolerance genes to absolutely describe the hop tolerance and overall ability of a BSR LAB to grow in and spoil beer. For example, there are reported cases of hop-tolerant isolates that do not possess any of the described hop tolerance genes (Behr *et al.*, 2006; Menz *et al.*, 2010), or isolates that harbour all three hop tolerance genes and are hop-sensitive (Back, 2005). In addition, the actual functionality and/or transcriptional activity of these genes in response to hop or beer stress have yet to be analysed broadly and in depth using current molecular methods. One study utilized RT-qPCR to show that a *L. brevis* strain possessing all four genes (*hitA*, *horA*, *horB*, *horC*) and a *P. claussenii* strain possessing only *horA* and *bsrA* did not utilize these genes to the same extent during mid-exponential growth in beer (Bergsveinson *et al.*, 2012). In fact, only *horC* showed significant expression in beer in *L. brevis*, while its transcriptional regulator *horB* was not similarly expressed, nor were *hitA* and *horA*. Comparatively, *P. claussenii* demonstrated a significant expression of the *horA* gene and to a lesser extent, *bsrA*. Given these two isolates differ in their beer spoilage virulence, the transcriptional data raises the following questions. Does the possession of more than one hop tolerance gene correlate with increased hop tolerance *and* beer spoilage virulence? When all genes are present, are they utilized or active to different extents and potentially at different times during an isolate's growth? Finally, what is the role of *horB*?

General hypotheses can be posed in response to these questions. First, there is evidence to suggest that the presence of more than one hop tolerance gene correlates with increased hop tolerance and potentially increased beer spoilage ability; however, these studies did not delineate whether all genes under analysis were complete, functional and/or active (Suzuki, 2011). Second, some results suggest that HorC is a major contributor to hop tolerance and is generally correlated with strong beer spoilage ability, and therefore might be the preferred mechanism of action (i.e. energy is spent transcribing this gene) in the face of hop or beer stress despite the presence of the other hop tolerance genes (Bergsveinson *et al.*, 2015a,c; Preissler *et al.*, 2010). Alternatively, hop tolerance genes may not be necessarily transcriptionally active simultaneously and instead are activated in some sequential fashion, perhaps to reduce the cell's energy burden. As the HorC transporter is dependent on the PMF and apparently active during mid-exponential growth in beer (in *L. brevis*), it could be that HorA and HitA, which act to either re-establish or maintain the PMF through removal of hop compounds or movement of divalent cations, could actually establish optimal conditions for HorC activity and thus facilitate growth.

The role of *horB* was previously unclear. Although *horB* is nearly always found in conjunction with *horC*, it does not exhibit parallel transcription (Bergsveinson *et al.*, 2012; Bokulich *et al.*, 2015). Recently, it has been confirmed via RNAseq and ddPCR analysis that *horB* is a likely transcriptional repressor of *horC* activity, with active transcription of *horB* in nutritive or non-stressful growth conditions resulting in repressed horC expression (Bergsveinson *et al.*, 2016a,b,c).

Hop tolerance genes and the brewing environment

Until a recent study by Bokulich *et al.* (2015), there was no previous analysis of hop tolerance gene dispersion in the brewery environment. This study made use of next-generation ddPCR to quantify the abundance of *hitA*, *horA*, *horB* and *horC*, in conjunction with associated microbial community profiles assessed by LAB-TRFLP on various brewery surfaces over time.

Firstly, this study determined that areas involved with sour beer production had the highest gene frequencies, specifically *horC* (Bokulich *et al.*, 2015). This gene was the most abundant gene in general and was found in a nearly equal ratio with that of its putative transcriptional regulator *horB*, enforcing the notion that *horC* is an important and prevalent hop tolerance gene selected for in the brewing environment. The *hitA* gene had the lowest frequencies throughout the brewery, corroborating a previous report of low detected frequencies in BSR LAB (Haakensen *et al.*, 2008). Indeed, *hitA* also had a lower correlation with the presence of the other hop tolerance genes, which shared amongst themselves high degrees of intercorrelation (Bokulich *et al.*, 2015). *horA* was the only gene correlated with *Pediococcus* in the brewery, supporting the previous observation that this gene is the primary known resistance gene for this genus (Haakensen *et al.*, 2008). Most interesting was the fact that no genes correlated with *L. brevis*, which is not only considered the most common brewery contaminant species, but has also been shown to be among the LAB most commonly positive for hop-resistance genes (Haakensen *et al.*, 2008). However, given that *L. brevis* was only a minor component of sour beer and processing surfaces, this finding is probably particular to the brewery under analysis (Bokulich *et al.*, 2015).

These results reveal the importance of tracking spoilage genes within the brewery environment in order to understand contamination risks and patterns, especially where more than one beer style is produced. For examples, barrel surfaces, fermenters and packaging-line surfaces (that all come into contact with beer) exhibited fairly high levels of hop-resistance genes with the highest detection on surfaces that contacted sour beers and in unsanitary areas such as the packaging-line sink and below the packaging belt (Bokulich *et al.*, 2015). These findings are highly illuminating for development of brewery best practices, in that equipment for the production of sour or specialty beer must be specifically dedicated and adequately separated from equipment used in conventional brewing. Further, contact with beer is strongly implicated in transmission of hop tolerance genes and BSR LAB between different areas of the brewery; thus protocol and human activities must limit this transfer. Tracking transmission of hop tolerance (and other important genes) within the brewing environment is an incredibly worthwhile undertaking given that it will add insight into the role of environmental versus raw material contamination, and into the propagation of hop tolerance genes and BSR LAB within the brewery (Bokulich *et al.*, 2015).

Utility of BSR LAB hop tolerance genes

Questions remain concerning the utility of hop tolerance genes in predicting beer spoilage ability; however, there is no denying these genes are relevant to BSR LAB and the brewing environment. The most notable feature of these genes is that they are not species-specific markers for hop-tolerance. For example, *horA* and *horC* (in addition to their flanking open reading frame (ORF) regions) are found to be well conserved in other BSR LAB isolates such as *L. backii*, *L. lindneri*, *L. paracollinoides*, *P. claussenii*, and *P. damnosus*, in addition to *L. brevis* (Iijima *et al.*, 2007; Pittet *et al.*, 2012; Suzuki *et al.*, 2004a, 2005, 2011). It has even been reported that these two genes are found at rates as high as 94% and 96% of BSR LAB tested and that all strains have at least one of the genes (Suzuki, 2011). Caution is required; however, since the full gene length of hop tolerance genes is rarely sequenced in brewery settings; therefore, the sequence similarity, let alone the functional integrity of these genes, is rarely guaranteed.

Though targeting hop tolerance genes currently remain the strongest discriminatory method to detect intra-species differences in beer spoilage ability (Iijima *et al.*, 2007; Sami *et al.*, 1997b), these genes are still unable to predict the beer spoilage capacity of the full spectrum of BSR LAB that have been described. Unfortunately, to date, there are few data suggesting alternative hop tolerance genes or mechanisms in the absence of any of *hitA*, *horA*, or *horC* (Menz *et al.*, 2010). This lack of compensatory theories is frustrating in light of the physiological variability (hop-tolerance, growth phenotype) between strains that have identical hop tolerance genes profiles; other uncharacterized hop tolerance mechanisms must exist (Haakensen *et al.*, 2008, 2009c; Menz *et al.*, 2010). Since the known hop tolerance functions are ABC transporters or efflux pumps, and since both types of transporter are common within LAB (Konings *et al.*, 1997; White *et al.*, 2012b), it is short-sighted to not conceive of other similar genes and proteins across the spectrum of BSR LAB that deal with hops directly or indirectly, or that deal with other stresses in beer. Indeed, given the multitude of stress in beer, the ability to grow in and mediate the damage of both beer and hops, is probably the result of a synergy of mechanisms and redundant genetic traits. Until more detailed and high-throughput analyses of these processes are conducted, we remain hindered in our capacity to screen for elements that describe true beer spoilage ability.

Stress tolerance and adaptation of BSR LAB

There is considerable literature that discusses general and specific stress responses of LAB in a variety of industries. Though stress tolerance can differ among isolates of the same species, LAB are highly adaptable to stressful environments and adaptation to one particular stress often affords LAB increased tolerance to the challenge of another stress, due to the cross-regulation and functions of stress-response pathways (de Angelis and Gobetti, 2011; Parente *et al.*, 2010). BSR LAB isolates exemplify complex stress-response regulation given that isolates must simultaneously employ tolerance mechanisms to a variety of stresses.

Stress tolerance to ethanol and low pH

Ethanol levels and pH differ among styles of beer worldwide, typically within the ranges of 0.5–14% (v/v) ethanol and pH 3.8–4.7 (Suzuki *et al.*, 2008a). As a consequence, LAB recovered from beer within or outside these ranges are typically well adapted to one or both of these stresses (Suzuki, 2011). Further, most BSR LAB produce either lactic or acetic acid due to their basic fermentation, which naturally lowers the pH of the surrounding environment. Indeed, it has been reported that decreased pH and increased ethanol in beer had little effect on the growth of LAB, and that there is no correlation between these two factors and contamination, though pH values near 4.0 or below had some inhibitory effect on LAB (Menz *et al.*, 2010). Nonetheless, adaptation to the acidity found in beer is necessary, as low pH can interfere with enzymatic reactions, protein folding and other intracellular processes of non-pH-tolerant organisms. LAB, and other pH-tolerant organisms are capable of regulating their intracellular pH in face of acidic conditions through means of proton transport across the cellular membrane (often coupled to cation transport) or through proton-translocating ATP synthase (de Angelis and Gobetti, 2011).

Ethanol, like hops, is an antimicrobial component of beer, easily crossing the bacterial membrane and then modifying activity of cytoplasmic processes such as protein folding and inhibiting enzymatic actions. Ethanol also increases cell membrane permeability through alteration of the polarity of aqueous and hydrophobic regions of the phospholipid membrane, causing leakage of small molecules from the cell and cell death (Ingram, 1990). Various tolerance mechanisms may combat these effects, such as membrane fortification through an increase in long-chain fatty acids (>20 carbons) (Uchida, 1974). Other general stress-response proteins such as the GroES chaperone, heat-shock proteins (HSP), and glutathione reductase (Fiocco *et al.*, 2007; Silveira *et al.*, 2004) confer increased survival during ethanol stress, as well as to other stresses (Angelis and Gobetti, 2011). For BSR LAB, it has been found that ethanol tolerance does not differ significantly between BSR and non-BSR LAB, and that overall LAB ethanol-tolerance levels were species-conserved, unlike beer spoilage capacity (Pittet *et al.*, 2011). Though BSR LAB adaptation to low pH and ethanol are important, it does not appear that either is necessarily predictive of the ability to tolerate hops, nor ability to spoil beer (Bergsveinson *et al.*, 2015a,b; Menz *et al.*, 2010; Pittet *et al.*, 2011).

Stress tolerance to low nutrient availability

Yeasts are used to consume and ferment nutrients in wort following the breakdown of grain starches during the malting and mashing processes, in order to produce ethanol. Yeasts can make use of the majority of sugars present in wort, in addition to using available amino acids as a source of nitrogen, in a sequence usually dependent on both the strain of yeast and conditions used (Lodolo *et al.*, 2008; Perpete *et al.*, 2005). Organic acids (acetic, citric, lactic, malic, pyruvic and succinic acid) are left behind by yeasts as metabolic by-products, in addition to unused compounds such as dextrins, arabinoxylans and β-glucans (Gupta *et al.*, 2010). Remaining nutrients in beer following fermentation are typically in low abundance and are often 'alternative' sources of carbon that can vary from brew to brew within and between breweries.

LAB naturally have an array of possible mechanisms to transfer nutrients into the cell, thus allowing them to inhabit various nutrient-rich or -poor environmental niches. In nutrient-depleted beer, primary nutrient transport via the use of ATP-binding-cassette (ABC) transporters is proposed to allow for advantageous growth (Konings, 1997). These transporters typically have high affinity for a given solute and use ATP-hydrolysis for high-rate transport. Further, secondary transport mechanisms, not requiring ATP but relying on the electrochemical ion gradient to transport molecules across the membrane, involves uniporters, antiporters and symporters for effective uptake of molecules (White *et al.*, 2012b). In some cases, this uptake can even contribute to the production of energy through contribution to the PMF gradient (White *et al.*, 2012b). Lastly, group translocation, a mechanism that chemically modifies a solute that has been internalized can also facilitate the uptake of a range of carbohydrates (White *et al.*, 2012b).

There is considerable evidence for the importance of each type of transport uptake mechanism for BSR LAB. First, there is a great number of ABC transporters among LAB in general, and the importance of these proteins for BSR LAB is probably not yet fully appreciated beyond hop tolerance mechanisms (Konings *et al.*, 1997). Second, recent transcriptomic work on *P. claussenii* ATCC-BAA344^{T} (Pc344) when grown in beer revealed the importance of both secondary transport systems (e.g. the arginine or agmatine deiminase pathways, citrate fermentation) and group translocation such as the phosphotransferase system (PTS) (Pittet *et al.*, 2013). Components of all these systems were actually found to be among the top 20 most significantly expressed transcripts in beer, suggesting the critical role of nutrient-acquisition pathways for survival in beer (Pittet *et al.*, 2013). Interestingly, the significantly expressed agmatine deiminase pathway (AgDI operon) in Pc344 is very similar to the arginine deiminase (ADI) pathway, which is not specific to nutrient-stress, but has been shown to be up-regulated in response to low pH and acid stress, low oxygen concentration, low arginine supply (6 mM) and cell adaptation to arginine in *L. sanfransciscensis* (de Angelis and Gobetti, 2004). Though the ADI operon is not found in Pc344, the similar AgDI operon was shown to be critically important for survival in the beer environment, and is a major example of the cross-specificity of LAB stress responses. Finally, another example of cross-resistance is the induction of stationary phase in LAB when faced with nutritional starvation, at which point the cells become more resistant to stresses such as heat and acid (Angelis and Gobetti, 2011; Gouesbet *et al.*, 2001).

Stress tolerance to low O_2 tension and dissolved CO_2

The low oxygen levels in beer selects for microbes capable of anaerobic respiration. LAB, specifically *Lactobacillus* and *Pediococcus* isolates, can produce energy in the absence of oxygen by using other electron acceptors to regenerate NAD^+ or by substrate-level phosphorylation during fermentation for the regeneration of NAD^+ (White *et al.*, 2012a). Fermentation capacities in anaerobic conditions are known to be different across subgroups, even genera, of LAB, nonetheless, the overall anaerobic nature of BSR LAB facilitates their resistance to the stress of low oxygen.

Recently, it was shown that the presence of headspace pressure and dissolved CO_2 (dCO_2) limits the ability of LAB to grow in and spoil beer (Bergsveinson *et al.* 2015b). Transcriptomic analysis of dCO_2-tolerant isolate *L. brevis* BSO 464 during growth in packaged beer has since revealed that this environment strongly induces modifications to cellular transcriptional regulation, and to both the cell membrane and wall, relative to what occurs in unpackaged beer (Bergsveinson *et al.*, 2016b). This finding strongly suggests that further analysis of changes in the cell membrane lipid profile of dCO_2 beer-tolerant BSR LAB may reveal signature alterations, or biomarkers of tolerance to the packaged beer environment and beer in general. Such information is important, since it is reasonable to expect brewery-adapted BSR LAB to be able to withstand the sudden, additional stress of high dCO_2 as a result of headspace flushing with CO_2 during packaging.

Viable, but not culturable, state

A general adaptation to the beer environment by BSR LAB is the modification of both cell size and morphology. Diminished cell size and a notable rounding or shrinking of bacilli or rod-shaped cells (i.e. taking on a coccoid appearance) have been noted for several BSR LAB isolates following beer-adaptation (Asano *et al.*, 2007). This phenomenon has been proposed to be an attempt by the cell to reduce surface area in contact with beer and to help membrane-associated tolerance genes (e.g. hop tolerance) deploy more efficiently (Suzuki *et al.*, 2006; Suzuki, 2011). Furthermore, increased time in beer results in induction of a viable, but not culturable (VNBC) state in LAB cells. Such cells are not detectable by routine non-beer culture media on which they would normally grow colonies, but are alive and capable of renewed metabolic activity through continued exposure to routine media (Deng *et al.*, 2014; Oliver, 2005). Understanding both the conditions inducing the VBNC state in a variety of different BSR LAB and how to retrieve the culturable phenotype more efficiently is of extreme utility to brewers in accurately detecting where BSR LAB exist in their brewery. Investigation into the genetics and transcriptional activity of BSR LAB throughout their VNBC cycle would also increase the understanding and be of upmost importance for a wide range of industries dependent on LAB.

Maintenance of BSR LAB and importance of biofilms

The mapping of BSR LAB and hop tolerance genes in the brewery by Bokulich *et al.* (2015) illustrates the risk of cross-contamination between different equipment surfaces, especially in environments where conventional and sour beer types are produced. Cross-contamination of surface areas supports the increase in diversity of the present microbial community, as well as the development of biofilms, thus probably driving spoilage incidence at various production stages (Matoulkova *et al.*, 2012; Timke *et al.*, 2005, 2008; Storgårds *et al.*, 2006).

The brewing industry has great concern about biofilms, given that they can be established not only in the brewery, but also in draft beer dispensing lines outside the brewery, which brewers usually do not monitor, nor control (Timke *et al.*, 2005; Thomas and Whitham, 1997). Though biofilms are typically comprised of a variety of microorganisms, they have a known correlation with product-spoiling bacteria and thus require prevention and attention (Timke *et al.*, 2005; Zottola and Sasahara, 1994). Increased analysis of brewery biofilms would be of great utility, especially given that biofilms may be highly individualistic and thus require specific or adapted control treatments. Though some specific strains of LAB have been shown to be able to form biofilms (Kubota *et al.*, 2008), in general, Gram-negative bacteria (and yeasts and moulds) are among the first to colonize surfaces in the brewery, while LAB are opportunistic colonizers that benefit from the multiple interactions within already established communities, especially if the biofilm provides reduced oxygen levels and an acidic environment (Stornsgard *et al.*, 2006).

Involvement in biofilms also increases the likelihood of acquiring genetic material advantageous for the brewing environment through HGT (Kubota *et al.*, 2008; Timke *et al.*, 2005). Biofilm-mediated transfer of beer spoilage virulence genes is evidenced by the finding of plasmid-harboured hop tolerance genes among many LAB species, with the interspecies nucleotide sequence identities of these genes and surrounding regions being highly conserved at approximately 99% (Suzuki *et al.*, 2005, 2008a; Suzuki, 2011). Indeed, a 5.6 kb region that contains *horA* was found to be 100% identical in *L. backii* and *P. inopinatus* strains isolated from the same brewery (Iijima *et al.*, 2007) and this same phenomenon is identified in other *horA*$^+$ isolates recovered from different sources (Pittet *et al.*, 2012a; J. Bergsveinson, unpublished data). As such, HGT among LAB in biofilms is believed to be how hop tolerance genes, and other putative plasmid-mediated beer resistance elements are spread within breweries (Suzuki, 2011).

Screening physiological capacities of BSR LAB: contaminant versus fermentor?

It is obvious that beer spoilage is not a binary phenotype mediated by the presence/absence of one or even a few genes, as antibiotic resistance or other phenotypes like motility most often are (Schwarz *et al.*, 2014). Instead, BSR LAB isolates exist on a scale of beer-spoiling virulence, just as LAB isolates in other industries sit on a continuum of capabilities and efficiency in performing a task (Mozzi *et al.*, 2013). Indeed the extent to which a LAB isolate can grow and metabolize is important for its classification as either a contaminant or fermentor. Contaminants are considered thus because their growth is at once unexpected, and *uncontrollable* by the environmental conditions posed by production of a given beer; however, the damage they impose on the beer product is relative across isolates as a result of their growth ability and metabolic by-products. On the other hand, ideal LAB participants in sour beer fermentations contribute some specific flavour component(s), but do not over-produce these compounds through cellular overgrowth, as this leads to flavour imbalance (i.e. spoilage). Thus, to prevent a helpful LAB isolate from being considered a contaminant, either the genetic makeup or surrounding environment of an isolate must limit its own growth and metabolism.

The relative ability for LAB to establish rapid growth in beer or produce flavour compounds in a moderate fashion is genetically based, which strongly points to subtle differences in the genetics and metabolism of BSR LAB and fermenting LAB. However, these changes are not readily interrogated using targeted analytical methods such as MLST. To distinguish contaminant from helpful isolate, the analytical method must be able to take into account the influence of the total beer environment (i.e. available fermentable sugars and other nutrients, ethanol levels, hop levels, pH, dCO_2) when describing the beer spoilage virulence of a given LAB. Thus, it is only through the use of meta-genomics, global transcriptomics and phenotype correlation, that researchers and the brewery industry will be able to effectively profile helpful, fermenting LAB for development of new beer product, as has been done in other non-brewing industries. For example, efforts are under way to preform en masse genome analysis of LAB *Oenococcus oeni* isolates relevant to the wine industry, in order to link genotype with an isolate's winemaking properties and wine characteristics (Bartowsky *et al.*, 2011, 2014; Borneman *et al.*, 2013a,b). Linkage analysis allows for the distinction between content diversity (specifically, gene presence/absence) and genome diversity (organization, regulation, plasmid and phage presence) and their link to overall isolate phenotype such as flavour profile produced. Importantly, this large-scale analysis ameliorates the potential bias that isolate-selection has on between-isolate comparisons (Cai *et al.*, 2009; Pfeiler and Klaenhammer, 2007; Sun *et al.*, 2014).

Within the brewing industry, non-academic institutions are beginning to conduct analyses similar to those of White Labs Ltd (San Diego, CA, USA) who are analysing brewing yeast genomes in relation to the flavour profile of beers that the yeasts impact (Herkewitz, 2014). It is reasonable to expect that comparable analyses could be performed not only for LAB in relation to the styles and composition of beer they are able to spoil, but also for the characterization of helpful, fermenting LAB for sour beer production and for defining LAB exhibiting beneficial interactions with yeast in specialty brews.

Conclusions and the future

The demonstrated variability of LAB as to genetics, niche-adaptations, and stress tolerance, together with the individuality of brewery environments leads to the conclusion that single episodes of BSR LAB contamination cannot be considered exemplary. Unfortunately, historical preoccupation with a select few *exemplary* isolates, brew styles, and physiological stresses means that the sum total of current BSR LAB research does not adequately describe the dynamic event that is LAB-related beer spoilage. Examples of shortsightedness or incomplete information about BSR LAB include the following:

1. Although *the total* beer environment is recognized as stressful for microbial growth, the effect of dissolved CO_2 on LAB growth has not been studied to the same extent as for the other described stresses in beer.
2. There has been virtually no investigation of the multiple beer stresses simultaneously as to their effect on BSR LAB physiology or gene transcription.
3. The presence and role that pervasive or common bacteriophages might have in establishing BSR LAB communities as well as patterns of bacteriophage resistance remain unknown.
4. No data are available on the role of redundant genetic mechanisms that may operate in BSR LAB to deal with the individual, let alone the simultaneous growth stresses found in beer.
5. Only minimal efforts have been made to describe *non-traditional* BSR LAB or to discover the full spectrum of microbial diversity in breweries.
6. There is limited understanding of how shifts in beer composition (nutrient levels, processing conditions, small changes in ethanol or hops) affect the survival of BSR LAB or microbial communities.

Further, when considering these examples of where research on BSR LAB is needed, it must be emphasized that the brewing industry lags behind other LAB-fermentation industries such as the dairy/cheese and wine industry in understanding both product-fermenting and product-spoiling LAB isolates, how they behave within a microbial community, and if these interactions are influenced by raw substrates as well as the brewing process.

To address these issues and enlarge the data available on BSR LAB, the brewing industry must align more closely with advances made in other LAB industries. Concurrently, omics-based technologies must be applied to the study of BSR LAB within both academia and the brewing industry. Not only is the power and versatility of these technologies well demonstrated for interrogating microbial processes, but the simple utilitarian perspective remains that these methods clearly provide *more* information through *fewer* experimental trials than other approaches. For example, as opposed to using classic methods (laborious processes with need for excessive replication) to investigate the presence or absence of a gene and the correlation to a specific physiological trait, the use of metatranscriptomics together with metagenomics can reveal patterns of gene activity in relation to growth environment and phenotype – and do so for multiple isolates with fewer procedural steps.

Support for application of such broad scale omics analyses to brewery LAB contamination has come from recent research, which has strengthened the call for a shift in research paradigm from targeted-analysis to community-analysis. To begin with, and most importantly, genomic and transcriptomic analyses of BSR LAB isolates have firmly called into question the utility of the few known hop tolerance genes as indicators of LAB beer spoilage potential, and have begun to indicate other potential genetic markers of beer spoilage or bacteriophage content that could be potentially exploited for control of BSR LAB. Further, microbial community analysis using next-generation PCR applications and sensitive molecular methods has indicated that this type of analysis is critical to perform for breweries producing multiple brew styles, particularly to determine and then modify personnel-driven contamination patterns with the brewery. Additionally, this analysis has begun to reveal the presence of unique microbes not yet described in brewery settings that must be explored for their potential to contribute via HGT beer spoilage resistance genes to LAB found in the brewery setting.

Concerns of accessibility of omics technologies for those who work within brewery settings expose an interesting developing niche within the industry – for commercial or academic laboratories that can provide omics services to breweries and assist with interpretation of omics data. The emergence of such an approach represents an important evolution in the intersection of academia and the brewing industry. Based on the established progression of omics accessibility (i.e. decreasing cost, with increasing ease of use) there is a foreseeable future wherein brewers readily rely on these technologies for investigating incidences of spoilage, as well as for obtaining information needed for product innovation.

Ultimately, the two main issues that remain endemic to BSR LAB research are the dogma that a few select genes are responsible for conferring beer spoilage ability and that a handful of BSR LAB species can be used as models for investigating all matters of LAB spoilage. Current omics results already point to the fallacy of these two dogmas and future research is expected to provide confirmation. In this context, it is important to note that omics approaches can be used to search answers to critical remaining questions. Specifically, are there are other conserved plasmid sequences transferred in brewery HGT events? How do BSR LAB interact with one another within microbial communities (metabolome, quorum sensing), and how do these interactions influence the relative proportions of the microbes present? How or what is the connection between genomic content and the good or the bad or ugly, physiological capacities of BSR LAB? Finally, how can omics technologies be exploited to screen for helpful brewing LAB, and help us understand as well as control their contribution to beer flavour and interaction with brewing yeasts? Applications of omics approaches to BSR LAB truly have the potential to quickly and exponentially expand our understanding of these bacteria.

The reality of the highly adaptable and variable nature of LAB, and the selective and individual environment of a given brewery, guarantee that LAB-contamination will continue to pose a threat to the global beer industry. Undoubtedly, improved brewery hygiene has increased the general stability of beer products; however, the current industry environment fosters competition and innovation, thus necessitating not only the production of new beers, but often of different types of beer simultaneously within a given physical plant. This means there are developing layers of complexity and new challenges to maintaining a known, stable, and controlled microbial brewing environment. Community-scale analysis together with increased application of omics approaches for troubleshooting, general research, and innovation is the only way to keep pace with these demands.

References

Asano, S., Iijima, K., Suzuki, K., Motoyama, Y., Ogata, T., and Kitagawa, Y. (2009). Rapid detection and identification of beer spoilage lactic acid bacteria by microcolony method. J. Biosci. Bioeng. *108*, 124–129. http://dx.doi.org/10.1016/j.jbiosc.2009.02.016

Asano, S., Suzuki, K., Iijima, K., Motoyama, Y., Kuriyama, H., and Kitagawa, Y. (2007). Effects of morphological changes in beer spoilage lactic acid bacteria on membrane filtration in breweries. J. Biosci. Bioeng. *104*, 334–338. http://dx.doi.org/10.1263/jbb.104.334

Back, W. (1987). New description of a type of *Lactobacillus* harmful to beer — *Lactobacillus brevisimilis* spec. nov. Monatsschr. Brauwiss. *40*, 484–488.

Back, W. (1994). Secondary contaminations in the filling area. Brauwelt Int. *4*, 326–333.

Back, W. (2005). Brewery. In Colour Atlas and Handbook of Beverage Biology, Back, W. ed. (Germany: Verlag Hans Carl: Nürnberg),,pp. 10–112.

Back, W., Leibhard, M., and Bohak, I. (1992). Flash pasteurization– membrane filtration. Comparative biological safety. Brauwelt Int. *1*, 42–49.

Bartowsky, E.J., and Borneman, A.R. (2011). Genomic variations of *Oenococcus oeni* strains and the potential to impact on malolactic fermentation and aroma compounds in wine. Appl. Microbiol. Biotechnol. *92*, 441–447. http://dx.doi.org/10.1007/s00253-011-3546-2

Bartowsky, E., Costello, J., Chambers, J., and Borneman, A.R. (2014). Presentation: Genomic analysis of 80 *Oenococcus oeni* strains and connecting the genome with winemaking properties. 11th Symposium on Lactic Acid Bacteria proceedings, Egmond aan Zee, Netherlands August 31–September 4.

Basanta, A., Sánchez, J., Gómez-Sala, B., Herranz, C., Hernández, P.E., and Cintas, L.M. (2008). Antimicrobial activity *of Enterococcus faecium* L50, a strain producing enterocins L50 (L50A and L50B), P and Q, against beer spoilage lactic acid bacteria in broth, wort (hopped and unhopped), and alcoholic and non-alcoholic lager beers. Int. J. Food Microbiol. 125, 293–307. http://dx.doi.org/10.1016/j.ijfoodmicro.2008.04.011

Baugher, J.L., and Klaenhammer, T.R. (2011). Invited review: Application of omics tools to understanding probiotic functionality. J. Dairy Sci. *94*, 4753–4765. http://dx.doi.org/10.3168/jds.2011-4384

Behr, J., and Vogel, R.F. (2010). Mechanisms of hop inhibition include the transmembrane redox reaction. Appl. Environ. Microbiol. *76*, 142–149. http://dx.doi.org/10.1128/AEM.01693-09

Behr, J., Gänzle, M.G., and Vogel, R.F. (2006). Characterization of a highly hop-resistant *Lactobacillus brevis* strain lacking hop transport. Appl. Environ. Microbiol. 72, 6483–6492. http://dx.doi.org/10.1128/AEM.00668-06

Behr, J., Israel, L., Gänzle, M.G., and Vogel, R.F. (2007). Proteomic approach for characterization of hop-inducible proteins in *Lactobacillus brevis*. Appl. Environ. Microbiol. 73, 3300–3306. http://dx.doi.org/10.1128/AEM.00124-07

Ben-Amor, K., Vaughan, E.E., and de Vos, W.M. (2007). Advanced molecular tools for the identification of lactic acid bacteria. J. Nutr. *137* (Suppl. 3), 741S–747S.

Bergsveinson, J., Baecker, N., Pittet, V., and Ziola, B. (2015a). Role of plasmids in *Lactobacillus brevis* BSO 464 hop tolerance and beer spoilage. Appl. Environ. Microbiol. *81*, 1234–1241. http://dx.doi.org/10.1128/AEM.02870-14

Bergsveinson, J., Ewen, E., Friesen, V., and Ziola, B. (2016a). Transcriptional activity and role of plasmids of *Lactobacillus brevis* BSO 464 and *Pediococcus claussenii* ATCC BAA-344T during growth in the presence of hops. AIMS Microbiol. *2*, 460–478. http://dx.doi.org/10.3934/microbiol.2016.4.460

Bergsveinson, J., Friesen, V., and Ziola, B. (2016b). Transcriptome analysis of beer-spoiling *Lactobacillus brevis* BSO 464 during growth in degassed and gassed beer. Int. J. Food Microbiol. *235*, 28–35. http://dx.doi.org/10.1016/j.ijfoodmicro.2016.06.041

Bergsveinson, J., Goerzen, S., Redekop, A., Zoerb, S., and Ziola, B. (2015b). Dissolved carbon dioxide selects for lactic acid bacteria able to grow in and spoil packaged beer. J. Am. Soc. Brew. Chem. 73, 331–338. http://dx.doi.org/10.1094/ASBCJ-2015-0726-01

Bergsveinson, J., Goerzen, S., Redekop, A., Zoerb, S., and Ziola, B. (2016c). Genetic variability in the hop-tolerance *horC* gene of beer-spoiling lactic acid bacteria. J Am Soc Brew Chem. *74*, 173–182. http://dx.doi.org/10.1094/ASBCJ-2016-3962-01

Bergsveinson, J., Pittet, V., and Ziola, B. (2012). RT-qPCR analysis of putative beer spoilage gene expression during growth of *Lactobacillus brevis* BSO 464 and *Pediococcus claussenii* ATCC BAA-344(T) in beer. Appl. Microbiol. Biotechnol. *96*, 461–470. http://dx.doi.org/10.1007/s00253-012-4334-3

Bergsveinson, J., Pittet, V., Ewen, E., Baecker, N., and Ziola, B. (2015c). Genome Sequence of Rapid Beer-Spoiling Isolate *Lactobacillus brevis* BSO 464. Genome Announc. 3, e01411–15. http://dx.doi.org/10.1128/genomeA.01411-15

Bohak, I., Back, W., Richter, L., Ehrmann, M., Ludwig, W., and Schleifer, K.H. (1998). *Lactobacillus amylolyticus* sp. nov., isolated from beer malt and beer wort. Syst. Appl. Microbiol. *21*, 360–364. http://dx.doi.org/10.1016/S0723-2020(98)80045-3

Bohak, I., Thelen, K., and Beimfohr, C. (2006). Description of *Lactobacillus backi* sp. nov., an obligate beer-spoiling bacterium. Monatsschr. Brauwiss. *59*, 78–82.

Bokulich, N.A., and Bamforth, C.W. (2013). The microbiology of malting and brewing. Microbiol. Mol. Biol. Rev. *77*, 157–172. http://dx.doi.org/10.1128/MMBR.00060-12

Bokulich, N.A., and Mills, D.A. (2012a). Differentiation of mixed lactic acid bacteria communities in beverage fermentations using targeted terminal restriction fragment length polymorphism. Food Microbiol. 31, 126–132. http://dx.doi.org/10.1016/j.fm.2012.02.007

Bokulich, N.A., and Mills, D.A. (2012b). Next-generation approaches to the microbial ecology of food fermentations. BMB Rep. *45*, 377–389. http://dx.doi.org/10.5483/BMBRep.2012.45.7.148

Bokulich, N.A., Bamforth, C.W., and Mills, D.A. (2012a). A review of molecular methods for microbial community profiling of beer and wine. J. Am. Soc. Brew. Chem. 70, 150–162. http://dx.doi.org/10.1094/ASBCJ-2012-0709-01.

Bokulich, N.A., Bergsveinson, J., Ziola, B., and Mills, D.A. (2015). Mapping microbial ecosystems and spoilage-gene flow in breweries highlights patterns of contamination and resistance. Elife 4, e04634. http://dx.doi.org/10.7554/eLife.04634

Bokulich, N.A., Joseph, C.M., Allen, G., Benson, A.K., and Mills, D.A. (2012b). Next-generation sequencing reveals significant bacterial diversity of botrytized wine. PLoS ONE 7, e36357. http://dx.doi.org/10.1371/journal.pone.0036357

Borneman, A.R., Pretorius, I.S., and Chambers, P.J. (2013a). Comparative genomics: a revolutionary tool for wine yeast strain development. Curr. Opin. Biotechnol. *24*, 192–199. http://dx.doi.org/10.1016/j.copbio.2012.08.006

Borneman, A.R., Schmidt, S.A., and Pretorius, I.S. (2013b). At the cutting-edge of grape and wine biotechnology. Trends Genet. *29*, 263–271. http://dx.doi.org/10.1016/j.tig.2012.10.014

Bottari, B., Ercolini, D., Gatti, M., and Neviani, E. (2006). Application of FISH technology for microbiological analysis: current state and prospects. Appl. Microbiol. Biotechnol. *73*, 485–494. http://dx.doi.org/10.1007/s00253-006-0615-z

Broadbent, J.R., Neeno-Eckwall, E.C., Stahl, B., Tandee, K., Cai, H., Morovic, W., Horvath, P., Heidenreich, J., Perna, N.T., Barrangou, R., *et al.* (2012). Analysis of the *Lactobacillus casei* supragenome and its influence in species evolution and lifestyle adaptation. BMC Genomics *13*, 533. http://dx.doi.org/10.1186/1471-2164-13-533

Bron, P.A., Wels, M., Bongers, R.S., van Bokhorst-van de Veen, H., Wiersma, A., Overmars, L., Marco, M.L., and Kleerebezem, M. (2012). Transcriptomes reveal genetic signatures underlying physiological variations imposed by different fermentation conditions in *Lactobacillus plantarum*. PLoS ONE. 7, e38720. http://dx.doi.org/10.1371/journal.pone.0038720

Cai, H., Thompson, R., Budinich, M.F., Broadbent, J.R., and Steele, J.L. (2009). Genome sequence and comparative genome analysis of *Lactobacillus casei*: insights into their niche-associated evolution. Genome Biol. Evol. 1, 239–257. http://dx.doi.org/10.1093/gbe/evp019

Cintas, L.M., Casaus, M.P., Herranz, C., Nes, I.F., and Hernández, P.E. (2001). Review: Bacteriocins of lactic acid bacteria. Food Sci. Tech. Int. 7, 281–305. http://dx.doi.org/10.1106/R8DE-P6HU-CLXP-5RY

Claesson, M.J., van Sinderen, D., and O'Toole, P.W. (2007). The genus *Lactobacillus* – a genomic basis for understanding its diversity. FEMS Microbiol. Lett. *269*, 22–28. http://dx.doi.org/10.1111/j.1574-6968.2006.00596.x

Cocolin, L., Manzano, M., Cantoni, C., and Comi, G. (2001). Denaturing gradient gel electrophoresis analysis of the 16S rRNA gene V1 region to monitor dynamic changes in the bacterial population during fermentation of Italian sausages. Appl. Environ. Microbiol. *67*, 5113–5121. http://dx.doi.org/10.1128/AEM.67.11.5113-5121.2001

Corsetti, A., Settanni, L., van Sinderen, D., Felis, G.E., Dellaglio, F., and Gobbetti, M. (2005). *Lactobacillus rossii* sp. nov., isolated from wheat sourdough. Int. J. Syst. Evol. Microbiol. *55*, 35–40. http://dx.doi.org/10.1099/ijs.0.63075-0

Cotter, P.D., Hill, C., and Ross, R.P. (2005). Bacteriocins: developing innate immunity for food. Nat. Rev. Microbiol. *3*, 777–788. http://dx.doi.org/10.1038/nrmicro1273.

Curk, M.C., Hubert, J.C., and Bringel, F. (1996). *Lactobacillus paraplantarum* sp. now., a new species related to *Lactobacillus plantarum*. Int. J. Syst. Bacteriol. *46*, 595–598. http://dx.doi.org/10.1099/00207713-46-2-595

De Angelis, M., and Gobbetti, M. (2004). Environmental stress responses in *Lactobacillus*: a review. Proteomics *4*, 106–122. http://dx.doi.org/10.1002/pmic.200300497

de Angelis, M., and Gobbetti, M. (2011). Chapter 11: Stress Response of Lactobacilli. In Stress Responses of Lactic Acid Bacteria, Tsakalidou, E., and Papadimitriou, K., eds. (Boston: Springer), pp. 219–249. http://dx.doi.org/10.1007/978-0-387-92771-8

de Bok, F.A., Janssen, P.W., Bayjanov, J.R., Sieuwerts, S., Lommen, A., van Hylckama Vlieg, J.E., and Molenaar, D. (2011). Volatile compound fingerprinting of mixed-culture fermentations. Appl. Environ. Microbiol. 77, 6233–6239. http://dx.doi.org/10.1128/AEM.00352-11

de Lipthay, J.R., Enzinger, C., Johnsen, K., Aamand, J., and Sørensen, S.J. (2004). Impact of DNA extraction method on bacterial community composition measured by denaturing gradient gel electrophoresis. Soil Biol. Biochem. 36, 1607–1614. http://dx.doi.org/10.1016/j.soilbio.2004.03.011

de Man, J.C., Rogosa, M., and Sharpe, M.E. (1960). A medium for the cultivation of lactobacilli. J. Appl. Bacteriol. 23, 130–135.

Deasy, T., Mahony, J., Neve, H., Heller, K.J., and van Sinderen, D. (2011). Isolation of a virulent *Lactobacillus brevis* phage and its application in the control of beer spoilage. J. Food Prot. *74*, 2157–2161. http://dx.doi.org/10.4315/0362-028X.JFP-11-262

Delves-Broughton, J., Blackburn, P., Evans, R.J., and Hugenholtz, J. (1996). Applications of the bacteriocin, nisin. Antonie Van Leeuwenhoek *69*, 193–202.

Deng, Y., Liu, J., Li, H., Li, L., Tu, J., Fang, H., Chen, J., and Qian, F. (2014). An improved plate culture procedure for the rapid detection of beer spoilage lactic acid bacteria. J. Inst. Brew. 120, 127–132. http://dx.doi.org/10.1002/jib.121

Dequin, S. (2001). The potential of genetic engineering for improving brewing, wine-making and baking yeasts.

Appl. Microbiol. Biotechnol. *56*, 577–588. http://dx.doi.org/10.1007/s002530100700
Di Cagno, R., De Angelis, M., Calasso, M., and Gobbetti, M. (2011). Proteomics of the bacterial cross-talk by quorum sensing. J. Proteomics *74*, 19–34. http://dx.doi.org/10.1016/j.jprot.2010.09.003
Dobson, C.M., Deneer, H., Lee, S., Hemmingsen, S., Glaze, S., and Ziola, B. (2002). Phylogenetic analysis of the genus *Pediococcus*, including *Pediococcus claussenii* sp. nov., a novel lactic acid bacterium isolated from beer. Int. J. Syst. Evol. Microbiol. *52*, 2003–2010. http://dx.doi.org/10.1099/00207713-52-6-2003
Douillard, F.P., Ribbera, A., Kant, R., Pietilä, T.E., Järvinen, H.M., Messing, M., Randazzo, C.L., Paulin, L., Laine, P., Ritari, J., *et al.* (2013). Comparative genomic and functional analysis of 100 *Lactobacillus rhamnosus* strains and their comparison with strain GG. PLoS Genet. *9*, e1003683. http://dx.doi.org/10.1371/journal.pgen.1003683
Dziewit, L., and Bartosik, D. (2014). Plasmids of psychrophilic and psychrotolerant bacteria and their role in adaptation to cold environments. Front. Microbiol. *5*, 596. http://dx.doi.org/10.3389/fmicb.2014.00596
Ehrmann, M.A., Preissler, P., Danne, M., and Vogel, R.F. (2010). *Lactobacillus paucivorans* sp. nov., isolated from a brewery environment. Int. J. Syst. Evol. Microbiol. *60*, 2353–2357. http://dx.doi.org/10.1099/ijs.0.018077-0
Endo, A., Futagawa-Endo, Y., and Dicks, L.M. (2011). Influence of carbohydrates on the isolation of lactic acid bacteria. J. Appl. Microbiol. *110*, 1085–1092. http://dx.doi.org/10.1111/j.1365-2672.2011.04966.x
Enright, M.C., and Spratt, B.G. (1999). Multilocus sequence typing. Trends Microbiol. *7*, 482–487.
Fiocco, D., Capozzi, V., Goffin, P., Hols, P., and Spano, G. (2007). Improved adaptation to heat, cold, and solvent tolerance in *Lactobacillus plantarum*. Appl. Microbiol. Biotechnol. *77*, 909–915. http://dx.doi.org/10.1007/s00253-007-1228-x
Fernandez, J.L., and Simpson, W.J. (1993). Aspects of the resistance of lactic acid bacteria to hop bitter acids. J. Appl. Microbiol. *75*, 315–319. http://dx.doi.org/10.1111/j.1365-2672.1993.tb02782.x
Franchi, M.A., Tribst, A.A.L., and Cristianini, M. (2011). Inactivation of *Lactobacillus brevis* in beer utilizing a combination of high-pressure homogenization and lysozyme treatment. J. Inst. Brew. *117*, 634–638. http://dx.doi.org/10.1002/j.2050-0416.2011.tb00515.x
Franchi, M.A., Tribst, A.A.L., and Cristianini, M. (2013). High-pressure homogenization: a non-thermal process applied for inactivation of spoilage microorganisms in beer. J. Inst. Brew. 119, 237–241. http://dx.doi.org/10.1002/jib.99
Fukao, M., Oshima, K., Morita, H., Toh, H., Suda, W., Kim, S.W., Suzuki, S., Yakabe, T., Hattori, M., and Yajima, N. (2013). Genomic analysis by deep sequencing of the probiotic *Lactobacillus brevis* KB290 harboring nine plasmids reveals genomic stability. PLoS ONE 8, e60521. http://dx.doi.org/10.1371/journal.pone.0060521
Geissler, A.J., Behr, J., von Kamp, K., and Vogel, R.F. (2016). Metabolic strategies of beer spoilage lactic acid bacteria in beer. Int. J. Food Microbiol. *216*, 60–68. http://dx.doi.org/10.1016/j.ijfoodmicro.2015.08.016
Gil, G., del Mónaco, S., Cerrutti, P., and Galvagno, M. (2004). Selective antimicrobial activity of chitosan on beer spoilage bacteria and brewing yeasts. Biotechnol. Lett *26*, 569–574. http://dx.doi.org/10.1023/B:BILE.0000021957.37426.9b
Gouesbet, G., Jan, G., and Boyaval, P. (2001). *Lactobacillus delbrueckii* ssp. *bulgaricus* thermotolerance. Lait. *81*, 301–309. http://dx.doi.org/10.1051/lait:2001133
Gupta, M., Abu-Ghannam, N., and Gallaghar, E. (2010). Barley for brewing: Characteristic changes during malting, brewing and applications of its by-products. Comp. Rev. Food Sci. Food Safety 9, 318–328. http://dx.doi.org/10.1111/j.1541-4337.2010.00112.x
Haakensen, M.C., Butt, L., Chaban, B., Deneer, H., and Ziola, B. (2007). *horA*-specific real- time PCR for detection of beer spoilage lactic acid bacteria. J. Am. Soc. Brew. Chem. *65*, 157–165. http://dx.doi.org/10.1094/ASBCJ-2007-0611-01
Haakensen, M., Dobson, C.M., Hill, J.E., and Ziola, B. (2009a). Reclassification of *Pediococcus dextrinicus* (Coster and White 1964) back 1978 (Approved Lists 1980) as *Lactobacillus dextrinicus* comb. nov., and emended description of the genus *Lactobacillus*. Int. J. Syst. Evol. Microbiol. *59*, 615–621. http://dx.doi.org/10.1099/ijs.0.65779-0
Haakensen, M., Schubert, A., and Ziola, B. (2008). Multiplex PCR for putative *Lactobacillus* and *Pediococcus* beer spoilage genes and ability of gene presence to predict growth in beer. J. Am. Soc. Brew. Chem. *66*, 63–70. http://dx.doi.org/10.1094/ASBCJ-2008-0314-01
Haakensen, M., Schubert, A., and Ziola, B. (2009b). Broth and agar hop-gradient plates used to evaluate the beer spoilage potential of *Lactobacillus* and *Pediococcus* isolates. Int. J. Food Microbiol. *130*, 56–60. http://dx.doi.org/10.1016/j.ijfoodmicro.2009.01.001
Haakensen, M., Vickers, D.M., and Ziola, B. (2009c). Susceptibility of *Pediococcus* isolates to antimicrobial compounds in relation to hop-resistance and beer spoilage. BMC Microbiol. *9*, 190. http://dx.doi.org/10.1186/1471-2180-9-190
Hammes, W.P., and Hertel, C. (2006). The Genera *Lactobacillus* and *Carnobacterium*. In The Prokaryotes. 3rd ed., Dworkin, M., Falkow, S., Rosenberg, E., Schleifer, K-H., and Stackebandt, E., eds. (New York: Springer-Verlag), pp. 319–403. http://dx.doi.org/10.1007/0-387-30743-5
Hammes, W.P., and Vogel, R.F. (1995). The Genus *Lactobacillus*. In The Genera of Lactic Acid Bacteria,, Holzapfel, W.H.N, and Wood, B.J.B., eds. (US: Springer), pp. 19–54. http://dx.doi.org/10.1007/978-1-4615-5817-0
Hayashi, N., Ito, M., Horiike, S., and Taguchi, H. (2001). Molecular cloning of a putative divalent-cation transporter gene as a new genetic marker for the identification of *Lactobacillus brevis* strains capable of growing in beer. Appl. Microbiol. Biotechnol. *55*, 596–603. http://dx.doi.org/10.1007/s002530100600
Herkewitz, W. (May 26th, 2014). Strange Brews: the Genes of Craft Beer. New York Times, http://www.nytimes.com/2014/05/27/science/craft-beer-at-the-genetic-level.html?_r=0.
Herrero, M., Quirós, C., García, L.A., and Díaz, M. (2006). Use of flow cytometry to follow the physiological states

of microorganisms in cider fermentation processes. Appl. Environ. Microbiol. *72*, 6725–6733. http://dx.doi.org/10.1128/AEM.01183-06

Heuer, H., and Smalla, K. (2007). Horizontal gene transfer between bacteria. Environ. Biosafety. Res. *6*, 3–13. http://dx.doi.org/10.1051/ebr:2007034

Hindson, B.J., Ness, K.D., Masquelier, D.A., Belgrader, P., Heredia, N.J., Makarewicz, A.J., Bright, I.J., Lucero, M.Y., Hiddessen, A.L., Legler, T.C., *et al.* (2011). High-throughput droplet digital PCR system for absolute quantitation of DNA copy number. Anal. Chem. 83, 8604–8610. http://dx.doi.org/10.1021/ac202028g

Holzapfel, W.H. (1992). Culture media for non-sporulating gram-positive food spoilage bacteria. Int. J. Food Microbiol. *17*, 113–133.

Holzapfel, W.H., and Wood, B.J.B. (2014). Chapter 1: Introduction to LAB. In Lactic Acid Bacteria: Biodiversity and Taxonomy, Holzapfel, W.H., and Wood, B.J.B., eds. (New Jersey:Wiley-Blackwell). pp.1–12. http://dx.doi.org/10.1002/9781118655252

Holzapfel, W.H., Franz, C.M.A.P., Ludwig,W., and Dicks, L.M.T. (2009). Genus III.. *Pediococcus..* In Bergey's Manual of Systematic Bacteriology. 2nd ed., Volume 3, De-Vos, P., Garrity, G., Jones, D., Krieg, N., Ludwig, W., Rainey, F.A., Schleifer, K-H., and Whitman, W.B.,eds. (New York: Springer) pp. 513–32. http://dx.doi.org/10.1007/978-0-387-68489-5

Horvath, P., Coûté-Monvoisin, A.C., Romero, D.A., Boyaval, P., Fremaux, C., and Barrangou, R. (2009). Comparative analysis of CRISPR loci in lactic acid bacteria genomes. Int. J. Food Microbiol. *131*, 62–70. http://dx.doi.org/10.1016/j.ijfoodmicro.2008.05.030

Iijima, K., Suzuki, K., Asano, S., Kuriyama, H., and Kitagawa, Y. (2007). Isolation and identification of potential beer spoilage *Pediococcus inopinatus* and beer spoilage *Lactobacillus backi* strains carrying the *horA* and *horC* gene clusters. J. Inst. Brew. *113*, 96–101. http://dx.doi.org/10.1002/j.2050-0416.2007.tb00262.x

Iijima, K., Suzuki, K., Ozaki, K., and Yamashita, H. (2006). *horC* confers beer spoilage ability on hop-sensitive *Lactobacillus brevis* ABBC45cc. J. Appl. Microbiol. *100*, 1282–1288. http://dx.doi.org/10.1111/j.1365-2672.2006.02869.x

Ingram, L.O. (1986). Microbial tolerance to alcohols: role of the cell membrane. Trends Biotechnol. *4*, 40–44. http://dx.doi.org/http://dx.doi.org/10.1016/0167-7799(86)90152-6

Izquierdo-Pulido, M., Hernández-Jover, T., Mariné-Font, A., and Vidal-Carou, M.C. (1996). Biogenic amines in European beers. J. Food Sci. *44*, 3159–3163. http://dx.doi.org/10.1021/jf960155j

James, T.C., Gallagher, L., Titze, J., Bourke, P., Kavanagh, J., Arendt, E., and Bond, U. (2014). *In situ* production of human β-defensin-3 in lager yeasts provides bactericidal activity against beer-spoiling bacteria under fermentation conditions. J. Appl. Microbiol. 116, 368–379. http://dx.doi.org/10.1111/jam.12382

Jespersen, L., and Jakobsen, M. (1996). Specific spoilage organisms in breweries and laboratory media for their detection. Int. J. Food Microbiol. 33, 139–155. http://dx.doi.org/10.1016/0168-1605(96)01154-3

Juvonen, R., Partanen, T., and Koivula, T. (2010). Evaluation of reverse-transcription PCR detection of 16S rRNA and *tuf* mRNA for viable/dead discrimination of beer spoilage lactic acid bacteria. J. Am. Soc. Brew. Chem. *68*, 101–106. http://dx.doi.org/10.1094/ASBCJ-2010-0416-01

Kalač, P., Šavel, J., Křížek, M. Pelikánová, T., and Prokopová, M. (2002). Biogenic amine formation in bottled beer. Food Chem. *79*, 431–434. http://dx.doi.org/10.1016/S0308-8146(02)00193-0

Kelly, D., Neve, H., McAuliffe, O., Ross, R.P., Arendt, E.K., and Coffey, A. (2011). Isolation and characterization of bacteriophages that inhibit strains of *Pediococcus damnosus*, *Lactobacillus brevis*, and *Lactobacillus paraplantarum* that cause beer spoilage. J. Am. Soc. Brew. Chem. *69*, 8–12. http://dx.doi.org/10.1094/ASBCJ-2010-1119-01

Kelly, D., O'Sullivan, O., Mills, S., McAuliffe, O., Ross, R.P., Neve, H., and Coffey, A. (2012). Genome sequence of the phage clP1, which infects the beer spoilage bacterium *Pediococcus damnosus*. Gene *504*, 53–63. http://dx.doi.org/10.1016/j.gene.2012.04.085

Kern, C.C., Vogel, R.F., and Behr, J. (2014). Differentiation of *Lactobacillus brevis* strains using matrix-assisted-laser-desorption-ionization-time-of-flight mass spectrometry with respect to their beer spoilage potential. Food Microbiol. *40*, 18–24. http://dx.doi.org/10.1016/j.fm.2013.11.015

Kheadr, E.E., Vachon, J.F., Paquin, P., and Fliss, I. (2002). Effect of dynamic high pressure on microbiological, rheological and microstructural quality of Cheddar cheese. Int. Dairy J. *12*, 435–446. http://dx.doi.org/10.1016/S0958-6946(01)00104-2

Klaenhammer, T.R. (1993). Genetics of bacteriocins produced by lactic acid bacteria. FEMS Microbiol. Rev. *12*, 39–85. http://dx.doi.org/10.1016/0168-6445(93)90057-G

Klaenhammer, T.R., Barrangou, R., Buck, B.L., Azcarate-Peril, M.A., and Altermann, E. (2005). Genomic features of lactic acid bacteria effecting bioprocessing and health. FEMS Microbiol. Rev. *29*, 393–409. http://dx.doi.org/10.1016/j.femsre.2005.04.007

Konings, W.N., Lolkema, J.S., Bolhuis, H., van Veen, H.W., Poolman, B., and Driessen, A..J. (1997). The role of transport processes in survival of lactic acid bacteria. Energy transduction and multidrug resistance. Antonie Van Leeuwenhoek. *71*, 117–128. http://dx.doi.org/10.1023/A:1000143525601

Kubota, H., Senda, S., Nomura, N., Tokuda, H., and Uchiyama, H. (2008). Biofilm formation by lactic acid bacteria and resistance to environmental stress. J. Biosci. Bioeng. *106*, 381–386. http://dx.doi.org/10.1263/jbb.106.381

Lewis, D. (1998). Biological mash and wort acidification. The New Brewer *15*, 36–45.

Liu, W.T., Marsh, T.L., Cheng, H., and Forney, L.J. (1997). Characterization of microbial diversity by determining terminal restriction fragment length polymorphisms of genes encoding 16S rRNA. Appl. Environ. Microbiol. 63, 4516–4522.

Lodolo, E.J., Kock, J.L., Axcell, B.C., and Brooks, M. (2008). The yeast *Saccharomyces cerevisiae* – the main character in beer brewing. FEMS Yeast Res. *8*, 1018–1036. http://dx.doi.org/10.1111/j.1567-1364.2008.00433.x

Mahony, J., Ainsworth, S., Stockdale, S., and van Sinderen, D. (2012). Phages of lactic acid bacteria: the role of genetics in understanding phage-host interactions and their co-evolutionary processes. Virology *434*, 143–150. http://dx.doi.org/10.1016/j.virol.2012.10.008

Mahony, J., McAuliffe, O., Ross, R.P., and van Sinderen, D. (2011). Bacteriophages as biocontrol agents of food pathogens. Curr. Opin. Biotechnol. *22*, 157–163. http://dx.doi.org/10.1016/j.copbio.2010.10.008

Maiden, M.C. (2006). Multilocus sequence typing of bacteria. Annu. Rev. Microbiol. *60*, 561–588. http://dx.doi.org/10.1146/annurev.micro.59.030804.121325

Maiden, M.C., Bygraves, J.A., Feil, E., Morelli, G., Russell, J.E., Urwin, R., Zhang, Q., Zhou, J., Zurth, K., Caugant, D.A., *et al.* (1998). Multilocus sequence typing: A portable approach to the identification of clones within populations of pathogenic microorganisms. Proc. Natl. Acad. Sci. U.S.A. *95*, 3140–3145. http://dx.doi.org/10.1073/pnas.95.6.3140

Makarova, K., Slesarev, A., Wolf, Y., Sorokin, A., Mirkin, B., Koonin, E., Pavlov, A., Pavlova, N., Karamychev, V., Polouchine, N., *et al.* (2006). Comparative genomics of the lactic acid bacteria. Proc. Natl. Acad. Sci. U.S.A. *103*, 15611–15616. http://dx.doi.org/10.1073/pnas.0607117103

Manzano, M., Giusto, C., Bartolomeoli, I., Buiatti, S., and Comi, G. (2005). Microbiological analyses of dry and slurry yeasts for brewing. J. Inst. Brew. *111*, 203–208. http://dx.doi.org/10.1002/j.2050-0416.2005.tb00667.x

Matoulková, D., Kosař, K., and Slabý, M. (2012). Occurrence and species distribution of strictly anaerobic bacterium *Pectinatus* in brewery bottling halls. J. Am. Soc. Brew. Chem. 70, 262–267. http://dx.doi.org/10.1094/ASBCJ-2012-0910-01

Mardis, E.R. (2011). A decade's perspective on DNA sequencing technology. Nature *470*, 198–203. http://dx.doi.org/10.1038/nature09796

Martens, H., Iserentant, D., and Verachtert, H. (1997). Microbiological aspects of a mixed yeast-bacterial fermentation in the production of a special Belgian acidic ale. J. Inst. Brew. *103*, 85–91. http://dx.doi.org/10.1002/j.2050-0416.1997.tb00939.x

Meng, S., Wang, D., Liu, X., Zhang, Y., Guo, L., Lin, Z., Jia, F., and Pavlovic, M. (2012). Rapid detection of *Lactobacillus brevis* in beer production by the combination of microcolony and Fluorescence In Situ Hybridization (FISH). J. Am. Soc. Brew. Chem. 70, 91–94. http://dx.doi.org/10.1094/ASBCJ-2012-0410-01

Menz, G., Andrighetto, C., Lombardi, A., Corich, V., Aldred, P., and Vriesekoop, F. (2010). Isolation, identification, and characterisation of beer spoilage lactic acid bacteria from microbrewed beer from Victoria, Australia. J. Inst. Brew. 116, 14–22. http://dx.doi.org/10.1002/j.2050-0416.2010.tb00393.x

Moneke, A.N., Okolo, B.N., Odo, G.C., and Ire, F.S. (2009). Screening of malting sorghum samples for lactic acid bacteria with potentials for antimicrobial activity. J. Biotechnol. *8*, 2821–2829. http://dx.doi.org/10.5897/AJB08.1158

Mozzi, F., Ortiz, M.E., Bleckwedel, J., De Vuyst, L., and Pescuma, M. (2013). Metabolomics as a tool for the comprehensive understanding of fermented and functional foods with lactic acid bacteria. Food Res. Int. 54, 1152–1161. http://dx.doi.org/10.1016/j.foodres.2012.11.010

Muyzer, G., de Waal, E.C., and Uitterlinden, A.G. (1993). Profiling of complex microbial populations by denaturing gradient gel electrophoresis analysis of polymerase chain reaction-amplified genes coding for 16S rRNA. Appl. Environ. Microbiol. 59, 695–700.

Nishikawa, N., and Kohgo, M. (1985). Microbial control in the brewery. MBAA TQ. *22*, 61–66.

Ogden, K., Waites, M.J., and Hammond, J.R.M. (1988). Nisin and brewing. J. Inst. Brew. *94*, 233–238. http://dx.doi.org/10.1002/j.2050-0416.1988.tb04578.x

Oliver, J.D. (2005). The viable but nonculturable state in bacteria. J. Microbiol. *43*, 93–100.

Papadimitriou, K., Anastasiou, R., Maistrou, E., Plakas, T., Papandreou, N.C., Hamodrakas, S.J., Ferreira, S., Supply, P., Renault, P., Pot, B., *et al.* (2015). Acquisition through horizontal gene transfer of plasmid pSMA198 by *Streptococcus macedonicus* ACA-DC 198 points towards the dairy origin of the species. PLoS ONE 10, e0116337. http://dx.doi.org/10.1371/journal.pone.0116337

Parente, E., Ciocia, F., Ricciardi, A., Zotta, T., Felis, G.E., and Torriani, S. (2010). Diversity of stress tolerance in *Lactobacillus plantarum, Lactobacillus pentosus* and *Lactobacillus paraplantarum*: A multivariate screening study. Int. J. Food Microbiol. 144, 270–279. http://dx.doi.org/10.1016/j.ijfoodmicro.2010.10.005

Park, E.J., Kim, K.H., Abell, G.C., Kim, M.S., Roh, S.W., and Bae, J.W. (2011). Metagenomic analysis of the viral communities in fermented foods. Appl. Environ. Microbiol. *77*, 1284–1291. http://dx.doi.org/10.1128/AEM.01859-10

Perpète, P., Santos, G., Bodart, E., and Collin, S. (2005). Uptake of amino acids during beer production: the concept of a critical time value. J. Am. Soc. Brew. Chem. 63, 23–27. http://dx.doi.org/10.1094/ASBCJ-63-0023

Petri, A., Pfannebecker, J., Fröhlich, J., and König, H. (2013). Fast identification of wine related lactic acid bacteria by multiplex PCR. Food Microbiol. 33, 48–54. http://dx.doi.org/10.1016/j.fm.2012.08.011

Pfannebecker, J., and Fröhlich, J. (2008). Use of a species-specific multiplex PCR for the identification of pediococci. Int. J. Food Microbiol. *128*, 288–296. http://dx.doi.org/10.1016/j.ijfoodmicro.2008.08.019

Pfeiler, E.A., and Klaenhammer, T.R. (2007). The genomics of lactic acid bacteria. Trends Microbiol. *15*, 546–553. http://dx.doi.org/10.1016/j.tim.2007.09.010

Picariello, G., Mamone, G., Addeo, F., and Ferranti, P. (2012). Novel mass spectrometry-based applications of the 'omic' sciences in food technology and biotechnology. Food Technol. Biotech. 50, 286–305.

Pinheiro, L.B., Coleman, V.A., Hindson, C.M., Herrmann, J., Hindson, B.J., Bhat, S., and Emslie, K.R. (2012). Evaluation of a droplet digital polymerase chain reaction format for DNA copy number quantification. Anal. Chem. *84*, 1003–1011. http://dx.doi.org/10.1021/ac202578x

Pittet, V., Abegunde, T., Marfleet, T., Haakensen, M., Morrow, K., Jayaprakash, T., Schroeder, K., Trost, B., Byrns, S., Bergsveinson, J., *et al.* (2012a). Complete genome sequence of the beer spoilage organism

Pediococcus claussenii ATCC BAA-344T. J. Bacteriol. *194*, 1271–1272. http://dx.doi.org/10.1128/JB.06759-11

Pittet, V., Ewen, E., Bushell, B.R., and Ziola, B. (2012b). Genome sequence of *Lactobacillus rhamnosus* ATCC 8530. J. Bacteriol. *194*, 726. http://dx.doi.org/10.1128/JB.06430-11

Pittet, V., Phister, T.G., and Ziola, B. (2013). Transcriptome sequence and plasmid copy number analysis of the brewery isolate *Pediococcus claussenii* ATCC BAA-344^T during growth in beer. PLoS ONE 8, e73627. http://dx.doi.org/10.1371/journal.pone.0073627

Pittet, V., Morrow, K., and Ziola, B. (2011). Ethanol tolerance of lactic acid bacteria, including relevance of the exopolysaccharide gene *gtf*. J. Am. Soc. Brew. Chem. 69, 57–61. http://dx.doi.org/10.1094/ASBCJ-2011-0124-01

Preissler, P., Behr, J., and Vogel, R.F. (2010). Detection of beer spoilage *Lactobacillus brevis* strains by reduction of resazurin. J. Inst. Brew. 116, 399–404. http://dx.doi.org/10.1002/j.2050-0416.2010.tb00790.x

Priest, F.G. (2003). Gram-positive Brewery Bacteria. In Brewing Microbiology 3rd edn, Priest, F.G., and Campbell, I., eds. (USA: Springer), pp. 181–217. http://dx.doi.org/10.1007/978-1-4419-9250-5

Rouse, S., and van Sinderen, D. (2008). Bioprotective potential of lactic acid bacteria in malting and brewing. J. Food Prot. *71*, 1724–1733.

Russell, C., and Walker, T.K. (1953). *Lactobacillus malefermentans* n.sp., isolated from beer. J. Gen. Microbiol. *8*, 160–162. http://dx.doi.org/10.1099/00221287-8-1-160

Sakamoto, K., and Konings, W.N. (2003). Beer spoilage bacteria and hop resistance. Int. J. Food Microbiol. *89*, 105–124.

Sakamoto, K., Margolles, A., van Veen, H.W., and Konings, W.N. (2001). Hop resistance in the beer spoilage bacterium *Lactobacillus brevis* is mediated by the ATP-binding cassette multidrug transporter HorA. J. Bacteriol. *183*, 5371–5375. http://dx.doi.org/10.1128/JB.183.18.5371-5375.2001

Salvetti, E., Torriani, S., and Felis, G.E. (2012). The genus *Lactobacillus*: A taxonomic update. Probiotics. Antimicrob. Proteins. *4*, 217–226. http://dx.doi.org/10.1007/s12602-012-9117-8

Sami, M., Yamashita, H., Hirono, T., Kadokura, H., Kitamoto, K., Yoda, K., and Yamasaki, M. (1997a). Hop-resistant *Lactobacillus brevis* contains a novel plasmid harboring a multidrug resistance-like gene. J. Ferment. Bioeng. 84, 1–6. http://dx.doi.org/10.1016/S0922-338X(97)82778-X

Sami, M., Yamashita, H., Kadokura, H., Kitamoto, K., Yoda, K., and Yamasaki, M. (1997b). A new and rapid method for determination of beer spoilage ability of lactobacilli. J. Am. Soc. Brew. Chem. 55, 137–140. http://dx.doi.org/10.1094/ASBCJ-55-0137

Sanders, J.W., Oomes, S.J., Membré, J.M., Wegkamp, A., and Wels, M. (2015). Biodiversity of spoilage lactobacilli: phenotypic characterisation. Food Microbiol. *45*, 34–44. http://dx.doi.org/10.1016/j.fm.2014.03.013

Schleifer, K.H., and Ludwig, W. (1995). Phylogenetic Relationships of Lactic Acid Bacteria. In The Genera of Lactic Acid Bacteria, Holzapfel, W.H.N., and Wood, B.J.B., eds. (US: Springer), pp. 7–18.

Schroeter, J., and Klaenhammer, T. (2009). Genomics of lactic acid bacteria. FEMS Microbiol. Lett. *292*, 1–6. http://dx.doi.org/10.1111/j.1574-6968.2008.01442.x

Schwarz, S., Shen, J., Wendlandt, S., Fessler, A.T., Wang, Y., Kadlec, K., and Wu, C.M. (2014). Plasmid-mediated antimicrobial resistance in *Staphylococci* and other *Firmicutes*. Microbiol. Spectr. 2, http://dx.doi.org/10.1128/microbiolspec.PLAS-0020-2014

Silveira, M.G., Baumgärtner, M., Rombouts, F.M., and Abee, T. (2004). Effect of adaptation to ethanol on cytoplasmic and membrane protein profiles of *Oenococcus oeni*. Appl. Environ. Microbiol. *70*, 2748–2755. http://dx.doi.org/10.1128/AEM.70.5.2748-2755.2004

Simpson, W.J. (1993a). Ionophoric action of trans-isohumulone on *Lactobacillus brevis*. J. Gen. Microbiol. *139*, 1041–1045.

Simpson, W.J. (1993b). Studies on the sensitivity of lactic acid bacteria to hop bitter acids. J. Inst. Brew. *99*, 405–411.

Simon, C., and Daniel, R. (2011). Metagenomic analyses: past and future trends. Appl. Environ. Microbiol. *77*, 1153–1161. http://dx.doi.org/10.1128/AEM.02345-10

Spitaels, F., Wieme, A.D., Janssens, M., Aerts, M., Daniel, H.M., Van Landschoot, A., De Vuyst, L., and Vandamme, P. (2014). The microbial diversity of traditional spontaneously fermented lambic beer. PLoS ONE 9, e95384. http://dx.doi.org/10.1371/journal.pone.0095384

Steenackers, B., De Cooman, L., and De Vos, D. (2015). Chemical transformations of characteristic hop secondary metabolites in relation to beer properties and the brewing process: a review. Food Chem. *172*, 742–756. http://dx.doi.org/10.1016/j.foodchem.2014.09.139

Storgårds, E., Tapani, K., Hartwall, P., Saleva, R., and Suihko, M. (2006). Microbial attachment and biofilm formation in brewery bottling plants. J. Am. Soc. Brew. Chem. 64, 8–15. http://dx.doi.org/10.1094/ASBCJ-64-0008

Sun, Z., Yu, J., Dan, T., Zhang, W., and Zhang, H. (2014). Chapter 1: Phylogenetics and Evolution of Lactic Acid Bacteria In Lactic Acid Bacteria: Fundamentals and Practice, Zhang, H., and Cai, Y. eds. (Netherlands: Springer), pp.1–101. http://dx.doi.org/10.1007/978-94-017-8841-0

Suzuki, K., Funahashi, W., Koyanagi, M., and Yamashita, H. (2004). *Lactobacillus paracollinoides* sp. nov., isolated from brewery environments. Int. J. Syst. Evol. Microbiol. *54*, 115–117. http://dx.doi.org/10.1099/ijs.0.02722-0

Suzuki, K. (2008). Beer spoilage lactic acid bacteria (supplementary chapter). In Beer in Health and Disease Prevention, Preedy, V.R., and Watson, R.R., eds. (San Diego: Elsevier Science) pp. 150–164. http://dx.doi.org/10.1016/B978-0-12-373891-2.50001-7

Suzuki, K. (2011). 125th Anniversary review: Microbiological instability of beer caused by spoilage bacteria. J. Inst. Brew. 117, 131–155. http://dx.doi.org/10.1002/j.2050-0416.2011.tb00454.x

Suzuki, K., Asano, S., Iijima, K., and Kitamoto, K. (2008a). Sake and beer spoilage lactic acid bacteria – A review. J. Inst. Brew. *114*, 209–223. http://dx.doi.org/10.1002/j.2050-0416.2008.tb00331.x

Suzuki, K., Asano, S., Iijima, K., Kuriyama, H., and Kitagawa, Y. (2008b). Development of detection medium for hard-to-culture beer spoilage lactic acid bacteria. J. Appl. Microbiol. *104*, 1458–1470. http://dx.doi.org/10.1111/j.1365-2672.2007.03669.x

Suzuki, K., Iijima, K., Asano, S., Kuriyama, H., and Kitagawa, Y. (2006). Induction of viable but nonculturable state in beer spoilage lactic acid bacteria. J. Inst. Brew. *112*, 295–301. http://dx.doi.org/10.1002/j.2050-0416.2006.tb00734.x

Suzuki, K., Iijima, K., Ozaki, K., and Yamashita, H. (2005). Isolation of a hop-sensitive variant of *Lactobacillus lindneri* and identification of genetic markers for beer spoilage ability of lactic acid bacteria. Appl. Environ. Microbiol. *71*, 5089–5097. http://dx.doi.org/10.1128/AEM.71.9.5089-5097.2005

Suzuki, K., Ozaki, K., and Yamashita, H. (2004a). Comparative analysis of conserved genetic markers and adjacent DNA regions identified in beer spoilage lactic acid bacteria. Lett. Appl. Microbiol. *39*, 240–245. http://dx.doi.org/10.1111/j.1472-765X.2004.01572.x

Suzuki, K., Ozaki, K., and Yamashita, H. (2004b). Genetic marker for differentiating beer spoilage ability of *Lactobacillus paracollinoides* strains. J. Appl. Microbiol. *97*, 712–718. http://dx.doi.org/10.1111/j.1365-2672.2004.02350.x

Taskila, S., Kronlöf, J., and Ojamo, H. (2011). Enrichment cultivation of beer-spoiling lactic acid bacteria. J. Inst. Brew. 117, 285–294. http://dx.doi.org/10.1002/j.2050-0416.2011.tb00473.x

Taskila, S., Tuomola, M., Kronlöf, J., and Neubauer, P. (2010). Comparison of enrichment media for routine detection of beer spoiling lactic acid bacteria and development of trouble-shooting medium for *Lactobacillus backii*. J. Inst. Brew. 116, 151–156. http://dx.doi.org/10.1002/j.2050-0416.2010.tb00411.x

Thayer, A.M. (August 18th, 2014). Next-gen sequencing is a numbers game. Chem. Eng. News. 92, 11–15.

Thelen, K., Beimfohr, C., and Snaidr, J. (2004). VIT-Bier: The rapid and easy detection method for beer-spoiling bacteria. MBAA TQ. *41*, 115–119.

Thelen, K., Beimfohr, C., and Snaidr, J. (2006). Evaluation study of the frequency of different beer-spoiling bacteria using the VIT analysis. MBAA TQ. 43, 31–35.

Thelen, K., Beimfohr, C., Munich, I., Bohak, I., and Back, W. (2002). Specific rapid detection method for beer spoilage bacteria using fluorescence-marked gene probes. Brauwelt Int. 3, 155–159.

Thomas, K., and Whitham, H. (1997). Improvement in beer line technology. European Brewery Convention Monograph 25, Symposium Draught beer, Packaging and Dispensing, Edinburgh September 1996, Nurnberg, Verlag Hans Carl, 124–137.

Timke, M., Wang-Lieu, N.Q., Altendorf, K., and Lipski, A. (2005). Community structure and diversity of biofilms from a beer bottling plant as revealed using 16S rRNA gene clone libraries. Appl. Environ. Microbiol. *71*, 6446–6452. http://dx.doi.org/10.1128/AEM.71.10.6446-6452.2005

Timke, M., Wang-Lieu, N.Q., Altendorf, K., and Lipski, A. (2008). Identity, beer spoiling and biofilm forming potential of yeasts from beer bottling plant associated biofilms. Antonie Van Leeuwenhoek *93*, 151–161. http://dx.doi.org/10.1007/s10482-007-9189-8

Tonsmeire, M. (2014). American Sour Beers: Innovative Techniques for Mixed Fermentations. (Boulder, CO: Brewers Publication).

Tribst, A.A.L., Augusto, P.E.D., and Cristianini, M. (2013). Multi-pass high-pressure homogenization of commercial enzymes: Effect on the activities of glucose oxidase, neutral protease and amyloglucosidase at different temperatures. Innov. Food Sci. Emerg. Techn. *18*, 83–88. http://dx.doi.org/10.1016/j.ifset.2013.01.002

Tribst, A.A.L., Franchi, M.A., and Cristianini, M. (2008). Ultra-high pressure homogenization treatment combined with lysozyme for controlling *Lactobacillus brevis* contamination in model system. Innov. Food Sci. Emerg. Technol. 9, 265–271. http://dx.doi.org/10.1016/j.ifset.2007.07.012

Tsuchiya, Y., Kano, T., and Koshino, S. (1994). Identification of lactic acid bacteria using temperature gradient gel electrophoresis for DNA fragments amplified by polymerase chain reaction. J. Am. Soc. Brew. Chem. 52, 95. http://dx.doi.org/10.1094/ASBCJ-52-0095

Uchida, K. (1974). Lipids of alcoholophilic lactobacilli. II. Occurrence of polar lipids with unusually long acyl chains in *Lactobacillus heterohiochii*. BBA. *369*, 146–155. http://dx.doi.org/10.1016/0005-2760(74)90248-3.

Ulmer, H.M., Herberhold, H., Fahsel, S., Gänzle, M.G., Winter, R., and Vogel, R.F. (2002). Effects of pressure-induced membrane phase transitions on inactivation of HorA, an ATP-dependent multidrug resistance transporter, in *Lactobacillus plantarum*. Appl. Environ. Microbiol. 68, 1088–1095. http://dx.doi.org/10.1128/AEM.68.3.1088-1095.2002

Van Reenen, C.A., Chikindas, M.L., Van Zyl, W.H., and Dicks, L.M.T. (2003). Characterization and heterologous expression of a class IIa bacteriocin, plantaricin 423 from *Lactobacillus plantarum* 423, in *Saccharomyces cerevisiae*. Int. J. Food Microbiol. *81*, 29–40. http://dx.doi.org/10.1016/S0168-1605(02)00164-2

Vaughan, A., O'Sullivan, T., and van Sinderen, D. (2005). Enhancing the microbiological stability of malt and beer—a review. J. Inst. Brew. *111*, 355–371. http://dx.doi.org/10.1002/j.2050-0416.2005.tb00221.x

Vaughan, A., Rouse, S., and van Sinderen, D. (2004). Investigating the antimicrobial efficacy of a lactococcal bacteriocin for the development of microbiologically stable beer. J. Inst. Brew. *110*, 181–188. http://dx.doi.org/10.1002/j.2050-0416.2004.tb00200.x

Vogel, R.F., Preissler, P., and Behr, J. (2010). Towards an understanding of hop tolerance in beer spoiling *Lactobacillus brevis*. Brew. Sci. *63*, 23–30.

Walling, E., Gindreau, E., and Lonvaud-Funel, A. (2005). A putative glucan synthase gene *dps* detected in exopolysaccharide-producing *Pediococcus damnosus* and *Oenococcus oeni* strains isolated from wine and cider. Int. J. Food Microbiol. 98, 53–62. http://dx.doi.org/10.1016/j.ijfoodmicro.2004.05.016

Warnecke, F., and Hess, M. (2009). A perspective: metatranscriptomics as a tool for the discovery of novel biocatalysts. J. Biotechnol. *142*, 91–95. http://dx.doi.org/10.1016/j.jbiotec.2009.03.022

Wiedenbeck, J., and Cohan, F.M. (2011). Origins of bacterial diversity through horizontal genetic transfer and adaptation to new ecological niches. FEMS Microbiol.

Rev. *35*, 957–976. http://dx.doi.org/10.1111/j.1574-6976.2011.00292.x

Wieme, A.D., Spitaels, F., Aerts, M., De Bruyne, K., Van Landschoot, A., and Vandamme, P. (2014). Identification of beer spoilage bacteria using matrix-assisted laser desorption/ionization time-of-flight mass spectrometry. Int. J. Food Microbiol. 185, 41–50. http://dx.doi.org/10.1016/j.ijfoodmicro.2014.05.003

White, D., Drummond, J., and Fuqua, C. (2012a). Chapter 5: Electron Transport. In The Physiology and Biochemistry of Prokaryotes, 4th Ed. White, D., Drummond, J., and Fuqua, C., eds. (New York: Oxford University Press). pp.146–174.

White, D., Drummond, J., and Fuqua, C. (2012b). Chapter 17: Solute Transport. In The Physiology and Biochemistry of Prokaryotes, 4th Ed. White, D., Drummond, J., and Fuqua, C., eds. (New York,: Oxford University Press). pp. 432–451.

Wuytack, E.Y., Diels, A.M.J., and Michiels, C.W. (2002). Bacterial inactivation by high-pressure homogenisation and high hydrostatic pressure. Int. J. Food Microbiol. *77*, 205–212. http://dx.doi.org/10.1016/S0168-1605(02)00054-5

Yasui, T., and Yoda, K. (1997). Purification and partial characterization of an antigen specific to *Lactobacillus brevis* strains with beer spoilage activity. FEMS Microbiol. Lett. *151*, 169–176. http://dx.doi.org/10.1111/j.1574-6968.1997.tb12566.x

Zhang, Z.G., Ye, Z.Q., Yu, L., and Shi, P. (2011). Phylogenomic reconstruction of lactic acid bacteria: an update. BMC Evol. Biol. *11*, 1. http://dx.doi.org/10.1186/1471-2148-11-1

Zheng, J., Ruan, L., Sun, M., and Gänzle, M. (2015). A genomic view of Lactobacilli and Pediococci demonstrates that phylogeny matches ecology and physiology. Appl. Environ. Microbiol. *81*, 7233–7243. http://dx.doi.org/10.1128/AEM.02116-15

Zottola, E.A., and Sasahara, K.C. (1994). Microbial biofilms in the food processing industry – should they be a concern? Int. J. Food Microbiol. *23*, 125–148. http://dx.doi.org/10.1016/0168-1605(94)90047-7

Brewery- and Beer Spoilage-related Gram-negative Bacteria: The Unpleasant, the Malodorous and the Outright Fetid

10

Barry Ziola* and Jordyn Bergsveinson

Department of Pathology and Laboratory Medicine, College of Medicine, University of Saskatchewan, Royal University Hospital, Saskatoon, SK, Canada.

*Correspondence: b.ziola@usask.ca

https://doi.org/10.21775/9781910190616.10

Abstract

As indicated by the chapter title, growth of selected Gram-negative bacteria in beer results in an unpalatable product, leading to consumer complaints, and loss of brewer and brand loyalty. Aerobic Gram-negative bacteria historically have been a major problem for brewers, but with improved packaging methods resulting in reduced oxygen levels in beer, these bacteria now are mostly found in improperly attended draft beer systems. Concurrently, reduced oxygen in packaged product resulted in emergence of the anaerobic Gram-negative spoilage bacteria, particularly those within the genera *Megasphaera* and *Pectinatus*. Little is known about the genetics of these anaerobic bacteria, given that minimal omics-based research has been done on them. This chapter presents the historical aspects of brewing-related Gram-negative bacteria, where the anaerobic brewing-related bacteria likely originate from and where they are found within breweries, and the evolution of molecular-based approaches for the rapid detection and identification of these bacteria for brewery quality control. Finally, application of metagenomics, genomics, and transcriptomics for improved understanding of brewing-related Gram-negative bacteria is discussed from the perspectives of bacterial persistence within breweries as well as growth in and spoilage of beer.

Introduction

Bacteria inevitably find their way into the brewing environment, whether via the raw materials used in making beer, environmental sources (e.g. air or water), or the individuals working in a brewery or where draft beer is offered. Regardless of its source and mode of introduction, once a bacterium is established within a brewery or draft beer distribution site, eradication is difficult, resulting in a high probability that the organism will periodically emerge and cause beer spoilage, e.g. as off-flavours, aromas, or turbidity. Although the Gram-negative bacteria (GNB) causing problems for brewers and beer comprise a number of bacterial genera and species, only two physiologically functional groups are involved. The first is comprised of bacteria growing under aerobic, or at least microaerophilic conditions, and the second consists of bacteria that are essentially strict anaerobes.

Selected aerobic GNB bacteria can be found in the brewery at all stages of beer production, except in packaged product where current-day packaging methods involve the removal of oxygen and creation of a restricted headspace, effectively resulting in an anaerobic environment (Jandreau and Hahn, 1978; Kuchel *et al.*, 2006). These bacteria can also be found growing at post-packaging points of opened kegs or draft beer dispensing lines if appropriate procedures to restrict oxygen entry and maintain hygiene are not observed. Aerobic GNB, by definition, are not a problem for packaged beer in which oxygen has been virtually eliminated. The situation for the anaerobic brewery-related GNB is the exact opposite, with these bacteria causing spoilage of oxygen-depleted packaged beer, but generally being unable to grow well at other points in the beer production process due to the presence of variable levels of oxygen. Although anaerobic GNB are more restricted in their brewery growth niche, brewers fear them much more than their aerobic counter-parts, as growth of anaerobic GNB is accompanied by stronger markers of beer spoilage: turbidity and strongly offensive off-flavours and aromas. Experiencing even one episode of such spoilage will emphatically affect a consumer's beer and brewery loyalty.

Aerobic GNB

Enterobacteria

Historically, numerous aerobic GNB have been found to spoil beer through production of off-flavours and turbidity. Included here is the coliform group of bacteria from the family *Enterobacteriaceae*, with beer spoilage isolates occasionally identified as *Citrobacter freundii*, *Enterobacter* spp. (e.g. *E. aerogenes*, *E. agglomerans*, *E. cloacae*), *Escherichia coli*, *Hafnia alvei*, *Hafnia protea* (formerly *Obesumbacterium proteus*), *Klebsiella* spp. (e.g. *K. aerogenes*, *K. pneumonia*) or *Rahnella aquatalis* (earlier often confused with *E. agglomerans*) (Greipsson and Priest, 1983; Hamze *et al.*, 1991; Martens *et al.*, 1991; Priest *et al.*, 1973, 1974; Van Vuuren *et al.*, 1978). These bacteria are considered to be wort spoilers (Priest *et al.*, 1974) and in standard beer production, growth of these aerobic bacteria in nutrient-rich wort at pH 5 to 6 can cause off-flavours through production of 2,3-butanediol, acetate, formate, and dimethyl sulfide, along with low levels of fusel alcohols (Priest and Hough, 1974; Priest *et al.*, 1974; Van Vuuren *et al.*, 1980). The result is sweet fruity or vegetable-like (celery or cooked cabbage) flavours. Ironically, many of the bacterial species viewed as wort-spoilers are obligatory for the initial open air-seeded fermentation process leading to production of traditional Lambic beer (Bokulich *et al.*, 2012; Martens *et al.*, 1991; Spitaels *et al.*, 2014; Van Oevelen *et al.*, 1977). In contrast, for current industrially produced lambic beer, the early *Enterbacteriaceae* phase is absent, with acetic acid bacteria instead playing a larger role (Spitaels *et al.*, 2015).

Most of the enterobacteria that have been described as wort-spoilers do not survive in the later stages of fermentation, as they are susceptible to the increased ethanol, and lower pH and oxygen levels. *E. agglomerans*, *E. cloacae*, *H. alvei*, *H. protea*, and *R. aquatalis* can better tolerate a higher ethanol, and lower pH and oxygen environment (Jespersen and Jakobsen, 1996), and thus are more problematic for brewers than other enterobacteria. This is primarily due to the possibility of these bacteria being carried into subsequent fermentations when yeast is pitched. Improved brewery hygiene, careful attention to sterile handling of wort through proper cleaning (decontamination) of transfer lines and tanks, and reduced yeast pitch rate (i.e. more frequent preparation of new fermentation yeast inoculant) have contributed to greatly reduced brewery problems with enterobacteria.

Acetic acid bacteria

Although aerobic (or microaerophilic) acetic acid bacteria are ubiquitous, it is specifically at the dispensing point for draught beer that the acetic acid bacteria *Acetobacter aceti*, *Acetobacter pasteurianus*, and *Gluconobacter oxydans* most often pose a problem (Bamforth, 2005). If air enters a keg or if beer stands on tap too long, sufficient oxygen may be present for growth of acetic acid bacteria. These bacteria produce acetic acid from the ethanol in beer and also can produce a ropy slime or surface pellicle. Isolates from the genus *Acetobacter* can further metabolize the acetic acid to CO_2 and water, while *Gluconobacter* isolates cannot. Because of their oxygen growth requirement, acetic acid bacteria, much like enterobacteria, have been eliminated almost entirely as a problem in modern breweries. Nonetheless, the recent description of a new acetic acid bacterium *Gluconobacter cerevisiae* isolated in a brewery environment indicates the necessity for continued vigilance by brewers for this group of GNB (Spitaels *et al.*, 2013).

Zymomonas mobilis

The remaining beer spoilage GNB that can grow in the presence of oxygen is *Z. mobilis*. In fact, this organism can grow aerobically or anaerobically, and actually shows increased ethanol tolerance if growth is occurring without oxygen present (Moreau *et al.*, 1997). *Z. mobilis* grows optimally at 25–30°C and is rarely found in larger breweries due to the 8–12°C fermentation temperatures used. Additionally, the organism has stringent carbohydrate requirements (sucrose, glucose, or fructose) from which it makes quantitative amounts of ethanol and CO_2. The main beer spoilage situation for *Z. mobilis* is cask-conditioned, primed beer with added sugar for achieving carbonation. Spoilage is caused by turbidity and the synthesis of variable levels of acetaldehyde and hydrogen sulfide, which together give a rotten apple or fruity smell, and an ester or sulfur flavour (Dadds *et al.*, 1971; Richards and Corbey, 1974). Interestingly, *Z. mobilis* is used in making alcoholic beverages in tropical locations (e.g. African palm wine and Mexican pulque), and is being actively investigated for potential usage in fuel alcohol production (Kosaric *et al.*, 1982; Rogers *et al.*, 2007).

Anaerobic GNB

The notion attributed to the ancient Greek philosopher Aristotle that *Nature abhors a vacuum* is exemplified by the emergence in the 1970s of beer spoilage by a group of related anaerobic bacteria. As brewers reduced oxygen levels in packaged beer to increase shelf life (Kuchel *et al.*, 2006), two things happened concurrently. First, as oxygen levels in beer went down, the role played by aerobic GNB decreased to the point that spoilage occurrences by these organisms became a rarity. At the same time, the increasingly anaerobic environment in beer allowed for certain oxygen-intolerant GNB to emerge and dominate as beer spoilers. Given the current importance of beer spoilage by anaerobic relative to aerobic GNB, the remainder of this chapter will focus on the anaerobic group of bacteria.

At present, brewers are concerned with possible spoilage of packaged beer by nine species of anaerobic bacteria in four genera, including *Megasphaera* (*cerevisiae*, *paucivorans*, and *sueciensis*), *Pectinatus* (*cerevisiiphilus*, *frisingensis*, and *haikarae*), *Propionispira* (genus formerly named *Zymophilus*; *paucivorans* and *raffinosivorans*), and *Selenomonas lacticifix* (Engelmann and Weiss, 1985; Juvonen and Suihko, 2006; Lee *et al.*, 1978, 1980; Schleifer *et al.*, 1990; Ueki *et al.*, 2014). The *Megasphaera* and *Pectinatus* spp. are reasonably well studied, while little research has been done on the *Propionispira* spp. and *Selenomonas lacticifix*.

Megasphaera and *Pectinatus*

The first of the anaerobic GNB to be associated with beer spoilage was *P. cerevisiiphilus* (Lee *et al.*, 1978, 1980). Subsequently, *P. frisingensis* and *P. haikarae* were also defined as beer spoilage organisms (Juvonen and Suihko, 2006; Schleifer *et al.*, 1990). A similar situation occurred for the three *Megasphaera* spp. now known to spoil packaged beer, with *M. cerevisiae* being described initially (Engelmann and Weiss, 1985), and *M. paucivorans* and *M. sueciensis* defined later as beer spoilage species (Juvonen and Suihko, 2006). Retrospectively, all six species appear to have emerged as anaerobic beer spoilers in the 1970s, with isolates early on often misidentified as being the initially described species in each genus (Haikara *et al.*, 1981; Weiss *et al.*, 1979).

When found to cause beer spoilage, *Megasphaera* and *Pectinatus* are generally isolated from non-pasteurized, packaged beer (Back, 1994) having lower alcohol levels and slightly higher pH. *Pectinatus* spp. appear to tolerate beer with ethanol up to 4.4% (w/v), while *Megasphaera* spp. are less ethanol tolerant, thought to grow only in beer with ethanol levels of 3.5% (w/v) or below (Haikara and Helander, 2006). Recently, however, an isolate of *M. cerevisiae* able to tolerate 5% (w/v) ethanol has been described (Bergsveinson *et al.*, 2017). *Pectinatus* isolates can tolerate beer with a pH of 4.0–4.5, while *Megasphaera* isolates start to show restricted growth at pH 4.5 (Haikara and Helander, 2006). Physiological growth data indicate *P. frisingensis* is more acidophilic than *P. cerevisiiphilus*, which makes this species more of a beer spoilage threat (Tholozan *et al.*, 1996, 1997). Finally, while low oxygen content is critical for the growth of *Pectinatus* and *Megasphaera* in beer, physiological differences exist between these bacteria, with *Pectinatus* isolates being more aero-tolerant than *Megasphaera* isolates (Chowdhury *et al.*, 1995; Haikara and Helander, 2006; B. Ziola, unpublished). Clearly, a combination of factors influence whether or not a particular packaged beer can support growth of *Pectinatus* or *Megasphaera*.

Beer spoilage by *Pectinatus* and *Megasphaera* invariably involves severe turbidity, particularly so for *Pectinatus* (Lee *et al.*, 1980). In addition, these bacteria cause offensive off-flavours and an obnoxious aroma due to the synthesis of variable levels of acetic, butyric, proprionic, caproic, isovaleric, and valeric acids; as well as acetoin, hydrogen sulfide, and methyl mercaptan (Foster and Andersen, 1999; Haikara *et al.*, 1981; Lee *et al.*, 1978, 1980; Membré and Tholozan, 1994). For beer with *Pectinatus* contamination, the resultant smell is most often that of rotten eggs, while the usual smell associated with *Megasphaera* growth is strikingly reminiscent of barnyard faecal waste slurries (particularly porcine), which is even more strikingly malodorous. Growth of these bacteria in beer occurs best at temperatures above 15°C (Haikara *et al.*, 1981), so occurs in packaged beer stored unrefrigerated. While the severe turbidity and terrible smell associated with spoilage is generally enough to cause consumer dissatisfaction, actually sampling the spoiled beer makes the experience much worse. The ultimate bad experience, never to be forgotten by a consumer, is the spilling of the spoiled beer on clothes or skin, as the horrible smell involved is difficult to remove even with repeated washing. Given the negative consumer outcome of experiencing spoilage by *Pectinatus* and *Megasphaera*, it is little wonder that brewers fear these bacteria. The good news for brewers globally is that the proportion of packaged beer spoilage due to these bacteria peaked in the later 1980s and early 1990s at values approaching 30% for *Pectinatus* and 5% for *Megasphaera* (Back, 1994, 2003; Paradh *et al.*, 2011). This is most likely due to improved brewery hygiene and increased vigilance for these bacteria in breweries.

Propionispira (formerly *Zymophilus*) and *Selenomonas*

When the three species for these two genera were first described, all isolates were from pitching yeast or undefined brewery waste; none were from contaminated beer (Schleifer *et al.*, 1990). In the intervening years, no definitive information has been forthcoming which associates *P. paucivorans*, *P. raffinosivorans*, or *S. lacticifix* with growth in packaged product. Of the three bacteria, *S. lacticifix* appears to be the most likely to be a beer spoiler, since laboratory modelling work has shown this bacterium can grow in beer with a pH of 4.3 to 4.6. In contrast, *P. paucivorans* and *P. raffinosivorans* require a higher pH of 6.0 or 5.0, respectively, for growth in beer (Seidel-Rüfer, 1990). As such, these bacteria are considered to be indicator microbes, whose detected presence within the brewery is indicative of a breakdown in brewery hygiene.

Accumulating evidence points to plants as the environmental origin for the nine currently defined anaerobic GNB species found in breweries. The initial indication that these bacteria are plant-associated came from analyses showing that their lipopolysaccharides contain unusual carbohydrate structures showing similarity with what is found in plant-associated *Rhizobium* spp. (Helander *et al.*, 2004). The first direct evidence for such plant association was provided by the finding of *P. cerevisiiphilus* in samples of mangrove sediment in Thailand (Saimmai *et al.*, 2012). About the same time, two new, albeit non-brewing-related *Pectinatus* species, *P. brassicae* (Zhang *et al.*, 2012) and *P. sotticetonis* (Caldwell *et al.*, 2013) were isolated from pickle waste water and spoiled pickles, respectively, Finally, a recent metagenomics analysis of a root sample from a sedge plant (*Carex* spp.) collected in a wetland at high altitude in the Tien Shan Mountains in the Issyk Kul region of Kyrgyzstan revealed that *Propionispira* spp. comprised some 25% of the bacterial population present (M. Haakensen and V. Friesen, Contango Strategies Ltd; personal communication). As sequencing the V3–V4 region of the 16S rRNA gene that was done on the sample does not allow *P. paucivorans* and *P. raffinosivorans* to be differentiated, identification to the species level could not be done.

Location in the brewery: biofilms and microbial communities

Anaerobic GNB most likely have long been brewery inhabitants and emerged as beer spoilers when the conditions in beer permitted anaerobic growth, rather than these bacteria coming into breweries coincidentally with the implementation of improved beer packaging which produced greatly reduced oxygen content. This hypothesis is indirectly supported by the same temporal emergence of spoilage by *Pectinatus* and *Megasphaera* in well-separated countries on different continents. *Pectinatus* spp. are widely distributed, perhaps due to their slightly higher oxygen tolerance than *Megasphaera* spp. *Pectinatus* spp. have been isolated in breweries in at least a dozen countries, including the USA; Japan; Finland, Norway, and Sweden; Belgium, France, Germany, Spain, the Netherlands, and the United Kingdom; and most recently, the Czech Republic (Lee *et al.*, 1978; Matoulková *et al.*, 2012b; Paradh *et al.*, 2011; Suikho and Haikara, 2001). In contrast, *Megasphaera* isolates have been reported in fewer, but still widely separated countries, including in Australia, Finland, Germany, Sweden, and the United Kingdom; and most recently, Canada (Bergsveinson *et al.*, 2017; Paradh *et al.*, 2011; Suikho and Haikara, 2001).

Additional evidence for *Megasphaera* and *Pectinatus* having a long-term association with breweries is based on the relative ease and frequency with which these bacteria are found. The recent papers by Paradh *et al.* (2011) and Matoulková *et al.* (2012b) document this, as isolation of *Megasphaera* and *Pectinatus* (particularly *P. frisingensis*) was readily achieved directly from beer, and even fermenting wort and yeast slurries, and from various sampling points within the brewery packaging environment. Both research groups emphasized that *Megasphaera* and *Pectinatus* are likely maintained within biofilms in the brewery and if stringent cleaning and hygiene are not maintained, such biofilms persist, continually acting as a potential source from which the bacteria can spread and contaminate yeast suspensions, fermenting wort, and most importantly, packaged beer. That at least *Pectinatus* isolates appear to be able to spread via aerosols within the brewery emphasizes the need for continued vigilance for the presence of biofilms and anaerobic GNB in all areas of a brewery, not just in the filling area (Back, 1994).

It has been suggested that brewery biofilms containing anaerobic GNB also include aerobic bacteria and yeast (to consume the oxygen in the immediate locale), as well as lactic acid bacteria (LAB; to provide lactate used as a carbon source for the anaerobic GNB growth) (Back, 1994). The frequent finding of both LAB and anaerobic GNB within a given brewery setting fits with this scenario (Back, 1994; Paradh *et al.*, 2011). In contrast, research specifically focused on microbes composing biofilms has not been as supportive, in that anaerobic GNB were found in only 3 of 78 samples taken from the bottling areas of two breweries in one study (Timke *et al.*, 2005a) and in none of the samples similarly collected from a single brewery in a second study (Timke *et al.*, 2005b). In contrast, a more recent analysis of 58 isolates obtained from sampling biofilms in a single brewery filling hall were analysed by MALDI-TOF MS, and *P. frisingensis* and *S. lacticifex* were identified 3 and 19 times, respectively (Vávrová *et al.*, 2014). This variability in frequency of anaerobic GNB isolation and identification is probably primarily due to differences in hygiene operating procedures in the different breweries sampled in the respective studies. Of secondary importance are the sampling and bacterial identification methods used. The limited and variable information available strongly suggest the need for more research on the role of biofilms in maintaining anaerobic GNB in the brewery setting.

Isolation, detection and identification

When *Megasphaera* and *Pectinatus* are isolated from brewery samples, it is invariably as a mixture with yeast, aerobic bacteria, and/or Gram-positive bacteria. In an effort to enhance selectivity for *Megasphaera* and *Pectinatus*, specific growth media have been developed, including SMMP medium, which contains lactate as the sole carbon source and reducing agents to promote the anaerobic growth by *Megasphaera* and *Pectinatus* (Lee, 1994). Added cycloheximide, alcohol, and crystal violet together with sodium fusidate inhibit growth of yeast, aerobic bacteria and Gram-positive bacteria, respectively. A more recently described selective *Pectinatus* growth medium consists of a modified de Man-Rogosa-Sharpe (MRS) medium with cysteine hydrochloride and sodium thioglycolate added to lower the redox potential to enable anaerobic growth, and a mixture of tetrahydroiso-alpha-acids and 2-phenyethanol to inhibit growth of other organisms. Bacteriological agar is added at a sub-solidification level to prevent medium oxidation during use over time (Matoulková *et al.*, 2012a). Although not tested, it is likely that *Megasphaera* would also grow in this medium. Once *Megasphaera* and *Pectinatus* spp. have been purified by plating on anaerobic MRS agar in Gas Pak chambers, isolates of both genera can readily be grown in MRS broth that has been autoclaved in tubes with a minimum headspace and the tube caps immediately tightened before the medium cools (SY Lee, personal communication; Pittet *et al.*, 2014).

Starting either with contaminated beer samples

or following growth of brewery samples in isolation medium, the question then is whether *Megasphaera* or *Pectinatus* spp. are present. If these bacteria are present in high levels, scent alone can indicate their presence, particularly if a *Megasphaera* isolate is involved. For breweries with gas chromatography capability, confirmation can be obtained of the obnoxious volatile compounds present due to the metabolism of *Megasphaera* or *Pectinatus* spp. (Foster and Andersen, 1999). More recently, a lipidomics approach to detecting *Megasphaera* or *Pectinatus* in brewing samples was described, with electrospray ionization tandem mass spectrometry used to detect plasmalogen phospholipids (Řezanka *et al.*, 2015). As plasmalogen phospholipids are found in anaerobic bacteria and not yeast or plant materials, this methodology allows detection of GNB in brewing samples, but without further identification to the genus level.

When considering not only detection but identification of *Megasphaera* or *Pectinatus* spp., from a brewery microbial quality control perspective it is important to know whether beer spoilage is the result of a newly introduced bacterium or re-emergence of a strain that already comprises part of the existing brewery microbial community. Also, when *Megasphaera* or *Pectinatus* are present at low levels in brewery samples, more sensitive and targeted approaches than just smell are required. However, the methods that have been developed to meet this criterion vary in both sensitivity and ability to answer whether a detected *Megasphaera* or *Pectinatus* isolate is new to or re-emerging in a brewery.

Both polyclonal and monoclonal antibody-based systems have been described for detection of *Megasphaera* and *Pectinatus* (Gares *et al.*, 1993; Haikara, 1985; Ziola *et al.*, 2000). There is some evidence that the age of the cells trapped by filtration influences subsequent detection by an immunofluorescence antibody test, at least if polyclonal antiserum is used (Haikara, 1985). Provided monoclonal antibodies specific for surface-exposed antigens of the bacterial cell are used, bacterial levels as low as 2–4 cells per 10 ml of contaminated beer were detectable by fluorescence immunoassays following filter concentration of the cells (Gares *et al.*, 1993). As long as the bacterial cells are intact, age of the cells is not a factor, as they are easily visible by microscope scanning. It should be noted that monoclonal antibodies which are strain or species subgroup specific are readily derived, making them useful in determining whether the *Megasphaera* or *Pectinatus* isolate in question is new to or re-emerging in a brewery. In contrast, monoclonal antibodies showing species- or genus-specific reactivity are much more difficult to obtain. Not surprisingly, this difficulty in obtaining monoclonal antibodies with broader binding specificity is reflected when polyclonal antiserum is used instead. Overall, polyclonal antibodies (whether mouse or rabbit) against a given *Megasphaera* or *Pectinatus* isolate generally do not work well in filter-based bacterial cell-surface immunofluorescence assays. This is because of a high assay background signal and the difficulty in generating antiserum showing isolate-, species subgroup-, species- or genus-specific reactivity (even following extensive adsorptions in attempts to remove different kinds of bacterial cross-reacting antibodies; B. Ziola, unpublished).

The first application of omics to the brewing-related anaerobic GNB involved sequencing of the 16S and 23S rRNA genes, and the rRNA gene interspacer DNA with the overall aim of developing DNA-based methods for detection with concurrent identification of these bacteria at the species level (Juvonen *et al.*, 1999; Motoyama and Ogata, 2000; Sakamoto *et al.*, 1997; Satokari *et al.*, 1998). These tests work well for cells following pre-enrichment growth in isolation medium, and recently have been extended to the detection of the two *Propionispira* (formerly *Zymophilus*) spp. and *S. lacticifix* (Felsberg *et al.*, 2014, 2015). In an attempt to apply this methodology to contaminated beer samples where *Megasphaera* or *Pectinatus* cells can be low in number, concentration of the bacteria by membrane filtration has been done prior to DNA extraction (Juvonen *et al.*, 1999; Juvonen and Haikara, 2009). As PCR inhibitors are found in beer, numerous pretreatments of the filter-trapped cells as well as PCR additives have been tested, with addition of bovine serum albumin and polyvinyl pyrrolidone proving useful in reducing the effect of PCR inhibitors in beer (Juvonen and Haikara, 2009).

In an effort to increase assay efficiency, from the perspective of potentially obtaining more data per reaction, multiplex PCR methods based on 16S rRNA gene sequences have been described for the brewing-related anaerobic GNB (Iijima *et al.*, 2008; Juvonen *et al.*, 2008). Such assays incorporate multiple primer sets that yield PCR amplicons of different sizes or melting curves, thus enabling a range of species to be identified with a single reaction. DNA sequences other than the 16S rRNA gene can be used in such a multiplex PCR approach as was recently demonstrated with the gene for the major outer membrane protein found in the three *Pectinatus* spp. used as the PCR target (Pittet *et al.*, 2014). Indeed, any gene with inherent variability involving deletions, or insertions, or even multiple nucleotide changes if clustered, can find utility in such a multiplex PCR. For *Pectinatus* spp., another gene that appears to meet the criteria for targeting via a multiplex PCR is the flagellin gene (Chaban *et al.*, 2005). In the case of *Megasphaera* spp., no gene has been sequenced for all three brewing-related species. Nonetheless, the readily extractable outer membrane protein from *M. cerevisiae* isolates exhibits size variability, suggesting the gene for this protein has the potential to fulfil the role as the target in a multiplex PCR for identifying *Megasphaera* spp. (Ziola *et al.*, 2000).

Ribosomal operon-based PCR methods also have been described that incorporate restriction enzyme-digestion of the reaction amplicons, followed by fragment size assessment in an approach termed ribotyping (Juvonen *et al.*, 2008; Motoyama *et al.*, 1998; Suihko and Haikara, 2001). Not only can identification of the brewing-related anaerobic GNB be done to the species level, but also to the subspecies and even in some cases to the strain level. Using conserved genes such as the rRNA genes or other so-called housekeeping genes are not ideal as PCR targets for determining whether a contaminant is a new strain *or* an existing strain of anaerobic GNB within a given brewery. This is because such genes, by definition, show only slow and thus limited change in genetic sequence. Instead, genes that show considerable sequence variability are better suited in this regard as exemplified by the *Pectinatus* flagellin and major outer membrane protein genes (Chaban *et al.*, 2005; Pittet *et al.*, 2014). Once either of these genes has been sequenced for a given *Pectinatus* isolate, a PCR; for example, a probe-based real-time PCR can be readily assembled for detecting re-emergence of that isolate in a brewery.

Viability of the bacterial cells being detected in a brewery setting is a consideration if the cells being detected have not been subjected to pre-enrichment growth in isolation medium. A DNA oligonucleotide microarray has been described for the detection and identification of viable beer spoilage bacteria, including anaerobic GNB (Weber *et al.*, 2008). Here the ribosomal interspacer DNA is the binding target, with determination of bacterial viability based on the correlation between growing cells and the presence of pre-rRNA that includes the rRNA interspacer region.

If the focus of microbial quality control within a brewery is on whether there has been a break in brewery hygiene, then the assays deployed need not be bacterial strain, species or even genus specific, and viability of any detected contaminate is not important. In this case, using a DNA-based rapid method for beer spoilage bacteria writ large will suffice. As the vast majority of brewing spoilage is caused by bacteria belonging to the *Firmicutes* phylum (i.e. both the Gram-positive bacteria dealt with in the previous chapter and the anaerobic GNB discussed here), detection of a brewery hygiene break can be accomplished by a 16S rRNA-based PCR designed to detect all *Firmicutes* (Haakensen *et al.*, 2008). Combining the *Firmicutes*-specific PCR primers with a primer that binds only to the 16S rRNA gene of Gram-negative *Firmicutes* allows concurrent determination of a detected *Firmicutes* bacterium as being either a Gram-positive or an anaerobic Gram-negative organism (Pittet *et al.*, 2010).

Common structural features, genes, and genomes

Despite their Gram-staining characteristics, the anaerobic brewing-related GNB are taxonomically placed with the low-GC Gram-positive bacteria in the *Firmicutes* phylum. This placement occurs whether the phylogeny is based on rRNA gene sequences (Juvonen and Suihko, 2006; Schleifer *et al.*, 1990; Stackebrandt *et al.*, 1985) or, for *Pectinatus* spp., when the much more variable flagellin protein sequence is used (Chaban *et al.*, 2005). While the brewing-related GNB were originally all placed in the class *Clostridia* within the *Firmicutes*, more recent taxonomic analysis has placed *Megasphaera* sp. in a new class called *Negativicutes* within the *Firmicutes* (Marchandin *et al.*, 2010). Since phylogenetic analyses place all the brewery-related anaerobic GNB close together taxonomically, it is expected that these bacteria will possess a wide range of common features.

From an omics perspective, comparing a group of bacteria, or even one bacterium to another, is best done starting with a comparison of global genome sequences. Unfortunately, for the nine species of brewery-related anaerobic GNB, only the two strains of *M. cerevisiae* (Bergsveinson *et al.*, 2017; Kutumbaka *et al.*, 2015) and one strain of *P. raffinosivorans* (Kyrpides *et al.*, 2013) have been genome sequenced, with the data publicly available through the National Centre for Biotechnology Information (NCBI). Thus far, no research has been published using these three genome sequences. At least four and seven non-brewing-related *Megasphaera* spp. and *Selenomonas* spp., respectively, have been genome-sequenced and the data deposited with the NCBI. Included here are four distinct isolates for each of *Megasphaera elsdenii* and *Selenomonas ruminatum*. As such, there already is a reasonable amount of comparative genomic data available for the non-beer spoilage immediate relatives of the brewing-related anaerobic GNB when additional genomes of the latter organisms do become sequenced.

Except for the ribosomal operon-related sequences used for establishment of the DNA-based detection and identification methods already described, no other DNA (i.e. genes) universal to all nine brewing-related anaerobic GNB have been sequenced. The only genes investigated in any detail for these bacteria include the *Pectinatus* flagellin gene (Chaban *et al.*, 2005; Hakalehto *et al.*, 1997) and the gene for the *Pectinatus* major outer membrane protein (Pittet *et al.*, 2014). Given the few brewing-related anaerobic GNB DNA sequences that have been analysed, it is not surprising that no gene expression (transcription) analysis has been performed on any of these bacteria.

That brewing-related anaerobic GNB ultimately will be shown to have common genetics is predicated on the anaerobic growth properties of these bacteria and what is known of their structural components. The research leading to the initial description of many of these bacteria determined that they possess an unusually thick layer of directly cross-linked *meso*-diaminopimelic acid peptidoglycan containing cadaverine and putrescine as major and minor components, respectively (Engelmann and Weiss, 1985; Haikara *et al.*, 1981; Schleifer *et al.*, 1990). Commonality of moieties in the peptidoglycan of these bacteria was later found through epitope mapping of peptidoglycan-specific monoclonal antibodies (Ziola *et al.*, 1999). A total of nineteen monoclonal antibodies were analysed and found to map to four distinct binding sites on the peptidoglycan. Monoclonal antibodies specific to each of the four sites were documented to bind to six of the nine currently described anaerobic brewing-related GNB, with the remaining species *M. paucivorans*, *M. sueciensis*, and *P. haikarae* not tested as these spp. were recognized subsequently. While it seems likely that cadaverine and putrescine would each be part of one of the four defined antibody-binding sites involved, this hypothesis remains untested.

Although the brewing-related anaerobic GNB are classified taxonomically in the *Firmicutes* phylum alongside the Gram-positive bacteria, these microbes contain an outer membrane and lipopolysaccharides (LPS) like other GNB. The LPS of *Pectinatus* isolates have been well characterized (Helander *et al.*, 1983, 1994, 2004) and shown to be rich in fatty acids and possess considerable heterogeneity, with several distinct molecules produced by an individual strain. *Pectinatus* LPS also contains unusual carbohydrate structures consisting of what is essentially a conserved core region carrying a large non-repetitive polysaccharide, which replaces the usual O-specific chain (Vinogradov *et al.*, 2003). Again, similar to the bacterial peptidoglycan, monoclonal antibodies have shown that the LPS from all brewing-related anaerobic GNB contain common structural features not found in other GNB. Using spleen cells from mice immunized and boosted with ethylenediaminetetraacetic acid extracts from different *M. cerevisiae* isolates, four monoclonal antibody hybridomas were isolated secreting antibodies that reacted with a surface-accessible antigen on all *M. cerevisiae* isolates. Further testing revealed that at least two distinct binding epitopes are recognized by the four monoclonal antibodies and that these epitopes are also present in *M. elsdenii* as well as *P. cerevisiiphilus*, *P. frisingensis*, *P. paucivorans*, *P. raffinosivorans* and *S. lacticifix*, (Ziola, 2016). Since all four monoclonal antibodies show reactivity with the bacterial surface in a surface immunofluorescence test with intact cells as the antibody binding target, the conserved antibody binding sites are most likely in the unusual large distal carbohydrate in the LPS. This commonality in a LPS-related structural feature among these bacteria points to yet another metabolic pathway and, hence, genetics, that they have in common.

Conclusions and the future

Beer spoilage by anaerobic GNB clearly is a problem for the brewing industry that waxes and wanes, but is not going away, meaning continuous attention to brewery hygiene is essential if these unwanted bacteria are to be kept out of beer. In contrast to the emerging use of omics to study the Gram-positive LAB that can spoil beer, as described in the preceding chapter, omics approaches have yet to be truly applied in studies of the anaerobic GNB of interest to brewers. Compared to the Gram-positive lactobacilli and pediococci beer spoilage bacteria, the brewing-related anaerobic GNB are the *poor country cousins* waiting their turn in the omics spotlight.

Genome sequencing and transcriptional analysis

Although only briefly touched upon here, physiological responses of the brewing-related anaerobic GNB have been studied broadly with summaries of the information available in review form (Haikara and Helander, 2006; Suzuki, 2011). To make better sense of the physiological data that has been accumulated, it is essential to have in-hand genome sequences for the bacteria involved. For example, having sequenced genomes for each of the nine anaerobic brewing-related GNB would allow elucidation of the genetics required for the synthesis of the common aspects of the peptidoglycan and LPS these bacteria exhibit. However, with only the genomes of two isolates of *M. cerevisiae* and one isolate of *P. raffinosivorans* so far available, the sequence cupboard is essentially bare. To remedy this, at least the type strains of the other seven brewing-related anaerobic GNB should have their genomes sequenced and assembled. Beyond that, several additional *Megasphaera* and *Pectinatus brewery* isolates should have their genomes sequenced because the majority of the available physiological data on the anaerobic brewing-related GNB has been done with isolates of these two genera. With multiple genome sequences available it would then be possible to start doing the genome comparisons that will underpin understanding of the different metabolic activities among the brewing-related anaerobic GNB. Since several isolates of closely related non-brewing *Megasphaera* spp. have had their genomes sequenced, including four strains of *M. elsdenii*, comparisons against genome-sequenced brewing-related anaerobic GNB would begin to reveal the genetic characteristics of these bacteria needed if growth in beer is to occur.

Genome sequencing for isolates representing the full spectrum of the brewing-related anaerobic GNB will also provide information pertinent to two other interesting areas of investigation. First, since these bacteria are classified as Firmicutes, what is it about the genetics of these low-GC Gram-positive bacteria that fits with both true Gram-positive organisms and true Gram-negative organisms? Genome sequence analyses in this regard have a good probability of providing additional insights into the evolution occurring among the existing spectrum of brewing-related bacteria. Second, genome sequencing of the anaerobic GNB will most certainly provide information on whether plasmids play a role in the growth of these bacteria in beer. It has long been suspected that plasmids obtained via horizontal gene transfer play a major role in the growth of lactobacilli and pediococci that are beer spoilers. Direct evidence for the importance of plasmids in beer-spoiling organisms was recently obtained by curing various plasmids from the rapidly growing beer spoiler *Lactobacillus brevis* BSO 464 and finding that growth in beer and hop resistance were dramatically reduced with the loss of particular plasmids (Bergsveinson *et al.*, 2015). Whether the brewing-related anaerobic GNB indeed even have plasmids must first be answered by genome-sequencing of different isolates; only then can the question be approached of whether plasmid-harboured genes contribute to the growth in and spoilage of beer by these bacteria.

Once genome sequences for the anaerobic brewing-related GNB become available, then the power of transcriptomic approaches can come into play. With the availability of different genomes of these bacteria and their close non-spoilage relatives, gene expression analysis of genes of interest can be performed. The alternative is to undertake global transcriptional analysis of given bacteria grown under specified growth conditions. In the initial instance, an enormous amount of data can be obtained by determining global gene transcription when a bacterium is growing in beer compared to growth in routine laboratory medium. Subsequently, the gene expression required to handle each of the growth stresses encountered by a bacterium when growing in beer can be individually determined. As described in the preceding chapter, such studies are beginning to take place for the LAB that spoil beer (Pittet *et al.*, 2013; Bergsveingson *et al.*, 2016a,b), but have yet to be undertaken for any of the brewing-related anaerobic GNB.

As the acidic pH, alcohol, and dissolved CO_2/pressure found in packaged beer are generally sufficient to inhibit the growth of most GNB, it would be of particular interest to determine the genetics allowing the brewing-related anaerobic GNB to handle these stresses when growing in beer. In fact, it would be of interest to compare gene transcription of the aerobic and anaerobic GNB in relation to growth in beer. Including hops as one of the growth stresses found in beer in such comparative analyses should also be done, because GNB as a whole are generally resistant to the effect of hops, while Gram-positive bacteria are generally sensitive. As the anaerobic brewing-related GNB possess features of Gram-positive bacteria, and beer spoilage Gram-positive bacteria have genetic mechanisms frequently acquired by horizontal gene transfer, for handling hop compounds, an important question is: what genes are specifically expressed in brewing-related GNB to counteract the antimicrobial effects of hops? The data obtained in this regard again will have relevance to our understanding of both bacterial evolution and bacterial adaptation to the beer environment.

Metagenomics and biofilms

Increasingly, metagenomic analysis of samples taken from a wide range of environments is filling in the large gaps in our knowledge of bacteria vis à vis where they are and what activities they are performing. Here, by way of example, it is worth repeating the findings of *P. cerevisiiphilus* in mangrove sediments in Thailand and *Propionispira* in a plant root sample collected in Kyrgyzstan. These metagenomics outcomes point to plant material or related soil being the source for these bacteria finding their way into breweries. As metagenomic analyses from a wide range of studies become deposited in public data bases, routine surveying of the information is likely to provide additional insights on the environmental sources for each of the nine brewing-related anaerobic GNB, as well as their close taxonomic relatives. Paying attention to sources and locations of bacteria that are closely related to the nine currently known brewing-related anaerobic GNB species could prove interesting in terms of understanding or even anticipating the emergence of yet other brewing-related GNB that have a high probability of belonging to the *Firmicutes* phylum.

Metagenomic analyses should also be undertaken using samples taken from materials coming into breweries and the brewing process, as well as using samples collected widely within breweries. Such analyses done in a systematic fashion on a periodic basis will provide not only baseline data for all bacteria moving into and establishing themselves with a brewery, but also for how different spoilage bacteria are able to relocate over time within a brewery. The metagenomics approach to sampling of biofilms within breweries is particularly relevant as we still do not clearly understand how both aerobic and anaerobic GNB establish and maintain themselves within a brewery. Provided the analysis is done such that eukaryotes (brewing and wild yeast) are also tested for along with all forms of bacteria, information should be forthcoming that clarifies exactly how GNB, particularly those that are anaerobes, are able to persist in a brewery. Such data collected over time will provide a better picture of what microbes co-exist within the same brewery locations. Moreover, this approach is the most detailed way to document whether brewery hygiene procedures are working or not and, if not, where remedial action in brewery operations are required.

Along with routine and systematic brewery biofilm assessment, experimental modelling of biofilms should be done, incorporating all the different types of microbes found in the brewery environment. The basis for this is the under-appreciated fact that brewing-related lactobacilli and pediococci, and aerobic or anaerobic GNB do not grow in isolation. There is an inherent need to determine how each microbial member found within a biofilm affects other members in the same location and how the microbial composition of biofilms evolves. As one example, using a quantitative PCR approach (e.g. droplet digital PCR), it should be possible to ascertain the relative quantities of different microbes within biofilms over time, with these data informing how best to inhibit biofilm development as well as decontaminate existing biofilms in the brewery setting.

Omics methodologies are continually being refined and increasingly cheaper to implement. As discussed in the preceding chapter, these experimental tools are already being applied to the Gram-positive LAB beer spoilers, driven in part because LAB are important in many food industries in addition to the brewing industry. In contrast, the brewing-related anaerobic GNB have no useful counterparts in other food or beverage industries; hence, the interest in these bacteria is essentially restricted to brewers, thus accounting for the type and limited amount of research done on these bacteria. Nonetheless, given the unusual features of these anaerobic GNB and given their ability to successfully invade and survive in the beer environment, detailed omics-based studies of these bacteria are certainly warranted.

References

Back, W. (1994). Secondary contaminations in the filling area. Brauwelt Int. 4, 326–333.

Back, W. (2003). Biofilme in der brauerei und getrankeindustrie– 15 jahre praxiserfahrung Brauwelt 24/25, 766 (4 pp.).

Bamforth, C.W. (2005). Beer. In Food, Fermentation and Microorganisms. (Ames, Iowa: Blackwell Publishing Professional), pp.40–88. http://dx.doi.org/10.1002/9780470995273.ch2

Bergsveinson, J., Baecker, N., Pittet, V., and Ziola, B. (2015). Role of plasmids in *Lactobacillus brevis* BSO 464 hop tolerance and beer spoilage. Appl. Environ. Microbiol. *81*, 1234–1241. http://dx.doi.org/10.1128/AEM.02870-14

Bergsveinson, J., Ewen, E., Friesen, V., and Ziola, B. (2016a). Transcriptional activity and role of plasmids of *Lactobacillus brevis* BSO 464 and *Pediococcus claussenii* ATCC BAA-344T during growth in the presence of hops. AIMS Microbiol. *2*, 460–478. http://dx.doi.org/10.3934/microbiol.2016.4.460

Bergsveinson, J., Friesen, V., and Ziola, B. (2016b). Transcriptome analysis of beer-spoiling *Lactobacillus brevis* BSO 464 during growth in degassed and gassed beer. Int. J. Food Microbiol. *235*, 28–35. http://dx.doi.org/10.1016/j.ijfoodmicro.2016.06.041

Bergsveinson, J., Thomson, E., Jacoby, D., Coady, Y., and Ziola, B. (2017). Genome sequence of *Megasphaera cerevisiae* NSB1, a bacterium isolated from a canning line and able to grow in beer with high alcohol content. Genome Announc. *5*, e01686-16. http://dx.doi.org/10.1128/genomeA.01686-16

Bokulich, N.A., Bamforth, C.W., and Mills, D.A. (2012). Brewhouse-resident microbiota are responsible for multi-stage fermentation of American coolship ale. PLoS ONE *7*, e35507. http://dx.doi.org/10.1371/journal.pone.0035507

Caldwell, J.M., Juvonen, R., Brown, J., and Breidt, F. (2013). *Pectinatus sottacetonis* sp. nov., isolated from a commercial pickle spoilage tank. Int. J. Syst. Evol. Microbiol. *63*, 3609–3616. http://dx.doi.org/10.1099/ijs.0.047886-0

Chaban, B., Deneer, H., Dowgiert, T., Hymers, J., and Ziola, B. (2005). The flagellin gene and protein from the brewing spoilage bacteria *Pectinatus cerevisiiphilus* and *Pectinatus frisingensis*. Can. J. Microbiol. *51*, 863–874. http://dx.doi.org/10.1139/w05-076

Chowdhury, I., Watier, D., and Hornez, J.-P. (1995). Variability in survival of *Pectinatus cerevisiiphilus*, a strictly anaerobic bacteria under different oxygen conditions. Anaerobe 1, 151–156. http://dx.doi.org/10.1006/anae.1995.1012

Dadds, M.J.S., Macpherson, A.L., and Sinclair, A. (1971). *Zymomonas* and acetaldehyde levels in beer. J. Inst. Brew. *77*, 453–456. http://dx.doi.org/10.1002/j.2050-0416.1971.tb03402.x

Engelmann, U., and Weiss, N. (1985). *Megasphaera cerevisiae* sp. nov.: A new gram-negative obligately anaerobic coccus isolated from spoiled beer. Syst. Appl. Microbiol. *6*, 287–290. http://dx.doi.org/10.1016/S0723-2020(85)80033-3

Felsberg, J., Jelinkova, M., Kubizniakova, P., and Matoulkova, D. (2014). Development of a species-specific PCR assay for identification of the strictly anaerobic bacterium *Selenomonas lacticifix* found in biofilm-covered surfaces in brewery bottling halls. J. Appl. Microbiol. *117*, 1328–1335. http://dx.doi.org/10.1111/jam.12610

Felsberg, J., Jelinkova, M., Kubizniakova, P., and Matoulkova, D. (2015). Development of a PCR assay based on the 16S-23S rDNA internal transcribed spacer for identification of strictly anerobic bacterium *Zymophilus*. Anaerobe *33*, 85–89. http://dx.doi.org/10.1016/j.anaerobe.2015.02.004

Foster, A., and Andersen, L.B. (1999). Gas chromatographic technique for the confirmation of *Megasphaera* and *Pectinatus* spp. J. Am. Soc. Brew. Chem. *57*, 91–93. http://dx.doi.org/10.1094/ASBCJ-57-0091

Gares, S., Whiting, M.S., Ingledew, W.M., and Ziola, B. (1993). Detection and identification of *Pectinatus cerevisiiphilus* suing surface-reactive monoclonal antibodies in a membrane filter-based fluoroimmunoassay. J. Am. Soc. Brew. Chem. *51*, 158–163. http://dx.doi.org/10.1094/ASBCJ-51-0158

Greipsson, S., and Priest, F.G. (1983). Numerical Taxonomy of *Hafnia alvei*. Int. J. Sys. Bacteriol. 33, 470–475. http://dx.doi.org/10.1099/00207713-33-3-470

Haakensen, M., Dobson, C.M., Deneer, H., and Ziola, B. (2008). Real-time PCR detection of bacteria belonging to the *Firmicutes* Phylum. Int. J. Food Microbiol. *125*, 236–241. http://dx.doi.org/10.1016/j.ijfoodmicro.2008.04.002

Haikara, A. (1985). Detection of *Pectinatus* contaminants in beer. J. Am. Soc. Brew. Chem. 43, 43–46. http://dx.doi.org/10.1094/ASBCJ-43-0043

Haikara, A., and Helander, I. (2006). *Pectinatus, Megasphaera* and *Zymophilus*. In The Prokaryotes, Volume 4, M. Dworkin, S. Falkow, E. Rosenberg, K.-H. Schleifer and E. Stackenbrandt, eds. (New York,

USA: Springer-Verlag), pp. 965–981. http://dx.doi.org/10.1007/0-387-30744-3

Haikara, A., Penttilä, L., Enari, T.M., and Lounatmaa, K. (1981). Microbiological, biochemical, and electron microscopic characterization of a pectinatus strain. Appl. Environ. Microbiol. *41*, 511–517.

Hakalehto, E., Santa, H., Vepsalainen, J., Laatikainen, R., and Finne, J. (1997). Identification of a common structural motif in the disordered N-terminal region of bacterial flagellins – Evidence for a new class of fibril-forming peptides. Eur. J. Biochem. 250, 19–29. http://dx.doi.org/10.1111/j.1432-1033.1997.00019.x

Hamze, M., Mergaert, J., van Vuuren, H.J., Gavini, F., Beji, A., Izard, D., and Kersters, K. (1991). *Rahnella aquatilis*, a potential contaminant in lager beer breweries. Int. J. Food Microbiol. *13*, 63–68. http://dx.doi.org/10.1016/0168-1605(91)90137-E

Helander, I.M., Haikara, A., Sadovskaya, I., Vinogradov, E., and Salkinoja-Salonen, M.S. (2004). Lipopolysaccharides of anaerobic beer spoilage bacteria of the genus *Pectinatus* – lipopolysaccharides of a Gram-positive genus. FEMS Microbiol. Rev. *28*, 543–552. http://dx.doi.org/10.1016/j.femsre.2004.05.001

Helander, I.M., Hakalehto, E., Ahvenainen, J., and Haikara, A. (1983). Characterization of lipopolysaccharides of *Pectinatus cerevisiiphilus*. FEMS Microbiol. Lett. 18, 223–226. http://dx.doi.org/10.1111/j.1574-6968.1983.tb00482.x

Helander, I.M., Kilpeläinen, I., Vaara, M., Moran, A.P., Lindner, B., and Seydel, U. (1994). Chemical structure of the lipid A component of lipopolysaccharides of the genus *Pectinatus*. Eur. J. Biochem. *224*, 63–70. http://dx.doi.org/10.1111/j.1432-1033.1994.tb19995.x

Iijima, K., Asano, S., Suzuki, K., Ogata, T., and Kitagawa, Y. (2008). Modified multiplex PCR methods for comprehensive detection of *Pectinatus* and beer spoilage cocci. Biosci. Biotechnol. Biochem. *72*, 2764–2766. http://dx.doi.org/10.1271/BBB.80297

Jandreau, G.L., and Hahn, C.W. (1978). Automated measurement of total oxygen in beer packages. J. Am. Soc. Brew. Chem. 36, 44–48. http://dx.doi.org/10.1094/ASBCJ-36-0044

Jespersen, L., and Jakobsen, M. (1996). Specific spoilage organisms in breweries and laboratory media for their detection. Int. J. Food Microbiol. *33*, 139–155. http://dx.doi.org/10.1016/0168-1605(96)01154-3

Juvonen, R., and Haikara, A. (2009). Amplification facilitators and pre-processing methods for PCR detection of strictly anaerobic beer spoilage bacteria of the Class *Clostridia* in brewery samples. J. Inst. Brew. 115, 167–176. http://dx.doi.org/10.1002/j.2050-0416.2009.tb00365.x

Juvonen, R., Koivula, T., and Haikara, A. (2008). Group-specific PCR-RFLP and real-time PCR methods for detection and tentative discrimination of strictly anaerobic beer spoilage bacteria of the class *Clostridia*. Int. J. Food Microbiol. *125*, 162–169. http://dx.doi.org/10.1016/j.ijfoodmicro.2008.03.042

Juvonen, R., Satokari, R., Mallison, K., and Haikara, A. (1999). Detection of spoilage bacteria in beer by polymerase chain reaction. J. Am. Soc. Brew. Chem. 57, 99–103. http://dx.doi.org/10.1094/ASBCJ-57-0099

Juvonen, R., and Suihko, M.L. (2006). *Megasphaera paucivorans* sp. nov., *Megasphaera sueciensis* sp. nov. and *Pectinatus haikarae* sp. nov., isolated from brewery samples, and emended description of the genus *Pectinatus*. Int. J. Syst. Evol. Microbiol. *56*, 695–702. http://dx.doi.org/10.1099/ijs.0.63699-0

Kosaric, N., Ong, S.L., and Duvnjak, Z. (1982). Fuel alcohol biosynthesis by *Zymomonas anaerobia*: Optimization studies. Biotechnol. Bioeng. *24*, 691–701. http://dx.doi.org/10.1002/bit.260240312

Kuchel, L., Brody, A.L., and Wicker, L. (2006). Oxygen and its reactions in beer. Packag. Technol. Sci. *19*, 25–32. http://dx.doi.org/10.1002/pts.705

Kutumbaka, K.K., Pasmowitz, J., Mategko, J., Reyes, D., Friedich, A., Ham, S., Martens-Habbena, W., Neal-McKinney, J., Janagama, H.K., Nadala, C., and Samadpour, M. (2015). Draft genome sequence of the beer spoilage bacterium *Megasphaera cerevisiae* strain PAT 1^T. Genome Announc. 3, e01045-15. http://dx.doi.org/10.1128/genomeA.01045-15

Kyrpides, N., Huntemann, M., Han, J., Chen, A., Mavromatis, K., Markowitz, V., Palaniappan, K., Ivanova, N., Schaumberg, A., Pati, A., Liolios, K., Nordberg, H.P., Cantor, M.N., Hua, S.X., and Woyke, T. (2013). *Propionispira raffinosivorans* DSM 20765^T whole geneome shotgun sequencing project. Direct submission; NCBI reference sequence NZ_ARLE00000000.1.

Lee, S.Y. (1994). SMMP – a medium for selective isolation of *Megasphaera* and *Pectinatus* from the brewery. J. Am. Soc. Brew. Chem. 52, 115–119. http://dx.doi.org/10.1094/ASBCJ-52-0115

Lee, S.Y., Mabee, M.S., and Jangaard, N.O. (1978). *Pectinatus*, a new genus of the family *Bacteriodaceae*. Int. J. Syst. Bacteriol. *28*, 582–594. http://dx.doi.org/10.1099/00207713-28-4-582

Lee, S.Y., Mabee, M.S., Jangaard, N.O., and Horiuchi, E.K. (1980). *Pectinatus*, a new genus of bacteria capable of growth in hopped beer. J. Inst. Brew. *86*, 28–30. http://dx.doi.org/10.1002/j.2050-0416.1980.tb03951.x

Marchandin, H., Teyssier, C., Campos, J., Jean-Pierre, H., Roger, F., Gay, B., Carlier, J., and Jumas-Bilak, E. (2010). *Negativicoccus succinicivorans* gen.nov., sp. Nov., isolated from human clinical samples, emended description of the family *Veillonellaceae* and description of *Negativicutes* classis nov., *Selenomondales* ord. nov. and *Adcidaminococcaceae* fam. Nov. in the bacterial phylum *Firmicutes*. Int. J. Syst. Evol. Microbiol. 60, 1271–1279. http://dx.doi.org/10.1099/ijs.0.013102-0

Martens, H., Dawoud, E., and Verachtert, H. (1991). Wort *Enterobacteria* and other microbial populations during the first month of Lambic fermentation. J. Inst. Brew. 97, 435–439. http://dx.doi.org/10.1002/j.2050-0416.1991.tb01082.x

Matoulková, D., Kosar, K., and Sigler, K. (2012a). Rapid, simple, and specific cultivation-based method for detection of *Pectinatus* spp. in brewery samples. J. Am. Soc. Brew. Chem. *70*, 29–34. http://dx.doi.org/10.1094/ASBCJ-2011-1214-01

Matoulková, D., Kosar, K., and Slaby, M. (2012b). Occurrence and species distribution of strictly anaerobic bacterium *Pectinatus* in brewery bottling halls. J. Am. Soc. Brew. Chem. *70*, 262–267. http://dx.doi.org/10.1094/ASBCJ-2012-0910-01

Membré, J.M., and Tholozan, J.L. (1994). Modeling growth and off-flavours production of spoiled beer bacteria *Pectinatus frisingensis*. J. Appl. Bacteriol. *77*, 456–460. http://dx.doi.org/10.1111/j.1365-2672.1994.tb03449.x

Moreau, R.A., Powell, M.J., Fett, W.F., and Whitaker, B.D. (1997). The effect of ethanol and oxygen on the growth of *Zymomonas mobilis* and the levels of hopanoids and other membrane lipids. Curr. Microbiol. *35*, 124–128. http://dx.doi.org/10.1007/s002849900224

Motoyama, Y., and Ogata, T. (2000). Detection of *Pectinatus* spp. by PCR using 16S-23S rDNA spacer regions. J. Am. Soc. Brew. Chem. *58*, 4–7. http://dx.doi.org/10.1094/ASBCJ-58-0004

Motoyama, Y., Ogata, T., and Sakai, K. (1998). Characterization of *Pectinatus cerevisiiphilus* and *P. frisingensis* by ribotyping. J. Am. Soc. Brew. Chem. *56*, 19–23. http://dx.doi.org/10.1094/ASBCJ-56-0019

Paradh, A.D., Mitchell, W.J., and Hill, A.E. (2011). Occurrence of *Pectinatus* and *Megasphaera* in the major UK breweries. J. Inst. Brew. *117*, 498–506. http://dx.doi.org/10.1002/j.2050-0416.2011.tb00497.x

Pittet, V., Haakensen, M., Chaban, B., and Ziola, B. (2014). Detection and identification of *Pectinatus* brewery contaminants based on the gene for the major outer membrane protein. J. Am. Soc. Brew. Chem. 72, 169–174. http://dx.doi.org/10.1094/ASBCJ-2014-0610-02

Pittet, V., Haakensen, M., and Ziola, B. (2010). Rapid-screening for gram-negative and gram-positive beer spoilage *Firmicutes* using a real-time multiplex PCR. J. Am. Soc. Brew. Chem. 68, 89–95. http://dx.doi.org/10.1094/ASBCJ-2010-0308-02

Pittet, V., Phister, T.G., and Ziola, B. (2013). Transcriptome sequence and plasmid copy number analysis of the brewery isolate *Pediococcus claussenii* ATCC BAA-344^T during growth in beer. PLoS ONE 8, e73627. http://dx.doi.org/10.1371/journal.pone.0073627

Priest, F.G., Cowbourne, M.A., and Hough, J.S. (1974). Wort Enterobacteria – a review. I. Inst. Brew. *80*, 342–356. http://dx.doi.org/10.1002/j.2050-0416.1974.tb03629.x

Priest, F.G., and Hough, J.S. (1974). The influence of *Hafnia protea (Obesumbacterium proteus)* on beer flavor. J. Inst. Brew. *80*, 370–376. http://dx.doi.org/10.1002/j.2050-0416.1974.tb03631.x

Priest, F.G., Somerville, H.J., Cole, J.A., and Hough, J.S. (1973). The taxonomic position of *Obesumbacterium proteus*, a common brewery contaminant. J. Gen. Microbiol. *75*, 295–307. http://dx.doi.org/10.1099/00221287-75-2-295

Řezanka, T., Matoulková, D., Benada, O., and Sigler, K. (2015). Lipidomics as an important key for the identification of beer spoilage bacteria. Lett. Appl. Microbiol. *60*, 536–543. http://dx.doi.org/10.1111/lam.12415

Richards, M., and Corbey, D.A. (1974). Isolation of *Zymomonas mobilis* from primed beer. I. Inst. Brew. 80, 241–244. http://dx.doi.org/10.1002/j.2050-0416.1974.tb03612.x

Rogers, P.L., Jeon, Y.J., Lee, K.J., and Lawford, H.G. (2007). *Zymomonas mobilis* for fuel ethanol and higher value products. Adv. Biochem. Eng. Biotechnol. *108*, 263–288. http://dx.doi.org/10.1007/10_2007_060

Sakamoto, K., Funahashi, W., Yamashita, H., and Eto, M. (1997). A reliable method for detection and identification of beer spoilage bacteria with internal positive control PCR (IPC-PCR). Proc. Eur. Brew. Conv. Congress 26, Maastricht (New York, USA: Oxford University Press), pp. 631–638.

Saimmai, A., Rukadee, O., Onlamool, T., Sobhon, V., and Maneerat, S. (2012). Characterization and phylogenetic analysis of microbial surface active compound-producing bacteria. Appl. Biochem. Biotechnol. *168*, 1003–1018. http://dx.doi.org/10.1007/s12010-012-9836-z

Satokari, R., Juvonen, R., Mallison, K., von Wright, A., and Haikara., A. (1998). Detection of beer spoilage bacteria *Megasphaera* and *Pectinatus* by polymerase chain reaction and colorimetric microplate hybridization. Int. J. Food Microbiol. 45, 119–127. http://dx.doi.org/10.1016/S0168-1605(98)00154-8

Schleifer, K.H., Leuteritz, M., Weiss, N., Ludwig, W., Kirchof, G., and Seidel-Rüfer, H. (1990). Taxonomic study of anaerobic, gram-negative, rod-shaped bacteria from breweries: emended description of *Pectinatus cerevisiiphilus*, and description of *Pectinatus frisingensis* sp. nov., *Selanomonas lacticifix* sp. nov. *Zymophilus raffinosivorans* gen. nov., sp. nov., and *Zymophilus paucivorans* sp. nov. Int. J. Syst. Bacteriol. *40*, 19–27. http://dx.doi.org/10.1099/00207713-40-1-19

Seidel-Rüfer, H. (1990). *Pectinatus* and other morphologically similar Gram-negative anaerobic bacilli from brewing. Monatssch. Brauwiss. *43*, 101–105.

Spitaels, F., Wieme, A., Balzarini, T., Cleenwerck, I., Van Landschoot, A., De Vuyst, L., and Vandamme, P. (2013). *Gluconobacter cerevisiae* sp. nov., isolated from the brewery environment. Int. J. Syst. Evol. Microbiol. 64, 1134–1141. http://dx.doi.org/10.1099/ijs.0.059311-0

Spitaels, F., Wieme, A.D., Janssens, M., Aerts, M., Daniel, H.M., Van Landschoot, A., De Vuyst, L., and Vandamme, P. (2014). The microbial diversity of traditional spontaneously fermented lambic beer. PLoS ONE 9, e95384. http://dx.doi.org/10.1371/journal.pone.0095384

Spitaels, F., Wieme, A.D., Janssens, M., Aerts, M., Van Landschoot, A., De Vuyst, L., and Vandamme, P. (2015). The microbial diversity of an industrially produced lambic beer shares members of a traditionally produced one and reveals a core microbiota for lambic beer fermentation. Food Microbiol. *49*, 23–32. http://dx.doi.org/10.1016/j.fm.2015.01.008

Stackebrandt, E., Pohla, H., Kroppenstedt, R., Hippe, H., and Woese, C.R. (1985). 16S rRNA analysis of *Sporomusa, Selenomonas,* and *Megasphaera:* on the phylogenetic origin of gram-positive eubacteria. Arch. Microbiol. 143, 270–276. http://dx.doi.org/10.1007/BF00411249

Suihko, M.-L., and Haikara, A. (2001). Characterization of *Pectinatus* and *Megasphaera* strains by automated ribotyping. J. Inst. Brew. 107, 175–184. http://dx.doi.org/10.1002/j.2050-0416.2001.tb00089.x

Suzuki, K. (2011). 125th anniversary review: microbiological instability of beer caused by spoilage bacteria. J. Inst. Brew. 117, 131–155. http://dx.doi.org/10.1002/j.2050-0416.2011.tb00454.x

Tholozan, J.L., Membré, J.M., and Grivet, J.P. (1997). Physiology and development of *Pectinatus cerevisiiphilus* and *Pectinatus frisingensis*, two strict anaerobic beer spoilage bacteria. Int. J. Food Microbiol. *35*, 29–39. http://dx.doi.org/10.1016/S0168-1605(96)01206-8

Tholozan, J.L., Membre, J.M., and Kubaczka, M. (1996). Effects of culture conditions on *Pectinatus cerevisiiphilus* and *Pectinatus frisingensis* metabolism: a physiological and statistical approach. J. Appl. Bacteriol. *80*, 418–424. http://dx.doi.org/10.1111/j.1365-2672.1996.tb03237.x

Timke, M., Wang-Lieu, N.Q., Altendorf, K., and Lipski, A. (2005a). Community structure and diversity of biofilms from a beer bottling plant as revealed using 16S rRNA gene clone libraries. Appl. Environ. Microbiol. *71*, 6446–6452. http://dx.doi.org/10.1128/AEM.71.10.6446-6452.2005

Timke, M., Wang-Lieu, N.Q., Altendorf, K., and Lipski, A. (2005b). Fatty acid analysis and spoilage potential of biofilms from two breweries. J. Appl. Microbiol. *99*, 1108–1122. http://dx.doi.org/10.1111/j.1365-2672.2005.02714.x

Ueki, A., Watanabe, M., Ohtaki, Y., Kaku, N., and Ueki, K. (2014). Description of *Propionispira arcuata* sp. nov., isolated from a methanogenic reactor of cattle waste, reclassification of *Zymophilus raffinosivorans* and *Zymophilus paucivorans* as *Propionispira raffinosivorans* comb. nov. and *Propionispira paucivorans* comb. nov. and emended description of the genus *Propionispira*. Int. J. Syst. Evol. Microbiol. *64*, 3571–3577. http://dx.doi.org/10.1099/ijs.0.063875-0

Van Oeleven, D., Spaepen, M., Timmermans, P., and Verachtert, H. (1977). Microbiological aspects of spontaneous wort fermentation in the production of lambic and gueuze. J. Inst. Brew. 83, 356–360. http://dx.doi.org/10.1002/j.2050-0416.1977.tb03825.x

Van Vuuren, H.J.J., Cosser, K., and Prior, B.A. (1980). The influence of *Enterobaccter agglomerans* on beer flavor. J. Inst. Brew. *86*, 31–33. http://dx.doi.org/10.1002/j.2050-0416.1980.tb03952.x

Van Vuuren, H.J.J., Kersters, K., De Ley, J., Toerien, D.F., and Meisel, R. (1978). *Enterobacter agglomerans* – a new bacterial contaminant isolated from lager beer breweries. J. Inst. Brew. 84, 315–317. http://dx.doi.org/10.1002/j.2050-0416.1978.tb03898.x

Vávrová, A., Matoulková, D., Balázová, T., and Sedo, O. (2014). MALDI-TOF MS analysis of anaerobic bacteria isolated from biofilm-covered surfaces in brewery bottling halls. J. Am. Soc. Brew. Chem. *72*, 95–101. http://dx.doi.org/10.1094/ASBCJ-2014-0324-01

Vinogradov, E., Petersen, B.O., Sadovskaya, I., Jabbouri, S., Duus, J.Ø., and Helander, I.M. (2003). Structure of the exceptionally large nonrepetitive carbohydrate backbone of the lipopolysaccharide of *Pectinatus frisingensis* strain VTT E-82164. Eur. J. Biochem. *270*, 3036–3046. http://dx.doi.org/10.1046/j.1432-1033.2003.03682.x

Weber, D.G., Sahm, K., Polen, T., Wendisch, V.F., and Antranikian, G. (2008). Oligonucleotide microarrays for the detection and identification of viable beer spoilage bacteria. J. Appl. Microbiol. *105*, 951–962. http://dx.doi.org/10.1111/j.1365-2672.2008.03799.x

Weiss, N., Seidel, H., and Back, W. (1979). Isolation and systematic classification of beer spoilage, Gram-negative bacteria. Part I. Gram-negative strictly anaerobic cocci. Monatssch. Brauwiss. 32, 189–194.

Zhang, W.W., Fang, M.X., Tan, H.Q., Zhang, X.Q., Wu, M., and Zhu, X.F. (2012). *Pectinatus brassicae* sp. nov., a Gram-negative, anaerobic bacterium isolated from salty wastewater. Int. J. Syst. Evol. Microbiol. *62*, 2145–2149. http://dx.doi.org/10.1099/ijs.0.032144-0

Ziola, B. (2016). Monoclonal antibodies binding to lipopolysaccharide from the beer spoilage bacterium *Megasphaera cerevisiae* exhibit pan-reactivity with the strictly anaerobic Gram-negative brewing-related bacteria. J. Am. Soc. Brew. Chem. *74*, 267–271. http://dx.doi.org/10.1094/ASBCJ-2016-4895-01

Ziola, B., Gares, S.L., Lorrain, B., Gee, L., Ingledew, W.M., and Lee, S.Y. (1999). Epitope mapping of monoclonal antibodies specific for the directly cross-linked *meso*-diaminopimelic acid peptidoglycan found in the anaerobic beer spoilage bacterium *Pectinatus cerevisiiphilus*. Can. J. Microbiol. 45, 779–785. http://dx.doi.org/10.1139/w99-071

Ziola, B., Gee, L., Berg, N.N., and Lee, S.Y. (2000). Serogroups of the beer spoilage bacterium *Megasphaera cerevisiae* correlate with the molecular weight of the major EDTA-extractable surface protein. Can. J. Microbiol. *46*, 95–100. http://dx.doi.org/10.1139/w99-123

Beer-spoiling Yeasts: Genomics, Detection, and Control

11

Chris D. Powell[1]* and Daniel W. M. Kerruish[2]

[1]School of Biosciences, University of Nottingham, Nottingham, UK.
[2]Diageo PLC, Science and Technology, St James Gate Brewery, Dublin, Ireland.

*Correspondence: chris.powell@nottingham.ac.uk

https://doi.org/10.21775/9781910190616.11

Abstract

Beer-spoiling yeasts comprise a diverse group of organisms that can have a variety of impacts on beer production. Invariably, contamination of wort or beer by these yeasts leads to inconsistencies within the process, and quality defects in packaged beer. Beer-spoiling yeasts can be broadly separated into non-fermentative (aerobic) and fermentative yeasts. The former typically exploit process steps associated with raw materials, and areas where oxygen ingress is difficult to prevent, such as unpasteurized cask beers or dispense. Fermentative yeasts are arguably more problematic due to their capacity to compete with production strains during fermentation. Major impacts include altered sugar utilization, flocculation and ethanol production, as well as the formation of phenolic compounds, acidity, estery off-flavours, and haze or turbidity. These effects occur primarily due to differences in the genetic, metabolic and physiological characteristics of the spoilage yeast and the production strain. In this chapter, we describe the characteristics and functionality of beer-spoiling yeasts, as well as methods for their isolation and identification.

Introduction

Yeasts are a group of organisms comprising unicellular fungi that are capable of dividing asexually by budding or fission, or sexually by the process of sporulation. Industrial or 'domesticated' yeasts typically belong to the *Saccharomyces* genus, and within the brewing industry two main species are encountered: *S. pastorianus* and *S. cerevisiae* (see Chapter 4). These organisms divide asexually and tend not to sporulate, due to their complex genetic make-up and unconventional ploidy. However, other yeasts can be found within brewing processes that fit the description above more completely. Such species are typically referred to as 'beer-spoiling' or 'wild' yeasts. Beer-spoiling yeasts are those that have a negative impact on the sensorial qualities of the final product, either by affecting process stages or by changing the character of the beer directly. Within the brewing industry, this definition often incorporates 'wild yeast', which is a generic term used historically to define any type of yeast not deliberately introduced into the brewery environment. It should be noted that, at the present time, and within certain sectors of the industry, there has been a growing interest in the potential applications of strains that do not belong to the traditional brewing yeast *Saccharomyces* genus (see Chapter 7). This is primarily due to the novel characteristics that non-*Saccharomyces* yeasts impart to the final product and has resulted in the phrase 'wild yeast' (isolated from 'the wild') taking on a slightly alternative meaning. Within this Chapter we use the term 'beer-spoiling yeast' to indicate unwanted organisms present within the brewing process or beer.

Beer-spoiling yeasts constitute a broad group of organisms that, although phylogenetically diverse, inevitably comprise individuals that share key physiological properties, which allow them to exploit sections of the brewing chain. Although most beer-spoiling yeast are isolated either directly or indirectly from areas of the brewery associated with fermentation, others may be found as contaminants of raw materials through to final pack (Table 11.1). Irrespective of their preferred environment, for simplicity beer-spoiling yeast are often characterized as *Saccharomyces* or non-*Saccharomyces* types. Of these, the *Saccharomyces* spoilage yeasts are often regarded as being the most hazardous since they are difficult to detect, and compete directly with the culture strain. However, there is also a range of non-*Saccharomyces* species that are associated with beer spoilage and these include members of the genera *Brettanomyces, Candida, Debaryomyces, Hanseniaspora* (*Kloeckera*), *Kluyveromyces, Pichia, Torulaspora,* and *Zygosaccharomyces.* There are inevitably incidences of contamination with other types of yeast, but species and strains belonging to these genera tend to predominate.

Taxonomy of beer-spoiling yeast species

As described by Sampaio *et al.* (see Chapter 4) and below (see 'Molecular methods for identification of beer-spoiling yeast species'), yeast species are increasingly classified according to DNA sequence homology. However, fungi were historically organized and named based primarily on structures associated with sexual reproduction (i.e. spore morphology), along with physiological characteristics, including cell size and shape, and nutritional requirements. While much of this information remains useful today, it has inevitably led to confusion in naming and identifying species, since the predominant mode of replication for many yeasts associated with industrial processes is asexual (i.e. non-spore forming). Furthermore, those species that do undergo sexual reproduction only do so under certain conditions and only one method of reproduction is typically observed at a specific point in time. Hence, in practice, many fungi have multiple names based on their sexual state when they were first isolated, a phenomenon particularly relevant to beer-spoiling yeasts. For example the genus names *Brettanomyces* and *Dekkera* refer to the same organism; *Brettanomyces* refers to the asexual (anamorph) form, while *Dekkera* is the name used to describe the sexual form (teleomorph). Current consensus within the world of yeast taxonomy is that fungal species should be referred to by a single name and that teleomorphic names cannot have priority over anamorphic ones (McNeill *et al.*, 2012; Gams, 2016). However, it is also acknowledged that the designated genus name will often be guided by common usage. Hence, within the brewing industry it is still acceptable (if not desirable) that certain nomenclatures will continue to be used. It is anticipated that this rule should lead to greater clarity in the future. Previously there has been a degree of confusion within the brewing industry, especially with regard to spoilage yeasts that were originally classified as separate species, or those that have undergone a number of re-classifications. To illustrate this point, *Pichia anomola* (previously known as *Hansenula anomola*) is a teleomorph of *Candida peliculosa;* each form has different characteristics and exerts potentially different impacts on the product and process. Other yeasts have been described as having a 'tortuous history' due to nomenclature confusion caused by repeated reclassification. One particular example is the yeast *Candida utilis,* a

Table 11.1 Typical points of contamination for beer-spoiling yeasts

Process stage	Species
Raw materials	*Candida, Cryptococcus, Debaryomyces, Pichia, Rhodotorula*
Pitching yeast	*Saccharomyces*, non-*Saccharomyces*
High sugars	*Zygosaccharomyces, Kluyveromyces*
Aerobic stages of fermentation	*Candida, Hanseniaspora* (*Kloeckera*), *Pichia*
Fermentation	*Saccharomyces*
Bottle conditioning	*Brettanomyces*
Draft beer	*Candida, Brettanomyces, Torulaspora, Hanseniaspora* (*Kloeckera*), *Pichia*
Various locations (non-spoilers)	*Rhodotorula, Cryptococcus*

potential beer-spoiler and an organism used in feed production, which has been known by five different names: *Torula utilis, Torulaspora utilis, Hansenula jadinii, Pichia jadinii* and *C. utilis* (Barnett, 2004). Within this chapter, the anamorphic names of species are utilized; where common usage dictates that the teleomorphic name is applied, the corresponding anamorphic term is added in parentheses.

The physiology of beer-spoiling yeast

Vegetative growth, cell structure and sexual division

When a yeast culture encounters favourable conditions, it is advantageous for the population to divide vegetatively, or asexually. This allows new individuals to be generated quickly through mitosis, producing cells that are theoretically identical barring random mutation events. The majority of yeasts found within brewery locations reproduce predominantly via this mode of replication. In most instances, vegetative growth occurs through budding, whereby a new cell is produced via localized expansion and extrusion of the mother cell wall. The precise mechanism can vary, with different yeasts employing multilateral or bipolar divisional patterns. It should be noted that other types of yeasts (including those not typically found in breweries) may divide vegetatively via a number of mechanisms, including binary fission (notably *Schizosaccharomyces pombe*), bud fission, and via stalks or outgrowths. In addition to asexual reproduction, some beer-spoiling yeasts have the ability to reproduce sexually via karyogamy (fusion of cells to form a zygote) and subsequently meiosis. These events result in the formation of spores contained within an ascus (essentially a sac derived from the original fusion of the two parental cells). Sexual reproduction is relatively common in many beer-spoiling yeasts and can be induced by sudden changes in environmental conditions or nutrient deficiency (starvation).

All yeasts exhibit characteristic cellular structures: a rigid outer wall and a fluid cell membrane envelope the cytoplasm, nucleus, mitochondria, Golgi apparatus, endoplasmic reticulum, vacuoles, and a variety of vesicles and microbodies. However, despite being virtually identical in terms of cellular constituents, yeasts as a group of organisms are morphologically diverse. A culture of brewing yeast typically comprises a population of uniform cells that are spherical or slightly ellipsoidal in shape, and between 6 and 10 µm in diameter (Fig. 11.1). In contrast, beer-spoiling yeast (as well as other non-domesticated species isolated from the wild) can show a wide variety of cell shapes and sizes (Fig. 11.2). Broadly speaking, vegetative cells belonging to different yeast species can be described as being spherical, ellipsoidal, ovoid (egg-shaped), apiculate (lemon-shaped), pointed, rectangle-like, bottle (or flask)-shaped, or elongated. Yeast cell morphology is influenced by a number of factors, often linked closely to budding or budding patterns. In general, cells that are oval usually exhibit either an axial (bud production occurs adjacent to previous site of division) or a bipolar (cells bud at either polar end of the cell) budding pattern, while cells that are elongated tend to produce buds in an almost exclusively bipolar fashion (Chant and Pringle, 1995). This relationship is true for most strains belonging to the species *Hanseniaspora* (*Kloeckera*), for example, which produce lemon-shaped cells and display a polar divisional pattern. In contract to brewing yeast strains, which show a high degree of morphological homogeneity, non-domesticated yeasts can be polymorphic; cells within a population are often visually diverse such that they may at first appear to belong to different species. For example, *Brettanomyces* populations often comprise individuals that are ellipsoidal, elongated, or even rectangular. Similarly, some *Hanseniaspora* species can show extensive polymorphisms with notable variations in size as well as shape. Differences in morphology within populations are further exacerbated by the pleomorphic nature of yeasts (Fig. 11.3). As described above, under certain conditions yeasts have the ability to form mating aggregates, which lead to the production of sexual spores. Much of our understanding of the process comes from analysis of *Saccharomyces* strains, where mating is triggered by specific pheromones that bind to receptor sites in two opposite 'mating types'. In *S. cerevisiae*, these are known as 'a' and 'α', while in *Sch. pombe* they are referred to as h+ and h– In other types of yeasts, these nomenclature may be used, although often they are referred to as simply plus (P) or minus (M). In many yeast species, cells may also switch mating types, allowing

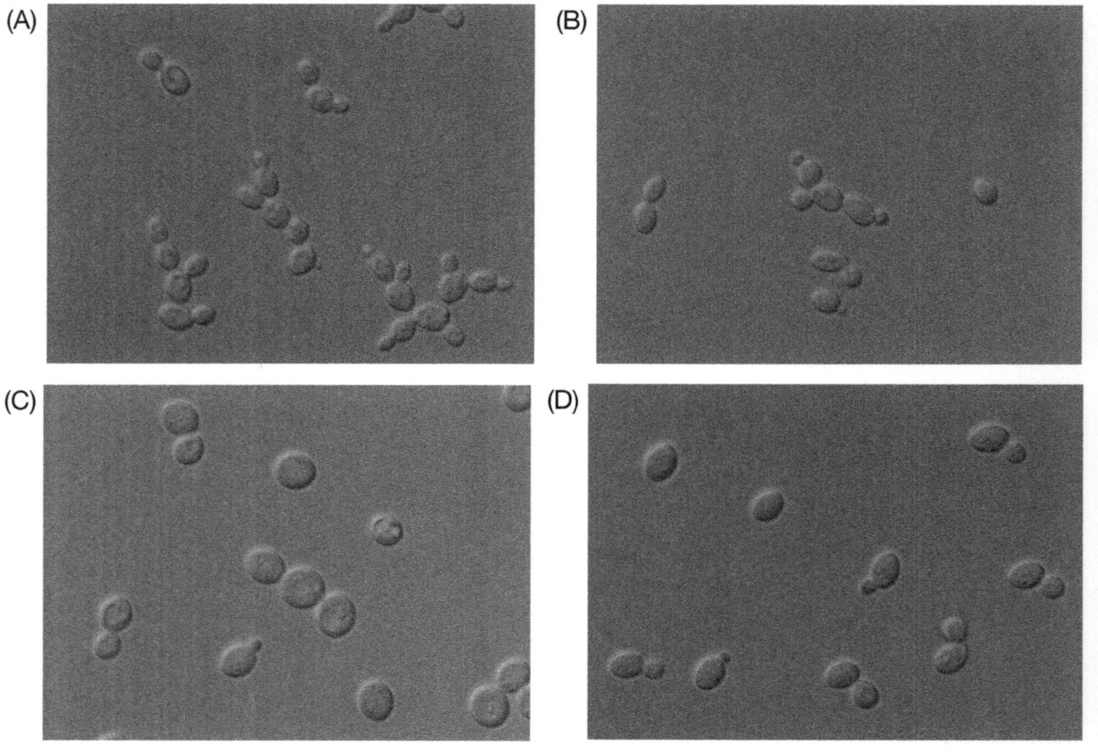

Figure 11.1 Types of cell morphology associated with *Saccharomyces* yeast species at ×400 magnification. All species typically show a similar uniform ovoid/spherical morphology. Note that the phenolic (A) and diastatic (B) wild *S. cerevisiae* strains exhibit a smaller cell size when compared to the lager and ale brewing yeast strains shown in C and D, respectively. Non-brewing strains and some ale-type strains may form chains (A), where individual cells are cytoplasmically discrete, but remain connected due to incomplete separation of mother and daughter at the cell wall.

colonies containing both mating types to develop and ensuring that there are always 'partners' available. Irrespective, the pheromones produced induce mating cells to develop a bottleneck-like projection that extends in the direction of a cell of the opposite mating type. This structure is known as a shmoo and allows for the eventual exchange of genetic material, culminating in spore formation. Typically 1–4 spores are formed within the ascus, the shape of which is highly variable between species. *Saccharomyces* yeasts tend to produce spores that are held in a tetrahedral formation, while other species have been described as having ellipsoidal, spherical, elongated, hat-shaped, Saturn-shaped, or kidney-shaped spores (Fig. 11.4). It should be noted that, in contrast to bacterial spores, yeast spores are not particularly stress tolerant; ascospores are only slightly more resistant to environmental challenges than vegetative cells.

Many yeast species are also able to adopt a variety of non-sexual structures, including chain formation and pseudohyphal (filamentous) growth. Although budding yeasts do not undergo true hyphal growth, they can exhibit a phenomenon in which cells fully separate by cytokinesis during division, but remain attached to each other due to the presence of specific proteins located in the cell wall. This is termed pseudohyphal growth and is closely linked to nutritional limitation, especially nitrogen deficiency (Gimeno *et al.*, 1992; Kron *et al.*, 1994). The majority of information on how this occurs is based on studies of *S. cerevisiae*, *Sch. pombe*, and *Candida* strains. In *S. cerevisiae*, it is known that immediately prior to the initiation of filamentous growth, a protein known as Ras2p (localized within the cell membrane) is activated, which in turn stimulates the synthesis of cAMP, an intracellular signalling molecule. This results in the activation

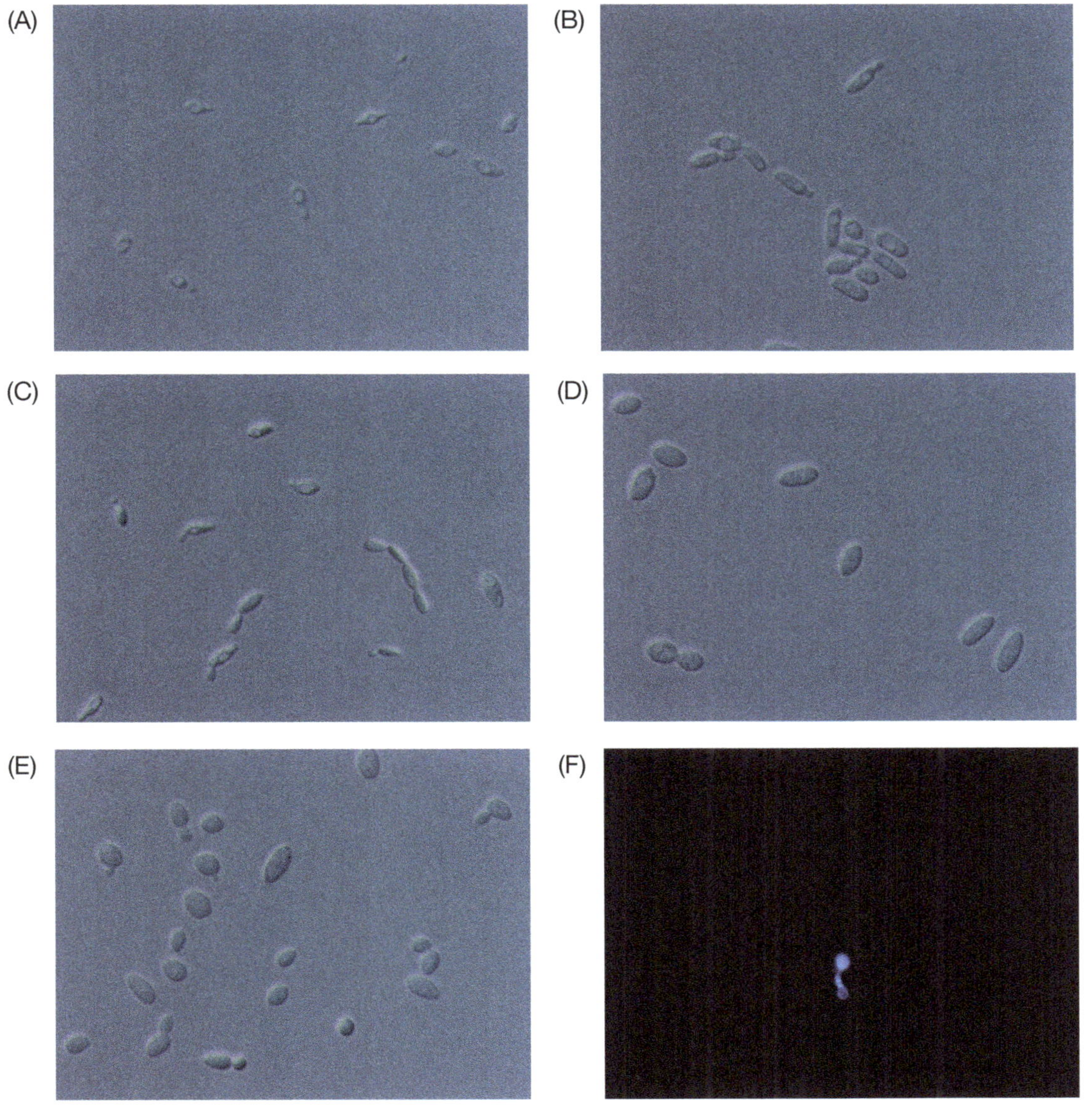

Figure 11.2 Types of cell morphology associated with non-*Saccharomyces* yeast species at ×400 magnification. Apiculate cells typical of *Hanseniaspora valbyensis* (A), rod-shaped cells of *Pichia membranifaciens* (B), elongated cells such as those associated with *Brettanomyces anomalus* (C) and ellipsoidal cells as seen in *Zygosaccharomyces bailii* (D). It should be noted that these images serve as examples and are not representative of all strains within each species. There is considerable variation within species and often individual strains will show several different morphological types, as seen in *Pichia anomola* (E). This can be compounded by the presence of sexual spores, formation of mating aggregates (shmoos), and the formation of pseudohyphae (F). For the latter, staining was conducted using calcofluor white to reveal the location of chitin deposits within the cell wall, clearing indicating two discrete cells.

of protein kinase A (PKA), which triggers a range of key transcription factors, including products of the *STE* gene family and Flo8p. These regulate the expression of a huge number of genes contributing to pseudohyphal growth (Jin *et al.*, 2008), the most well characterized being *FLO11*. The production of pseudohyphae is distinct from invasive hyphal growth observed in other fungi, but the appearance can be similar with an increase in cell length and enhanced cell–cell adhesion. Pseudohyphal

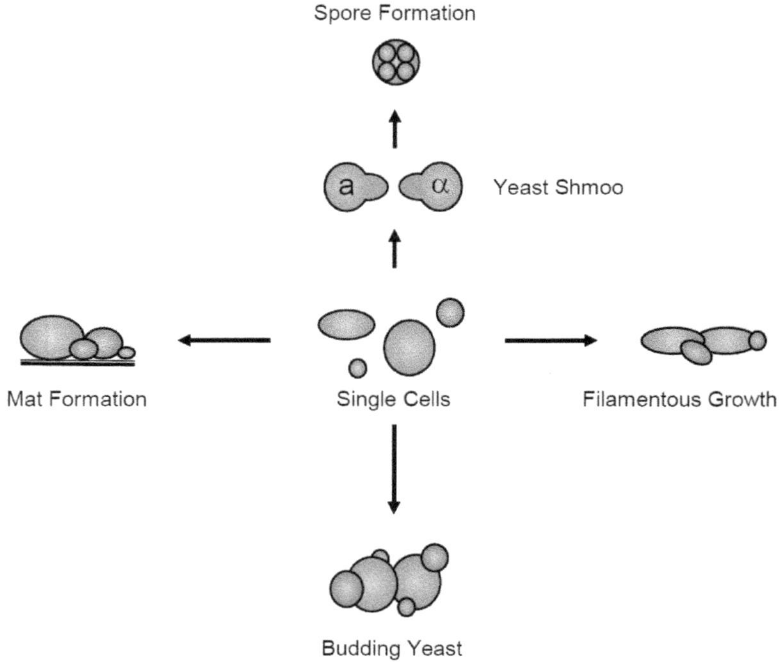

Figure 11.3 Forms of yeast growth. Yeast typically divide asexually to produce discrete cells, but failure to separate through budding can lead to chain formation. Cells can also undergo filamentous (pseudohyphal) growth, form biofilms through mat formation, and undergo 'shmoo' formation as a means of exchanging DNA during sexual division, which culminates in spore formation.

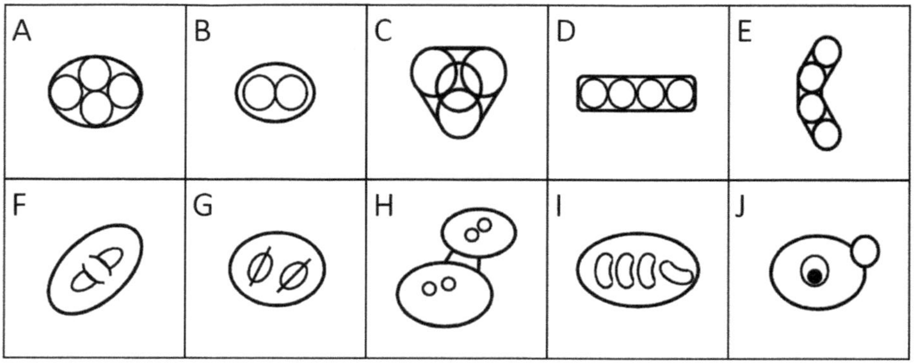

Figure 11.4 Spore formation in yeasts as observed under a light microscope at approximately ×400 magnification. Spore morphology is highly variable and dependent on genus and species. Typically between 1 and 4 globose spores will be formed, enclosed within an ellipsoidal ascus (A and B). *Saccharomyces* spores can be tetrahedral shaped (C) due to the tight rigidity of the ascus, although some may contain just 2 or 3 spores. Some yeasts such as the fission yeast *Schizosaccharomyces pombe* form linear asci (D). *Pichia* species are diverse, showing elongated asci (E), as well as producing spores that can be 'hat-shaped' (helmet-shaped) (F) or Saturn-shaped (G). The latter can also be observed in *Lindnera saturnus*, the genus originally designated for yeast forming such spores, while *Hanseniaspora* (*Kloeckera*) and *Dekkera* (*Brettanomyces*) form hat-shaped spores. In *Zygosaccharomyces* cells, a conjugation tube links mating cells together, forming a characteristic dumbbell-shaped zygote (H), while *Kluyveromyces marxianus* forms kidney- or bean-shaped spores (I). It is common for some species, such as *Debaryomyces*, to produce lipid bodies as a component of the spore (J), which can be clearly seen under a light microscope. Finally, it should be noted that ascospores can be smooth, warty, or ridged, although this level of detail is not always obvious without the use of a high-powered microscope.

cells are usually elongated (sometimes 20–50μm in length) or ellipsoidal and have constrictions at the septal junctions that connect adjacent cells (Fig. 11.2F). In some instances, complex aggregates of filamentous cells may be observed, including individuals in the process of developing pseudohyphae, which show a variety of intermediate morphologies. The formation of such structures is important from a beer spoilage perspective since elongated or pseudohyphal cells are often those that are able to form a pellicle and float on the surface of liquid media.

Pseudohyphae should not be confused with chain formation (Brown, 1970), which occurs in many yeasts, including some ale strains. In both instances, individual cells are cytoplasmically discrete, but remain connected due to incomplete separation of mother and daughter at the cell wall. However, in chain formation this typically occurs due to a deficiency in *CTS1* activity, which codes for chitinase (Kuranda and Robbins, 1991), responsible for digesting chitin scar tissue and allowing separation of a daughter cell from its mother. The impact of chain formation can be similar to pseudohyphae in that large clumps of cells can form; however, the significance is arguably of greater impact during fermentation, since chains can act as nucleation points for floc formation, resulting in changes to the flocculation properties of the culture yeast.

Some yeast species will also undergo mat formation in response to nutrient limitation, characterized by an increased ability to 'stick' to surfaces (Wood *et al.*, 1992; Reynolds *et al.*, 2008). This is also believed to be due to activation of the *FLO11* gene, which causes changes in the cell wall that enhance cell–cell adhesion. Interestingly, the mat itself is subject to nutrient and pH gradients, which is believed to impact on the activity of Flo11p rather than gene expression. Cells located at the 'rim' of the mat structure show decreased adherence properties as a result of higher pH, which may enable the colony to spread more easily (Reynolds, 2008). The ability of cells to stick together in this fashion can be desirable in production yeast strains that are used for bottle conditioning, but can be problematic in beer-spoiling yeasts since this can encourage the production and development of biofilms. Biofilms are aggregates of cells of one or more species (including in some circumstances both yeasts and bacteria together) that adhere to one another and form a community separated from the external environment by the development of a glycoprotein-polysaccharide layer known as a glycocalyx. This serves to protect organisms located within the biofilm structure, with the result that they will often display enhanced resistance to removal through conventional cleaning mechanisms. Biofilm formation in *Saccharomyces* yeast is known to be an intricate process with the activation of at least 71 genes (Andersen *et al.*, 2014), a phenomenon likely to be equally complex in other yeast species. Once formed, the biofilm is a relatively stable unit and will continue to grow in size until it becomes physically too large to remain attached, or is disrupted in some way, leading to a portion of the biofilm breaking away and cells being released. During biofilm formation, certain yeasts can act as primary colonisers, which then provide a mechanism for the attachment and inclusion of other species, which may not have the capacity to stick to surfaces on their own. Within the brewery environment, this can lead to the development of complex and potentially serious biofilms. For example, it has been shown that non-*Saccharomyces* yeasts such as *Pichia anomola* can provide an initial surface attachment for the subsequent colonization by *Saccharomyces* strains, the presence of which then increases the spoilage potential of the biofilm (Timke *et al.*, 2008).

Cell size and colour in beer-spoiling yeasts

While sexual state, cell shape, and the development of aggregates can often clearly differentiate between a production strain and a contaminant, there are other differences that may also be evident. Although beer-spoiling yeasts can be similar in size to industrial strains, typically they are smaller, with a diameter of 2–6 μm. The reason for the difference in size between domesticated and non-domesticated yeasts is unknown, although it may be an artefact of the greater ploidy of industrial strains, which tends to lead to an increase in cell volume (Müller, 1971). There can also be considerable variation in size within a population, which is often immediately evident in yeasts showing an elongated morphology, or in those strains that produce pseudohyphae. This intra-population variability is difficult to rationalize, as cell size is typically related to cell cycle regulation. In particular, there is a requirement to achieve a critical size prior to passing through 'Start' within the cell cycle, a checkpoint that commits the cell to division (Hartwell, 1974; Johnston *et al.*, 1977). This results in haploid cells (and the majority of industrial strains) typically having a similar size dispersion at the point of division. From a functional perspective this make sense; many cellular functions are dependent on cell size, and changes to cell volume can have a major impact on nutrient uptake, metabolic flux, and the capacity to biosynthesise important cellular compounds. Furthermore, the basic machinery of cell transport and division in eukaryotes relies on the cytoskeletal network, which plays an important role in forming and positioning the mitotic spindle (Marshall *et al.*, 2012). Microtubules have a limited range of lengths, and if cell structure and size become changed or abnormal, then the mitotic apparatus may have difficulty working, resulting in a reduced 'fitness' within the population. Consequently, the reason for size discrepancy within species, strains, and populations is unknown, although it is possible that a less tightly regulated control mechanism in non-domesticated yeasts may allow for greater functional plasticity as opposed to in brewing strains, where growth conditions are controlled within tightly defined parameters.

The colour of yeast cells belonging to different species may also be variable. Although most types of beer-spoiling yeasts produce colonies that are cream or white (using standard media that does not influence colour directly), other yeasts can be a variety of colours such as pink, orange, or yellow. The reason why there is a divergence in colour is yet to be fully explored, although many organisms (including plants and animals) produce a variety of pigments that can act to absorb light, such as carotenoid compounds. This can serve to benefit the organism in a number of ways, including as a form of protection, as vitamin precursors, or as antioxidants (Mata-Gómez *et al.*, 2014). Although not a beer spoiler, evidence indicates that this is almost certainly the case for the black yeast *Hortaea werneckii*, in which melanin is the primary cause of the colouration observed (Kejžar *et al.*, 2013). Of the beer-spoiling yeasts, *Rhodotorula* exhibits a characteristic red colouration derived from a variety of carotenoids (Mrak *et al.*, 1949), although the precise benefits of this to the cell have yet to be ascertained.

Origins and impacts of beer-spoiling yeasts

The natural environment of yeasts is known to be diverse, with species isolated from a huge range of locations and ecosystems. Interestingly, the natural habitat of *S. cerevisiae* yeasts (encompassing many industrial, and beer-spoiling strains) has been the subject of much debate, although current consensus is that they are primarily associated with the bark and litter of trees, specifically oak (Sampaio and Gonçalves, 2008; see Chapter 4). It is unclear whether oak trees are genuinely the primary habitat for *S. cerevisiae* yeasts, or whether this association is due to biased sampling; the more we look for this species of yeast, the more we find. Indeed, although the majority of yeasts exist in habitats associated with plant material, detritus, and soils, they can also be found in water (salt and fresh), on animals and insects, and can be dispersed through air. Most beer-spoiling yeasts are likely to originate from these sources and become introduced to the process opportunistically. Unlike certain species of bacteria (e.g. *Pediococcus damnosus*), spoilage yeasts are typically not unique to industrial locations, but are associated with raw materials such as hops, priming sugars and adjunct syrups, and casks. It is accepted that specific materials are often not the direct cause of contamination but act as a source of entry into the brewery, leading to contamination across different stages of the brewing chain (Table 11.1). Consequently, brewing equipment, surfaces, water supplies, and pitching yeast can also be considered to be potential sources of infection and should attract particular attention from a hygiene perspective, especially if a repeat contamination is observed.

As mentioned above, beer-spoiling yeasts are often described as being *Saccharomyces* or non-*Saccharomyces*. This classification is based primarily on the fact that different methods of detection are generally employed for each group. However, there are also broad differences in terms of spoilage potential; the majority of non-*Saccharomyces* yeasts do not typically compete with production strains during fermentation and cannot establish themselves within the process. This is primarily because many non-*Saccharomyces* yeasts depend on oxygen for growth, and their influence is therefore limited due to tight oxygen control within the brewing process. In addition, non-brewing strains are generally inefficient at metabolizing maltose, the predominant sugar associated with wort, and consequently many beer-spoiling yeast strains are at a competitive disadvantage during brewing fermentations. Spoilage by these yeasts is therefore restricted to raw materials and the early stages of fermentation, where simple sugars and some oxygen may be present. Exceptions to this can be found in breweries that employ open fermentation vessels or foeders (barrel fermentations), where oxygen ingress may act to stimulate growth.

Ethanol tolerance and the ability to grow at low pH are central properties that enable yeasts to spoil fermenting wort and beer, and this is where the most serious impacts can be found. During fermentation, use of an inappropriate production strain or the presence of a killer yeast (see section 'Beer-spoiling yeasts and killer toxins') can have particularly negative effects. Typically, spoilage is through the production of inappropriate levels of esters, higher alcohols and vicinal diketone (VDK), which cause flavour imbalance during fermentation. However, certain species of yeasts can also produce specific off-flavours including organic acids, sulfur-containing compounds, and phenolics. An additional impact is related to the performance of the culture yeast; the majority of beer-spoiling yeast strains do not sediment in the same fashion as production strains, often displaying a weaker flocculation potential (see 'Beer-spoiling yeasts and flocculation'). Many non-production yeasts also do not interact with finings, since they do not exhibit a strong negative charge. The result of this is to create cloudy beer with associated off-flavour production due to cell lysis.

Packaged beer that has been filtered and pasteurized rarely undergoes spoilage by yeasts. Although strains show variability in heat tolerance, their ability to withstand temperatures associated with pasteurization are such that their survival is a relatively rare occurrence (Tsang and Ingledew, 1982). When spoilage does occur, this is generally due to carry-over of culture yeast or spoilage yeasts that are smaller and less subject to fining. Some yeasts can also be an issue in traditional cask-conditioned beers, giving rise to quality defects. Typically, the result of contamination post-fermentation is that residual sugars are utilized and yeasty or phenolic off-flavours are produced. Growth can also result in the formation of haze and sediments and in some instances the development of pellicles or surface films that act to ensure the proximity of yeast to headspace oxygen.

Anaerobic (fermentative) beer-spoiling yeasts

Yeast contaminants that exhibit fermentative properties represent a serious problem within the brewery since they can compete directly with production strains. Furthermore, the similarity of these yeasts to brewing strains can make them difficult to detect. The characteristics and effects of fermentative beer-spoiling yeasts on the product and process are variable. Yeast species belonging to the genera *Kluyveromyces*, *Saccharomyces*, *Torulaspora*, and *Zygosaccharomyces* can show an enhanced growth rate such that they have the capacity to displace the culture yeast over the course of serial re-pitching. Given that a brewing yeast slurry may be used from as few as 3–4 times to as many as >100 serial re-pitchings, it is clear that under certain circumstances this could lead to significant issues. With the exception of strains that exhibit killer activity (see section 'Beer-spoiling yeasts and killer toxins'), fermentative yeasts do not harm the production strain but may compete for resources, reduce ethanol yield, and generate off-flavours. Although major changes to the end product can indicate their presence, often much less apparent and more subtle defects are caused. Fermentative beer-spoiling yeast strains often exhibit differences in sugar preference and patterns of nutrient utilization that can lead to variations in the concentrations of esters, higher alcohols, and VDK. In addition, many spoilage yeasts do not flocculate well and do not interact with finings, with the result that they can pass into conditioning where they can have negative sensorial effects on post-fermentation beer, as well as causing haze and turbidity. A summary of the characteristics and spoilage potential of aerobic beer-spoiling yeasts can be found in Table 11.2.

Table 11.2 Characteristics of typical anaerobic (fermentative) beer-spoiling yeasts

Genus	Common species	Characteristics	Beer spoilage potential	Additional information
Hanseniaspora (Kloeckera)	*H. uvarum (K. apiculata)* *H. valbyensis* *H. vineae*	Apiculate (lemon-shaped) yeast Fermentative yeast that prefers anaerobic conditions	Fermentation, turbidity, and off-flavours Not typically associated with beer, but may be present due to cross-contamination, for example from wine barrels if used for conditioning	Ascomycete Hanseniaspora is the teleomorph, producing spores *Kloeckera* is the anamorphic form. Can be found in wine production and in certain lambic beers
Kluyveromyces	*K. marxianus*	Various cell morphologies but typically ovoid/ellipsoidal Fermentative yeast	Fermentation, turbidity and off-flavours	Ascomycete Found as a contaminant throughout the food industry, particularly in soft drinks and dairy products
Saccharomyces	*S. bayanus* *S. cerevisiae* *S. pastorianus* *S. unisporus*	Spherical-shaped cells Includes non-production strains and variants *S. bayanus* cells may be elongated Fermentative yeasts	Fermentation, turbidity, and off-flavours Some strains have varying flocculation, which can interact with brewing yeast Some strains may be phenolic Diastatic yeasts can yield especially low attenuation and reduced mouthfeel	Ascomycete Cells often more irregular than brewing strains and size is typically smaller, which can result in haze/filtration issues Some non-brewing strains will produce tetrahedral-shaped spores
Schizosaccharomyces	*S. pombe*	Rod-shaped cells Fermentative yeast	Fermentation, turbidity, and off-flavours	Ascomycete Divides by fission and can be readily distinguished from budding yeasts
Torulaspora	*T. delbreuckii*	Spherical/ellipsoidal cells Fermentative yeast (some species are obligate fermenters) Grows poorly under anaerobic conditions	Fermentation, turbidity, and off-flavours Associated with pitching yeast Can also spoil unpasteurized beer	Ascomycete *Saccharomyces delbreuckii*, *Saccharomyces fermentati*, and *Saccharomyces rosei* are obsolete synonyms *T. delbreuckii* is a teleomorph of *Candida colliculosa* Highly osmotolerant organism
Zygosaccharomyces	*Z. bailii* *Z. bisporus* *Z. rouxii*	Oval-shaped cells Fermentative yeast High sugar tolerance	Fermentation, turbidity, and off-flavours	Ascomycete Stress-tolerant yeast found as a contaminant throughout the food industry Particularly associated with high-sugar products

Hanseniaspora (Kloeckera)

Hanseniaspora (anamorph: *Kloeckera*) species reproduce via bipolar budding, usually resulting in cells that display a characteristic apiculate shape. However, *Hanseniaspora* can produce a variety of cell morphologies, with pairs of cells being relatively common (with a skittle-like appearance), as well as elongated and bottle-shaped individuals. Short pseudohyphae structures can also be formed, leading to surface film formation, although this is typically not as strong as pellicles formed by other genera. Interestingly, *Hanseniaspora* species have been reported to be one of the predominant yeast genera found on grapes and can often be found during the early stages of natural wine fermentations (Heard and Fleet, 1985; Prakitchaiwattana *et al.*, 2004). There is some evidence to suggest that due to their association with fruits, certain *Hanseniaspora* species such as *H. uvarum* and *H. apiculata* can be carried by drosophila flies (Miller and Phaff, 1962), hence insect vectors are likely to form a major route into the brewery, as well as raw materials. Within the brewing process, *Hanseniaspora* spp. are mainly found during the aerobic stages of fermentation and are sometimes associated with draft beer (Wiles, 1950; Hemmons, 1954).

Kluyveromyces

Kluyveromyces cells are generally ovoid or ellipsoidal, although other morphologies can be observed, including elongated cells and the production of pseudohyphae. *Kluyveromyces* strains are homothallic and are therefore able to self-fertilize, producing heat-resistant ascospores. Yeast belonging to this genus are industrially significant and can have a positive impact in fermented milk products, natural wine fermentations and traditional African beers (Maoura *et al.*, 2005; Jolly *et al.*, 2014; Misihairabgwi *et al.*, 2015; Prado *et al.*, 2015). *Kluyveromyces* species such as *K. lactis* and *K. marxianus* (see also section on Candida, below) are also able to ferment a range of sugars that many other yeasts cannot, including lactose, inulin, and the pentose sugar xylose. Hence, these yeasts have been explored for the conversion of waste products (including whey) for bioethanol production. Isolates found within the brewery may be derived from malt and other raw materials. General spoilage effects include the production of turbidity and off-flavours as a result of vigorous fermentation. However, some strains contain killer plasmids (see section 'Beer-spoiling yeasts and killer toxins'), which can have a severely negative impact on a pitching yeast culture and hence overall fermentation progression (Stark and Boyd, 1986; Rodriguez-Cousino *et al.*, 2011).

Saccharomyces

The *Saccharomyces* genus is perhaps the most well described of all the yeasts, due to the widespread use of *S. cerevisiae* for industrial purposes, and the availability of many full genome sequences for analysis. Cells are typically globose or ellipsoidal, and vegetative reproduction is by multilateral budding. Most species are capable of forming pseudohyphae, although this rarely happens in *S. pastorianus* (lager) yeasts or in *S. cerevisiae* (ale) brewing strains. Within the genus, most species are able to replicate sexually, producing characteristic tetraploid asci (Fig. 11.4). However, brewing yeasts belonging to *S. cerevisiae* or *S. pastorianus* only do so rarely, due to their genetic complexity and hybrid status. As beer-spoiling yeasts, *Saccharomyces* strains can be found in various locations within the brewery, but are predominantly associated with the fermentation stages of the process.

From a brewing perspective, the term '*Saccharomyces* beer-spoiling yeast' invariably refers to *S. cerevisiae* strains. However, most *Saccharomyces* species are capable of contaminating wort, including strains belonging to *S. bayanus*, *S. kudriavzevii*, and *S. mikatae*. In general, *Saccharomyces* yeasts display properties that are relatively similar to brewing strains, and are therefore difficult to detect, while representing a very serious threat to the brewing process. It is also important to reiterate that the definition of beer-spoiling yeast includes production strains and variants that have not been directly introduced to the process by the brewer. Consequently, this can include the accidental mixing of different types of brewing yeasts (ale/lager) or the use of an incorrect strain within either category. The mixing of production strains used for ale- and lager-type products can be especially problematic due to their intrinsic differences in fermentation properties. This is apparent when considering their response to fermentation temperature (lager products are generally fermented at colder temperatures), the typical flavour profiles associated with these two styles of beer (ale strains produce more fruity notes), and their flocculation characteristics (lager strains are traditionally classified as bottom-fermenting, while lagers are top-fermenting). However, the precise effects of mixing equivalent 'types' of production strains are difficult to predict and are largely dependent on strain phenotype. The most likely results include variations in flavour, attenuation rate, flocculation and cropping patterns. However, in reality these may be relatively subtle changes, especially if the level of contamination is low, or if there is a high degree of strain similarity. In the latter instance, differences can sometimes be offset by blending of the product, although this is not desirable or recommended, especially for core brands or premium products.

A similar range of effects can be observed when mutants derived from production strains are used. These can be particularly problematic since they are by nature very difficult to detect, especially if there is a gradual change in the concentration of variants with successive fermentations. Arguably the two most commonly encountered types of mutants are flocculation variants (showing either decreased, or more commonly increased flocculation potential) and respiratory-deficient (RD) mutants with defective mitochondria. Changes to the flocculation characteristics of a culture can have consequences with regard to overall production consistency. For example, an unexpectedly early and heavy crop can result in stuck fermentations or can create issues with regard to mechanically removing the yeast crop. If mitochondrial deficient cells, otherwise known as 'petite' mutants (due to the production of small colonies on solid agar), are present in significant numbers, this tends to influence beer flavour production. As petite cells are slow to grow and divide, this can lead to a sluggish fermentation with an inappropriate balance of flavours. For example, the presence of petite cells is widely associated with unacceptable levels of diacetyl.

More conspicuous effects can arise when non-production *Saccharomyces* strains are encountered. These often display certain properties that are different and/or undesirable to those found in production yeasts (Jespersen *et al.*, 2000). The most noticeable are those that are capable of producing phenolic off-flavour (POF) compounds (see section 'Production of phenolic compounds') or that have diastatic activity. Most brewing strains are unable to utilize long-chain sugars (dextrins), often claimed to contribute to mouthfeel in beer. Some *S. cerevisiae* strains (previously classified separately as *S. diastaticus*) are amylolytic and possess the *STA* genes, responsible for glucoamylase production (Tamaki, 1978; Adam *et al.*, 2004). This allows the breakdown of dextrins, leading to super-attenuation of wort and resulting in a beer that has an unusually low final gravity and low residual extract. Diastatic yeasts can have more disastrous consequences when associated with unpasteurized bottled beer. The production of abnormally high concentrations of carbon dioxide can increase the risk of exploding bottles. Furthermore, many diastatic strains are also POF^+ and the use of residual dextrins for growth can therefore result in phenolic flavour production, as well as haze formation and other off-flavours.

Schizosaccharomyces

Schizosaccharomyces, along with *Saccharomyces* species, represent one of the most widely studied yeast genera. This is partly due to the ease at which cell structure and cell cycle events can be visualized; cells are rod-shaped and do not divide by budding, but through lateral fission via cross-wall formation, which causes individuals to take on a 'v'-like configuration during vegetative growth. Due to their distinctive budding pattern, *Schizosaccharomyces* yeasts look considerably different to other yeasts encountered within the brewery. Cells are typically uniform and although pseudohyphae may be present, pellicles are not generally formed. Similar to *Saccharomyces* yeasts, *Schizosaccharomyces* are primarily heterothallic (although mating type switching can occur), producing linear or dumbbell-shaped asci. By far the most common species, *Sch. pombe*, is employed in traditional African beers (*pombe* being the Swahili word for beer), some fermented tea products, and can be found naturally associated with soil and plants including barley. *Sch. pombe* is a Crabtree-positive yeast (see section 'Sugar uptake and metabolism of beer-spoiling yeasts') capable of fermentation and shows osmophilic properties and resistance to some chemical preservatives. Although it is not a widely reported beer-spoiling yeast, the presence of this organism can lead to off-flavour production and inconsistent fermentations.

Torulaspora

Torulaspora cells are spherical and homogeneous in structure, with less intra-population variation than seen with some other yeast genera. Cells are haploid during vegetative growth, but due to their homothallic nature, mating can occur between the nuclei of a mother and daughter cell. The resulting spores are formed in the husk of the two cellular structures, which remain closely attached, although sometimes a small conjugation tube may be observed. Genetically, *Torulaspora* species are closely related to *Zygosaccharomyces* and *Saccharomyces*; however, they are physiologically distinct and are particularly well adapted to tolerate osmotically challenging environments. Consequently, species such as *T. delbrueckii* can often be found in habitats that are associated with high sugar concentrations, including food products such as molasses, honey and, importantly, sugar-based syrups or adjuncts. This genus of yeast is capable of good growth under anaerobic conditions and is suited to conditions associated with fermentation, as well as contamination of raw materials such as priming sugars. The main impact of this organism is to cause turbidity and off-flavour production.

Zygosaccharomyces

Zygosaccharomyces yeasts form round or ellipsoidal cells and sporulate rarely, producing characteristic dumbbell-shaped asci due to spores being located in each of the two parental cell structures. Pseudohyphae may be formed, though these are typically short and do not lead to robust pellicle formation. *Zygosaccharomyces* species are arguably one of the most important spoilage yeasts within the food industry in general, with the potential to contaminate a wide range of food products. This is primarily because they are resistant to many traditional means of preserving foods, exemplified by their ability to withstand desiccation and resist high levels of ethanol and low pH (including tolerance to weak acids, such as acetic) (Stratford *et al.*, 2013). In addition, they are particularly osmotolerant (Dakal *et al.*, 2014) and can thrive under high sugar concentrations, potentially contaminating syrups and adjuncts. There are several species that can be found in brewery locations, the most common being *Z. bailii* and *Z. rouxii*. These organisms can cause issues with beer clarity and flavour as they ferment strongly, producing a range of higher alcohols as well as typical yeasty off-flavours.

Aerobic (non-fermentative) beer-spoiling yeasts

The use of the phrase 'non-fermentative' is slightly misleading, as many species of yeast within this category can ferment (albeit weakly) under certain conditions. However, others are unable to ferment complex sugars or only show limited replication capacity under anaerobic conditions; the fact that these yeasts require oxygen for robust growth suggests that the term 'aerobic' may be a more appropriate descriptor. Aerobic yeasts are more likely to be found within areas of the brewery that are not associated with the fermentation step, for example raw materials, surfaces and equipment, and in final pack. However, it is important to note that even aerobic species are often able to survive at low concentrations during fermentations and persist through serial re-pitching. Generally their presence during the fermentation stage is unlikely to significantly affect the final product since they will most likely remain below a threshold cell number or be out-competed by the production strain. Of the aerobic beer-spoiling yeasts, *Pichia* and *Candida* species are arguably the most important since they are relatively prevalent (Jespersen and Jakobsen, 1996); however, *Kluyveromyces, Torulaspora* and *Brettanomyces* can also act as opportunistic spoilers during the process, particularly during the aerobic phase of fermentation and in situations where oxygen ingress is difficult to prevent, such as in unpasteurized cask beers. A summary of the characteristics and spoilage potential of aerobic beer-spoiling yeast can be found in Table 11.3.

Brettanomyces (Dekkera)

Of the non-*Saccharomyces* beer-spoiling yeasts, *Brettanomyces* spp. have arguably the greatest potential impact on beer and have been the subject of detailed scientific investigation. This is partly because they are desirable for the production of certain beer types, including lambic, guez, and some saison-style products, but also because they are considered to be amongst the most dangerous spoilage microbes in alcoholic beverages, particularly in wine production (Schifferdecker *et al.*, 2014). *Brettanomyces* cells are typically sausage- or bullet-shaped, but overall morphology can be variable. They are able to produce pseudohyphae and can sporulate; as mentioned previously (see section 'Taxonomy of beer-spoiling yeast species'), the sexual forms of *Brettanomyces* were historically classified as belonging to the species *Dekkera*. Reclassification of many yeasts within the genus has resulted in five species: *B. bruxellensis, B. anomalus, B. custerianus, B. naardenensis*, and *B. nanus*. All of these have been isolated from beers, but arguably the first two are the most frequently reported. In the wild, *Brettanomyces* yeasts are associated with trees and the surfaces of fruit; it is thought that *Brettanomyces* strains may be introduced to breweries by insect vectors such as fruit flies (Christiaens *et al.*, 2014).

Yeasts belonging to this species are often described as 'survivalists' since they are ethanol tolerant and resistant to low pH which means that they are well adapted to withstand the fermentation process (Steensels *et al.*, 2012). As well as being recognized contaminants of beer and wine, they can also cause issues in the manufacture of

Table 11.3 Characteristics of typical aerobic (non-fermentative) beer-spoiling yeasts

Genus	Common species	Characteristics	Beer spoilage potential	Additional information
Brettanomyces (*Dekkera*)	*B. anomalus* (*D. anomola*) *B. bruxellensis* (*D. bruxellensis*) *B. lambicus*	Elongated cell structure can form short chains Fermentative yeast in presence of oxygen due to the Custers effect Cannot ferment sucrose and limited fermentation with maltose	Produces acetic acid and 4-ethyl phenol (horse blanket/barnyard character) Sometimes forms a pellicle Causes off-flavours especially if present in bottle-conditioned beers Can be found in unpasteurized draught beer	Ascomycete *Brettanomyces* is the anamorphic form *Dekkera* is the teleomorph, producing spores Sometimes used as the primary yeast or as a secondary culture in lambic style and 'Brett' beers, or in some traditional UK cask-fermented ales
Candida	*C. boidinii* *C. stellata* *C. tropicalis* *C. vini*	Spherical-shaped cells Fermentation of glucose and sometimes maltose Grows poorly under anaerobic conditions	Fermentation, turbidity, and off-flavours Can form films (pellicles) Spoilage often limited to aerobic production stages	Ascomycete Some teleomorphs of *Candida* are *Pichia* species Other species may be associated with the human microbiome
Debaryomyces	*D. hansenii*	Small spherical cells Weak or no fermentation	Turbidity and yeasty off-flavours Sometimes forms a pellicle or deposit Spoilage often limited to aerobic production stages	Ascomycete Osmotolerant yeast found as a contaminant throughout the food industry
Lindnera (*Williopsis*)	*L. saturnus*	Spherical-shaped cells Weak or no fermentation	Produces strong estery flavours Often associated with killer activity	Ascomycete Has been studied as a means of producing banana (isoamyl acetate) flavouring for the food industry
Pichia (*Hansenula*)	*P. anomola* *P. fermentans* *P. membranifaciens*	Typically ovoid/ ellipsoidal or rod-shaped cells Prefers aerobic conditions Fermentation weak or absent	Turbidity and yeasty off-flavours Can cause elevated ester production Sometimes forms a pellicle or deposit	Ascomycete *Hansenula* is an obsolete synonym of *Pichia* Pichia are teleomorphs and produce spores. Some *Candida* species are anamorphs of *Pichia* species
Rhodotorula	*R. glutinis* *R. mucilaginosa*	Typically ovoid/ ellipsoidal May produce pseudohyphae Fermentation absent	Can assimilate sugars, reducing fermentation efficiency Survives in pitching yeast but typically does not spoil beer	Basidiomycete Can act as a nitrate reducer, potentially contributing to apparent total N-nitroso compounds (ATNC) in beer

soft beverages and dairy products. Although *Brettanomyces* yeasts are facultative anaerobes, due to the Custers effect (see section, 'The Custers effect', below) they readily convert sugar into ethanol and acetic acid in the presence of oxygen. Furthermore, they are able to degrade and ferment complex sugars (dextrins) that are not readily utilized by the culture yeast. Consequently, they pose a significant threat to unpasteurized beers in which oxygen ingress may occur and spoilage of beer is characterized by the formation of particularly high concentrations of acetic acid. In addition, these yeasts exhibit an enhanced capacity for biomass generation, causing turbidity, while in the case of extreme infections a surface film may also be visible. *Brettanomyces* are also responsible for the production of volatile phenolic compounds (see section, 'Production of phenolic compounds', below) such as 4-ethylphenol (often described as barnyard, horsey or medicinal) and 4-ethylguaiacol (bacon, spice, cloves, smoky), and volatile fatty acids such as isovaleric acid (sweaty saddle, cheese, rancid). As can be imagined, many of these flavours are undesirable in the vast majority of beers produced worldwide. For a comprehensive review of positive characteristics associated with the use of *Brettanomyces* strains for beverage production, the reader is directed to Steensels *et al.*, 2015; see also Chapters 4, 6 and 7.

Candida

The *Candida* genus is often described as being 'oversized', since it comprises species with a wide spectrum of characteristics. Diversity within the genus is an artefact of the original classification of yeasts, during which many anamorphic species were placed into this group. Since that time the genus has undergone many changes, with some *Candida* yeasts being reclassified and others, such as *Torulopsis*, becoming incorporated. Despite this, many *Candida* species remain genetically and phenotypically close to those within other genera. Further ambiguity remains due to the fact that some species are anamorphic forms of other yeasts (including *Pichia* and some *Kluyveromyces* species).

Candida strains produce vegetative cells that are typically ovoid and small in size. Although they are often associated with human disease, some have a purely commensal relationship with humans and the majority are not pathogenic. Others, such as *C. kefyr* (*Kluyveromyces marxianus*), can be used to produce fermented milk products (e.g. kefir), while *C. tropicalis* is a major spoilage organism with respect to a variety of food types, including beer. *Candida* species found within the brewery can only grow aerobically and are not a major threat during active fermentation. However, they may cause issues during the initial stages of the fermentation process and, since cells are typically small in size, they may ultimately be slow to sediment and cause haze and filtration issues. *Candida* species can also oxidize ethanol to produce acetic acid in the presence of air, which can be particularly problematic in the draft beer trade.

Debaryomyces

Debaryomyces yeasts are spheroidal to ellipsoidal and reproduce vegetatively by lateral budding. Sexual reproduction involves the fusion of two cells, typically mother and daughter, and often via a short conjugation tube. *Debaryomyces* species are weakly fermentative organisms that exhibit poor growth in the absence of oxygen. However, *D. hansenii* is one of the most prevalent foodborne yeasts since it is tolerant to a variety of stresses including those associated with cold, salt, and changes in osmolality. Due to these characteristics, it is a common organism related to the production of surface-ripened cheeses and varieties of dried sausage. Within the brewing chain, this species can be found associated with raw materials, particularly malt. In addition, it is able to form a film on the surface of liquids due to hydrophobic cell wall components that cause cells to congregate at the liquid–gas interface, as well as having the potential to produce acetic acid and esters when associated directly with wort or beer.

Pichia

This genus is a large and diverse group of yeasts that represent one of the more common non-fermentative spoilage organisms found within the brewery. The genus has grown due to gene sequence analysis and now incorporates a number of species previously classified elsewhere, including all *Issatchenkia* species (Kurtzman *et al.*, 2008) and some *Candida* species. *Pichia* yeasts tend to reproduce via lateral budding, although pseudohyphal formation and sexual reproduction is relatively common. The spores produced are highly variable depending on the species and can be round, hat-shaped, Saturn-shaped, or elongated. *Pichia* spp. survive best under aerobic conditions and cells are typically found associated with draft beer, raw materials, or the early stages of fermentation. Although they are able to grow anaerobically, they can only ferment glucose, and cell function is impaired by alcohol and low pH. However, they are capable of surviving through fermentation and can represent a recurring issue within the brewery. *Pichia* species typically produce turbidity and can form films or pellicles on the surface of liquid media. In doing so, *Pichia* strains increase their opportunity to utilize oxygen that may be present in the head space above the beer. Generally, *Pichia* yeast also produce a range of estery off-flavours, particularly ethyl acetate, as well as yeasty notes, and are capable of oxidizing ethanol to acetic acid. There is a range of species that can be encountered, typically more than one of which may be present at a given time. This includes *P. membranifaciens*, *P. anomola*, and *P. fermentans*, although others such as *P. guilliermondii* and *P. kudriavzevii* (previously *C. krusei*) have also been reported.

Lindnera (previously *Williopsis*)

Lindnera species are ellipsoidal yeasts that grow primarily under aerobic conditions, although they may show limited fermentation giving rise to ester production. Yields are extremely high, such that *Lindnera* species have been explored for the manufacture of isoamyl acetate as a natural flavouring compound (Yilmaztekin *et al.*, 2008), and for the production of low alcohol but intensely fruity wines (Erten and Campbell, 2001). *Lindnera* yeasts are primarily associated with soil, and the most predominant species encountered in a brewing context is *L. saturnus* (previously *W. saturnus*), which has been isolated from early fermentation. This species produces characteristic Saturn-shaped spores, and is also closely associated with killer yeast activity (see 'Beer-spoiling yeasts and killer toxins', below). It should be noted that the species *Williopsis californica*, which has previously been isolated from barley in the field and during the germination stages of malting (Laitila *et al.*, 2006), is currently classified as *Barnettozyma californica* (Kurtzman *et al.*, 2008).

Other yeast contaminants

There are inevitably yeast species that have been found within breweries that have not been mentioned within this chapter. Furthermore, there may be species of which we are not yet aware, or incidences where current yeasts have been misidentified. Furthermore, there are also some types of yeast that can be considered to be 'non spoilers' in the sense that they do not significantly contribute to off-flavour production, aromas, or turbidity. Unless conditions are grossly atypical, these species are unable to survive in hopped wort or spoil beer; however, their presence indicates a poor level of hygiene at some point within the brewery chain. In such instances, it is important to treat contamination with the same degree of gravity as with other more 'harmful' yeast or bacterial species. Yeasts that fall under this category include members of the genera *Cryptococcus* and *Rhodotorula*, both of which are basidiomycete yeasts.

Rhodotorula species are non-fermenters, but are capable of thriving under cold conditions and in aerobic environments. Colonies form a characteristic orange, pink, or red colour on solid media. *Rhodotorula* is found as a contaminant of fish, poultry, and dairy products, but is relatively rare in brewing environments. Although several studies have shown that *Rhodotorula* species may be recovered from surfaces and air samples around the brewery, this species is primarily associated with contamination of water and with raw materials. Malt may be a route of entry into the brewery since *Rhodotorula* yeasts are frequently found on both barley and in finished malt (Flannigan, 1974). They have been isolated from cask conditioned beers and from pitching yeast samples (Brady, 1958) and cause minor sensorial effects.

Cryptococcus species produce a characteristically thick polysaccharide capsule that surrounds the cell. As a group of organisms, they are primarily associated with soil and guano, and some species are serious human pathogens. *Cryptococcus* spp. have been detected in wheat and barley, and also during the malting process. Although these organisms are not normally associated with fermentation, *C. diffluens* has been found in a variety of brewery locations including pipework and surfaces, while *C. albidus* has been isolated from German wines and traditional African beer fermentations. In such instances, they pose minimal threat to human health and do not have a serious impact on the sensorial quality of beer or production efficiency.

Genomics and metabolomics of beer-spoiling yeast

The haploid *S. cerevisiae* strain S288C was the first eukaryotic organism to be fully sequenced in 1996 (Goffeau *et al.*, 1996). Since this time, developments to next-generation sequencing have enabled the process to be undertaken cheaply and much more rapidly. Consequently, the genomes of over 40 different yeast species, including many capable of beer spoilage, have been published (Dujon, 2010). The vast majority of functional genomic studies in yeasts have focused on *S. cerevisiae*, largely due to the industrial applications of this species. However, with an increasing number of alternative sequences available for a variety of type strains and industrially significant yeasts, it is anticipated that the field of comparative genomics will develop significantly in the near future.

Currently, analysis has focused on the Saccharomycetales yeasts, which include industrial strains and the majority of beer-spoiling species, including *B. bruxellensis*, *C. tropicalis*, *D. hansenii*, *K. lactis*, *K. marxianus*, *P. guilliermondii*, *Sch. pombe* and *Z. Rouxii*. All of these organisms are relatively closely related (Fig. 11.5) and have genomes that range in size from 9 to 20 Mb (for haploids) and contain ≈4700–6500 protein-coding genes located on between 4 and 16 chromosomes (Dujon, 2010). Chromosome number is highly variable between species and sometimes between strains, while there is also strong evidence to suggest that heterospecific hybridization between yeasts is relatively common. Interestingly, all of the yeast genomes analysed to date appear to contain a large number of paralogous gene copies, i.e. genes with shared ancestry due to duplication events. In beer-spoiling organisms, this may either impact on how 'robust' strains are, since gene duplications offer a degree of genetic and evolutionary plasticity (Gu *et al.*, 2003), or expand functionality due to overlapping metabolic roles (Kuepfer *et al.*, 2005). Analysis of gene variation between species indicates that there is a trend towards expansion of tandem gene arrays. This is significant from a beer-spoiling perspective since genes known to impact on flocculation contain highly repeated sequences that influence the 'strength' of flocculation (Verstrepen *et al.*, 2005). Similarly, repeat elements are associated with the *CUP1* gene (Zhao *et al.*, 2014), which dictates copper resistance, one of the major means of differentiating between culture yeast and non-*Saccharomyces* strains (see section 'Detection and identification', below). Finally, there is also

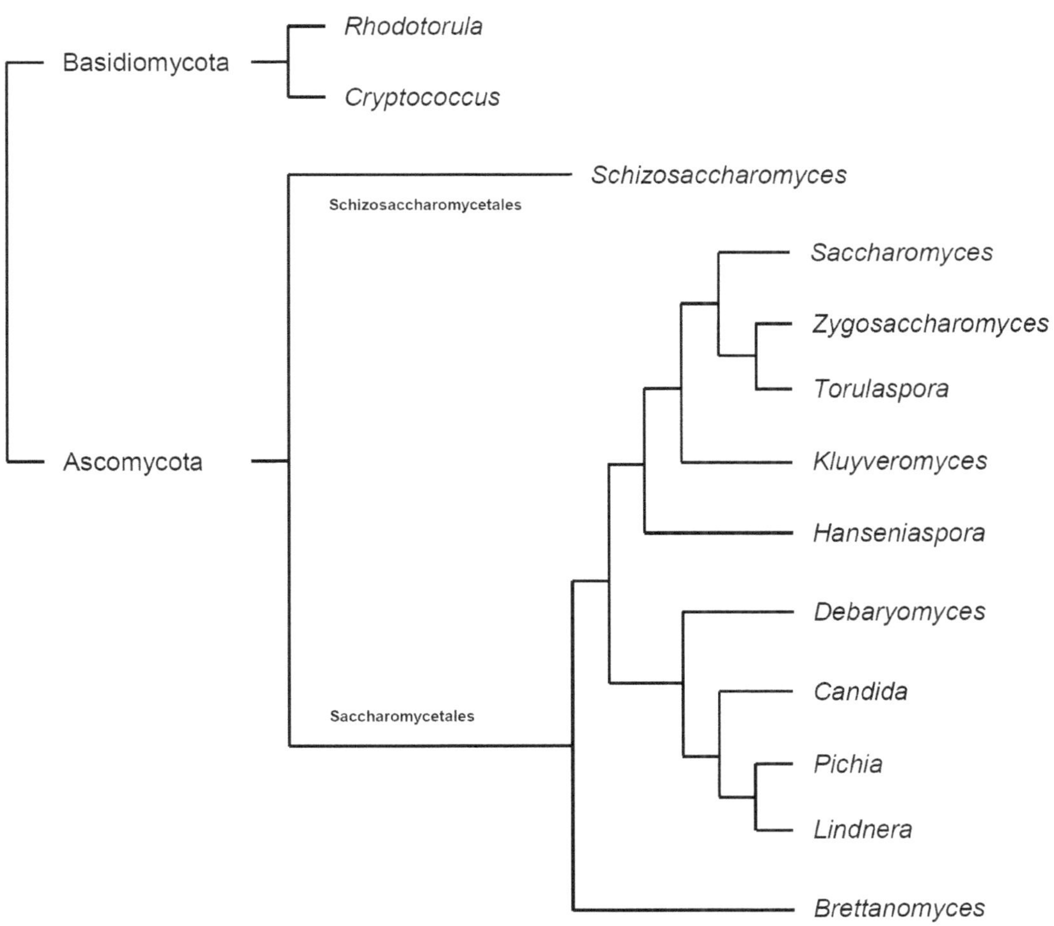

Figure 11.5 Phylogenetic tree of beer-spoiling yeasts, providing an indication of the evolutionary relationship between species based on a variety of DNA analyses (see Suh *et al.*, 2006; Kurtzman *et al.*, 2008; Kurtzman and Robnett, 2013, for more details). Branch lengths are not indicative of evolutionary distance. Note that although *Brettanomyces* is often described as being a member of the *Pichiaceae* family, this assignment is uncertain. According to certain measures of analysis, *Brettanomyces* and other *Pichiaceae* yeasts are only distantly related (Kurtzman and Boekhout, 2011), hence it is placed separately here.

evidence of horizontal gene transfer from bacterial species in certain yeasts and the presence of autonomous plasmids or viral elements that could have a significant impact on both the functionality and the threat to industrial fermentation systems (Keeling and Palmer, 2008; Coelho *et al.*, 2013; Lacroix and Citovsky, 2016). When comparing the genomic impact of production brewing strains and beer-spoiling yeasts directly, there are also some more obvious differences that have a major influence on spoilage potential. These are related to flavour development, flocculation, killer activity, and general metabolic activity as described below.

Production of phenolic compounds

One of the major spoilage characteristics of *Saccharomyces* beer-spoiling yeasts and some non-*Saccharomyces* species (including *Brettanomyces*), is the production of phenolic compounds during fermentation. Although these represent desirable flavour attributes in many wheat beers, in the majority of instances the presence of these compounds signifies a serious quality defect. In *Saccharomyces* yeasts, the production of phenolic compounds is largely influenced by the *PAD1* gene (often referred to as the phenolic-off-flavour or *POF1* gene) which encodes for phenylacrylic acid decarboxylase (Clausen *et al.*, 1994), while in *Brettanomyces* a similar function is performed by *dbPAD* (Godoy *et al.*, 2014). This essentially regulates the decarboxylation of hydroxycinnamic acids such as ferulic acid, *p*-coumaric acid, and cinnamic acid to create 4-vinylguaiacol (4-VG), 4-vinylphenol (4-VP), and styrene, respectively (Schwarz *et al.*, 2012). These impart characteristic phenolic (4-VG), clove-like (4-VP), and plastic aromas (styrene) to the beer. At the cellular level, the role of these enzymes is not fully understood. However, it is interesting that they may function to detoxify the cell, since overexpression of *PAD1* in *S. cerevisiae* results in enhanced growth rate and ethanol productivity in the presence of the hydroxycinnamic acids (Larsson *et al.*, 2001). In *Brettanomyces* species, vinylphenol reductase (VPR) further converts 4-VG and 4-VP to 4-ethylguaiacol and 4-ethylphenol (Heresztyn, 1986), which impart the smoky, leathery, medicinal, and clove like flavours often summarized as being 'barnyard' or simply 'Brett' characteristics. It has been argued that the physiological impact of VPR is to contribute towards redox balance within the cell, since NADH is used as a cofactor in the reduction of 4-VG and 4-VP to their ethyl derivatives. This hypothesis is supported by observations that VPR activity is enhanced under oxygen-limited conditions, increasing NAD^+ availability (Curtin *et al.*, 2013).

Beer-spoiling yeasts and killer toxins

Some beer-spoiling yeast strains are capable of producing proteins (killer factors) that are toxic and are able to kill a wide range of other (sensitive) yeasts. Toxin-secreting strains are referred to as killer yeasts and their presence within the brewery can have devastating effects. Typically, a killer yeast will rapidly replace the production strain and become the dominant organism in the fermentation (Young, 1987). Although killer yeasts are not particularly frequent contaminants within the brewing industry, killer activity is surprisingly common across many yeast species. A survey of 148 species comprising 964 strains from the NCYC collection indicated that 59 produced killer factors, the majority of which were *Saccharomyces* strains (Philliskirk and Young, 1975). However, killer activity can also be observed in other species and genera, including *Barnettozyma californica*, *Hanseniaspora uvarum*, *Kluyveromyces lactis*, *Pichia* spp., and *Zygosaccharomyces bailii*.

The killer phenotype is caused by several different mechanisms. In *Saccharomyces* strains and some other species including *Z. bailii*, it is believed to arise due to cytoplasmic infection by a double-stranded RNA (dsRNA) virus, while in other yeasts (including *K. lactis*) it is encoded for by a linear dsDNA plasmid, also located in the cytoplasm (Stark *et al.*, 1990). A third variation is seen in yeast species including *L. saturnus*, *P. kluyveri*, and *H. uvarum*, whereby the toxin is encoded chromosomally within the nucleus (Kimura *et al.*, 1993; Radler *et al.*, 1985). The best categorized killer strains belong to the first type, the virally induced killer phenotypes. These are caused by an infection with dsRNA viruses belonging to the *Totiviridae* family, which are widely distributed amongst yeast and higher fungi. In the majority of instances, the virus is inherited cytoplasmically, spreading horizontally during sexual reproduction. In *Saccharomyces* strains there are at least four major types of killer virus known as ScV-M1, ScV-M2, ScV-M28, and ScV-Mlus. Each encodes a specific killer toxin referred to as K1, K2, K28 and Klus, respectively, as well as a self-protective immunity component (Schmitt and Breinig, 2006; Rodriguez-Cousino *et al.*, 2011). The result of this is that members of each group are able to kill sensitive strains (as well as killer yeasts belonging to any of the other types), but are immune to toxins produced by strains of the same killer type. Other yeasts can give rise to different toxins, for example in *Lindnera* species four toxins have been observed known as HMK, K500, WmKT, and wicaltin (Theisen *et al.*, 2000; Yamamoto *et al.*, 1986), while *Z. bailii* killer yeast can produce an antifungal called zygocin (Radler *et al.*, 1993). There is also considerable variation within species; *Saccharomyces* yeasts can also produce toxins designated KHR and KHS, which are encoded on the chromosomal DNA. Irrespective of their origin, viral toxins typically kill sensitive cells using one of several different modes of action, the most common of which appear to target either cell membrane function or DNA synthesis. This is certainly the case for K28, which causes inhibition of DNA synthesis, and K1, which causes disruption of the plasma membrane, resulting in the formation of ion channels and thus ion leakage. Other effects include cell cycle arrest in G1 as seen in response to zygocin and inhibition of β-1–3-glucan synthesis, which is associated with the WmKT toxin.

Beer-spoiling yeasts and flocculation

One of the major effects of beer-spoiling yeasts on the fermentation process is related to the impact of contamination on the flocculation potential of the culture strain. Flocculation is a form of non-sexual aggregation and can be described as a reversible, calcium-mediated process that is characterized by the adhesion of cells within a population to form aggregates known as flocs (see Chapter 1). These sediment rapidly from the medium in which they are suspended, facilitating beer clarification and providing a cost-effective means of collecting yeast for re-pitching into a successive fermentation. Within the brewing industry, it is desirable that this occurs in a regular and predictable fashion, and preferentially towards the end of fermentation once fermentable sugars have been utilized. Consequently, the flocculation properties of the culture yeast are of great importance, since they have an impact on the consistency of the process, and the quality of the yeast and the final product.

In *Saccharomyces* yeasts, flocculation is governed by a group of closely related genes, referred to as the *FLO* gene family. This family incorporates several groups of genes, the first of which facilitate cell–cell adhesion (*FLO1*, *FLO5*, *FLO9*, and *FLO10*). *FLO8* encodes a transcriptional activator of *FLO1* which itself is responsible for the structural protein directly involved in the flocculation process. The remaining gene of interest, *FLO11*, induces cell-substrate adhesion associated with invasive growth or pseudohyphae formation as described above (see section, 'Vegetative growth, cell structure and sexual division'). The flocculation process is largely driven by lectin-like proteins (flocculins) coded for by the *FLO1* gene (Teunissen *et al.*, 1993). These extend out of the cell wall and bind to mannan receptor sites on adjacent cells (Miki *et al.*, 1982; Kobayashi *et al.*, 1998). Less is known about the other genes within the family; although they are structurally similar, the main difference is the degree of flocculation induced by their expression. Genome analysis of non-*Saccharomyces* yeast species have indicated that they also possess genes that code for cell wall lectins, suggesting that flocculation may be broadly similar in nature across the yeasts. Functional analysis of beer-spoiling yeasts including *Debaryomyces* (Cubells Martinez *et al.*, 1996), *Candida* (Bauer and Wendland, 2007), *Hanseniaspora* (Suzzi *et al.*, 1996), *Pichia* (Mbawala *et al.*, 1990), *Kluyveromyces* (El-Behhari *et al.*, 2000), *Brettanomyces* (Steensels *et al.*, 2015), *Torulaspora* (Canonico *et al.*, 2016), and *Zygosaccharomyces* (Suzzi *et al.*, 1992), have indicated that each species demonstrates flocculation phenotypes of varying degrees. However, analysis of the mechanism of flocculation in *K. marxianus* has indicated that the specific structure and spatial arrangement of the cell wall groups involved in flocculation may be species- or genus-specific (Sousa *et al.*, 1992), indicating that the underpinning mechanisms may differ.

Irrespective of the precise mechanism of aggregation, at the very basic level brewing strains are highly flocculent compared to most other types of yeast. Hence, it would be expected that the presence of beer-spoiling species would lead to a reduction in flocculation potential, causing issues for post-fermentation processing. However, this is simplistic since flocculation potential in beer-spoiling yeasts is dependent on the species, its physical properties, and the level of contamination. Furthermore, it is known that mixing yeast cultures can give rise to co-flocculation, a phenomenon first described by Eddy (1958), based on observations that non-flocculent strains became flocculent when mixed together. This definition has been extended to describe variations in flocculation observed when two different strains or species are mixed together, irrespective of their individual intrinsic flocculation capacity (Nishihara *et al.*, 2000). Intra- and inter-specific co-flocculation has been reported in *Kluyveromyces* species (El-Behhari *et al.*, 2000; Sosa *et al.*, 2008), as well as in *D. hansenii* (previously *C. famata*) and *Sch. pombe* (Martinez *et al.*, 1996), and is likely to be a prevalent phenotype across the yeasts. In fact, co-flocculation has also been observed between yeast and beer-spoiling bacteria (Peng *et al.*, 2001), suggesting that such interactions may be important in developing microbial ecosystems. This hypothesis is supported by specific analysis into the genetic regulation of co-flocculation in yeast. It has been reported that individual *FLO* genes may impact differently on cell–cell adhesion phenotypes, favouring adhesion between some species while excluding others in mixed flocs (Rossouw *et al.*, 2015). This type of interaction was observed between *Hanseniaspora* strains and *S. cerevisiae* wine yeasts, as well as between other species, including *P. kudriavzevii*. In the same study, *FLO* gene-specific differences in co-flocculation behaviour were observed between non-*Saccharomyces* strains analysed, with *FLO1* overexpression consistently leading to increased co-flocculation. Analysis of the remaining *FLO* genes revealed a complex pattern of results dependent on the combination of strains investigated, indicating that co-flocculation is a multifaceted process. However, allowing organisms in co-culture to respond differently to one another may help explain the evolutionary persistence of the *FLO* gene family, comprising a number of genes that exhibit apparently similar function.

Sugar uptake and metabolism of beer-spoiling yeasts

All yeasts are heterotrophic organisms that require a variety of nutrients in order to be able to produce energy, maintain cellular function, grow, and reproduce. A carbohydrate source is essential for the production of ATP, and to fulfil structural carbon requirements, while nitrogen is important for protein and amino acid synthesis. Phosphorus is required for energy transduction, and as a major component of membrane phospholipids and nucleic acid (DNA and RNA) synthesis. Metal ions, vitamins, growth factors, and other trace elements may also be required for a variety of structural and functional roles within the cell. The majority of beer-spoiling yeast share broadly the same nutritional requirements as production strains; however, there can be significant differences with regard to the preferred sources of individual nutrients and the ways in which they are assimilated and utilized. Many of these differences form the basis of tests for detection of specific groups of yeast (see section 'Detection and identification'). There are also variations in the extent to which specific metabolic end products are formed; in some species carbon may be directed primarily towards biomass production rather than ethanol. In others, there may be a more diverse range of products, including a higher propensity to form glycerol as a mechanism for redox balance. Furthermore, nutritional requirements may lead to certain metabolic pathways being preferentially employed, leading to the production of compounds, such as diacetyl, that are produced as by-products of amino acid metabolism.

The mechanisms of sugar assimilation and breakdown form a major difference between brewing and beer-spoiling yeasts. At the basic level, some beer-spoiling yeasts may not be able to transport certain sugars into the cell, while others may efficiently use those that brewing strains cannot, one example being the utilization of the disaccharide lactose by *Kluyveromyces*. Some yeasts cannot (or prefer not to) undergo fermentation (i.e. non-fermentative yeast strains), while others may only utilize fermentation pathways in the presence or absence of certain sugars. Yeasts such as *Zygosaccharomyces* are highly effective at fermenting simple carbon sources, while some species (especially *S. cerevisiae* strains with diastatic properties) are able to metabolize complex long-chain sugars. The pattern of sugar assimilation and utilization displayed by individual beer-spoiling yeasts can therefore have an impact not only on ethanol yield, but in some instances on mouthfeel, due to the removal of dextrins from beer. The causative reasons for differences in sugar utilization are unknown but are likely to be evolutionarily driven, perhaps related to primary habitat or, in the case of brewing strains, artificial selection. Irrespective, the process of sugar assimilation is controlled both genetically and metabolically, allowing the cell to produce and regulate a series of enzymes involved in uptake and metabolic pathways. In *Saccharomyces* yeasts, glucose uptake principally occurs via facilitated diffusion with no expense of metabolic energy (ATP). However, in other types of yeast the transport mechanism varies; for example, *Candida, Kluyveromyces* and *Pichia* strains employ active transport as the primary mechanism for glucose uptake. In such systems ATP is required to expel hydrogen ions, which functions to create an electrochemical gradient allowing for the transport of sugars into the cell via proton symport. This is an effective method when glucose is in short supply since it allows for transport of sugar against a concentration gradient, but under optimum conditions the benefit of expending energy for the uptake of sugar is questionable. A potential reason for this divergence in uptake strategy is that for respiratory (i.e. non-fermentative) yeasts the 'cost' is relatively insubstantial given that a theoretical yield of 38 ATP can be achieved from a single glucose molecule through aerobic metabolism. However, for yeasts that rely primarily on the fermentation pathway, yielding 2 ATP per glucose, this is a considerably higher proportion of cellular energy, hence other more cost-effective strategies are preferred (Griffin, 1994).

When yeasts are cultivated on sugar, a flux occurs through central carbon metabolism (Fig. 11.6). As alluded to above, yeasts have the potential to convert this sugar into energy through fermentation (glycolysis) or via respiration (Krebs cycle and the electron transport chain). Simplistically, it might be expected that under aerobic conditions yeast would preferentially utilize the respiratory pathway to produce ATP, carbon dioxide, and water. Likewise,

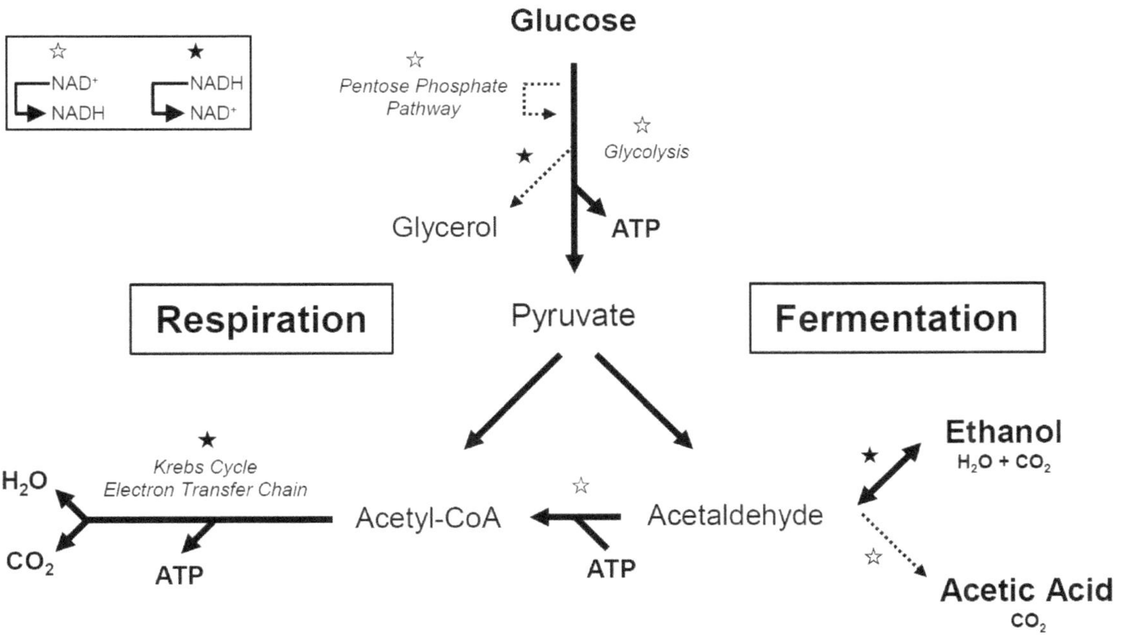

Figure 11.6 Yeast carbohydrate metabolism. Yeasts can employ two major pathways for ATP production from glucose: respiration and fermentation. Glycolysis forms the initial stage of each pathway, which involves the conversion of glucose to pyruvate with the net gain of two units of ATP. During fermentation, pyruvate is subsequently converted into ethanol. This process does not produce additional ATP but is favourable due to the recycling of NAD^+, which regenerates the cellular pool of NADH for re-use in glycolysis. During respiration, pyruvate is oxidized to H_2O and CO_2 through the process of oxidative phosphorylation. This yields ATP and regenerates NAD^+, but has an absolute requirement for oxygen. Ethanol (and acetic acid in some yeasts) are believed to be formed by yeast as part of a MAC strategy and can be recycled for ATP production if needed. This process yields less energy than the direct oxidation of pyruvate because the synthesis of Acetyl-CoA (through acetaldehyde for ethanol, but directly for acetic acid) requires ATP. Pathways are often employed as a means of redox balance (NAD^+/NADH ratio) and are partially regulated by oxygen (Pasteur and Custers effects), and the type and concentration of sugar (Kluyver and Crabtree effects).

under anaerobic conditions it might be expected that the fermentation pathway would be employed with acetaldehyde acting as the final electron acceptor to produce ATP, ethanol, and carbon dioxide. However, as described above, this is not always the case, with some yeasts preferentially exploiting one or other pathway (i.e. being fermentative or non-fermentative). In reality, many yeasts are respiro-fermentative and have evolved to utilize both pathways virtually simultaneously (Hagman and Piskur, 2015). The extent to which this happens is variable and the direction of carbon flux is influenced by a variety of parameters, including the presence/absence of oxygen, the presence/absence of sugars, as well as lesser effects based on the concentrations of inorganic phosphate, glycolytic intermediates, ammonium ions, and intracellular pH. Glucose in particular (although other sugars may be significant, albeit to a lesser extent) and oxygen act as metabolic triggers, causing pathways to be preferentially employed (Rolland *et al.*, 2002). These regulatory mechanisms are summarized in Table 11.4 and include the Pasteur, Crabtree, Kluyver and Custers effects. With the exception of the Kluyver effect, adoption of these mechanisms is typically driven by the ability to metabolize faster under specific conditions, or as a consequence of the generation of specific end-products, as described below. Both of these strategies are likely to have evolved to provide strains with some form of competitive advantage in 'natural' environments.

The Pasteur effect

The Pasteur effect describes the phenomenon whereby fermentation is suppressed by the presence of oxygen. In reality, what this means from the perspective of the yeast cell is that the rate of glycolysis is *slower* under aerobic conditions than anaerobic conditions (due to regulation of phosphofructokinase within the pathway). This ensures that bottlenecks are not encountered and that all carbon can be metabolized via the Krebs cycle and through oxidative phosphorylation, which is an efficient means of maximizing ATP yield. Arguably of equal or greater significance is that this also means that glycolysis proceeds at a *faster* rate under anaerobic conditions. This is necessitated due to the reduction in ATP yield which is encountered through switching from respiration to fermentation; the cell must increase glycolytic rate to counter this effect and to ensure that ATP demands are met. One caveat for this phenomenon is that the level of glucose must be relatively low (Lagunas, 1979) and that other nutrients (including nitrogen) are also limiting. Interestingly, these criteria are relatively flexible which means that yeasts can differ in their capacity to implement the Pasteur effect. *Saccharomyces* yeasts only exhibit the Pasteur effect under quite stringent conditions and when cells are in a 'resting' state (i.e. not actively growing). In contrast, yeasts such as *Candida* and *Pichia* are highly susceptible to this effect. The reason why there is such discrepancy is that the Pasteur effect can

Table 11.4 Regulatory mechanisms for sugar metabolism in yeast

Metabolic effect	Description	Notable yeast species
Pasteur effect	Inhibition of fermentation pathways in the presence of oxygen *Oxygen restricts ethanol production*	*Saccharomyces* yeasts are not affected due to the Crabtree effect
Crabtree effect	Suppression of respiration pathways in the presence of glucose *When a threshold of glucose is reached, yeast will only utilize the fermentation pathway*	Positive: *Saccharomyces*, *Schizosaccharomyces*, *Zygosaccharomyces*, *Brettanomyces* Negative: *Pichia*, *Kluyveromyces*, *Debaryomyces*, *Torulaspora* Species-dependent: *Candida*, *Hanseniaspora*
Kluyver effect	Inhibition of fermentation pathway in the presence of certain sugars *Yeast can only utilize certain sugars aerobically*	*Saccharomyces* strains are Kluyver effect negative. Most non-*Saccharomyces* beer-spoiling yeasts are Kluyver effect positive for various sugars
Custers effect	Anaerobic conditions cause fermentation to be suppressed *Yeast utilize the fermentation pathway when oxygen is present*	*Brettanomyces*

be influenced by the presence of other metabolic regulatory phenomena. For example, in some yeast (e.g. brewing strains) the Pasteur effect is rendered insignificant due to the Crabtree effect (see below). Irrespective, increasing the fermentation rate in the absence of oxygen leads to the rapid production of ethanol that can procure a competitive advantage in the 'natural' environment. This type of approach is often referred to as the 'make–accumulate–consume' (MAC) strategy, whereby organisms produce a compound that accumulates within the environment, and which can be re-assimilated and used for energy production at a later point in time (Fig. 11.6). This strategy may provide a significant benefit to a population since competing organisms, including bacterial species, are typically less tolerant to ethanol than is yeast.

The Crabtree effect (glucose repression)

Arguably one of the most important mechanisms for metabolic regulation in yeast species is the Crabtree effect, often referred to as simply 'glucose repression' (Barnett and Entian, 2005). The Crabtree effect occurs irrespective of the presence of oxygen and specifies that if the concentration of sugar is above a certain threshold, cells will metabolize exclusively via fermentation. As the name suggests, this is primarily due to repression of the respiratory pathway caused by the presence of glucose, and occurs to the extent that the Pasteur effect is overridden (i.e. the rate of glycolysis is not restricted by oxygen). It should be noted that other sugars such as fructose may also play a similar role to glucose, since many yeasts are able to conduct aerobic fermentation in their presence. However, glucose is certainly the primary sugar involved. The Crabtree effect is observed in most fermentative beer-spoiling yeasts, including *Saccharomyces*, *Brettanomyces* and *Zygosaccharomyces*. In these organisms, the Crabtree effect can be induced by as little as 0.2% glucose, although this limit can vary upwards. The extent to which the Crabtree effect is induced can also be species-, and to a lesser extent strain-dependent, as indicated by analysis of the ratio of glucose that is fermented/respired under defined conditions (De Deken, 1966; Hagman *et al.*, 2014).

In yeasts, the Crabtree effect can be observed in the form of a 'short-term' or 'long-term' response. The short-term Crabtree effect occurs when the cell encounters a sudden increase in glucose concentration (Van Urk *et al.*, 1989), which causes the respiratory pathway to become saturated at pyruvate, resulting in carbon being directed towards ethanol formation (overflow metabolism) (Fig. 11.6). In contrast, the long-term Crabtree effect, characterized by the response to steady-state conditions, indicates that glucose (or a product of glucose metabolism) actively functions to repress the synthesis of respiratory enzymes. It is known that the yeast cell is able to respond to the presence of glucose by a series of signal transduction pathways that are regulated based on the concentrations of extracellular and intracellular glucose, related metabolites, and flux through key glycolytic enzymes. This includes activation of a range of genes that function as transcriptional regulators, and those that are involved in glucose sensing and sugar uptake (for a comprehensive review see Conrad *et al.*, 2014). Although many of the regulatory activities are believed to operate at the transcription regulation level, some may also act directly on specific respiratory enzymes, including those involved in gluconeogenesis, the Krebs cycle, and mitochondrial function and maintenance (Käppeli, 1986). It should be noted that despite effort in this direction, at the present time much is still unknown about glucose sensing and its regulatory mechanisms in *S. cerevisiae*, while in other yeast species it remains largely unexplored. However, it is likely that the underlying principles behind glucose repression are similar among yeasts, even if the precise mechanisms differ.

Although many yeasts are subject to the long-term Crabtree effect, the reason why this form of metabolism has evolved is difficult to explain, especially since fermentation has a significantly lower ATP yield than respiration. However, it has been suggested that the long-term Crabtree effect is essentially an evolutionary extension of overflow metabolism (Hagman and Piskur, 2015), while the acceleration of glycolysis and the preference for alcohol production is also likely to function to create a hostile environment for competitors as part of a MAC strategy. In addition, it is possible that rapid sugar uptake can be considered to be a form of glucose scavenging, inhibiting growth of other microbes by starvation, while others have argued that the Crabtree effect evolved simply due to the benefits associated with an overall increase in the rate of ATP production (Pfeiffer and Morley,

2014). In the context of beer-spoiling yeasts, the fact that many species are not Crabtree-positive, or that they exhibit either a short-term response or a 'weaker' long-term response overall is significant. For example, purely respiring yeast species such as *Kluyveromyces marxianus* or *Candida utilis* may need to adopt other strategies to limit the glycolytic rate if a metabolic overflow is encountered. This may manifest in regulation of sugar uptake, formation of glycerol, or in the production of intracellular reserve carbohydrates (glycogen/trehalose). There is also likely to be an impact based on niche development, which may ultimately determine which species are likely to compete with production stains during fermentation. Crabtree-negative yeasts will inevitably be less competitive than production strains during fermentation and will only be able to survive at low concentrations, hence one of the reasons why they are typically associated with alternative processing stages.

The Custers effect

The Custers effect was first observed during analysis of *Brettanomyces* strains isolated from lambic-style beers (for a review, see Barnett and Entian, 2005). In contrast to the Pasteur effect, Custers-positive yeast ferment glucose to ethanol faster when oxygen is present than when anaerobic conditions are applied. The reason for this is related to the capacity of cells to convert acetaldehyde into acetic acid (Scheffers, 1961, 1979). This acid production has also been implicated in a MAC strategy, whereby the population is able to create a temporary pH-hostile environment that eliminates or restricts the growth of competitors, before being re-assimilated for an energetic gain (Fig. 11.6). However, the initial process of generating acetic acid requires the reduction of NAD^+ to NADH, causing a cellular redox balance disparity that must be restored. This imbalance is compounded by the fact that Custers-positive *Brettanomyces* strains lack the cellular machinery to synthesize glycerol, which can act as an important mechanism for NAD^+ regeneration (Wijsman *et al*., 1984). In the absence of oxygen, the cell cannot readily re-oxidize NADH and so the glycolytic pathway is restricted and the rate of fermentation decreases. Conversely, under aerobic conditions the yeasts circumvent this issue by increasing the glycolytic rate and diverting a proportion of carbon flux towards the production of ethanol in order to regenerate NAD^+ rapidly (van Dijken and Scheffers, 1986), although it should be noted that cellular oxygen (i.e. the respiratory chain) can also perform this function. With regard to the Custers effect, there is also considerable variation between strains, which may also be influenced by media composition and other regulatory mechanisms. For example, nitrate assimilation is believed to prevent the Custers effect from occurring since cells can replenish the NADH pool through reduction of nitrate to ammonium (Steensels *et al*., 2015). Furthermore, Crabtree-negative yeasts only respire under aerobic conditions and consequently would not be subject to the Custer effect at all.

The Kluyver effect

Although brewing production strains can metabolize sugars efficiently under aerobic or anaerobic conditions, some yeast species cannot utilize the fermentative pathway efficiently in the presence of certain sugars. This phenomenon is known as the Klyuver effect and describes the process whereby specific sugars can only be metabolized aerobically. Yeast species that are subject to this effect are often referred to as being 'respiration dependent' or 'Kluyver effect positive' for a specific sugar, the nature of which can vary between organisms. For example, the Kluyver effect can be observed in *Kluyveromyces wickerhamii* in response to lactose, in *Debaryomyces yamadae* in the presence of sucrose, and in *Pichia heimeii* and *Candida utilis* in response to maltose. Consequently, it can be seen that, in contrast to the other metabolic effects described above, the Kluyver effect has broad specifications and wide-ranging consequences; a yeast species may be Kluyver effect positive for some sugars but not others. Furthermore, many Kluyver-positive yeasts are often able to ferment glucose efficiently, as well as some of the component parts (i.e. hexoses) of oligosaccharides. To simplify things slightly, the Kluyver effect is not observed in the majority of species that have a predominantly fermentative metabolism, including *Saccharomyces* yeasts. However, it is frequently observed in facultatively fermentative yeasts, especially those displaying poor or weak fermentation properties (Sims and Barnett, 1978).

The precise reasons why the Kluyver effect occurs are unknown, and indeed it is perplexing that a particular yeast may be able to utilize glucose under anaerobic conditions but not maltose, which is a disaccharide comprising 2 glucose units. However, there is strong evidence to suggest that the Kluyver effect is linked to sugar transporters within the cell membrane. The uptake of certain sugars may be restricted due to the general reduction in ATP synthesis associated with anaerobic metabolism. This acts to limit proton pump activity (since this is an energy-dependent transport mechanism) and consequently cells are unable to efficiently assimilate these sugars (Barnett, 1992). Essentially, the low level of sugar transporter activity observed in Kluyver effect positive yeast is not sufficient to sustain the high substrate flow necessary for robust fermentative growth (Goffrini *et al.*, 2002). Hence, it is also likely that the level of sugar transport activity under anaerobic conditions may at least partially determine whether a yeast is Kluyver-positive or -negative for a specific sugar (Fukuhara, 2003). If this is indeed the case then this has implications for interpreting the Kluyver effect, especially since it is known that the transport activity for a given sugar can vary significantly within a species. It is also possible that other factors may be involved, albeit to a lesser extent. This may include the effects of redox imbalance (in highly aerobic yeasts), the relative decrease in activity of pyruvate decarboxylase, product inhibition (specifically ethanol), and the relationship between anaerobiosis and sugar carrier properties, as reviewed by Fukuhara (2003) and Barnett and Entian (2005). From the perspective of beer-spoiling yeasts, the Kluyver effect predominantly influences the capacity of a strain to compete with the production yeast. For example, yeasts that are Kluyver effect positive for maltose are unlikely to be a serious threat during fermentation. However, it is important to note that such yeasts may survive at low concentrations or be able to thrive at different stages of the process.

Frequency and control of beer-spoiling yeasts

Incidences of beer-spoiling yeasts in breweries

In comparison to reports of bacterial contamination, there are relatively few surveys documenting the occurrence of yeast contaminants within breweries. It is debatable why this is the case; either they are genuinely less problematic, or their presence goes largely unnoticed owing to difficulties in isolating and detecting beer-spoiling yeasts. It is also worth noting that contamination is often not as pronounced as with bacterial species and therefore yeast contaminants may remain unnoticed since only minor changes to fermentation performance or flavour profiles are observed, and these may be erroneously attributed to other sources.

In one major study, van der Aa Kühle and Jespersen (1998) analysed 101 cropped yeast samples obtained from 45 lager breweries. These were assessed for yeast contamination by plating samples of yeast collected from fermentation vessels onto selective media. In total, 41 of the samples analysed (representing 24 breweries) were identified as containing beer-spoiling yeasts. In total, 126 beer-spoiling yeasts were isolated and identified to the species level, a frequency that was considerably higher than expected. *Saccharomyces* species accounted for more than 57% of the infections detected, with *Pichia* and *Candida* representing 28% and 15% of the contaminants isolated, respectively. In the same study, the capacity of the isolated yeast strains to grow in de-carbonated beer and wort were also examined. Most strains investigated were capable of growth in both wort (99%) and beer (98%). Further analysis revealed that inoculation of 26 of the isolated strains into bottled lager beer resulted in significantly reduced survival and growth. Eight of these isolates were not able to grow in bottled beer even when incubated for 17 days at 21°C. For those isolates that were able to grow, most demonstrated moderate replication potential and a few showed rapid growth. In a separate study, Pham *et al.* (2011) analysed yeast contaminants isolated from conditioning tanks and fermentation vessels. Analysis of fermentation samples indicated that 90% of isolates were *Saccharomyces* species, while 61% of samples from conditioning tanks belonged to the *Pichia* genus (either *P. fermentans* or *P. membranifaciens*). This was perhaps surprising, given that *Pichia* are primarily aerobic yeasts. However, these data highlight the possible effects of oxygen in beer post fermentation; in this instance, it is possible that air was either the direct cause of the contamination, or more likely that it facilitated the growth of the *Pichia* yeast, which had survived through fermentation at subdetection levels.

Irrespective of the source and nature of yeast contamination, there is a range of consequences. These include having a direct negative impact on product quality and consistency, while also increasing product wastage through disposal of contaminated beer. If a contamination issue is not addressed then this could also lead to complaints from the market, bruised consumer confidence, decreased brand loyalty, and potentially a trade recall.

Control of beer-spoiling yeasts

The potential for contamination by beer-spoiling yeasts is largely determined by the level of hygiene applied throughout the brewing chain. The possibility of introducing beer-spoiling yeasts should be eliminated by ensuring that equipment is sterile, and that the pitching culture is free of contaminants. With respect to raw materials arriving from an external source, confidence should be assured by accreditation systems such as ISO 9001. However, brewers should still be aware of the microbiological risk associated with water, barley, hops, adjuncts, and culture yeast. For some raw materials, this may necessitate extra analyses; for example, where water is sourced from wells or boreholes the risk of contamination should be considered and an appropriate general microbiological sampling plan generated. Similar strategies should be implemented to check for the presence of beer-spoiling yeasts within the brewery on a routine basis. On a practical level, this inevitably involves sampling wherever there is a high risk of microbial contamination, including raw materials such as primings or dry hops. This should also include areas associated with yeast propagation, fermentation and conditioning tanks, yeast storage vessels (brinks), and the packaging line.

Arguably the easiest way that a yeast contaminant can be introduced into the brewery is via the pitching yeast. At the simplest level, breweries that regularly use several yeast strains have a greater chance of observing cross-contamination. However, in all breweries where yeast is propagated in-house (see Chapter 3), contaminated slopes or ampoules pose a significant microbiological risk as any undesirable yeasts that are present will be cultivated as part of the yeast propagation process. Consequently, the routine yeast supply process should be designed such that there is confidence that the yeast is contamination free. Key stages include long-term storage of yeast, removal of yeast stocks from storage, and the initial cultivation of biomass. Third-party suppliers are often used to store yeast strains and to provide working cultures to individual breweries on request. These companies typically perform a series of quality-assurance checks to ensure that the culture is pure, free of mutations, and of the correct strain. Once the yeast propagation process is initiated, human error represents the greatest threat; poor aseptic technique, failure to follow the correct procedures, or a lack of understanding about the potential risks of beer-spoiling yeast can all lead to contamination. Within the laboratory, each step of the propagation process should ideally be performed in a laminar flow hood, or alternatively close to a Bunsen burner using appropriate aseptic technique. The yeast growth media should be sterile before inoculation, preferably by autoclaving at 121°C and 15 psi for 15 minutes, and the sterility of any air or oxygen introduced should be assured by passing through a sterile filter. Once the biomass is sufficient to seed the brewery propagation vessel, transferral of yeast should be carefully monitored. Correct propagation management at the plant level can also assist in ensuring the absence of beer-spoiling yeasts and should involve regular microbiological checks on equipment, cooled wort and sterile air lines, as well as equipment used for the transport and storage of pitching yeast. If contamination is found at any point, it is best practice to immediately discard the culture, re-initiate the process of yeast supply, and implement a rigorous cleaning regime. It should be noted that although acid washing may be used to control spoilage bacterial contamination, this has no effect on beer-spoiling yeast.

For smaller brewing companies, yeast cultures may be obtained from a variety of sources, including manufacturers of active dried yeast (ADY), liquid yeast suppliers, or cooperative neighbouring breweries. While the microbiological quality of the latter is largely dependent on the brewer, suppliers of yeast cultures in pitchable quantities (ADY or liquid yeast) will typically perform a check for yeast contaminants as part of their routine quality assurance prior to sale. However it is also advisable to conduct analyses in-house to confirm purity of the culture, even if this only provides retrospective information.

Although brewers should strive for absolute

sterility, in reality many will accept slightly lower standards (often unknowingly) that are sufficient to ensure that the final product is not compromised. Historically, the critical level was considered to be around 4–10 cells per bottle (Brumstead and Glenister, 1962), since yeast would not be competitive at lower concentrations. However, current consensus is that for most yeasts, there is no 'safe' level and best practice is to maintain a high degree of hygiene throughout the brewing process (van der Aa Kühle and Jespersen, 1998). With regard to general brewery hygiene, the reader is directed to a thorough recent review (Davies *et al.*, 2015); however, it is important to note that a well-designed plant greatly mitigates microbial contamination. Internal surfaces should be smooth and dead-legs in pipework should be avoided. Areas that cannot be effectively sterilized using cleaning in place (CIP) pose a considerable contamination risk. The CIP system employed within a plant should be effective with spray balls or rotating spray heads able to deliver the appropriate cleaning agents at the correct pressure and for the required time. It should be noted that many beer-spoiling yeasts can be airborne; surfaces that may have become contaminated should not be 'hosed down' as this can create an aerosol that could potentially act as a conduit for further contamination.

When considering beer quality overall, it is also important to recognize that opportunities for spoilage also occur at the point of sale (Quain, 2016). In particular, ale-type products in cask, and dispense systems in general, can provide a semi-aerobic environment in which many beer-spoiling yeasts are able to proliferate. Typically, these yeasts are opportunistic contaminants that are able to thrive where a combination of poor hygiene and bad practice provides the opportunity for proliferation. Unfortunately for the brewer, once the beer leaves the brewery they often have little influence on the product or the standards of hygiene applied. However, adhering to best practice for the sanitation of dispense equipment and using inert gas to prevent air ingress into casks can minimize risk at this stage.

Detection and identification

At critical stages of the brewing process it is good microbiological practice to conduct tests to detect the presence of unwanted microorganisms. Unfortunately, while many brewers implement strategies for monitoring bacterial species, many do not test for beer-spoiling yeasts, or do so only infrequently. If a beer-spoiling yeast is detected, it is often useful to identify the species since this information can assist in determining if a recurring issue has presented itself. Furthermore, identifying the species may give clues as to the source of the contamination and the best course of action for elimination. Traditional methods based on growth properties are still widely employed for the detection of yeasts, despite a growing range of alternative rapid methods. The reason for this is that using culture media is relatively simple, cheap, and provides robust information on the presence of viable organisms. Alternative methods for detection may not meet all of these criteria and hence are typically used alongside traditional methods, perhaps to speed up an initial diagnosis. In contrast, novel methods for the identification of beer-spoiling yeast have been broadly accepted and are implemented in most large brewing companies. There are now many techniques available to identify yeasts to the genus, species, or strain level, and although some of these can be easily conducted in-house, third-party laboratories can also perform such analysis relatively quickly and without huge cost.

Isolating beer-spoiling yeasts

Traditional techniques used to isolate yeast contaminants in breweries are largely based on the cultivation and growth requirements of the genus and species involved. Such techniques are popular since they are relatively easy to implement; however, results are typically available only several days or even weeks after beer has been packaged. These techniques therefore only offer a retrospective view of final product quality and do not allow for proactive process control.

The most common means of detecting beer-spoiling yeast within the brewery is to cultivate unknown samples (collected either via direct sampling or by filtration of liquids) on selective nutrient media. These media types are designed to either promote the growth of spoilage yeasts or to restrict growth (e.g. of brewing strains or bacterial species). Given that organisms frequently exhibit different growth requirements, variations in carbon or nitrogen source, nutrient composition, and aerobic conditions are invariably used in media for the isolation of beer-spoiling yeasts. At the most basic level, Wallerstein Laboratory Nutrient (WLN) media can be employed as this supports the growth of a range of microbes; by incorporating tetracycline or chloramphenicol, it is possible to prevent bacterial growth, providing specificity for yeast cultivation. More specific media for yeast detection include CLEN (cadaverine, lysine, ethylamine, and nitrate medium), Lins wild yeast medium (LWYM), lysine medium, MYGP + copper, and XMACS (Table 11.5). The use of these media types can facilitate the detection of either *Saccharomyces* or non-*Saccharomyces* strains, or both. However, in reality it is good practice to employ more than one type of media to detect a broad range of yeasts. For example, many breweries will conduct testing using MYGP + copper to isolate non-production strains in conjunction with lysine media to test for non-*Saccharomyces* yeasts.

Traditional methods for characterizing beer-spoiling yeasts

Identifying yeasts using traditional methods is a complex and inexact process. Direct microscopy is useful in determining cell structure, budding patterns, the presence of pseudohyphae, and the shape of spores (if present). This can be supplemented with analysis of phenotypic traits including fermentation of sugars, production of acids, and a range of biochemical tests to determine nutritional dependencies. The latter can be facilitated by the use of commercially available API strips (Biomerieux, France). However, although they have been widely used for analysis of clinical isolates, their use within the brewing industry to date has been limited. Other methods that are more accurate but have yet to be fully translated to the brewing industry include matrix assisted laser desorption/ionization time-of-flight mass spectrometry (MALDI-TOF MS) (Usbeck *et al.*, 2014; see Chapter 7), as well as pyrolysis mass spectroscopy and Fourier transform infrared spectroscopy (FTIR) (Timmins *et al.*, 1998), which allow species differentiation based on mass fingerprinting. These offer interesting opportunities since production strain data can be stored in databases, allowing quick and precise comparisons between unknown samples to be made. Other methods include profiling of total fatty acids based on determination of fatty acid methyl ester (FAME) compounds (Timke *et al.*, 2008), protein fingerprinting (Kobi *et al.*, 2004), and immunological-based methods, which employ species-specific monoclonal antibodies to detect organisms by enzyme-linked immunosorbent assays (ELISA) (Kuniyuki *et al.*, 1984). In addition, techniques can be applied to characterize unknown yeasts by determining spoilage characteristics based on phenotype analysis. Examples include testing for acetic acid or

Table 11.5 Media for the detection of wild yeasts. For the majority of media types, it should be noted that other yeasts may show limited growth over an extended period of time

Method	Specificity and details	Reference
CLEN	Non-*Saccharomyces* wild yeast Multinitrogen media containing cadavarine, lysine, ethylamine, and nitrate. Supports the growth of multiple wild yeasts	Martin and Siebert (1992)
Lins Wild Yeast Medium (LWYM)	*Saccharomyces* wild yeasts Contains crystal violet and fuschin-sulfite to suppress brewing yeast growth	Lin (1981)
Lysine	Non-*Saccharomyces* wild yeasts *Saccharomyces* strains are unable to use lysine as a sole nitrogen source	Walters and Thiselton (1953)
MYGP + Copper	Range of wild yeasts Production strains are inhibited by copper in the range of 100–200 mg/l	Taylor and Marsh (1984)
XMACS	Range of wild yeasts Multicarbon media containing xylose, mannitol, adonitol, cellobiose and sorbitol	De Angelo and Siebert (1987)

phenolic compounds, both of which are obvious by performing a basic sniff test. Although care should always be taken when sniffing 'unknown' laboratory samples, by inoculating a suspected beer-spoiling yeast into media comprising ferulic acid, POF^+ yeast can be detected due to the production of a medicinal-like (phenolic/clove) aroma, typical of 4-vinylguiacol (Cowley *et al.*, 2016).

Traditional methods for differentiation of production strains and variants

Ale and lager yeast strains can be readily differentiated based on key physiological characteristics. For example, ale strains are able to grow on solid media and form colonies at 37°C while lager strains are not. In addition, the ability of lager strains to utilize melibiose can be exploited using X-α-Gal based medium. This medium determines the capacity of yeast to cleave the melibiose homologue X-α-gal (5-bromo-4-chloro-3-indolyl-α-D-galactopyranoside), resulting in the development of a blue/green coloration derived from indol. While this is traditionally conducted using solid media, more rapid results can be obtained using a revised protocol where cells are suspended in liquid media and results obtained within one hour (Box *et al.*, 2012). Differentiation of production strains of the same type is more challenging, particularly if the strains in question are both *S. pastorianus*. Most ale strains show differences in giant colony morphology when grown for an extended period of time on WLN or gelatin-based media (Morris and Hough, 1958). For differentiation of lager strains, a number of methods have been used with varying degrees of success. Ultimately, without undertaking molecular analyses, often the best means of determining strain identity is to conduct some form of test to examine known fermentation characteristics, such as flocculation capacity or sugar utilization patterns.

Variants create similar problems and typically necessitate precise phenotypic analyses, as is the case for cultures that display enhanced or reduced flocculation patterns. In this instance, flocculation tests such as the modified Helm assay (Bendiak *et al.*, 1996) can be applied. Respiratory-deficient petite mutants (see Chapter 2) can be detected by growth on glycerol-based media since petites are unable to grow only on media containing only non-fermentable carbon sources. Alternatively, these can also be identified using the TTC (triphenyl tetrazolium chloride) overlay technique. In this method, the colourless TTC stain is poured over colonies present on the surface of an agar plate. The TTC dye is reduced by respiratory-sufficient yeasts to form a pinkish red colour. Respiratory-deficient yeast colonies cannot reduce the dye and remain white (Ogur *et al.*, 1957).

Alternative methods for yeast detection and identification

Although the detection methods described above provide accurate and reliable information on the presence of culturable (live) cells, these tests can be labour-intensive and require anywhere from 3–14 days to obtain results. This has prompted alternative techniques to be developed that are less laborious and provide results more rapidly, resulting in a more immediate, rather than retrospective, process analysis. Although there are many methods available for detection of beer-spoiling yeasts, the most widely employed are those that utilize polymerase chain reaction (PCR) technology. This technique permits the amplification of DNA fragments (corresponding to specific organisms) by several orders of magnitude over a few hours. Consequently, using PCR it is possible to amplify extremely small amounts of DNA to levels that can be used for detection and identification purposes; an approach that lends itself to forensic science, medical microbiology, food microbiology, as well as brewing quality assurance. The PCR procedure consists of repeated cycles of DNA denaturation, primer annealing, and extension by DNA polymerase, and relies on the selection of a DNA sequence that adequately differentiates the genotype of one organism from another. The amplified DNA is then visualized either by gel electrophoresis in conjunction with a DNA stain such as ethidium bromide (standard PCR), or by using primers in conjunction with a probe to provide a quantitative indication of PCR amplification through the production of fluorescence (quantitative real-time PCR, or qPCR). One basic example is to employ primers designed for detection of the *STA1* gene responsible for diastatic activity in *Saccharomyces* strains (Yamauchi *et al.*, 1998), although a range of DNA sequences that are unique to different species can be exploited.

It is relatively common for larger brewing companies to use commercially available PCR kits, as these provide a rapid and reliable means of both detecting and identifying beer-spoiling yeasts in a single reaction. Implementation of PCR technology has gained momentum since the first commercially available kits became available in the early 2000s. These have gained in popularity, partly because of the development of equivalent methodology for bacterial detection, which allows a range of brewing microbes to be investigated in parallel. However, the primary reason is that such methods are easy to perform on a routine basis and provide results that are often supported by data available on-line, allowing for increased confidence in the data obtained. Despite this, it should be recognized that such tools are not amenable to many brewery laboratories since the level of expertise required is relatively high and costs are often prohibitive. In addition, there are still issues with regard to inhibition of the PCR reaction by beer components, and the fact that PCR methods do not easily discriminate between live and dead cells. Finally, the level of sensitivity remains low (around 1×10^2 – 1 10^3 cells/ml), which often necessitate a period of 'pre-enrichment' to ensure that false negatives are eliminated. As a result of these factors, many breweries continue to perform traditional testing as a back-up measure, indicating that there may still be an element of caution in relying solely on PCR. Despite this, due to the advances in next-generation sequencing, it is likely that DNA-based techniques will continue to develop. The capacity to identify beer-spoiling yeast species by mechanisms such as fluorescence *in situ* hybridization (FisH) (Xufre *et al.*, 2006), or PCR analysis based on single nucleotide polymorphisms (SNPs) (Wilkening *et al.*, 2013) or ribosomal DNA sequence divergence offer interesting opportunities for the future.

Molecular methods for identification of beer-spoiling yeast species

As described above (see 'Detection and identification'), methods for identification of yeasts were historically based on physiological characteristics, including sexual reproduction (where present), cell morphology, growth characteristics, and biochemical features. Some of these criteria are still utilized today for characterizing yeasts, although for identification purposes analysis of DNA homology is increasingly the most widely accepted method. Once a beer-spoiling yeast has been isolated, it can be identified to the genus and species level by analysing specific regions of DNA. These regions are typically selected based on specific criteria and should demonstrate a high degree of interspecific polymorphism and a low or non-existent intraspecific polymorphism. This enables the sequence to be used to accurately differentiate between species, but also provides a robust means of ensuring that all strains within a species yield identical results. There are several regions of the genome that meet these criteria, with the most widely used being those located within the ribosomal DNA (rDNA) sequence, coding for the ribosomal RNA. The rDNA contains tracts designated as 26S, 5S, 18S, 5.8S and 26S, which are highly conserved and arranged in tandem units (Fig. 11.7). These genes are separated by internal transcribed spacer (ITS) regions and non-transcribed spacer (NTS) regions, both of which are non-coding, highly variable, and match the criteria outlined above for classification purposes. Studies have shown that the ITS region is particularly useful for identification of yeasts to the genus and species level and this has recently been proposed for adoption as the primary fungal barcode marker (Schoch *et al.*, 2012). In addition, a small region of the 26S subunit known as the D1/D2 domain can be employed to accurately identify yeast species. Analysis of this region is important from a phylogenetic perspective as it has previously acted as one of the predominant means of confirming existing taxonomic groupings and in identifying novel yeast species (Kurtzman and Robnett, 1998; Weiss *et al.*, 2013).

D1/D2 sequencing

The D1/D2 region of the 26S subunit was previously adopted as a convenient mechanism for characterizing yeast to the species level. By analysing this sequence, Kurtzman and Robnett (1998) were able to identify a large number of yeast species based on sequence divergence within the D1/D2 domain. These authors demonstrated that strains belonging to a species exhibited less than 1% difference within the D1/D2 sequence, while distinct species had a much greater variation. Analysis of the D1/D2 domain can be conducted using a

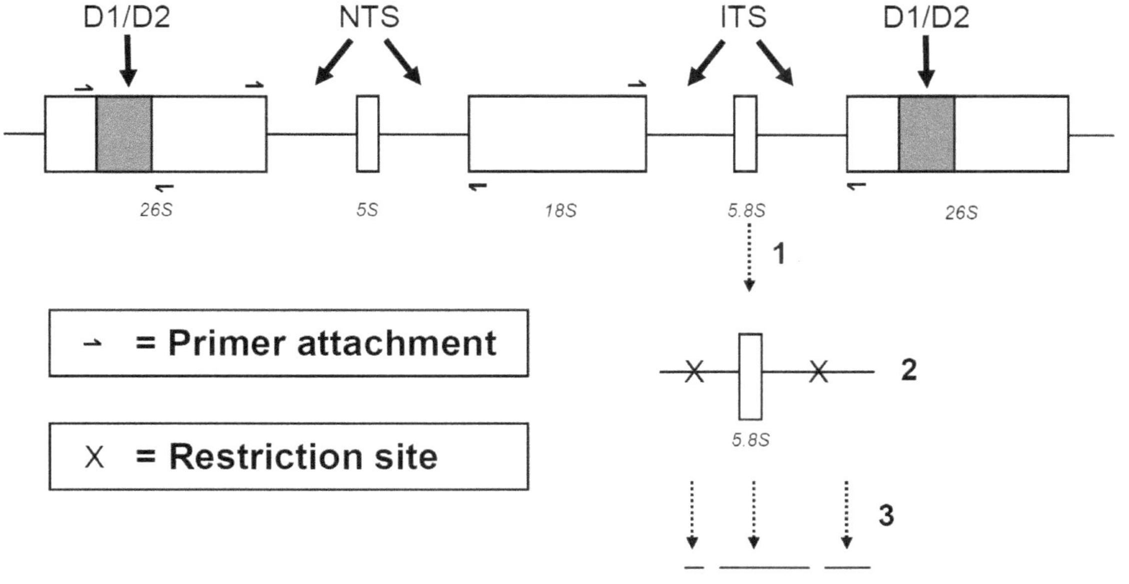

Figure 11.7 Identification of yeast species based on analysis of ribosomal RNA gene sequences. Several sections of the rRNA gene region are variable between species but conserved between strains, including the D1/D2 domain of the 26S subunit, and the ITS and NTS regions. The D1/D2 domain is typically analysed by PCR followed by sequencing of the ≈600 bp fragment. The ITS region (including the 5.8S region) can be amplified in its entirety by primers designed to target the ends of the 18S (small subunit) and 26S (large subunit) ribosomal RNA genes (1). The PCR product is then cut using restriction enzymes (2) to yield a number of DNA fragments that can be resolved using gel electrophoresis (3). The sizes of each resulting fragment (along with that of the original PCR product) are determined by comparison to a DNA ladder (molecular weight marker). For analysis of species using ITS or NTS RFLP, or D1/D2 sequencing, identification is based on comparison to the literature, or to an internal or external database.

Table 11.6 Primer sequences employed for the identification of yeast species

Primers	Sequence	Target	Reference
ITS1	5′ TCC GTA GGT GAA CCT GCG G 3′	ITS region	White *et al.* (1990), Kurtzman and Robnett (1991)
ITS4	5′ TCC TCC GCT TAT TGA TAT GC 3′		
ITS4	5′ TCC TCC GCT TAT TGA TAT GC 3′	ITS region	White *et al.* (1990), Scoch *et al.* (2012)
ITS5	5′ GGA AGT AAA AGT CGT AAC AAG G 3′		
r-1234	5′ AAC GGT GCT TTC TGG TAG 3′	NTS2 region	Nguyen and Gaillardin (1997)
r-2516	5′ TGT CTT CAA CTG CTT T 3′		
NL1	5′ GCA TAT CAA TAA GCG GAG GAA AA 3′	D1/D2 Domain	O'Donnell (1993), Kurtzman and Robnett (1998)
NL4	5′ GGT GCG TGT TTC AAG ACG G 3′		

combination of PCR and DNA sequencing. PCR is required to amplify the region of interest, generating an amplicon of ≈600 bp based on targeted primers (Table 11.6). This region is then sequenced to allow for the generation of qualitative data that can be analysed using sequence alignment software or compared directly to a database. There are several sources of reference that can be used, many of which are freely available online. This is most easily performed by conducting a BLAST (Basic Local Alignment Search Tool) analysis of deposited sequences that are maintained in databases such as GenBank or the YeastIP gene database for molecular taxonomy and phylogeny of yeasts (Weiss *et al.*,

2013). Either approach can be used to calculate phylogenetic relationships, as well as for the identification of unknown species.

ITS/NTS PCR

Analysis of the ITS region has gained significant traction and is currently preferred to other methods for yeast identification (Schoch *et al.*, 2012). However, in some instances the NTS region can also be used to good effect, being useful for the differentiation of *Saccharomyces sensu stricto* yeasts (Pulvirenti *et al.*, 2000) and *Kluyveromyces* species (Nguyen *et al.*, 2000). Analysis of these sequences typically involves a preliminary PCR reaction using specific primers (Table 11.6) to amplify the designated region, and to create a sufficient quantity of DNA, which can then be analysed by alternative methods. For analysis of the ITS region, this can include analysis for total length polymorphism, restriction fragment length polymorphism (RFLP), DNA probe hybridization, and DNA sequencing. Of these, DNA sequencing is the most precise due to the vast number of full-length ITS sequences that have been deposited in databases such as GenBank. However, for brewery troubleshooting purposes RFLP tends to be used preferentially, simply due to the speed and ease at which the technique can be conducted. RFLP is a term used to describe the analysis of DNA fragments, obtained by the use of specific enzymes that cut at pre-designated locations within a DNA sequence. Restriction enzymes, originally isolated from bacteria, cut at recognition sites within the DNA giving rise to multiple fragments that can then be separated by electrophoresis (Fig. 11.7). The application of these enzymes provides a useful means of identifying yeast species, since the enzymes utilized can be selected based on their capacity to cut at a location known to highlight polymorphisms between species and genera. Since the development of the technique, a series of studies have been performed, primarily to identify important yeast strains isolated from wine (Guillamón *et al.*, 1998; Esteve-Zarzoso *et al.*, 1999), but also from clinical sources (Trost *et al.*, 2004; Leaw *et al.*, 2006) and a range of industrial environments including breweries (Pham *et al.*, 2011).

The primers used for amplification of the ITS region are known to have near universal function with ascomycete yeasts (Kurtzman and Robnett, 1991) and the majority of studies employ the restriction enzymes CfoI, HinfI and HaeIII for increasing specificity. This is primarily because they have been proven to have highly suitable restriction sites. The size of the ITS region varies between species from approximately 300 bp to 1000 bp, which does allow for some differentiation, particularly at the genus level. However, individual species can only be reliably differentiated based on either RFLP or sequence-based analysis as described above. Due to a combination of the size of the ITS fragment and the restriction enzyme cutting site, the size and number of restriction fragments yielded by RFLP analysis differs according to species (Table 11.7). It should be noted that strains belonging to the *Saccharomyces sensu stricto* group exhibit ITS sequences in the region of 840–880 bp in size, a feature which can be used to immediately differentiate them from non-*Saccharomyces* contaminants. Within the sensu stricto group there are less obvious differences that can cause issues, especially if contamination or a mix-up between production yeasts is suspected. In such instances, further identification is required. There are a large number of molecular-based techniques that can be used to differentiate yeasts to the strain level. While this list is by no means exhaustive, methods that are known to be successful include analysis of yeast transposons (Wightman *et al.*, 1996), micro- and mini- satellites (Baleiras Couto *et al.*, 1996; Pérez *et al.*, 2001; Schuller *et al.*, 2004), chromosomal karyotyping (Schwartz and Cantor, 1984; Casey *et al.*, 1990; Oakley-Gutowski *et al.*, 1991; Boekhout *et al.*, 1993; Casey, 1996), inter-delta sequence fingerprinting (Ness *et al.*, 1993; Legras and Karst, 2003), mitochondrial DNA profiling (Aigle *et al.*, 1984; López *et al.*, 2001), and analysis of single nucleotide polymorphisms (SNPs) (Ben-Ari *et al.*, 2005; Schacherer *et al.*, 2007, 2009).

Conclusions

Irrespective of their characteristics and threat to the process, the presence of beer-spoiling yeasts within the brewery is often an indication of poor general hygiene. If left to proliferate, unwanted yeasts can lead to a range of issues, including spoilage of raw materials and the production of off-flavours, as well as fermentation inconsistencies and poor ethanol yields. The occurrence of spoilage yeasts can also lead to confusion within breweries, since many

Table 11.7 Size (base pairs) of PCR products and restriction fragments derived from analysis of the ITS region of ribosomal DNA isolated from beer-spoiling species of yeast. Fragment sizes may not be absolute due to the use of standard gel electrophoresis, which can be subjective. Furthermore, fragments of less than 30–40 bp in size are difficult to visualize and may account for discrepancies between restriction fragment sizes and the PCR product length

Species	Reference	Size of PCR product (bp)	Size of restriction fragments (bp)		
			CfoI	HaeIII	HinfI
B. bruxellensis	Guillamón *et al.*, 1998	500	230, 130, 80	375, 105	265, 215
C. stellata	Guillamón *et al.*, 1998	500	220, 130	480	260, 240
D. hansenii	Esteve-Zarzoso *et al.*, 1999	650	300, 300, 50	420, 150, 90	325, 325
H. uvarum	Guillamón *et al.*, 1998	760	320, 315, 105	760	360, 200, 180
K. marxianus	Esteve-Zarzoso *et al.*, 1999	740	285, 185, 140, 100	655, 80	240, 185, 120, 80, 65, 50
L. saturnus	Jeyaram *et al.*, 2008	630	470, 70	310, 130, 90, 60	300
P. anomola	Esteve-Zarzoso *et al.*, 1999	650	575	600, 50	310, 310
R. glutinis	Esteve-Zarzoso *et al.*, 1999	640	320, 240, 80	430, 210	340, 225, 75
S. cerevisiae	Guillamón *et al.*, 1998	880	385, 365	32, 220, 180, 145	365, 155
S. pombe	Esteve-Zarzoso *et al.*, 1999	1050	600, 400	1050	600, 450
T. delbreuckii	Esteve-Zarzoso *et al.*, 1999	800	330, 220, 150, 100	800	410, 380
Z. bailii	Esteve-Zarzoso *et al.*, 1999	790	320, 270, 95, 95	690, 90	340, 225, 160, 55

strains are able to survive within the culture yeast population at relatively low levels and without any noticeable impact on product quality. For example, there have been occasions where brewing yeast populations have shown a cyclical deterioration in quality towards the end of fermentation and in storage vessels (in the region of a 3–5% reduction in overall viability). This can be linked to the physiological characteristics of non-*Saccharomyces* yeasts, which are often pH and ethanol sensitive. These individuals are less able to tolerate conditions encountered at the end of fermentation and dead cells are therefore a function of the presence of physiologically 'weaker' yeast contaminants. Hence, while there are certain broader aspects of contamination that are simple to define, it is important to be aware that there may also be a range of additional associated affects.

Although hygiene in the brewery is usually attended to assiduously, contamination of process stages or the final product with yeasts may be more frequent than the majority of brewers would acknowledge. *Saccharomyces* beer-spoiling yeasts, which include other production strains and variants, pose the greatest threat to the fermentation process since they are able to compete directly with the culture yeast. Non-*Saccharomyces* yeasts such as *Brettanomyces, Candida, Kluyveromyces, Pichia* and *Torulaspora* are opportunistic spoilers that can be found in the initial aerobic stages of fermentation and in unpasteurized cask beers. *Brettanomyces* is a particular threat in cask beer due to its ability to form high concentrations of acetic acid. It should be noted that although breweries may encounter repeat problems with specific yeast strains, it is not common for isolated outbreaks involving different species to occur.

Currently, the preferred methodology in breweries for the detection of yeast contaminants remains the use of nutrient media. Although generally slow, the technique is robust and allows for the detection of small quantities of beer-spoiling yeast, even when large numbers of brewing yeast are present. However, cultivation-based methods are increasingly used in conjunction with more rapid techniques. Real-time PCR analysis in particular offers an alternative means of detecting and identifying beer-spoiling yeasts, leading to faster decision time and more efficient quality-control procedures. If the presence of a beer-spoiling yeast is confirmed, identification can be an aid to fully understanding the root causes of the problem. Eradication should be based on a systematic procedure of investigation and sterilization to eliminate the source of the issue.

References

Adam, A.C., Latorre-Garcia, L., and Polaina, J. (2004). Structural analysis of glucoamylase encoded by the *STA1* gene of *Saccharomyces cerevisiae* (var. *diastaticus*). Yeast *21*, 379–388.

Aigle, M., Erbs, D., and Moll, M. (1984). Some molecular structures in the genome of lager brewing yeasts. J. Am. Soc. Brew. Chem. *42*, 1–7.

Andersen, K.S., Bojsen, R.K., Gro Rejkjaer Sorensen, L., Weiss Nielsen, M., Lisby, M., Folkesson, S.A., and Regenberg, B. (2014). Genetic basis for *Saccharomyces cerevisiae* biofilm in liquid medium. Genes, Genomes, Genet. *4*, 1671–1680.

Baleiras Couto, M.M., Eijsma, B., Hofstra, H., Huis in't Veld, J.H.J., and van der Vossen, J.M.B.M. (1996). Evaluation of molecular typing techniques to assign genetic diversity among *Saccharomyces cerevisiae* strains. Appl. Environ. Microbiol. *62*, 41–46.

Barnett, J.A., and Entian, K.-D. (2005). A history of research on yeasts 9: regulation of sugar metabolism. Yeast *22*, 835–894.

Barnett, J.A. (1992). Some controls on oligosaccharide utilization by yeasts: the physiological basis of the Kluyver effect. FEMS Microbiol. Lett. *100*, 371–378.

Barnett, J.A. (2004). A history of research on yeasts 8: taxonomy. Yeast *21*, 1141–1193.

Bauer, J., and Wendland, J. (2007). *Candida albicans* Sfl1 suppresses flocculation and filamentation. Eukaryot. Cell *6*, 1736–1744.

Ben-Ari, G., Zenvirth, D., Sherman, A., Simchen, G., Lavi, U., and Hillel, J. (2005). Application of SNPs for assessing biodiversity and phylogeny among yeast strains. Heredity *95*, 493–501.

Bendiak, D., Van der Aar, P., Barbero, F., Benzing, P., Berndt, R., Carrick, K., Dull, C., Dunn, S., Eto, M., Gonzalez, M., Hayashi, N., Lawrence, D., Miller, J., Pugh, T., Rashel, L., Rossmore, K., Smart, K., Sobczak, J., Speers, A., and Casey, G. (1996). Yeast flocculation by absorbance. J. Am. Soc. Brew. Chem. *54*, 245–248.

Boekhout, T., Renting, M., Scheffers, W.A., and Bosboom, R. (1993). The use of karyotyping in the systematics of yeasts. Antonie Van Leeuwenhoek *63*, 157–163.

Box, W., Bendiak, D., Castonguay, L., Feliciano, S., Fischborn, T., Gibson, B., Kuenker, M., Lodolo, B., Miller, M., Nicholls, S., Thiele, F., White, L., and Powell, C. (2012). Differentiation of ale and lager yeast strains by rapid X-α-Gal analysis. J. Am. Soc. Brew. Chem. *70*, 313–315.

Brady, B.L. (1958). Wild yeasts in pitching yeast. J. Inst Brew. *64*, 304–307.

Brown, M.L. (1970). Development of chains of cells of *Saccharomyces cerevisiae* during fermentation. J. Inst. Brew. *76*, 61–65.

Brumstead, D.D., and Glenister, P.R. (1962). The viability of minimal populations of a wild yeast in beer. Proc. Am. Soc. Brew. Chem. 72.

Canonico, L., Agarbati, A., Comitini, F., and Ciani, M. (2016). *Torulaspora delbrueckii* in the brewing process: A new approach to enhance bioflavour and to reduce ethanol content. Food Microbiol. *56*, 45–51. http://dx.doi.org/10.1016/j.fm.2015.12.005

Casey, G.P. (1996). Practical applications of pulsed field electrophoresis and yeast chromosome fingerprinting in brewing QA and R&D. MBAA Tech. Quart. *33*, 1–10.

Casey, G.P., Pringle, A.T., and Erdmann, P.A. (1990). Evaluation of recent techniques used to identify individual strains of *Saccharomyces* yeasts. J. Am. Soc. Brew. Chem. *48*, 100–106.

Chant, J., and Pringle, J.R. (1995). Patterns of bud-site selection in the yeast *Saccharomyces cerevisiae*. J. Cell Biol. *129*, 751–765.

Christiaens, J.F., Franco, L.M., Cools, T.L., De Meester, L., Michiels, J., Wenseleers, T., Hassan, B.A., Yaksi, E., and Verstrepen, K.J. (2014). The fungal aroma gene ATF1 promotes dispersal of yeast cells through insect vectors. Cell Rep. *9*, 425–432. http://dx.doi.org/10.1016/j.celrep.2014.09.009

Clausen, M., Lamb, C.J., Megnet, R., and Doerner, P.W. (1994). PAD1 encodes phenylacrylic acid decarboxylase which confers resistance to cinnamic acid in *Saccharomyces cerevisiae*. Gene *142*, 107–112.

Coelho, M.A., Gonçalves, C., Sampaio, J.P., and Gonçalves, P. (2013). Extensive intra-kingdom horizontal gene transfer converging on a fungal fructose transporter gene. PLoS Genet. *9*, e1003587. http://dx.doi.org/10.1371/journal.pgen.1003587

Conrad, M., Schothorst, J., Kankipati, H.N., Van Zeebroeck, G., Rubio-Texeira, M., and Thevelein, J.M. (2014). Nutrient sensing and signaling in the yeast *Saccharomyces cerevisiae*. FEMS Microbiol. Rev. *38*, 254–299. http://dx.doi.org/10.1111/1574-6976.12065

Cowley, T., Belden, E., Besora, N.F., Orive Camprubi, M., Carlucci, C., Cheung, A., Coony, A., Crisafulli, V., Davis, M., Depalma, S., Dornberg, M., Eideman, R., Emeline Gless, E., Firli, R., Fisher, G., Gibson, B., Gotsch, D., Grandell, J., Gutierrez, C., Hutzler, M., Juanes, J.C., Krogerus, K., Lemar, H., Marchesiani, L., Mather, S., Monsour, B., Norwood, R., Palausky, K., Pawlowsky, K., Perez, B., Radke, B., Rice, C., Rojano, V., Schurmeyer, E., Tafoya Gibbs, K., Tretter, K., Tynan, S., Vinkmeier, K., Vogeser, G., Voigt, J., and Pachello, C. (2016). Phenolic yeast detection. J. Am. Soc. Brew. Chem. In Press.

Cubells Martinez, X.C., Narbad, A., Carter, A.C., and Stratford, M. (1996). Flocculation of the yeast *Candida famata* (*Debaryomyces hansenii*): an essential role for peptone. Yeast *12*, 415–423.

Curtin, C.D., Langhans, G., Henschke, P.A., and Grbin, P.R. (2013). Impact of Australian *Dekkera bruxellensis* strains grown under oxygen-limited conditions on model wine composition and aroma. Food Microbiol. *36*, 241–247. http://dx.doi.org/10.1016/j.fm.2013.06.008

Dakal, T.C., Solieri, L., and Giudici, P. (2014). Adaptive response and tolerance to sugar and salt stress in the food yeast *Zygosaccharomyces rouxii*. Int. J. Food Microbiol. *185*, 140–157. http://dx.doi.org/10.1016/j.ijfoodmicro.2014.05.015

Davies, S., Sykes, T., Philips, M., and Hancock, J. (2015). Hygienic design and cleaning-in-place (CIP) systems in breweries. In Brewing Microbiology: Managing microbes, ensuring quality and valorising waste. A.E. Hill ed (Cambridge, UK; Woodhead Publishing), pp. 221–239.

De Angelo, J., and Siebert, K.J. (1987). A new medium for the detection of wild yeast in brewing culture yeast. J. Am. Soc. Brew. Chem. *45*, 135–140.

De Deken, R.H. (1966). The Crabtree effect: a regulatory system in yeast. J. Gen. Microbiol. *44*, 149–156. http://dx.doi.org/10.1099/00221287-44-2-149

Dujon, B. (2010). Yeast evolutionary genomics. Nat. Rev. Genet. *11*, 512–524. http://dx.doi.org/10.1038/nrg2811

Eddy, A.A. (1958). Composite nature of the flocculation process of top and bottom strains of *Saccharomyces*. J. Inst. Brew. *64*, 143–151.

El-Behhari, M., Géhin, G., Coulon, J., and Bonaly, R. (2000). Evidence for a lectin in *Kluyveromyces* sp. that is involved in co-flocculation with *Schizosaccharomyces pombe*. FEMS Microbiol. Lett. *184*, 41–46.

Erten, H., and Campbell, I. (2001). The production of low-alcohol wines by aerobic yeasts. J. Inst. Brew. *107*, 207–215.

Esteve-Zarzoso, B., Belloch, C., Uruburu, F., and Querol, A. (1999). Identification of yeasts by RFLP analysis of the 5.8S rRNA gene and the two ribosomal internal transcribed spacers. Int. J. Syst. Bacteriol. *49*, 329–337. http://dx.doi.org/10.1099/00207713-49-1-329

Flannigan, B. (1974). Distribution of seed-borne micro-organisms in naked barley and wheat before harvest. Trans. Br. Mycol. Soc. *62*, 51–58.

Fukuhara, H. (2003). The Kluyver effect revisited. FEMS Yeast Res. *3*, 327–331.

Gams, W. (2016). Recent changes in fungal nomenclature and their impact on naming of microfungi. In Biology of Microfungi. D-W Lei ed. (New York, USA; Springer), pp. 7–24.

Gimeno, C.J., Ljungdahl, P.O., Styles, C.A., and Fink, G.R. (1992). Unipolar cell divisions in the yeast *S. cerevisiae* lead to filamentous growth: regulation by starvation and *RAS*. Cell *68*, 1077–1090.

Godoy, L., Garcia, V., Pena, R., Martinez, C., and Ganga, M.A. (2014). Identification of the *Dekkera bruxellensis* phenolic acid decarboxylase (PAD) gene responsible for wine spoilage. Food Control *45*, 81–86.

Goffeau, A., Barrell, B.G., Bussey, H., Davis, R.W., Dujon, B., Feldmann, H., Galibert, F., Hoheisel, J.D., Jacq, C., Johnston, M., Louis, E.J., Mewes, H.W., Murakami, Y., Philippsen, P., Tettelin, H., and Oliver, S.G. (1996). Life with 6000 genes. Science *274*, 546–567.

Goffrini, P., Ferrero, I., and Donnini, C. (2002). Respiration-dependent utilization of sugars in yeasts: a determinant role for sugar transporters. J. Bacteriol. *184*, 427–432.

Griffin, D.H. (1994). Chemical requirements for growth. In Fungal Physiology. D.H. Griffin ed (New York, USA; Wiley-Liss), pp. 130–157.

Gu, Z., Steinmetz, L.M., Gu, X., Scharfe, C., Davis, R.W., and Li, W.H. (2003). Role of duplicate genes in genetic robustness against null mutations. Nature *421*, 63–66. http://dx.doi.org/10.1038/nature01198

Guillamón, J.M., Sabaté, J., Barrio, E., Cano, J., and Querol, A. (1998). Rapid identification of wine yeast species based on RFLP analysis of the ribosomal internal transcribed spacer (ITS) region. Arch. Microbiol. *169*, 387–392.

Hagman, A., and Piškur, J. (2015). A study on the fundamental mechanism and the evolutionary driving forces behind aerobic fermentation in yeast. PLoS ONE *10*, e0116942. http://dx.doi.org/10.1371/journal.pone.0116942

Hagman, A., Säll, T., and Piškur, J. (2014). Analysis of the yeast short-term Crabtree effect and its origin. FEBS. J. *281*, 4805–4814. http://dx.doi.org/10.1111/febs.13019

Hartwell, L.H. (1974). *Saccharomyces cerevisiae* cell cycle. Bacteriol. Rev. *38*, 164–198.

Heard, G.M., and Fleet, G.H. (1985). Growth of Natural Yeast Flora during the Fermentation of Inoculated Wines. Appl. Environ. Microbiol. *50*, 727–728.

Hemmons, L.M. (1954). Wild yeasts in draught beer I. An exploratory study. J. Inst. Brew. *60*, 288–291.

Heresztyn, T. (1986). Metabolism of volatile phenolic compounds from hydroxycinnamic acids by *Brettanomyces* yeast. Arch. Microbiol. *146*, 96–98.

Jespersen, L., and Jakobsen, M. (1996). Specific spoilage organisms in breweries and laboratory media for their detection. Int. J. Food Microbiol. 33, 139–155.

Jespersen, L., van der Aa Kühle, A., and Petersen, K.M. (2000). Phenotypic and genetic diversity of *Saccharomyces* contaminants isolated from lager breweries and their phylogenetic relationship with brewing yeasts. Int. J. Food Microbiol. *60*, 43–53.

Jeyaram, K., Mohendro Singh, W., Capece, A., and Romano, P. (2008). Molecular identification of yeast species associated with 'Hamei'– A traditional starter used for rice wine production in Manipur, India. Int. J. Food Microbiol. *124*, 115–125.

Jin, R., Dobry, C.J., McCown, P.J., and Kumar, A. (2008). Large-scale analysis of yeast filamentous growth by systematic gene disruption and overexpression. Mol. Biol. Cell *19*, 284–296.

Johnston, G.C., Pringle, J.R., and Hartwell, L.H. (1977). Coordination of growth with cell division in *Saccharomyces cerevisiae*. Exp. Cell. Res. *105*, 79–98.

Jolly, N.P., Varela, C., and Pretorius, I.S. (2014). Not your ordinary yeast: non-*Saccharomyces* yeasts in wine production uncovered. FEMS Yeast Res. *14*, 215–237. http://dx.doi.org/10.1111/1567-1364.12111

Käppeli, O. (1986). Regulation of carbon metabolism in *Saccharomyces cerevisiae* and related yeasts. Adv. Microb. Physiol. *28*, 181–209.

Keeling, P.J., and Palmer, J.D. (2008). Horizontal gene transfer in eukaryotic evolution. Nat. Rev. Genet. *9*, 605–618. http://dx.doi.org/10.1038/nrg2386

Kejžar, A., Gobec, S., Plemenitaš, A., and Lenassi, M. (2013). Melanin is crucial for growth of the black yeast *Hortaea werneckii* in its natural hypersaline environment. Fungal. Biol. *117*, 368–379. http://dx.doi.org/10.1016/j.funbio.2013.03.006

Kimura, T., Kitamoto, N., Matsuoka, K., Nakamura, K., Iimura, Y., and Kito, Y. (1993). Isolation and nucleotide sequences of the genes encoding killer toxins from *Hansenula mrakii* and *H. saturnus*. Gene *137*, 265–270.

Kobayashi, O., Hayashi, N., Kuroki, R., and Sone, H. (1998). Region of FLO1 proteins responsible for sugar recognition. J. Bacteriol. *180*, 6503–6510.

Kobi, D., Zugmeyer, S., Potier, S., and Jaquet-Gutfreund, L. (2004). Two-dimensional protein map of an 'ale'-brewing yeast strain: proteome dynamics during fermentation. FEMS Yeast Res. *5*, 213–230.

Kron, S.J., Styles, C.A., and Fink, G.R. (1994). Symmetric cell division in pseudohyphae of the yeast *Saccharomyces cerevisiae*. Mol. Biol. Cell *5*, 1003–1022.

Kuepfer, L., Sauer, U., and Blank, L.M. (2005). Metabolic functions of duplicate genes in *Saccharomyces cerevisiae*. Genome Res. *15*, 1421–1430.

Kuniyuki, A.H., Rous, C., and Sanderson, J.L. (1984). Enzyme-linked immunosorbent assay (ELISA) detection of *Brettanomyces* contaminants in wine production. Am. J. Enol. Vitic. *35*, 143–145.

Kuranda, M.J., and Robbins, P.W. (1991). Chitinase is required for cell separation during growth of *Saccharomyces cerevisiae*. J. Biol. Chem. *266*, 19758–19767.

Kurtzman, C., Fell, J.W., and Boekhout, T. (eds) (2011). Pichia and associated genera. In The yeasts: a taxonomic study fifth edition, Part IVA. (Amsterdam, Netherlands; Elsevier). pp. 298.

Kurtzman, C.P., and Robnett, C.J. (1991). Phylogenetic relationships among species of *Saccharomyces*, *Schizosaccharomyces*, *Debaryomyces* and *Schwanniomyces* determined from partial ribosomal RNA sequences. Yeast *7*, 61–72.

Kurtzman, C.P., and Robnett, C.J. (1998). Identification and phylogeny of ascomycetous yeasts from analysis of nuclear large subunit (26S) ribosomal DNA partial sequences. Antonie Van Leeuwenhoek *73*, 331–371.

Kurtzman, C.P., and Robnett, C.J. (2013). Relationships among genera of the *Saccharomycotina* (*Ascomycota*) from multigene phylogenetic analysis of type species. FEMS Yeast Res. *13*, 23–33. http://dx.doi.org/10.1111/1567-1364.12006

Kurtzman, C.P., Robnett, C.J., and Basehoar-Powers, E. (2008). Phylogenetic relationships among species of *Pichia*, *Issatchenkia* and *Williopsis* determined from multigene sequence analysis, and the proposal of *Barnettozyma* gen.nov., *Lindnera* gen.nov. and *Wickerhamomyces* gen.nov. FEMS Yeast Res. *8*, 939–954.

Lacroix, B., and Citovsky, V. (2016). Transfer of DNA from bacteria to eukaryotes. MBio *7*, e00863–16. http://dx.doi.org/10.1128/mBio.00863-16

Lagunas, R. (1979). Energetic irrelevance of aerobiosis for *S. cerevisiae* growing on sugars. Mol. Cell. Biochem. *27*, 139–146.

Laitila, A., Wilhelmson, A., Kotaviita, E., Olkku, J., Home, S., and Juvonen, R. (2006). Yeasts in an industrial malting ecosystem. J. Ind. Microbiol. Biotechnol. *33*, 953–966. http://dx.doi.org/10.1007/s10295-006-0150-z

Larsson, S., Nilvebrant, N.O., and Jonsson, L.J. (2001). Effect of overexpression of *Saccharomyces cerevisiae* Pad1p on the resistance to phenylacrylic acids and lignocellulose hydrolysates under aerobic and oxygen-limited conditions. Appl. Microbiol. Biotechnol. *57*, 167–174.

Leaw, S.N., Chang, H.C., Sun, H.F., Barton, R., Bouchara, J.P., and Chang, T.C. (2006). Identification of medically important yeast species by sequence analysis of the internal transcribed spacer regions. J. Clin. Microbiol. *44*, 693–699.

Legras, J.L., and Karst, F. (2003). Optimisation of interdelta analysis for *Saccharomyces cerevisiae* strain characterisation. FEMS Microbiol. Lett. *221*, 249–255.

Lin, Y. (1981). Formulation and testing of cupric sulphate medium for wild yeast detection. J. Inst. Brew. *87*, 151–154.

López, V., Querol, A., Ramón, D., and Fernández-Espinar, M.T. (2001). A simplified procedure to analyse mitochondrial DNA from industrial yeasts. Int. J. Food Microbiol. *68*, 75–81.

Maoura, N., Mbaiguinam, M., Nguyen, H.V., Gaillardin, C., and Pourquie, J. (2005). Identification and typing of the yeast strains isolated from bili bili, a traditional sorghum beer of Chad. African J. Biotech. *4*, 646–656.

Marshall, W.F., Young, K.D., Swaffer, M., Wood, E., Nurse, P., Kimura, A., Frankel, J., Wallingford, J., Walbot, V., Qu, X.. and Roeder, A.H.K. (2012). What determines cell size? BMC Biol. *10*, 101. http://dx.doi.org/10.1186/1741-7007-10-101

Martin, C.P., and Siebert, K.J. (1992). Evaluation of multinitrogen source media for wild yeast detection in brewing culture yeast. J. Am. Soc. Brew. Chem. *50*, 134–138.

Mata-Gómez, L.C., Montañez, J.C., Méndez-Zavala, A., and Aguilar, C.N. (2014). Biotechnological production of carotenoids by yeasts: an overview. Microb. Cell Fact. *13*, 12. http://dx.doi.org/10.1186/1475-2859-13-12

Mbawala, A., al Mahmood, S., Loppinet, V., and Bonaly, R. (1990). Acetolysis and 1H NMR studies on mannans isolated from very flocculent and weakly flocculent cells of *Pichia pastoris* IFP 206. J. Gen. Microbiol. *136*, 1279–1284. http://dx.doi.org/10.1099/00221287-136-7-1279

McNeill, J., Barrie, F.R., Buck, W.R., Demoulin, V., Greuter, W., Hawksworth, D.L., Herendeen, P.S., Knapp, S., Marhold, K., Prado, J., Prud'homme van Reine, W.F., Smith, G.E., Wiersema, J.H., and Turland, N.J. (eds.) (2012). International code of nomenclature for algae, fungi, and plants (Melbourne code) adopted by the eighteenth international botanical congress, Melbourne, Australia, July 2011. [Regnum Vegetabile no. 154.] Ruggell: A.R.G. Gantner Verlag.

Miki, B.L., Poon, N.H., James, A.P., and Seligy, V.L. (1982). Possible mechanism for flocculation interactions governed by gene FLO1 in *Saccharomyces cerevisiae*. J. Bacteriol. *150*, 878–889.

Miller, M.W., and Phaff, H.J. (1962). Successive Microbial Populations in Calimyrna Figs. Appl. Microbiol. *10*, 394–400.

Misihairabgwi, J.M., Kock, L., Pretorius, E., Pohl, C., and Zvauya, R. (2015). Characterisation of yeasts isolated from traditional opaque beer beverages brewed in Zimbabwean households. African J. Microbiol. Res. *9*, 549–556.

Morris, E.O., and Hough, J.S. (1958). Some aspects of the giant colony characters of *Saccharomyces cerevisiae*. J. Inst. Brew. *62*, 466–468.

Mrak, E.M., Phaff, H.J., and Mackinney, G. (1949). A Simple test for carotenoid pigments in yeasts. J. Bacteriol. *57*, 409–411.

Müller, I. (1971). Experiments on ageing in single cells of *Saccharomyces cerevisiae*. Arch. Mikrobiol. *77*, 20–25.

Ness, F., Lavallee, F., Dubourdieu, D., Aigle, M., and Dulau, L. (1993). Identification of yeast strains using the polymerase chain reaction. J. Sci. Food Agric. *62*, 89–94.

Nguyen, H.-V., and Gaillardin, C. (1997). Two subgroups within the *Saccharomyces bayanus* species evidenced by PCR amplification and restriction polymorphism of the non-transcribed spacer 2 in the Ribosomal DNA unit. System. Appl. Microbiol. *20*, 286–294.

Nguyen, H.-V., Pulvirenti, A., and Gaillardin, C. (2000). Rapid differentiation of the closely related *Kluyveromyces lactis* var. *lactis* and *K. marxianus* strains isolated from dairy products using selective media and PCR/RFLP of the rDNA non transcribed spacer 2. Can. J. Microbiol. *46*, 1115–1122.

Nishihara, H., Kio, K., and Imamura, M. (2000). Possible mechanism of co-flocculation between non-flocculent yeasts. J. Inst. Brew. *106*, 7–10.

O'Donnell, K. (1993). *Fusarium* and its near relatives. In The Fungal Holomorph: Mitotic, meiotic and pleomorphic speciation in fungal systematics. D.R Reynolds and J.W Taylor eds (Wallingford, UK; CAB International), pp. 225–233.

Oakley-Gutowski, K.M., Hawthorne, D.B., and Kavanagh, T.E. (1991). Application of chromosome fingerprinting to the differentiation of brewing yeasts. J. Am. Soc. Brew. Chem. *50*, 48–52.

Ogur, M., St. John, R., and Nagal, S. (1957). Tetrazolium overlay technique for population studies of respiration deficiency in yeast. Science *125*, 928–929.

Peng, X., Sun, J., Iserentant, D., Michiels, C., and Verachtert, H. (2001). Flocculation and coflocculation of bacteria by yeasts. Appl. Microbiol. Biotechnol. *55*, 777–781.

Pérez, M.A., Gallego, F.J., Martínez, I., and Hidalgo, P. (2001). Detection, distribution and selection of microsatellites (SSRs) in the genome of the yeast *Saccharomyces cerevisiae* as molecular markers. Lett. Appl. Microbiol. *33*, 461–466.

Pfeiffer, T., and Morley, A. (2014). An evolutionary perspective on the Crabtree effect. Frontiers Mol. Biosci. *1*, 1–6. http://dx.doi.org/10.3389/fmolb.2014.00017

Pham, T., Wimalasena, T., Box, W.G., Koivuranta, K., Storgards, E., Smart, K.A., and Gibson, B.R. (Evaluation of ITS PCR and RFLP for differentiation and identification of brewing yeast and brewery 'wild' yeast contaminants. (2011). J. Inst. Brew. *117*, 556–568.

Philliskirk, G., and Young, T.W. (1975). The occurrence of killer character in yeasts of various genera. Antonie Van Leeuwenhoek *41*, 147–151.

Prado, M.R., Blandon, L.M., Vandenberghe, L.P.S., Rodrigues, C., Castro, G.R., Thomaz-Soccol, V., and Soccol, C.R. (2015). Milk kefir: composition, microbial cultures, biological activities, and related products. Frontiers Microbiol. *6*, 1–10. http://dx.doi.org/10.3389/fmicb.2015.01177

Prakitchaiwattana, C.J., Fleet, G.H., and Heard, G.M. (2004). Application and evaluation of denaturing gradient gel electrophoresis to analyse the yeast ecology of wine grapes. FEMS Yeast Res. *4*, 865–877. http://dx.doi.org/10.1016/j.femsyr.2004.05.004

Pulvirenti, A., Nguyen, H., Caggia, C., Giudici, P., Rainieri, S., and Zambonelli, C. (2000). *Saccharomyces uvarum*, a proper species within *Saccharomyces sensu stricto*. FEMS Microbiol. Lett. *192*, 191–196.

Quain, D.E. (2016). Draught beer hygiene: cleaning of dispense tap nozzles. J. Inst. Brew. http://dx.doi.org/10.1002/jib.335

Radler, F., Pfeiffer, P., and Dennert, M. (1985). Killer toxins in new isolates of the yeasts *Hanseniaspora uvarum* and *Pichia kluyveri*. FEMS Microbiol. Letts. *29*, 269–272.

Radler, F., Herzberger, S., Schönig, I., and Schwarz, P. (1993). Investigation of a killer strain of *Zygosaccharomyces bailii*. J. Gen. Microbiol. *139*, 495–500. http://dx.doi.org/10.1099/00221287-139-3-495

Reynolds, T.B., Jansen, A., Peng, X., and Fink, G.R. (2008). Mat formation in *Saccharomyces cerevisiae* requires nutrient and pH gradients. Eukaryot. Cell *7*, 122–130.

Rodriguez-Cousino, N., Maqueda, M., Ambrona, J., Zamora, E., Esteban, R., and Ramirez, M. (2011). A new wine *Saccharomyces cerevisiae* killer toxin (Klus), encoded by a double-stranded RNA virus, with broad antifungal activity is evolutionarily related to a chromosomal host gene. Appl. Environ. Microbiol. *77*, 1822–1832.

Rolland, F., Winderickx, J., and Thevelein, J.M. (2002). Glucose-sensing and -signalling mechanisms in yeast. FEMS Yeast Res. *2*, 183–201.

Rossouw, D., Bagheri, B., Setati, M.E., and Bauer, F.F. (2015). Co-flocculation of yeast species, a new mechanism to govern population dynamics in microbial ecosystems. PLoS ONE *10*, e0136249. http://dx.doi.org/10.1371/journal.pone.0136249

Sampaio, J.P., and Gonçalves, P. (2008). Natural populations of *Saccharomyces kudriavzevii* in Portugal are associated with oak bark and are sympatric with *S. cerevisiae* and *S. paradoxus*. Appl. Environ. Microbiol. *74*, 2144–2152. http://dx.doi.org/10.1128/AEM.02396-07

Schacherer, J., Ruderfer, D.M., Gresham, D., Dolinski, K., Botstein, D., and Kruglyak, L. (2007). Genome-wide analysis of nucleotide-level variation in commonly used *Saccharomyces cerevisiae* strains. PLoS ONE *2*, e322. http://dx.doi.org/10.1371/journal.pone.0000322

Schacherer, J., Shapiro, J.A., Ruderfer, D.M., and Kruglyak, L. (2009). Comprehensive polymorphism survey elucidates population structure of *Saccharomyces cerevisiae*. Nature *458*, 342–345. http://dx.doi.org/10.1038/nature07670

Scheffers, W.A. (1961). On the inhibition of alcoholic fermentation of *Brettanomyces* yeasts under anaerobic conditions. Experientia *17*, 40–42.

Scheffers, W.A. (1979). Anaerobic inhibition in yeasts (Custers effect). Antonie Van Leeuwenhoek *45*, 150.

Schifferdecker, A.J., Dashko, S., Ishchuk, O.P., and Piskur, J. (2014). The wine and beer yeast *Dekkera bruxellensis*. Yeast *31*, 323–332.

Schmitt, M.J., and Breinig, F. (2006). Yeast viral killer toxins: lethality and self-protection. Nat. Rev. Microbiol. *4*, 212–221.

Schoch, C.L., Seifert, K.A., Huhndorf, S., Robert, V., Spouge, J.L., Levesque, C.A., and Chen, W., and Fungal Barcoding Consortium. (2012). Nuclear ribosomal internal transcribed spacer (ITS) region as a universal DNA barcode marker for Fungi. Proc. Natl. Acad. Sci. U.S.A. *109*, 6241–6246. http://dx.doi.org/10.1073/pnas.1117018109

Schuller, D., Valero, E., Dequin, S., and Casal, M. (2004). Survey of molecular methods for the typing of wine yeast strains. FEMS Microbiol. Lett. *231*, 19–26. http://dx.doi.org/10.1016/S0378-1097(03)00928-5

Schwarz, K.J., Boitz, L.I., and Methner, F.-J. (2012). Enzymatic formation of styrene during wheat beer fermentation is dependent on pitching rate and cinnamic acid content. J. Inst. Brew. *118*, 280–284.

Schwartz, D.C., and Cantor, C.R. (1984). Separation of yeast chromosome-sized DNAs by pulsed field gradient gel electrophoresis. Cell 37, 67–75.

Sims, A.P., and Barnett, J.A. (1978). The requirement of oxygen for the utilization of maltose, cellobiose and D-galactose by certain anaerobically-fermenting yeasts (Kluyver effect). J. Gen. Microbiol. *106*, 277–288.

Sosa, O.A., Manca de Nadra, M.C., and Farias, M.E. (2008). Behaviour of *Kloeckera apiculata* flocculent strain in coculture with *Saccharomyces cerevisiae*. Food Technol. Biotechnol. *46*, 413–418.

Sousa, M.J., Teixeira, J.A., and Mota, M. (1992). Differences in the flocculation mechanism of *Kluyveromyces marxianus* and *Saccharomyces cerevisiae*. Biotechnol. Letts. *14*, 213–218.

Stark, M.J.R., and Boyd, A. (1986). The killer toxin of *Kluyveromyces lactis*: characterization of the toxin subunits and identification of the genes which encode them. EMBO J. *5*, 1995–2002.

Stark, M.J., Boyd, A., Mileham, A.J., and Romanos, M.A. (1990). The plasmid-encoded killer system of *Kluyveromyces lactis*: a review. Yeast *6*, 1–29.

Steensels, J., Daenen, L., Malcorps, P., Derdelinckx, G., Verachtert, H., and Verstrepen, K.J. (2015). *Brettanomyces* yeasts – From spoilage organisms to valuable contributors to industrial fermentations. Int. J. Food Microbiol. *206*, 24–38. http://dx.doi.org/10.1016/j.ijfoodmicro.2015.04.005

Stratford, M., Steels, H., Nebe-von-Caron, G., Novodvorska, M., Hayer, K., and Archer, D.B. (2013). Extreme resistance to weak-acid preservatives in the spoilage yeast *Zygosaccharomyces bailii*. Int. J. Food Microbiol. *166*, 126–134. http://dx.doi.org/10.1016/j.ijfoodmicro.2013.06.025

Suh, S.O., Blackwell, M., Kurtzman, C.P., and Lachance, M.A. (2006). Phylogenetics of Saccharomycetales, the ascomycete yeasts. Mycologia 98, 1006–1017.

Suzzi, G., Romano, P., and Benevelli, M. (1992). The flocculation of wine yeasts: biochemical and morphological characteristics in *Zygosaccharomyces*–flocculation in *Zygosaccharomyces*. Antonie Van Leeuwenhoek. *61*, 317–322.

Suzzi, G., Romano, P., Westall, F., and Vannini, L. (1996). The flocculation of wine yeasts: biochemical and morphological characteristics in *Kloeckera apiculata*. Antonie Van Leeuwenhoek *69*, 273–277.

Tamaki, H. (1978). Genetic studies of ability to ferment starch in *Saccharomyces*: gene polymorphism. Mol. Gen. Genet. *164*, 205–209.

Taylor, G.T., and Marsh, A.S. (1984). MYGP+Copper, a medium that detects both *Saccharomyces* and non-*Saccharomyces* wild yeast in the presence of culture yeast. J. Inst. Brew. *90*, 134–145.

Teunissen, A.W.R.H., van den Berg, J.A., and Steensma, H.Y. (1993). Physical localization of the flocculation gene *FLO1* on Chromosome I of *Saccharomyces cerevisiae*. Yeast *9*, 1–10.

Theisen, S., Molkenau, E., and Schmitt, M.J. (2000). Wicaltin, a new protein toxin secreted by the yeast *Williopsis californica* and its broad-spectrum antimycotic potential. J. Microbiol. Biotechnol. *10*, 547–550.

Timke, M., Wang-Lieu, N.Q., Altendorf, K., and Lipski, A. (2008). Identity, beer spoiling and biofilm forming potential of yeasts from beer bottling plant associated biofilms. Antonie Van Leeuwenhoek *93*, 151–161. http://dx.doi.org/10.1007/s10482-007-9189-8

Timmins, E.M., Quain, D.E., and Goodacre, R. (1998). Differentiation of brewing yeast strains by pyrolysis mass spectrometry and Fourier transform infrared spectroscopy. Yeast *14*, 885–893.

Trost, A., Graf, B., Eucker, J., Sezer, O., Possinger, K., Göbel, U.B., and Adam, T. (2004). Identification of clinically relevant yeasts by PCR/RFLP. J. Microbiol. Methods *56*, 201–211.

Tsang, E.W.T., and Ingledew, W.M. (1982). Studies on the heat resistance of wild yeasts and bacteria in beer. J. Am. Soc. Brew. Chem. *40*, 1–8.

Usbeck, J.C., Wilde, C., Bertrand, D., Behr, J., and Vogel, R.F. (2014). Wine yeast typing by MALDI-TOF MS. Appl. Microbiol. Biotechnol. *98*, 3737–3752. http://dx.doi.org/10.1007/s00253-014-5586-x

van der Aa Kühle, A., and Jespersen, L. (1998). Detection and identification of wild yeasts in lager breweries. Int. J. Food Microbiol. *43*, 205–213.

van Dijken, J.P., and Scheffers, W.A. (1986). Redox balances in the metabolism of sugars by yeasts. FEMS Microbiol Rev. *32*, 199–224.

van Urk, H., Postma, E., Scheffers, W.A., and van Dijken, J.P. (1989). Glucose transport in crabtree-positive and crabtree-negative yeasts. J. Gen. Microbiol. *135*, 2399–2406. http://dx.doi.org/10.1099/00221287-135-9-2399

Verstrepen, K.J., Jansen, A., Lewitter, F., and Fink, G.R. (2005). Intragenic tandem repeats generate functional variability. Nat. Genet. 37, 986–990.

Walters, L.S., and Thiselton, M.R. (1953). Utilization of lysine by yeasts. J. Inst. Brew. *59*, 401–404.

Weiss, S., Samson, F., Navarro, D., and Casaregola, S. (2013). YeastIP: a database for identification and phylogeny of *Saccharomycotina* yeasts. FEMS Yeast Res. *13*, 117–125. http://dx.doi.org/10.1111/1567-1364.12017

White, T.J., Bruns, T., Less, S., and Taylor, J. (1990). Amplification and direct sequencing of fungal ribosomal RNA genes for phylogenetics. In PCR Protocols: A Guide to Methods and Applications. M.A Innis, D.H. Gelfand, J.J. Sninsky and T.J. White eds. (San Diego, USA; Academic Press), pp. 315–322.

Wightman, P., Quain, D.E., and Meaden, P.G. (1996). Analysis of production brewing strains of yeast by DNA fingerprinting. Lett. Appl. Microbiol. *22*, 90–94.

Wijsman, M.R., van Dijken, J.P., van Kleef, B.H., and Scheffers, W.A. (1984). Inhibition of fermentation and growth in batch cultures of the yeast *Brettanomyces intermedius* upon a shift from aerobic to anaerobic conditions (Custers effect). Antonie Van Leeuwenhoek *50*, 183–192.

Wiles, A.E. (1950). Studies of some yeasts causing spoilage of draught beers. J. Inst. Brew. *56*, 183–193.

Wilkening, S., Tekkedil, M.M., Lin, G., Fritsch, E.S., Wei, W., Gagneur, J., Lazinski, D.W., Camilli, A., and Steinmetz,

L.M. (2013). Genotyping 1000 yeast strains by next-generation sequencing. BMC Genomics *14*, 90. http://dx.doi.org/10.1186/1471-2164-14-90

Wood, K.A., Quain, D.E., and Hinchcliffe, E. (1992). The attachment of brewing yeast to glass. J. Inst. Brew. *98*, 325–327.

Xufre, A., Albergaria, H., Inácio, J., Spencer-Martins, I., and Gírio, F. (2006). Application of fluorescence in situ hybridisation (FISH) to the analysis of yeast population dynamics in winery and laboratory grape must fermentations. Int. J. Food Microbiol. *108*, 376–384.

Yamamoto, T., Hiratani, T., Hirata, H., Imai, M., and Yamaguchi, H. (1986). Killer toxin from *Hansenula mrakii* selectively inhibits cell wall synthesis in a sensitive yeast. FEBS Lett. *197*, 50–54.

Yamauchi, H., Yamamoto, H., Shibano, Y., Amaya, N., and Saeki, T. (1998). Rapid methods for detecting *Saccharomyces diastaticus*, a beer spoilage yeast, using the polymerase chain reaction. J. Am. Soc. Brew. Chem. *56*, 58–63.

Yilmaztekin, M., Erten, H., and Cabaroglu, T. (2008). Production of isoamyl acetate from sugar beet molasses by *Williopsis saturnus* var. *saturnus*. J. Inst. Brew. *114*, 34–38.

Young, T.W. (1987). Killer yeasts. In The Yeasts, Vol 2: Yeasts and the environment. A.H. Rose and J.S. Harrison eds (London, UK; Academic Press), pp. 131–164.

Zhao, Y., Strope, P.K., Kozmin, S.G., McCusker, J.H., Dietrich, F.S., Kokoska, R.J., and Petes, T.D. (2014). Structures of naturally evolved CUP1 tandem arrays in yeast indicate that these arrays are generated by unequal nonhomologous recombination. Genes, Genomes, Genet. *4*, 2259–2269.

Index

www.ingramcontent.com/pod-product-compliance
Ingram Content Group UK Ltd.
Pitfield, Milton Keynes, MK11 3LW, UK
UKHW062000041026
15591UKWH00016B/1099

9 781910 190616